Modern Selective Fungicides

Modern Selective Fungicides

— Properties, Applications, Mechanisms of Action —

Editor

Professor Dr. Horst Lyr

In cooperation with 34 scientists

Dr. P. Braun (FRG), Prof. Dr. H. Buchenauer (FRG), Dr. L. C. Davidse (NL),
Prof. Dr. J. Dekker (NL), Dr. C. J. Delp (USA), Prof. Dr. M. A. de Waard (NL),
Dr. W. Edlich (FRG), Dr. M. Gasztonyi (Hungaria),
Prof. Dr. S. G. Georgopoulos (Greece), Dr. U. Gisi (Switzerland),
Dr. D. W. Hollomon (UK), Dr. H. Ishii (Japan), Prof. Dr. F. Jacob (FRG),
Dr. A. Kerkenaar (NL), Dr. D. Kleier (USA), Dr. K. H. Kuck (FRG),
Dr. M. Kulka (Canada), Dr. G. Lorenz (FRG), Prof. Dr. H. Lyr (FRG),
Dr. St. Neumann (FRG), Dr. E.-H. Pommer (FRG), Dr. R. Pontzen (FRG),
Dr. N. Ragsdale (USA), Prof. Dr. H. Scheinpflug (FRG),
Prof. Dr. T. Schewe (FRG), B. von Schmeling (USA), Dr. H. H. Schmidt (FRG),
Dr. B. Schreiber (FRG), Prof. Dr. F. Schwinn (Switzerland),
Prof. Dr. H. D. Sisler (USA), Dr. T. Staub (Switzerland), Dr. J. Steffens (USA),
Dr. H. Vanden Bossche (Belgium), Dr. J. Yamaguchi (Japan)

2nd revised and enlarged edition

70 Figures and 93 Tables

Gustav Fischer Verlag · Jena · Stuttgart · New York

1st edition 1987

Die Deutsche Bibliothek – CIP-Einheitsaufnahme

Modern selective fungicides : properties, applications,
mechanisms of action ; 93 tables / ed. H. Lyr. In coop. with P.
Braun ... – 2., rev. and enl. ed. – Jena ; Stuttgart ; New York :
Fischer ; New York : VCH, 1995
  ISBN 3-334-60455-1 (Fischer)
  ISBN 1-56081-403-9 (VCH)
NE: Lyr, Horst [Hrsg.]; Braun, Peter

Gedruckt mit Unterstützung des Förderungs- und Beihilfefonds Wissenschaft der VG Wort

---

Published jointly by:
Gustav Fischer Verlag
220 East 23rd Street, Suite 909, New York 10010, USA
Gustav Fischer Verlag
Villengang 2, D-07745 Jena, Federal Republic of Germany
For orders from USA and Canada:
VCH Publishers Inc., 303 N.W. 12th Avenue, Deerfield Beach, Florida 33442-1705, USA
© Gustav Fischer Verlag · Jena · Stuttgart · New York · 1995
This work with all its parts is protected by copyright. Any use beyond the strict limits of the copyright law without the consent of the publisher is inadmissible and punishable. This refers especially to reproduction of figures and/or text in print or xerography, translations, microforms and the data storage and processing in electronic systems.
Printed: Druckhaus Köthen GmbH
Bound: Druckerei zu Altenburg GmbH
Printed in Germany

List of contributors

Dr. P. Braun
AgrEvo GmbH
Products Development/Ecology,
Building G 836
D-65926 Frankfurt/Main
Germany

Prof. Dr. H. Buchenauer
Institute of Phytomedizin
University Stuttgart-Hohenheim
Otto-Sander-Str. 3
D-70599 Stuttgart
Germany

Dr. L. C. Davidse
Melkbon
1602 JD Enkhuizen
The Netherlands

Prof. Dr. J. Dekker
Department of Phytopathology
Agricultural University Wageningen
Binnenhaven 9
6709 PD Wageningen
The Netherlands

Dr. C. J. Delp
2041 Misty Sunrise Trail
Sarasota, FL 34240-9686
USA

Prof. Dr. M. A. de Waard
Department of Phytopathology
Agricultural University Wageningen
Binnenhaven 9
6709 PD Wageningen
The Netherlands

Dr. W. Edlich
Karl-Marx-Str. 123
D-14532 Kleinmachnow
Germany

Dr. M. Gasztonyi
Plant Protection Institute
Hungarian Academy of Sciences
Herman Otto Ut. 15
H-1525 Budapest
Hungaria

Prof. Dr. S. G. Georgopoulos
Lapapharm Inc.
Trade and Distribution Comp.
73, Menadrou str.
104 37 Athens
Greece

Dr. U. Gisi
Sandoz Agro Ltd.
Agrobiological Research Station
CH-4108 Witterswil
Switzerland

Dr. D. W. HOLLOMON
Department Agricultural Sciences
University of Bristol
Long Ashton, Research Station
Long Ashton, Bristol, BS18 9AF
UK

Dr. H. ISHII
Ministry Agriculture, Forestry, Fisheries
Fruit Tree Res. Station
2-1 Fujimoto, Tsukaba, Ibaraki,
305 Japan

Prof. Dr. F. JACOB
Department of Biosciences
Botanical Institute
Martin-Luther-University
Am Kirchtor 1
D-06108 Halle/S.
Germany

Dr. A. KERKENAAR
Plaggewagen 12
1261 KG Blaricum
The Netherlands

Dr. D. KLEIER
DuPont Agricultural Products
Experimental Station E402/4229
P.O.B. 80402
Wilmington, DE 19880-0402
USA

Dr. K. H. KUCK
Institute for Plant Diseases
Pflanzenschutzzentrum Monheim
Bayer AG
D-51368 Leverkusen – Bayerwerke
Germany

Dr. M. KULKA
Uniroyal Ltd. Research Laboratories
120 Huron Str.
Guelph Ontar.
Canada

Dr. G. LORENZ
BASF AG
Agricultural Research Station
D-67114 Limburgerhof
Germany

Prof. Dr. H. LYR
Georg-Herwegh-Str. 8
D-16225 Eberswalde
Germany

Dr. S. NEUMANN
Department of Biosciences
Botanical Institute
Martin-Luther-University
Am Kirchtor 1
D-06108 Halle/S.
Germany

Dr. E.-H. POMMER
Berliner Platz 7
D-67117 Limburgerhof
Germany

Dr. R. PONTZEN
Institute for Plant Diseases
Pflanzenschutzzentrum Monheim
Bayer AG
D-51368 Leverkusen – Bayerwerke
Germany

Dr. N. RAGSDALE
NAPIAP/USDA
Rn 324-A Admn. Bldg.
Washington DC 20250-0100
USA

Prof. Dr. H. SCHEINPFLUG
Am Thelenhof 15
D-51377 Leverkusen
Germany

Prof. Dr. T. SCHEWE
Institute of Biochemistry
Humboldt-University
Hessische Straße 3–4
D-10115 Berlin
Germany

Boe von Schmeling
Uniroyal Ltd.
World Headquaters
Middlebury CT 06749
USA

Dr. H. H. Schmidt
Biologische Bundesanstalt
Stahnsdorfer Damm 81
D-14532 Kleinmachnow
Germany

Dr. B. Schreiber
AgrEvo GmbH
Building K 607
D-65926 Frankfurt/Main
Germany

Prof. Dr. F. Schwinn
Rebgasse 10
CH-4132 Muttenz
Switzerland

Prof. Dr. H. D. Sisler
University of Maryland
Department of Botany
College Park
Maryland 20742
USA

Dr. T. Staub
Agricultural Div. Geigy
CH-4002 Basel
Switzerland

Dr. J. Steffens
DuPont Agricultural Products
Experimental Station E402/4229
P.O.B. 80402
Wilmington, DE 19880-0402
USA

Dr. H. van den Bossche
Department of Comparative Biochemistry
Janssen Research Foundation
B-2340 Beerse
Belgium

Dr. J. Yamaguchi
Institute of Physical and Chemical Research
Microbiology, Toxicology Laboratory
Hirosawa 2-1, Wako-shi
Saitama, 351-01
Japan

Table of Contents

	Foreword H. Lyr	11
1	Selectivity in modern fungicides and its basis. H. Lyr	13–22
2	Development of resistance to modern fungicidies and strategies for its avoidance J. Dekker	23–38
3	The genetics of fungicide resistance. S. G. Georgopoulos	39–52
4	Principles of uptake and systemic transport of fungicides within the plant. St. Neumann and F. Jacob	53–73
5	Aromatic hydrocarbon fungicides and their mechanism of action. H. Lyr	75–98
6	Dicarboximide fungicides. E.-H. Pommer and G. Lorenz	99–118
7	Mechanism of action of dicarboximide fungicides. W. Edlich and H. Lyr	119–131
8	Carboxin fungicides and related compounds. M. Kulka and B. v. Schmeling	133–147
9	Mechanism of action of carboxin fungicides and related compounds T. Schewe and H. Lyr	149–161
10	Morpholine fungicides and related compounds. E.-H. Pommer	163–183
11	Mechanism of action of cyclic amine fungicides: morpholines and piperidines. A. Kerkenaar	185–204
12	DMI fungicides. K. H. Kuck, H. Scheinpflug and R. Pontzen	205–258
13	DMI-fungicides – side effects on the plant and problems of resistance H. Buchenauer	259–290

13.1	Mode of action of N-heterocyclic plant growth retardants and side effects of structurally related fungicides in plants	259–279
13.2	Resistance of fungi to sterol demethylation inhibitors	280–290
14	Benzimidazole and related fungicides . . . . C. J. Delp	291–303
15	Biochemical and molecular aspects of the mechanisms of action of benzimidazoles, N-phenylcarbamates and N-phenylformamidoximes and the mechanisms of resistance to these compounds in fungi . . . L. C. Davidse and H. Ishii	305–322
16	Oomycetes fungicides	323–354
16.1	Phenylamides and other fungicides against Oomycetes . . . . F. Schwinn and T. Staub	323–346
16.2	Phenylamide fungicides – biochemical action and resistance . . . . L. C. Davidse	347–354
17	2-Aminopyrimidine fungicides . . . . D. W. Hollomon and H.-H. Schmidt	355–371
18	Organophosphorus fungicides . . . . P. Braun and N. Schreiber	373–388
19	Miscellaneous fungicides . . . . M. Gasztonyi and H. Lyr	389–414
20	Antibiotics as antifungal agents . . . . I. Yamaguchi	415–429
21	Chemotherapy of human fungal infections . . . . H. Vanden Bossche	431–484
22	Fungicides for wood preservation . . . . E.-H. Pommer	485–502
23	Fungicides for the preservation of materials . . . . E.-H. Pommer	503–515
24	Computer aided discovery and optimization of fungicides . . . . . J. J. Steffens and D. A. Kleier	517–542
25	Disease control by nonfungitoxic compounds . . . . H. D. Sisler and N. N. Ragsdale	543–564
26	Synergism and antagonism in fungicides . . . . M. A. de Waard and U. Gisi	565–578
27	Outlooks . . . . H. Lyr	579–584
	Subject Index	585–595

Foreword

Stimulated by the worldwide positive acceptance of the first edition of this book (1987), Gustav Fischer Verlag Jena and the editor in cooperation with the authors have decided to issue a second revised and enlarged edition.

Meanwhile several new fungicides, or even new groups of fungicides, have been developed and have reached the market. In spite of the efforts of plant breeders to produce more resistant cultivars, protection of plants by application of modern fungicides is unavoidable in agro- and horticulture. Of course new demands for still better protection of the environment by avoidance of undesirable side effects must be met.

In medicine, the need to use fungicides to control human fungal diseases has become more urgent because an affected immune system favours the spreading of fungal diseases. We decided to add some new chapters such as "Chemotherapy of human fungal infections", "Fungicides for wood preservation", "Fungicides for material preservation" and "Antibiotics as antifungal agents". This allows for a broad survey of nearly all fields related to the practical use of fungicides. Other chapters, which are still of high relevance, have been updated or in part newly written.

Progress has been made towards a better understanding of the mechanisms of action of various fungicides, which in turn allows a better understanding of the function of fungal cells as well as progress towards a more rational design of new fungicidal compounds.

This book now describes all the main selective fungicides which have reached the market, their properties, spectrum of activity and the mode of action.

I wish to thank all the contributors of this book, who are all internationally well known experts in this field, for their cooperation and stimulating suggestions for improving the content. The authors hope that it will fulfill the demands of readers in many parts of the world to become familiar with the great number of useful compounds and their proper usage.

We are indebted to Gustav Fischer Verlag Jena for contribution to the high-quality realization of this project.

H. LYR

Chapter 1

Selectivity in modern fungicides and its basis

H. LYR

Biologische Bundesanstalt für Land- und Forstwirtschaft, Institut für Integrierten Pflanzenschutz, Kleinmachnow, Germany

Introduction

The word "fungicides" suggests that such compounds kill all fungi more or less unselectively. But this is not the cases even with classic "multisite inhibitors" such as dithiocarbamates, phthalimides, or quinones and similar compounds. Although multisite inhibitors have a nonspecific mechanism of action (TORGESON 1969), surprisingly they act in some cases rather selectively (Tab. 1.1), and cannot be used to control all fungal diseases. For these types of fungicides, a certain "tolerance" has been described for some fungal strains but no true resistance (see chapter 2). This is based on their unspecific mechanism of action. Most "fungicides" act fungistatically in principle and a fungicidal effect results only after long lasting interactions with the fungus, or by support of host defence mechanisms. Some strong biocidal compounds, such as mercurials, several disinfectants (such as phenol, cresol, peracetic acid, quaternary-N-compounds, Na-hypochlorite) and other biocidal agents act more or less fungicidal, but they are also phytotoxic (cf. chapters 22, 23).

Modern fungicides must be selective in order to avoid phytotoxic effects. They must have a low toxicitiy for mammals, fish, bees and other harmless organisms in the ecosystem. Although not all demands are optimally combined in one compound, modern fungicides increasingly meet the demands for protection of plants and humans, as well as of the environment.

Table 1.1 Spectrum of activity of some fungicides with an unspecific mechanism of action (ED_{50} values for mycelial radial growth on malt medium, in mg/l)

Species	Thiram	Maneb	Captan	Dichlone
Phytophthora cactorum	20	10	30	5
Pythium ultimum	20	8	20	5
Mucor mucedo	30	100	100	100
Botrytis cinerea	3	40	20	3
Fusarium oxysporum	20	100	40	100
Gloeosp. fructigenum	20	40	–	50
Mycosph. pinodes	4	30	–	30
Trametes versicolor	20	30	30	100
Verticillium albo-atrum	15	10	–	10
Rhizoctonia solani	10	35	30	4

In spite of considerable efforts by plant breeders, the usage of fungicides in agriculture, and also in medicine, is necessary to keep plants and humans healthy. Tabs. 1.2 – 1.4 give an overview of the economic value of the fungicide world market.

Table 1.2 Crop Protection World Market 1991. Industrial Countries Market Share in %

	North America	Western-Europe	Japan
Herbicides	36	27	7
Fungicides	11	47	14
Insecticides	19	17	14
Others	15	42	9

Table 1.3 Fungicides World Market 1991. Classification by Chemical Type

Market	US $ approx. 5 Mrd
Inorganics	11%
Dithiocarbamates	16%
Organophosphorus	7%
Phtalimides	7%
Other non Systemics	10%
Triazoles	18%
Benzimidazoles	10%
Other Systemics	21%

Table 1.4 Crop Protection World Market 1991. Major Crops and Product Groups Mrd US $

Culture	Herbicides 10.0 Mrd US $ Share in %	Insecticides 6.4 Mrd US $ Share in %	Fungicides approx. 5 Mrd US $ Share in %
Cereals	20	3	27
Maize	18	8	1
Rice	9	13	12
Soybeans	15	3	2
Cotton	3	21	1
Sugarbeet	5	3	2
Fruit/Vegetable (incl. Vine)	15	26	41
Others	15	23	14

Facts and problems

With the introduction of modern, highly efficient fungicides which produce no, or tolerable side effects, practice and science have been more urgently confronted with the problem of mycological selectivity.

The positive side is a more directed control of pathogens without severe disturbances of the harmless or useful fungal flora on foliage or in the soil.

For example, a selective control of *Pythium* in the soil may favor its antagonists, such as *Trichoderma* species, which can support the suppression of the pathogen (BETH et al. 1984).

On the other hand, a very selective control of *Pythium* can lead to increased severity of other diseases, such as those caused by soil-borne *Phytophthora* species, *Fusarium* or *Rhizoctonia* and others. It has been observed that selective control of *Pseudocercosporella herpotrichoides* by benomyl favors the attack of *Rhizoctonia* in wheat. Still other examples in this respect have been described.

However, very little is known about the positive or negative effects of selective fungicides on the fungal phyllosphere. A selective control of mildew in cereals may favor the attack of rusts and other pathogens. *Alternaria*, *Epicoccum* and *Trichoderma* significantly antagonize the infection of lettuce plants by *Sclerotinia* (MERCIER and REEFELDER 1984). Other antagonistic effects are described by HINSDORF and KETTERER (1984), LEINHOS and BUCHENAUER (1984), and STENZEL (1984).

But the danger to beneficial fungi under practical conditions seems not to be too great. This is due to the high stability of microbial associations in nature and the short lasting effects of selective fungicides.

Advantages and disadvantages must be considered when using modern fungicides. To compensate for gaps in the spectrum of activity of some fungicides, combinations (mixtures of compatible fungicides with different mechanisms of action) are very often used. This also avoids, or slows down, the built up of resistant fungal populations.

The same is true for insecticides and herbicides.

Spectra of activity

Modern fungicides have indeed distinct spectra of activity, which demonstrate their selectivity even within fungal taxa.

Some examples of fungicides selected from different groups with different mechanisms of action are given in Tab. 1.5. Similarities or differences are not simply related to the mechanism of action. Even within such a group quantitative and qualitative differences can be found. They may be based on the potential for net-uptake or degradation or other features. This is even true for very specific fungicides with a narrow spectrum of activity, for example those acting against Oomycetes (Tab. 1.6). Within the genus *Phytophthora*, considerable differences in the sensitivity to a special fungicide may exist. *Pythium* species, although related to *Phytophthora* species, may behave very differently from the latter regarding their fungicide sensitivity. This points to still unknown biochemical differences among these fungi.

It should be kept in mind that also within one species differences do exist in the sensitivity to a fungicide. This is demonstrated in Tab. 1.7 for various pathotypes of *Phytophthora infestans*, which had no previous contact with the fungicide metalaxyl. This fact must be kept in mind in discussions about shifting of the level of resistance in a fungal population under selection pressure by fungicides. The data in Tab. 1.7 reflect

Table 1.5 Comparison of the spectra of activity of some modern fungicides (malt-agar-growth test) ED_{50} values in mg/l (* barley leaf-test)

	Tridemorph	Triforine	Triadimefon	Etridiazole	Chloroneb	Dicloran	Tolclophos-methyl	Vinclozolin	Benomyl	Thiophanate-methyl	Carboxin	Dichlofluanid	BAS 490 F Kresoxim-methyl
Phytophthora cactorum	300	20	200	0.5	2	200	100	200	50	100	100	4	>500
Ph. infestans	–	–	–	1	10	–	–	500	–	–	>500	10	<1
Pythium ultimum	60	15	200	0.5	10	200	100	50	100	100	100	2	>500
Mucor mucedo	300	20	200	8	2	0.5	–	10	100	100	50	40	(>500)
*Erysiphe graminis**	0.02	1	0.006	20	20	–	–	3	7	100	20	3	–
Penicillium chrysogenum	150	200	5	90	500	0.5	1.6	2	0.2	50	50	2	<1
Botrytis cinerea	0.4	20	40	20	5	1	1.9	0.5	0.4	3	10	0.5	>500
Fusarium oxysporum	300	20	50	70	500	200	100	40	2	70	200	50	<50
Gloeosp. fructigenum	10	50	200	80	500	200	–	200	0.2	0.2	300	6	–
Verticillium alb.-atrum	0.3	80	3	40	10	8	100	40	3	3	5	9	<10
Rhizoctonia solani	40	60	100	70	15	200	0.1	20	200	200	5	3	–
Trametes versicolor	6	30	0.5	30	2	200	–	100	200	200	5	2	–

Table 1.6 Selectivity of action of some fungicides with relative specificity for the genus *Phytophthora* or Oomycetes (inhibition of radial mycelial growth as % of control on agar medium, concentrations in mg/l) (Andoprim is a anilinopyrimidine fungicide)

Species Conc.	Meta- laxyl (0.1)	Al-phos- ethyl (100)	Cymo- xanil (10)	Chlo- roneb (10)	Etri- diazole (1)	Hyme- xazole (1)	Ando- prim (3)
P. cactorum	100	57	3	51	67	30	57
P. cinnamomi	28	74	11	52	66	17	90
P. citricola	17	65	0	33	47	18	100
P. citrophthora	15	65	7	39	73	28	50
P. cryptogaea	68	34	0	54	80	32	40
P. infestans	85	22	100	47	56	98	48
P. megasperma	75	48	76	73	89	75	58
P. nicotianae	38	71	30	64	77	42	32
P. palmivora	77	12	34	50	73	19	88
P. parasitica	72	65	52	65	66	17	97
Pyth. ultimum	56	–	24	0	100	100	8
Botrytis cinerea	0	–	5	50	2	–	0
Fusarium oxyspor.	0	–	0	0	0	–	3
Rhizoctonia solani	0	–	3	30	0	–	0

Table 1.7 Variability in sensitivity of fungal strains to the fungicides metalaxyl and tolclophosmethyl. A: Sensitivity of strains of *Phytophthora infestans* (various pathotypes collected in the GDR) to metalaxyl. B: Sensitivity of strains of *Rhizoctonia solani* to tolclophosmethyl (results mainly from information of Sumitomo Chemical Co.). The inhibitory concentrations are expressed as ED_{50} values in mg/l (AG means Ogoshi's hyphal anastomosis groups)

A: *Phytophthora infestans* Pathotypes:	ED_{50}	B: *Rhizoctonia solani* strains:	ED_{50}
1.2.3.4.	0.004	coll. IPF	0.6
1.5.	0.007	AG-1	0.15
1.3.4.	0.012	AG-2	0.10
1.3.4.5.7.10.11.	0.015	AG-2	0.12
1.2.3.4.5.7.10.	0.056	AG-3	0.43
4.1.	0.022	AG-4	0.01
4.5.	0.008	AG-5	0.39

the natural variance of sensitivity within a species excluding any problems of resistance. A similar type of variation could be demonstrated with tolclofos-methyl for strains of *Rhizoctonia solani* (Tab. 1.7). Tab. 1.8 demonstrates the large variability of 10 anastomosis groups of *Rhizoctonia solani* in the sensitivity to 11 fungicides (KATARIA et al. 1991).

LYR and CASPERSON (1982) explored a method to quantify similarities and dissimilarities in the spectra or activities of some fungicides. Those with the same

Table 1.8 *In vitro* activity of fungicides (EC90: mg a.i./l) against ten anastomosis groups of *Rhizoctonia solani* (KATARIA *et al.* 1991)

Fungicide	R. solani anastomosis groups									
	AG1	AG2-1	AG2-2	AG3	AG4	AG5	AG6	AG7	AG8	AG9
	EC90: mg a.i./l									
Carboxin	19.0	9.3	42.0	140	11.0	70.0	7.5	1.7	0.6	2.4
Furmecyclox	11.0	2.0	2.1	40.0	2.6	2.5	1.4	1.3	0.2	0.3
Tiabendazole	3.2	40.0	33.0	2.5	5.2	14.0	5.4	7.0	9.0	11.0
Pencycuron	>500	300	0.2	0.2	>500	>500	0.6	>500	>500	3.6
Tolclofos-methyl	<0.1	0.3	0.4	0.3	0.5	0.4	0.4	0.3	1.1	0.4
Iprodione	1.8	3.4	2.7	1.9	2.7	2.5	2.1	4.0	3.0	4.6
Vinclozolin	7.5	37.5	45.0	25.0	16.0	10.0	10.0	7.8	49.0	48.0
Cyproconazole	0.5	8.5	11.0	10.0	17.0	17.0	1.0	0.9	4.6	7.9
Fenpropimorph	6.0	2.2	5.0	4.7	30.0	40.0	4.0	5.1	22.0	7.0
Fenarimol	2.5	>500	>500	>500	>500	>500	170	14.0	>500	>500
Imazalil	23.0	>500	>500	400	>500	>500	110	>500	190	>500

L.S.D. ($P = 0.01$): Isolates 3.6; Fungicides 3.8; Isolate × fungicide interactions 12.0

mechanism of action have a greater similarity in their spectra than those with quite different mechanisms of action (for example: chloroneb and triadimefon).

Some fungicides have a specificity for taxonomical groups of fungi. Well known is the selectivity of carboxins for basidiomycetes (EDGINGTON and BARRON 1967), or that of benzimidazole fungicides for the *Porosporaceae* group within the Ascomycetes (EDGINGTON *et al.* 1971). Very often large differences in response to a fungicide occur within one fungal family or even within one genus.

This situation points to the fact that fungi are not only very different in their phylogeny, but have distinct differences in their biochemistry. Unfortunately, progress in the knowledge of the differential biochemistry of fungi (LYR 1977) is slow.

Basis of selectivity

The reasons for selectivity of modern fungicides are very poorly understood. From a theoretical point of view and from experimental data, it can be deduced that one or more of the following possibilities account for the differences in sensitivity of various fungi to a fungicide.

1. Differences in the accumulation of a fungicide in the cell
2. Different structures of the receptor or target systems
3. Differences in ability to toxify (activate) a compound
4. Differences in ability to detoxify a compound
5. Different degrees of importance of a receptor or target system for survival of the fungus.

Differences in the accumulation of fungicides in the fungal cell

Numerous investigations have revealed that there are a few cases of selective action based on differences in fungicide accumulation, although this has been very frequently claimed as an explanation of selectivity. This mechanism of selectivity has been suggested for cycloheximide (WESTCOTT and SISLER), pentachloronitrobenzene (NAKANISHI and OKU 1969), and some others compounds mentioned by RICHMOND (1977).

A new mechanism for avoiding fungicide accumulation in the fungal cell was described by DE WAARD and VAN NISTELROY (1982 and earlier papers). They demonstrated that some fungi have the ability to extrude fungicides, such as fenarimol, out of the cell by an active, energy dependent process whose mechanism is still unknown. This lowers the fungicide concentration in the cell and protects the fungus. It seems that such a mechanism is not widely distributed among fungi. No further examples have been described.

Different structures of the receptor or target

A prerequisite for the involvement of this principle as a basis for selectivity is the binding of a fungicide to a specific receptor. A well-known example is the selective action of carboxins and related compounds against basidiomycetes, which results from their binding to a specifically configured receptor at site II (see chapter 9). The exchange of a single amino acid (leucine for histidine) hinders the binding of the fungicide (BROOMFIELD and HARGREAVES 1992), probably by a change of the tertiary structure (configuration) of the binding site of the receptor. On the other hand, the introduction of a phenyl group at the ortho-position of the carboxin molecule (compound F427) causes a dramatic change of selectivity towards Ascomycetes and even Phycomycetes (EDGINGTON and BARRON 1967). This indicates that the receptor structures in various fungal groups differ, perhaps only in a few amino acid sequences.

The same seems true in the selectivity of benzimidazole fungicides of the carbendazim type. The exchange of one amino acid by mutation, (glutamic acid for glycin) prevents the binding of the fungicides to the tubulin structure and shifts the fungal response from sensitive to resistant (FUJIMURA et al. 1990; ISHII et al. 1992).

In contrast, phenylcarbamates act exclusively on highly benzimidazole resistant strains (SUMITOMO patent EP 51 871 ff., KATO et al. 1984). For details see chapter 15.

The most advanced description of the interaction of fungicides with a receptor exists for the DMI fungicides (chapters 13, 21). Here cytochrome P450 (perhaps a special species of the cytochrome P450 family) represents the receptor structure. In spite of targeting the same receptor, various members of the DMI fungicides differ remarkably in their spectrum of activity. This may be based on small differences in receptor structures of various fungal species (Gozzo et al. 1990). Details remain to be elucidated.

An interaction with specific receptors, although of a still unknown structure, can also be postulated for morpholine-, dicarboximide-, aromatic hydrocarbon-, strobilurin-, phenylpyrrole- and anilinopyrimidine-fungicides.

The target for strobilurins is well defined (see chapter 19), but the exact receptor structure remains to be elucidated. In most of the new fungicide groups quite new targets seem to be involved, but their nature is still unexplored.

Differences in the ability to toxify (activate) a compound

Some well known fungicides seem to be only prefungicides, which must be converted to the biochemically active toxicant. This was described by DE WAARD (1974) for pyrazophos (chapter 19), a rather selective fungicide, which was said to be converted to the real toxic products (PO-pyrazophos and PP). Although this conversion occurs in several fungi, it may be that Pyrazophos itself is the toxic principle which accumulates in high amounts in the cell, and the conversion products are only accompanying effects (LYR et al. 1993). Very surprising was the fact that triadimefon, a highly active DMI-fungicide must be enzymatically reduced to triadimenol, which is the actual fungicide (see chapter 13).

This phenomenon may occur more often than generally assumed, and can be a basis of selectivity.

Differences in the ability to detoxify a compound

Metabolic degradation can be a cause of selectivity for some compounds. The enzymes involved in detoxication are of different types, and not well described (LYR 1984).

Distribution of such enzymes and their level of activity in various species can be responsible for selective effects of some fungicides. Such enzymes are: oxido-reductases (flavin enzymes), esterases, phenol-oxidases, peroxidases, cytochrome P-450-monoxygenases, glutathione-S-aryltransferases, -acyltransferases, -reductases, and others. Also defense mechanisms within the cell can contribute to selective actions. Such mechanisms are: high superoxide dismutase levels, high glutathione reductase and -peroxidase activites within the cell.

The degree of saturation of membrane lipids or sterol content of the membranes also brings differences in the sensitivity for some fungicides (sensitivity to polyenic antibiotics, induced oxidative attacks).

A new interesting example are fungicides of the group of strobilurins (BAS 490 F) (cf. chapter 19). The receptor sensitivity seems to be similar in all organisms, but a high selectivity is caused by different degradation (and accumulation?) rates (KÖHLE et al. 1994).

Different degrees of importance of a receptor or target system

Very little is known at present about the natural function of some receptor or target systems and their importance for survival or fitness of a fungus. A variation among species regarding the essential role of such system or possible mechanisms of circumvention of a block in the metabolism can affect the selectivity of fungicides. However, more information is needed in this field.

This survey should give the impression that the phenomenon of fungicide selectivity is very complex in nature and that the knowledge of the biochemical processes involved, is still scanty. A better understanding of this phenomenon is not only of theoretical value but of great importance from a practical point of view e.g. for possibilities to enlarge the spectrum of activity of given antifungal substances. Advances in the chemical control of fungal diseases will depend to a high degree on a better understanding of the factors which control the selective action of a compound.

References

Anonymus: Sumitomo Patent EP 51 871.
BETH, H., DENZER, H., and SCHLÖSSER, H.: Antagonistische Wirkungen von *Trichoderma harzianum* auf einige phytopathogene Pilze. Mitt. Biol. Bundesanstalt f. Land- und Forstwirtsch. **223** (1984): 255.
BROOMFIELD, P. L. E., and HARGREAVES, J. A.: A single amino acid change in the iron-sulphur protein subunit of succinate dehydrogenase confers resistance to carboxin in *Ustilago maydis*. Curr. Genet. **22** (1992): 117–121.
DE WAARD, M. A.: Mechanism of action of the organophosphorus fungicide pyrazophos. Ph. D. Dissertation. Wageningen 1974.
— and VAN NISTELROY, J. G. M.: An energy dependent efflux mechanism for fenarimol in a wild-type strain and fenarimol resistant mutants of *Aspergillus nidulans*. Pesticide Biochem. Physiol. **13** (1980): 255–266.
EDGINGTON, L. V., and BARRON, G. L.: Fungitoxic spectrum of oxathiin compounds. Phytopathology **57** (1967): 1256–1257.
— KHEW, K. L., and BARRON, G. L.: Fungitoxic spectrum of benzimidazole compounds. Phytopathology **61** (1971): 42–44.
FUJIMURA, M., OEDA, K., INOUE, H., and KATO, T.: Mechanism of action of N-phenylcarbamates in benzimidazole-resistant *Neurospora* strains. In: GREEN, M. B., LEBARON, H. M., and MOBERG, K. (Eds.): Managing resistance to Agrochemicals. Americ. Chem. Soc. (1990): 224–236.
GOZZO, F., GARAVAGLIA, C., and MIRENNA, L.: Enantiomers and racemate of M 14360: Investigation on their antifungal activity. In: LYR, H., and POLTER, C. (Eds.): Systemic Fungicides and Antifungal Compounds, Tagungsbericht Nr. 291 Akademie Landwirtschaftswiss. DDR (Berlin) (1990): 87–94.
HINDORF, H., and KETTERER, N.: Einfluß von Stoffwechselprodukten pilzlicher Antagonisten auf *Sclerotium rolfsii* und *Verticillium fungicola*. Mitt. Biol. Bundesanstalt f. Land- und Forstwirtsch. **223** (1984): 258–259.
ISHII, H., VAN RAAK, M., INOUE, I., and TOMIKAWA, A.: Limitations in the exploitation of N-phenylcarbamates and N-phenylformamidoximes to control benzimidazole-resistant *Venturia nashicola* on Japanese pear. Plant Pathol. **41** (1992): 543–553.
KATARIA, H. R., VERMA, P. R., and GISI, U.: Variability in the sensitivity of *Rhizoctonia solani* anastomosis groups to fungicides. J. Phytopathol. **133** (1991): 121–133.
KATO, T., SUZUKI, K., TAKANASHI, J., and RAMOSHITA, K.: Negatively correlated cross resistance between benzimidazole fungicides and methyl-N-(3,5-dichlorophenyl)carbamate. J. Pesticide Sci. **9** (1984): 489–495.
KÖHLE, H., RADEMACHER, W., and RETZLAFF, G.: Wirkungsmechanismus von Strobilurinen bei

Pilzen und Pflanzen. Mitt. Biol. Bundesanstalt **301**, 49. Deutsche Pflanzenschutztagung Heidelberg (1994): 396.

LEINHOS, G., and BUCHENAUER, H.: Untersuchungen zu Antagonismus und Hyperparasitismus einiger ausgewählter Pilze gegenüber Getreiderostpilzen. Mitt. Biol. Bundesanstalt f. Land- u. Forstwirtsch. **223** (1984): 255.

LYR, H.: Effect of fungicides on energy production and intermediary metabolism. In: SIEGEL, M. R., and SISLER, H. D. (Eds.): Antifungal Compounds. Vol. 2. Marcel Dekker Inc., New York 1977, pp. 301–332.

LYR, H.: Biochemical mechanisms of fungi for toxification and detoxification of fungicides. In: Tagungsbericht Akad. Landwirtschaftswiss. DDR **222**, 1984.

— and CASPERSON, G.: On the mechanism of action and the phenomenon of cross resistance of aromatic hydrocarbon fungicides and dicarboximide fungicides. Acta Phytopath. Acad. Sci. Hungary **17** (1982): 317–326.

— POLTER, C., and BRAUN, P.: The mechanism of action of pyrazophos – a reinvestigation. In: LYR, H., and POLTER, C. (Eds.): Proceed. 10th Intern. Symposium of Systemic Fungicides and Antifungal Compounds. Ulmer Verlag, Stuttgart 1993, pp. 141–149.

MERCIER, J., and REEFELDER, R.: A method for screening phylloplane antagonists to *Sclerotinia sclerotiorum* on lettuce. Phytopathology **74** (1984): 804 and personal communication 1984.

NAKANISHI, T., and OKU, H.: Metabolism and accumulation of pentachloronitrobenzene by phytopathogenic fungi in relation to selectice toxicity. Phytopathology **59** (1969): 1761–1762.

RICHMOND, D. V.: Permeation and migration of fungicides in fungal cells. In: SIEGEL, M. R., and SISLER, H. D. (Eds.): Antifugal Compounds. Vol. 2. Marcel Dekker Incorp., New York 1977, pp. 251–276.

STENZEL, K.: Untersuchungen zur verminderten Ausbreitung Echter Mehltaupilze auf induziert resistenten Pflanzen. Mitt. Biol. Bundenanstalt f. Land- u. Forstwirtsch. **223** (1984): 238.

TORGESON, D. C.: Fungicides, an advanced treatise. Vol. II, Academic Press Inc., New York – London 1969.

WESTCOTT, E. W., and SISLER, H. D.: Uptake of cycloheximide by a sensitive and a resistant yeast. Phytopathology **54** (1964): 1261–1264.

ZOLLFRANK, G., and LYR, H.: Die Variabilität der Metalaxyl-Empfindlichkeit bei verschiedenen Pathotypen von *Phytophthora infestans*. Arch. Phytopath. u. Pflanzenschutz **23** (1987), 351–356.

Chapter 2

Development of resistance to modern fungicides and strategies for its avoidance

J. Dekker

Wageningen Agricultural University, The Netherlands

Introduction

Like many other living organisms, fungi may become resistant to toxicants. In a population of a fungal plant pathogen, sensitive to a particular fungicide, cells may emerge by mutation or otherwise which are significantly less sensitive. The frequency of mutation towards resistance, usually between $10^{-4}-10^{-10}$, may increase considerably by exposure of the fungus to mutagenic agents. Administration of the fungicide concerned will favour the resistant cells by elimination of competition from the sensitive cells. The fungicide itself does not induce resistance, but acts as a selecting agent. If a particular fungicide would appear to be also mutagenic, it should not be used on agricultural crops which are grown for consumption. When under selection pressure by the fungicide build-up of a resistant pathogen population occurs, it may result in failure of disease control. This has happened many times, especially after the introduction in the nineteen sixties of specific-site fungicides, among which were many systemic fungicides. In a review, compiled in 1979 (Ogawa et al. 1983), more than 100 plant pathogen species are listed, which have become resistant to fungicides under field conditions, a number which has grown considerably in recent years. Failure of disease control due to acquired resistance has been reported, among others, for acylalanines, benzimidazole fungicides, thiophanates, carboxanilides, hydroxypyrimidines, organophosphates (see chapters 14, 16, 18, 19) and for antibiotics such as kasugamicin, used against *Pyricularia oryzae* on rice (Miura et al. 1975) and polyoxin B, used against *Alternaria kikuchiana* on Japanese pear (Nishimura et al. 1973). In other cases a moderate decrease in sensitivity has been observed without a rapid loss of disease control, e.g. with some inhibitors of sterol biosynthesis (triazoles, pyrimidines, imidazoles) and with dicarboximides, which will be discussed below (see also chapter 12, 6).

Development of fungicide-resistance is now one of the major problems in plant disease control. It causes unexpected crop losses for the grower, and may put him in a difficult position if no adequate substitutes are available. It may reduce the profits of the manufacturer, which has developed the fungicide at high cost, and may even lead to costly lawsuits. Should this occur repeatedly, the agrochemical industry might become hesitant to invest in the costly development of modern fungicides, which in the long run would reduce the arsenal of available disease control agents, again to the disadvantage of the farmer. The development of resistance may further have consequences for the extension officers, the regulatory authorities, and in some cases even for the consumer and the national economy (Schwinn 1982).

The problems with fungicide-resistance have given impetus to genetic, biochemical and epidemiological studies, in order to clarify the origin of resistance and to understand the factors which govern the build-up of a resistant pathogen population in the field. In this chapter the principles underlying the fungicide-resistance phenomenon will be treated, illustrated by examples. The genetic aspects are discussed in chapter 3 by GEORGOPOULOS.

Attention will further be paid to strategies to delay or avoid the occurrence of fungicide-resistance. No attempt has been made to prepare an exhaustive review of resistance to all fungicides concerned. For additional information various reviews are available, among others by GEORGOPOULOS (1977), OGAWA et al. (1983), DEKKER (1985) and a book on "Fungicide-resistance in crop protection" by various authors, edited by DEKKER and GEORGOPOULOS (1982).

Definitions

Fungicide-resistance may be defined as the stable, inheritable adjustment by a fungus to a fungicide, resulting in a less than normal sensitivity to that fungicide. The term is used for strains of a sensitive fungal species, which have become, usually by mutation, significantly less sentitive to the fungicide.

Resistance to fungicides with specific-site action is often easily obtained on agar, containing a fungicide concentration, which is lethal to the wild-type fungus. However, the appearance of so-called "laboratory-resistance" does not necessarily imply that resistance problems will arise in the field. This will depend on the level and frequency of resistance. Further, it is obvious that failure of disease control should not be easily attributed to resistance before appropriate tests have shown that it is indeed caused by the presence of resistant strains.

Quite often the term **tolerance** is used instead of or interchangeably with the term resistance, and sometimes also in addition to resistance to indicate a less severe type of resistance. The latter view has no grammatical basis. Moreover, the concept tolerance is already used in another way, namely as the maximum amount of a pesticide residue that may lawfully remain on or in food. In view of this the FAO panel of experts on pest resistance to pesticides, meeting in 1978, "agreed that resistance should continue to apply to hereditable resistance in fungi and bacteria, and recommended that the word tolerance should not be used, since it may be ambiguous" (Anonymous 1979).

Some plant breeders prefer the term insensitivity instead of resistance, in order to avoid confusion with resistance of the plant to diseases. A disadvantage of the word insensitivity, however, is that it suggests a complete loss of sensitivity, which will seldom occur.

The level of resistance, or **resistance factor**, may be expressed as the ratio: ED_{50} or EC_{50} resistant strain/ED_{50} or EC_{50} wild-type fungus. This term should not be confused with frequency of resistance, which is the percentage of resistant isolates in the pathogen population.

The term (positive) **cross-resistance** designates resistance to two or more toxicants, mediated by the same genetic factor. When such a factor mediates resistance to one fungicide and at the same time increases sensitivity to a second fungicide, the term negative cross-resistance is used. Cross-resistance should not be confused with multiple resistance, which means resistance to two or more toxicants, mediated by different genetic factors.

Mechanisms of resistance

General

Nowadays several fungicides are known which interfere at specific sites with biosynthetic processes in fungi (e.g. synthesis of nucleic acids, proteins, ergosterol, chitins), with respiration, membrane structure or nuclear function. Development of resistance to such specific-site inhibitors is often due to single gene mutation, resulting in a slightly changed target site, with reduced affinity to the fungicide. It is obvious that resistance to multi-site inhibitors, comprising most conventional fungicides cannot be achieved in this way, as it would require simultaneous changes at many sites. This does not imply, however, that reduction of sensitivity to these fungicides will not occur. It may happen when a change in the fungal cell occurs, which prevents the fungicide from reaching the site of action. This may happen by binding at other places, conversion into non fungicidal compounds, or reduced uptake. Experience during a century of fungicide use in agriculture has learned, however, that this does not occur easily. In fact, with conventional fungicides only rarely resistance problems arose.

The mechanism of resistance is not only of scientific interest, but also of practical significance, as it may influence the fitness of resistant strains. This will be discussed on page 28. Resistance to specific-site fungicides is mostly caused by a change at the site of action or by reduced uptake, whereas with insecticides conversion into an inactive compound is the predominant mechanism.

Change at site of action

There is an increasing number of fungicides, where resistance can be attributed to a change at the site of action, which reduces its affinity to the fungicide. This may be illustrated by a few examples. For more extensive information the reader is referred to a review by DEKKER (1985) and other chapters in this book.

Carbendazim, the toxic principle of benomyl and thiophanate-methyl, binds to tubulin, the major constituent of microtubules, which constitute the spindle. The assembly of microtubules is prevented, and as a consequence mitosis and other cellular processes in which microtubules are involved, are inhibited. Resistance to carbendazim is caused by single gene mutation, resulting in slightly changed tubulin with a reduced affinity to carbendazim (DAVIDSE 1982, see also chapter 15).

The antibiotic kasugamycin, used for control of rice blast, inhibits protein synthesis in some bacteria and fungi by attaching to one of the subunits of the ribosome. After several years of application in Japan, failure of disease control occurred, due to development of resistance. This appeared to be caused by single mutation in each of three different loci for resistance in the causal organism, *Pyricularia oryzae* (perfect stage *Magnaporthe grisea*) (TAGA et al. 1978). Studies with cell free systems learned that resistance was due to a change in the ribosome, resulting in reduced affinity to the antiobiotic (MISATO and KO 1975).

Also metalaxyl, an acylalanine active against Oomycetes, has met serious resistance problems in the field. In studies with *Phytophthora megasperma* f. sp. *medicaginis* DAVIDSE et al. (1983) showed that metalaxyl inhibits RNA synthesis by specific interference with template bound RNA polymerase activity. Endogenous polymerase activity of nuclei isolated from a metalaxyl resistant mutant was not inhibited by the fungicide, which indicates a modification of the target site.

Changes in the respiratory chain may be responsible for resistance to fungicides, which specifically act at certain steps in this chain. Carboxin, a carboxamide and mainly active against Basidiomycetes, inhibits fungal respiration by specific binding to the succinate-ubiquinone reductase complex (complex II), thus blocking the electron flow (MOWERY et al. 1977). In laboratory experiments carboxin resistant mutants of *Ustilago* spp. emerged readily. Evidence has been obtained that at least some of these mutants have a slightly changed complex II, resulting in a decreased afficity for carboxin (GEORGOPOULOS 1982b).

Reduced uptake, detoxification

The fungal cell may become less sensitive to a fungicide by changes, which keep the fungicide from reaching the site of action in sufficient quantity. These changes may hamper entrance by the fungicide through the membrane, or they may lead to an increased efflux immediately after entrance, preventing accumulation. Further there may be changes which increase the capacity of the cell to detoxify the fungicide. This may happen by conversion of the fungicide into non fungitoxic compounds or by binding to other cell constituents before the sites of action have been reached. A few examples may illustrate this.

Polyoxin antibiotics, which interfere with chitin synthesis in fungi, are used for control of black spot in Japanese pear. Strains of *Alternaria kikuchiana*, resistant to polyoxin B, appeared in orchards treated with the antibiotic, resulting in failure of disease control. Resistance was not due to a change at the target site, chitin synthetase, as this enzyme was equally inhibited in cell free systems of resistant and sensitive strains, but it appeared to be caused by a change in the fungal membrane, resulting in reduced uptake (MISATO et al. 1977). Lack of accumulation may also be due to the presence of a constitutive, energy dependent efflux mechanism, as has been shown in studies with fenarimol and *Aspergillus nidulans* (DE WAARD and VAN NISTELROOY 1980a) (cf. chapter 26).

Fungicide-resistance due to conversion into non fungitoxicants seems to occur only rarely. Kitazin-P (5-benzyl 0,0-diisopropyl phosphorothiolate), which among others is used for control of rice blast, interferes with the biosynthesis of phosphatidylcholine. Strains of the causal organism, *Pyricularia oryzae*, were obtained with moderate, others with a high level of resistance to this fungicide. The resistance mechanism of the latter has not yet been elucidated, but that of the former appears to be due to cleavage of the S—C bond of the molecule by the pathogen, which gives non-fungitoxic derivatives (UESUGI and SISLER 1978). It is interesting that resistance can also be based on a lack of conversion of an itself non-fungitoxic compound into a fungicide. Resistance to pyrazophos in *P. oryzae* due to the inability of resistant strains to convert pyrazophos into the actual toxic principle: 2-hydroxy-5-methyl-6-ethoxy-carbonylpyrazolo(1,5a)pyrimidine (DE WAARD and VAN NISTELROOY 1980b).

Other resistance mechanisms

In principle, resistance could be caused also by metabolic changes in the fungal cell which result in a compensation for the inhibiting effect, for example by an increase in the production of an inhibited enzyme, or by changes which result in circumvention of the blocked site by an alternative pathway. No clear cut examples of such a mechanism seem available.

Build-up of a resistant pathogen population

General

The emergence of fungicide-resistant strains on agar medium in laboratory experiments does not necessarily mean that failure of disease control has to be expected after application of this fungicide in the field. In the first place the frequency of resistant strains may stay low, because their properties and the environmental conditions are not conducive to the build-up of a resistant pathogen population, and, secondly, the level of resistance may stay relatively moderate, so that the fungicide continues to provide control. Failure of disease control will not arise before a considerable proportion of the pathogen population has become resistant. In some cases this happened shortly after introduction, e.g. with benomyl and metalaxyl, but in other cases it took many years, e.g. Kitazin-P resistance in *P. oryzae* (UESUGI 1982) and dodine resistance in *Venturia inaequalis* (GILPATRICK 1982). Factors which influence the speed of build-up of a resistant pathogen population are the genetic basis of resistance, the fitness of resistant strains in the presence or absence of the fungicide, the nature of the pathogen and the disease, the selection pressure by the fungicide and others.

Genetic basis of resistance

Fungicide resistance may be monogenic or polygenic. By the former a high level of resistance may be achieved in only one step. In a particular fungus this may happen at more than one locus, but there is no positive interaction between mutant genes at different loci. High selection pressure by the fungicide may then cause a rapid increase of the frequency of resistant cells, and resistance problems in practice may arise within one or only a few seasons, as occurred e.g. with benomyl and metalaxyl.

When fungicide resistance is polygenic, many mutant genes may contribute to resistance, with each of them responsible for only a small resistance step. In such a case build-up of resistance to a dangerous level will take more time, and may last several of even many years. An example is the resistance to fungicides which inhibit biosynthesis of sterols (s.b.i.'s). Although such compounds are considered to belong to the low resistance category, build-up of resistance still may occur after a period of continuous high selection pressure by the fungicide. This was shown in laboratory experiments with *Penicillium italicum*, where the resistance level to imazalil could be increased, step by step, with a factor of almost 200 (DE WAARD and VAN NISTELROOY 1990). This may, under continued selection pressure, also cause control problems with s.b.i.'s in practice. Moreover it has been reported recently that with an s.b.i., terbinafine, also monogenic resistance may occur (LASSERON-DE FALANDRE et al. 1991).

Fitness of resistant strains

Fitness is a comparative concept: a strain may be more or less fit than another one in a particular situation. It is further a rather complex property with many parameters involved, including infection chance, speed of colonization of the host tissue and degree of sporulation. A change in the fungal cell, which is responsible for resistance, may appear to be disadvantageous in absence of the fungicide. In this way there may be a link between resistance and reduced fitness. This may be illustrated by studies with the antibiotic pimaricin, an experimental fungicide active against bulb rot in narcissus, caused by *Fusarium oxysporum* f. sp. *narcissi* (DEKKER and GIELINK 1979a). This fungicide complexes with ergosterol in the fungal membrane, thus causing leakage and cell death. In mutants of yeasts with resistance to pimaricin or the closely related nystatin, ergosterol appeared to be replaced by one of its precursors. This resulted in two effects: resistance, because of lower affinity of this precursor to pimaricin, and at the same time to reduced fitness, because the more primitive precursor functions less well in the membrane stabilization than ergosterol.

There are indications that a link between resistance and reduced fitness might exist also for certain commercial fungicides, such as fenarimol (DE WAARD and VAN NISTELROOY 1982). Reduced fitness of fenarimol resistant strains can possibly be explained by the assumption that the constant presence of an energy dependent efflux mechanism is an extra burden for the cell, which is disadvantagenous in absence of the fungicide. Reduced fitness of resistant strains has also been reported for other sterol biosynthesis inhibitors, such as triforine (FUCHS et al. 1977). Another example, may be, is dicarboximide resistance in *Botrytis cinerea*. From vineyards, attacked by grey mold, some strains of *B. cinerea* were isolated with a low level and others with a higher level of resistance. The latter strains appeared to possess a higher osmotic sensitivity than the wild-type pathogen. This might be the cause of reduced virulence of such strains in view of the high sugar content of the grapes (BEEVER and BYRDE 1982) (see also chapter 7). Decreased fitness has further been reported for ethirimol resistant strains of *Erysiphe graminis* (SHEPHARD et al. 1975) and pyrazophos resistant strains of *Sphaerotheca fuliginea* (DEKKER and GIELINK 1979b). It is obvious that reduced fitness will hamper the build-up of a resistant pathogen population, because the less fit strains will lose ground to the sensitive pathogen population in absence of the fungicide, e.g. on plant parts not hit by the fungicide, or at the end of the spray-interval. Moreover, when resistance is inversely related to fitness, the level of resistance may be limited. This may also have consequences for disease control.

There are, however, also fungicides where resistance is not linked to reduced fitness. This means that also resistant strains may emerge with a fitness which does not significantly differ from that of sensitive strains. Examples are benzimidazoles and acylalanine fungicides. In such cases there are, apparently, changes possible at the sites of action (namely the tubulin monomers of the microtubules and the RNA polymerase-template complex, respectively) which are not disadvantageous to the cell. The emergence of such strains will favour the shift towards a resistant pathogen population.

It may be concluded that the fitness of fungicide-resistant strains seems related to the mechanism by which the pathogen becomes resistant to that fungicide. Some metabolic changes, conferring resistance, are linked to lower fitness in absence of the fungicide, others are not. On this basis some fungicides seem more risky with respect to development of resistance than others. However, the possibility may exist that strains, in which originally resistance seems linked to lower fitness in the absence of the fungicide, may increase their fitness in the course of time. It should further be stressed that development of resistance in the field will depend also on other factors.

Nature of pathogen and disease

Resistance will build up more rapidly in an abundantly sporulating pathogen on aerial plant parts than in a pathogen which sporulates scarcely and spreads slowly, such as with certain soil born root or foot diseases. For example, resistance to metalaxyl developed rapidly in *Phytophthora infestans* in potatoes (DAVIDSE et al. 1981) so that control of this disease by the single compound even had to be abandoned, but problems with control of various *Phytophthora* root diseases in other crops, seem to occur less readily. However, also here problems may arise. Indications were obtained that reduced control of *Phytophthora cinnamomi* in avocado by metalaxyl was due to development of fungicide-resistance (KOTZÉ 1983).

The influence which the fungal life cycle may have was shown in apple orchards, treated with benomyl for control of scab and rust. After some time failure of scab control occurred but there were no problems with control of rust. The explanation may be that *Venturia inaequalis* does have a repetitive summer cycle on apple, which favours build-up of resistance, but *Gymnosporangium juniperi viriginianae* does not (GILPATRICK 1982).

Thus in calculating the risk for development of resistance, also the type of disease should be taken into consideration.

Selection pressure by fungicide

The degree of selection pressure will mainly be determined by the doses applied, the frequency of application, the persistence of the fungicide on the crop or in the soil, and by the thoroughness of the treatment. Also the method of application may play a role, as, for example, treatment of the seed or the soil by a systemic compound may make a long lasting uptake of the chemical possible. Further, the size of the area treated with a particular fungicide may contribute to selection pressure, since this influences the influx of sensitive strains from outside. Insight in these factors is important for the design of counter measures to avoid resistance (DEKKER 1982).

Environmental and other factors

In addition to the above-mentioned factors, there may be other phenomena, which influence the build-up of a resistant pathogen population. SAMOUCHA and COHEN (1984) made the interesting observation that synergism occurs between metalaxyl-sensitive and -resistant strains of *Pseudoperonospora cubensis*. They found that the sensitive strains stimulated the release of zoospores and the infection by the resistant strains. WILD and ECKERT (1982) found synergism between a benzimidazole-sensitive and benzimidazole-resistant isolate of *Penicillium digitatum*. The former, although unable to infect citrus fruit in the presence of benomyl, germinated and produced pectolytic enzymes. This caused breakdown of cell wall material, followed by release of nutrients, which increased infection by resistant isolates.

Further, environmental conditions, among others the weather, may play a role. For example, failure of disease control due to resistance will sooner occur under conditions which favour the outbreak of an epidemic.

Level of resistance in relation to disease control

In addition to the proportion of resistant strains in the pathogen population, also the level of resistance is important. If it is only low or moderate, fungicide applications may continue to provide satisfactory disease control, even when the majority of the pathogen population has become less sensitive to the fungicide. It is obvious that the level of resistance will be limited in those cases where an inverse relation exists between resistance and fitness as has been suggested for resistance to dicarboximides. Although resistant strains of *Botrytis cinerea* have been found in many areas, where dicarboximides had been frequently used for several years, loss of field control has not become a major field problem (POMMER and LORENZ 1982). They attributed this to a loss of vigour in dicarboximide resistant strains. LORENZ and EICHHORN (1982) found that the proportion of resistant strains increases during periods of high selection pressure, but decreases again in between two growing seasons. Lower fitness of resistant strains, however, is not a guarantee that no problems will arise in the future. Under continuous selection pressure by the fungicide there may be an evolution towards higher fitness among resistant strains, by acquiring the ability to compensate for the reduced fitness in some way or another. Recently indeed there have already been reports about reduced disease control, due to decreased dicarboximide sensitivity (LEROUX and BESSELAT 1984). Further there may be exceptions to the rule that high resistance is connected with lower fitness, as shown by GRINDLE and TEMPLE (1985) for one out of a number of vinclozolin-resistant isolates of the non-pathogen *Neurospora crassa*. However, comparable variants of fungal pathogens have, so far, not been recovered from field populations. If they would occur, they might compete with the normal wild type pathogen and become a problem in practice.

Reduced fitness of resistant strains has also been reported for sterol biosynthesis inhibitors as fenarimol, triforine and triadimefon. This may explain why the use of triadimefon against barley powdery mildew in the U.K has not resulted in clear cut failure of disease control, in spite of a considerable increase in the frequency of less sensitive strains (WOLFE et al. 1984). Several surveys are being carried out to investigate whether the level of resistance to sterol biosynthesis inhibitors increases in the course of the years. An example is the one carried out from 1982–1984 in commercial cucumber greenhouses in The Netherlands, where several biosynthesis inhibitors have been in use for powdery mildew control during a number of years (SCHEPERS 1985). Although a significant shift towards fungicide resistance was observed with fenarimol, imazalil and triforine, complete failure of disease control did not yet occur. It still has to be awaited whether the level of resistance will continue to rise in the years to come.

Tactics to avoid development of resistance

General

On the basis of the knowledge obtained, measures may be taken to counteract the development of resistance. This should be done before a build-up of resistance has taken place. Once the pathogen population has become resistant, the only possibility is to change to other fungicides with a different mechanism of action, or to non-chemical control measures, if available.

Firstly, it is desirable to obtain some information about the resistance risks of a new fungicide. When a fungicide has to be used which is resistance prone, it is imperative to take stringent measures to avoid or at least delay the development of resistance. In cases where resistance may reach a high level beyond control by the fungicide, a reduction of selection pressure by the fungicide concerned should be considered, when necessary and possible in combination or alternation with other fungicides. On the other hand, when only a moderate level of resistance is likely to occur, an increase of the selection pressure, for as far technically and economically feasible, may be considered in order to kill also the moderate-level resistant strains. However, monitoring is desirable to keep track of the level of resistance, which may increase in the course of years, and the frequency of resistance.

For more information on tactics to avoid resistance the reader is referred to DEKKER (1982), DELP (1980), STAUB and SOZZI (1984).

Prediction

Experiments on an artificial medium, with or without mutagenic agents, may inform us whether emergence of resistant cells is possible, by mutation or otherwise. When no resistant mutants emerge in this way, in the pathogen concerned or in related species, it is unlikely that they will appear in the field. But if they do, or when they are already present at low frequency in a wild-type population, their fitness should be tested, in vitro and on the plant. This information may provide an indication of the resistance risk of the fungicide concerned. Further, genetic studies may reveal whether monogenic or polygenic resistance is involved, which is also important for risk estimation. Nevertheless it will not be possible to give an exact prediction of what is going to happen in practice. Conditions in the greenhouse are never exactly the same as in the field, and even field experiments may not yield the results that can be obtained in large-scale application in practice, as also the size of the area may play a role. From experience it is known that fungicides, classified in the low-risk category, still may give resistance problems after many years of use. Therefore, after introduction of a new chemical, detection and monitoring of resistance development should be advised or lcast considered.

Detection and monitoring

In order to detect build-up of resistance in an early stage, so that counter measures might still be possible, it is often advised to monitor for resistance in the field. In view of this, data should be available on the sensitivity of the wild-type pathogen population before the introduction of the fungicide in the field. However, if development of resistance occurs very fast, the information may come too late for adequate counter measures. Timely detection of resistance might offer perspectives in those cases, where build-up of resistance occurs only slowly, and requires more than one growing season.

Another difficulty is that detection of resistance in an early stage, at the 1% level, would require a fairly large scale and costly sampling procedure. It is easier and less costly to monitor for performance of the fungicide, since in that case only few samples have to be analyzed from spots where failure of disease control is suspected to be caused by resistance. Although it will then be too late for counter measures in the field concerned, the information will provide a warning to take measures in other fields.

Reduction of selection pressure

It should be clear that a continuous selection pressure by one particular fungicide or by fungicides which show cross-resistance, will increase the chance for build-up of resistance, if not on short notice, than anyway in the long run. It should be avoided to carry out unnecessary applications and to apply larger quantities of the chemical than needed, not only to avoid resistance, but also for economical and environmental reasons. It is not advisable to use the same type of chemical for treatment of seed or plant material, spraying of the crop and post harvest treatment, since development of resistance in the earlier treatments may jeopardize the effect of the later applications. Also the treatment of a crop in a whole region or country with the same type of chemical should be avoided, as it increases the selection pressure. It should further be realized that also a very thorough treatment of the crop will increase the selection pressure and therefore favour the build-up of resistance. It will be obvious, however, that the possibilities to reduce the selection pressure, without use of other fungicides, are limited.

Combination and alternation of fungicides

A prerequisite for the use of two or more fungicides to avoid resistance, is that these fungicides do not show cross-resistance. In a mixture the companion should preferably be a conventional and anyway a low risk fungicide, as a combination of two risky fungicides may lead to multiple resistance. Considering the usually extremely large populations of fungal cells, this may happen rather easily (WILD 1983).

With respect to the effect of a **mixture** on development of resistance, there is much difference of opinion in the literature. Some authors claim that the use of a mixture counteracts resistance, but others do not observe a delay in the build-up of resistance. Resistance to the risky chemical in such a mixture may build-up in that part of the pathogen population which has not been killed by the companion compound. The speed with which this occurs depends on various factors, among which, according to a model by KABLE and JEFFERY (1980), the "escape" is the single most important factor. When there are only few fungal cells which escape from being hit by the mixture and when selection pressure is continuous, the speed of build-up of resistance of the risky compound, expressed in percentage of the pathogen population, will be the same for the mixture as when the risky compound alone is applied. In that case resistant strains face hardly any competition from sensitive strains during spray intervals. Such a situation will not likely occur in the field, and in most cases a mixture will at least delay the build-up of resistance. When the duration of decreased selection pressure between treatments, and reduced fitness of resistant strains allow a sufficient drop in the proportion of resistant cells, resistance problems might even be avoided indefinitely.

With **alternating use** of a risky and a non risky fungicide, there will only be selection pressure towards resistance during the periods that the risky chemical is present in the plants. During the periods that only the non risky chemical is present, the proportion of resistant cells may remain the same, or it may decrease, depending on the fitness of resistant strains in absence of the fungicide at risk. If it remains the same, and increases only in presence of the risky compound, there will be a build-up of resistance in steps, which eventually may result in failure of disease control. If the proportion of resistant cells decreases during periods that the risky compound is absent, the overall step-wise increase of resistance will be slower than in the former case, or not happen at all. It is not possible to make a general statement, whether combination or alternation is a better tactic to avoid resistance. The mixture may be more effective under certain conditions and the alternations under other conditions, depending on the values of the parameters involved.

In addition to combinations or simple alternations more complex application sequences may be followed. A mixture has the disadvantage that the compound at risk is continuously applied, and with alternation the application of the non risky compound is interrupted for no good reason. Better results might therefore be obtained by a sequential scheme in which mixtures and rotation are combined in such a way that the non risky compound is constantly present and that only the application of the risky compound is interrupted. Calculations, based on the model by KABLE and JEFFERY, support this assumption (Tab. 3.1).

Table 3.1 The effect of combined or alternating use of two fungicides on buildup of a resistant pathogen population. Stimulation model, assuming specific values for spray coverage[1]), efficacies[2]) and initial resistance level[3]) (DEKKER 1982, modified after KABLE and JEFFERY 1980)

Fungicide	Percentage of population resistant after 5—40 sprayings				
S − S − S − S	0.0	82.6	100.0	100.0	100.0
(S + S) − (S + C) − (S + C) − (S + C)	0.0	0.0	99.6	100.0	100.0
S − C − S − C	0.0	0.0	82.6	100.0	100.0
(S + C) − C − (S + C) − C	0.0	0.0	0.0	26.1	99.6

[1]) E (escape) = proportion of population escaping fungicide contact, set at 5%.
[2]) S (systemic fungicide) with efficacy against sensitive and resistant subpopulations set at 90% and 10%, respectively.
C (conventional fungicide) with efficacy against both S-sensitive and S-resistant subpopulations set at 80%.
[3]) Initial resistance frequency set at 10^{-9}.

In addition to the valuable but rather limited model of KABLE and JEFFERY, other authors have presented models with more parameters (SKYLAKALIS 1982; LEVY et al. 1983). Although these models cannot yet be used for application in practice, they certainly help to increase our knowledge about the influence which particular parameters have on the build-up of resistance.

Integrated control

For avoidance of resistance a combination of chemical with non-chemical methods may be considered, e.g. natural host resistance, cultural measures, biological control. WOLFE (1984) advocates an approach in which different fungicides are used in combination with several varieties, where resistance is based on different resistance genes. Exerting selection pressure on the pathogen in repeatedly changing directions should 'confuse' the pathogen.

Long term strategies

Availability of a broad fungicide arsenal

To be able to avoid the development of fungicide resistance in the future it is important that a broad arsenal of fungicides is available, so that flexibility exists in the design of counter measures. Therefore the industry should continue to search for and develop new fungicides, especially fungicides with yet unknown mechanisms of action. Special attention should be given to site-specific inhibitors that show a low risk for development of resistance. In view of this the search for new disease control agents should include chemicals which are not fungitoxic in itself, but which interfere in other ways with the relation between plant and parasite. Such compounds may induce or increase the natural resistance of the host, or may decrease the capability of the pathogen to attack the plant. The compounds which have shown activity along these lines have been reviewed earlier (DEKKER 1983; WADE 1984). Some of these might not or might less readily encounter resistance. In addition to strengthening our research efforts to discover new chemicals, we must ensure that the number of conventional fungicides is not needlessly decreased by regulatory agencies. They may be needed to fall back upon in case resistance to specific-site fungicides occurs, and they may be used in combinations or rotations to avoid resistance.

Negative cross-resistance

A quite interesting and potentially promising phenomenon is negatively correlated cross-resistance between two fungicides. When one fungicide is especially active against strains which have become resistant to a second fungicide, and when this second fungicide is more than normally active against strains resistant to the first fungicide, the development of resistance will in principle be precluded. However, this will only be successful when negative cross-resistance holds for all strains, and when no other resistance mechanisms become apparent. Negative cross-resistance has been observed for benomyl and thiabendazole (VAN TUYL et al. 1974), for some carboxamide fungicides (GEORGOPOULOS 1982a) and for phosphoramidate and phosphorothiolate fungicides (UESUGI 1982). The problem with these cases of negative cross-resistance was that it did not hold for all resistant strains. Remarkable is the negative cross-resistance, shown by *Botrytis cinerea* and three other plant pathogens for benomyl and methyl N-(3,5-dichlorophenol) carbamate (MDPC) which seems to hold for all resistant isolates (KATO et al. 1984). In experiments with *Venturia nashicola*, however, ISHII et al. (1984) found negative cross-resistance to these compounds only for the highly carbendazim resistant strains, but not for the strains with intermediate or low carbendazim resistance. ROSENBERGER and MEYER (1985) observed that benomyl-resistant strains of *Penicillium expansum* were more sensitive to diphenylamine than the wild type pathogen, and that this effect was temperature dependent.

Negative cross-resistance has further been reported for two sterol biosynthesis inhibitors. Fenarimol-resistant mutants of *Penicillium italicum*, which cause post harvest rot in citrus, appeared more sensitive to fenpropimorph than the wild type strain (DE WAARD and VAN NISTELROOY 1982), but negative cross-resistance with respect to these compounds was not observed for *Aspergillus nidulans*. The use of negative cross-resistance as a tool to avoid resistance should further be explored (DE WAARD 1984).

Synergism

Several reports mention synergistic action between two fungicides. This phenomenon might appear to have also value for avoidance of resistance if the second fungicide interferes with the resistance mechanism towards the first fungicide. As an example the synergism between fenarimol and captan might be mentioned. As has been mentioned above, resistance to fenarimol depends on an energy-dependent efflux. This efflux is inhibited by addition of captan or other compounds, which exert an inhibitory effect on respiration (DE WAARD 1984). It is desirable that such possibilities of the phenomenon of synergism are further investigated (see chapter 24).

Implementation

Any long-term strategy should include the creation of possibilities for the implementation of tactics to prevent or manage fungicide resistance. For this reason it is necessary to establish and maintain an efficient communication system among growers, extension officers, teachers, research workers, manufacturers, salesmen, the press, regulatory agencies and the government.

Concluding remarks

Fungicide-resistance has become one of the major problems for control of fungal plant diseases. Tactics to prolong the life of badly needed fungicides and long term strategies to ensure efficient disease control are therefore much necessary. For the design of such tactics and strategies knowledge about the underlying genetical and biochemical principles of the fungicide-resistance phenomenon is needed, which can only be obtained by research. Equally important is insight in the behaviour of resistant strains in the field, to which the use of epidemiological models may contribute.

A prerequisite for the design of tactics to avoid resistance is the availability of a varied aresenal of fungicides, conventional as well as systemic fungucides, which act at different sites in the fungal cell, and against which different mechanisms of resistance operate. This requires continued research efforts and a prudent and conservative policy by regulatory agencies.

References

Anonymus: Pest resistance to pesticides and crop loss assessment. Report on the 2nd session of the FAO panel of experts, held in Rome, 28 August – 1 September 1979. FAO Plant Product. and Protect. Paper **6**, 2 (1979): 1–41.

BEEVER, R. E., and BYRDE, R. J. W.: Resistance to the dicarboximide fungicides. In: DEKKER, J., and GEORGOPOULOS, S. G. (Eds.): Fungicide Resistance in Crop Protection. Pudoc, Wageningen 1982, pp. 101–117.

DAVIDSE, L. C.: Benzimidazole compounds: selectivity and resistance. In: DEKKER, J., and GEORGOPOULOS, S. G. (Eds.): Fungicide Resistance in Crop Protection. Pudoc, Wageningen 1982, pp. 66–70.

– HOFMAN, A. E., and VELTHUIS, G. C. M.: Specific interference of metalaxyl with endogenous RNA polymerase activity in isolated nuclei from *Phytophthora megasperma* f. sp. *medicaginis*. Experiment. Mycology **7** (1983): 344–361.

– LOOYEN, D., TURKENSTEEN, L. J., and VAN DER WAL, D.: Occurrence of metalaxyl resistant strains of *Phytophthora infestans* in Dutch potato fields. Netherl. J. Plant Pathol. **87** (1981): 65–68.

DEKKER, J.: Counter measures for avoiding fungicide resistance. In: DEKKER, J., and GEORGOPOULOS, S. G. (Eds.): Fungicide Resistance in Crop Protection. Pudoc, Wageningen 1982, pp. 177–176.

– Non fungicidal compounds, which prevent disease development. Proc. Brit. Crop Protect. Conf. Brighton **1** (1983): 237–248.

– The development of resistance to fungicides. Progress in Pesticide Biochem. Toxicology **4** (1985): 165–218.

– and GEORGOPOULOS, S. G. (Eds.): Fungicide Resistance in Crop Protection. Pudoc, Wageningen 1982, pp. 273.

– and GIELINK, A. J.: Acquired resistance to pimaricin in *Cladosporium cucumerinum* and *Fusarium oxysporum* f. sp. *narcissi* associated with decreased virulence. Netherl. J. Plant Pathol. **85** (1979a): 67–73.

– – Decreased sensitivity to pyrazophos of cucumber and gherkin powdery mildew. Netherl. J. Plant Pathol. **85** (1979): 137–142.

DELP, C. J.: Coping with resistance to plant disease control agents. Plant Disease **64** (1980): 652–657.

DE WAARD, M. A.: Negatively correlated cross-resistance and synergism as strategies in coping with fungicide resistance. Proc. Brit. Crop Protect. Conf. **2** (1984): 573–584.

– and van NISTELROOY, J. G. M.: An anergy-dependent efflux mechanism for fenarimol in a wild-type strain and fenarimol-resistant mutants of *Aspergillus nidulans*. Pestic. Biochem. Physiol. **13** (1980a): 255–266.

- – Mechanism of resistance to pyrazophos in *Pyricularia orycae*. Netherl. J. Plant Pathol. **86** (1980b): 251–258.
- – Laboratory resistance to fungicides which inhibit ergosterol biosynthesis in *Penicillium italicum*. Netherl. J. Plant Pathol. **88** (1982): 99–112.
- – Stepwise development of laboratory resistance to DMI fungicides in *Penicillium italicum*. Netherl. J. Plant Pathol. **96** (1990): 321–329.
- FUCHS, A., DE RUIG, S. P., VAN TUYL, J. M., and DE VRIES, F. W.: Resistance to triforin: a non existent problem? Netherl. J. Plant Pathol. **83** (1977) Suppl. 1: 189–205.
- GEORGOPOULOS, S. G.: Development of fungal resistance to fungicides. In: SIEGEL, M. R., and SISLER, H. D. (Eds.): Antifungal Compounds. Vol. 2. Marcel Dekker Inc., New York–Basel 1977, pp. 409–495.
- – Cross-resistance. In: DEKKER, J., and GEORGOPOULOS, S. G. (Eds.): Fungicide Resistance in Crop Protection. Pudoc, Wageningen 1982a, pp. 53–59.
- – Genetical and biochemical background of fungicide resistance. In: DEKKER, J., and GEORGOPOULOS, S. G. (Eds.): Fungicide Resistance in Crop Protection. Pudoc, Wageningen 1982b, pp. 46–52.
- GILPATRICK, J. D.: Case study 2: *Venturia* of pome fruits and *Monilinia* of stone fruits. In: DEKKER, J., and GEORGOPOULOS, S. G. (Eds.): Fungicide Resistance in Crop Protection. Pudoc, Wageningen 1982, pp. 195–206.
- GRINDLE, M., and TEMPLE, W.: Sporulation and osmotic sensitivity of dicarboximide resistant mutants of *Neurospora crassa*. Transactions Brit. Mycolog. Soc. **94** (1985): 369–372.
- ISHII, H., YANASE, H., and DEKKER, J.: Resistance of *Venturia nashicola* to benzimidazole fungicides. Meded. Fac. Landbouwwet., Rijksuniv. Gent **49/2** (1984): 163–172.
- KABLE, P. F., and JEFFERY, H.: Selection for tolerance in organisms exposed to sprays of biocide mixtures: a theoretical model. Phytopathology **70** (1980): 163–172.
- KATO, T., SUZUKI, K., TAKANASHI, J., and KAMOSHITA, K.: Negatively correlated cross-resistance between benzimidazole fungicides and methyl N-(3,5dichlorophenyl) carbamate. J. Pesticide Sci. **9** (1984): 489–495.
- KOTZÉ, J. M.: Integrated protection in subtropical crops. Proc. 10th Intern. Congr. Plant Protect. November 20–25 (1983): 984–989.
- LASSERON-DE FALANDRE, A., DABOUISSI, M. J., and LEROUX, P.: Inheritance of resistance to fenpropimorph and terbinafine, two sterol biosynthesis inhibitors, in *Nectria haematococca*. Phytopathology **81** (1991): 1432–1438.
- LEROUX, P., and BESSEAT, B.: Pourriture grise: La résistance aux fongicides de *Botrytis cinerea*. Phytoma 359 (1984): 25–31.
- LEVY, Y., LEVI, R., and COHEN, Y.: Build up of a pathogen subpopulation resistant to a systemic fungicide under various control strategies: a flexible simulation model. Phytopathology **73** (1983): 1475–1480.
- LORENZ, P. H., and EICHHORN, K. W.: *Botrytis cinerea* and its resistance to dicarboximide fungicides. EPPO Bull. **12** (1982): 125–129.
- MISATO, T., KAKIKI, K., and HORI, M.: Mechanism of polyoxin resistance. Neth. J. Plant Pathol. **83** (1977), Suppl. 1: 253–260.
- – and Ko, K.: The development of resistance to agricultural antibiotics. Environmental quality and safety, suppl. **3** (1975): 437–440.
- MIURA, H., ITO, H., and TAKAHASHI, S.: Occurrence of resistant strains of *Pyricularia oryzae* to kasugamycin as a cause of the diminished fungicidal activity to rice blast. Ann. Phytopathol. Soc. Japan **41** (1975): 415–427.
- MOWEY, P. C., STEENKAMP, D. J., ACKRELL, B. A. C., SINGER, T. P., and WHITE, G. A.: Inhibition of Mammalian Succinate Dehydrogenase by Carboxins. Arch. Biochem. and Biophys. **178** (1977): 495–506.

NISHIMURA, S., KOHMOTO, K., and UDAGAWA, H.: Field emergence of fungicide-tolerant strains in *Alternaria kikuchiana* Tanaki. Rep. Tottori Mocylog. Inst., Jap. **10** (1973): 677–686.

OGAWA, J. M., MANJI, B. T., HEATON, C. R., PETRIE, J., and SONODA, R. M.: Methods for detection and monnitoring the resistance of plant pathogens to chemicals. In: GEORGIOU, G. P., and SAITO, T. (Eds.): Pest Resistance to Pesticides. Plenum Press, New York – London 1983, pp. 117–162.

POMMER, E. H., and LORENTZ, G.: Resistance to *Botrytis cinerea* to dicarboximide fungicides – a literature review. Crop Protect. **1** (1982): 221–230.

ROSENBERGER, D. A., and MEYER, F. W.: Negatively correlated cross-resistance to diphenylamine in benomyl-resistant *Penicillium expansum*. Phytopathology **75** (1985): 74–79

SAMOUCHA, Y., and COHEN, Y.: Synergy between metalaxyl-sensitive and metalaxyl-resistant strains of *Pseudoperonospora cubensis*. Phytopathology **74** (1984): 376–378.

SCHEPERS, H. T. A. M.: Changes during a three-year period in the sensitivity to ergosterol biosynthesis inhibitors of *Sphaerotheca fuliginea* in The Netherlands. Netherl. J. Plant Pathol. **91** (1985): 105–118.

SCHWINN, F. J.: Socio-economic impact of fungicide-resistance. In: DEKKER, J., and GEORGOPOULOS, S. G. (Eds.): Fungicide Resistance in Crop Protection. Pudoc, Wageningen 1982, pp. 16–23.

SHEPHARD, M. C., BENT, K. J., WOOLNER, M., and COLE, M. A.: Sensitivity to ethirimol of powdery mildew from UK barley crops. Proc. Brit. Insecticide Fungicide Conf., Brighton **1** (1975): 59–65.

SKYLAKAKIS, G.: The development and use of models describing outbreaks of fungicide resistance. Crop Protect. **1** (1982): 249–262.

STAUB, T., and SOZZI, D.: Fungicide resistance: a continuing challenge. Plant Disease **68** (1984): 1026–1031.

TAGA, M., NAKAGAWA, H., TSUDA, M., and UEGAMA, A.: Ascospore analysis of kasugamycin resistance in the perfect stage of *Pyricularia oryzae*. Phytopathology **68** (1978): 815–817.

UESUGI, Y.: *Pyricularia oryzae* in rice. In: DEKKER, J., and GEORGOPOULOS, S. G. (Eds.): Fungicide resistance in Crop Protection. Pudoc, Wageningen 1982, pp. 207–218.

– and SISLER, H. D.: Metabolism of a phosphoramidate by *Pyricularia oryzae* in relation to tolerance and synergism by a phosphorothiolate and isoprothiolane. Pesticide Biochem. Physiol. **9** (1978): 247–254.

VAN TUYL, J. M., DAVIDSE, L. C., and DEKKER, J.: Lack of cross-resistance to benomyl and thiabendazole in some strains of *Aspergillus nidulans*. Netherl. J. Pl. Pathol. **80** (1974): 165–168.

WADE, M.: Antifungal agents with an indirect mode of action. In: TRINCI, A. P. J., and RYLEY, J. F,. (Eds.): Mode of Action of Antifungal Agents. Symp. Brit. Mycolog. Soc. Manchester, UK, September 1983. Cambridge Univ. Press, Cambridge 1984, pp. 283–298.

WILD, B. L.: Double resistance by citrus green mould *Penicillium digitatum* to the fungicides guazatine and benomyl. Ann. appl. Biol. **103** (1983): 237–241.

– and ECKERT, J. W.: Synergy between a benzimidazole-sensitive and benzimidazole-resistant isolate of *Penicillium digitatum*. Phytopathology **72** (1982): 1329–1332.

WOLFE, M. A.: Trying to understand and control powdery mildew. Plant Pathol. **33** (1984): 451–466.

– MINCHIN, P. N., and SLATER, S. E.: Dynamics of triazole sensitivity in barley mildew, nationally and locally. Brit. Crop Protect. Conf. **2** (1984): 465–470.

Chapter 3

The genetics of fungicide resistance

S. G. Georgopoulos

Lapapharm Inc., 73 Menandrou Str., Athens 104 37, Greece

Introduction

All properties of living organisms are subject to genetic control and sensitivity to toxic compounds is certainly no exception. Evolution of resistance to such compounds should be regarded as a normal part of the overall evolutionary process which has enabled life on earth to be sustained and to achieve levels of amazing complexity. An organism may become resistant to toxicants either by changes in its own genetic material or by acquisition of additional genes from an external source. This latter way of acquiring resistance is very common in bacteria where the additional genes that are needed are carried on plasmids: cytoplasmic, circular DNA molecules which are transmissible between bacterial cells of the same or of different, even unrelated, species. As a rule, plasmid-borne resistance genes code for detoxifying enzymes, while modification of the molecular target of the inhibitor is usually the result of chromosomal mutations (Gale et al. 1981).

Cytoplasmic, plasmid-like DNAs are also known to exist in fungi (Esser et al. 1983; Gunge 1983), but no evidence is so far available that such elements may play a role in transmiting resistance to antifungal compounds of agricultural importance. Sensitivity to one type of agricultural fungicide (fentin) has been shown to be controlled by a mitochondrial gene, though not in a pathogenic fungus (Lancashire and Griffiths 1971). In all other cases where heritable variation for sensitivity to an agricultural fungicide has been demonstrated, the phenotypic differences were shown to result from differences in chromosomal genes.

To understand the work on the genetics of resistance in the fungi, some knowledge of the genetic features of this group of organisms is essential (Fincham et al. 1979). Fungi are eukaryotic organisms with well-defined nuclei, each bounded by an envelope which remains intact during division. In many fungi, two or more genetically different nuclei can be carried in the same cytoplasm. The balance of nuclear types in such a heterokaryon may vary from the very stable dikaryons of many heterothallic Basidiomycetes to the multinucleate cells of many Ascomycetes in which the proportions of different nuclei may change in response to selection (Davis 1966). If only some nuclei in a heterokaryon carry genes for resistance to a particular fungicide, their proportion may be increased by exposure to the chemical (Summers et al. 1984). To what extent such a change will affect the degree of resistance of the heterokaryon will also depend on the dominance of the genes involved.

In the Fungi Imperfecti, a sexual reproductive cycle is not known. In these organisms, genes can be identified only by the analysis of vegetative (mitotic) segregation: The fungus is normally haploid, but occasional fusion of two nuclei in somatic cells permits the selection of diploid strains by appropriate techniques. If such diploids are heterozygous, they tend to produce sectors showing segregation of markers originally present in the heterozygous condition and resulting mainly from occasional crossing-over between homologous mitotic chromatids or from non-disjunction. This sectoring can be exploited for purposes of genetic mapping.

In members of all other groups, the sexual reproductive cycle consists, as in higher organisms, of a regular alternation of a haploid phase, in which the nuclei contain a single set of chromosomes, and a diploid phase, in which a double set of chromosomes is present in each nucleus. The transition from haploidy to diploidy results from fusion (karyogamy) of two haploid gamete nuclei which may be contributed by the same strain in the homothallic but have to be produced by two compatible strains in the heterothallic fungi. The converse transition is accomplished through meiosis which reduces the chromosomes from a double to a single set.

In the Basidiomycetes, the filamentous Ascomycetes and the Fungi Imperfecti the vegetative pathogenic phase is entirely haploid, with meiosis following immediately after karyogamy. In the Peronosporales, the most important order of the Oomycetes, the situation is quite different: karyogamy occurs immediately or soon after meiosis so that the organism is predominantly diploid.

If the pathogenic phase is comprised by a haploid homokaryon, as in Ascomycetes and most Fungi Imperfecti, it is irrelevant whether a resistance gene is dominant or semidominant or recessive. It is also sufficient in such a case to only examine the phenotypes in the F_1 from a sensitive × resistant cross, in order to identify the genes involved. In diploids, such as *Phytophthora* or dikaryons, such as *Ustilago*, a recessive gene for resistance will not affect the phenotype until it becomes homozygous. A semidominant or dominant gene, however, will be expressed in the heterozygote, so that resistance will be recognised earlier in the laboratory and probably also in the field. On the other hand, if the vegetative phase is diploid or dikaryotic, recognition of a Mendelian ratio requires selfing of the F_1 from a sensitive × resistant cross and examination of the phenotypes of the F_2 generation.

As is the case with other genetic traits, fungicide resistance may be **qualitative** or **quantitative** (GEORGOPOULOS 1985; GRINDLE 1987). Qualitative resistance is controlled by genes of large effects which cause distinct discontinuity in sensitivity distribution, while the continuous variation characteristic of quantitative resistance is caused by the simultaneous segregation of many genes, each of a small individual effect. It cannot be excluded, of course, that both types of genetic control may occur in a particular fungus-fungicide combination (GEORGOPOULOS and SKYLAKAKIS 1986). One known example is that of *Nectria haematococca* var. *cucurbitae* in which resistance to all fungicides acting as C-14 demethylation inhibitors (DMIs) in the sterol biosynthetic pathway is under polygenic control, but a major gene has also been recognised which gives high resistance to triazoles, particularly triadimenol (KALAMARAKIS et al. 1989, 1991). This, however, appears to be the exception rather than the rule.

Major-gene control of fungicide resistance

With many fungicides, mutation of a single gene is all that is required, in order for the highest possible level of resistance to be achieved and this appears to be independent of the organism which is used to study resistance. If such a highly resistant strain is crossed to a sensitive wild type, a Mendelian ratio of distinct phenotypes will be obtained in the progeny. It is not excluded that mutation at a different locus may also affect sensitivity, but what distinguishes this type of resistance from the polygenic (quantitative) one is the absence of additivity of gene effects. In other words, a major gene is epistatic over other genes which may affect sensitivity to the same type of fungicide. With this type of genetic control of sensitivity, field populations of sensitive species give a discontinuous distribution, i.e. each population consists of at least two distinct subpopulations, one sensitive and one resistant. Often, fungicide rates which are 100% effective on the sensitive have no effect on the resistant subpopulation. Because of this lack of overlap, each isolate can be easily classified as unequivocally resistant or sensitive (GEORGOPOULOS 1987).

Benzimidazoles

The genetics of resistance to benzimidazole fungicides has been studied in several organisms, including *Aspergillus nidulans* (HASTIE and GEORGOPOULOS 1971; VAN TUYL 1977), *A. niger* (VAN TUYL 1977), *Botryotinia fuckeliana* (FARETRA and POLLASTRO 1992), *Ceratocystis ulmi* (BRASIER and GIBBS 1975; WEBBER et al. 1986), *Neurospora crassa* (BORCK and BRAYMER 1974), *Physarum polycephalum* (BURLAND et al. 1984), *Saccharomyces cerevisiae* (THOMAS et al. 1985), *Schizosaccharomyces pombe* (YAMAMOTO 1980; UMEZONO et al. 1983), *Talaromyces flavus* (KATAN et al. 1984), *Ustilago maydis* (ZIOGAS et al. 1993), *Venturia inaequalis* (KATAN et al. 1983), *V. nashicola* (ISHII et al. 1984), and *V. pyrina* (SHABI and KATAN 1979). In all of these organisms, high resistance to benzimidazoles can be obtained by mutation of a single gene. Only exceptionally, e.g. in *A. nidulans* or *S. pombe*, mutation at one to a few other loci may give smaller decreases in sensitivity. In such a case, crosses of strains differing in the degree of resistance, yield some sensitive recombinants, but the recombinants carrying two mutant genes are not more resistant than the major-gene mutant parent, indicating absence of additivity. A different situation has been claimed for *Penicillium italicum* (BERAHA and GARBER 1980) and *Fusarium oxysporum* (HORNOK et al. 1988) where interaction of two mutant genes appears to be required for a highly resistant phenotype.

More frequently, only one locus for resistance to benzimidazoles is recognised in any given species. Even in cases where large numbers of resistant mutants were studied, e.g. in *S. cerevisiae* (THOMAS et al. 1985), they were all found to be allelic. In this situation, different alleles of this major gene may give different levels of resistance to benzimidazoles and may or may not affect sensitivity to compounds which, like benzimidazoles, bind to tubulin (see ISHII 1992 for a review and also JOSEPOVITS and GASTONYI 1992; ZIOGAS et al. 1993). In diploid strains of *A. nidulans*, heterozygous for benomyl resistance, the sensitive allele was shown to be dominant over the resistant one (HASTIE and GEORGOPOULOS 1971). Similarly, in the heterozygous plasmodia of *P. polycephalum* carbendazim caused enlarged nuclei as in the case of sensitive homozygotes, indicating recessiveness of the resistant allelomorph (BURLAND et al. 1984). In *N. crassa* heterokaryons, however, it was the sensitive gene that was recessive (BORCK and BRAYMER 1974). A more detailed study was done with *S. cerevisiae*. A total of 65 resistant mutants were crossed to an appropriate wild type to synthesize diploids. Of these, 15 showed partial or complete dominance and the remaining 50 showed partial or nearly complete recessiveness of the resistant phenotype (THOMAS et al. 1985).

As indicated above, benzimidazole fungicides exert their antifungal activity by binding to tubulin. It has been shown with a number of organisms that mutations in the structural gene for β-tubulin are responsible for benzimidazole resistance. In a few cases, this gene has been molecularly characterised. In the case of *S. cerevisiae*, a G to A transversion at the second base of codon 241 was shown to determine resistance by changing arginine to histidine (THOMAS et al. 1985). In a benzimidazole resistant mutant of *N. crassa* the mutation caused a phenylalanine-to-tyrosin change at position 167 (ORBACH et al. 1986). A very interesting study has recently been made with field strains of *V. inaequalis* and other plant pathogenic fungi (KOENRAADT and JONES 1992; KOENRAADT et al. 1992). Four types of strains with widely different levels of resistance to benzimidazoles and with or without change in sensitivity to N-phenylcarbamates were compared to wild type. It was shown that each phenotypic class, except the one with low resistance to benomyl, was associated with a unique amino acid substitution at position 198 or 200 of the β-tubulin molecule. This indicates a key role of the amino acids at these positions in the action of benzimidazole fungicides in these organisms. A study of 10 field strains of *B. cinerea* also showed substitutions at amino acids 198 and 200 associated with various benzimidazole-resistant phenotypes (YARDEN and KATAN 1993). It appears there may be a difference between field and laboratory strains regarding the amino acids which are important for benzimidazole sensitivity.

Aromatic hydrocarbon group

Genes for resistance to the various members of this group, including the dicarboximides, tolclofos methyl and, apparently, also the new phenylpyrrole fungicides (FARETRA and POLLASTRO 1993; LEROUX and FRITZ 1993) have been recognised in *A. nidulans* (THRELFALL 1968; GRINDLE and ZHOU 1988), *Botryotinia fuckeliana* (FARETRA and POLLASTRO 1993; KHAI and BOMPEIX 1990), *Nectria haematococca* (syn. *Hypomyces solani* – GEORGOPOULOS and PANOPOULOS 1986), *Neurospora crassa* (GRINDLE 1984), *N. sitophila* (WHITTINGHAM 1962), *Penicillium expansum* (BERAHA and GARBER 1966), *P. italicum* (BERAHA and GARBER 1980), *Phytophthora parasitica* (CHANG and KO 1990), *U. maydis* (TILLMAN and SISLER 1973), and *U. violacea* (GARBER et al. 1982). In all cases, resistance was qualitative. In *N. haematococca*, a mutant gene at any one of a number of different loci gives apparently the same degree of resistance to chlorinated nitrobenzenes, and this is not affected by the presence of mutant or wild type genes at other loci (GEORGOPOULOS 1963). In *B. fuckeliana*, only one locus for resistance to dicarboximides has been recognised and the genes for low and high resistance are allelic, the latter usually causing hypersensitivity to high osmotic pressure (FARETRA and POLLASTRO 1991).

The intraallelic interactions appear to vary depending on the organism. Thus, in diploid *A. nidulans* quintozene resistance behaves as a recessive character (THRELFALL 1968), while in *N. crassa* heterokaryons with approximately equal proportions of wild type and mutant nuclei, vinclozolin resistance appeared semidominant (GRINDLE 1984). In heterokaryotic thalli of *B. fuckeliana*, high resistance of dicarboximides was dominant, while the associated phenotype of osmotic hypersensitivity was partially recessive (FARETRA and POLLASTRO 1993). Resistance to sodium orthophenylphenolate was recessive in *P. expansum* (BERAHA and GARBER 1966) and dominant in *P. italicum* (BERAHA and GARBER 1980).

Carboxamides

Resistance to the 1,4 oxathiin carboxamide fungicide carboxin in *A. nidulans* results from mutation of one of at least three, freely recombining genes (GUNATILLEKE *et al.* 1976). Depending on the carbon source utilised, the three types of mutants may be distinguished by the degree of resistance to the fungicide. In *U. maydis*, low resistance to carboxin is conferred by the *ants* mutation which is more easily recognisable by its effect on the cyanide insensitive respiration (ZIOGAS and GEORGOPOULOS 1984). Intermediate and high resistance, however, are the result of two allelic mutations at the *oxr*-1 locus (GEORGOPOULOS and ZIOGAS 1977). In both *A. nidulans* and *U. maydis*, genes for carboxin resistance affect also sensitivity to other carboxamides (WHITE *et al.* 1978; WHITE and THORN 1980; WHITE and GEORGOPOULOS, 1986). Different cross-resistance relationships have also been observed with *U. nuda* (LEROUX and BERTHIER 1988). With respect to intraallelic interaction, it appears that resistant alleles are semidominant, apparently because heterozygous diploid or dikaryotic cells contain a mixture of sentive and resistant mitochondrial succinate dehydrogenase complexes (GEORGOPOULOS *et al.* 1975).

The *U. maydis* gene for high resistance to carboxin (*oxr*-1B) has recently been isolated using gene transfer techniques (KEON *et al.* 1991). The sequence of this gene shows a high degree of homology to succinate dehydrogenase iron sulfur protein subunit genes from a number of other organisms. Comparison of this sequence with that of the gene encoding the same subunit from a carboxin sensitive strain identified a two-base difference between the sequences. This causes a single amino acid change with a leucine residue being substituted for a histidine in the carboxamide resistant form (BROOMFIELD and HARGREAVES 1992).

Phenylamides

These fungicides are active against mainly the Peronosporales which are more difficult to manipulate genetically than the haploid heterothallic Ascomycetes and Basidiomycetes used to study the genetics of resistance to fungicides of other groups. However, genetic analyses of crosses involving phenylamide resistant strains have been performed in *Bremia lactucae* (CRUTE and HARRISON 1988), *Phytophthora infestans* (SHATTOCK 1988), and *P. parasitica* (CHANG and KO 1990). The results of these analyses support the view that phenylamide resistance in each species is controlled by a major gene. In *B. lactucae* and *P. infestans* the gene exhibits incomplete dominance, so that the diploid F_1 progeny from crosses between highly resistant and sensitive isolates are intermediate and a 1:2:1 ratio of sensitive:intermediate:resistant phenotypes is obtained in the F_2 generation. In *P. parasitica* the major gene for phenylamide resistance is apparently fully dominant.

Kasugamycin

It is only in *Pyricularia oryzae* among the fungi that kasugamycin resistance has been studied. Genetic work has shown the existence of three unlinked loci for resistance to the antibiotic in this organism. Mutation of only the one of the three genes, however, gives cross resistance to blasticidin S (TAGA *et al.* 1978). Double mutant recombinants from didybrid crosses did not appear to exhibit higher resistance to kasugamycin than the parental strains, showing no additivity of gene effects.

Kasugamycin is also active against bacteria and several studies on bacterial resistance to this antibiotic have been conducted (GALE et al. 1981). In *Escherichia coli*, all three resistance genes identified are chromosomal. In merodiploids heterozygous for the *Ksg*A gene resistance was recessive.

Streptomycin

Streptomycin is known as an antibacterial antibiotic, but it inhibits the growth of some fungi particularly Oomycetes. Several strains of *Phytophthora cactorum* resistant to this antibiotic were obtained by SHAW and ELLIOT (1968). One of these strains required streptomycin for growth. Resistance and dependence were stable through asexual and sexual reproduction, but whether chromosomal or cytoplasmic determinants were involved was not shown. A recent study with *P. parasitica* (CHANG and KO 1990) suggests that streptomycin resistance in this organism is controlled by a cytoplasmic gene and is inherited solely through the maternal parent.

In bacteria, high level resistance to streptomycin is easily obtained in the laboratory through mutation of one chromosomal gene (GALE et al. 1981). The three phenotypic responses to the antibiotic (sensitivity, dependence, and high-level resistance) appear to involve multiple alleles of the same gene. In merodiploid heterozygous strains of *E. coli* sensitivity to streptomycin appears to be dominant over resistance. In field strains of bacteria that are resistant to streptomycin, genes for antibiotic inactivating enzymes are usually carried on plasmids. The genetics of resistance of plant pathogenic bacteria to bactericides has recently been reviewed (COOKSEY 1990). Streptomycin resistant strains of *Pseudomonas syringae* pv. *papulans* contained a plasmid that was not present in sensitive strains. In *Xanthomonas campestris* pv. *vesicatoria* streptomycin resistance genes may reside on plasmids of different sizes and perhaps also on the chromosome.

Copper

Although genes for resistance to copper have not been recognised in filamentous fungi and no failures of copper fungicides against fungal diseases of plants have been assigned to resistance (GEORGOPOULOS and SKYLAKAKIS 1986), it has been known for many years that copper resistance in the yeast *S. cerevisiae* is controlled by the *CUP* 1 chromosomal locus (BRENES-POMALES et al. 1955). Further increases of the resistance of mutant cells are the result not of mutation of additional genes, but of gene amplification: the locus copy number is positively correlated with the resistance level of a yeast strain (FOGEL and WELCH 1982; KARIN et al. 1984).

Copper compounds are also used against bacterial plant pathogens and in this case several failures have been assigned to resistance. Two types of genetic determinants of bacterial resistance to copper have been identified (COOKSEY 1990). One is known only in *X. compestris* pv. *vesicatoria* and is associated with pXcv-type plasmids. The other determinant is the *cop* operon from *P. syringae* pv. *tomato* and related genes in other *Pseudomonas* species and in *X. campestris* pv. *vesicatoria*.

Others

As stated earlier, when stable resistance to a fungicide has been recognised but not studied genetically, involvement of major genes can be suspected if the sensitivity distributions of field populations show distinct, non-overlapping subpopulations. This appears to be the case with the polyoxins (GEORGOPOULOS 1987). High level resistance to these antibiotics in *Alternaria kikuchiana* appeared suddenly and caused control failures in Japan. Resistant and sensitive isolates do not seem to overlap which indicates involvement of a major gene.

Polygenic control of fungicide resistance

High resistance to some fungicides requires many mutant genes acting additively. The effect of single-gene mutations may be measurable, at least in the laboratory and Mendelian segregation is obtained from crosses of single-gene mutants to wild type strains. Though the number of genes identified in any fungus-fungicide combination may be rather small, this may be due only to the limited number of resistant strains that have been studied genetically. That the genes involved are many is often indicated by the fact that even first-step selection may yield some strains carrying more than one gene for resistance. Higher resistance levels may be achieved through recombination and/or stepwise selection. If a highly resistant strain is crossed to a highly sensitive one, the progeny will give the rather continuous distribution characteristic of quantitative inheritance. In nature, the various combinations of resistance genes and often modifiers also result in a continuous distribution of sensitivities, so that distinct subpopulations cannot be recognised even after long exposures to the selector.

Dodine

The genetics of dodine resistance was first studied in *N. haematococca* var. *cucurbitae*. Single-gene mutations caused variable but generally small reductions in sensitivity. Though only a small number of single-gene mutants were studied, four loci for resistance were recognised and in recombinants mutant alleles acted additively (KAPPAS and GEORGOPOULOS 1970). Dodine is used mainly against *V. inaequalis* and the work on fungicide resistance in this organism has recently been reviewed (CRUTE 1992). Polygenic control of resistance to dodine is indicated by both the genetic analyses and the absence of discontinuity in sensitivity distributions of field populations of *V. inaequalis*.

Sterol C-14 demethylation inhibitors

Most of the agricultural fungicides which inhibit the biosynthesis of sterols (SBIs) act as C-14 demethylation inhibitors (DMIs). Polygenic control of resistance to these fungicides was first indicated by work with *A. nidulans*. In this organism, eight genes for resistance to the DMI imazalil were recognised (VAN TUYL 1977). Single gene mutations caused a relatively low level of resistance with a maximum increase of the minimal inhibitory concentration by a factor of 10. This factor could reach 1 000 by combining proper genes. Modifiers were also recognised. A study of 51 fenarimol-selected strains of *N. haematococca* var. *cucurbitae* recognised a similar polygenic system for resistance to DMI fungicides with at least nine chromosomal loci involved (KALAMARAKIS *et al.* 1991). Additivity of effects was shown with regard not only to growth inhibition but also to accumulation of fenarimol in the mycelium. Polygenic control of DMI resistance has also been suggested by other genetic studies e.g. with *Erysiphe graminis* f. sp. *hordei* (HOLLOMON *et al.* 1984) and *U. maydis* (WELLMAN and SCHAUZ 1992) or by the stepwise increase in resistance, as in *P. italicum* (DE WAARD and VAN NISTELROOY 1990).

In some recent publications, major-gene control of resistance to at least some DMI fungicides has been claimed (BROWN *et al.* 1992; PEEVER and MILGROOM 1992; STANIS and JONES 1985). Though with some organisms it is difficult to study a large number of mutants, it should be remembered that polygenic control cannot be excluded because one or a few resistant × resistant crosses did not yield wild type recombinants. A large enough number (200) of such crosses were analysed, however, in the case of triadimenol resistance of *N. haematococca* var. *cucurbitae*. These analyses showed that all of 30 mutants studied carry the same gene (KALAMARAKIS *et al.* 1989). This gene gives high resistance, particularly to the highly active diastereoisomer of triadimenol (DEMOPOULOS *et al.* 1993) and some resistance to other triazoles. Sensitivity to the pyrimidine fenarimol and the piperazine triforine is not affected, while sensitivity to some imidazoles is increased rather than decreased by this mutation. This major gene is thus distinct from the minor, additively acting genes recognised by selection on fenarimol which give resistance to all DMIs, including triadimenol (KALAMARAKIS *et al.* 1991).

With the knowledge so far available it seems justified to say that resistance to DMIs in general is polygenic but in some fungi there may be a major gene for high resistance to one or more of these fungicides. In some crosses of field isolates of *Pyrenophora teres*, in adition to a major gene for triadimenol resistance, a number of minor genes were, in fact, observed to be segregating (PEEVER and MILGROOM 1992). The importance of the polygenic system for DMI resistance is indicated by the fact that no discontinuity of sensitivity distributions has been observed in field populations of pathogens against which DMI fungicides have been used for several years, e.g. *E. graminis* f. sp. *hordei* (HEANEY 1988). Populations collected from treated fields are of lower sensitivity than those from untreated ones, but distinct subpopulations are not found.

Morpholines and piperidines

Mutations for resistance to DMIs do not generally affect sensitivity to the morpholine derivatives fenpropimorph and tridemorph and the piperidine derivative fenpropidin, agriculturally important SBIs which do not act on the C-14 demethylation step. Fenpropimorph resistant mutants have been studied in the cucurbit and the pea pathogens *N. haematococca* var. *cucurbitae* (DEMAKOPOULOU *et al.* 1989) and var. *pisi* (DE FALANDRE *et al.* 1991). Though only a small number of such mutants were studied, three genes for resistance were recognised in each organism and the additivity of effects led to the conclusion that fenpropimorph resistance is polygenic. It is characteristic that in each study one homokaryotic strain carrying two genes for resistance was recognised, which indicates that there are many such genes in the genome. In *N. haematococca* var. *cucurbitae*, evidence was obtained that one of the fenpropimorph resistance genes does not affect sensitivity to tridemorph and fenpropidin (DEMAKOPOULOU et al. 1989).

Cycloheximide

Six chromosomal genes for resistance to cycloheximide have been recognised in *S. cerevisiae* (WILKIE and LEE 1965). While the level of first-step mutants did not exceed 20 µg of the antibiotic per ml, this level could be increased to 1,000 µg/ml by crossing first-step mutants among themselves or by stepwise selection, indicating quantitative nature of the trait. Of the six genes, two were recessive, three were semidominant and one was dominant. Polygenic control of cycloheximide resistance has also been observed in *A. nidulans* (VAN TUYL 1977) and *N. crassa* (VOMVOYANNI 1974).

Others

The use of ethirimol against powdery mildew of barley was associated with slightly reduced sensitivity in the surviving populations of the pathogen (BRENT 1982). The sensitivity distribution, however, remained unimodal in all cases, which favors a polygenic control of resistance. Progenies from crosses of field isolates of differing sensitivities made by HOLLOMON (1981) gave continuous frequency distributions, also indicating polygenic control. This, however, was not the case in experiments carried out by BROWN et al. (1992) who concluded that variation in ethirimol resistance might always be resolvable into components controlled by single genes. Polygenic control cannot be excluded, of course, unless it is shown that such genes do not act additively.

In the control of *Cercospora beticola* with fentin fungicides, mean sensitivity may decrease slowly, but the distribution remains continuous with considerable overlapping between treated and untreated populions (GIANNOPOLITIS 1978). A similar response of field populations of *P. oryzae* to phosphorothiolates has been observed (UESUGI 1982). It is likely that polygenes are involved in these two cases also.

Practical implications

To recognise whether genetic variation for sensitivity to a particular fungicide is available to target organisms and what is the type of genetic control of such variation is very important in order to make some prediction regarding the likely useful life of the respective product (GEORGOPOULOS 1987). Theoretically, the genes which would mutate to give resistance to a chemical may not be available to a given species or may be lethal in their mutated form. From what evidence is so far available, it seems that plant pathogenic fungi are not capable of developing resistance to some of the protectant fungicides. Mutants of pathogens resistant to these fungicides have not been isolated, even under laboratory conditions, and there is no data showing selection of less sensitive forms in the field, in spite of long use.

If genetic variation is recognised, it is important to determine whether it is qualitative or quantitative, in order to understand the type of selection that will be caused by fungicide treatments (GEORGOPOULOS 1988). Sudden and complete loss of fungicidal effectiveness has so far taken place only in cases of major-gene resistance. Of course, major genes do not always lead to complete failures, particularly if the mutations substantially lower fitness. This seems to have been the case, for example, with kasugamycin (ITO and YAMAGUCHI 1979). In many cases (e.g. with benzimidazoles of phenylamides), however, fitness is not seriously affected and because the response is qualitative, selection proceeds unnoticed in terms of efficacy, until highly resistant strains reach a proportion of $10^{-2}-10^{-1}$ (BRENT 1992). This is only one or two treatments away from complete failure.

The quantitative changes which are based on polygenic systems are apparently less likely to lead to sudden and complete failures. Treatments cause a gradual shifting of the population towards decreased sensitivity and satisfactory control cannot suddenly be followed by failure. Indications of less satisfactory performance will be obtained much earlier and the control strategies may be modified accordingly. In addition, it may be anticipated that with the large number of mutations required for high resistance, there may be an increased likelihood for substantial loss of fitness.

References

BERAHA, L., and GARBER, E. D.: Genetics of phytopathogenic fungi. XV. A genetic resistance to sodium orthophenylphenate and sodium dihydroacetate in *Penicillium expansum*. Amer. J. Bot. **53** (1966): 1041–1047.
– – A genetic study of resistance to thiabendazole and sodium o-phenylphenate in *Penicillium italicum* by the parasexual cycle. Bot. Gaz. **141** (1980): 204–209.
BORCK, K., and BRAYMER, H. D.: The genetic analysis of resistance to benomyl in *Neurospora crassa*. J. Gen. Microbiol. **85** (1974): 51–56.
BRASIER, C. M., and GIBBS, J. N.: MBC tolerance in aggressive and non-aggressive isolates of *Ceratocystis ulmi*. Ann. appl. Biol. **80** (1975): 231–235.
BRENES-POMALES, A., LINDEGREN, G., and LINDEGREN, C. C.: Gene control of copper sensitivity in *Saccharomyces*. Nature **176** (1955): 841–842.
BRENT, K. J.: Case study 4: Powdery mildews of barley and cucumber. In: DEKKER, J., and GEORGOPOULOS, S. G. (Eds.): Fungicide Resistance in Crop Protection. Pudoc, Wageningen 1982, pp. 219–230.
– Monitoring fungicide resistance: purposes, procedures and progress. In: DENHOLM, I., DEVONSHIRE, A. L., and HOLLOMON, D. W. (Eds.): Resistance '91: Achievements and Developments in Compating Pesticide Resistance. Elsevier, London 1992, pp. 1–18.
BROOMFIELD, P. L. E., and HARGREAVES, J. A.: A single amino-acid change in the iron-sulfur protein subunit of succinate dehydrogenase controls resistance to carboxin in *Ustilago maydis*. Curr. Genet. **22** (1992): 117–121.
BROWN, J. K. M., JESSOP, A. C., THOMAS, S., and BEZANOOR, H. N.: Genetic control of the response of *Erysiphe graminis* f. sp. *hordei* to ethirimol and triadimenol. Plant Pathol. **41** (1992): 126–135.
BURLAND, T. G., SCHERL, T., GULL, K., and DOVE, W. F.: Genetic analysis of resistance to benzimidazoles in *Physarum*: differential expression of β-tubulin genes. Genetics **108** (1984): 123–141.

CHANG, T. T., and KO, W. H.: Resistance to fungicides and antibiotics in *Phytophthora parasitica*: Genetic nature and use in hybrid determination. Phytopathology **80** (1990): 1414–1421.

COOKSEY, D. A.: Genetics of bactericide resistance in plant pathogenic bacteria. Annu. Rev. Phytopathol. **28** (1990): 201–219.

CRUTE, I. R.: The contribution of genetic studies to understanding fungicide resistance. In: DENHOLM, I., DEVONSHIRE, A. L., and HOLLOMON, D. W. (Eds.): Resistance '91: Achievements and Developments in Compating Pesticide Resistance. Elsevier, London 1991, pp. 190–202.

– and HARRISON, J. M.: Studies on the inheritance of resistance to metalaxyl in *Bremia lactucae* and on the stability and fitness of field isolates. Plant Pathol. **37** (1988): 231–250.

DAVIS, R. H.: Mechanisms of inheritance. 2. Heterokaryosis. In: AINSWORTH, G. C., and SUSSMAN, A. S. (Eds.): The Fungi, Vol. 2. Academic Press, New York 1966, pp. 567–588.

DE FALANDRE, A., DABOUSSI, M., and LEROUX, P.: Inheritance of resistance to fenpropimorph and terbinafine, two sterol biosynthesis inhibitors, in *Nectria haematococca*. Phytopathology **81** (1991): 1432–1438.

DEMAKOPOULOU, M. G., ZIOGAS, B. N., and GEORGOPOULOS, S. G.: Evidence for polygenic control of fenpropimorph resistance in laboratory mutants of *Nectria haematococca* var. *cucurbitae*. ISPP Chemical Control Newsletter, No. 12 (1989): 34–35.

DEMOPOULOS, V. P., ZIOGAS, B. N., and GEORGOPOULOS, S. G.: Stereospecificity of a triadimenol resistance mutation in *Nectria haematococca* var. *cucurbitae* and its effect on sensitivity to triadimefon. In: LYR, H., and POLTER, C. (Eds.): Systemic Fungicides and Antifungal Compounds. Eugen Ulmer, Stuttgart 1993, pp. 393–398.

DE WAARD, M. A., and NISTLELROOY, J. G. M. VAN: Stepwise development of laboratory resistance to DMI-fungicides in *Penicillium italicum*. Neth. J. Plant Pathol. **96** (1990): 321–329.

ESSER, K., KUCK, U., STAHL, U., and TUDZYNSKI, P.: Cloning vectors of mitochondrial origin for eukaryotes: a new concept in genetic engineering. Curr. Genet. **7** (1983): 239–243.

FARETRA, F., and POLLASTRO, S.: Genetic basis of resistance to benzimidazole and dicarboximide fungicides, in *Botryotinia fuckeliana* (*Botrytis cinerea*). Mycol. Res. **95** (1991): 943–951.

– – Genetics of sexual compatibility and resistance to benzimidazole and dicarboximide fungicides in isolates of *Botryotinia fuckeliana* (*Botrytis cinerea*) from nine countries. Plant Pathol. **42** (1993): 48–57.

– – Genetic basis of resistance to the phenylpyrrole fungicide CGA 173506 in *Botryotinia fuckeliana* (*Botrytis cinerea*). In: LYR, H., and POLTER, G. (Eds.): Systemic Fungicides and Antifungal Compounds. Eugen Ulmer, Stuttgart 1993, pp. 405–409.

FINCHMAN, J. R. S., DAY, P. R., and RADFORD, A.: Fungal Genetics. 4th edit. Blackwell, Oxford 1979.

FOGEL, S., and WELCH, J. W.: Tandem gene amplification mediates copper resistance in yeast. Proc. Natl. Acad. Sci. (U.S.A.) **79** (1982): 5342–5346.

GALE, E. F., CUNDLIFE, E., REYNOLDS, P. E., RICHMONT, M. H., and WARING, M. J.: The Molecular Basis of Antibiotic Action. 2nd edit. Wiley Inters., London 1981.

GARBER, E. D., ENG, C., PUSCHECK, E. E., WELL, M., and WARD, S.: Genetics of *Ustilago violacea*. XII. Half tetrad analysis and double selection. Bot. Gazette **143** (1982): 524–529.

GEORGOPOULOS, S. G.: Tolerance to chlorinated nitrobenzenes in *Hypomyces solani* f. *cucurbitae* and its mode of inheritance. Phytopathology **53** (1963): 1086–1093.

– The genetic basis of classification of fungicides according to resistance risk. EPPO Bull. **15** (1985): 513–517.

– The development of fungicide resistance. In: WOLFE, M. S., and CATEN, C. E. (Eds.): Populations of Plant Pathogens: their Dynamics and Genetics. Blackwell Sci. Publ., Oxford 1987, pp. 239–251.

– Genetics and population dynamics. In: DELP, C. J. (Ed.): Fungicide Resistance in North America, APS Press, St. Paul, Minn. 1988, pp. 12–13.

– and PANOPOULOS, N. J.: The relative mutability of the *cnb* loci in *Hypomyces*. Can. J. Genet. Cytol. **8** (1966): 347–349.

- and SKYLAKAKIS, G.: Genetic variability in the fungi and the problem of fungicide resistance. Crop Prot. **5** (1986): 299–305.
- and ZIOGAS, B. N.: A new class of carboxin resistant mutants of *Ustilago maydis*. Netherl. J. Plant Pathol. **83** (1977): 235–242.
- CHRYSAYI, M., and WHITE, G. A.: Carboxin resistance in the haploid, the heterozygous diploid, and the plant parasitic dikaryotic phase of *Ustilago maydis*. Pestic. Biochem. Physiol. **5** (1975): 543–551.

GIANNOPOLITIS, C. N.: Occurence of strains of *Cercospora beticola* resistant to triphenyltin fungicides in Greece. Plant Dis. Reprt. **62** (1978): 205–208.

GRINDLE, M.: Isolation and characterisation of vinclozolin resistant mutants of *Neurospora crassa*. Trans. Br. mycol. Soc. **82** (1984): 635–643.
- Genetic basis of fungicide resistance. In: FORD, M. G., HOLLOMON, D. W., KHAMBAY, B. P. S., and SAWICKI, R. M. (Eds.): Compating Resistance to Xenobiotics. Ellis Horwood, Chichester, U.K. 1987, pp. 74–93.
- and ZHOU, Y.: Isolation, characterisation and genetic analysis of mutants of *Aspergillus nidulans* resistant to tolclofos-methyl. Proc. 1988 Brighton Crop Prot. Conf. – Pests and Diseases: 409–414.

GUNATILLEKE, I. A. U. N., ARST, H. N., and SCAZZOCHIO, C.: Three genes determine the carboxin sensitivity of mitochondrial succinate oxidation in *Aspergillus nidulans*. Genet. Res., Camb. **26** (1976): 297–305.

GUNGE, N.: Yeast DNA plasmids. Ann. Rev. Microbiol. **37** (1983): 253–275.

HASTIE, A. C., and GEORGOPOULOS, S. G.: Mutational resistance to fungitoxic benzimidazole derivatives in *Aspergillus nidulans*. J. Gen. Microbiol. **67** (1971): 371–374.

HEANEY, S. P.: Population dynamics in DMI fungicide sensitivity changes in barley powdery mildew. In: DELP, C. E. (Ed.): Fungicide Resistance in North America, APS Press, St. Paul, Minn. 1988, pp. 89–92.

HOLLOMON, D. W.: Genetic control of ethirimol resistance in a natural population of *Erysiphe graminis* f. sp. *hordei*. Phytopathology **71** (1981): 536–540.
- BUTTERS, J., and CLARK, J.: Genetic control of triadimenol resistance in barley powdery mildew. Proc. 1984 Br. Crop. Prot. Conf. – Pests and Diseases: 477–482.

HORNOK, L., MOLNAR, A., and OROS, G.: Variations in sensitivity to benzimidazole and non-benzimidazole fungicides of genetically different benomyl resistant *Fusarium oxysporum* strains. Acta Phytop. et Entom. Hungarica **23** (1988): 3–10.

ISHII, H.: Target sites of tubulin-binding fungicides. In: KOLLER, W. (Ed.): Target Sites of Fungicide Action, CRC Press, Boca Raton, Fl. 1992, pp. 43–52.
- Yanase, H., and DEKKER, J.: Resistance of *Venturia nashicola* to benzimidazole fungicides. Meded. Fac. Landbouwwet., Rijksuniv. Gent. **49** (1984): 163–172.

ITO, I., and YAMAGUCHI, T.: Competition between sensitive and resistant strains of *Pyricularia oryzae* Cav. against kasugamycin. Ann. Phytopathol. Soc. Japan **45** (1979): 40–46.

JOSEPOVITS, G., and GASTONYI, M.: Negative cross-resistance to N-phenylanilines in benzimidazole-resistant strains of *Botrytis cinerea*, *Venturia nashicola* and *Venturia inaequalis*. Pestic. Sci. **35** (1992): 237–242.

KALAMARAKIS, A. E., DEMOPOULOS, V. P., ZIOGAS, B. N., and GEORGOPOULOS, S. G.: A highly mutable major gene for triadimenol resistance in *Nectria haematococca* var. *cucurbitae*. Neth. J. Plant Pathol. **95** (1989), Suppl. 1: 109–120.
- DE WAARD, M. A., ZIOGAS, B. N., and GEORGOPOULOS, S. G.: Resistance to fenarimol in *Nectria haematoccoca* var. *cucurbitae*. Pestic. Biochem. Physiol. **40** (1991): 212–220.

KAPPAS, A., and GEORGOPOULOS, S. G.: Genetic analysis of dodine resistance in *Nectria haematococca* (Syn. *Hypomyces solani*). Genetics **66** (1970): 617–622.

KARIN, M., NAJARIAN, R., HASLINGER, A., VALENZUELLA, P., and WELCH, J.: Pirmary structure and transcription of an amplified genetic locus: The *CUP* 1 locus in yeast. Proc. Natl. Acad. Sci. (U.S.A.) **81** (1984): 337–341.

KATAN, T., DUNN, M. T., and PAPAVIZAS, G. C.: Genetics of fungicide resistance in *Talaromyces flavus*. Can. J. Microbiol. **30** (1984): 1079–1087.
– SHABI, E., and GILPATRICK, J. D.: Genetics of resistance to benomyl in *Venturia inaequalis* from Israel and New York. Phytopathology **73** (1983): 600–603.
KEON, J. P. R., WHITE, G. A., and HARGREAVES, J. A.: Isolation, characterization and sequence of a gene conferring resistance to the systemic fungicide carboxin from the maize smut pathogen *Ustilago maydis*. Vurr. Genet. **19** (1991): 475–481.
KHAI, M., and BOMPEIX, G.: Inheritance of resistance/sensitivity to iprodione in a cross of two isolates of *Botryotinia fuckeliana* (de Bary) Whetz. C. R. Acad. Sci. Paris, t. 311, Serie III (1990): 163–168.
KOENRAADT, H., and JONES, A. L.: The use of allele-specific oligonucleotide probes to characterize resistance to benomyl in field strains of *Venturia inaequalis*. Phytopathology **82** (1992): 1354–1358.
– SOMERVILLE, S. G., and JONES, A. L.: Characterization of mutations in the beta-tubulin gene of benomyl resistant field strains of *Venturia inaequalis* and other plant pathogenic fungi. Phytopathology **82** (1992): 1348–1354.
LANCASHIRE, W. E., and GRIFFITHS, D. E.: Biocide resistance in yeast: isolation and general properties of trialkyltin resistant mutants. Fed. Europ. Biochem. Soc. Letters **17** (1971): 209–214.
LEROUX, P., and BERTHIER, G.: Resistance to carboxin and fenfuran in *Ustilago nuda* (Jens.) Rostr., the causal agent of barley loose smut. Crop Prot. **17** (1988): 16–19.
– and FRITZ, R.: Similarities between the antifungal activity of fenpiclonil, iprodione and tolclofos methyl. In: LYR, H., and POLTER, C. (Eds.): Systemic Fungicides and Antifungal Compounds. Eugen Ulmer, Stuttgart, 1993, pp. 91–97.
ORBACH, M. J., PORRO, E. B., and YANOFSKY, C.: Cloning and characterization of the gene for β-tubulin from a benomyl-resistant mutant of *Neurospora crassa* and its use as a dominant selectable marker. Mol. and Cell. Biol. **6** (1986): 2452–2461.
PEEVER, T. L., and MILGROOM, M. G.: Inheritance of triadimenol resistance in *Pyrenophora teres*. Phytopathology **82** (1992): 821–828.
SHABI, E., and KATAN, T.: Genetics, pathogenicity and stability of carbendazim-resistant isolates of *Venturia pyrina*. Phytopathology **69** (1979): 267–269.
SHATTOCK, R. C.: Studies on the inheritance of resistance to metalaxyl in *Phytophthora infestans*. Plant Pathol. **37** (1988): 4–11.
SHAW, D. S., and ELLIOT, C. G.: Streptomycin resistance and morphological variation in *Phytophthora cactorum*. J. Gen. Microbiol. **51** (1968): 75–84.
STANIS, V. F., and JONES, A. L.: Reduced sensitivity to sterol-inhibiting fungicides in field isolates of *Venturia inaequalis*. Phytopathology **75** (1985): 1098–1101.
SUMMERS, R. W., HEANY, S. P., and GRINDLE, M.: Studies of a dicarboximide resistant heterokaryon of *Botrytis cinerea*. Proc. 1984 Br. Crop Prot. Conf. – Pests and Diseases: 453–458.
TAGA, M., NAKAGAWA, H., TSUDA, M., and UEYAMA, A.: Ascospore analysis of kasugamycin resistance in the perfect stage of *Pyricularia oryzae*. Phytopathology **68** (1978): 815–817.
THOMAS, J. H., NEFF, N. F., and BOTSTEIN, D.: Isolation and characterisation of mutations in the β-tubulin gene of *Saccharomyces cerevisiae*. Genetics **112** (1985): 715–734.
THRELFALL, R. J.: The genetics and biochemistry of mutants of *Aspergillus nidulans* resistant to chlorinated nitrobenzenes. J. Gen. Microbiol. **52** (1968): 35–44.
TILLMAN, R. W., and SISLER, H. D.: Effect of chloroneb on the metabolism and growth of *Ustilago maydis*. Phytopathology **63** (1973): 219–225.
TUYL, J. M. VAN: Genetics of fungicide resistance. Meded. Landbouwhogeschool, Wageningen **77-2** (1977).
UESUGI, Y.: *Pyricularia oryzae* in rice. In: DEKKER, J., and GEORGOPOULOS, S. G. (Eds.): Fungicide Resistance in Crop Protection. Pudoc, Wageningen 1982, pp. 207–218.

UMEZONO, K., TAKASHI, T., HAYASHI, S., and YANAGIDA, M.: Two cell division cycle genes *ADA2* and *NDA3* of the fission yeast *Schizosaccharomyces pombe*: control of microtubular organisation and sensitivity to antimitotic benzimidazole compounds. J. Mol. Biol. **168** (1983): 271–284.

VOMVOYANNI, V.: Multigenic control of ribosomal properties associated with cycloheximide sensitivity in *Neurospora crassa*. Nature **248** (1974): 508–510.

WEBBER, J., MITCHELL, A. G., and SMITH, F.: Linkage of the genes determining mating type and fungicide tolerance in *Ophiostoma ulmi*. Plant Pathol. **35** (1986): 512–516.

WELLMANN, H., and SCHAUZ, K.: DMI-resistance in *Ustilago maydis*. I. Characterization and genetic analysis of triadimefon-resistant laboratory mutants. Pestic. Biochem. Physiol. **43** (1992): 171–181.

WHITE, G. A., THORN, G. D., and GEORGOPOULOS, S. G.: Oxathiin carboxamides highly active against carboxin-resistant succinic dehydrogenase complexes from carboxin selected mutants of *Ustilago maydis* and *Aspergillus nidulans*. Pestic. Biochem. Physiol. **9** (1978): 165–182.

– – Thiophene carboxamide fungicides: Structure activity relationships with succinate dehydrogenase complex from wild-type and carboxin-resistant mutant strains of *Ustilago maydis*. Pestic. Biochem. Physiol. **14** (1980): 26–40.

– and GEORGOPOULOS, S. G.: Thiophene carboxamide fungicides: Structure-activity relationships with the succinate dehydrogenase complex from wild-type and carboxin-resistant mutant strains of *Aspergillus nidulans*. Pestic. Biochem. Physiol. **25** (1986): 188–204.

WHITTINGHAM, W. F.: The inheritance of acenaphthene tolerance in *Neurospora sitophila*. Amer. J. Bot. **49** (1962): 866–869.

WILKIE, D., and LEE, B. K.: Genetic analysis of actidione resistance in *Saccharomyces cerevisiae*. Genet. Res. Camb. **6** (1965): 130–138.

YAMAMOTO, M.: Genetic analysis of resistant mutants to antimitotic benzimidazole compounds in *Schizosaccharomyces pombe*. Molec. Gen. Genet. **180** (1980): 231–234.

YARDEN, O., and KATAN, T.: Mutations leading to substitutions at amino acids 198 and 200 of beta-tubulin correlate with benomyl-resistance phenotypes of field strains of *Botrytis cinerea*. Phytopathology **83** (1994): 1478–1483.

ZIOGAS, B. N., and GEORGOPOULOS, S. G.: Mitochondrial electron transport in a carboxin-resistant, antimycin A – sensitive mutant of *Ustilago maydis*. Pestic. Biochem. Physiol. **22** (1984): 24–31.

– GIRGIS, M., and GEORGOPOULOS, S. G.: Allelic mutations for benzimidazole resistance with and without change in sensitivity to N-phenylcarbamates in *Ustilago maydis*. In: LYR, H., and POLTER, C. (Eds.): Systemic Fungicides and Antifungal Compounds. Eugen Ulmer, Stuttgart 1993, pp. 45–50.

Chapter 4

Principles of uptake and systemic transport of fungicides within the plant

St. Neumann and F. Jacob

Fachbereich Biologie, Martin-Luther-Universität Halle – Wittenberg, FRG

Introduction

Although several protective, non-penetrating fungicides have a stable place on the fungicide market, most of the modern, selective fungicides exhibit systemic properties or are at least locally mobile (Shephard 1985; Cohen and Coffey 1986; Kuck and Scheinpflug 1986). A necessary prerequisite is a sufficient selectivity which allows control of the fungal pathogen without damaging the host organism. Therefore, the main principles of penetration and transport of xenobiotics, especially of fungicides, are summarised in this chapter. The specific behaviour of the various fungicides is described in other chapters.

Reviews and general descriptions

In the past twenty years results of studies on uptake and translocation of xenobiotic substances in higher plants were repeatedly summarised (Crafts and Crisp 1971; Crisp 1972; Jacob et al. 1973; Crowdy 1973; Shephard 1973, 1985; Price 1977, 1979; Christ 1979; Hartley and Graham-Bryce 1980; Jacob and Neumann 1983; Neumann et al. 1985a, 1988; Lichtner 1986; Kleier 1988; Devin 1989; Grayson and Kleier 1990; Bromilow et al. 1991). A special interest was taken in herbicides, because their systemic distribution is performed not only in the xylem but also partly in the phloem. Some reports dealt with the systemic behaviour and pattern of distribution of fungicides: Evans (1971), van de Kerk (1971), Crowdy (1972), Erwin (1973), Grossmann (1974), Fehrmann (1976), Marsh (1977), Peterson and Edgington (1981), Hassall (1982), Neumann et al. (1985b), Shephard (1985), and Cohen and Coffey (1986).

Terminology of transport processes

Transport in Tissues

1. apoplastic	– transport in the coherent network of free space, cell walls and non-living cells (= apoplast)
1.1. euapoplastic	– apoplastic movement without any passage through protoplasts

1.2. pseudoapoplastic — apoplastic movement with occasional passage through or retention in protoplasts
2. symplastic — transport in the coherent network of protoplasts connected by plasmodesmata (= symplast)

Long-distance transport

1. xylem-mobile — apoplastic transport in vessels and tracheids of xylem by means of the transpiration stream (xylem-systemic)*)
2. phloem-mobile — symplastic transport in the sieve tubes of the phloem by means of the mass flow from source to sink (phloem-systemic)
3. ambimobile — transport in the xylem and phloem (ambisystemic)
4. locally mobile — transport within the organ of application (locosystemic)
5. amobile — no long-distance transport from the site of application (non-systemic)

General aspects

The chemical control of phytopathogenic fungi has to be adapted to the biology of the pathogen and the sites of its occurrence. The protection of plant tissues requires systemic fungicides to be taken up by leaves, roots, seeds or fruits and to be translocated a short distance in the parenchyma (locosystemic) or a long distance in the flow of solutions through the xylem (xylem-systemic) or the phloem (phloem-systemic) or both (ambisystemic). The two long-distance transport systems differ (i) in the anatomy, (ii) in the driving force, and (iii) in the factors regulating strength and direction of the transport. For practical reason is important to know to what extent a given exogenously applied substance could be translocated in the both streams and if it is able to reach special target organs, e.g. a fungicide which is expected to protect young leaves or fruits should be translocated by the phloem after the application to a fully expanded exporting leaf.

It is difficult to obtain an equal distribution of substances on the whole leaf surface. Therefore, systemic fungicides which undergo also some redistribution ensure a better disease control. For several reasons (costs, resistance problems etc.) both, systemic and non-systemic antifungal compounds will be used in future. Obviously, the application of systemic fungicides has advantages as well as disadvantages, as evidenced by increased rates of decomposition and phytotoxic or growth-regulating side effects. The increase of resistance (cf. chapter 2) of the fungi, e.g. to benomyl (DEKKER 1977) and thiabendazole (GEORGOPOULOS 1977), can lead to ineffectiveness of some compounds for a practical disease control (monosite inhibitors).

Locosystemic fungicides show a limited spread from the site of application and the mobility is intermediate between systemic and non-systemic. It is very often unclear,

---

*) The term "xylem-systemic" characterises the ability of an absorbed pesticide to furnish the whole plant with the pesticide activity by means of the xylem transport; "xylem-mobile" indicates the transport in the xylem of any substance.

whether the penetration takes place apoplastically and/or symplastically. Systemic fungicide movement comprises uptake, translocation, and distribution in the plant to reach the pytopathogenic fungus. Therefore, a successful application requires certain properties of mobility as well as a strong fungicidal activity. The passage through outer cell wall layers, cells and intercellular spaces could influence the strength of the fungicide effect by chemical or physical interactions with cellular constituents. A reduced activity can be the consequence of adsorption, chemical binding or biotransformation, whereas an enhanced activity may be due to the transport in the plant, an accumulation in target organs or the release of active metabolites.

Till now the absorption of fungicides by pathogenic fungi in host-parasite combinations has not been investigated intensively in a quantitative respect. Undoubtedly, non-polar substances, bases as well as acids in non-ionic forms are favoured in penetration of the lipoproteinic phase of the plasmalemma. Xenobiotic compounds are taken up more easily by germ hyphae than by spores (HASSALL 1982). Pesticides as xenobiotics move passively through the cell membrane by diffusion. A small number of compounds with structural similarities to biotic compounds was found to be absorbed by a carrier-mediated membrane transport, e.g. the herbicide glufosinate (ULLRICH et al. 1990) and phosphonate, the fungitoxic metabolite of the fungicide Aliette (OUIMETTE and COFFEY 1990).

Differences in chemical structures result in different physico-chemical properties, such e.g. lipo-hydrophilicity, solubility, mol volume, steric parameter, ionisation, and distribution of charge density. These characteristic features influence uptake and transport in the plant. Furthermore, metabolic and non-metabolic changes of the compounds could modify these properties and by this, often the mobility. There is a close connection between mobility, stability and strength of the fungicide effect. The relevant concentration of a chemical in the tissue of higher plants and in the cells of parasitic fungi is determined by the equilibrium between the rate of absorption and the rate of destruction.

The use of formulations could increase the amounts of fungicide which are taken up into the plant and subsequently submitted the long distance transport (KUCK 1987). However, recent experiments show that mobility of the compound is not changed.

Translocation in the xylem

Xylem translocation takes place in tracheids and vessels forming a continuous network of dead elongated cells with more or less lignified secondary cell walls. The distribution pattern and the extent of movement in the xylem are determined by the gradient in the water potential between soil and air. Thus, the transport is usually directed from the root to the transpiring areas, especially to the leaves. Factors controlling the intensity of transpiration: relative humidity, temperature, light, and phytohormones (especially ABA) could influence the translocation rate and the distribution of xenobiotics dissolved in the xylem sap.

The distribution of xylem-mobile xenobiotics is characterised by the following peculiarities (PETERSON and EDGINGTON 1975a):

1. Substances are accumulated at the sites of high transpiration, e.g. at the tips and margins of leaves.
2. The transport into plant organs with negligible transpiration (fruits and young leaves) is very limited.

3. Xylem-mobile substances do not undergo any downward movement from expanded leaves. If they are applied to the basis of a leaf, they are mainly transported to the tip. Transport in the opposite direction is extremely rare.

Xylem-mobile xenobiotics are absorbed together with water, mainly in the root hair zone, which is 5–50 mm from the root tip. In some cases the absorption of xenobiotics seems to be more intensive than that of water (Shone and Wood 1974). In contrast to the leaves the rhizodermis is not covered by a well developed cuticle. Therefore, it could not expected to be a barrier for the uptake of xenobiotics into the roots (Esau 1965). Radial movement through the cortex zone occurs either

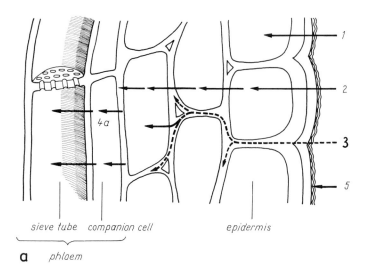

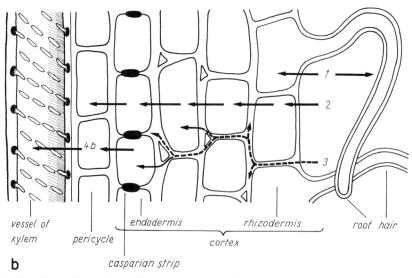

Fig. 4.1 The pathway of xenobiotic substances in plant tissue by penetrating (a) the leaf or (b) the root surface. 1 uptake in outer cells, 2 symplastic transport, 3 apoplastic transport, 4a absorption in sieve tubes, 4b transfer to vessels, 5 absorption in cuticular layers.

symplastically or apoplastically (Fig. 4.1). Symplastically moving substances cross the plasmalemma and are transported *via* protoplasts and plasmodesmata of cortex cells and through the endodermis cells to the vessels. Transfer in the apoplast is supposed to occur in the free space of the cell walls. Since this route is blocked by lipophilic incrustations of the Casparian strips at the endodermis level, xenobiotics must enter the symplast at this site to be transferred to the xylem. Entrance into the symplast requires lipophilic properties. Therefore, extremely hydrophilic compounds like the fluorescent dye tri-sodium-3-hydroxy-5,8,10-pyrenetrisulfonate (PTS) are retained in the free space of the cortex. Despite of one exception they are not translocated in the long distance stream of the xylem to the leaves. In a certain stage the endodermis continuity is disturbed by emerging lateral branches of the root (BONNET 1969; DUMBROFF and PEIRSON 1971; KARAS and MCCULLY 1973) and apoplastic connections seem to be responsible for the appearance of highly hydrophilic substances in the xylem in very low concentrations (PETERSON and EDGINGTON 1975b; PETERSON et al. 1981).

Uptake of xenobiotics by roots

Time courses of uptake of fungicides (CROWDY and RUDD-JONES 1956a; LEROUX and GREDT 1983; DOBE et al. 1986) and herbicides (MOODY et al. 1970; SHONE et al. 1974; DONALDSON et al. 1973; MORRISON and VAN DEN BORN 1975; BALKE and PRICE 1988) were found to be initially rapid, followed by a phase of declining rate of absorption. The initial part represents the uptake into the apoplast, the second is correlated with the translocation of the absorbed substances from the root to the leaves depending on the extent of transpiration (MINSHALL 1954; CROWDY and RUDD-JONES 1956b; SHONE and WOOD 1972; WALKER and FEATHERSTONE 1973). Linear concentration response of absorption, a low temperature coefficient, and the insensibility to metabolic inhibitors suggest an uptake by diffusion (SHONE and WOOD 1974; DONALDSON et al. 1973; LEROUX and GREDT 1983; DOBE et al. 1986).

In hydroponic culture the applied amounts of xenobiotics are completely available for absorption by the plants. Some physical and chemical interactions between xenobiotic compounds and the soil matter could affect the quantities available for uptake by the root (BRIGGS 1973; HANCE 1983; THIELERT et al. 1986; TOPP et al. 1986). Due to the pH-value of the soil solution the cationic moieties of dissociating fungicides like carbendazim could be adsorbed to a certain extent on negatively charged soil matter. Consequently, smaller amounts of molecules are accessible for root uptake (AHARONSON and KAFKAFI 1975; AUSTIN and BRIGGS 1976). An analogous effect is attributed to hydrophobic interaction between non-ionised lipophilic molecules and soil components. The absorption of non-ionised compounds was shown to be positively correlated to the hydrophilic-lipophilic balance of substances characterised by the log K_{OW}-values (octanol-water partition coefficient) (BRIGGS 1973; BRIGGS et al. 1977). The derivatives of the fungicide isoprothiolane are adsorbed better, the longer the alkyl side chain is. Accordingly the actual amount in the soil solution available for the uptake is decreased (UCHIDA and KASAI 1980; UCHIDA and SUZUKI 1982). From studies with oxymecarbamate and phenylureas a log K_{OW} of 0.5 was found to be optimal for root uptake following soil application. In hydroponic culture an optimum uptake can be expected for compounds with log K_{OW} 1.7 (BRIGGS et al. 1977).

The uptake of xenobiotics by roots can be characterised by the root concentration factor (RCF) defined by SHONE and WOOD (1974) as

$$RCF = \frac{\text{concentration of the xenobiotic in the root}}{\text{concentration of the xenobiotic in the external solution}}$$

The RCF is usually independent on the xenobiotic concentration in the external solution, supporting the view that the entry is accomplished by diffusion. Furthermore log K_{OW} and RCF are positively correlated: Increasing lipophilicity is linked with an increasing absorption (SHONE and WOOD 1974; BRIGGS et al. 1977).

This is in accordance with findings by LEROUX and GREDT (1983) who demonstrated the pH-dependence of uptake of benzimidazoles and thiophanates. The same refers to the uptake of ethirimol investigated by SHONE and WOOD (1974). RCF-values >1 indicate accumulation. It is due to the association of lipophilic uncharged xenobiotics with lipophilic cell components or due to the ion trapping of acidic compounds. Cationic fungicides like carbendazim (pK_a = 4,1), thiabendazole (pK_a = 4.7) and ethirimol (pK_a = 4.8) become protonated in physiological pH ranges. Independently of lipophilic interactions, they are accumulated by adsorption to negative charges of the cell wall. Uptake of these compounds is obviously pH dependent and was inhibited by addition of bivalent cations (Ca^{2+} and Mg^{2+}) (LEROUX and GREDT 1983; SHONE et al. 1974). BRIGGS et al. (1982) suggested that there exists a general relationship between RCF and log K_{OW} expressed by the function $\log (RCF - 0.82) = 0.77 \log K_{OW} - 1.52$, which is valid for young roots without lignified cell walls. Subsequently, the lipophilic nature of root constituents involved in adsorption of xenobiotics, seems to be similar among different plant species, at least during young stages of root development.

Properties of xylem-systemic substances

According to their physico-chemical properties, xenobiotics attain a characteristic equilibrium of adsorbed moieties, and those which are freely mobile in the root apoplast and symplast. This equilibrium is responsible for the extent of translocation from the root to the shoot. Since xylem translocation is preferably determined by transpiration, its efficiency can be described by the so-called "transpiration stream concentration factor" (TSCF) (SHONE and WOOD 1974):

$$TSCF = \frac{\text{concentration of the xenobiotic in the transpiration stream}}{\text{concentration of the xenobiotic in the external solution}}$$

TSCF values are independent on the xenobiotic concentration in the external solution. They are always <1 and those of non-ionised compounds are correlated to log K_{OW}.

From investigations by CROWDY et al. (1959) on mobility of griseofulvin derivatives after root application in *Vicia faba*, it was concluded that substances with an hexan water partition coefficient of 0.3 are favourably and of 0.8 are sufficiently transported in xylem. K values <6 indicate amobility or retention in the root. Transport of oxymecarbamates and phenylurea derivatives in *Hordeum* (BRIGGS et al. 1977) as well as dialkyl-1,3-dithiolan-2-ylidenemalonates (derivatives of the fungicide isoprothiolane) in rice (UCHIDA 1980) exhibit a similar correlation between partition coefficient and transport in xylem. BRIGGS et al. (1977, 1982 and 1983) demonstrated that the rates of translocation, expressed as TSCF-values, against partition coefficient yield a Gaussian curve: A sharp increase of TSCF was found, when log K_{OW} exceeds -0.5, a decline was noted above 3 and the optimum lipophilicity-hydrophilicity balance for translocation is around 1.7. UCHIDA (1980) also showed a log K_{OW} dependent increase of transport in xylem up to 2.9, substances with log K_{OW} > 4 are not translocated. Furthermore similar results derived from studies on transport of atrazine and linuron in lettuce, parsnip, carrots and turnip (WALKER and FEATHERSTONE 1973). Retention of lipophilic xenobiotics occurs by adsorption to cell constituents in the root and additionally in the lower and middle parts of the stem. Substances of strong lipophilicity never appear in transpiring leaves. Lignin is generally accepted to adsorb lipophilic compounds and to cause the retention in the root and in the stem. BARAK et al. (1983a, 1983b) found a linear correlation between log K_{OW} and adsorption of the fungicides carbendazim, triarimol, triadimefon, nuarimol, fenarimol (cf. chapter 12) and of the herbicide fluometuron to "ground stems" as well as to purified lignin of *Phaseolus vulgaris*, *Capsicum annuum* and *Gossypium hirsutum*. Association to cellulose, polygalacturonic acids and proteins seems to be negligible. Nevertheless, methylation of "ground stems" resulted in a two to threefold increase of adsorption. Both, hydrophilic interaction of undissociated carbendazim molecules with lignin and the association of their protonated moieties with negative charges of the cell wall are pH-dependent.

Ca^{2+} inhibited adsorption to *Pinus* lignin. The latter phenomenon also became obvious from experiments with cationic dyes (VAN BELL 1978), cationic xenobiotics (EDGINGTON and DIMOND 1964; WILHELM and KNÖSEL 1976), and positively charged amino acids (HILL-COTTINGHAM and LLOYD-JONES 1968, 1973). Since translocation in the xylem is limited by adsorptive binding to cell wall charges it could be promoted by the addition of Ca^{2+} and or Mg^{2+}. Due to proceeding lignification, adsorption of xenobiotics on lignin increases with the age of stems, which could be the reason for differences in transport of ethirimol and diethirimol in herbaceous and woody plants (SHEPHARD 1973).

Practical consequences

Absorption of fungicides or other xenobiotics by the root and the subsequent transport in the xylem result from different consecutive stages of partition between hydrophilic and lipophilic compartments. From the particular knowledge of these processes information becomes available about optimum partition coefficients, pH values of the application medium. Considering the distribution of xenobiotics in soil and plants, BRIGGS et al. (1977) concluded that xylem-systemic fungicides should exhibit log K_{OW} in the range between -0.5 to 3, if responses are expected in leaves. On the other hand, more lipophilic compounds (log K_{OW} 3 – 5) seem to be favourable for acting in roots because of a confined movement. Generally speaking, hydrophilic compounds are transferred through the soil and rapidly absorbed by the root, whereas lipophilic chemicals undergo a retarded movement in soil and long-lasting uptake into the plant. Furthermore the transport limitation of lipophilic substances in the xylem due to the age of the plants and the actually effective proportion of applied substances should be taken into account.

Translocation in the phloem

Phloem transport occurs in the continuous network of sieve tubes, consisting of elongated living cells (sieve elements) with characteristic cell wall perforations, the so-called sieve areas or sieve plates. Both, the nucleus and the tonoplast have disappeared in completely differentiated sieve elements. The pH of the sieve tube protoplasm amounts to 7.2 – 8.7 which is markedly higher than in surrounding parenchyma cells. This pH is maintained by the activity of plasmalemma-bound ATPases pumping protons into the apoplast. The apoplast pH is about 5 (GIAQUINTA 1983). Sieve elements and companion cells form an anatomical and physiological unit: the sieve element-companion cell complex (se-cc-complex). Both types of cells seem to be involved in the loading of assimilates, whereas the longitudinal movement takes place in the sieve tubes exclusively. Phloem transport is accomplished between source and sink. Fully expanded green leaves producing and exporting assimilates are sources, whereas growing or storing organs with photosynthate demand are sinks.

The distribution of phloem- and ambimobile xenobiotics are characterised by the following peculiarities (CRAFTS and CRISP 1971; MÜLLER 1976).

1. The movement of xenobiotics in the phloem is inevitably connected with transport of photoassimilates. Thus, it depends on the source-sink-relationship due to a given developmental stage. There is no export of xenobiotics from very young leaves. When the leaves reach two third of their final size, they become source leaves, which are capable of phloem loading and export of externally applied xenobiotics too. An the other hand a movement into fully expanded exporting leaves was not found.

2. When the assimilate transport was ceased the movement of xenobiotics was also stopped. Xenobiotics which interfere with the sucrose synthesis, with the loading process or which damage cell membranes will stop the phloem translocation from the source leaf. Thus, they interrupt their own transport. This phenomenon is known as "self-limiting" of mobility by phytotoxic action and was demonstrated by compounds such as chlorsulfuron and glyphosate (CHALEFF and MAUVAIS 1984; GEIGER et al. 1986, 1987).

3. Following root application only small amounts of phloemmobile compounds are transported to upper parts, whereas ambimobile compounds show the characteristic distribution patterns of xylem-systemic xenobiotics.

4. In contrast to xylem-mobile substances, which do not exhibit common structural features, xenobiotics mobile in the phloem (phloem- and ambimobile) are often acids. In many cases the acidity depends on carboxylic groups (phenoxy acids), on phenolic hydroxyl groups (maleic hydrazide) or on the arrangement of $-SO_2-NH-$ in the so-called nitrogen acids (mefluidid, asulam) (JACOB et al. 1973). It was shown that the phloem mobility disappears with the elimination of $-COOH$ groups. Conversely, the introduction of carboxylic groups into xylem-mobile chemicals resulted in ambimobile derivatives (MITCHELL et al. 1954; JACOB et al. 1973; CRISP and LOOK 1979; JACOB and NEUMANN 1983; NEUMANN et al. 1985).

Uptake of xenobiotics into the leaf

Unlike the rhizodermis of the root, the epidermis of the leaves is not adapted for absorption of dissolved substances. The leaf epidermis with stomata and cuticle enables the regulation of the gaseous exchange avoiding an appreciable loss of water. The cuticle covering leaves as a $0.5-14\,\mu m$ thick layer consists of cutin, intracuticular lipids, polysaccharides, polypeptides, and phenols (MARTIN and JUNIPER 1970; NORRIS 1974; SCHÖNHERR and HUBER 1977; HUNT and BAKER 1980; WATTENDORF and HOLLOWAY 1980; HOLLOWAY 1982). Epicuticular waxes impede the adhesion of compounds in aqueous solution on the surface of leaves. Adhesion can be quantified by the so-called contact angle (HOLLOWAY 1970; RENTSCHLER 1971; NETTING and WETTSTEIN-KNOWLES 1973) (Fig. 4.2).

Principles of uptake and systemic transport of fungicides within the plant 61

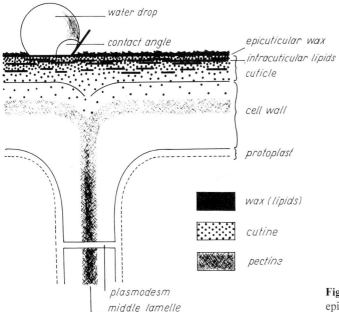

Fig. 4.2 Structure of a leaf epidermis.

Nevertheless the main barrier for the penetration of polar substances through the cuticle are the intracuticular lipids. Thus, penetration rates vary greatly according to the proportion of intracuticular lipids. In contrast to charged moieties, undissociated molecules of acidic compounds penetrate preferably. Additionally, intracuticular lipids retain lipophilic xenobiotics, in some cases a depot effect could be observed (RIEDERER and SCHÖNHERR 1984, 1985). Entrance *via* the abaxial epidermis was found to be more efficient than through the adaxial one, facilitated by a lower content of intracuticular lipids and a higher number of stomata (NORRIS and BUKOVAC 1968). The promoting impact of stomata seems to be due to the absence of epicuticular waxes in the stomata chamber (DYBING and CURRIER 1961; PRASSAD *et al.* 1967; MANSFIELD *et al.* 1983). Furthermore a high humidity in the stomata chamber and other environmental factors which cause opening of stomata enhance absorption (SHARMA and VAN DEN BORN 1973; GREENE and BUKOVAC 1974).

Uptake of xenobiotics by the phloem

The transfer of sucrose from the mesophyll cells to the vascular bundle and loading into the sieve tube could be either symplastically or at least in the final step apoplastically (VAN BEL 1992). The symplastic transfer and loading proceed *via* plasmodesmata. The mechanism which is responsible for the accumulation of transport sugars in the sieve tube is unknown still. The apoplastic loading of sucrose into the sieve tubes was found to be mediated by proton-cotransport (GIAQUINTA 1983). In contrast to this the entrance of xenobiotics into the se-cc-complex is accomplished by diffusion (GRIMM et al. 1983). The velocity of uptake was found proportional to log K_{OW}. Both, phloem- and xylem-mobile substances enter the sieve tube. Nevertheless, the essential prerequisite for translocation in the phloem is the retention of the xenobiotics in the sieve tubes during the translocation along the path. Outside of the application zone xylem-mobile substances are leaked out from the phloem very quickly. Most of the phloem- and ambimobile compounds are acidic ones which can be retained by an ion trapping mechanism (JACOB and NEUMANN 1983; NEUMANN 1985; NEUMANN et al. 1985 a). According to the pH regime in the phloem and the pK_a-values, acidic subtances dissociate to different extents in the apoplast (pH 5), and in the symplast (pH 8). The proportion of dissociated moieties is higher in the alkaline medium (sieve tube) than in the acidic one (apoplast). Since the more hydrophilic anions penetrate membrane barriers at lower velocities and the undissociated molecules are usually lipophilic enough to cross the plasmalemma by diffusion with sufficient velocity, the latter represent the principal form of xenobiotics which enter the sieve tube. Due to their pK_a-values which are usually smaller than 7, acidic phloem- and ambimobile xenobiotics undergo a nearly complete dissociation in the sieve tube. The relatively lipophilic, undissociated molecules are leaked out very quickly from the sieve tube along the path, while the relatively hydrophilic anions are escaped to a lesser extent and translocated together with the assimilates to sink organs (Fig. 4.3). The accumulation of xenobiotics in the symplast of the sieve tube can be calculated by the following equation (RAVEN and RUBERY 1982):

$$\frac{[HA]_i + [A]_i}{[HA]_o + [A]_o} = \frac{(1 + 10^{pHi-pKa}) [P_{HA}/P_A + \{(FE/RT)/(1 - e^{-FE/RT})\} \cdot 10^{pHo-pKa}]}{(1 + 10^{pHo-pKa}) [P_{HA}/P_A + \{(FE/RT)/(1 - e^{-FE/RT})\} \cdot 10^{pHo-pKa} \cdot e^{-FE/RT}]}$$

where the subscripts i and o refer to the concentration in the symplast and in the apoplast, respectively. P_{HA} and P_A are the permeability coefficients of the undissociated and the dissociated moieties, F is the Faraday constant, E is the charge on the membrane, T is the temperature in Kelvin, and R is the universal gas constant.

There are some experimental proofs for the ion trapping mechanism (NEUMANN et al. 1985 a and GRIMM et al. 1985). Following leaf application in *Sinapis alba* 98.5% of the absorbed xylem-mobile defenuron is retained in the treated cotyledon. By introduction of a —COOH group defenuron is converted to the ambimobile N-(p-carboxyphenyl)-N'-methylurea (CPMU). Both compounds are taken up by isolated conducting tissue of *Cyclamen persicum* in the linear proportion to the external concentration without any accumulation.

However in efflux experiments defenuron, as well as other xylem-mobile xenobiotics (SHONE and WOOD 1974; SHONE et al. 1974; BOULWARE and CAMPER 1973; EL IBAOUI et al. 1986) was leaked out very quickly in contrast to the moderate efflux of CPMU. The total concentration in the tissue after leaching for 2 h still amounted to 32% of the ambimobile compound and to 2,4% of the xylem-mobile one. 2,4-D, maleic hydrazide and amitrole were also found to be leaked out moderately from *Cyclamen* conducting tissue (GRIMM et al. 1985). Similarly experiments with potato tuber disks revealed a markedly slower escape of phloem-mobile substances compared with xylem-mobile ones (EDGINGTON and PETERSON 1977; MARTIN and EDGINGTON 1981).

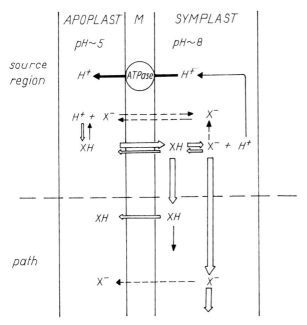

Fig. 4.3 Diagram of phloem transport of xenobiotics. Uptake of phloem- and ambimobile, as well as xylem-mobile xenobiotics into se-cc complex by diffusion of the undissociated molecules (XH). Retention of the relatively hydrophilic anions (X$^-$) by ion trap mechanism. Along the pathway the relatively lipophilic, undissociated molecules XH are leaked out very quickly from the sieve tube, while the relatively hydrophilic anions X$^-$ escaping to a lesser extent are translocated together with the assimilates to sink regions.

A number of compounds without acidic properties is retained by means of the so-called "intermediate permeability" (TYREE et al. 1979), e.g. the nematicide oxamyl (PETERSON et al. 1978), the herbicides amitrole (LICHTNER 1983) and glyphosate (GOUGLER and GEIGER 1981), and the fungicide pyroxychlor (SMALL et al. 1970). These substances are characterised by a high degree of water solubility (oxamyl 60 g/l; amitrole 280 g/l; glyphosate 10 g/l, pyroxychlor 26 g/l) and permeate very slowly through biological membranes. In contrast to this the velocity of the transport in the sieve tubes is relatively high. Thus, the longitudinal transfer along the path occurs without a substantial lateral efflux from the sieve tube into the apoplast and noticeable amounts could arrive in sink organs.

Xenobiotics without any structural prerequisites to be submitted either ion trapping or retention in the phloem sap by optimum permeability never appear in the sink organs. Although xylem-mobile compounds are submitted a rapid leakage along the phloem path, a very limited translocation in the sieve tube can be observed. The covered distance is determined by the hydrophilicity-lipophilicity balance (expressed as log K_{OW}) and the applied concentration of the compound. From an increased concentration at the source leaves arose an elongated translocation path of the xenobiotic (SOLEL et al. 1973; PRICE 1979). Thus, highly effective xylem-mobile pesticides can be expected to show some response in basipetal parts after leaf application.

Metabolism and mobility

Generally speaking, the fungitoxic quality of any substance is dependent on its chemical structure, which usually does not offer favourable conditions for absorption into higher plants and long distance transport. Metabolic changes and decomposition of active substance in plants mainly result in detoxification and in a loss of effectiveness. Slightly water soluble substances often become more polar. This could be connected with changes of the mobility. In some cases species specific delayed decomposition combined with depot effects were observed (WAIN and CARTER 1967; GASZTONYI and JOSEPOVITS 1979). On the other hand effective derivatives with favourable conditions for translocation could arise from metabolism; e.g. the esters chlorfenprop-methyl (BOLDT and PUTNAM 1980) and dichlofop-methyl (SHIMABUKURO et al. 1979) are transformed metabolically to acids which are translocated in the phloem. Among the fungicides benomyl is rapidly conversed after absorption. The remove of the butylisocyanat side chain (CLEMONS and SISLER 1969) leads to carbendazim, which is the real fungitoxic compound with decreased lipophilicity. Both, benomyl and carbendazime are xylem-mobile compounds. They are absorbed in equal amounts by pumpkin leaves (YOUNG and HAMMET 1973). N-butylisocyanat has been identified as an inhibitor of cutinase. It could cause a protective effect against fungi which infect the cuticle (KÖHLER et al. 1982) (c.f. chapter 25).

The piperazin derivative triforine (chapter 12) has only a small water solubility and it is xylem-mobile. However, slight transport from the leaves to the roots could be ascertained in light (MARS 1977; GREEN et al. 1977), whereas it is unclear, whether it is due to the formation of phloem-mobile metabolites (FUCHS et al. 1976; HASSALL 1982):

After application of triadimefon (chapter 12), two diastereomere forms (triadimenol I and II) appear (BUCHENAUER 1979; BUCHENAUER and RÖHRER 1982), whereas triadimenol I is more fungitoxic. Both triadimenoles are four times more water soluble than the initial substance (PERKOW 1985). Nevertheless, the biotransformation seems to be without any change of the pseudoapoplastic transport. The highly absorbed carboxin is oxidised quickly to the non fungitoxic sulphoxide (BRIGGS et al. 1974). The carboxin sulphon (oxycarboxin) has a higher persistence, but it is absorbed to a smaller extent (SNEL and EDGINGTON 1970). It can be supposed that the high carboxin uptake is due to the strong lipophilicity of the compound. Nevertheless, carboxin is submitted a fast decomposition, which prevent a longlasting fungicide effect.

The phloem-systemic properties of the anti-oomycete fungicide Aliette (aluminium tris-O-ethyl phosphonate) bases on the fungitoxic metabolite phosphonate (FENN and COFFEY 1989). For the membrane transport of phosphonate the use of a phosphate carrier is supposed. Its physicochemical properties allow the retention in the sieve tube according an ion trapping mechanism (OUIMETTE and COFFEY 1990).

Altogether, metabolic processes could be connected with changes in mobility of the absorbed substances which can be advantageous for the distribution in the plants as far as the pesticide properties are preserved. The knowledge of the fungicide biotransformation and the activity and mobility of metabolits is still insufficient. Thus, it seems to be useful to consider the interaction between metabolism and transport.

Practical conclusions

Since the introduction of carboxins (cf. chapter 8) for the control of the loose smut pathogen *Ustilago nuda* by SCHMELING and KULKA (1966) systemic fungicides have very quickly gained in importance (EVANS 1971; GROSSMANN 1974; HASSALL 1982). However, the molecular configurations of the most of the present available systemic fungicides allow a distribution by the xylem stream. They are xylem-systemic; only very few of them are phloem-systemic to a limited extent. Therefore, a protection of the transpiring parts of the plant can be expected after treatment of the seeds and an application to the soil. The latter seems to be economically acceptable and ecologically harmless when granula or incrustations are used. The important disadvantage of xylem-systemic fungicides consists in the lack of the protection of the young leaves, the growing tip of the stem, flowers and fruits. Nevertheless, a continuous release of xylem-mobile fungicides which adsorbed on soil matter might be advantageous, since a depot effect could counteract a rapid decomposition in the plant (GRAHAM-BRYCE and COUTTS 1971). The uptake into the root seems to be more effective than into the leaf because the lack of lipophilic cell wall barriers. In some cases a protection of the cortex could be expected after lateral transfer of xylem-mobile fungicides (carbendazim, PETERSON and EDGINGTON 1970). From this consideration the demand for new phloem-systemic fungicides which will reach the growing and storing plant organs, young leaves, flower and fruits, becomes obvious. Due to a high thinning of phloem-systemic compounds along their route through the plant a high effectiveness and persistence is a decisive requirement.

Tests for uptake and translocation

Uptake tests

Since the cuticle is the main barrier for the entrance of exogenously applied substances a number of penetration tests are described (KERLER *et al.* 1984). Using either isolated cuticles or leaf disks the penetration through or the adsorption on the cuticle is determined (CURLER and SCHÖNHERR 1988a, 1988b; SCHREIBER 1990).

Translocation tests

The general scheme of translocation tests comprises the application of substances under investigation to roots or fully expanded leaves and the analysis in organs which are specifically supplied by the transpiration stream (leaves) or by the assimilate stream (young leaves, roots, flowers and fruits). In some cases the test system was reduced to a single detached leaf. Compounds could move in the xylem to the leaf tip or basipetally in the phloem (NEUMANN 1982; KUCK and THIELERT 1987; SHEPHARD 1987). Furthermore seeds were treated with fungicides. In a defined growth stage of the developing plant the distribution pattern of the applied substances are determined.

Detection of translocated fungicides is carried out by different methods (NEUMANN *et al.* 1988): (i) Bioassays on the basis of a protective or a curative disease control. The fungicides were applied before or after inoculation with the pathogenous fungus.

(ii) The application of radioactive labelled or unlabelled compounds and their detection by means of autoradiography, liquid scintillation counting, chromatographic methods (GC, TLC, HPLC) or immunoassay after extraction of the tissue and purification. The tests according to (i) show the biological activity of the applied compound in the target organs meeting directly the needs of practical use. Quantification of the translocated amounts are hardly possible. The application of radioactive labelled substances allow a semiquantitative (autoradiography) or quantitative (liquid scintillation counting) analysis of the transported label. Apart from chromatographic and immunological analysis a general disadvantage of both testing approaches consists in the lack of information on biotransformation of the applied substances.

A simple method to quantify translocation of a given substance in the long distance streams is the *Sinapis* seedlings mobility test (JACOB and NEUMANN 1983; NEUMANN et al. 1985a). The transport is characterised by the translocation quotient (Q_{tr}) which represents the proportion between phloem and xylem mobility. It is calculated from the ratio of radioactivity in hypocotyl and root after leaf application to the radioactivity in cotyledons and apex after root application.

Phloem-mobile substances are characterised by $Qtr > 1$, ambimobile by values between 1.0 and 0.2, xylemmobile by values <0.2. In this way it is possible to compare the translocation of different compounds and to determine the influence of surfactants and environmental conditions on the translocation quantitatively.

Some special tests of phloem transport based on the collection and analysis of phloem exudate from intact *Ricinus* plants (CHAMBERLAIN et al. 1984), detached inflorescence stalks from *Yucca* (NEUMANN 1982; NEUMANN et al. 1985a), from detached leaves (KING and ZEEVAART 1974; GROUSSOL et al. 1986), from excised hypocotyls (HANSON and COHEN 1985) and from flower pods (HOAD 1980).

Conclusions

1. Systemic fungicides differ from non-systemic ones by their ability to be redistributed in the plant. The long distance transport systems of xylem and phloem, however, are essential for distribution and the chemotherapeutic effect. Untill now most of the highly effective systemic fungicides are mainly xylem-systemic.
2. The absorption of fungicides by roots, leaves, seeds or fruits occurs passively by diffusion. The chemical structure and metabolite changes decide whether the tissue transport takes place on an apoplastic or symplastic route, and whether any compound is subjected to the long distance transport or not. The outer cell wall layers of the leaves play an important role in the process for absorption.
3. The relationship between hydrophilic and lipophilic molecular parameters of the applied compounds is responsible for used transport system. In the case of dissociating molecules the degree of dissociation and the pH value of the application solution play a decisive role. Substances of high lipophilicity or hydrophilicity are not or poorly transported.
4. The accumulation of substances might be due to the loss of water in transpiring leaves, due to the adsorption on molecular constituents, due to the biotransformation and due to the ion trapping.
5. The knowledge on relationship between the chemical and physical properties of xenobiotics and their mobility, should also be used for the recent control of fungal diseases and for the development of new agents with improved fungitoxic activity.

The authors thank Dr. Eckhard Grimm for helpful discussion and critical reading of the manuscript.

References

AHARONSON, N., and KAFKAFI, U.: Adsorption, mobility and persistence of thiabendazole and methyl-2-benzimidazolecarbamate in soils. J. Agric. Food Chem. **23** (1975): 720–724.

AUSTIN, D. J., and BRIGGS, G. G.: A new extraction method for benomyl residues in soil and its application in movement and persistence studies. Pesticide Sci. **7** (1976): 201–210.

BALKE, N. E., and PRICE, T. P.: Relationship of lipophilicity to influx and efflux of triazine herbicides in oat roots. Pestic. Biochem. Physiol. **30** (1988): 228–237.

BARAK, E., DINOOR, A., and JACOBY, B.: Adsorption of systemic fungicide and a herbicide by some components of plant tissues, in relation to some physico-chemical properties of the pesticide. Pesticide Sci. **14** (1983): 213–219.

– JACOBY, B., and DINOOR, A.: Adsorption of systemic pesticides on ground stems and in the apoplastic pathway of stems, as related to lignification and lipophilicity of the pesticides. Pesticide Biochem. Physiol. **20** (1983): 194–202.

BOLDT, P. F., and PUTNAM, A. R.: Selectivity mechanisms for foliar application of diclofop-methyl. I. Retention, absorption, translocation and volatility. Weed Sci. **28** (1980): 474–477.

BONNETT, H. T.: Cortical cell death during lateral root formation. J. Cell Biol. **40** (1969): 144–159.

BOULWARE, M. A., and CAMPER, N. D.: Sorption of some ^{14}C-herbicides by isolated plant cells and protoplasts. Weed. Sci. **21** (1973): 145–149.

BRIGGS, D. E., WARING, R. W., and HACKETT, A. M.: The metabolism of carboxin in growing barley. Pesticide Sci. **5** (1974): 599–607.

BRIGGS, G. G.: A simple relationship between soil adsorption of organic chemicals and their octanol/water partition coefficients. Proc. 7th Brit. Insecticide Fungicide Conf. (1973): 83–86.

– BROMILOW, R. H., EDMUNDSON, R., and JOHNSTON, M.: Distribution coefficients and systemic activity. In: MCFARLANE, N. R. (Ed.): Herbicides and Fungicides – Factors Affecting Their Activity. Chem. Soc., London 1977, pp. 129–134.

– – and EVANS, A. A.: Relationship between lipophilicity and root uptake of non-ionised chemicals by barley. Pesticide Sci. **13** (1982): 495–504.

– – and WILLIAMS, M.: Relationship between lipophilicity and the distribution of non-ionised chemicals in the barley shoot following uptake by roots. Pesticide Sci. **14** (1983): 492–500.

BROMILOW, R. H., CHAMBERLAIN, K., and EVANS, A. A.: Molecular structure and properties of xenobiotics in relation to phloem translocation. In: BONNEMAIN, J.-L., DELROT, S., LUCAS, W. J., and DAINTY, J. (Eds.): Recent advances in phloem transport and assimilate compartmentation. Ouest Editions, Nantes 1991, pp. 332–340.

BUCHENAUER, H.: Conversion of triadimefon into two diastereomers, triadimenol-I and triadimenol-II by fungi and plants. IX. Congr. Plant Prot. Washington (1979): Abstract 939.

– and RÖHRER, E.: Aufnahme, Translokation und Transformation von Triadimefon in Kulturpflanzen. Z. Pflanzenkrankh. Pflanzenschutz **89** (1982): 385–398.

CHALEFF, R. S., and MAUVAIS, C. J.: Acetolactat synthase is the site of action of two sulfonylurea herbicides in higher plants. Science **224** (1984): 1443–1445.

CHAMBERLAIN, K., BURREL, M. M., BUTCHER, D. N., and WHITE, J. C.: Phloem transport of xenobiotics in *Ricinus communis* var. *Gibsonii.* Pestic. Sci. **15** (1984): 1–8.

CHRIST, R. A.: Physiological and physiochemical requisites for the transport of xenobiotics in plants. In: GEISSBÜHLER, H. (Ed.): Adv. Pesticide Sci. Vol. 3. Pergamon Press, Oxford 1979, pp. 420–429.

CLEMONS, G. P., and SISLER, H. D.: Formation of a phytotoxic derivative from Benlate. Phytopathology **59** (1969): 705–706.

COHEN, Y., and COFFEY, M. D.: Systemic fungicides and the control of oomcyetes. Ann. Rev. Phytopathol. **24** (1986): 311–338.

CRAFTS, A. S., and CRISP, C. E.: Movement of exogenous substances. In: Phloem transport in Plants. Freeman and Comp., San Francisco 1971, pp. 168–264.

CRISP, C. E.: The molecular design of systemic insecticides and organic functional groups in translocation. In: TAHORI, A. S. (Ed.): Proc. 2nd Congr. Pesticide Chemistry **1** (1972): 211–264.
- and LOOK, M.: Phloem loading and transport of weak acids. In: GEISSBÜHLER, H. (Ed.): Adv. Pesticide Sci. Vol. 3 Pergamon Press, Oxford 1979, pp. 430–437.

CROWDY, S. H.: Translocation. In: MARSH, R. W. (Ed.): Systemic Fungicides. Longman, London 1972, pp. 92–115.
- Patterns and processes of movement of chemicals in higher plants. Proc. 7th Brit. Insecticide Fungicide Conf. (1973): 831–835.
- GROVE, J. F., and MCCLOSKEY, P.: The translocation of antibiotics in plants. 4. Systemic fungicidal activity and chemical structure in griseofulvin relatives. Biochem. J. **72** (1959): 241–249.
- and RUDD-JONES, D.: The translocation of sulphonamides in higher plants. I. Uptake and translocation in broad beans. J. exp. Bot. **7** (1956a): 335–346.
- – Partition of sulphonamides in plant roots: a factor of their translocation. Nature **178** (1956b): 1165–1167.

DEKKER, J.: Resistance. In: MARSH, R. W. (Ed.): Systemic Fungicides. 2nd edit. Longman, London 1977, pp. 176–197.

DEVINE, M. D.: Phloem translocation of herbicides. Rev. Weed Sci. **4** (1989): 191–213.

DOBE, H., SIEBER, K., NEUMANN, St., and JACOB, F.: Untersuchungen zur Aufnahme und Translokation von Aldimorph in ausgewählten Pflanzenspecies. Arch. Phytopathol. Pflanzensch. **21** (1986): 415–425.

DONALDSON, T. W., BAYER, D. E., and LEONARD, O. A.: Absorption of 2,4-dichlorophenoxyacetic acid and 3-(p-chlorophenyl)-1,1-dimethylurea (monuron) by barley roots. Plant Physiol. **52** (1973): 638–645.

DUMBROFF, E. B., and PEIRSON, D. R.: Probable sites for passive movement of ions across the endodermis. Can. J. Bot. **49** (1971): 35–38.

DYBING, C. D., and CURRIER, H. B.: Foliar penetration by chemicals. Plant Physiol. **36** (1961): 169–174.

EDGINGTON, L. V.: Structural requirements of systemic fungicides. Ann. Rev. Phytopathol. **19** (1981): 107–124.
- and DIMOND, A. E.: The effect of adsorption of organic cations to plant tissue on their use as systemic fungicides. Phytopathology **54** (1964): 1193–1197.
- and PETERSON, C. A.: Systemic fungicides: Theory, uptake, and translocation. In: SIEGEL, M. R., and SISLER, H. D. (Eds.): Antifungal Compounds. Vol. 2, Dekker, New York 1977, pp. 51–89.

EL IBAOUI, H., DELROT, S., BESSON, J., and BONNEMAIN, J.-L.: Uptake and release of a phloem-mobile (glyphosate) and of a non-phloem-mobile (iprodione) xenobiotic by broad bean leaf tissues. Physiol. Vég. **24** (1986): 431–442.

ERWIN, E. C.: Systemic fungicides: Disease control, translocation and mode of action. Ann. Rev. Phytopath. **11** (1973): 389–422.

Esau, K.: Plant Anatomy. 2nd edit. John Wiley and Sons, New York–London–Sydney 1965.

EVANS, E.: Problems and progress in the use of systemic fungicides. Proc. 6th Brit. Insecticide Fungicide Conf. (1971): 758–764.

FENN, M. E., and COFFEY, M. D.: Quantification of phosphonate and ethyl phosphonate in tobacco and tomato tissues and significance for the mode of action of two phosphonate fungicides. Phytopathology **79** (1989): 76–83.

FEHRMANN, H.: Systemische Fungizide – ein Überblick. Phytopath. Z. **86** (1976): 67–89.

FUCHS, A., DE VRIES, F. W., and AALBERS, A. M. J.: Uptake, distribution and metabolic fate of ^{3}H-triforine in plants. I. Short-term experiments. Pesticide Sci. **7** (1976): 115–126.

GASTONYI, M., and JOSEPOVITS, G.: The activation of triadimefon and its role in the selectivity of fungicide action. Pesticide Sci. **10** (1979): 57–65.

GEIGER, D. R., KAPITAN, S. W., and TUCCI, M. A.: Glyphosate inhibits photosynthesis and allocation of carbon to starch in sugar beet leaves. Plant Physiol. **82** (1986): 468–472.
– TUCCI, M. A., and SERVAITES, J. C.: Glyphosate effects on carbon assimilation and gas exchange in sugar beet leaves. Plant Physiol. **85** (1987): 365–369.
GEORGOPOULOS, G.: Development of fungal resistance to fungicides. In: SIEGEL, M. R., and SISLER, H. D. (Eds.): Antifungal Compounds. Vol. 2, Dekker, New York 1977, pp. 409–495.
GIAQUINTA, R. T.: Phloem loading of sucrose. Ann. Rev. Plant Physiol. **34** (1983): 347–387.
GOUGLER, J., and GEIGER, D. R.: Uptake and distribution of N-phosphonomethylglycine in sugar beet plants. Plant Physiol. **68** (1981): 668–672.
GRAHAM-BRYCE, I. J., and COUTTS, J.: Interaction of pyrimidine fungicides with soil and their influence on uptake by plants. Proc. 6th Brit. Insecticide Fungicide Conf. (1971): 419.
GRAYSON, B. T., and KLEIER, D. A.: Phloem mobility of xenobiotics. IV. Modelling of pesticide movement in plants. Pestic. Sci. **30** (1990): 67–79.
GREEN, M. B., HARTLEY, G. S., and WEST, T. F.: Chemicals for Crop Protection and Pest Control. Pergamon Press, Oxford 1977.
GREENE, D. W., and BUKOVAC, M. J.: Stomatal penetration, effect of surfactans and role in foliar absorption. Amer. J. Bot. **61** (1974): 100–106.
GRIMM, E., NEUMANN, St., and JACOB, F.: Aufnahme von Saccharose und Xenobiotika in isoliertes Leitgewebe von *Cyclamen*. Biochem. Physiol. Pflanzen **178** (1983): 29–42.
GRIMM, E., NEUMANN, St., and JACOB, F.: Transport of xenobiotics in higher plants. II. Absorption of defenuron, carboxyphenylmethylurea and maleic hydrazide by isolated conducting tissue of *Cyclamen*. Biochem. Physiol. Pflanzen **180** (1985): 383–392.
GROSSMANN, F.: Möglichkeiten und Grenzen des Einsatzes systemischer Fungizide. Z. Pflanzenkrankh. Pflanzensch. **81** (1974): 670–674.
GROUSSOL, J., DELROT, S., CARUHEL, P., and BONNEMAIN, J.-L.: Design of an improved exudation method for phloem sap collection and its use for the study of phloem mobility of pesticides. Physiol. Veg. **24** (1986): 123–133.
HANCE, R. J.: Processes in soil which control the availability of pesticides. Proc. 10th Int. Congr. Plant Prot. Brighton (1983): 537–544.
HANSON, S. D., and COHEN, J. D.: A technique for collection of exudate from pea seedlings. Plant Physiol. **78** (1985): 734–738.
HARTLEY, G. S., and GRAHAM-BRYCE, I. J.: Physical Principles of Pesticide Behaviour. Vol. 2, Academic Press, London–New York–Toronto–Sydney–San Francisco 1980.
HASSALL, K. A.: The Chemistry of Pesticides, Their Metabolism, Mode of Action and Uses in Crop Protection. Verlag Chemie, Weinheim–Deerfield–Beach–Florida–Basel 1982.
HILL-COTTINGHAM, D. G., and LLOYD-JONES, C. P.: Relative mobility of some nitrogenous compounds in the xylem of apple shoots. Nature **220** (1968): 389–390.
– – A technique studying the adsorption and metabolism of amino acids in apple stem tissue. Physiol. Plant. **28** (1973): 443–446.
HOAD, G. V.: A simple system for determining the phloem mobility of compounds using excises pods of lupin (*Lupinus albus*). Planta **150** (1980): 175–178.
HOLLOWAY, P. J.: Surface factors affecting the wetting of leaves. Pesticide Sci. **1** (1970): 156–163.
– The chemical constitution of plant cutin. In: CUTLER, D. F., ALVIN, K. L., and PRICE, C. E. (Eds.): The Plant Cuticle. Academic Press, New York–London 1982, pp. 45–85.
HUNT, G. M., and BAKER, E. A.: Phenolic constituents of tomato fruit cuticles. Phytochemistry **19** (1980): 1415–1419.
JACOB, F., and NEUMANN, St.: Quantitative determination of mobility of xenobiotics and criteria of their phloem and xylem mobility. In: MIYAMOTO, J., and KEARNEY, P. C. (Eds.): IUPAC Pesticide Chemistry – Human Welfare and the Environment. Vol. 1. Pergamon Press, Oxford–New York–Toronto–Sydney–Paris–Frankfurt 1983, pp. 357–362.
– – and STROBEL, U.: Studies on mobility of exogen applied substances in plants. In: Transac-

tions of the 3rd Symp. on Accumulation of Nutrients and Regulators in Plant Organism. Warsaw. Proc. Res. Inst. Pomology Skierniewice Poland, Series E (1973): 315–330.

KARAS, I., and MCCULLY, M. E.: Further studies on the histology of lateral root development in *Zea mays*. Protoplasma **77** (1973): 243–269.

KERLER, F., RIEDERER, M., and SCHÖNHERR, J.: Non-electrolyte permeability of plant cuticles: a critical evaluation of experimental methods. Physiol. Plant. **62** (1984): 599–602.

— and SCHÖNHERR, J.: Accumulation of lipophilic chemicals in plant cuticles: prediction from octanol/water partition coefficient. Arch. Environ. Contam. Toxicol. **17** (1988a): 1–6.

— — Permeation of lipophilic chemicals across plant cuticles: prediction from partition coefficients and molar volumes. Arch. Environ. Contam. Toxicol. **17** (1988b): 7–12.

KING, R. W., and ZEEVAART, J. A. D.: Enhancement of phloem exudation from cut petioles by chelating agents. Plant Physiol. **53** (1974): 96–103.

KLEIER, D. A.: Phloem mobility of xenobiotics. I. Mathematical model unifying the weak acid and intermediate permeability theories. Plant Physiol. **86** (1988): 803–810.

KÖLLER, W., ALLAN, C. R., and KOLATTUKUDY, P. E.: Inhibition of cutinase and prevention of fungal penetration into plant by benomyl-a possible protective mode of action. Pesticide Biochem. Physiol. **18** (1982): 15–25.

KUCK, K. H.: Studies on uptake and translocation of ®Bayleton in wheat plants. Pflanzenschutz-Nachrichten Bayer **40** (1987): 1–28.

— and SCHEINPFLUG, H.: Biology of sterol-biosynthesis inhibiting fungicides. In: HANG, G., and HOFFMANN, H. (Eds.): Chemistry of plant protection. Vol. 1: Sterol biosynthesis, inhibitors and anti-feeding compounds. Springer-Verlag, Berlin–Heidelberg–New York–Tokyo 1986, pp. 65–96

— and THIELERT, W.: On the systemic properties of HWG 1608 the active ingredient of the fungicides ®Folicur and ®Raxil. Pflanzenschutz-Nachrichten Bayer **40** (1987): 133–152.

LEROUX, P., and GREDT, M.: Uptake of systemic fungicides by maize roots. Netherl. J. Plant Path. **83**, Suppl. 1 (1977): 51–61.

LICHTNER, F. T.: Amitrole absorption by bean. (*Phaseolus vulgaris* L. c. v. Red Kidney) roots. Plant Physiol. **71** (1983): 64–72.

— Phloem transport of agricultural chemicals. In: CRONSHAW, J., LUCAS, W. J., and GIAQUINTA, R. T. (Eds.): Phloem transport. Alan R. Liss Inc., New York 1986, pp. 601–608.

MANSFIELD, T. A., PEMASADA, M. A., and SNAITH, P. J.: New possibilities for controlling foliar absorption *via* stomata. Pesticide Sci. **14** (1983): 294–298.

MARSH, R. W. (Ed.): Systemic fungicides. 2nd edit. Longman, London 1977.

MARTIN, J. T., and JUNIPER, B. E.: The Cuticles of Plants. Arnold, London 1970.

MARTIN, R. H., and EDGINGTON, L. V.: Comparative systemic translocation of several xenobiotics and sucrose. Pesticide Biochem. Physiol. **16** (1981): 87–96.

MINSHALL, W. H.: Translocation path and place of action of 3-(4-chlorophenyl)-1,1-dimethylurea in bean and tomato. Can. J. Bot. **32** (1954): 765–798.

MITCHELL, J. W., MARTH, P. C., and PRESTON, W. H.: Structural modification that increases translocability of some plant-regulating carbamates. Science **120** (1954): 263–265.

MOODY, K., KUST, C. A., and BUCHHOLZ, K. P.: Uptake of herbicides by soybean roots in culture solution. Weed Sci. **18** (1970): 642–647.

MORRISON, I. N., and VAN DEN BORN, W. H.: Uptake of picloram by roots of alfalfa and barley. Can. J. Bot. **53** (1975): 1774–1785.

MÜLLER, F.: Translokation von ^{14}C-markiertem MCPA in verschiedenen Entwicklungsstadien einiger mehrjähriger Unkräuter. Acta Phytomedica (Suppl. J. Phytopath.) **4** (1976): 5–160.

NETTING, A. G., and WETTSTEIN-KNOWLES, P.: The physico-chemical basis of leaf wettability in wheat. Planta **114** (1973): 289–309.

NEUMANN, St.: Untersuchungen zur Mobilität von Xenobiotika in höheren Pflanzen. Promotion B,

Mathematisch-Naturwissenschaftliche Fakultät der Martin-Luther-Universität Halle-Wittenberg, 1982.
- Die Ionenfalle als Akkumulations- und Retentionsmechanismus bei der Aufnahme von xenobiotischen Substanzen in das Phloem. Colloquia Pflanzenphysiologie Humboldt-Univ. Berlin **8** (1985): 117–130.
- GRIMM, E., and JACOB, F.: Transport of xenobiotics in higher plants. I. Structural prerequisites for translocation in the phloem. Biochem. Physiol. Pflanzen **180** (1985 a): 257–268.
- PETZOLD, U., and JACOB, F.: Mobilität von Xenobiotika in höheren Pflanzen – eine tabellarische Übersicht. I. Allgemeine Prinzipien der Aufnahme und des Transports. II. Fungizide. Wiss. Z. Univ. Halle **XXXIV** '85M (1985 b): 56–70.
- SCHWEINGEL, M., and JACOB, F.: Mobilität von Xenobiotika in höheren Pflanzen – eine tabellarische Übersicht. III. Mobilität von Herbiziden mit Carbonsäureesterstruktur. Wiss. Z. Univ. Halle **XXXVII** '88M (1988): 68–71.

NORRIS, R. F.: Penetration of 2,4-D in relation to the cuticle thickness. Amer. J. Bot. **61** (1974): 74–79.
- and BUKOVAC, M. J.: Structure of pear leaf cuticle with special reference to cuticular penetration. Amer. J. Bot. **55** (1986): 975–983.

OUIMETTE, D. G., and COFFEY, M. D.: Symplastic entry and phloem translocation of phosphonate. Pesticide Biochem. Physiol. **38** (1990): 18–25.

PERKOW, W.: Wirksubstanzen der Pflanzenschutz- und Schädlingsbekämpfungsmittel. 2. Aufl. Parey, Berlin–Hamburg 1983.

PETERSON, C. A., DE WILDT, P. Q., and EDGINGTON, L. V.: A rationale for the ambimobile translocation of the nematicide oxamyl in plants. Pesticide Biochem. Physiol. **8** (1978): 1–9.
- and EDGINGTON, L. V.: Transport of systemic fungicide, benomyl, in bean plants. Phytopathology **60** (1970): 475–478.
- – Factors influencing apoplastic transport in plants. In: LYR, H., and POLTER, C. (Eds.): Systemic Fungicides (Int. Symp. Reinhardsbrunn 1974). Akademie-Verlag, Berlin 1975, pp. 287–299.
- – Uptake of the systemic fungicide methyl-2-benzimidazolecarbamate and the fluorescent dye PTS by the onion roots. Phytopathology **65** (1975): 1254–1259.
- EMANUEL, M. E., and HUMPHREYS, G. B.: Pathway of movement of apoplastic fluorescent dye tracers through the endodermis at the site of secondary root formation in corn (*Zea mays*) and broad bean (*Vicia faba*). Can. J. Bot. **59** (1981): 618–625.

PRASAD, R., FOY, C. L., and CRAFTS, A. S.: Effects of relative humidity on absorption and translocation of foliary applied dalapon. Weed Sci. **15** (1967): 149–156.

PRICE, C. E.: Penetration and translocation of herbicides and fungicides in plants. In: MCFARLANE, N. R. (Ed.): Herbicides and Fungicides – Factors affecting Their Activity. Chem. Soc., London 1977, pp. 42–66.
- Movement of xenobiotics in plants – perspectives. In: GEISSBÜHLER, H. (Ed.): Adv. Pesticide Sci. Vol. 3, Pergamon Press, Oxford 1979, pp. 401–409.

RAVEN, J. A., and RUBERY, P. H.: In: SMITH, H., and GRIERSON, D. (Eds.): The molecular biology of plant development. Blackwell Sci Publ., Oxford. U.K. 1982, pp. 28–48.

RENTSCHLER, I.: Die Wasserbenetzbarkeit von Blattoberflächen und ihre submikroskopische Wachsstruktur. Planta **96** (1971): 119–135.

RIEDERER, M., and SCHÖNHERR, J.: Accumulation and transport of (2,4-dichlorophenoxy)acetic acid in plant cuticles. I. Sorption in the cuticular membrane and its components. Ecotoxicol. Environ. Safety **8** (1984): 236–247.
- – Accumulation and transport of (2,4-dichlorophenoxy)acetic acid in plant cuticles. II. Permeability of the cuticular membrane. Ecotoxicol. Environ. Safety **9** (1985): 196–208.

SCHMELING, B. v., and KULKA, M.: Systemic fungicidal activity of 1,4 oxathiin derivatives. Science **152** (1966): 659–660.

SCHÖNHERR, J., and HUBER, R.: Plant cuticles are polyelectrolytes with isoelectric points around three. Plant Physiol. **59** (1977): 145–150.

SCHREIBER, L.: Untersuchungen zur Schadstoffaufnahme in Blätter. Dissertation, Technische Universität München, 1990.

SHARMA, M. P., and VAN DEN BORN, W. H.: Uptake, cellular distribution and metabolism of ^{14}C-picloram by excised plant tissues. Physiol. Plant. **29** (1973): 10–16.

SHEPHARD, M. C.: Barriers to the uptake and translocation of chemicals in herbaceous and woody plants. Proc. 7th Brit. Insecticide Fungicide Conf. (1973): 841–850.

– Factors which influence the biological performance of pesticides. Proc. Brit. Crop. Conf. (1981): 711–721.

– Fungicide behaviour in the plant systemicity. In: BRENT, K. J. (Ed.): One hundred years of fungicide use. Crop Prot. BCPC Monogr. No. 31, Thornton Heath. Br. Crop Prot. Counc. Publ. 1985: pp. 99–106.

– Screening for fungicides. Ann. Rev. Phytopathol. **25** (1987): 189–206.

SHIMABUKURO, R. H., WALSH, W. C., and HOERAUF, R. A.: Metabolism and selectivity of diclofop-methyl in wild oat and wheat. J. Agric. Food Chem. **27** (1979): 615–623.

SHONE, M. G. T., BARTLETT, B. O., and WOOD, A. V.: A comparison of the uptake and translocation of some herbicide and a systemic fungicide by barley. II. Relationship between uptake by roots and translocation to shoots. J. exp. Bot. **25** (1974): 401–409.

– and WOOD, A. V.: Factors affecting absorption and translocation of simazine by barley. J. exp. Bot. **23** (1972): 141–151.

– – A comparison of the uptake and the translocation of some organic herbicides and a systemic fungicide by barley. I. Absorption in relation to physico-chemical properties. J. exp. Bot. **25** (1974): 390–400.

SMALL, L. W., MARTIN, R. A., and EDGINGTON, L. V.: A comparison of the translocation within plants of pyroxychlor and a 6-amino analogue. Netherl. J. Plant Path. **83**, Suppl. 1 (1977): 63–70.

SNEL, M., and EDGINGTON, L. V.: Uptake, translocation and decomposition of systemic oxathiin fungicides in beans. Phytopathology **60** (1970): 1708–1716.

SOLEL, Z., SCHOOLEY, J. M., and EDGINGTON, L. V.: Uptake and translocation of benomyl and carbendazim (methylbenzimidazole-2-ylcarbamate) in the symplast. Pesticide Sci. **4** (1973): 713–718.

TOPP, E., SCHEUNERT, I., ATTAR, A., and KORTE, F.: Factors affecting the uptake of ^{14}C-labelled organic chemicals by plants from soil. Ecotoxicol. Environ. Safety **11** (1986): 219–228.

TYREE, M. T., PETERSON, C. A., and EDGINGTON, L. V.: A simple theory regarding ambimobility of xenobiotics with special reference to the nematicide oxamyl. Plant Physiol. **63** (1979): 367–374.

UCHIDA, M.: Affinity and mobility of fungicidal diakyl dithioanylidenemalonates in rice plants. Pesticide Biochem. Physiol. **14** (1980): 249–255.

– and KASAI, T.: Adsorption and mobility of fungicidal dialkyl dithioanylidenemalonates and their analogs in soil. J. Pesticide Sci. **5** (1980): 553–559.

– and SUZUKI, T.: Affinity and mobility of fungicides in soils and rice plants. 5th Int. Congr. Pesticide Chem. (IUPAC) Kyoto (1982): Abstract IIe-3.

ULLRICH, W. R., ULLRICH-EBERIUS, C. I., and KÖCHER, H.: Uptake of glufosinate and concomitant membrane potential changes in *Lemna gibba* G1. Pesticide Biochem. Physiol. **37** (1990): 1–11.

VAN BEL, A. J. E.: Behaviour of differently charged amino acids towards tomato wood powder. Z. Pflanzenphysiol. **89** (1978): 331–320.

– Different phloem loading machineries correlated with the climate. Acta Bot. Neerl. **41** (1992): 121–141.

VAN DER KERK, G. J. M.: Systemic fungicides new solutions and new problems. Proc. 6th Brit. Insecticide Fungicide Conf (1971): 791–802.

— Evolution of the chemical control of plant diseases an evaluation. Netherl. J. Plant Path. **83**, Suppl. 1 (1977): 3–13.

WAIN, R. L., and CARTER, G. A.: Uptake, translocation and transformation by higher plants. In: TORGESON, D. C. (Ed.): Fungicides. Vol. 1. Academic Press, New York–London 1967, pp. 561–611.

WALKER, A., and FEATHERSTONE, R. M.: Absorption and translocation of atrazine and linuron by plants with implications concerning linuron selectivity. J. exp. Bot. **24** (1973): 450–458.

WATTENDORF, J., and HOLLOWAY, P. J.: Studies on the ultrastructure and histochemistry of plant cuticles: The cuticular membrane of *Agave americana* L. in situ. Ann. Bot. **46** (1980): 13–28.

WILHELM, H., and KNÖSEL, H.: Penetration and Translokation von ^{3}H-Tetracyclinhydrochlorid in pflanzlichem Gewebe. Z. Pflanzenkrankh. Pflanzenschutz **83** (1976): 241–251.

YOUNG, H., and HAMMET, K. R. W.: Comparative uptake of benomyl-2-^{14}C and methylbenzimidazole-2-ylcarbamate-2-^{14}C by cucumber leaves. New Zealand J. Sci. **16** (1973): 535–541.

Chapter 5

Aromatic hydrocarbon fungicides and their mechanism of action

H. LYR

Biologische Bundesanstalt für Land- und Forstwirtschaft, Institut für Integrierten Pflanzenschutz, Kleinmachnow, Germany

Introduction

"Aromatic hydrocarbon fungicides" (AHF) represent an old and heterogeneous group of fungicides, which have been in use for a longer time, but lost now their importance. Most of them could be replaced by new compounds with a broader antifungal spectrum, better physical properties and higher activity. Although chemically different, they are connected by fungal cross resistance and a very similar mechanism of action. Therefore, also etridiazole and tolclofos-methyl are mentioned here (Fig. 5.1). Some properties are summerized in Tab. 5.1. Because of the high volatility many compounds were used only as soil fungicides. Tab. 5.2 demonstrates that they exhibit very selective effects against some important fungi, but are inactive for other ones.

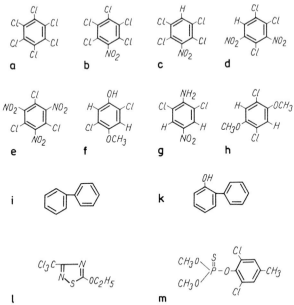

Fig. 5.1 Structure formulae for compounds of the group of aromatic hydrocarbon fungicides. a) Hexachlorobenzene, b) Pentachloronitrobenzene (Quintozene), c) Tetrachloronitrobenzene (Tectacene), d) 1,2,4-trichloro-3,5-dinitrobenzene (Olpisan), e) 1,3,5-trichloro-2,4,6-trinitrobenzene (Phomasan), f) 2,4-Dichloro-3-methoxy-phenol (DCMP), g) Dicloran, h) Chloroneb, i) Diphenyl, k) o-Phenylphenol, l) Etridiazole, m) Tolclophosmethyl.

Table 5.1 Aromatic hydrocarbon fungicides compounds, trade-names, properties and origins

Compound	Trade-names	Usage	Acute oral doses LD_{50} (rats) mg/kg	Vapour pressure mm Hg	Patents	Introduction
Hexachlorobenzene		Seed dressing Soil fungicide	10,000	1.09×10^{-5}	–	1945
Quintozene (Pentachloronitrobenzene)	Brassicol Tritisan Folosan	Soil fungicide Seed dressing	12,000	13.3×10^{-3}	IG Farben AG (DRP 682048)	1930
Tecnazene (Tetrachloronitrobenzene TCNB)	Fosolan Fusarex	Soil fungicide	57	volatile	Bayer AG USP 2615801	1946
Trichlorodinitrobenzene	Olpisan	Soil fungicide (*Plasmodiophora brassicae*, a. o. fungi)		moderate volatile		1950
Trichlorotrinitrobenzene	Phomasan	Soil fungicide		moderate volatile		1953
Chloroneb	Demosan	Soil fungicide	11,000	3×10^{-5}	Du Pont de Nemours & Co. (Inc.) USP 3265564	1967
Dicloran	Allisan Bortran	Fruit storage diseases, ornamental *Botrytis* and *Sclerotinia*, Rhizopus diseases	1,500 – 4,000	1.2×10^{-6}	Boots Co. Ltd. (BP 845916)	1930
Etridiazole	Terrazol	Soil fungicide Seed dressing	2,000	1×10^{-4}	Olin Chemicals USP 3260588 USP 3260588 USP 3260725	1969
Tolclofosmethyl	Rizolex	Soil, foliar fungicide	5,000		Sumitomo Chemical Co. Ltd.	
2-Phenylphenol (OPP)	Dowicide Nectryl	Fruit storage diseases, Disinfectants	2,480	moderate volatile		1936
Biphenyl		Citrus Citrus storage	3,280	volatile		1944

Table 5.2 Antifungal spectrum of some aromatic hydrocarbon fungicides (ED_{50} values in mg/l for inhibition of radial mycelial growth on malt agar medium). Values for Tolclophosmethyl according to Technical Information "Rizolex" Sumitomo Chem. Corp. Osaka, Japan

Fungus	Chloroneb	Dicloran	Quintozene	Olpisan	Diphenyl	Tolclofosmethyl	Etridiazole
Phytophthora cactorum	2	200	200	35	100	100	0.4
Pythium ultimum	1	200	150	4	100	100	0.3
Mucor mucedo	2	0.5	5	30	30	100	8
Botrytis cinerea	3	1	0.5	1	30	1	20
Sclerotinia sclerotiorum	3	2	5	2	30	1.4	10
Penicillium chrysogenum		0.5	20	–	100	–	90
Penicillium italicum	2	0.4	–	–	1	1.6	–
Aspergillus niger		100	200	–	100	–	100
Fusarium oxysporum	500	200	500	10	100	100	70
Ophiobolus graminearum	70	–	30	–	100	1.4	3
Pseudocercosp. herpotrich.	200	10	60	–	100	–	100
Rhizoctonia solani	15	200	200	20	30	0.1	70
Verticill. albo-atrum	10	8	20	–	–	100	40
Schizophyllum commune	3	100	10	–	100	–	40

The value of this group was the low costs for synthesis, low mammal toxicity (if no toxic impurities from the technical synthesis were present), and their specificity in the control of fungi. Very surprising was the fungal cross resistance with the structurally unrelated dicarboximides (compare chapter 6) (LEROUX et al. 1977). This pointed to a common mechanism of action, and showed the pathway for the elucidation of the mode of action. The biological conversion of some of these fungicides is summarized by KAARS-SIJPESTEIJN et al. 1977.

Some members of this group will be briefly characterized here.

Hexachlorobenzene (HCB)

HCB (Fig. 5.1a) was an effective fungicide of the halogenated benzene class. It was discovered in France (YERSIN et al. 1946) and exhibits a high degree of vapour-phase activity against some important soil-borne fungi and seed-borne diseases. HCB controls common bunt (*Tilletia foetida* and *T. caries*) and dwarf bunt of wheat (*T. controversa*) (HOLTON and PURDY 1954; PURDY 1965). It is also effective against the seed borne fungus *Urocystis agropyri* (initiant of flag smut), but is inactive against *Ustilago* spp.

HCB had been used for more than 20 years as seed dressing and soil fungicide because of its low mammalian toxicity and low price. In Greece it was used since 1958 as a seed dressing agent as an alternative to the organic mercury compounds. However, in 1973 a marked decrease in effectivity against *Tilletia* spp. was observed. SCORDA (1977) reported that races of *T. foetida* were cross resistant to HCB, PCNB imazalil and hydantoin. Similar observations were made in Australia (KUIPER 1965). The use of HCB as a seed dressing has creased because it has been replaced by other fungicide combinations with broader spectra of activity against seed-borne pathogens.

Chlorinated nitrobenzenes

The most widely used compound in this group is pentachloronitrobenzene (Fig. 5.1 b) (PCNB or quintozene). Quintozene exhibits an antifungal spectrum that is characteristic of this group, combined with low toxicity for mammals (Tab. 5.1 and 5.2). It was extensively used as a soil fungicide to control diseases caused by *Rhizoctonia solani*, *Botrytis* spp., *Sclerotinia* spp., *Sclerotium rolfsii*. However quintozene is completely ineffective against *Phytium*, *Phytophthora* and *Fusarium* (Tab. 5.2). The related fungicide TCNB (Tectacene) (Fig. 5.1 c) controls *Fusarium coeruleum*, the cause of dry rot in potatoes (BROOK and CHESTERS 1957). PCNB and other chlorinated nitrobenzenes act fungistatically decreasing the mycelium growth of sensitive fungi but without hindering spore germination (ESURUOSO et al. 1968). Isolates of *Rhizoctonia solani* from diseased cotton plants, differed significantly in their sensitivity to PCNB (SHATLA and SINCLAIR 1963). PCNB stimulated sclerotia production and this phenomenon may result in the selection of resistant strains of *Rhizoctonia* (SHATLA and SINCLAIR 1965). The 2,3,4,6- and 2,3,4,5-isomers of tetrachloronitrobenzene were more effective against *Rhizoctonia solani* than the 2,3,5,6-isomer (ECKERT 1962). Strains resistant to these fungicides can be isolated after culturing sensitive fungi on sublethal concentrations (KATARIA and GROVER 1974). The strains are cross resistant to all members of this group of compounds (PRIEST and WOOD 1961). The effectiveness of various compounds of this group depends partly on the conditions of testing because very often vapour phase action contributes to overall fungistatic activity. Fungal species may vary in sensitivity to the chloronitrobenzenes. 1,2,4-Trichloro-3,5-dinitrobenzene (Fig. 5.1 d) (trade name olpisan) has been used against *Plasmodiophora brassicae* (cause of club root disease) in the soil. A 1,3,5-trichloro-2,4,6-trinitrobenzene (trade name Phomasan) (Fig. 5.1 e) had been used against diseases of cucumber and tomato (THIELECKE 1963).

The chlorinated nitrobenzenes affect the growth of higher plants to some degree (BROWN 1947; ECKERT 1962) and some of these compounds exhibit a systemic activity against *Fusarium* wilt of tomato (GROSSMANN 1958). Compared to modern benzimidazoles however their innertherapeutic activity is weak.

The chemical and microbial stability of the chlorinated nitrobenzenes is a major reason for their use as seed and soil fungicides over several decades. Also, the acute mammalian toxicity is quite low, but long term effects have been noted in recent studies. Development of resistance of practical importance in soil fungi seems to require a longer time period compared with leaf pathogens which have a higher sporulation capacity. Therefore the problems of resistance has never become a widespread problem. But the moderate activity of the chloronitrobenzenes against other important soil fungi has limited their practical use (further literature is cited by CORDEN 1969).

Chloroneb

Chloroneb was developed by Du Pont de Nemours Co. in 1967. It differs from the compounds described above in that it does not contain nitro-groups (Fig. 5.1 h). In common with other chlorobenzene fungicides, chloroneb has a low mammalian toxicity (Tab. 5.1) and a significant vapour pressure so that it was used as a soil fungicide, in the culture of beans, cucumber, cotton (FIELDING and RHODES 1967). In contrast to the nitrobenzene compounds, *Phytophthora* spp. are rather sensitive to chloroneb (Tab. 1.6, Tab. 5.2). *Pythium* spp. vary in sensitivity to chloroneb. Because of its low water solubility chloroneb is only very weakly systemic (GRAY and SINCLAIR 1970; SINCLAIR 1975).

The main targets of chloroneb were *Rhizoctonia solani*, *Pythium* spp., *Ustilago maydis*, *Typhula* spp. (VARGAS and DEARO 1970), and soil inhabiting *Phytophthora* spp. Chloroneb is not effective against *Fusarium*, but has a relatively broad spectrum of activity compared with other compounds which are specifically active against Oomycetes (KLUGE 1978). By controlling *Rhizoctonia solani*, by seed-piece or infurrow applications, chloroneb (LIPE and THOMAS 1979) increased potato yields in Texas.

Like other fungicides of this group, resistant strains can obtained easily in the laboratory and differences in sensitivity seem to exist also in the natural population.

Dicloran

Dicloran (Botran®) (Fig. 5.1 g) was discovered in 1959 by Boots Co. Ltd. Because of its high activity against *Botrytis*, *Mucor*, *Rhizopus*, some *Penicillium* spp., *Monilia fructicola* and its low mammalian toxicity dicloran is used against fruit and vegetable storage diseases, especially in peaches, sweet cherries, grapes, tomatoes, lettuce and cabbage (OGAWA et al. 1963; CHASTAGNER and OGAWA 1979; ECKERT 1969, 1979).

Tab. 5.2 shows that dicloran has a selective spectrum of antifungal activity similar to that of PCNB. Both fungicides are highly active against *Mucor*, *Rhizopus* and *Botrytis*, but not active against *Phytophthora*, *Pythium*, *Fusarium*. Dicloran may be applied as spray or as fungicide-wax formulation (CHASTAGNER and OGAWA 1979). It has been used as a seed treatment against *Sclerotium cepivorum* on onions (LOCKE 1965), *Sclerotinia sclerotiorum* on beans (BECKMAN and PARSONS 1965) or on sunflowers. The chemical stability and low volatility of dicloran results in a prolonged disease control on leaves, fruits and in the soil.

Similar to other compounds of this group, dicloran does not inhibit germination of spores but inhibits mycelium growth. Distortion and bursting of the germ tubes have been described.

Resistant isolates were obtained that are cross resistant to other chloronitrobenzene fungicides and to dicarboximide fungicides (RITCHIE 1982; LEROUX and GREDT 1984), but not to benzimidazol- or triadimefon-fungicides (BOLTON 1976). Some properties of dicloran resistant isolates of *Monilinia fructicola* were similar to dicarboximide-resistant isolates (RITCHIE 1983). In cross resistant isolates, mycelial growth on agar medium was inhibited more by dicloran than by the dicarboximides. For some practical applications dicloran has been replaced by the more effective dicarboximide fungicides.

Etridiazole (Terrazol®)

Etridiazole produced by OLIN Corp. USA, although of dissimilar chemical structure (Fig. 5.11), it is included into this group of compounds because of its similar mechanism of action and its cross resistance to aromatic hydrocarbon fungicides (BISCHOFF and LYR

1980) with which, it is sharing some properties. As Table 5.2 demonstrates, this compound has a pretty broad spectrum of activity and is highly effective against *Phytophthora* and *Pythium* species which can also be seen from Tab. 1.6. Because this compound is pretty sensitive against UV-irradiation it is mainly used as soil fungicide. It is used in various cultures for treatment of seed beds in nurseries, soil treatment for flower bulb crops and other plants which are attacked by *Pythium*, *Phytophthora* spp., *Botrytis* spp. *Sclerotinia* and similar pathogens (BAKKEREN and OLWEHAND 1970; WHEELER *et al.* 1970).

An unique side effect is its ability to retard soil nitrification (TURNER 1979). It has weak insecticidal effects (USP 4,057,639) and can be used as synergist for some inecticides (DP WP A O1N/226475).

o-Phenylphenol (OPP)

This compound (Fig. 5.1 k) is unique among the phenol group in that it has retained a limited use in plant protection. Its main application is the protection of stored fruits, especially citrus and the disinfection of storage material (ROSE *et al.* 1951; WOLF 1956). By vapour action it can protect packed fruits against decay by *Penicillium italicum* and *P. digitatum*, *Diplodia natalenis*, *Botrytis cinerea* and other species.

OPP is more selective than other free phenols but does produce phytotoxic effects. Sodium o-phenylphenate (SOPP) is less toxic and is used in practice because it is much less phytotoxic to fruits and has a greater water solubility. TOMKINS (1963) tried to diminish the phytotoxic effect by esterification of the phenolic group. Acetate and butyrate esters gave efficient control of *Botrytis* on grapes and *Monilinia* on peaches, but

Table 5.3 Activity of some chlorinated anisoles and phenols against certain fungi on malt agar medium and their phytotoxicity (measured by the TTC-Test with young tissue of maize and bean hypocotyles). ED_{50}-values in mg/l of mycelial growth inhibition or reduction of TTC, values related to controls) DCMP = Dichloromethoxyphenol

	Chloroneb	DCMP	2,4,6-trichloroanisole	Tetrachloroanisole	2,4,6-trichlorophenol	Tetrachlorophenol	Pentachlorophenol
Phytophthora cactorum	2	3	2	80	2	3	1
Pythium debaryanum	10	40	60	100	2	2	3
Mucor mucedo	2	3	40	25	8	20	2
Botrytis cinerea	5	10	4	6	3	1	0.4
Fusarium oxysporum	100	100	100	100	3	3	5
Colletotrichum lindem.	2	–	30	–	2	0.5	1.5
Cochliobol. carboneum	5	–	30	–	4	2	1.5
Rhizoctonia solani	2	2	20	15	3	0.5	1
Trametes versicolor	2	–	20	–	10	10	2
Phytotoxicity:							
Maize	150	100	1.000	300	30	15	3
Bean	300	200	100	–	3	3	2

esters of o-phenylphenol are easily hydrolyzed into the free phenol (ECKERT 1979). The methyl ether was less active against *Diplodia*, and virtually inactive against *Penicillia*. ECKERT (1979) demonstrated that *Diplodia* can demethylate about 10% of the anisole, whereas *Penicillium digitatum* does not metabolize this compound. However it is not likely that this small difference can account for the differences in the growth inhibition of these two fungi. Apparently biphenyl anisole which is rapidly accumulated, is itself active against *Diplodia*, but not against *Penicillium*. This resembles strongly the situation with chloroneb, where the first demethylation product (DCMP, dichloromethoxyphenol), has the same spectrum of activity as chloroneb itself and does only partly behave as a phenol (see Tab. 5.3).

Cross resistance of SOPP and diphenyl against *Penicillia* became of practical importance in California in lemon packing houses after continuous use (ECKERT et al. 1981; ECKERT and WILD 1983; ECKERT and OGAWA 1985).

Diphenyl (biphenyl)

Diphenyl (Fig. 5.1i) is a vapour-phase fungistat which is used to control postharvest diseases in citrus caused by *Penicillium digitatum* and *P. italicum*, inhibiting mycelial growth and sporulation (NAGY et al. 1982). It is impregnated into paper sheets which are added to storage and transportation cartons. Citrus fruits absorb this very stable compound in proportions to the vapour concentration in the surrounding atmosphere which is related to storage temperature and storage time. The official tolerance limit for this compound on citrus fruits is 110 ppm (mg/kg) in the United States, and 70 ppm in Europe and Japan.

Diphenyl, one of the weakest fungicides of this group, because it is highly hydrophobic, is still a part of the protection strategy of citrus shippers.

Tolclophos-methyl

This compound, developed in the recent years by Sumitomo Chemical Co., is the newest member of the AHF group (Fig. 5.1m), and the first organo-phosphorus compound (KATO 1983). Its spectrum of activity has been published by the Sumitomo Co. ("A new Fungicide Rizolex"). Some data have been incorporated into Tab. 1.7 and 5.2. This compound is highly active against *Corticium rolfsii*, *Typhula* spp., *Ustilago maydis*, *Botrytis*, *Sclerotinia*, *Penicillium italicum* and especially *Rhizoctonia solani*. *Phytophthora*, *Pythium*, *Fusarium* and *Verticillium albo-atrum* are rather insensitive. Table 5.2 shows the spectrum of activity of this compound. It differs remarkably from other fungicides of this group. Tolclophos-methyl is mainly recommended for control of all diseases where *Rhizoctonia* is involved (seed potatoes, control of stem canker and black scurf). This fungicide is highly effective in controlling this fungus on ornamental crops and vegetables (lettuce). *Sclerotia* of *Rhizoctonia* on potatoes are killed within 30 min after contact with this compound. Such an action cannot be achieved with thiram, mancozeb or iprodione (BARNES et al. 1984) (compare chapter 9).

Mechanism of action of aromatic hydrocarbon fungicides

Several investigators have attempted to clarify the mechanism of action of this rather old group of selective fungicides. Although many effects in very different sites of the metabolism have been reported, no clear picture of the primary mechanism of action of

Table 5.4 Some effects of aromatic hydrocarbon fungicides on sensitive fungi which have been described in the literature

	Compounds	Fungus	Author(s)
Inhibition of motility of *Phytophthora* zoospores	biphenyl, chloroneb, dicloran, PCNB, tolclophosmethyl	*Phytophthora capsi*	KATO (1983)
Inhibition of respiration	chloroneb PCNB etridiazole PCNB	*Rhizoctonia Pythium Mucor*	HALOS and HUISMAN (1976b); RADZUHN (1978); WERNER (1980); KATARIA and GROVER (1975)
Thickening of cell walls or abnormal cell wall synthesis	biphenyl chloroneb dicloran PCNB	*Mucor mucedo Actinomucor Mucor*	THRELFALL (1968, 1972), LYR and CASPERSON (1982); FISHER (1977);
Yeast like growth	2,4-dichloro-aniline	*Mucorales*	LYR and CASPERSON (1982)
Effects on cellular or nuclear division	tolclophos-methyl, chloroneb	*Ustilago*	TILLMAN and SISLER (1973); KATO (1983)
Genetic (mutagenic) effects	chloroneb PCNB dicloran and others	*Asp. nidulans*	THRELFALL (1968); GEORGOPOULOS et al. (1976) AZEVEDO et al. (1977); KAPPAS (1978)
Mitochondrial destruction	etridiazole, PCNB chloroneb DCMP	*Mucor, Botrytis*	CASPERSON and LYR (1982); CASPERSON and LYR (1975)
Protein and nucleic acid synthesis	many compounds	*Ustilago Rhizopus Rhizoctonia Asp. nidulans*	WEBER and OGAWA (1965); TILLMAN and SISLER (1973); THRELFALL (1968); HOCK and SISLER (1969); CRAIG and PEBERDY (1983); FRITZ et al. (1977)
Lipid synthesis	dicloran	*Asp. nidulans*	CRAIG and PEPERDY (1983) further literature see LEROUX and FRITZ (1984)

these fungicides had emerged (KAARS SIJPESTIJN 1982; LEROUX and FRITZ 1984). Some of these effects are summarized in Table 5.4.

All authors agree that there should be a common mechanism of action in spite of the very different chemical structures (Fig. 5.1), because fungi which developed resistance for one member of this group of fungicides are resistant to other members as well, and also to dicarboximide fungicides, a more recently discovered group of fungicides (cf. chapter 6).

Results on the mechanism of action

With regard to the mechanism of action, the most thoroughly investigated compounds are chloroneb, quintozene (PCNB) and etridiazole, but there is at present no doubt that the other fungicides of this group act by a similar or identical mechanism. The reason for the different antifungal spectra is another problem that is not yet elucidated, but this difference may be due to mechanisms such as detoxification or other processes (chapter 1).

The mechanism of action of chloroneb has been analyzed in detail in *Mucor mucedo* by WERNER (1980). Some aspects of these studies have been described by LYR and WERNER (1982), WERNER et al. (1978).

The ultrastructural effects of chloroneb (Fig. 5.2 and 5.3) are nearly identical with those of PCNB (Fig. 5.4), which has a similar molecular space configuration compared to chloroneb. Therefore, they shall be characterized together. Both compounds induce a lysis of the inner mitochondrial membranes, beginning with a swelling of the cristae. The nuclear envelope vacuolizes, and the cell wall thickness increases dramatically within a few hours in *Mucor mucedo* (WERNER et al. 1978; CASPERSON and LYR 1982; LYR and CASPERSON 1982) (Fig. 5.5, 5.8). This stops the hyphal tip (radial) growth and thus hinders the spread of the fungus and a pathogenic attack. Simultaneously the respiration decreases, but it is not specifically inhibited, because the respiratory quotient of mitochondria remains nearly constant. Dry matter accumulation also decreases.

Chloroneb binds to mitochondrial proteins of sensitive, but not of resistant strains in *Mucor*. The latter do not show changes in the ultrastructure when compared to the control. Comparing the amino acid composition of the proteins of isolated mitochondria of R- and S-strains of *Mucor mucedo* WERNER (1980) found nearly an equal percentage of all amino acids except for tyrosine, which was only 0.2% in resistant strains compared to 2.9% in S-strains. This means, that a tyrosine rich protein (enzyme) has decreased in content or its composition has changed. It may be, that a tyrosine molecule located in the active centre of the enzyme has something to do with the binding capacity for chloroneb. *Neurospora crassa*, a moderately sensitive species, binds significantly more chloroneb in its mitochondria than the insensitive *N. sitophila*. The tyrosine content of mitochondrial proteins of the former species is higher than in the latter (WERNER 1980).

Several biochemical investigations have established that tyrosine and tryptophane play an important role in the binding of flavin coenzymes to the apoenzyme. Tyrosine can form an intermolecular complex with charge-transfer interaction, in contrast to hydroquinones (INOUE et al. 1980). Addition of tyrosine to the culture medium did not change the sensitivity of *Mucor* towards chloroneb (in contrast to results of KATARIA and GROVER (1978) in *Rhizoctonia solani*). This indicates that tyrosine synthesis is not blocked but

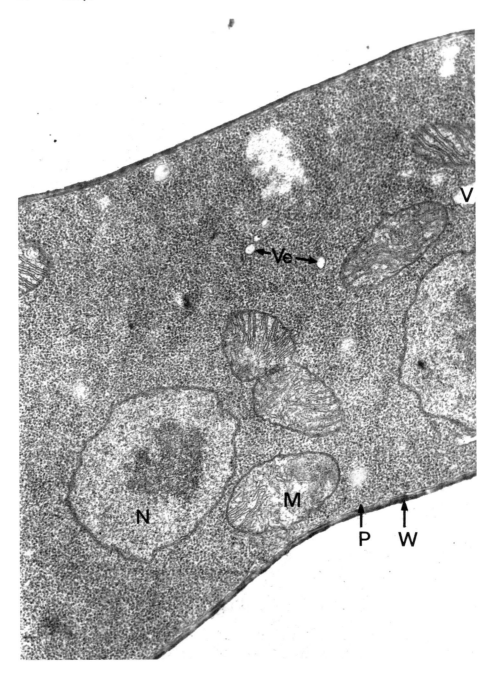

Fig. 5.2 Ultrastructure of *Mucor mucedo*, control. N = nucleus; M = mitochondrium; V = vacuole; Ve = vesicle; P = plasmalemma; W = cell wall. 36,000:1, phot.: CASPERSON.

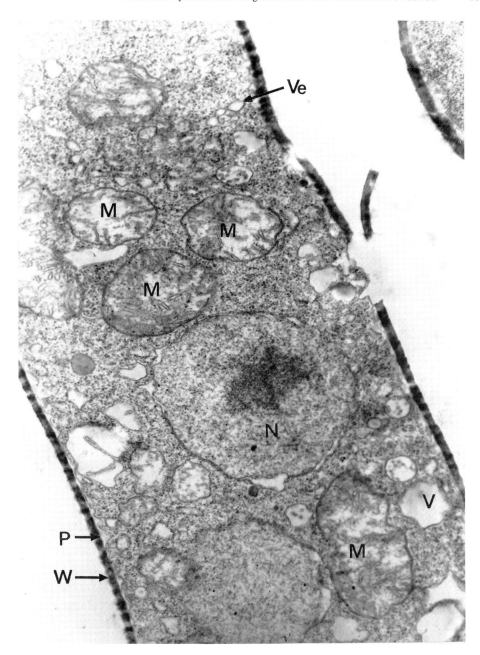

Fig. 5.3 Ultrastructural changes in *Mucor mucedo* 2 hours after application of 20 ppm Chloroneb. Lysis of the cristae in the swollen mitochondria, number of vacuoles and vesicles increased. The perinuclear space is partially enlarged, the plasmalemma is loosened from the thickened cell wall. 36,000:1, phot.: CASPERSON.

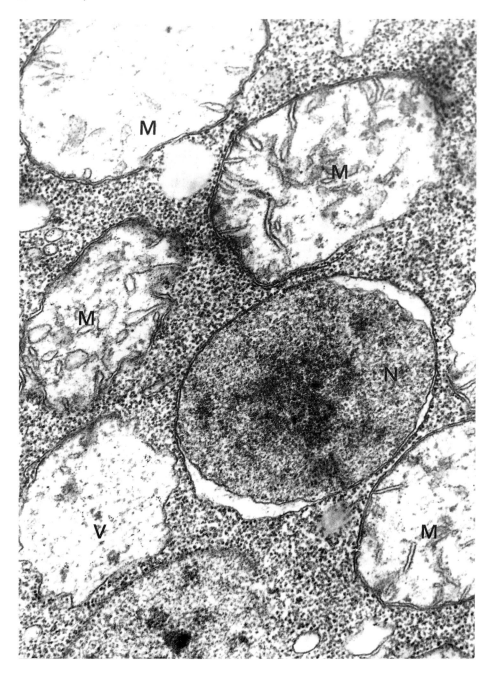

Fig. 5.4 Ultrastructural changes in *Mucor mucedo* 2 hours after application of 20 ppm Pentachloronitrobenzene (PCNB). The perinuclear space is enlarged, in the swollen mitochondria the lysis of the inner membrane is visible, 60,000:1, phot.: CASPERSON.

Fig. 5.5 Cell wall of *Mucor mucedo* 2 hours after application of 20 ppm Pentachloronitrobenzene (PCNB). The anomalous thickened cell wall shows a lamellation after PAS-reaction. Dark lamellae indicate mannan and light lamellae glucan and chitin. 60,000:1 (According to LYR and CASPERSON 1982).

that a specific mutation reducing tyrosine rich sequences seems to have occured in R-strains. According to RUCKPAUL and REIN (1984) tyrosine is also an essential part of the active centre of cytochrome P 450. Therefore, a point mutation at the active centre could abolish the enzyme activity and by this the production of toxic radicals under the influence of AHF. AHF seem to interact with tyrosine sequences in monooxygenases which are capable of hydroxylation of benzene rings. AHF could play the role of substrate analogs which increase the rate of oxidation of NADPH but do not serve as substrates which can be hydroxylated (effector role) (MASSEY et al. 1982). Resistance can involve less effective binding or low enzymatic activity.

It should be mentioned here that AESCHBACH et al. (1976) described the formation of diphenyl-bridge bonding between hydrophobic tyrosine sequences by a peroxidative action. This could be an interesting receptor structure for diphenyl or o-phenylphenol. DCMP, the demethylation product of chloroneb (Fig. 5.1 g) behaves nearly identical as chloroneb, but has an additional weak uncoupling effect in isolated mitochondria, as can be expected because of its phenolic structure. But chlorinated phenols to which chloroneb resistant mutants do not show cross resistance (Tab. 5.5) have a much stronger uncoupling activity and never produce a cell wall thickening.

Newer investigations by LYR and EDLICH (1984) and EDLICH and LYR (1984) revealed, that chloroneb as well as other members of AHF induce a lipid peroxidation of mitochondrial and endoplasmic membranes which can be nearly nullified by addition of α-tocopherole-acetate, piperonylbutoxide or SKF-525 A, but perhaps merely by physical reasons. Simultaneously the growth inhibition decreased or is counteracted (Tab. 5.6). The basis of this phenomenon is the intereraction of AHF with an enzyme which was preliminarily characterized as cytochrome c-reductase, or with similar NADPH depend-

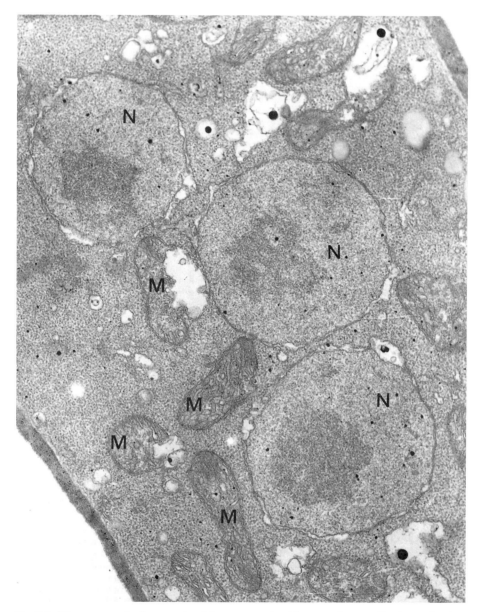

Fig. 5.6 Ultrastructural changes in *Mucor mucedo* 2 hours after application of 10 ppm Ethridiazole. Mitochondria are attacked, the outer membrane is enlarged, the inner membrane shows a partially, local lysis, the nuclear envelope has an enlarged perinuclear space. The number of vesicles is increased, the cell wall is abnormally thickened. 36,000:1, phot.: CASPERSON.

Table 5.5 Cross resistance of *Mucor mucedo* comparing a wild (S-)strain with a resistant (R-strain) selected on chloroneb amended medium. Inhibition of radial growth on malt agar (in percent of controls) by various concentrations of some fungicides (According to BISCHOFF and LYR 1980)

		Fungicide concentrations (mg/l)				
		1	3	10	30	100
Chloroneb	S	–	31	70	72	67
	R	–	0	0	0	0
PCNB	S	46	91	96	98	100
	R	0	18	32	47	49
Biphenyl	S	–	2	8	29	70
	R	–	0	0	3	8
Etridiazole	S	–	5	28	73	100
	R	–	4	9	24	71
Dicloran	S	69	85	100	–	–
	R	0	0	0	–	–
Pentachlorophenyl-	S	0	43	83	91	100
methylether	R	0	17	34	52	66
Pentachlorophenol	S	23	42	64	97	100
	R	22	45	65	96	100

Table 5.6 Growth inhibition and lipid peroxidation in *Mucor mucedo* (LYR and EDLICH, 1984) caused by AHF after 16 hours incubation time. Growth inhibition (fresh-weight) is in % related to controls without fungicide. Lipid peroxidation was measured by the method of TERAO and MATSUSHITA 1981

		growth inhibit. %	Lipid peroxide content in %	Lipid peroxide per/g fresh weight µmol
Control	1% DMSO	0	100	6.44
Etridiazole	5 ppm	65	160	10.32
Chloroneb	2 ppm	72	210	13.66
DCMP	2 ppm	64	210	13.66
PCNB	5 ppm	68	146	9.54

ent flavin enzymes (chapter 7). Cytochrome c-reductase is totally inhibited by chloroneb and other fungicides of this group, which are the most effective inhibitors of this enzyme presently known. The primary toxic side effect of this inhibition is a stimulated NADPH-oxidation probably *via* a flavin peroxide (ZIEGLER et al. 1980), which results in a lipid peroxidation of the phospholipide of the enzyme. This starts a cascade process in peroxidation of unsaturated fatty acids in the membranes of sensitive fungi.

According to KATARIA and GROVER (1978) cysteine, glutathione, hydroquinone and tyrosine decreased or nullify the growth inhibition by chloroneb and PCNB in *Rhizoctonia solani*.

This mechanism of action can sufficiently explain all effects by AHF described until now. A lipid peroxidation by monooxygenases within mitochondria must destroy

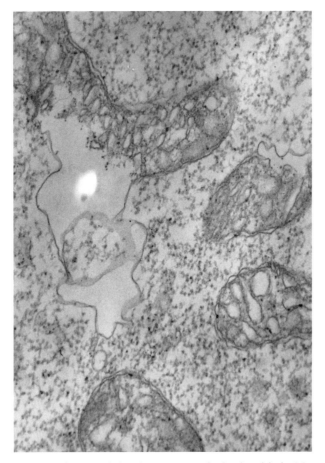

Fig. 5.7 Changes of the ultrastructure of mitochondria in *Mucor mucedo* 2 hours after application of 10 ppm Ethridiazole. The outer membrane is enlarged and the inner membrane shows a beginning destruction. 50,000:1 (According to Casperson and Lyr 1975).

the structure of the inner membrane system (Fig. 5.3, 5.4) and by this, decrease overall respiration without a specific inhibition of the respiratory chain. It may be that the monooxygenase is normally necessary for the synthesis of ubiquinone (Knoell 1981). This could explain the partial relief of etridiazole inhibition of growth and respiration after addition of ubiquinone or menadione, as observed by Halos and Huisman (1976a).

A production of free radicals and lipid peroxidation within the nuclear envelope [the occurrence of a "cytochrome c reductase" was demonstrated in nuclear membranes of mammals by Stasiecki et al. (1980)] not only impairs the membrane function and transport of RNA but can affect DNA, leading to strand scissions and chromosome aberrations (Baird et al. 1980; Brawn and Fridovich 1981; Ames et al. 1982). Desoxynucleosides are very sensitive to singlet oxygen or hydroxyl radicals. This could explain the genetic effects in fungi, observed by Georgopoulos et al. (1976).

Cytochrome c reductase is very often located in the endoplasmic reticulum therefore after induction of lipid peroxidation by AHF it is not surprising that protein synthesis at the ribosomal site is impaired.

Membrane bound lipid synthesis can also be disturbed in many ways by lipid peroxidation.

There still remains the phenomenon of pathological cell wall thickening under the influence of AHF as a typical effect (Fig. 5.8), which contributes to the fungistatic effect of AHF.

According to ULANE and CABIB (1974) chitin and glucan synthetase are present behind the hyphal tip in the cytoplasmic membrane as a dormant form under the influence of an inhibitor protein, and can be activated by proteinases decomposing the inhibitor proteins. This was demonstrated also in vitro experiments. AHF also seem to activate chitin synthetase probable allosterically (LYR and SEYD 1979). WATANABE and KONDO (1983) found an activation of proteinases by interaction of lipid peroxidation with a proteinase inhibitor protein. An activations of proteinase by chloroneb was described by WERNER (1980) in *Mucor mucedo*. Therefore it is possible, that under the influence of a moderate lipid peroxidation in the cytoplasmic membrane, chitin- and glucan synthetases are directly or indirectly activated which results in the observed pathological cell wall thickening (Fig. 5.8) and yeast-like growth. Other processes could support this effect (LYR and CASPERSON 1982).

The cytoplasmic membrane itself seems not to be very sensitive to AHF, because it is not lysed and no extreme leakage of cell constituents occurs. But this must not mean that it is not affected at all.

The sensitivity of a membrane to lipid peroxidation is dependent on the presence of

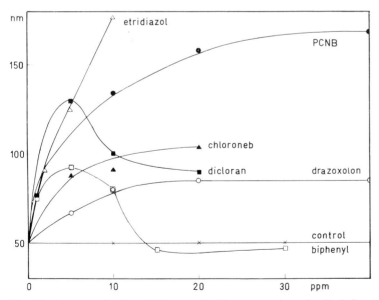

Fig. 5.8 Increase of cell wall thickness in *Mucor mucedo* under the influence of some AHF in dependance of fungicide concentration. (According to CASPERSON and LYR 1982.)

radical forming enzymes or radicals, on the content of unsaturated phospholipids, and on the ratio of phospholipids to sterols within the membrane.

In *Mucor mucedo* the rank of sensitivity, appears to occur in the following order: inner mitochondrial membrane > outer membrane > nuclear membrane > EPR > cytoplasmic membrane.

The inner mitochondrial membrane seems to be most sensitive perhaps because of the abundance of radical generating flavin enzymes and/or because of a relatively high content of unsaturated fatty acids combined with a relative low sterol content especially in some fungi (BOTTEMA and PARKS 1980).

In the case of etridiazole some differences in the mechanism of action exist, when comparisons are made with other AHF, which make it necessary to discuss this compound separately.

This is demonstrated in the cross resistance values. Etridiazole selected R-strains of *Mucor mucedo* have a high factor of cross resistance with most other AHF, but strains selected on chloroneb exhibit only a cross resistance factor of 3,3 against etridiazole (Tab. 5.5). The ultrastructural changes are somewhat different from those caused by chloroneb or PCNB in *Mucor* (Fig. 5.5, 5.6), especially in regard to the destruction of the inner mitochondrial membrane (CASPERSON and LYR 1975). Here a localized lysis of the membrane is typical and this enlarges to a total destruction of the mitochondrial inner membranes (RADZUHN and CASPERSON 1979). The overall effects on respiration, cell wall thickening, nuclear envelopment and other parameters are rather similar to those of other AHF (RADZUHN and LYR 1984). The same is true for lipid peroxidation (LYR and EDLICH 1984) (Tab. 5.6).

With etridiazole, in contrast to chloroneb and PCNB, the effect on growth and destruction of mitochondrial membranes can be counteracted by procaine or high doses of calcium ions (RADZUHN 1978). This means, that an activation of a phospholipase is involved in the total process of interaction.

A possible explanation of these deviations can be that a different, but similar enzyme is involved in this process. The molecular mechanism of action seems to differ in some details. The CCl_3-group of etridiazole is essential for its biological activity, replacement by CH_3- or CH_2Cl- results in inactive compounds as our experiments demonstrated. Therefore, one can speculate, that a monoxygenase, such as cytochrome P 450 oxygenase or a cytochrome c reductase can activate the CCl_3-group in a manner described for the insecticide DDT by BAKER *et al.* (1984). The produced intermediary R-C$^\bullet$·Cl_2 radicals beside initiating a lipid peroxidation can bind to phospholipids and hydrophobic proteins, yielding covalent bondings. This perhaps could explain the lower degree of resistance in etridiazole because by this it would have a multiside effect by its reactivity.

Resistance and cross resistance in AHF has been described by several authors for various fungi (GEORGOPOULOS 1977, 1982; LEROUX *et al.* 1984). An open question is, whether the AHF inhibited monooxygenases (target enzymes) are essential for normal growth and are still operating in the R-strains, though perhaps with small structural changes. Resistant strains in most cases have similar growth rates as S-strains, and sporulation can be quite normal.

VAN TUYL (1977) observed, that a chloroneb resistant strain of *Penicillium expansum* was unable to grow and to sporulate normally and could not produce the typical green colour of mature spores. Addition of chloroneb or PCNB restored growth and sporulation to normal. WERNER (1980) confirmed theses results and found, that dichlorohydro-

Table 5.7 ED_{50} values (in mg/l), and factors of resistance (RF) in *Mucor mucedo* (S- and R-strains) (from strain P3 of our collection) for various fungicides. The R_{Chl}-strain was selected on chloroneb medium (300 ppm), the R_{ET}-strain on etridiazole medium (according to BISCHOFF and LYR 1980)

Fungicides	S-strain	R_{Chl}-strain	R_{ET}-strain	RF
Chloroneb	3	>300	>300	>100
PCNB	1	100	120	>100
Pentachlorophenol	5	5	5	0
Biphenyl	50	>300	>300	>6
Hexachlorobenzene	>300	>300	–	–
Drazoxolon	2	2.5	–	0
Etridiazole	20	65	70	3.3
Fenarimol	180	260	–	1.4

quinone, p-dichlorobenzene, 2,4,5-trichlorophenol, 2,4,5-trichloroanisole, hexachlorobenzene, dicloran or diphenyl did not reverse this defect, whereas chloroneb, DCMP and PCNB at 100 and 1,000 ppm induced normal sporulation and a green colouring of the spores. This means that the R-mutant needs chloroneb for induction of the target enzymes or for some peroxidative reactions necessary for normal development. On the other hand, these experiments could reflect differences in the receptor structure for the various AHF. The basis of this phenomenon is not yet elucidated.

References

AESCHBACH, R., AMADO, R., and NEUKOM, H.: Formation of dityrosine cross-links in proteins by oxidation of tyrosine residues. Biochem. Biophys. Acta **439** (1976): 292–299.

AMES, B. N., HOLLSTEIN, M. C., and CATHCART, R.: Lipid peroxidation damage to DNA. In: YAGI, K. (Ed.): Lipid Peroxides in Biology and Medicine. Academ. Press Inc., New York 1982, pp. 339–351.

AZEVEDO, J. L., SANTANA, E. P., and BONATELLI, R.: Resistance and mitotic instability to chloroneb and 1,4-oxathiin in *Aspergillus nidulans*. Mutation Res. **48** (1977): 163–172.

BAIRED, M. B., BIRNBAUM, L. S., and SPEIR, G. T.: NADPH-driven lipid peroxidation in rat liver nuclei and nuclear membranes. Arch. Biochem. Biophys. **200** (1980): 108–115.

BAKER, M. T., and VAN DYKE, R. A.: Metabolism-dependent binding of the chlorinated insecticide DDT and its metabolite, DDD, to microsomal protein and lipids. Biochem. Pharmacology **33** (1984): 255–260.

BAKKEREN, M., and OUWEHAND, I.: Some experiments for control of *Pythium*-caused diseases in flower bulb-crops with 5-ethoxy-3-trichloromethyl-1,2,4-thiadizol. Brit. Crop. Protect. Conf. 70-B 3/2 25 (1970): 241–242.

BARNES, G., HARRIS, R. I., and RUSSEL, P. E.: Tolclophosmethyl, a new fungicide for the control of *Rhizotonia solani*. Brit. Crop. Protect. Conf. 2 C–S 23 (1984): 460–464.

BECKMANN, K. M., and PARSONS, J. E.: Fungicidal control of *Sclerotinia* wilt in green beans. Plant Disease Rep. **49** (1965): 357–358.

BEEVER, R. E., and BYRDE, J. W.: Resistance of the dicarboximide fungicides. In: DEKKER, J., and GEORGOPOULOS, S. G. (Eds.): Fungicide Resistance in Crop Protection. Centre for Agricult. Publ. and Document., Wageningen 1982, pp. 101–117.

BISCHOFF, G., and LYR, H.: Das Resistenzverhalten von *Mucor mucedo* (L.) Fres. gegenüber Chloroneb und anderen Fungiziden. Arch Phytopathol. u. Pflanzenschutz, Berlin **16** (1980): 111–118.

BOLTON, A. T.: Fungicide resistance in *Botrytis cinerea*, the result of selective pressure on resistant strains already present in nature. Can. J. Plant Sci. **56** (1976): 861–864.

BOTTEMA, C. K., and PARKS, L. W.: Sterol analysis of inner and outer mitochondrial membranes in yeast. Lipids **15** (1980): 987–992.

BRAWN, K., and FRIDOVICH, I.: DNA strand scission by enzymatically generated oxygen redicals. Arch. Biochem. Biophys. **206** (1981): 414–419.

BROWN, W.: Experiments on the effect of chlorinated nitrobenzenes on the sprouting potato tubers. Ann. appl. Biol. **34** (1947): 422–429.

CASPERSON, G., and LYR, H.: Wirkung von Terrazol auf die Ultrastruktur von *Mucor mucedo*. Z. Allg. Mikrobiol. **15** (1975): 481–488.

– and LYR, H.: Wirkung von Pentachlornitrobenzol (PCNB) auf die Ultrastruktur von *Mucor mucedo* und *Phytophthora cactorum*. Z. Allg. Mikrobiol. **22** (1982): 219–226.

CHASTAGNER, G. A., and OGAWA, J. M.: A fungicide-wax treatment to suppress *Botrytis cinerea* and protect fresh-market tomatoes. Phytopathology **69** (1979): 59–63.

CORDEN, M. E.: Aromatic compounds. In: TORGESON, D. C. (Ed.): Fungicides, an Advanced Treatise. Vol. II. Acad. Press Inc., New York–London 1969.

CRAIG, G. D., and PEPERDY, J. F.: The mode of action of s-benzyl-0,0-di-isopropyl phosphorothionate and of dicloran on *Aspergillus nidulans*. Pesticide Sci. **14** (1983): 17–24.

ECKERT, J. W.: Fungistatic and phytotoxic properties of some derivatives of nitrobenzene. Phytopathology **52** (1962): 642–649.

– Chemical treatment for control of postharvest diseases. World Rev. Pest Control Vol. **8** (1969): 116–137.

– Interaction of derivatives of o-phenylphenol with sensitive and tolerant fungi. In: LYR, H., and POLTER, C. (Eds.): Systemic Fungicides. Akademie-Verlag, Berlin 1979.

– Fungicidal and fungistatic agents: Control of pathogenic microorganisms on fresh fruits and vegetables after harvest. In: Food Mycology, G. K. Hall Publ. Co., Boston 1979, Chapt. 14.

– BRETSCHNEIDER, B. F., and RATNAYAKE, M.: Investigations on new post harvest fungicides for citrus fruits in California. Proc. Int. Soc. Citriculture **2** (1981): 804–810.

– and WILD, B. L.: Problems of fungicide resistance in *Penicillium* rot of citrus fruits. In: GEORGHIOU, G. P., and SAITO, T. (Eds.): Pest Resistance to Pesticides. Plenum Publ. Corp. 1983.

– and OGAWA, J. M.: The chemical control of postharvest diseases: subtropical and tropical fruits. Ann. Rev. Phytopathol. **23** (1985): 421–454.

EDLICH, W., and LYR, H.: Occurrence and properties of a Cytochrome c-reductase in *Mucor* and its interaction with some fungicides. In: LYR, H., and POLTER, C. (Eds.): Systemic Fungicides and Antifungal Compounds. Tagungsbericht Akad. Landwirtschaftswiss. DDR 222 (1984): 203–206.

ESURUOSO, O. F., PRICE, T. V., and WOOD, R. K. S.: Germination of *Botrytis cinerea* conidia in the presence of quintocene, tecnazene and dichloran. Trans. Br. Mycol. Soc. **51** (1968): 405 to 410.

FIELDING, M. J., and RHODES, R. C.: Studies with ^{14}C-labeled chloroneb fungicide in plants. Cotton Disease Council Proc. **27** (1967): 56–60.

FISHER, D. J.: Induction of yeast-like growth in *Mucorales* by systemic fungicides and other compounds. Transact. Brit. Mykol. Soc. **68** (1977): 397–402.

FRITZ, R., LEROUX, P., and GREDT, M.: Mécanisme de l'action fongitoxique de la promidone (260 19 RP ou glycophène), de la vinclozoline et du dichloran sur *Botrytis cinerea* Pers. Phytopatholog. Z. **90** (1977): 152–163.

GEORGOPOULOS, S. G.: Development of fungal resistance to fungicides. In: SIEGEL, M. R., and SISLER, H. D. (Eds.): Antifungal Compounds. Vol. 2, Marcel Dekker Inc., New York 1977, pp. 439–495.

- Cross-resistance. In: DEKKER, J., and GEORGOPOULOS, S. G. (Eds.): Fungicide Resistance in Crop Protection. Centre for Agricult. Publ. and Document., Wageningen 1982, pp. 53–59.
- KAPPAS, A., and HASTIE, A. C.: Induced sectoring in diploid *Aspergillus nidulans* as a criterion of fungitoxicity by interference with hereditary processes. Phytopathology **66** (1976): 217–220.
- and ZARACOVITIS, C.: Tolerance of fungi to organic fungicides. Ann Rev. Phytopathol. **5** (1967): 109–130.

GRAY, L. F., and SINCLAIR, J. B.: Uptake and translocation of systemic fungicides by soybean seedlings. Phytopathology **60** (1970): 1486–1488.

GROSSMANN, F.: Untersuchungen über die innertherapeutische Wirkung organischer Fungizide. II. Chlornitrobenzole. Z. Pflanzenkrankh. Pflanzenschutz **65** (1958): 594–599.

HALOS, P. M.: Inhibition of respiration in *Pythium* species by ethazol. Phytopathology **66** (1976b): 158–164.
- and HUISMAN, O. C.: Mechanism of tolerance of *Pythium* species to ethazol. Phytopathology **66** (1976a): 152–157.

HOCK, W. K., and SISLER, H. D.: Specificity and mechanism of antifungal action of chloroneb. Phytopathology **59** (1969): 627–632.

HOLTON, C. S., and PURDY, L. H.: Control of soil-borne common bunt of winter wheat in the Pacific Northwest by seed treatment. Plant Disease Rep. **38** (1954): 753–754.

INOUE, M., SHIBATA, M., and ISHIDA, T.: X-ray cristal structure of 7,8-dimethyl-isoalloxazine-10-acetic acid: tyramine (1:1) tetrahydrate complex. Biochem. Biophys. Res. Commun. **93** (1980): 415–419.

KAARS SIPESTEIJN, A.: Mechanism of action of fungicides. In: DEKKER, I., and GEORGOPOULOS, S. G. (Eds.): Fungicide Resistance in Crop Protection. Centre for Agricult. Publ. and Document. Wageningen 1982, pp. 32–45.
- DEKHUIZEN, A., and VONK, J. W.: Biological conversion of fungicides in plants and microorganisms. In: SIEGEL, M. R., and SISLER, H. D. (Eds.): Antifungal Compounds. Vol. 2. Marcel Dekker Inc., New York 1977.

KAPPAS, A.: On the mechanisms of induced somatic recombination by certain fungicides in *Aspergillus nidulans*. Mutation Res. **51** (1978): 189–197.

KATARIA, H. R., and GROVER, R. K.: Adaption of *Rhizoctonia solani* to systemic and nonsystemic fungitoxicants. Z. Pflanzenkr. Pflanzenschutz **81** (1974): 472–478.
- – Effect of chloroneb (1,4-dichloro-2,5-dimethoxybenzene) and pentachloronitrobenzene on metabolic activities of *Rhizoctonia solani* Kühn. Ind. J. Exp. Biology **13** (1975): 281–285.
- – Reversal of toxicity of some systemic and non-systemic fungitoxicants by chemicals. Z. Pflanzenkrankh. Pflanzenschutz **85** (1978): 76–83.

KATO, T.: Mode of action of a new fungicide, tolclophos-methyl. In: MIYAMOTO, I., and KEARNEY, P. C. (Eds.): Pesticide Chemistry. Vol. 3. Pergamon Press, Oxford 1983.

KLUGE, E.: Vergleichende Untersuchungen über die Wirksamkeit von Systemfungiziden gegen Oomyzeten. Arch. Phytopath. u. Pflanzensch. **14** (1978): 115–122.

KNOELL, H.-E.: On the nature of the monooxygenase system involved in ubiquinone-8 synthesis. FEMS Microbiol. Letters **10** (1981): 63–65.

KUIPER, J.: Failure of hexachlorobenzene to control common bunt of wheat. Nature **206** (1965): 1219–1220.

LEROUX, P., and FRITZ, R.: Antifungal activity of dicarboximides and aromatic hydrocarbons and resistane to these fungicides. In: TRINCI, A. P. J., and RYLEY, J. F. (Eds.): Mode of Action of Antifungal Agents. Brit. Mycol. Soc. (1984): 207–237.
- – and GREDT, M.: Etudes en laboratoire de souches de *Botrytis cinerea* Pers. resistantes à la dichlozoline, au dicloran, au quintocene, à la vinchlozoline et au 26019 RP (ou glycophene). Phytopath. Z. **89** (1977): 347–358.
- and GREDT, M.: Resistance of *Botrytis cinerea* Pers. to fungicides. In: LYR, H., and POLTER, C.

(Eds.). Systemic Fungicides and Antifungal Compounds. Tagungsber. Akad. Landwirtschaftswiss. DDR **222** (1984): 329–333.

LIPE, W. N., and THOMAS, D. G.: Effects of seed-piece and in-furrow fungicide treatments on grade and yield of potatoes in Texas high plains. Texas Agricult. Experim. Station MP-1430 (1979): 1–11.

LOCKE, S. B.: An improved laboratory assay for testing effectiveness of soil chemicals in preventing white rot of onion. Plant Disease Rep. **49** (1965): 546–549.

LYR, H.: Effect of fungicides on energy production and intermediary metabolism. In: SIEGEL, M. R., and SISLER, H. D. (Eds.): Antifungal Compounds. Vol. 2, Marcel Dekker Inc., New York 1977, pp. 301–332.

— and CASPERSON, G.: Anomalous cell wall synthesis in *Mucor mucedo* (L.) Fres. induced by some fungicides and other compounds related to the problem of dimorphism. Z. Allg. Mikrobiol. **22** (1982): 245–254.

— and EDLICH, W.: On the molecular mechanism of action of the fungicides chloroneb and etridiazole. In: LYR, H., and POLTER, C. (Eds.): Systemic Fungicides and Antifungal Compounds. Tagungsbericht Akad. Landwirtschaftswiss. DDR 222 (1984): 59–64.

— and SEYD, W.: Die Wirkung einiger Fungizide und anderer Verbindungen auf die Chitin-Synthese *in vitro*. In: LYR, H., and POLTER, C. (Eds.): Systemic Fungicides. Abh. Akad. Wiss. DDR N 2 (1979): 151–157.

— and WERNER, P.: On the mode of action of the fungicide Chloroneb. Pesticide Biochem. Physiol. **18** (1982): 69–78.

MASSEY, V., CLAIBORNE, AL., DETMER, K., and SCHOPFER, L. M.: Comparative aspects of flavo protein monooxygenases. In: Oxygenases and oxygen metabolism. Acad. Press Inc., New York 1982, pp. 185–195.

NAGY, ST., and WARDOWSKI, W. F.: Diphenyl absorption by honey tangerines: the effects of washing and waxing and time and temperature of storage. J. Agr. Food Chem. **29** (1981): 760–763.

— — and HEARN, C. J.: Diphenyl absorption and decay in "Dancy" and "Sunburst" tangerine fruit. J. Am. Hortic. Sci. **107** (1982): 154–157.

OGAWA, J. M., MATHRE, J. H., WEBER, D. J., and LYDA, ST. D.: Effects of 2,6-dichloro-4-nitroaniline on *Rhizopus* species and its comparison with other fungicides on control of *Rhizopus* rot of peaches. Phytopathology **53** (1963): 950–955.

PURDY, L. H.: Common and dwarf bunts, their chemical control in the Pacific Northwest. Plant Disease Rep. **49** (1965): 42–46.

RADZUHN, B.: Zum Wirkungsmechanismus des systemischen Fungizides Terrazol. Diss. Humboldt-Univ., Berlin 1978, S. 1–81.

— and CASPERSON, G.: Zum Wirkungsmechanismus von Terrazol. In: LYR, H., and POLTER, C. (Eds.): Systemic Fungicides. V. Intern. Symp., Abh. Akad. Wiss. DDR N 2. Akademie-Verlag, Berlin 1979, pp. 195–206.

— and LYR, H.: On the mode of action of the fungicide etridiazole. Pesticide Biochem. Physiol. **22** (1984): 14–23.

RENNER, G., and RUCKDESCHEL, G.: Effects of pentachloronitrobenzene and some of its known and possible metabolites on fungi. Appl. Environm. Microbiol. **46** (1983): 765–768.

RITCHIE, D. F.: Mycelial growth, peach fruit-rotting capability, and sporulation of strains on *Monilinia fructicola* resistant to dicloran, iprodione, procymidone and vinclozolin. Phytopathology **73** (1983): 44–47.

— Effect of dicloran, iprodione, procymidone, and vinclozolin on the mycelial growth, sporulation, and isolation of resistant strains of *Monilinia fructicola*. Plant Disease Rep. **66** (1982): 484 to 486.

ROSE, D. H., COOK, H. T., and REDIT, W. H.: Harvesting, handling, and transportation of citrus fruits. U.S. Dept. Agr. Bibliogr. Bull. **13** (1951): 1–178.

RUCKPAUL, K., and REIN, H.: Cytochrome P 450. Akademie-Verlag, Berlin 1984.

SCORDA, E. A.: Insensitivity of wheat bunt to hexachlorobenzene and quintocene (Pentachloronitrobenzene) in Greece. Proc. Brit. Crop. Prot. Conf. (1977): 67–71.

SHATLA, M. N., and SINCLAIR, J. B.: Tolerance to pentachloronitrobenzene among cotton isolates of *Rhizoctonia solani*. Phytopathology **53** (1963): 1407–1411.

— — Effect of pentachloronitrobenzene on *Rhizoctonia solani* under field conditions. Plant Disease Rep. **49** (1965): 21–23.

SINCLAIR, J. B.: Uptake and translocation of systemic fungicides by soybean, creeping bentgras and strawberry. In: LYR, H., and POLTER, C. (Eds.): Systemic Fungicides. Int. Symp. Reinhardsbrunn 1974. Akademie-Verlag, Berlin 1975.

STASIECKI, P., OESCH, F., BRUDER, G., JARASCH, E.-D., and FRANKE, W. W.: Distribution of enzymes involved in metabolism of polycyclic aromatic hydrocarbons among rat liver endomembranes and plasma membranes. Eur. J. Cell Biol. **21** (1980): 79–92.

STOTA, Z., and TOMAN, M.: Effect of the constitution of isomers and homologs of benzene on their biological activity. II. Phytotoxicity and fungicidal activity of Cl-substituted derivatives of benzene. Biologia **12** (1957): 683–692.

TERAO, J., and MATSUSHITA, S.: Thiobarbituric acid reaction of methyl arachidonate monohydroperoxide isomers. LIPIDS **16** (1981): 98–101.

THIELECKE, H.: Pflanzenschutzmittel. Wiss. Taschenbücher, Akademie-Verlag, Berlin 1963.

THRELFALL, R. J.: The genetics and biochemistry of mutants of *Aspergillus nidulans* resistant to chlorinated nitrobenzenes. J. Gen. Microbiol. **52** (1968): 35–44.

— Effect of pentachloronitrobenzene (PCNB) and other chemicals of sensitive and PCNB resistant strains of *Aspergillus nidulans*. J. Gen. Microbiol. **71** (1972): 173–180.

TILLMAN, R. W., and SISLER, H. D.: Effect of chloroneb on the growth and metabolism of *Ustilago maydis*. Phytopathology **63** (1973): 219–225.

TOMKINS, R. G.: Use of paper impregnated with esters of o-phenylphenol to reduce the rotting of stored fruit. Nature **199** (1963): 669–670.

TURNER, F. T.: Soil nitrification retardation by rice pesticides. J. Soil Sci. Soc. Amer. **43** (1979): 955–957.

ULANE, R. E., and CABIB, E.: The activating system of chitin synthetase from *Saccharomyces cerevisiae*. J. Biol. Chem. **249** (1974): 3418–3422.

VAN BRUGGEN, A. H. C., and ARNESON, P. A.: Resistance in *Rhizoctonia solani* to tolclofosmethyl. Neth. J. Plant Path. **90** (1984): 95–106.

VAN TUYL, J. M.: Genetics of fungal resistance to systemic fungicides. Mededelingen Landbouwhogeschool (Wageningen) **77** (1977): 1–136.

VARGAS, J. M., and BEARD, J. B.: Chloroneb, a new fungicide for the control of *Typhula* blight. Plant Disease Rep. **54** (1970): 1075–1080.

WATANABE, T., and KONDO, N.: The change in leaf protease and protease inhibitors activities after supplying various chemicals. Biol. Plant **25** (1983): 100–109.

WEBER, J. R., and OGAWA, J. M.: The mode of action of 2,6-dichloro-4-nitroaniline in *Rhizopus arrhizus*. Phytopathology **55** (1965): 159–165.

WERNER, P.: Zum Wirkungsmechanismus des systemischen Fungicides Chloroneb und zu den möglichen Ursachen erzeugter Resistenz gegenüber *Mucor mucedo*. L. Fres. Diss. Martin-Luther-Univ., Halle–Wittenberg 1980. S. 1–142.

— LYR, H., and CASPERSON, G.: Die Wirkung von Chloroneb, seinen Abbauprodukten sowie von chlorierten Phenolen auf das Wachstum und die Ultrastruktur verschiedener Pilzarten. Arch. Phytopathol. Pflanzenschutz, Berlin **14** (1978): 301–312.

WHEELER, J. E., HINE, R. B., and BOYLE, A. M.: Comparative activity of dexon and terrazole against *Phytophtora* and *Pythium*. Phytopathology **60** (1970): 561–562.

WOLF, P. A.: Sodium orthophenyl phenate (Dowicide A) combats microbiological losses of fruits and vegetables. Down Earth **12** (1956): 16–19.

YERSIN, H., CHOMETTE, H., BAUMANN, G., and LHOSTE, J.: Hexachlorobenzene, an organic synthetic used to combat wheat smut. Compt. Rend. Acad. Agric. France **31** (1974): 5247 to 5251.

ZIEGLER, D. M., POULSEN, L. L., and DUFFEL, M. W.: Kinetic studies on mechanism and substrate specificity of the microsomal flavin-containing monooxygenases. In: Microsomes, Drug oxidations, and Chemical Carcinogenesis. Acad. Press Inc., New York 1980, pp. 637 to 645.

Chapter 6

Dicarboximide fungicides

E.-H. POMMER and Gisela LORENZ

BASF Aktiengesellschaft, Limburgerhof, FRG

Introduction

The antimicrobial activity of 3-phenyl-oxazolidine-2,4-diones and related compounds was reported by FUJINAMI, OZAKI and YAMAMOTO in 1971. In their studies on structure-activity relationships, these authors were able to show the effect on the fungicidal activity of, firstly, substituents in the benzene ring and secondly, substitutes in the 5-position of the oxazolidine ring. A powerful fungicidal effect was only produced by a dichloro-substitution in the 3,5-position of the benzene ring; a dimethylsubstitution in the 5-position of the oxazolidine ring gave the best activity. Their studies were carried out using *Sclerotinia sclerotiorum* as the test fungus. The first active compound to emerge from this class of compounds, now designated dicarboximides, was dichlozoline (Fig. 6.1 f). On account of toxicological problems (NAKAYA et al. 1969), dichlozoline was not pursued further, although it possessed excellent fungicidal activity. A particular feature of dichlozoline is its strong activity against *Botrytis* and *Sclerotinia* species (MENAGER et al. 1971). Dimethachlor (Fig. 6.1 g) resultet from investigations into substituted 1-phenylpyrrolidine-2,5-diones (FUJINAMI et al. 1972). However, this compound found no wide application and is currently of only minor importance.

The search for further cyclic imides of high fungicidal activity and substituted on the nitrogen atom by a 3,5-dichlorophenyl group was successful. Within three years, three new dicarboximide fungicides were introduced onto the market: iprodione (LACROIX et al. 1974), vinclozolin (POMMER and MANGOLD 1975) and procymidone (HISADA et al. 1976). These compounds are used worldwide. Although the interest in new compounds from this group has been greatly reduced on account of the resistance of *Botrytis cinerea* to dicarboximides since the end of the 70's, two further compounds have been developed into marketed products; chlozolinate (DI TORO et al. 1980) and metomeclan (VULIE et al. 1984) (Fig. 6.1).

Among the development products that have emerged from this group of compounds, mention should be made of myclozoline, experimental code BAS 436 F, in which the active ingredient is 3-(3,5-dichlorophenyl)-5-(methoxymethyl)-5-methyl-2,4-oxazolidinedione (POMMER and ZEEH 1982).

Dicarboximides which have established themselves on the market are listed in Table 6.1. Table 6.2 lists those dicarboximides that are currently in use, with details of the active compounds that they contain; only limited information is currently available for chlozolinate and metomeclan.

In ornamental plant cultivation, dicarboximides are also used in bulb flower and other flower crops for desinfection against *Botrytis* (bulb fire) in tulips; *Sclerotinia bulborum* (black slime) in hyacinths; *Botrytis, Curvularia* (dry rot) and *Stromatinia* (dry rot) in gladioli. The fungicides are applied by dipping.

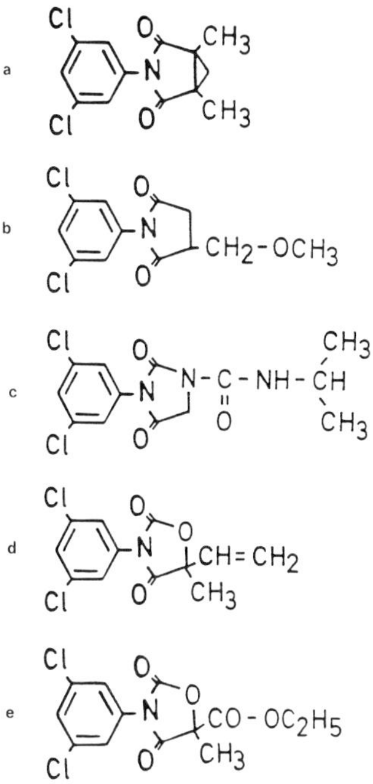

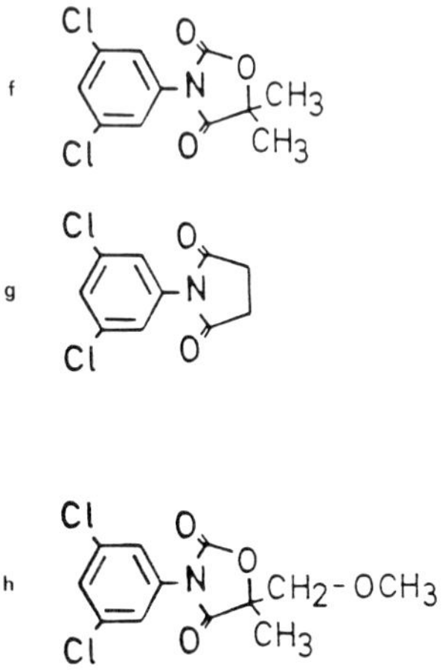

Fig. 6.1 Structural formulae of dicarboximide fungicides. a) Procymidone, b) Metomeclan, c) Iprodione, d) Vinclozolin, e) Chlozolinate, f) Dichlozoline, g) Dimethachlor, h) Myclozoline.

Table 6.1 Development of dicarboximides as fungicides for commercial use

Common name Code number	Product name on introduction	Year of introduction by
Dichlozoline CS 8890	Sclex	1967 Sumitomo/Hokko withdrawn
Dimethachlor S-47127	Ohric	1969 Sumitomo withdrawn
Iprodione 26019 RP	Rovral	1974 Rhone-Poulenc
Vinclozolin BAS 352 F	Ronilan	1975 BASF
Procymidone S-7131	Sumilex Sumisclex	1976 Sumitomo
Chlozolinate M 8164	Serinal	1980 Montedison/ Farmoplant
Metomeclan Co-6054	Drawifol	1984 Wacker withdrawn 1986

Table 6.2 Dicarboximides in commercial use as fungicides

Structure	Chemical name	Patent No.	LD_{50} rat p.o. mg/kg
Procymidone	N-(3,5-dichlorophenyl)-1,2-dimethyl-cyclopropane-1,2-dicarboximide	US 3 903 090 (Sumitomo)	6 800
Iprodione	3-(3,5-dichlorophenyl)-N-isopropyl-imidazolidine-2,4-dione-1-dicarboximide	FR 2 120 222 (Rhone-Poulenc)	3 500
Vinclozolin	3-(3,5-dichlorophenyl)-5-ethenyl-5-methyl-oxazolidine-2,4-dione	DE 2 207 576 (BASF)	>10 000
Chlozolinate	ethyl (±)-3-(3,5-dichlorophenyl)-5-methyl-2,4-dioxolidine-5-carboxylate	GB 874 406 (Montedison)	>4 500

There are no indications in the literature of serious problems with regard to plant tolerance when dicarboximides are used for horticultural or agricultural crops; however, the manufacturers' recommendations should be followed.

In order to broaden the spectrum of activity of dicarboximides, or in connection with anti-resistance strategies, mixtures containing other fungicides in addition have been developed into commercial products:

iprodione + carbendazim
iprodione + carbendazim + dinicazole
iprodione + dinicazole
iprodione + imazalil
iprodione + maneb + sulphur
iprodione + metalaxyl + thiabendazol

procymidone + copper oxychloride
procymidone + chlorothalonil
procymidone + maneb + zineb
procymidone + thiram

vinclozolin + carbendazim
vinclozolin + chlorothalonil
vinclozolin + metiram
vinclozolin + thiram
iprodione + thiophanate-methyl
iprodione + thiram

Fungicidal spectrum of activity and the use of dicarboximides in horticultural and agricultural crops

A general feature of dicarboximides is their essentially protective activity against representatives of the following genera of fungi: *Botrytis, Sclerotinia, Monilinia, Alternaria, Sclerotium* and *Phoma. Helminthosporium, Rhizoctonia* and *Corticium* should also be mentioned. Since the cyclic amide component in the dixarboximides can be an oxazolidine-dione (vinclozolin or chlozolinate), a succinimide (procymidone or metomeclan) or a hydantoin (iprodione), there are understandable differences in their degree of activity against the organisms listed or shifts in the spectrum of activity. Dicarboximides are mainly used for the control of *Botrytis cinerea* and *Sclerotinia sclerotiorum* or *Sclerotinia minor* in agricultural crops. Controllable plant diseases and the most important crops are listed in Table 6.3.

Table 6.3 The plant diseases of important crops which can be controlled by dicarboximide fungicides

Organism	Disease	Main crops
Botrytis cinerea	grey mould	vines, strawberries, lettuce,
Botrytis spp.		capsicums, aubergines, beans, onions, tomatoes, ornamentals
Sclerotinia sclerotiorum	*Scelerotinia* rot *Sclerotinia* wilt *Sclerotinia* stem rot	lettuce, endives, chicory, cucurbits, celery, peanuts, beans, rape, turf, lawn
Sclerotinia minor	*Sclerotinia* rot	lettuce
Sclerotinia homoeocarpa	dollar spot	turf, lawn
Monilia laxa	blossom wilt	peaches, apricots, cherries
Monilia fructigena	brown rot	apples, pears, peaches, apricots
Monilia fructicola		
Alternaria brassicae	dark leafspot	cabbage, rape
Alternaria radicina	black rot	celery
Sclerotium bataticola	stem rot	sunflower
Phoma betae	black rot	beet
Phoma lingam	black leg	cabbage, rape
Didymella lycopersici	stem and fruit rot	tomatoes
Didymella bryoniae	black rot	cucumber
Rhizoctonia solani	wirestem, bottom rot, head rot	radish, black radish, kohlrabi, potatoes
Laetisaria fuciformis (*Corticium fuciforme*)	black rot, red thread	turf, law

Effect of the development stages of fungi

Chapter 7 is concerned with the biochemical aspects of the mechanism of activity of dicarboximides. This section, therefore, will deal with the extent of our knowledge of the effects of the active compounds on the various development stages of fungi:

Depending on the concentration of active compound present, in *Botrytis cinerea* conidial germination is inhibited less strongly than mycelial development (BUCHENAUER 1976; HISADA and KAWASE 1977; HISADA et al. 1978; PAPPAS and FISHER 1979). Should conidial germination occur, the germ tubes remain short and stumpy, swell up and burst; in the case of an already developed mycelium, the cell wall structure is altered in the region of the hyphal tips that are still growing. Disturbances in cell wall synthesis occur, which is associated with an outflow of cytoplasm (BUCHENAUER 1976; HISADA and KAWASE 1977; EICHHORN and LORENZ 1978; ALBERT 1979). Using ^{14}C-labelled procymidone, HISADA and KAWASE (1977) established that the compound was rapidly bound to the hyphal cell wall. The binding process was reversible, and more than 95% of the bound procymidone was removed from the cell walls through washing. Hyphal growth, which was completely suppressed under the influence of the compound, recovered after washing. HAGAN and LARSEN (1979) reported a slight effect on the germination of conidia after using iprodione for the control of *Bipolaris* (*Drechslera*) *sorokiniana* on *Poa pratensis*, but the growth of germ tubes and the formation of appressoria was substantially suppressed. Conidia of *Drechslera sorokiniana* germinated in the vitro tests in the presence of iprodione, but they proceeded to swell and burst (DANNENBERGER and VARGAS 1982). Mycelial growth of *Corticium rolfsii* was strongly inhibited by iprodione but not by procymidone or vinclozolin, and all three fungicides suppressed the formation of sclerotia. GEORGOPOULOS et al. (1979) observed in *Aspergillus nidulans* that procymidone, iprodione and vinclozolin increased the frequency of mitotic recombination in diploid colonies. This effect was dependent on concentration.

Although the dicarboximides are classed among the contact fungicides which are generally applied prophylactically, various papers are devoted to the translocation of these compounds. The significance of these results for practical use is, however, still not clear, as these were essentially model experiments. COOKE et al. (1979) improved the activity of procymidone against *Botrytis cinerea* on strawberries by treating both the blossoms and the soil rather than blossom treatment alone. In tests with ^{14}C-labelled procymidone, MIKAMI et al. (1984) were not able to detect any uptake of the compound via cucumber leaves (in contrast to HISADA et al. (1977); in bean seedlings, uptake of the compound from a nutrient solution occurs via the roots. HISADA et al. (1977) also worked with a ^{14}C-labelled compound. In cucumbers they observed uptake via the leaf, and migration to the stem, with subsequent translocation both upwards and slightly downwards. Iprodione was taken up by potato plants growing in different soils, but the amount of fungicide detected in the aerial parts was dependent on the soil type (CAYLEY and HIDE (1980). After inoculation with *Phoma exigua* var. *foveata*, the development of lesions was prevented. Application of iprodione to the sprouts of seed potato tubers prior to planting decreased the incidence and severity of infection by *Rhizoctonia solani* and *Polycytalum pustulans* on stem bases.

Conversely, POMMER and MANGOLD (1975) established that the uptake of vinclozolin via the roots of beans and green peppers was slight, and that transfer of the compound in the stem tissue takes place over only short distances.

Effect on yeast flora and the course of fermentation

The treatment of vines against grape-*Botrytis* is an important field for the use of dicarboximides. The effect that dicarboximides have on the yeast flora and the course of fermentation is consequently the subject of numerous papers. Considerable attention has also been given to the behaviour of residues in wine.

In *in vitro* tests, the addition of 0.025 – 0.075% vinclozolin and procymidone to the nutrient agar did not inhibit the growth of *Kloeckera apiculata, Saccharomyces oviformis* and *S. ellipsoideus* (STOJANOVIC and VUKMIROVIC 1979). Similar results were obtained by VOJTEKOVA *et al.* (1983) with *Saccharomyces oviformis* and *S. cerevisiae* after adding vinclozolin to grape must. In field trials with vinclozolin (BENDA 1978; BARBERO and GAIA 1979) no changes in the composition of the natural yeast flora of the grape or the must resulted. Non-sulphited, non-defecated grape juice treated with iprodione or vinclozolin was recovered and allowed to ferment spontaneously. Yeast development during fermentation was identical to that in juice from grapes which had not been treated with these fungicides (SAPIS-DOMERCQ *et al.* 1977). In contrast, in France in 1977, a dry year, SAPIS-DOMERCQ *et al.* (1978) observed a slight effect on the course of fermentation after the use of vinclozolin and procymidone in viticulture. The authors attributed this to the residual effect of the fungicides applied to the grapes. According to BENDA (1983), the auxiliary substances used for formulation of the active compounds iprodione and procymidone can inhibit the fermentation process: the auxiliary substances in the vinclozolin formulation did not affect fermentation.

The degradation of the dicarboximides chlozolinate, iprodione, procymidone and vinclozolin has been studied in wine at pH 3.0 and 4.0 at 30 °C (CABRAS *et al.* 1984). The authors reached the conclusion that chlozolinate is degraded very rapidly. Procymidone and vinclozolin, which have greater stability than chlozolinate, showed degradation times that were shorter than the regular maturation time of wines. Iprodione exhibited high stability even after 92 days. Metabolites found were 3-(3,5-dichlorophenyl)-5-methyloxazolidine-2,4-dione (from chlozolinate) and 3′,5′-dichloro-2-hydroxy-2-methylbut-3-enanilide (from vinclozolin). The second of these probably results from hydrolytic opening of the heterocyclic part of the molecule. 3,5-dichloroaniline was not detected as a degradation product of any of the compounds.

PIRISI *et al.* (1986) investigated the process of degradation of chlozolinate, vinclozolin and procymidone in wine and in a hydroalcoholic medium (water + 10% ethanol + tartaric acid/sodium hydroxide buffer; pH 4.0). The fungicides were added to the wine after fermentation. Degradation was dependent on the nature of the heterocyclic ring. The oxazoline-2,4-diones underwent hydrolysis, producing carbamic acids as intermediates, and these underwent a loss of carbon dioxide, generating the corresponding anilides. The imidic ring of procymidone also underwent acid-catalysed ring opening; in wine the product of this reaction was 3,5-dichloroaniline, which was degraded to unidentified products.

FLORI and ZORINI (1984) were able to demonstrate that residues of vinclozoline, iprodione and procymidone were partially removed by cold clarification of the must or by treatment of the wine with active charcoal. Grapes treated with vinclozolin, iprodione and procymidone were also examined for the presence of residues. The residues detected were considerably below the legally tolerated limits (FIMA and WOMASTEK 1983).

Behaviour in the soil and plants

Several lines of investigation have been followed with regard to the behaviour of dicarboximides in soils. Only limited information is available on the degradation and migration of the compounds in soil. WALKER et al. (1984) reported enhanced degradation of iprodione in soils previously treated with this fungicide. ^{14}C-labelled iprodione was incubated in two soils which had not been treated previously. After 45 days' incubation, 18–20% of the iprodione applied initially was recovered from the previously treated soils, whereas recovery from one of the soils not treated previously was 40% of the initial dose, from the other it was 80%. In a second experiment with four paired soil samples, iprodione degraded at least twice as rapidly in the previously treated sample compared with the previously untreated one.

The effects of soil pH on the degradation rates of iprodione and vinclozolin were measured by WALKER (1987a) in a silky clay loam soil. There was little degradation of either fungicide at pH 4.3 or 5.0, and degradation at pH 5.7 was slower than at pH 6.5. In both of the higher-pH soils, the rate of loss of a second application of either fungicide was faster than that of the first, and a third application degraded even more rapidly. Studies of iprodione degradation in 33 soils from commercial fields demonstrated a clear trend towards faster rates of loss in soils with an extensive history of iprodione use. The time for 90% loss from previously untreated soils varied from 22 to 93 days. In soils treated once previously it varied from 16 to 28 days, and in those treated twice previously from 6 to 23 days. In soils that had received three or more previous applications, the time to 90% degradation varied from 3.8 to 15 days. WALKER (1987b) carried out measurements with gas-liquid and high performance liquid chromatography, confirming that 3,5-dichloroaniline was an important degradation product of iprodione in soil. A colorimetric test based on the production of a diazo colour complex was shown to differentiate the amounts of the aniline. Studies with vinclozolin indicated that the test could also be used to identify soils which rapidly degrade this fungicide.

In a comparative study of dicarboximide persistence in soil, FLORY et al. (1982) stated that procymidone was the slowest to degrade and showed the longest persistence. According to these authors, chlozolinate is the dicarboximide with the fastest rate of degradation. It was no longer detectable in field soils after 1 week, or in glasshouse soils after 1 month.

The migration and degradation of iprodione and vinclozolin in Woodstown loamy sand (a peanut soil) and in Lod. loam (not a peanut soil) have been investigated by ELMER and STIPES (1985) using a bioassay. They found that both fungicides were washed into the soil to a depth of 25–35 cm, the fungicide mobility being greatly affected by the soil type. Both fungicides disappeared slightly faster after incubation at 28 °C than at 21 °C. The time required to reduce the fungicides to trace levels depended on the initial concentration.

Evidence suggests that dicarboximides have only a slight effect on the microbial flora and microbiological activity of the soil.

10 and 100 mg iprodione per kg soil had little effect on soil respiration. Similarly, 1 and 10 mg/kg soil did not affect ammonification or nitrification; a slight retardation was observed with the addition of 100 ppm (HELWEG 1983).

Vinclozolin added at a concentration of 300 ppm to a humus sandy soil produced no adverse effects on the soil microflora or respiration; there was a slight negative effect on the activity of various enzymes, such as dehydrogenase, amylase and protease, and also on nitrification (POMMER and MANGOLD 1975).

In a comparative study of the decay of vinclozolin on four different grapevines in four widely differing areas of Italy, GENNARI et al. (1985) found that the theoretical half-life of the original residue varied quite considerably. The minimum half-life was 1.2 days; the maximum was 4.9 days. The fastest decay was observed in a region with a greater daily temperature range and longer hours of sunshine. Under laboratory conditions, SZETO et al. (1989) treated pea leaflets with a wettable powder formulation of vinclozolin; the fungicide persisted for 21−46 days. The dissipation of vinclozolin on the leaves was linear, and the calculated half-life was 31.1 days for the formulated fungicide. No translocation was observed in the plants after application of the fungicide to one of the leaflets. None of the products of hydrolytic degradation was detected in treated plants. The half-life of vinclozolin varies considerably under natural conditions, depending on plant species, frequency of application, and weather conditions. ZENON-ROLAND and GILLES (1978) found a vinclozolin half-life in strawberries of between 12 and 14 days in a "normal" year, and 22 days in a very dry year. In grapes, DEL RE et al. (1980) found a vinclozolin half-life of between 7.5 and 9 days.

The chemical photolysis and hydrolysis of procymidone were studied by MIKAMI et al. (1984) in buffered distilled water, and river and sea water. The fungicide was degraded mainly by hydrolysis, whereas sunlight had a limited effect on breakdown. Product cleaved at the cyclic linkage was predominant in neutral and basic water, whereas product cleaved further at the amide linkage was predominant in acidic water.

In a study of the photodegradation of pesticides, CLARK and WATKINS (1984) investigated the effect of UV light on vinclozolin in methanol and benzene. Among other things, from photolysis in methanol, they detected and isolated the mono-chloro analogue of vinclozolin, which was found to be ineffective against *Botrytis cinerea*.

SCHWACK and BOURGEOIS (1989, 1990), BOURGEOIS et al. (1991) and BOURGEOIS (1991) worked on the photodegradation of iprodione, procymidone and vinclozolin dissolved in isopropanol and cyclohexene respectively. Isopropanol was used as a model of cutin hydroxy fatty acids, and cyclohexene was used to represent the olefinic structure of cuticle constituents. In model photoreactions at 280 nm they noted a marked degradation of these fungicides. Both iprodione and procymidone were preferably dechlorinated, forming the corresponding mono-chloro derivatives. With vinclozolin, photo-addition to the vinyl group occurred and was followed by dechlorination. Whether photo-addition can lead to the formation of "bound residue" in plants if there are reactions with components of the cuticles is still an open question.

Only very limited information is available about the effect of dicarboximides on plant mycorrhizas. According to RHODES and LARSEN (1981), iprodione has a reducing activity. Iprodione reduced mycorrhizal development on creeping bentgrass when it was applied in the spring to golf course greens or 4−8 weeks after bentgrass was seeded and inoculated with *Glomus fasciculatus* in a glasshouse.

In connection with the use of dicarboximides on lawns and turf or in soils, important observations were made by ROARK and DALE (1979), according to which iprodione did not cause a reduction in the longevity of earthworms.

Resistance to dicarboximides

Several reviews on resistance to dicarboximides have been published in the past few years which have in part laid emphasis on different aspects (BEEVER and BYRDE 1982; POMMER and LORENZ 1982; LEROUX and FRITZ 1983).

The following section deals essentially with the resistance situation in vitro and in vivo, the biological properties of resistant strains, their population dynamics and the consequences arising therefrom (resistance management) for the practical use of this group of fungicides.

Induction of resistance in vitro

The relevant literature indicates that it is comparatively easy to produce resistant strains of fungi in appropriate laboratory experiments, even without pretreatment with mutagens (Table 6.4). Naturally occurring and very considerable fluctuations in mutation rates are found, these depending on the fungus and pretreatment. In general the values range from 1×10^{-6} to 1×10^{-8} and for *Botrytis cinerea*, the fungus most frequently and most thoroughly investigated on account of its practical importance, lie between 1×10^{-6} an 1×10^{-7} (see references in BEEVER and BYRDE 1982; POMMER and LORENZ 1982; LEROUX and FRITZ 1983).

Using mycelium as the starting material, resistant strains of fungi can be obtained either by adaptation to slowly increasing concentrations of fungicide or by the further propagation of resistant sectors that can develop spontaneously at fairly high concentrations of fungicide (LORENZ and EICHHORN 1978; SCHUEPP and KÜNG 1978).

Table 6.4 List of dicarboximide-resistant species of fungi, as determined by in vitro induction of resistance

Fungus	Source
Alternaria alternata	MCPHEE 1980
Alternaria kikuckiana	KATO et al. 1979
Aspergillus nidulans	LECROUX et al. 1977
	BEEVER 1983
Botrytis cinerea	LEROUX et al. 1977
	LORENZ and EICHHORN 1978
	SCHUEPP and KÜNG 1978
	ALBERT 1979
	GULLINO and GARIBALDI 1979
	HISADA et al. 1979
	MARAITE et al. 1980
	DAVIS and DENNIS 1981 (a)
Botrytis squamosa	PRESLEY et al. 1980
Botrytis tulipae	CHASTAGNER and VASSEY 1979
Cochliobolus miyabeanus	KATO et al. 1979
Monilinia fructicola	SZTEJNBERG and JONES 1978
	RITCHIE 1983
Monilinia laxa	KATAN and SHABI 1982
Neurospora crassa	GRINDLE 1984
Penicillium expansum	LECROUX et al. 1978
	ROSENBERGER et al. 1979
	BEEVER 1983
Rhizoctonia solani	KATO et al. 1979
Rhizopus nigricans	LEROUX et al. 1979
Sclerotinia minor	BRENNEMANN and STIPES 1983
	PORTER and PHIPPS 1985
Sclerotinia sclerotiorum	KATO et al. 1979
Sclerotium cepivorum	LITTLEY and RACKE 1984
Ustilago maydis	LECROUX et al. 1978

The occurrence of cross-resitance has been investigated in particular detail for *Botrytis cinerea*. As a rule, cross resistance covers all dicarboximides introduced (see Table 6.2), irrespective of the compound used to induce resistance (see references in LEROUX and FRITZ 1983; KATAN 1982).

Cross-resistance can, however, also appear to fungicides from the aromatic hydrocarbon group (dichlorane or quintozene) and other aromatic ring compounds (see references in LEROUX and FRITZ 1983). It is interesting to note that resistant strains of fungi selected on nutrient media containing ergosterol biosynthesis inhibitors are occasionally also resistant to dicarboximides, although similar mechanisms of action can be excluded (FUCHS et al. 1984; LEROUX and FRITZ 1983; LORENZ, unpublished results).

No cross-resistance exists between MBC fungicides and dicarboximides. The fact that in the majority of cases field isolates of *Botrytis cinerea* that are resistant to dicarboximides are also MBC-resistant (double resistance) is due to the fact that in virtually all cases MBC fungicides were used for *Botrytis* control before dicarboximides. MBC-resistant strains of *Botrytis* are thus widely distributed; in addition. MBC-resistance is extremely stable.

The development of resistance under practical conditions

The intensive observation of tests plots (vine/*Botrytis*) treated with dicarboximides in the period 1973 – 1978, and also the deliberate attempts to induce resistance in the field or in a glasshouse have shown that the selection of resistant (*Botrytis*) strains under practical conditions does not proceed with the rapidity that would perhaps have been expected from the laboratory results described above. Thus for example, is an experimental plot of vines that had been treated with dicarboximides up to 5 times a year since 1973 (as specified in the official recommendations for use), it was still not possible to isolate any resistant *Botrytis* stains ever in 1978 (SPENGLER et al. 1979). Similarly, no resistant strains could be isolated after two years in a field trial (vine/*Botrytis*) in which vinclozolin and iprodione had been used in gradually increasing concentrations (5 – 1000 ppm) up to eleven times a year (LORENZ and EICHHORN 1978).

Nor could HISADA et al. (1979) achieve the selection of resistant *Botrytis* isolates on glasshouse roses in spite of applying procymidone nineteen times in the course of three years.

The first dicarboximide-resistant *Botrytis* field isolates were found in a vineyard in the Mosel growing area at the end of 1978 (HOLZ 1979). In 1979 isolates of this type were found in all other wine-growing areas of Germany (SPENGLER et al. 1979; LORENZ and EICHHORN 1980), and in 1980 they were also found in vines in Switzerland (SCHUEPP et al. 1982) and in France (LEROUX et al. 1982). Since due to the lack of good alternative fungicides, dicarboximide use was continued for *Botrytis* control in the wine-growing areas and countries mentioned, the proportion of resistant strains increased further. Since 1981 the proportion of resistant strains has settled to a level of about 60 – 80% in West Germany (LÖCHER et al. 1987). Despite this relatively high percentage, total losses of activity, such as occurred very rapidly following the initial appearance of MBC resistance, has not yet been observed. A similar situation was reported after the appearance of dicarboximide-resistant *Botrytis* strains in vineyards in Canada (NORTHOVER and MATTEONI 1986) and New Zealand (BEEVER et al. 1989).

Further reports relating to the appearance of dicarboximide-resistant *Botrytis* strains are mainly concerned with strawberries (DAVIS and DENNIS 1979; HUNTER et al. 1979; MARAITE et al. 1981) and glasshouse crops of a wide variety of types (KATAN 1981, 1982; PANAGIOTAKU and MALATHRAKIS 1981; PAPPAS 1982; BEAVER and BRIEN 1983; HARTILL et al. 1983; NORTHOVER and MATTEONI 1986).

In the case of other fungal species as for example *Fusarium nivale* and *Sclerotinia homoeocarpa* in lawns (CHASTAGNER and WASSEY 1982; DETWEITER et al. 1983), *Monilinia fructicola* on stone fruit (PENROSE et al. 1985; RITCHIE 1982; ELMER and GAUNT 1988) *Penicilium expansum* on apples (ROSENBERGER and MEYER 1981), *Sclerotium cepivorum* (LITTLEY and RAHE 1984) and *Alternaria alternata (*McPHEE 1980; HUTTON 1988), resistance to dicarboximides has so far been found only sporadically without causing practical problems.

Despite the presence of resistant *Botrytis* strains in field crops dicarboximides can still be used to a limited extent following the approved recommendations for their use in resistant situations (LEROUX and FRITZ 1983; LÖCHER et al. 1985; LORENZ and LÖCHER 1988).

With glasshouse crops however, the occurrence of resistance is more critically assessed (PANAGIOTAKU and MALATHRAKIS 1981; KATAN 1982). This applies particularly to the Mediterranean area where vegetables, such as cucumbers and tomatoes, are cultivated under plastic film during the winter months. The conditions which prevail under the plastic film are extremely favourable for the mass propagation of *Botrytis* and it is necessary to employ botryticides in a frequency appropriate to the situation. This can have the result that populations consisting solely of resistant individuals develop and also maintain themselves, as a result of the high infestation and selection pressure, and also with no entry of conidia from outside. In such cases severe losses of activity can result (KATAN 1981).

The biological characteristics of dicarboximide-resistant strains of fungi

The vast majority of investigations to characterise dicarboximide-resistant fungal isolates have been carried out with *Botrytis cinerea* (see references in BEEVER and BYRDE 1982; POMMER and LORENZ 1982). Hence the results discussed here relate to this fungus, especially as investigations on other fungi, for example *Sclerotinia homoeocarpa* (DETWEITER et al. 1983) *Monilinia fructicola* (SZTEJNBERG and JONES 1979; RITCHIE 1983). *Monilinia laxa* (KATAN and SHABI 1982), *Sclerotium cepivorum* (LITTLEY and RAHE 1984), *Botrytis squamosa* (PRESLEY et al. 1980) and *Neurospora crassa* (GRINDLE 1984; GRINDLE and TEMPLE 1985) have not led to fundamentally different conclusions.

In general, dicarboximide-resistant strains of fungi are less vigorous than sensitive ones. This manifests itself particularly when adapted isolates are compared with the appropriate sensitive strains (LORENZ and POMMER 1985).

However it is possible for overlapping to occur due to the range of variation that exists in all characteristics, which is especially pronounced in *Botrytis cinerea* and which is to be found in both sensitive and resistant strains. This natural variation, in addition to the method chosen, probably also accounts for the contradictory results obtained in some cases by various authors (see references in POMMER and LORENZ 1982).

Mycelium growth rate and sporulation are, as a rule, lower in resistant than in sensitive strains; it has been repeatedly pointed out, however, that resistant isolates can also be found which are comparable to sensitive strains in these respects (see references in POMMER and LORENZ 1982). Our investigations with sensitive strains and the appropriate adapted strains have indicated that mycelium growth rates are insignificantly different between the two groups; on the other hand, conidial formation in resistant strains was, with only one exception, considerably less than that of the sensitive strains. In the one case mentioned the original isolate itself showed a very weak rate of sporulation (LORENZ and POMMER 1985).

It is very evident that the method chosen to test the pathogenicity of resistant strains has a pronounced effect on the results obtained.

Thus, for example, where mycelium is used as inoculum and/or separated plant parts are used as the host material, differences in pathogenicity between resistant and sensitive strains are, if at all present, only relatively slight (LORENZ and EICHHORN 1978; PAPPAS et al. 1979; HARTILL et al. 1983). On the other hand, when conidial suspensions and entire plants are used, dicarboximide-resistant isolates of *Botrytis* are, as a rule, less pathogenic than sensitive strains (HISADA et al. 1979; GULLINO and CARIBALDI 1979, 1981; LORENZ and POMMER 1982, 1985).

An increased osmotic sensitivity of resistant strains is frequently quoted as an explanation for their lower pathogenicity (BEEVER 1983; BEEVER and BRIEN 1983; LEROUX and FRITZ 1983).

The mycelial growth of such isolates on nutrient media which contain an increased concentration of NaCl (0,68 M – BEEVER 1983) or various sugars (10 mg/ml – LEROUX and FRITZ 1983) is strongly inhibited. LEROUX and GREDT (1982a) were able to establish a direct correlation between osmotic sensivity and the degree of resistance in the strains they used. Similarly, BEEVER and BRIEN (1983) found that all their resistant isolates were more sensitive to increased osmotic values than sensitive strains; this was, however, substantially independent from the degree of resistance. In investigations with *Neurospora crassa*, GRINDLE and TEMPLE (1984) were able to show, however, that dicarboximide resistance and osmotic sensitivity are not necessarily correlated. This agrees with our own investigations in which only one of five isolates had a higher osmotic sensitivity after adaptation than the appropriate sensitive strain (LORENZ and POMMER 1985). In addition to this the comparison of 50 sensitive and 70 resistant field isolates (Table 6.5) yielded no decisive differences in the range of variation of osmotic sensitivity between the two groups. The proportion of isolates having an average or higher osmotic sensitivity was the same in both groups.

Table 6.5 Osmotic sensitivity of sensitive and dicarboximide-resistant field-isolates of *Botrytis cinerea* grown on media amended with 0.68 M and 1.0 M (figures in brackets) respectively

	Total number of Isolates tested	% Isolates with ... osmotic sensitivity			
		no	low	medium	high
sensitive Isolates	50	86% (20%)	10% (67%)	2% (3%)	2% (10%)
resistant Isolates	70	77% (8%)	19% (77%)	3% (11%)	1% (4%)

no osmotic sensitivity — >60% control
low osmotic sensitivity — 30–60% control
medium osmotic sensitivity — 11–29% control
high osmotic sensitivity — <11% control

The degree of resistance to dicarboximides exhibited by *Botrytis* strains is generally relatively low with ED_{50} values of $< 5 - < 10$ ppm a.i. This applies to both field isolates (MARAITE et al. 1981; LEROUX and GREDT 1982b; LORENZ and POMMER 1985) as well as to isolates from glasshouse crops (PANAGIOTAKU and MALATHRAKIS 1981; KATAN 1982; PAPPAS 1982). Under laboratory conditions, on the other hand, it is possible to obtain strains which exhibit much higher degrees of resistance (ED_{50} values $> 100 - > 500$ ppm a.i.: LEROUX et al. 1977). In laboratory experiments of this kind the isolates used have generally been in culture for a prolonged period of time, which would suggest that there is a correlation between progressive homokaryotisation and degree of resistance.

In our own investigations we too could only obtain daughter isolates having a high degree of resistance ($ED_5 > 500$ ppm a.i.) by adaptation from a very homogeneous, stable laboratory strain, whereas with relatively freshly isolated field strains the low degree of resistance characteristic of such isolates was retained (LORENZ and POMMER 1985).

It is very evident that the question whether mainly heterokaryotic or homokaryotic strains have been used is of importance in interpreting the various results obtained on the stability of resistance to dicarboximides. With respect to the characteristic "resistance to dicarboximides", homokaryotic isolates only produce resistant conidia whereas heterokaryotic strains produce both sensitive and resistant conidia. In the case of the latter, the proportion of sensitive or resistant conidia, respectively, increases very rapidly under certain circumstances depending on whether selection pressure caused by the appropriate fungicide is absent or present (SUMMERS et al. 1984).

The continued application of selection pressure in relatively closed systems, which exist, for example, under laboratory conditions and also to a certain extent under glasshouse conditions, can finally result in the overwhelming homokaryotisation of individual isolates, and also entire populations (KATAN 1982; SUMMERS et al. 1984). In the field, on the other hand, as a result of constant mixing of populations, predominantly heterokaryotic populations can be expected. If the results obtained by various authors are considered from this aspect, then a clear correlation between the origin of the isolates and the stability of resistance does in fact manifest itself. In laboratory isolates and strains from glasshouse crops, resistance to dicarboximides has generally proved stable (LEROUX et al. 1977; LORENZ and EICHHORN 1978; KATAN 1982). With field isolates that have not been cultivated on nutrient media containing fungicides for any length of time, the proportion of resistant conidia decreased very rapidly; occasionally such isolates regained complete sensitivity (DENNIS and DAVIS 1979; LORENZ and EICHHORN 1980, 1982; MARAITE et al. 1981; LORENZ and POMMER 1982).

The latter does not, however, represent the rule. Testing isolates from vine plots in which no further applications of dicarboximides had been carried out for four years, has shown that a certain, if only small proportion of resistant strains is retained in the population (LORENZ and POMMER 1985). A similar result was obtained from infection tests (Tab. 6.6) with resistant strains over several passages in a glasshouse (LORENZ and POMMER 1985).

This, and also the studies by SUMMERS et al. (1984) on heterokaryotic and homokaryotic *Botrytis* isolates support the hypothesis formulated by DEKKER (1976) in which the heterokaryotic state which exists in multinucleate fungi "may serve as a mechanism in maintaining low levels of fungicide tolerance in populations in the presence and absence of selection pressure from the fungicide".

Table 6.6 Competitive ability of a resistant isolate of *Botrytis cinerea* Nr. 920) in mixture with conidia from the corresponding sensitive isolates

Mixture sensitive : resistant	% Resistant isolates after reisolation			
	I	II	III	IV
100% sensitive	0	0	0	0
80% sensitive : 20% resistant	44%	0	0	0
50% sensitive : 50% resistant	50%	0	10%	0
20% sensitive : 80% resistant	33%	30%	0	10%
100% resistant	80%	37%	10%	14%

The competitive ability of resistant *Botrytis* isolates is markedly inferior to that of sensitive isolates. It has been repeatedly possible to demonstrate this in a very wide variety of test situations, both under glasshouse conditions (HISADO et al. 1979; LACROIX and GOUOT 1981; LORENZ and POMMER 1985) and in the field (DAVIS and DENNIS 1981; GULLINO and GARIBALDI 1981). Field observations also indicate a low competitive ability. LEROUX and GREDT (1982a) observed a considerable reduction in resistant strains in vine plots when the selection pressure was absent during the vegetation period. The decline of resistant strains was particularly pronounced at high infection pressure. In our own tests (LÖCHER and LORENZ 1985) the proportion of resistant strains was, as a rule, highest after the use of dicarboximides in the autumn (at harvest), but decreased decisively during the winter and early spring (Fig. 6.2). The explanation for this is surely to be found in the interaction of various factors, namely the weaker production of conidia, the reduced pathogenicity and the, at least under field conditions, low stability of resistance.

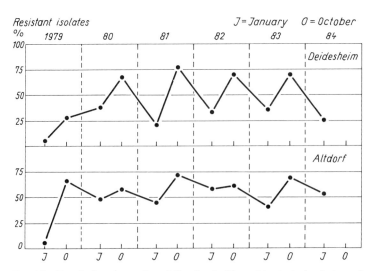

Fig. 6.2 Population dynamics of dicarboximide-resistant strains in two vineyards of the Rhenish-Palatinate area (Deidesheim and Altdorf) during the years 1979–1984 (dicarboximides were used continuously at least three times per season).

References

ALBERT, G.: Wirkungsmechanismus und Wirksamkeit von Vinclozolin bei *Botrytis cinerea* Pers. Diss. Rheinische Friedrich-Wilhelms-Univ., Bonn 1979.
— Sphäroplastenbildung bei *Botrytis cinerea*, hervorgerufen durch Vinclozolin. Z. Pflanzenkrankh. u. Pflanzenschutz **88** (1981): 337–342.
BARBERO, L., and GAIA, P.: Enological consequence of the use of Ronilan in grape cultivation. Vini Ital. **21** (119) (1979): 95–100.
BEEVER, R. E.: Osmotic sensitivity of fungal variants resistant to dicarboximide fungicides. Transactions of the Brit. Mycolog. Soc. **80** (1983): 327–331.
— and BRIEN, H. M. R.: A survey of resistance to the dicarboximide fungicides in *Botrytis cinerea*. New Zealand J. Agric. Res. **26** (1983): 391–400.
— and BYRDE, R. J. W.: Resistance to the dicarboximide fungicides. In: DEKKER, J., and GEORGOPOULOS, S. G. (Eds.): Fungicide Resistance in Crop Protection. Centre for Agricult. Publ. and Document. Pudoc, Wageningen 1982, pp. 101–117.
— LARACY, E. P., and PAK, H. A.: Strains of *Botrytis cinerea* resistant to dicarboximide and benzimidazole fungicides in New Zealand vinegards. Plant Pathology **38** (1989): 427–437.
BENDA, J.: Mikroskopische Untersuchungen über den Einfluß des Fungizides Ronilan auf die Hefeflora der Trauben und des Weines. Weinwissenschaft **33** (1978): 153–158.
— Botrytizide, ihre Wirkstoffe und Formulierungsmittel in der mikrobiologischen Prüfung. Weinwissenschaft **38** (1983): 41–50.
BOURGEOIS, B.: Untersuchungen zur Photochemie der Dicarboximid-Fungizide Vinclozolin, Procymidon und Iprodion. Diss. Universität Karlsruhe 1991.
— FLÖSSER-MÜLLER, H., and SCHWACK, W.: Modellreaktionen zum photochemischen Verhalten von Fungiziden auf Pflanzenoberflächen am Beispiel der Dicarboximide und N-Trichlormethylthioimide. Lebensmittelchemie **45** (1991): 43.
BRENNEMAN, T. P., and STIPES, R. J.: Sensitivity of *Sclerotinia minor* from peanut to dicloran, iprodione and vinclozolin. Phytophathology **73** (1983): 964.
BUCHENAUER, H.: Preliminary studies on the mode of action of vinclozolin. Meded. Fak. Landbouwwet. Rijksuniv. Gent **41** (1976): 1509–1519.
CABRAS, P., MELONI, M., PIRISI, F. P., and PIRISI, M.: Degradation of dicarboximide fungicides in wine. Pesticide Sci. **15** (1984): 247–252.
CAYLEY, G. R., and HIDE, G. A.: Uptake of iprodione and control of diseases on potato stems. Pesticide Sci. **11** (1980): 15–19.
CHASTAGNER, G. A., and VASSEY, W. E.: Tolerance of *Botrytis tulipae* to glycophene and vinclozolin. Phytopathology **69** (1979): 914.
— — Occurrence of iprodione tolerant *Fusarium nivale* under field conditions. Plant Dis. **66** (1983): 112–114.
CLARK, T., and WATKINS, D.: Photolysis of vinclozolin. Chemosphere **13** (1984): 1391–1396.
COOKE, B. K., PAPPAS, A. C., JORDAN, V., and WESTERN, N. M.: Translocation of benomyl prochloraz and procymidone in relation to control of *Botrytis cinerea* in strawberries. Pesticide Sci. **10** (1979): 467–472.
DANNEBERGER, T. K., and VARGAS, J. M.: Systemic activity of iprodione in *Poa annua* and postinfection activity for *Drechslera sorokiniana*. Plant Dis. **66** (1982): 914–915.
DAVIS, R. P., and DENNIS, C.: Use of dicarboximides on strawberries and potential problems of resistance in *Botrytis cinerea*. Proc., 1979 Brit. Crop Protect. Conf., Brighton, Pests and Diseases **1** (1979): 193–201.
— — Properties of dicarboximide-resistant strains of *Botrytis cinerea*. Pesticide Sci. **12** (1981): 521–535.
— — Studies on the survival and infective ability of dicarboximide-resistant strains of *Botrytis cinerea*. Ann. appl. Biol. **98** (1981): 395–402.

DEKKER, J.: Acquired resistance to fungicides. Ann. Rev. of Phytopathol. **14** (1976): 405–428.

DEL RE, A., FONTANA, P., MELONI, G. P., and NATALI, P.: Residues of Vinclozolin in Grapevine Musts after Grey Moulds Control. Ann. Fac. Agrar. Univ. Cattol. Sacro. Cucre. **20** (1980): 171–181.

DENNIS, C., and DAVIS, R. P.: Tolerance of *Botrytis cinerea* to iprodione and vinclozolin. Plant Pathol. **28** (1979): 131–133.

DETWEITER, A. R., VARGAS, J. M., and DANNEBERG, T. K.: Resistance of *Sclerotinia homoeocarpa* to iprodione and benomyl. Plant Dis. **67** (1983): 617–630.

DI TORO et al.: Atti Giornate Fitopatologico. 1980.

EICHHORN, K. W., and LORENZ, D. H.: Untersuchungen über die Wirkung von Vinclozolin gegenüber *Botrytis cinerea* in vitro. Z. für Pflanzenkrankh. u. Pflanzenschutz **85** (1978): 449–460.

ELMER, P. A. G., and GAUNT, R. E.: A survey of fungicide insensitivity in *Monilinia fructicola*. Proc. 41st. N.Z. Weed and Pest Control Conf. (1988): 167–169.

– – Dicarboximide resistance and disease control in brown rot of stonefruit. Proc. 41st. N.Z. Weed and Pest Control Conf. (1988): 271–274.

ELMER, W. H., and STIPES, R. J.: Movement and disappearance of dicloran, iprodione and vinclozolin in peanut and nonpeanut soils. Plant Dis. **69** (1985): 292–294.

FIMA, P., and WOMASTEK, R.: Residues in/on grapes and in wine. Mitt. Klosterneuburg **33** (1983): 253–256.

FLORI, P., PANCALDI, D., TENTONI, P., and MUSACCI, P.: Persistenza nel terreno dei fungicidi Dicloran, Vinclozolin, Iprodione, Procymidone e Dichlozolinate. La difesa delle piante **4** (1982): 221–230.

– and ZORINI, R.: Anti-botrytis fungicide residues and ecological technology. Vignevini **11** (1984): 47–50.

FUCHS, A., DE VRIES, F. W., and DE WAARD, M. A.: Simultaneous resistance in fungi to ergosterol biosynthesis inhibitors and dicarboximides. Netherl. J. Plant Pathol. **90** (1984): 3–11.

FUJINAMI, A., OZAKI, K., NODERA, T., and TANAKA, K.: Studies on biological activity of cyclic imide compounds. Part II. Antimicrobial activity of 1-phenylpyrolidine-2,5-diones and related compounds. Agric. Biol. Chem. **36** (1972): 318.

– – and YAMAMOTO, S.: Studies on biological activity of cyclic amide compounds. Part I. Antimicrobial activity of 3-phenyloxazolidine-2,4-diones and related compounds. Agric. Biol. Chem. **35** (1971): 1707–1719.

GENNARI, M., ZANINI, E., CIGNETTI, A., BICCHI, C., D'AMATO, A., TACCHEO, M., SPESSOTO, C., DE PAOLI, M., FLORI, P., IMBROGLINI, G., LEANDRI, A., and CONTE, E.: Vinclozolin Decay on Different Grapevines in Four Differing Italian Areas. J. Agric. Food Chem. **32** (1985): 1232–1237.

GEORGOPOULUS, S. G., SARRIS, M., and ZIOGAS, B. N.: Mitotic instability in *Aspergillus nidulans* caused by the fungicides iprodione, procymidone and vinclozolin. Pesticide Sci. **10** (1979): 389–392.

GRINDLE, M.: Isolation and characterization of vinclozolin resistant mutants of *Neurospora crassa*. Transactions of the Brit. mycolog. Soc. **82** (1984): 635–643.

– and TEMPLE, W.: Sporulation and osmotic sensitivity of dicarboximide-resistant mutants of *Neurospora crassa*. Transactions of the Brit. mycolog. Soc. **84** (1985): 369–372.

GULLINO, M. L., and GARIBALDI, A.: Osservazioni sperimentali sulla resistenza di isolamenti Italiani di *Botrytis cinerea* a vinclozolin. La difesa delle piante **6** (1979): 341–350.

– – Biological properties of dicarboximide-resistant strains of *Botrytis cinerea* Pers. Phytopathol. Mediterranea **20** (1981): 117–122.

HAGAN, A., and LARSEN, P. O.: Effect of fungicides on conidium germination, germ tube elongation, and appressorium formation by *Bipolaris sorokiniana* on Kentucky bluegrass. Plant Dis. Rep. **63** (1979): 474–478.

HARTILL, W. F. T., TOMPKINS, G. R., and KLEINSMAN, P. J.: Development in New Zealand of resistance to dicarboximide fungicides in *Botrytis cinerea*, to acylalanines in *Phytophthora infestans* and to guazatine in *Penicillium italicum*. New Zealand J. Agric. Res. **26** (1983): 261–269.

HELWEG, A.: Influence of the fungicide iprodione on respiration, ammonification and nitrification in soil. Pedobiologia **25** (1983): 87–92.

HISADA, Y., KATO, T., and KAWASE, Y.: Systemic movements in cucumber plants and control of cucumber gray mould by a new fungicide, S-7131. Netherl. J. Plant Pathol. **83** (1977): 71–78.

– – – Mechanism of antifungal action of procymidone in *Botrytis cinerea*. Ann. Phytopathol. Soc. Japan **44** (1978): 509–518.

– and KAWASE, Y.: Morphological studies of antifungal action of N-(3′5′-dichlorophenyl)-1,2-dimethylcyclopropane-1,2′-dicarboximide on *Botrytis cinerea*. Ann. Phytopathol. Soc. Japan **43** (1977): 151–158.

– – Reversible binding of procymidone to a sensitive fungus *Botrytis cinerea*. J. Pesticide Sci. **5** (1980): 559–564.

– MAEDA, K., TOTTORI, N., and KAWASE, Y.: Plant disease control by N-(3,5-dichlorophenyl)-1,2-dimethyl-cyclopropane-1,2-dicarboximide. J. Pesticide Sci. **1** (1976): 145–149.

– TAKAKI, H., KAWASE, J., and OZAKI, T.: Difference in the potential of *Botrytis cinerea* to develop resistance to procymidone *in vitro* and in the field. Ann. Phytopathol. Soc. Japan **45** (1979): 283–290.

HOLZ, B.: Über eine Resistenzerscheinung von *Botrytis cinerea* an Reben gegen die neuen Kontaktbotrytizide im Gebiet der Mittelmosel. Weinberg u. Keller **26** (1979): 18–25.

HUNTER, T., JORDAN, V. W. L., and PAPPAS, A. C.: Control of strawberry fruit rots caused by *Botrytis cinerea* and *Phytophthora cactorum*. Proc., 1979 Brit. Crop. Protect. Conf., Brighton-Pests and Dis. **1** (1979): 177–183.

HUTTON, D. G.: The appearance of dicarboximide resistance in *Alternaria alternata* in passionfruit in south-east Queensland. Australasian Plant Pathology **17** (1988): 34–36.

KATAN, T.: Resistance to dicarboximide fungicides in *Botrytis cinerea* from cucumber, tomatoes, strawberries and roses. Netherl. J. Plant Pathol. **87** (1981): 244.

– Resistance to 3,5-dichlorophenyl-N-cyclic imide (dicarboximide) fungicides in the grey mould pathogen *Botrytis cinerea* on protected crops. Plant Pathol. **31** (1982): 133–141.

– and SHABI, E.: Characterization of a dicarboximide-resistant laboratory isolate of *Monilinia laxa*. Phytoparasitica **10** (1982): 241–245.

KATO, T., HISADA, I., and KAWASE, I.: Nature of procymidone-tolerant *Botrytis cinerea* strains obtained in vitro. In: Proc. of Seminar on Pest resistance in Pesticides, Palm Springs, USA 1979.

LACROIX, L., BIC, C., BURGAUD, L., GUILLOT, M., LEBLANC, R., RIOTTOT, R., and SAULI, M.: Etude des propriétés antifongiques d'une nouvelle famille des dérivés de l'hydantoine et en particulier du 26019 R.P.: Phytiatrie-Phytopharmacie **23** (1974): 165–174.

– and GOUOT, J. M.: Investigations on the resistance of *Botrytis cinerea* to iprodion in greenhouses. Meded. Fak. Landbouwwet. Rijksuniv., Gent **46** (1981): 979–989.

LEBOUX, P., and FRITZ, R.: Antifungal activity of dicarboximides and aromatic hydrocarbons and resistance to these fungicides. In: TRINCI, A. P. J., and RYLEY, J. F. (Eds.): Mode of Action of Antifungal Agents. Cambridge Univ. Press, Cambridge 1983, pp. 207–237.

– – and GREDT, M.: Etudes en laboratoire de souches de *Botrytis cinerea* Pers. résistance à la Dichlozoline, au Dicloran, au Quintozène, à Vinclozoline et au 26019 RP (Glycophéne). Phytopathol. Z. **89** (1977): 347–358.

– – – Cross-resistance between 3,5-dichlorophenyl cyclic imide fungicides and various aromatic compounds. In: LYR, H., and POLTER, C. (Eds.). Systematische Fungicide und Antifungale Verbindungen. VI. Internationales Symposium. Akademie-Verlag, Berlin 1982, S. 79–88.

- and GREDT, M.: Effets d'alcools primaires, de polyols, de sels minéraux et de sucres sur des souches de *Botrytis cinerea* sensibles ou résistantes à l'iprodione et à la procymidone. Comptes rendus de l'Académie des Sci., Paris, série III, **294** (1982a): 53–56.
- — Phénomènes de resistance de *Botrytis cinerea* aux fongicides. La Défense des Végétaux **213** (1982b): 3–17.
- — and FRITZ, R.: Etudes en laboratoire de souches de quelques champignons phytopathogènes, résistantes à la vinclozoline et à divers fongicides aromatique. Meded. Fak. Landbouwwet, Rijksuniv., Gent **43** (1978): 881–889.
- — — Resistance to 3,5-dichlorophenyl-N-cyclic imide fungicides. Netherl. J. Plant Pathol. **87** (1981): 242.
- LAFON, R., and GREDT, M.: La resistance de *Botrytis cinerea* aux benzimidazoles et aux imides cycliques: situation dans les vignobles alsaciens, bordelais et champenois. OEPP/EPPO Bull. **12** (1982): 137–143.
- LITTLEY, E. R., and RAHE, J. E.: Specific tolerance of *Sclerotium cepivorum* to dicarboximide fungicides. Plant Dis. **68** (1984): 371–374.
- LÖCHER, F. J., BRANDES, W., LORENZ, G., HUBER, W., SCHILLER, R., and SCHREIBER, B.: Development of a strategy to maintain the efficacy of the dicarboximides in the presence of resistant strains of *Botrytis* in grapes. Gesunde Pflanzen **37** (1985): 3–8.
- LORENZ, G., and BEETZ, K. J.: Resistance management strategies for dicarboximide fungicides in grapes: results of six years' trial work. Crop Protect. **6** (1987): 139–147.
- LORENZ, D. H., and EICHHORN, K. W.: Untersuchungen zur möglichen Resistenzbildung von *Botrytis cinerea* an Reben gegen die Wirkstoffe Vinclozolin und Iprodione. Wein-Wissensch. **33** (1978): 2–10.
- — — Vorkommen und Verbreitung der Resistenz von *Botrytis cinerea* gegen Dicarboximid-Fungizide im Anbaugebiet der Rheinpfalz. Wein-Wissensch. **35** (1980): 199–210.
- — — *Botrytis* and its resistance against dicarboximide fungicides. OEPP/EPPO Bull. **12** (1982): 125–129.
- LORENZ, G., and LÖCHER, F.: Strategies to control dicarboximide resistant strains of *Botrytis cinerea*. Proc. Brit. Crop. Protect. Conf., Pests and Diseases (1988): 1107–1115.
- and POMMER, E.-H.: Studies on pectolytic and cellulolytic enzymes of dicarboximide-sensitive and -resistant strains of *Botrytis cinerea*. OEPP/EPPO Bull. **12** (1982): 145–149.
- — — Morphological and physiological characteristics of dicarboximide-sensitive and -resistant isolates of *Botrytis cinerea*. OEPP/EPPO Bull. **15** (1985): 353–360.
- MARAITE, H., GILLES, G., MEUNIER, S., WEYNS, J., and BAL, E.: Resistance of *Botrytis cinerea* Pers. ex Pers. to dicarboximide fungicides in strawberry fields. Parasitica **36** (1981): 90–101.
- — MEUNIER, S., POURTOIS, A., and MEYER, J. A.: Emergence *in vitro* and fitness of stains of *Botrytis cinerea* resistant to fungicides. Meded. Fak. Landbouwwet. Rijksuniv. Gent **45** (1980): 150–167.
- MCPHEE, W. J.: Some characteristics of *Alternaria alternata* strains resistant to iprodione. Plant Dis. **64** (1980): 847–849.
- MENAGER, TISSIER, COMELLI, and ALLUIS: La dichlozoline. Phytiatrie-Phytopharmacie **20** (1971): 169–172.
- MIKAMI, M., IMANISHI, K., YAMADO, H., and MIYAMOTO, J.: Photolysis and hydrolysis of the fungicide procymidone in Water. J. Pesticide Sci. **9** (1984): 223–228.
- MIKAMI, N., YOSHIMURA, J., YAMADA, H., and MIYAMOTO, J.: Translocation and metabolism of procymidone in cucumber and bean plants. Nippon Noyaku Gakkaishi **9** (1984): 131–136.
- NAKAYA, S., KITAYAMA, H., UEDA, M., KITAYAMA, M., SAITO, M., and KITSUTAKA, T.: Subacute toxicity of a new pesticide, 3-(3,5-dichlorophenyl)-5,5-dimethyl-2,4-oxazolidinedione (DDOD). Hokkaidoritsu Eisei KENKYUSHOHO **19** (1969): 132–138.
- NORTHOVER, J., and MATTEONI, J. A.: Resistance of *Botrytis cinerea* to benomyl and iprodione in

vinegards and greenhouses after exporate to the fungicides alone or mixed with captan. Plant Disease **70** (1986): 398–402.

PANAGIOTAKU, M. G., and MALATHRAKIS, N. E.: Resistance of *Botrytis cinerea* Pers. to dicarboximide fungicides. Netherland J. Plant Pathol. **87** (1981): 242.

PAPPAS, A. C.: Unzureichende Bekämpfung des Grauschimmels auf Alpenveilchen mit Dicarboximid-Fungiziden in Griechenland. Z. Pflanzenkrankh. u. Pflanzenschutz **89** (1982): 52–58.

COOKE, B. K., and JORDAN, V. W. L.: Insensitivity of *Botrytis cinerea* to iprodione, procymidone and vinclozolin and their uptake by the fungus. Plant Pathol. **28** (1979): 71–76.

— and FISHER, D. J.: A comparison of the mechanismen of action of vinclozolin, procymidone and prochloraz against *Botrytis cinerea*. Pesticide Sci. **10** (1979): 239–246.

PENROSE, L. J., HOFFMANN, W., and NICHOLLS, M. R.: Field occurrence of vinclozolin resistance in *Monilinia fructicola*. Plant Pathol. **34** (1985): 228–234.

PIRISI, F., MELONI, M., CABRAS, P., BIONDUCCI, M., and SERRA, A.: Degradation of dicarboximide fungicides in wine. Part II: Isolation and identification of the major breakdown products of chlozolinate, vinclozolin and procymidone. Pestic. Sci. **17** (1986): 109–118.

POMMER, E.-H., and LORENZ, G.: Resistance of *Botrytis cinerea* Pers. to dicarboximide fungicides – a literature review. Crop Protect. **1** (1982): 221–230.

— and MANGOLD, D.: Vinclozolin (BAS 352 F), ein neuer Wirkstoff zur Bekämpfung von *Botrytis cinerea*. Meded. Fak. Landbouwwet. Rijksuniv. Gent **40** (1975): 713–722.

— and ZEEH, B.: Myclozolin, ein neuer Wirkstoff aus der Klasse der Dicarboximide. Meded. Fak. Landbouwwet. Rijksuniv. Gent (1982): 935–942.

PORTER, D. M., and PHIPPS, P. M.: Effects of three fungicides on mycelial growth, *sclerotium* production and development of fungicide-tolerant isolates of *Sclerotinia minor*. Plant Dis. **69** (1985): 143–146.

PRESLEY, A. H., and MAUDE, R. B.: Tolerance in *Botrytis squamosa* to iprodione. Ann. appl. Biol. **100** (1982): 117–127.

RHODES, L. H., and LARSEN, P. O.: Effects of fungicides on mycorrhizal developmenmt of creeping bentgrass. Plant Dis. **65** (1981): 145–147.

RITCHIE, D. F.: Effect of dicloran, iprodione, procymidone, and vinclozolin on the mycelial growth, sporulation, and isolation of resistant strains of *Monilinia fructicola*. Plant Disease **66** (1982): 484–486.

— Mycelial growth, peach fruit-rotting capability and sporulation of strains of *Monilinia fructicola* resistant to dichloran, iprodione, procymidone and vinclozolin. Phytopathology **73** (1983): 44–47.

ROARK, J. H., and DALE, J. L.: The effect of turf fungicides on earthworms. Proc. Arkansas Academy of Sci. **33** (1979): 71–74.

ROSENBERGER, D. A., and MEYER, F. W.: Postharvest fungicides for apples: development of resistance to benomyl, vinclozolin and iprodione. Plant Diseases **65** (1981): 1010–1013.

— — and CECILIA, C. V.: Fungicide strategies for control of benomyl-tolerant *Penicillium expansum* in apple storage. Plant Dis. Rep. **63** (1979): 1033–1037.

SAPIS-DOMERCQ, S., BERTRAND, A., MUR, Fr., and SARRE, C.: Effects of fungicides residues from grapevines on fermentating microflora, 1976 Experiments. Connaissance Vigne Vin **11** (1977): 227–242.

— — JOYEUX, A., LUCMARET, V., and SARBE, C.: Study of the effect of vine treatment products on grape and vine microflora. 1977 Experiments. Connaissance Vigne Vin **12** (1978): 245–275.

SCHÜEPP, H., and KÜNG, M.: Gegenüber Dicarboximid-Fungiziden tolerante Stämme von *Botrytis cinerea* Pers. Ber. Schweizer. Bot. Ges. **88** (1978): 63–71.

— — and SIEGFRIED, W.: Dévelopment des souches de *Botrytis cinerea* résistentes aux dicarboximides dans les vignes de la Suisse alémanique OEPP/EPPO Bull. **12** (1982): 157–161.

SCHWACK, W., and BOURGEOIS, B.: Fungicides and photochemistry: iprodione, procymidone, vinclozolin. Z. Lebensm. Unters. Forsch. **188** (1989): 346–347.

- — Fungicides and photochemistry: iprodione, procymidone, vinclozolin. Model reactions with isopropanol and cyclohexene. 7th Intern. Congr. of Pestic. Chem., Book of Abstracts, Vol. III, 1990: 38.
- SPENGLER, G., SCHERER, M., and POMMER, E. H.: Untersuchungen über das Resistenzverhalten von *Botrytis cinerea* gegenüber Vinclozolin. Mitt. Biol. Bundesanstalt für Land- und Forstwirtschaft, Berlin-Dahlem, **191** (1979): 236.
- STOJANOVIC, M., and VUKMIROVIC, M.: Effect of some fungicides on fermentation yeasts. Mikrobiologija **16** (1979): 39–49.
- SUMMERS, R. W., HEANEY, S. P., and GRINDLE, M.: Studies on a dicarboximide resistant heterokaryon of *Botrytis cinerea*. Proc. Brit. Crop Protect. Conf. 1984, Pests and Diseases Vol. 2 (1984): 453–458.
- SZETO, S., BURLINSON, N., RAHE, J., and OLOFFS, P.: Persistence of the fungicide vinclozolin on pea leaves under laboratory conditions. J. Agric. Food Chem. **37** (1989): 529–534.
- SZTEJNBERG, A., and JONERS, A. L.: Resistance of the brown rot fungus *Monilinia fructicola* to iprodione, vinclozolin and procymidone. Phytoparasitica **7** (1979): 46.
- VOJTEKOVA, G., VODOVA, J., and JURKACKOVA, M.: Effect of several fungicides on the fermentation of grape must. Vinohrad (Bratislava) **21** (1983): 207–210.
- VULIE, M., EBERLE, O., and HÄBERLE, N.: Metomeclan, ein neuer Dicarboximid-Wirkstoff mit breitem fungizidem Wirkungsspektrum. Meded. Fak. Landbouwwet. Rijksuniv. Gent **49** (1984): 293–301.
- WALKER, A.: Further observations on the enhanced degradation of iprodione and vinclozolin in soil. Pestic. Sci. **21** (1987a): 219–236.
- — Enhanced degradation of iprodione and vinclozolin in soil: a simple colorimetric test for identification of rapid-degrading soils. Pestic. Sci. **21** (1987b): 233–240.
- — ENTWISTLE, A. R., and DEARNALEY, N. J.: Evidence for enhanced degradation of iprodione in soils treated previously with this fungicide. Monograph-British Crop Protect. Council **27** (Soils Crop Protect. Chemistry) (1984): 117–123.
- ZENON-ROLAND, L., and GILLES, G.: Evolution des résidues de triadimefon et de vinclozoline sur fraises. Med. Fac. Landbouww. Rijksuniv. Gent **43/2** (1978): 1269–1282.

Chapter 7

Mechanism of action of dicarboximide fungicides

W. EDLICH and H. LYR*)

*) Institut für Integrierten Pflanzenschutz der Biologischen Bundesanstalt für Land- und Forstwirtschaft Kleinmachnow, Germany

Introduction

Dicarboximide fungicides (DCOF) represent a group of highly active and selective fungicides as described in chapter 6.
Surprising was the detection of a general cross resistance between the dicarboximide and the Aromatic Hydrocarbon Fungicides (AHF) (LEROUX et al. 1978).

Table 7.1 Some effects of dicarboximides on fungal cell structure and metabolism

Affected Process	Symptoms	References
cell divison	mitotic instability	GEORGOPOULOS et al. 1979
DNA synthesis	somatic segregation	LEROUX et al. 1978
	low inhibition	FRITZ et al. 1977
	strong inhibition	HISADA and KAWASE 1978
	without effect	BUCHENAUER 1976
RNA and protein synthesis	low inhibition	PAPPAS and FISHER 1979
cell wall synthesis	without effect	BUCHENAUER 1976
	increased precursor insertion	HISADA and KAWASE 1977
	low inhibition	BUCHENAUER 1976
	strong inhibition	ALBERT 1981
metabolism of sterols	low effects	BUCHENAUER 1976
		PAPPAS and FISHER 1979
metabolism of lipids	increased level of free fatty acids	PAPPAS and FISHER 1979
		BUCHENAUER 1976
		FRITZ et al. 1977
K^+-efflux	without effect	HISADA and KAWASE 1977
respiration	without effect	BUCHENAUER 1976
		HISADA and KAWASE 1977
		FRITZ et al. 1977
		PAPPAS and FISHER 1977
cell structure	swelling of mitochondria and their lysis; vesiculation of ER	MÜLLER 1981

Morphological effects

DCOF inhibit spore germination and growth of mycelia, but inhibit the latter much more efficiently (Buchenauer 1976; Fritz et al. 1977; Parker and Fisher 1979). In liquid nutrient media germ tubes of *Botrytis cinerea* swell rapidly after treatment with DCOF and may burst when certain concentrations of DCOF are used (Hisada and Kawase 1977). At sublethal doses, the growth of the hyphal tips is disturbed and anomalous branching occurs (chapter 6) similar to that which occurs with some AHF (chapter 5).

But morphological changes are rather unspecific in relation to the mechanism of action of a fungicide, because very different agents (inhibitors of protein synthesis, antitubulins and others) can induce analogous abnormalities. Such effects have been described for example, with AHF (Sharples 1961; Georgopoulos et al. 1967; Macris

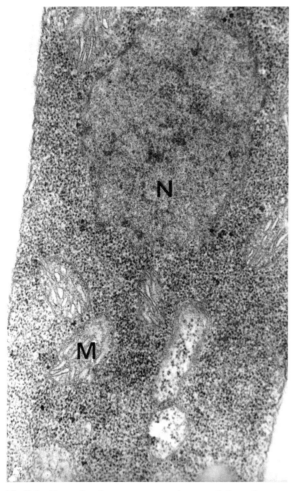

Fig. 7.1 *Botrytis cinerea*, untreated control in submers culture 8 h after spore germination. ×35,000 (phot. Dr. H. M. Müller). M = mitochondrium, N = nucleus.

and GEORGOPOULOS 1973; THRELFALL 1972; TILLMANN and SISLER 1973), with sorbose (MOORE 1981), or with cytochalasin (BETTINA et al. 1972; ALLEN et al. 1980).

HISADA and KAWASE (1977) working with fungal protoplasts did not find an influence of DCOF on cell wall synthesis, whereas ALBERT (1979, 1981) in similar investigations could inhibit cell wall synthesis with DCOF. But this may have been favoured by the rather high osmotic pressure (up to 2.3 M) in the culture medium. The strong pathological cell wall thickening induced by chloroneb and other AHF (chapter 5) is not produced by DCOF.

Physiological and biochemical effects

As with the AHF the main metabolic pathways are not, or only to a small degree disturbed by DCOF. Some authors found minor changes of the synthesis of proteins, nucleic acids and lipids, whereas in other experiments no severe influence on catabolic or anabolic pathways was observed (BUCHENAUER 1979; FRITZ et al. 1977; HISADA and KAWASE 1978; PAPPAS and FISHER 1979).

DNA-synthesis seems to be relatively strongly impaired by iprodione (PAPPAS and FISHER 1979). But all authors agree that the observed effects are only the consequence of an unknown primary action. GEORGOPOULOS et al. (1979) and FRITZ et al. (1977) described irregularities in the course of the cell division under the influence of DCOF.

Stimulated by our results with Aromatic Hydrocarbon Fungicides, we investigated the effect of DCOF on the lipid peroxidation ability in sensitive fungi, mainly in *Mucor mucedo* and *Botrytis cinerea*. Electron pictures (MÜLLER 1981) demonstrated damage in fungal cells in the mitochondrial inner membrane and in membranes of the endoplasmic reticulum (ERP) caused by DCOF (Fig. 7.1 and 7.2). There exist clear differences between the effects produced by DCOF and those produced by AHF (chapter 5). The cell wall is not thickened in cells treated with DCOF, and lysis of mitochondrial membranes is a bit different in those treated with AHF. It seems possible however, that DCOF initiate effects similar to AHF by inducing pathological oxidative processes in sensitive fungal cells. Indeed we found a correlation between inhibition of hyphal growth in *Mucor mucedo* and *Botrytis* and the level of intracellular lipid peroxidation caused by DCOF application (Fig. 7.3). Addition of α-tocopherol acetate not only antagonized the growth inhibition but also decreased the level of peroxides to the level of the untreated controls. Among other compounds which we investigated, only piperonylbutoxide and SKF 525 A significantly inhibited the lipid peroxidation which is induced by DCOF. These compounds were active at concentrations of $5 \times 10^{-6}\,mol \times l^{-1}$ (1 – 2 ppm). Other fungicides such as tridemorph, carbendazim, or chlorothalonil had little or no effect on this reaction, whereas the effect of chloroneb, dichloran and PCNB was similar to that of DCOF.

The results of ORTH et al. (1992a) indicate that the reversal of lipid peroxidation and growth inhibition by external application of α-tocopherol to the aquaeous medium (and perhaps also of other hydrophobic substances such as piperonylbutoxide) must be cautiously interpreted. Such compounds could bring by their tendency to form hydrophobic micellar structure in the medium

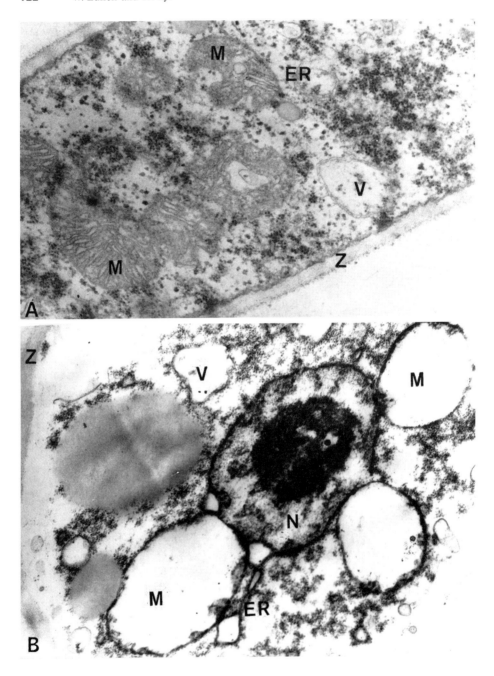

Fig. 7.2 *Botrytis cinerea*, treated by $1\,\text{mg}\cdot\text{l}^{-1}$ vinclozolin. A − 2 h after treatment, B − 4 h after treatment. Figures show swelling of mitochondria, loss of their matrix and strong reduction of christae, vesiculation of cytoplasma ×30,000 (phot. H. M. MÜLLER). M = mitochondria, ER = endoplasmatic reticulum, V = vesicles, N = nucleus, Z = cell wall.

a redistribution of the fungicide into these physical structures. This means that the actual concentration of the fungicide within the cell is reduced, which could be the reason of the observed effects. But this does not exclude a lipid peroxidative activity of DCOF. ORTH et al. (1992 b) did not find a lipid peroxidation in purified microsomal membranes of *Ustilago maydis* under the influence of AHF and DCOF. EDLICH and LYR (1984, 1986) used raw or only partly purified fractions of *Mucor mucedo*, therefore a cofactor could have been lost, or other membrane systems of *Ustilago* are bearing the target enzyme, which must not be identical with the cytochrome P-450 reductase. Only further investigations can solve this problem.

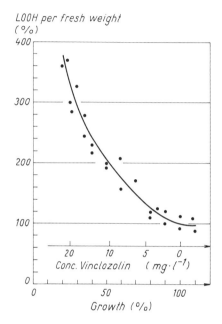

Fig. 7.3 Correlation between growth of mycel and level of lipid peroxides in cells of *Mucor mucedo* in dependence of concentration of vinclozolin. Growth is expressed as fresh weight of mycel in relationship to a control without fungicide. Level of lipid peroxides (LOOH) were measured by using the thiobarbituric acid method.

Mechanism of action

There is a question, of course, concerning the mechanism by which DCOF initiates lipid peroxidation. Experiments with *Mucor mucedo* and *Botrytis cinerea* revealed, that a membrane bound NADPH dependent flavin enzyme of the type of a "cytochrome-c-reductase" (EC 1.6.2.) (as well as some other, but not all flavin enzymes (Fig. 7.4)) are specifically inhibited by vinclozolin and other DCOF *in vitro*. The inhibition begins at concentrations of 5×10^{-6} mol$\times$l^{-1} (1–2 ppm). This reaction is not sensitive to CO, therefore a participation of a cytochrome P-450 dependent enzyme can be excluded. The involvement of a cytochrome P-450 dependent enzyme has been proposed by LEROUX et al. (1983) and is discussed by GULLINO and SISLER. They concluded this indirectly from an antagonism of some cytochrome P-450 inhibitors like piperonylbutoxide, SKF 525 A, metyrapone a. o.

Indeed, according to our results the effect of DCOF on lipid peroxidation can be counteracted *in vivo* and also *in vitro* by α-tocopherol acetate, piperonylbutoxide and partly by SKF 525 A. Piperonylbutoxide is in practical use as an insecticide synergist and the assumption is that it inhibits cytochrome P-450 enzymes which detoxify insecticides. Apparently it interacts with cytochrome c reductase which can, but must not be a part of the P-450 enzyme complex, i.e. it can also exist as a distinct enzyme. SKF 525 A has also been described as an inhibitor of cytochrome P-450 enzymes. In contrast, α-

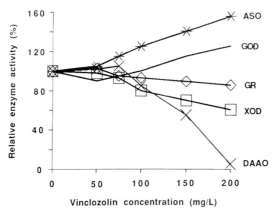

Fig. 7.4 Effect of vinclozoline on the relative activities of some flavin-containing enzymes (commercial sources). XOD, xantine oxidase; GR, oxidized glutathione reductase; GOD, glucose-1-oxidase; ASO, ascorbate oxidase; DAAO, D-amino acid oxidase.

Table 7.2 Effects of some antidotes acting against vinclozolin induced growth inhibition of *Botrytis cinerea* (S-strain) in nutrition solution in relation to the level of lipid peroxides

Compounds	Conc. $mg \times l^{-1}$	Dry weight decrease %	Level of lipid*) peroxides %
Without	–	0	100
Vinclozolin	0.1	67	180
	0.5	97	295
	10	100	–
α-Tocopherol acetate	100	0	96
	500	0	98
Piperonyl butoxide	0.1	4	103
	1.0	8	107
SKF 525A	1	1	93
	10	14	99
Vinclozolin +	0.5 + 100	77	210
α-Tocopherol acetate	0.5 + 500	50	175
	1.0 + 100	82	235
	1.0 + 500	67	180
Vinclozolin +	0.5 + 0.1	82	175
Piperonyl butoxide	0.5 + 1.0	76	155
Vinclozolin +	0.5 + 1.0	88	220
SKF 525A	0.5 + 10	87	215

*) Lipid peroxide content was measured by using the thiobarbituric acid method.

tocopherol acetate acts as general radical scavenger in membranes and by this mechanism, suppresses lipid peroxidation.

Therefore the antagonizing effects of these two types of aforementioned substances, which can be demonstrated *in vitro* as well as *in vivo* in mycelial growth tests (Tab. 7.2. and Fig. 7.3), are realized by different mechanisms.

DCOF even at relative high concentrations do not very strongly inhibit NADPH independent endogenous monoxygenases such as xanthine oxidase, whereas some AHF, such as dicloran, PCNB and chloroneb are more inhibitory towards this enzyme (Tab. 7.3). This demonstrates that the two groups of fungicides, AHF and DCOF, have many common features in their mechanism of action but do differ in some properties.

There is at present no doubt, that the pathological oxidative processes observed *in vitro* are identical with those occurring *in vivo*. DCOF seems to interact with the flavin enzyme cytochrome-c-reductase in such a manner that the normal electron flow from NADPH to cytochrome c is blocked: As a consequence NADPH as well as the essential phospholipids surrounding the active centre of the enzyme are oxidized probably by a peroxide intermediary product of the flavin enzyme (ZIEGLER *et al.* 1980) and/or by free radicals.

Table 7.3 Inhibitory effects of some fungicides on xanthine oxidase and D-amino acid oxidase

Fungicide	Xanthine oxidase I_{50} $mol \times l^{-1}$	D-Amino acid oxidase I_{50} $mol \times l^{-1}$
Iprodione	2×10^{-4}	3×10^{-4}
Procymidone	10^{-3}	4×10^{-4}
Vinclozolin	10^{-3}	3×10^{-4}
Chloroneb	10^{-4}	7×10^{-4}
Dicloran	4×10^{-4}	9×10^{-4}
Etridiazole	10^{-3}	8×10^{-4}
PCNB	3×10^{-4}	4×10^{-4}
Tolclophosmethyl	10^{-3}	7×10^{-4}

*) I_{50} — Concentration for 50% inhibition of enzyme activity

Such a pathological side reaction has been described for monoxygenases under the influence of pseudosubstrates which cannot be hydroxylated (MASSEY *et al.* 1982). The lipid peroxidation can spread in a cascade process if the natural cellular defence systems are overrun by an overproduction of free radicals. Their nature in connection with the action of DCOF is not yet established, but they could be single oxygen, hydroxyl ($\cdot$OH) or superoxide ($O_2 \cdot$) radicals which by their very good solubility in lipids oxidatively attack membranes and other cell structures. That has been described for several systems in the literature (Tab. 7.4). Protective systems within the cells are enzymes, such as superoxide dismutase (SOD), catalase and peroxidases, or scavengers of various kinds (tocopherols and mannitol). When there is an overproduction of aggressive radicals, these systems are inactivated and a collapse of cell structures or damage of DNA results, as has been demonstrated in numerous biological systems (FLAMINGINI *et al.* 1982; MITCHELL and MORRISON 1983; MELLO FILHO *et al.* 1984; IWATA *et al.* 1984; HALL *et al.* 1984; KLEBANOFF *et al.* 1984; NORKUS *et al.* 1983; REINER and KAZURA 1982).

Table 7.4 Actions induced by active kinds of oxygen

	References
Peroxidation of polyunsaturated fatty acids	BRYAN et al. 1982
Disintegration of membranes	HALLIWELL 1984
Inactivation of enzymes	HALLIWELL and GUTTERIDGE 1984
Lignin degradation	BES et al. 1983
Single strain breaks and disintegration of DNA	MELLO FILHO et al. 1984
Destruction of protease inhibitors	HALLIWELL and GUTTERIDGE 1984
Oxidation of sulfhydryl groups	Hall et al. 1984
Oxidation of cytosine	HAZRA and STEENKEN 1983
Destruction of chromosomes	HASSAN and MOODY 1984
Inhibition of respiration	HALLIWELL and GUTTERIDGE 1984

Due to their high relativity active oxygen radicals can attack various structures. Unsaturated fatty acids are peroxidized, SH-groups in enzymes are oxidized (BHUYAN et al. 1982; HALL et al. 1984) and biopolymers such as nucleic acids (GUTTERIDGE and HALLIWELL 1982; MITCHELL and MORRISON 1983; MELLO FILHO et al. 1984; IWATA et al. 1984), proteins (BRUNORI and ROTILO 1984; HALLIWELL and GUTTERIDGE 1984) and cell wall components (BES et al. 1983; GOLD et al. 1983) are attacked. The consequences are numerous, such as an inactivation of enzymes (KARAGEZYAN 1982), activation of proteolysis (WATANABE and KONDO 1983) strand scission of nucleic acids (HASSAN and MOODY 1984; HAZRA and STEENKEN 1983), destruction of cell wall material (GREEN and GOULD 1983) and destruction of membranes (ESTERBAUER 1982).

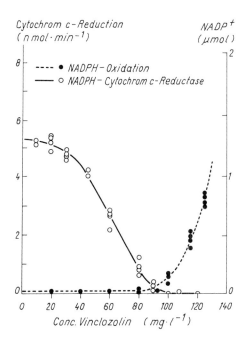

Fig. 7.5 Influence of vinclozolin on both NADPH dependent reduction of cytochrome c and NADPH oxidation *in vitro*. Solubilized 100,000 g pellets of homogenates of *Botrytis cinerea* serve as enzyme sources. Results represent from 4 independent experiments.

Several of these effects could to be consequence of the action of dicarboximide fungicides. The inner membranes of mitochrondria (Fig. 7.2) are especially sensitive which is probably due to their heigh content of unsaturated fatty acids. The plasmalemma, in contrast is much more stable, therefore, it does not lyse or allow cell constituents to leave the cell at an early stage of cell damage. But there may be a specific impairment, as inhibition of uptake of amino acids or osmotic lability suggests.

It should be mentioned that cytochrome P-540 produces free radicals also under certain circumstances (RUCKPAUL and REIN 1984). It seems, that piperonylbutoxide inhibits specifically the cytochrome-c-reductase and as a consequence of course, the cytochrome P-450 enzymes (mfo).

Selectivity

The reason for the selectivity of dicarboximide fungicides as in many other groups of fungicides is not yet clear, because the mechanism of action described here could be a common biocidal principle. *Botrytis cinerea*, *Sclerotinia* and *Monilia* are especially sensitive. They could be distinguished by a high activity of the toxifying enzyme, by an especially sensitive membrane composition, by weak protection mechanisms, or by a combination of these properties. This remains to be investigated.

Resistance

Another problem is that of resistance to DCOF. The practical observations regarding *Botrytis cinerea* are contradictory (compare chapter 6). But more or less resistant strains can be isolated, either in the laboratory or from the field and from glasshouse. Sporulation and growth rates are often weaker compared to wild strains (POMMER and LORENZ 1982). But there exist also R-strains with normal or high pathogenicity (LORENZ and POMMER 1982), which demonstrates the great genetic variability of *Botrytis cinerea* and indicates, that resistance is not firmly linked genetically to other properties as fitness and pathogenicity (chapter 6). STEEL and NAIR (1991) found an increased level of activity of catalase, but not of SOD, in strains of *Alternaria alternata* and *Botrytis cinerea* resistant to iprodione. This means that antioxidative enzymes (and compounds) within the fungal cell may contribute to resistance to DCOF. ORTH (1995) suggests that a ser/thr proteinkinase could be a target site.

The infection biochemistry of *Botrytis* is not very clear. Active oxygen species derived from a cell wall bound glucose oxidase seem to play, among other factors, an important role in the virulence (RIST and LORBEER 1984; WEIGEND and LYR 1995). In some investigations an increased osmolability in R-strains has been observed (BEEVER and BRIEN 1983; EDLICH and LYR 1986) which indicates changes in the membrane or cell wall structure. However, investigation of more than 100 resistant isolates showed no clear correlation between these parameters (LORENZ (chaper 6)). The surprising observation of a general cross resistance between dicarboximide and AHF (LEROUX *et al*. 1978), that there exist equal resistance mechanisms. Their nature is at present not identified. It can be concluded from the results of BEEVER and BYRDE (1982) that there is not an absolute identity, because some strains isolated from glycophen media are resistant to DCOF, but not to dicloran and PCNB. A reverse effect was found with strains isolated from dicloran media which are resistant to dicloran and PCNB, but sensitive to DCOF.

This demonstrates that in spite of many similarities in their general mechanism of action (inducing an increased, pathological lipid peroxidation) DCOF and AHF differ in some respects regarding their detailed mechanism of action. Perhaps there are also some differences in fungal mechanisms of resistance to them. This is shown for a special case in Table 7.5. This point remains to be elucidated.

LEROUX (1991) demonstrated with laboratory strains of various fungi also a cross resistance of DCOF and phenylpyrrole fungicides. This unexpected feature could be a hint for a similar mechanism of action. HILBER et al. (l.c. chapter 19) confirmed this fact, which is strengthened by the results of genetic investigations. Both R-genes could not be separated, which means that they are tightly coupled or identical.

On the other hand the authors failed to isolate resistant strains from the field, which seems to indicate that R-strains have a lower fitness than sensitive ones.

JESPERS and DEWAARD (1992) found as main effect of fenpiclonil a rapid block of uptake of amino acids and desoxyglucose and an increased accumulation of tetraphenylphosphonium bromide in a sensitive strain of *Fusarium sulphureum*. This suggests that the mechanisms of action of fenpiclonil may be related to membrane dependent active transport processes. A transport associated phosphorylation of glucose seems to be the site of action of fenpiclonil (JESPERS and DEWAARD 1995). It may be that the cytoplasmic membrane is affected too in a specific manner by DCOF, as the osmolability of most (but not all) R-strains indicates. However, the correlations of

Table 7.5 Cross resistance of *Botrytis cinerea*

Fungicide	Strain	Concentration [mg$\times$l^{-1}]		
		0.1	1.0	10
Iprodione	S	25	75	100
	R	5	7	12
Vinclozolin	S	40	97	100
	R	5	5	14
Carbendazim	S	77	100	100
	R	66	100	100
Chloroneb	S	–	10	70
	R	–	2	10
Dicloran	S	2	30	100
	R	0	5	24
PCNB	S	–	43	82
	R	–	8	20
Tridemorph	S	22	57	92
	R	19	68	90

Cross resistance, measured by the inhibition of radial growth on malt agar petri dishes. Results are expressed in relationship to untreated control.
S – dicarboximide sensitive strain
R – dicarboximide resistant strain

the findings in *Botrytis* and *Mucor* for the mechanism of action of DCOF and those for phenylpyrroles in *Fusarium sulphureum* need further investigations.

Acknowledgements: We are indebted to Dr. POMMER, Dr. MANGOLD and Dr. LORENZ for stimulating discussions and Prof. SISLER for his kind help in linguistic respect.

References

ALBERT, G.: Wirkungsmechanismus und Wirksamkeit von Vinclozolin bei *Botrytis cinerea*. Inaug. Diss. Landw. Fak. Rhein. Friedr. Wilh. Univ., Bonn 1979.
— Sphäroblastenbildung bei *Botrytis cinerea*, hervorgerufen durch Vinclozolin. Z. Pflanzenkr. Pflanzenschutz **88** (1981): 337–342.
ALLEN, G. R., AIUTO, R., and SUSSMAN, A. S.: Effect of cytochalasin on *Neurospora crassa*. Protoplasma **102** (1980): 63–75.
BEEVER, R. E., and BRIEN, H. M. R.: A survey of resistance to the dicarboximide fungicides in *Botrytis cinerea*. N. Z. J. Agric. Res. **26** (1983): 391–400.
— and BYRDE, R. J. W.: Resistance to the dicarboximide fungicides. In: DEKKER, J., and
— GEORGOPOULOS, S. G. (Eds.): Fungicide Resistance in Crop Protection. Centre for Agricult. Publ. and Document. Pudoc, Wageningen 1982.
BES, B., RANJEVA, R., and BOUDET, A. M.: Evidence for the involvement of activated oxygen in fungal degradation of lignocellulose. Biochemie (Paris), **65** (1983): 283–290.
BETINA, V., MICEKOVA, D., and NEMEC, P.: Antimicrobial properties of cytochalasin and their alternation of fungal morphology. J. Gen. Microbiol. **71** (1972): 343–349.
BHUYAN, K. C., BHUYAN, D. K., KUCK, J. F. R., KUCK, K. D., and KERN, H. L.: Increased lipid peroxidation and altered membrane functions in binory mouse cataract. Curr. Eye Res. **2** (1982/1983): 597–606.
BRUNORI, M., and ROTILO, G.: Biochemistry of oxygen radical species. In: COLOWICK, S. P., and KAPLAN, N. O. (Eds.): Methods of Enzymology. Vol. 105. Academic Press, New York 1984.
BUCHENAUER, H.: Preliminary studies on the mode of action of vinclozolin. Meded. Fac. Landbouwwet. Rijksuniv. Gent **41** (1976): 1509–1519.
EDLICH, W., and LYR, H.: Occurrence and properties of a cytochrome c-reductase in *Mucor* and its interaction with some fungicides. Tagungsber. Akad. Landwirtschaftswiss. DDR **222** (1984): 37–40.
— — Effectors of various flavin containing enzymes initiating a lipid peroxidation in fungi. Hoppe Seyler Z. physiol. Chemie **367** (1986): 257–263.
— — Target sites of fungicides with primary effects on lipid peroxidation. In: KÖLLER, W. (Ed.): Target Sites of Fungicide Action. CRC Press, Boca Raton 1992, pp. 53–68.
ESTERBAUER, H.: Aldehydic products of lipid peroxidation. In: MCBRIEN, D. C. H., and SLATER, T. F. (Eds.): Free Radicals, Lipid Peroxidation and Cancer. Academic Press, London 1982, pp. 101–108.
FLAMINGHINI, F., GUANIERI, C., TONI, R., and CALDARERA, C. M.: Effect of oxygen radicals on heart mitochrondrial function in tocopherol deficient rabbits. Int. J. Vitam. Nutr. Res. **52** (1982): 402–406.
FRITZ, R., LEROUX, P., and GREDT, M.: Mecánisme de l'action fongitoxique de la promidione (26019 RP ou glycophene), de la vinclozolin et du dicloran sur *Botrytis cinerea*. Phytopathol. Z. **90** (1977): 152–163.
GEORGOPOULOS, S. G., ZAFIRATOS, C., and GEORGIADIS, E.: Membrane functions and tolerance to aromatic hydrocarbon fungitoxicants in conidia of *Fusarium solani*. Physiol. Plant **20** (1967): 373–381.

- SARRIS, M., and ZIOGAS, B.: Mitotic instability in *Aspergillus nidulans* caused by the fungicides iprodione, procymidone and vinclozolin. Pesticide Sci. **10** (1979): 389–392.
- GOLD, M. H., KUTSUKI, H., and MORGAN, M. A.: Oxidative degradation of lignin by photochemical and chemical radical generating systems. Photochem. Photobiol. **38** (1983): 647–652.
- GREEN, R. V., and GOULD, J. M.: Substrate-induced hydrogen peroxide production in mycelia from the lignin-degrading fungus *Phaenerochaete chrysosporium*. Biochem. Biophys. Res. Commun. **117** (1983): 275–281.
- GUTTERIDGE, J. M. C., and HALLIWELL, B.: The role of superoxide and hydroxyl radicals in the degradation of DNA and deoxyribose induced by a copper-phenanthroline complex. Biochem. Pharmacol. **31** (1982): 2801–2806.
- HALL, N. D., MASLEN, C. L., and BLAKE, D. R.: The oxidation of serum sulfhydryl groups by hydrogen peroxide secreted by stimulated phagocytic cells in rheumatoid arthritis. Rheumatol. Int. **4** (1984): 35–38.
- HALLIWELL, B., and GUTTERIDGE, J. M. C.: Role of iron in oxygen radical reactions. In: COLOWICK, S. P., and KAPLAN, N. O. (Eds.): Methods in Enzymology. Vol. 105. Academic Press, New York 1984.
- HASSAN, H. M., and MOODY, C. S.: Determination of the mutagenity of oxygen free radicals using microbial systems. In: COLOWICK, S. P., and KAPLAN, N. O. (Eds.): Methods in Enzymology. Academic Press, New York 1984.
- HAZRA, D. K., and STEENKEN, S.: Pattern of hydroxyl radical addition to cytosine and 1-, 3-, 5- and 6 substituted cytosines. J. Am. Chem. Soc. **105** (1983): 4380–4386.
- HISADA, Y., and KAWASE, Y.: Morphological studies on antifungal action of N-(3′,5′-dichlorophenyl)-1,2-dimethyl-cyclopropane-1,2-dicarboximide on *Botrytis cinerea*. Ann. Phytopathol. Soc. Jap. **43** (1977): 151–158.
- IWATA, K., SHIBUYA, H., OHKAWA, V., and INUI, N.: Chromosomal aberrations in V79 cells induced by superoxide radical generated by the hypoxanthine-xanthine oxidase system. Toxicol. Letters **22** (1984): 75–82.
- JESPERS, A. B. K., and DEWAARD, M. A.: Biochemical effects of the phenylpyrrole fungicide fenpiclonil in *Fusarium sulphureum*. In: LYR, H., and POLTER, C. (Eds.): Systematic Fungicides and Antifungal Compounds. Ulmer Verl., Stuttgart 1993, pp. 99–103.
- — — Mode of action of the phenylpyrrole fungicide fenpiclonil in *Fusarium sulphureum*. 8th IUPAC Congress Washington D. C. (1994).
- KARAGEZYAN, K. G.: Phospholipid-phospholipid correlations and the dynamics of free radical oxidation of lipids in biological membranes in alloxan diabetes. Voprosij medicinoi khimii **28** (1982): 56–60.
- KLEBANOFF, S. J., WALTERSDORPH. A. M., and ROSEN, H.: Antimicrobial activity of myeloperoxidase. In: COLOWICK, S. P., and KAPLAN, N. O. (Eds.): Vol. 105 Academic Press, New York 1984.
- LEROUX, P.: Mise en évidence d'une similitude d'action fongicide entre le fenpiclonil, l'iprodione et le tolclofos-methyl. Agronomie **11** (1991): 115–117.
- FRITZ, R., and GREDT, M.: Cross resistance between 3,5-dichlorophenyl cyclic imidfungicides (Dichlozolin, Iprodion, Procymidone, Vinclozolin) and various aromatic compounds. In: LYR, H., and POLTER, C. (Eds.): Systematische Fungizide und Antifungale Verbindungen. Abh. Akad. d. Wiss. DDR, Akademie-Verlag, Berlin 1983, pp. 79–88.
- GREDT, M., and FRITZ, R.: Etudes en laboratoire de souches de quelques champions phytopathogenes resistantes a la dichlozoline, a la dicyclidine, a l'iprodione, a la vinclozolin et a divers fondicides aromatiques. Med. Fac. Landbouw. Rijksuniv. Gent **43** (1978): 881–889.
- — — — Resistance to 3,5-dichlorophenyl-N-cyclic imide fungicides. Neth. J. Plant Pathol. **87** (1981): 244–249.
- LORENZ, G., and POMMER, E.-H.: Pectolytic and cellolytic enzymes of dicarboximide sensitive and resistant strains of *Botrytis cinerea*. EPPO Bull. **12** (1982): 145–149.

MACRIS, B., and GEORGOPOULOS, S. G.: Reduced hexosamine content of fungal cell wall due to the fungicide pentachloronitrobenzene. Z. Allg. Mikrobiol. **13** (1973): 415–423.

MASSEY, V., PALMER, G., and BALLOU, D.: On the reaction of reduced flavins and flavoproteins with molecular oxygen. In: SLATER, E. C. (Ed.): Flavins and Flavoproteins. BBA Library, Vol. 8, 1966.

MELLO FILHO, A. C., HOFFMANN, M. E., and MENEGHINI, R.: Cell killing and DNA damage by hydrogen peroxide are mediated by intracellular iron. Biochem. J. **218** (1984): 273–276.

MITCHELL, R. E., and MORRISON, D. P.: A comparison between rates of cell death and DNA damage during irradiation of *Saccharomyces cerevisiae* nitrogen and nitrous oxide. Radiat. Res. **96** (1983): 374–379.

MOORE, D.: Effects of hexose analogues on fungi, mechanism of inhibition and of resistance. New Phytol. **87** (1981): 487–515.

MÜLLER, H. M.: Zytopathologische Veränderungen bei *Botrytis cinerea* unter dem Einfluß verschiedener Fungicide. Tag.-Ber. 10. Tagung "Elektronenmikroskopie", Ges. Topochem. und Elektronenmikroskopie DDR, Leipzig 1981.

NORKUS, E. P., KUENZIG, W., and CONNEY, A. H.: The mutagenic activity of ascorbic acid in vitro and in vivo. Mutat. Res. **117** (1983): 183–191.

ORTH, A. B.: Studies on the mode of action of dicarboximide and aromatic hydrocarbon fungicides in *Ustilago maydis*. 8th IUPAC Congress, Washington D. C. (1994).

–, SFARRA, A., PELL, E. J. and TIEN, M.: Assessing the involvement of free radicals in fungicide toxicity using α-tocopherol analogs. Pesticide Biochem. and Physiol. **44** (1992a): 19–23.

– – – – An Investigation into the role of lipid peroxidation in the mode of action of aromatic hydrocarbon and dicarboximide fungicides. Pesticide Biochem. and Physiol. **44** (1992b): 91–100.

PAPPAS, A. C., and FISHER, D. H.: A comparison of the mechanism of action of vinclozolin, procymidone, iprodione and prochloraz against *Botrytis cinerea*. Pesticide Sci. **10** (1979): 239–246.

POMMER, E.-H., and LORENZ, G.: Resistance of *Botrytis cinerea* Pers. to dicarboximide fungicides – literature review. Crop Protect. **1** (1982): 221–230.

REINER, N. E., and KAZURA, J. W.: Oxidant-mediated damage of *Leishmania donovani* promastigotes. Infect. Immun. **36** (1982): 1023–1027.

RIST, D. L., and LORBEER, J. W.: Moderate dosages of ozone enhance infection of onion (*Allium cepa*) leaves by *Botrytis cinerea* but not by *Botrytis squamosa*. Phytopathology **74** (1984): 1217–1220.

RUCKPAUL, K., and REIN, H.: Cytochrome P-450. Akademie-Verlag, Berlin 1984.

SHARPLES, R. O.: The fungitoxic effects of dicloran on *Botrytis cinerea*. Proc. Brit. Insect. Fung. Conf. London 1961: 327–336.

STEEL, CH. C., and (TAN)NAIR, N. G.: Role of free radical enzymes in resistance to the dicarboxamide fungicides. Biochem. Soc. Transactions **21** (1991): 254.

THRELFALL, R. J.: Effect of pentachloronitrobenzene (PCNB) and other chemicals on sensitive and PCNB-resistant strains of *Aspergillus nidulans*. J. Gen. Microbiol. **71** (1972): 173–180.

TILLMAN, R. W., and SISLER, H. D.: Effect of chloroneb on the growth and metabolism of *Ustilago maydis*. Phytopathology **63** (1973): 219–225.

WATANABE, T., and KONDO, N.: The change in leaf protease and protease inhibitor activities after supplying various chemicals. Biol. Plant **25** (1983): 100–109.

WEIGEND, M., and LYR, H.: The involvement of oxidative stress in pathogenesis of *Botrytis cinerea* on *Vicia faba* plants. (1995): in preparation.

ZIEGLER, D. M., POULSEN, L. L., and DUPPEL, M. W.: Kinetic studies on mechanism of the microsomal flavin-containing monoxygenase. In: Microsomes, Drug Oxidations and Chemical Carcinogenesis (1980): 637–645.

Chapter 8

Carboxin fungicides and related compounds

M. KULKA*) and B. VON SCHMELING**)

*) Uniroyal Ltd. Research Laboratories Guelph/Ontario, Canada
**) Uniroyal Chemical Company, World Headquaters, Middleburry CT 06749, USA

Introduction

The protectant and eradicant fungicides, which have been available of more than a century, are effective only in controlling fungal pathogens on the surface of the plant. To control internal pathogens of plants it was necessary to find a systemic fungicide which

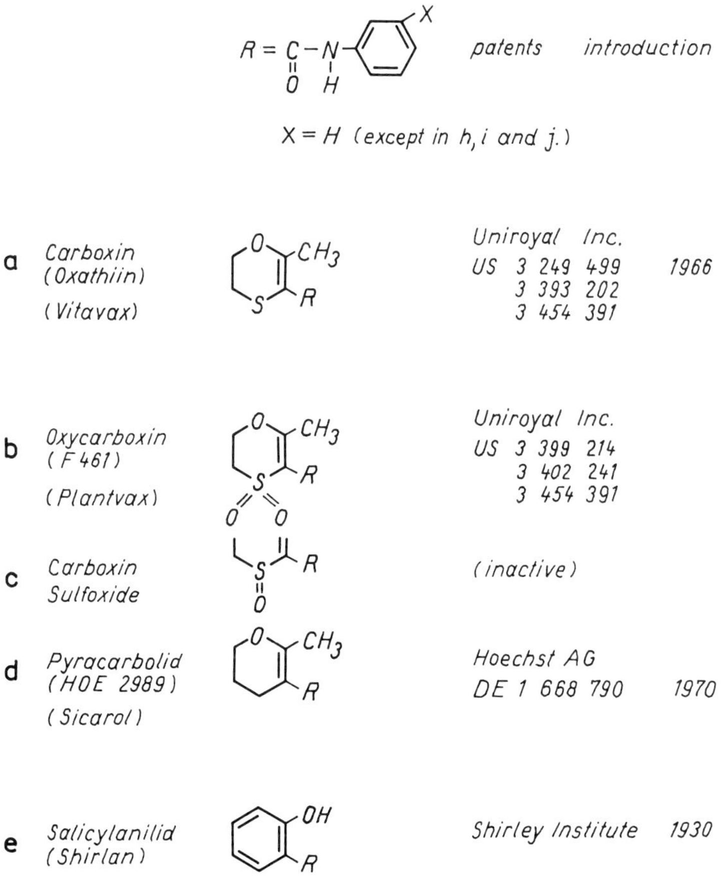

$$R = \underset{\underset{O}{\|}}{C} - \underset{H}{N} -\text{\large{\textbigcircle}}\!\!-X \qquad \text{patents} \qquad \text{introduction}$$

$$X = H \text{ (except in h, i and j.)}$$

a	Carboxin (Oxathiin) (Vitavax)	[structure]	Uniroyal Inc. US 3 249 499 3 393 202 3 454 391	1966
b	Oxycarboxin (F 461) (Plantvax)	[structure]	Uniroyal Inc. US 3 399 214 3 402 241 3 454 391	
c	Carboxin Sulfoxide	[structure]	(inactive)	
d	Pyracarbolid (HOE 2989) (Sicarol)	[structure]	Hoechst AG DE 1 668 790	1970
e	Salicylanilid (Shirlan)	[structure]	Shirley Institute	1930

Fig. 8.1 (1) Structure relationships of carboxins and related carboxamides.

could be absorbed by the host plant and be transported via the xylem or phloem systems to the site of the pathogen where eradication could take place. Although a search for such fungicides continued for many years, it was not until the nineteen sixties that a great deal of progress was made.

In 1966 von SCHMELING and KULKA reported the discovery of systemic fungicides carboxin (5,6-dihydro-2-methyl-N-phenyl-1,4-oxathiin-3-carboxamide) (Vitavax) (carbathiin in Canada) and oxycarbox in (5,5-dihydro-2-methyl-N-phenyl-1,4-oxathiin-3-carboxamide, 4,4-dioxide) (Plantvax) (Fig. 8.1). They announced that these chemicals were effective in controlling plant pathogenic fungi such as wheat leaf rust, *Puccinia rubigo-vera tritici* (Eriks) Carleton; bean rust, *Uromyces phaseoli typica* Arth.; loose smut of barley, *Ustilago nuda* (Jens) Rostr. and damping off, *Rhizoctonia solani* Kühn.

In the initial experiments, bean seeds (*Phaseolus vulgaris*) were treated with carboxin and oxycarboxin at the rate of 0.25% chemical per weight of seed. The seeds were planted in the greenhouse and the resulting plants were inoculated at 1 and 2 week intervals with uredospores of bean rust *Uromyces phaseoli*. Both chemicals controlled the development of rust symptoms on the primary leaves when inoculated 7 days after planting. In the experiment of the 2 week interval between planting and inoculation, oxycarboxin gave complete control whereas carboxin failed to control the rust. Barley seed

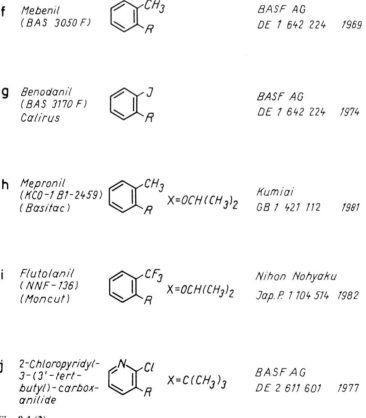

Fig. 8.1 (2)

k	Fenfuram (WL 22 361) (Panoram)			Shell Res. Ltd. GB 1 215 066 Uniroyal Inc. Canada P. 893,676	1974 1972
l	Methfuroxam (UBI H 719) Furavax			Uniroyal Inc. Canada P. 893 676 DE 2 006 471 US 3,959,481 4,054,585	1972
m	Furcarbanil (BAS 3191 F)			BASF AG DE 1 768 686	1970
n	2,4-Dimethyl-thiazole-5-carboxanilide G 696, F849	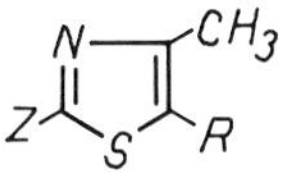	Z=CH$_3$ Z=NH$_2$	Uniroyal Inc. US 3,547,917 3,505,055 3,709,992 3,725,427	1970
o	3-Methylthiophene-2-carboxanilide				1975
p	1,3,5-Trimethyl-pyrazole-4-carboxanilide				1976

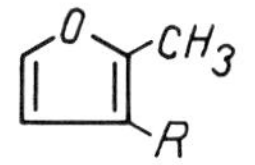

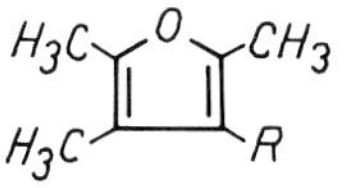

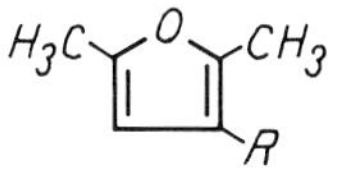

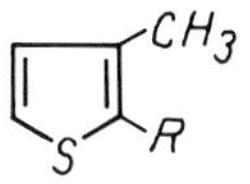

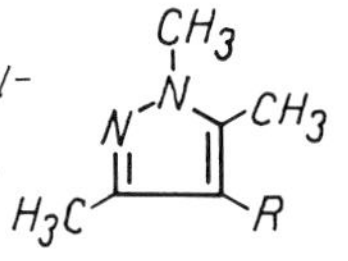

Fig. 8.1 (3)

(*Hordeum vulgare*) infected with loose smut, *Ustilago nuda,* when treated with carboxin (0.125% chemical per weight of seed) and planted in the field, yielded two months later smut free barley plants. Oxycarboxin in a similar experiment was only moderately effective.

Systemic nature of carboxin and oxycarboxin

The systemic nature of carboxin and oxycarboxin was demonstrated by a number of methods. The control of rust disease on the leaves of bean plants following soil treatment as well as control of the systemic loose smut fungus by seed treatment (v. SCHMELING and KULKA 1966; v. SCHMELING *et al.* 1966, 1968, 1969, 1970; MATHRE 1968) is evidence that the fungitoxic compound entered the plant and was transported to the site of the pathogen in sufficient quantity to cause eradication.

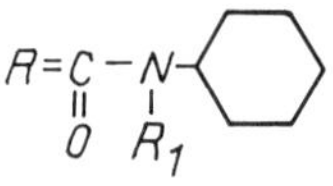

		$R = C - N -\bigcirc$ $\overset{\|}{O}\overset{\|}{R_1}$			
q	Furmecyclox (BAS 389 F) (Campogran) (Xyligen B)	(furan ring with CH₃, R, $R_1 = OCH_3$)	BASF AG DE 2 455 082		1977
r	Cyclafuramid (BAS 3270 F)	(H₃C, furan ring with CH₃, R, $R_1 = OCH_3$)	BASF AG DE 1 914 954		1971
s		(H₃C, furan ring with CH₃, R, $R_1 = H$)			
t		R_1, $C-N-R_2$, $\overset{\|}{O}H$	$R_1 = CH_3, CF_3, Cl, J$ $R_2 = $ Phenyl, Cyclohexyl		
u	F427	(dihydrothiazine ring with CH₃, S, C-N, biphenyl)	Uniroyal Inc. US- 3,249,499 3,393,202		1966

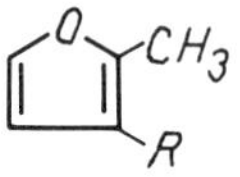

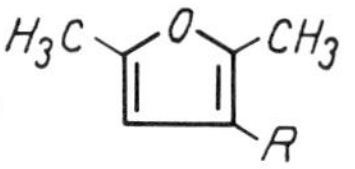

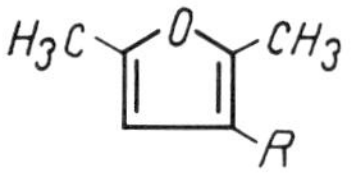

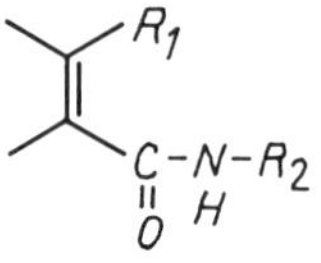

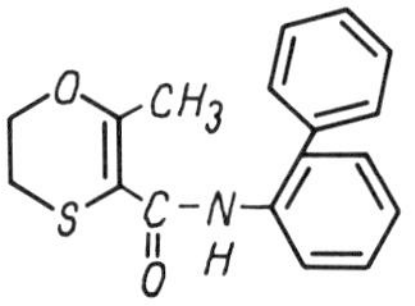

Fig. 8.1 (4)

More direct evidence for the systemicity of carboxin and oxycarboxin was obtained (SNEL and EDGINGTON 1968) by labelling these chemicals with ^{14}C and then applying them and tracing their movements in the plant by autoradiography and liquid scintillation. SNEL and EDGINGTON (1968) found that these chemicals were translocated in bean plants upwards from roots and shoots and accumulated in the margins of the leaves. Similarly, it was found (KIRK et al. 1969) that the ^{14}C labelled carboxin translocated from treated seeds or soil to stems and leaves in cotton plants (Gossypium hirsutum).

Applications in agriculture

Great interest was shown in carboxin and oxycarboxin following the diclosure of their systemic properties. They were evaluated all over the world in different climates and over seven hundred papers were published confirming the results of the origin disclosure and extending applications to other fungal pathogens. This extensive research also showed that carboxin and oxycarboxin are specific systemic fungicides controlling mainly pathogens of the Basidiomycetes — a class of fungi which include such important pathogens as smuts, bunts and rusts of cereal grains and the soil fungi *Rhizoctonia solani*. The economic benefits of carboxin and oxycarboxin became evident. The treatment of seed or plants with these chemicals results in higher yields of crops not only because of disease control but also because of growth stimulation. Varieties of grain seed that yield well but could not be used in agriculture because of their high susceptibility to diseases, could now be used after treatment with these systemic fungicides. ERWIN (1970) reviewed the development of the systemic fungicides up to 1970. An exhaustive review of the many publications which followed 1970 is beyond the scope of this chapter.

Carboxin was first registered for use on grains in France in 1969 and in Canada in 1970. Other countries followed and today carboxin and oxycarboxin are used in many countries of the world to control such diseases as smuts, bunts and rusts of cereal grains and damping-off diseases on cotton. These systemic fungicides may be applied to seed, soil or plants but seed application is the most practical. Carboxin is applied as a seed dressing at the rate of 0.4 g to 1 g per kg of cereal seed and 1 to 2 g per kg of cotton seed and these applications are sufficient to control the diseases for the growing season. For the control of rust diseases of grain, in addition to seed treatment, one or more foliar applications of oxycarboxin usually as 0.1% spray is required. Some pathogens such as onion smut require considerably heavier dosages of carboxin for complete control (EDGINGTON and KELLY 1966). The diseases of rust on ornamentals are controlled by a few 0.1% spray applications of oxycarboxin to the growing plants.

Carboxin is used either alone or in combinations with other fungicides such as thiram and copper 8-hydroxyquinolinate. The purpose of this is to increase the efficacy of carboxin and to broaden the spectrum of fungal control. Also synergistic effects of one fungicide on another can increase efficacy. A synergistic action between carboxin and copper 8-hydroxy quinolinate has been reported (RICHARD and VALLIER 1969).

Another method used for increasing the efficacy of the systemic fungicides is through formulation techniques. Thus oil dispersible formulations are better than wettable powder formulations mainly because oil formulations penetrate the host and pathogen better. In comparing the efficacy of different formulations in the control of oat leaf rust (*Puccinia coronata avenae*) with oxycarboxin, it was found that the emulsifiable concentrates and oil dispersible concentrates had lower ED values than wettable powder formulations (CHIN et al. 1975).

Tables 8.1 and Table 8.2 list the pathogens which are controlled by carboxin and oxycarboxin.

Structure-activity relationships of carboxin and related carboxamides

Diligent search by the inventors and by others, in the area of structure-activity relations followed soon after the discovery of carboxin. The aim was to find out what structural modifications could be made to carboxin without losing its unique biological properties.

Table 8.1 Fungal pathogens controlled by carboxin

Fungal pathogen	Host	References
Rhizoctonia solani (damping off)	cotton, bean	AL-BELDAWI and PINCĆARD (1970); SHARMA and SOHI (1982)
	conifer seeds	BELCHER and CARLSON (1968)
Ustilago nuda (loose smut)	barley	MAUDE and SHURING (1969); REINBERGS et al. (1968)
Ustilago nigra (black loose smut)	barley	BATALOVA et al. (1978)
Ustilago tritici (loose smut)	wheat	MAUDE and SHURING (1969); TYLER (1969)
Ustilago avenae (loose smut)	oats	TOLLENAAR et al. (1969)
Ustilago kolleri (covered smut)	oats	PATHAK et al. (1971); RICHARD and VALLIER (1969); WALLACE (1969)
Ustilago hordei (covered smut)	barley	PATHAK et al. (1971); WALLACE (1969)
Tilletia foetida (bunt)	wheat	PATHAK et al. (1971); WALLACE (1969)
Tilletia caries (bunt)	wheat	PATHAK et al. (1971); WALLACE (1969)
Sphacelotheca panici-miliacei (smut)	millet	KOISHIBAEV (1974)
Sphacelotheca sorghi (covered kernel smut)	sorghum	DUSCHANOV (1983); POPOV and SILAEV (1978)
Sphacelotheca reilianum (head smut)	sorghum, corn	SIMPSON and FENWICK (1971); POPOV and SILAEV (1978)
Urocystis agropyri (flag smut)	wheat	METCALFE and BROWN (1969)
Urocystis cepulae (onion smut)	onion	EDGINGTON and KELLY (1969)
Exobasidium vexans (blister blight)	tea	VENKATA RAM (1969)
Exobasidium vaccinii (red leaf disease)	blueberry	LOCKHARDT (1969)
Sclerotium rolfsii (white mold)	peanuts	DIOMANDE and BEUTE (1977)
Tolyposporium penicillariae (kernel smut)	pearl millet	WELLS (1967)
Helminthosporium gramineum (leaf stripe)	barley	KINGSLAND (1969); NAVUSHTANOV (1978)
Helminthosporium sativum (root rot)	barley	SMIRNOVA et al. (1977)
Cochliobolus sativus (seedling blight)	barley	MILLS and WALLACE (1970); WALLACE (1969)
Drechslera sorokiniana	barley	HAMPTON (1978)
Corticium sasakii (sheath blight)	rice	LAKSHMANAN et al. (1980)
Ustilago maydis (smut)	corn	MATHRE (1968)

Controlled oxidation of carboxin (Fig. 8.1 a) yielded the sulfoxide (Fig. 8.1 c) (also formed in plants and animals) and this is many times less active than the parent compound. Further oxidation of carboxin (Fig. 8.1 a) or (Fig. 8.1 c) yielded oxycarboxin (Fig. 8.1 b) which was found to be less active on smut diseases but more active on the rust diseases. Of the many carboxins substituted in the phenyl ring which were prepared, only a few had significant fungicidal activity. The 3'-methyl- and the 3'-methoxyderivatives of carboxin were at least as active as carboxin.

The 2'-phenyl-derivative of carboxin (F427) (Fig. 8.1 u) is unique in that it exhibits a broader spectrum of activity than carboxin, controlling some species of Deuteromycetes

Table 8.2 Fungal pathogens controlled by oxycarboxin

Fungal pathogen	Host	References
Uromyces phaseoli typica (rust)	bean	ROLIM *et al.* (1981)
Puccinia recondita (leaf rust)	wheat	HAGBORG (1971); ROWELL (1967)
Puccinia graminis (stem rust)	wheat	HAGBORG (1971); ROWELL (1967)
Puccinia coranata avenae (leaf rust)	oats	CHIN *et al.* (1975)
Puccinia hordei Otth. (brown rust)	barley	UDEOGALANYA and CLIFFORD (1982)
Puccinia striiformis	wheat	POWELSON and SHANER (1966)
(stripe, yellow rust)	K. bluegrass	HARDISON (1967, 1971)
Puccinia arenariae (rust)	dianthus	UMGELTER (1969)
Puccinia antirrhini (rust)	snapdragon	UMGELTER (1969)
Puccinia malvacearum (rust)	mallow	UMGELTER (1969)
Puccinia horiana (white rust)	chrysanthemum	UMGELTER (1969); ZADOKS *et al.* (1969)
Puccinia obscura (rust)	daisy	UMGELTER (1969)
Puccinia carthami (rust)	safflower	ZIMMER (1967)
Puccinia helianthi (rust)	sunflower	BHOWMIK and SINGH (1979)
Urocystis agropyri (flag smut)	wheat, K. bluegrass	HARDISON (1967, 1971); METCALFE and BROWN (1969)
Ustilago striiformis (stripe smut)	cr. bent grass c. orchard grass K. bluegrass	HARDISON (1967, 1971)
Hemileia vastatrix (rust)	coffee	FIGUEIREDO *et al.* (1981)
Phragmidium mucronatum (rust)	rose	UMGELTER (1969); SHATTOCK and RAHBAR BHATTI (1983)
Coleosporium campanulae (rust)	bluebell	UMGELTER (1969)
Uromyces transveralis (rust)	gladiolus	LOPES *et al.* (1983)
Uromyces dianthi (rust)	carnation	SPENCER (1979)
Puccinia pelargonii-zonalis (rust)	geranium	HARWOOD and RAABE (1979)
Melampsora lini (flax rust)	flax	FROILAND and LITTLEFIELD (1972)

and of Phycomycetes as well as those of the Basidiomycetes (EDGINGTON and BARRON 1967).

The existence of the 1,4-oxathiin ring in carboxin does not appear to be important since its replacement by other rings does not cause a loss of the systemic fungicidal activity. Thus two thiazolecarboxamides, 2-amino-4-methyl-, (Fig. 8.1 n), and 2,4-dimethyl- (Fig. 8.1 n) N-phenyl-5-thiazole-carboxamide are systemic fungicides with biological properties similar to that of carboxin (HARRISON *et al.* 1970, 1971, 1973). The thiophenecarboxamide, (Fig. 8.1 o) also possesses fungicidal activity (WHITE and THORN 1980). The replacement of the 1,4-oxathin ring in carboxin with a furan ring creates little change in biological activity. Thus 2,5-dimethyl-N-phenyl- and 2,4,5-trimethyl-N-phenyl- (Fig. 8.1 l, m) 3-furan-carboxamide (DAVIS *et al.* 1972, 1973, 1976, 1977) are systemic fungicides with activity similar to that of carboxin and oxycarboxin, but (Fig. 8.1 l) is somewhat more effective than carboxin. The replacement of sulfur in carboxin with a CH_2 group resulted in 5,6-dihydro-2-methyl-N-phenyl-4H-pyran 3-carboxamide (Fig. 8.1 d) and this is also active against smuts, rusts and *Rhizoctonia* spp. although (Fig. 8.1 d) is somewhat more phytotoxic than carboxin (JANK and GROSSMANN 1971).

The simple compound mebenil (2-methyl-N-phenylbenzamide) (Fig. 8.1 f) which contains the 2-butenamide moiety (Fig. 8.1 t) common to all the systemic fungicides (Fig. 8.1 a) to (Fig. 8.1 u) above controls rust in cereals (POMMER and KRADEL 1969). Thus it appears that the main structural feature responsible for systemic fungicidal activity is N-phenyl-2-butenamide (Fig. 8.1 t). The contributions of the methyl group, the double bond, the carboxanilide function and stereoisomerism to systemic fungicidal activity are discussed in detail by TEN HAKEN and DUNN (1971).

Among the benzoic acid anilides substituted in position 2, the 2-hydroxy-benzoic acid anilide (salicylic acid anilide) (Fig. 8.1 e) introduced in 1930 is worth mentioning. This was used at first for material and textile protection (FARGHER *et al.* 1930). It is still used in limited amounts in plant protection for the control of *Cladosporium fulvum* in tomatoes. The 2-methyl derivative Mebenil (Fig. 8.1 f), was not used commercially in spite of good activity against cereal rusts and *Rhizoctonia solani*. By replacement of the methyl-group by halogen, especially iodine, new compounds with a specific activity against Basidiomycetes were found (POMMER *et al.* 1974). It should be mentioned, that 2-jodobenzoic acid anilide (Fig. 8.1 g) (benodanil) has excellent photostability and can be used as fungicide for the control of *Puccinia* species in cereals (POMMER *et al.* 1973; FROST *et al.* 1973). Benodanil exhibits good activity against diseases caused by Basidiomycetes in turf and lawn, such as *Corticium fuciforme*, *Puccinia* spp. and *Marasmius oreades*. Among the fungicidally active benzanilides substituted in the aniline moiety, two are distinguished by their usefulness for controlling sheath blight in rice: mepronil (Fig. 8.1 h) (DOI 1981) and flutolanil (Fig. 8.1 i) (AKARI and YABUTANI 1981; KURONO 1985). Both compounds are also active as are the other benzoic acid anilides against *Rhizoctonia solani* in potatoes (black scurf) and *Typhula* snow blight in wheat.

In regard to the substituted furanilides and furamides apparently none came to marked as leaf fungicides, in spite of good results in the greenhouses. The main reasons seem to be the insufficient UV-stability and the correlated low persistence (BUCHENAUER 1975). Fenfuram (Fig. 8.1 k) (TEN HAKEN and DUNN 1971; JORDOW 1978), methfuroxam (Fig. 8.1 l) (ALCOCK 1978) and Furmecyclox (Fig. 8.1 q) are used mainly in seed dressing formulations for *Tilletia*- and *Ustilago* spp. in cereals. For enlarging the fungicidal spectrum of activity, i.e. for *Drechslera graminea* and *Fusarium nivale,* various seed dressing combinations have been developed:

Fenfuram + Guazatine
Fenfuram + Guazatine + Imazalil
Fenfuram + Quintocene + Thiabendazol
Methfuroxam + Thiabendazol
Furmecyclox + Imazalil

Methfuroxam (Fig. 8.1 l) and Furmecyclox (Fig. 8.1 q) are quite suitable for the control of *Rhizoctonia solani* in cotton, potatoes and flower bulbs. Furmecyclox is formulated as a soluble preservative. This compound gives excellent control of the economically important wood destroying *basidiomycetes* (POMMER and REUTHER 1978). Pyracarbolid (Fig. 8.1 d) described by STINGL *et al.* (1970) was experimentally used against *Ustilago* and *Tilletia* spp. and *Rhizoctonia solani*. It also exhibits activity against *Hemileia vastatrix* in coffee.

Animal toxicology

Carboxin has a low toxicity to animals. Albino rats suffered no detectable symptoms when fed 600 ppm of carboxin in their diets for two years. The acute oral LD_{50} values for white rats, quail and partridge are 3,200, 5,600 and 2,000 mg per kg body weight, respectively. The acute dermal LD_{50} for rabbits is >8,000 mg per kg. The formulation known as Vitavax 250 which consists of 25.3% carboxin and 47.7% diluent, antifreeze and dispersants has the following toxicity values: acute oral LD_{50} (rats) >5,000 mg/kg; acute dermal LD_{50} (rabbits) >23,600 mg/kg; inhalation LC_{50} (rat) >20 mg/L. The acute oral LD_{50} for hens is 24,400 mg/kg with cumulation coefficient >5. However when administered repeatedly to hens at 1% LD_{50} it caused some symptoms of poisoning (PADALKIN 1978).

The formulation Plantvax 75W which consists of 75% oxycarboxin and 25% carrier and surfactants has an oral LD_{50} to rats of 2,570 mg/kg body weight; dermal LD_{50} to rabbits of >8,000 mg/kg and inhalation LC_{50} to rats of >6.5 mg/L.

Carboxin is not toxic to bees. When carboxin was blown into the holes of the nests of alfalfa leafcutting bees once a week, the chalkbrood disease of *Megachile rodundata* was reduced and the fungicide had no effect on the nesting or mortality of the bees (PARKER 1984).

Fate of carboxin and oxycarboxin in soil, plants and animals

Carboxin is rapidly oxidized to the sulfoxide, in soil, plants and animals (CHIN et al. 1969, 1970). It loses its fungicidal effectiveness in soil and plants three weeks after application mainly because of conversion to the sulfoxide, which is much less active than the parent compound. Eventually the sulfoxide, disappears and is believed to be bound to lignin by complex formation. No residue were present in wheat (*Triticum aestivum*), barley or cotton seed harvested from ^{14}C-carboxin-treated seed. The residues of carboxin and its sulfoxide, were also determined in wheat, barley, oats (*Avena*), peanuts (*Arachis hypogaea*), sorghum (*Holcus gramineae*), flax (*Phormium tenax*), and cotton seed (each harvested from carboxin-treated seeds), by extraction, base hydrolysis and chromatographic determination of the cleavage product aniline (SISKEN and NEWELL 1971). No residues were detected. This easily met the tolerance level requirement of 0.2 ppm set by the U.S. Environmental Protection Agency.

More recent metabolic studies (LARSON and LAMOUREUX 1984) of carboxin have shown that in peanuts not only oxidation takes place but also some amide cleavage occurs and in barley plants the hydroxylation product of carboxin, namely, 5,6-dihydro-2-methyl-N-(4-hydroxyphenyl)-1,4-oxathiin-3-carboxamide forms and was detected and identified (BRIGGS et al. 1974). This phenol is believed to be bound to lignin.

Dogs fed carboxin did not accumulate it in the body. Instead they excreted it in the faeces and urine together with the sulfoxide (Fig. 8.1c) (CHIN et al. 1969, 1970).

Carboxin and oxycarboxin are also degraded by bacteria. The bacterium *Pseudomonas aeruginosa* isolated from red sandy loam soil, when perfused with a solution of carboxin, first oxidized carboxin to oxycarboxin. Then the oxycarboxin was converted to 2-aminophenol and 2-(2-hydroxyethylsulfonyl) acetic acid (BALASUBRAMANYA et al. 1980). The latter was apparently formed via the intermediate 2-(vinylsulfony) acetanilide with hydroxylation of the phenyl group and addition of water to the vinyl group occurring at some stage. Somewhat similar degradation of carboxin occurs in cultures of *Rhizopus japonicus* (WALLNOEFER et al. 1972).

Chemical degradation of oxycarboxin in base follows the same pattern and 2-(vinylsulfonyl) acetanilide has been isolated. It forms from the intermediate 3-oxo-N-phenyl-2-(vinylsulfonyl) butanamide through de-acetylation. 2-(Vinylsulfonyl) acetanilide is a reactive compound and readily undergoes intra-molecular cyclization to form N-phenyl-3-thiomorpholinone, 1,1-dioxide but in the presence of nucleophilic reagents, addition to the vinyl group occurs to form 2-(2-substituted-ethylsulfonyl) acetanilide.

Development of resistance

Resistance development of fungal organisms to specific systemic fungicides would be expected to take place rapidly. However, so far, no widespread development of resistance of fungal pathogens to carboxin and oxycarboxin has been observed in the field. Strains of *Ustilago maydis* resistant to carboxin were first obtained in 1970 (GEORGOPOULOS and SISLER 1970) by UV irradiation of *Ustilago* sporidia placed on carboxin-containing media. In 1981 (GROUET *et al.*) in France a *Puccinia horiana* strain resistant to oxycarboxin was isolated in the greenhouse from chrysanthemums treated repeatedly for one year with oxycarboxin sprays. In the absence of further treatment with oxycarboxin, the resistance decreased but was still noticeable three months later after the last treatment. At about the same time, similar cases of resistance development of the same fungal pathogen were observed in the Netherlands (DIRKSE *et al.* 1982) and in Taiwan (PAI and SUN 1981). The LC_{50} for the resistant strains of *P. horiana* was >100 ppm while the LC_{50} for the sensitive strain was only 1.4 ppm.

In Israel (BEN-YEPHET *et al.* 1974) the tolerance of UV-induced mutants of *Ustilago hordei* to carboxin was demonstrated in smut inoculated and fungicide treated barley plants. These mutants were also resistant to oxycarboxin.

BOCHOW *et al.* (1971) found that when cultivated in repeated passages on agar in the presence of the systemic fungicides, the pathogens *Sclerotinia sclerotiorum* and *Fusarium solani pisi* developed resistance to carboxin and oxycarboxin but *Rhizoctonia solani* did not.

In recent years some observations indicated that in some winter barley varieties, but not in others, resistant strains of *Ustilago nuda* against carboxin and fenfuram have developed. Surprising was their greater susceptibility for mepronil and flutolanil (LEROUX and BERTHIER 1988). This negative cross resistance within the carboxins is correlated to a specific mutation of the receptor site (cf. chapter 9).

Growth stimulation and other side effects

Carboxin and oxycarboxin when applied to seeds or plants not only control diseases but they also stimulate growth of the plants (v. SCHMELING and KULKA 1969). Thus when pinto beans (*Phaseolus vulgaris*) with or without disease were sprayed with a 125 ppm suspension of carboxin and then allowed to grow in the greenhouse, growth stimulation was evidenced by such effects as increased height of the plants, increased number of leaves, increased length of internodes, increased weight and dark green colour of the leaves as compared to controls. Seed treatment of barley seed with carboxin and oxycarboxin resulted in production of greener plants and in an increase of the number of seed heads as compared to controls.

Concern was expressed that these systemic fungicides might interfere with nitrogen fixation bacteria in the soil in view of the fact that carboxin and oxycarboxin are bactericides effective against *Staphylococcus aureus* Rosenbach on Petri plates. However, while high concentrations of carboxin and oxycarboxin in the soil significantly inhibited nitrogen fixation on white clover, there was no effect on growth or nodulation at lower concentrations, such concentrations still being much greater than would be normally encountered in the field (FISHER and HAYES 1981). There could be no accumulation of these fungicides in the soil through repeated seasonal applications because of rapid degradation.

Carboxin and oxycarboxin also protect plants against air pollutants (HAGER 1973). Thus pinto beans growing in soil treated with 12 ppm carboxin showed no ozone damage as compared to controls. Tobacco, cotton, soybean and tomato plants were also protected by carboxin and oxycarboxin against injury by ozone in the air.

Conclusion

Carboxin and oxycarboxin are systemic fungicides effective in the control of such fungal pathogens as smuts, bunts and rusts of cereal grains, ornamentals and vegetables and the soil fungi *Rhizoctonia solani* of cotton. They have low animal toxicity and are quickly degraded in the soil, plants and animals and leave no residue in crops. These systemic fungicides provide an added benefit in that they also stimulate the growth of plants resulting in crop yield increases. While carboxin and oxycarboxin are effective against some bacteria they do not interfere with the nitrogen fixation bacteria in the soil when applied at the recommended rates.

Acknowledgement: We are indebted to Dr. E.-H. POMMER for contributing the last two paragraphs to the section on structure-activity relationships of carboxin and related carboxamides.

References

AL-BELDAWI, A. S., and PINCKARD, J. A.: Control of *Rhizoctonia solani* on cotton seedings by means of a derivative of 1,4-oxathiin. Plant. Dis. Rep. **54** (1970): 524–528.

ALCOCK, K. T.: Field evaluation of 2,4,5-trimethyl-N-phenyl-3-furancarboxamid (UBI-H 719) against cereal smut diseases in Australia. Plant Dis. Rep. **62** (1978): 854–858.

ARAKI, F., and YABUTANI, K.: α,α,α-Trifluoro-3'-isopropoxy-o-toluanilide (NNF-136), a new fungicide for the control of diseases caused by Basidiomycetes. Proc. Brit. Crop Protect. Conf. – Pests and Diseases (1981): 3–10.

BALASUBRAMANYA, R. H., PATIL, R. B., BHAT, M. V., and NEGENDRAPPA, G.: Degradation of carboxin and oxycarboxin by *Pseudomonas aeruginosa* isolated from soil. J. Environ. Sci. Health, Part B B **15** (1980): 485–505.

BATALOVA, T. S., TYULINA, L. R., and MAL'TSEVA, A. I.: Disinfection of barley seeds for the elimination of black loose smut. Khim. Sel'sk Khoz. **16** (1978): 33–34.

BELCHER, J., and CARLSON, L. W.: Seed treatment fungicides for control of conifer damping off. Can. Plant Dis. Surv. **48** (1968): 47–52.

BEN-YEPHET, Y., HENIS, Y., and DINOOR, A.: Genetic studies on tolerance of carboxin and benomyl at the asexual phase of *Ustilago hordei*. Phytopathol. **64** (1974): 51–56.

BHOWMIK, T. P., and SINGH, A.: Evaluation of certain fungitoxicants for the control of sunflower rust. Indian Phytopathol. **32** (1979): 443–444.

BOCHOW, H., LUC, L. H., and SUNG, PH. O.: Development of tolerance by phytopathogenic fungi to systemic fungicides. Acta Phytopathol. **6** (1971): 399–414.

BRIGGS, D. E., WARING, R. H., and HACKETT, A. M.: Metabolism of carboxin in growing barley. Pesticide Sci. **5** (1974): 599–607.

BUCHENAUER, H.: Differences in light stability of some carboxylic acid anilide fungicides in relation to their applicability for seed and foliar treatment. Pesticide Sci. **6** (1975): 525–535.

CHIN, M. Y., EDGINGTON, L. V., BRUIN, G. C. A., and REINBURGS, E.: Influence of formulations on efficacy of three systemic fungicides for control of oat leaf rust. Can. J. Plant Sci. **55** (1975): 911–917.

CHIN, W. T., STONE, G. M., SMITH, A. E., and VON SCHMELING, B.: Fate of carboxin in soil, plants and animals. Proc. 5th Brit. Insecticide Fungicide Conf. (1969): 322–327; J. Agric. Food Chem. **18** (1970): 709–712; 731–732.

DAVIS, R. A., VON SCHMELING, B., KULKA, M., and FELAUER, E.: Furan-3-carboxamides. U.S. Patent 3,959,481 (1976); U.S. Patent 4,054,585 (1977); Can. Patent 893,676 (1972); Can. Patent 932,334 (1973).

DIOMANDE, M., and BEUTE, M. K.: Comparison of soil plate fungicide screening and field efficacy in control of *Sclerotium rolfsii* on peanuts. Plant Dis. Rep. **6** (1977): 408–412.

DIRKSE, F. B., DIL, M., LINDERS, R., and RIETSTRA, I.: Resistance in white rust (*Puccinia horiana* P. Hennings) of chrysanthemum to oxycarboxin and benodanil in the Netherlands. Meded. Fac. Landbouwwet., Rijksuniv. Gent, **47** (1982): 793–800.

DOI, S.: Basitac (Mepronil). Japan Pesticide Inf. **38** (1981): 17–20.

DUSCHANOV, I. D.: Characteristics of spore germination by kernel smut of sorghum. Khim. Sel'sk. Khoz. (1983): 32–34.

EDGINGTON, L. V., and BARRON, G. L.: Relation of structure to fungitoxic spectrums of oxathiin fungicides. Can. Phytopathol. Soc. Proc. **33** (1967): 18–19.

– and KELLY, C. B.: Chemotherapy of onion smut with oxathiin systemic fungicides. Phytopathology **56** (1966): 876.

ERWIN, D. C.: Progress in the development of systemic fungitoxic chemicals for control of plant diseases. FAO Plant Protect. Bull. **18** (1970): 73–82.

FARGHER, R. G., GALLOWAY, L. D., and PROBERT, M. E.: The inhibitory action of certain substances on the growth of mould fungi. Mem. Shirley Inst. **9** (1930): 37.

FIGUEIREDO, P., MARIOTTO, P. R., BONINI, R., DE OLIVEIRA, F. N. L., and OLIVEIRA, D. A.: Effect of pyrocarbolid and oxycarboxin applied alone, in mixture, and alternated with copper fungicide in the control of coffee rust (*Hemileia vastatrix*). Biologico (Brazil) **47** (1981): 239–244.

FISHER, D. J., and HAYES, A. L.: Effects of some fungicides used against cereal pathogens on the growth of *Rhizobium trifolii* and its capacity to fix nitrogen in white clover. Ann. appl. Biol. **98** (1981): 101–108.

FROILAND, G. E., and LITTLEFIELD, L. J.: Systemic protectant and eradicant chemical control of flax rust. Plant Dis. Rep. **56** (1972): 737–739.

FROST, A. J. P., JUNG, K. V., and BEDFORD, J. L.: The timing of application of benodanil (BAS 3170 F) for the control of careal rust diseases. Proc. 7th Brit. Insecticide and Fungicide Conf. **2** (1973): 105–110.

GEORGOPOULOS, S. G., and SISLER, H. D.: Gene mutation eliminating antimycin A-tolerant electron transport in *Ustilago maydis*. J. Bacteriol. **103** (1970): 745–750.

GROUET, D., MONTFORT, F., and LEROUX, P.: Presence of a strain of *Puccinia horiana* resistant to oxycarboxin in France. Phytiatr-Phytopharm. **30** (1981): 3–12.
HAGBORG, W. A. F.: Oxycarboxin emulsifiable concentrate in the control of leaf and stem rusts in wheat. Can. J. Plant Sci. **51** (1971): 239–241.
HAGER, F. M.: 5,6-Dihydro-2-methyl-1,4-oxathiin-3-carboxamide as plant protectant against air pollution. Ger. Offen. 2,238,053 (1973).
HAMPTON, J. G.: Seed treatments for the control of *Drechslera sorokiniana* in barley. N.Z.J. Exp. Agric. **6** (1978): 85–89.
HARDISON, J. R.: Chemotherapeutic control of stripe smut (*Ustilago striiformis*) in grasses by two derivatives of 1,4-oxathiin. Phytopathol. **57** (1967): 242–245; **61** (1971): 731–735.
HARRISON, W. A., KULKA, M., and VON SCHMELING, B.: Thiazolecarboxamides U.S. Patent 3,547,917 (1970); Can. Patent 873,888 (1971); U.S. Patent 3,505,055 (1970); Can. Patent 837,517 (1970); U.S. Patent 3,709,992 (1973); Can. Patent 985,901 (1970); U.S. Patent 3,725,427 (1973).
HARWOOD, C. A., and RAABE, R. D.: The disease cycle and control of geranium rust. Phytopathology **69** (1979): 923–927.
JANK, B., and GROSSMANN, F.: 2-Methyl-5,6-dihydro-4H-pyran-3-carboxylic acid anilide, a new systemic fungicide against smut diseases. Pesticide Sci. **2** (1971): 43–44.
JORDOW, E.: Panoram-chemical for fungus control on cereals. Vaextskyddsrapporter (Vaextskyddskonferensen) **4** (1978): 118–123.
KINGSLAND, G. C.: Barley leaf stripe control and *in vitro* inhibition of *Helminthosporium sorokinianum* obtained with carboxin in South Carolina. Phytopathology **59** (1969): 115.
KIRK, B. T., SINCLAIR, J. R., and LAMBREMONT, E. N.: Translocation of C_{14}-labelled chloroneb and DMOC in cotton seedlings. Phytopathology **59** (1969): 1473–1476.
KOISHIBAEV, M.: Effectiveness of new fungicides in controlling millet smut. Khim. Sel'sk. Khoz. **12** (1974): 765–766.
KURONO, H.: Steps to Moncut, a new systemic fungicide. Jap. Pestic. Infor. **46** (1985): 6–10.
LAKSHMANAN, P., NAIR, M. C., and MENNON, M. R.: Comparative efficacy of certain fungicides on the control of sheath blight of rice. Pesticides (India) **14** (1980): 31–32.
LARSON, J. D., and LAMOUREUX, G. L.: Comparison of the metabolism of carboxin in peanut plants and peanut cell suspension cultures. J. Agric. Food Chem. **32** (1984): 177–182.
LEROUX, P., and BERTHIER, G.: Resistance to carboxin and fenfuram in *Ustilago nuda* (jens.) Rostr., the causal agent of barley loose smut. Crop Protect. **7** (1988): 16–19.
LOCKHARDT, C. L.: Control of red leaf in lowbush blueberry. Pesticide Res. Rep. (1969): 235–236.
LOPES, L. C., BARBOSA, J. G., and FILHO, J.: Effects of fungicides on the control of rust *Uromyces transversalis* (Thueman, Winter) on gladiolus. Rev. Ceres. **30** (1983): 366–374.
MATHRE, D. E.: Uptake and binding of oxathiin systemic fungicides by resistant and sensitive fungi. Phytopathology **58** (1968): 1464–1469.
MAUDE, R. B., and SHURING, C. G.: Seed treatments with carboxin for the control of loose smut of wheat and barley. Ann. appl. Biol. **64** (1969): 259–263.
METCALFE, P. B., and BROWN, J. F.: Evaluation of nine fungicides in controlling flag smut of wheat. Plant. Dis. Rep. **53** (1969): 631–633.
MILLS, J. T., and WALLACE, H. A. H.: Effect of fungicides on *Cochliobolus sativus* and other fungi on barley seed in soil. Can. J. Plant Sci. **49** (1969): 543–548; **50** (1970): 129–136.
NAVUSHTANOV, S.: Results of tests with some fungicides for control of barley stripe disease (*Helminthosporium gramineum*). Rastenievud. Nauki **15** (1978): 134–140 (Bulgaria).
PADALKIN, I. YA.: Toxicity of Vitavax to hens. Veterinariya (Moscow) **6** (1978): 85–86.
PARKER, F. D.: Effect of fungicide treatments on incidence of chalkbrood disease in nests of the alfalfa leaf cutting bee (*Hymenoptera: Megachilidae*). J. Econom. Entomol. **77** (1984): 113–117.
PATHAK, K. D., SHARMA, R. C., and JOSHI, L. M.: Seed treatment for control of some important cereal smuts. Plant Dis. Rep. **55** (1971): 544–545.

PEI, C. L., and SUN, S. K.: Study on fungicide-tolerant strain of pathogenic fungi in Taiwan. Oxycarboxin-resistance of *Puccinia horiana* P. Hennings, the white rust of chrysanthemum. Chih Wu Pao Hu Hsueh Hui Hui K'an **23** (1981): 221–227.

POMMER, E.-H., GIRGENSOHN, B., KÖNIG, K.-H., OSIEKA, H., and ZEEH, B.: Development of new systemic fungicides with carboxanilide structure. Kemia-Kemi **1** (1974): 617–618.

– JUNG, K., HAMPEL, M., and LOECHER, F.: BAS 3170F (2-Jodbenzoesäureanilid), ein neues Fungizid zur Bekämpfung von Rostpilzen in Getreide. Mitt. Biol. Bundesanstalt f. Land- und Forstwirtschaft **151** (1973): 204.

– and KRADEL, J.: Mebenil (BAS 3050 F) new compound with specific action against some Basidiomycetes. Proc. 5th Brit. Insecticide Fungicide Conf. **2** (1969): 563–568.

– and REUTHER, W.: Furmetamid, a new active ingredient for the control of wood-destroying Basidiomycetes. Proc. 4th Intern. Biodeterioration Symp. (1978): 67–70.

POPOV, V. I., and SILAEV, A. I.: Effectiveness of dressing sorghum seed against two types of smut. Nauchn. Tr. Leningrad S-kh. Inst. **351** (1978): 85–87.

POWELSON, R. L., and SHANER, G. E.: An effective chemical seed treatment for systemic control of seedling infection of wheat by stripe rust (*Puccinia striiformis*). Plant. Dis. Rep. **50** (1966): 806–807.

REINBERGS, E., EDGINGTON, L. V., METCALFE, D. R., and BENDELOW, V. M.: Field control of loose smut in barley with systemic fungicides Vitavax and Plantvax. Can. J. Plant Sci. **48** (1968): 31–35.

RICHARD, G., and VALLIER, J. P.: Treatment of cereal seed by combination of carboxin with copper quinolinate. Proc. 5th Brit. Insecticide Fungicide Conf. (1969): 45–53.

ROLIM, P. R. R., NETO, F. B., ROSTON, A. J., and OLIVEIRA, D. A.: Chemical controls of bean (*Phaseolus vulgarus* L.) diseases Biologico (Brazil) **47** (1981): 201–205.

Rowell, J. B.: Control of leaf and stem rust on wheat by an 1,4-oxathiin derivative. Plant Dis. Rep. **51** (1967): 336–339; **52** (1968): 856–858.

SCHMELING, B. VON, and KULKA, M.: Systemic fungicidal activity of 1,4-oxathiin derivatives. Science **152**, No. 3722 (1966): 659–660.

– – Regulation of plant growth. Can. Patent 828,771 (1969); U.S. Patent 3,454,391 (1969).

– – THIARA, D. S., and HARRISON, W. A.: U.S. Patents 3,249,499 (1966); 3,393,202 (1968); 3,399,214 (1968); and 3,402,241 (1968). Canadian Patents 787,893 (1968); 791,151 (1968); 842,243 (1970) and 825,665 (1969).

SHARMA, S. R., and SOHI, H. S.: Effect of different fungicides against *Rhizoctonia* root rot of French bean (*Phaseolus vulgaris*), Indian. J. of Mycology and Plant Pathol. (1982): 216–220.

SHATTOCK, R. C., and RAHBAR, BHATTI, M. H.: Fungicides for control of *Phragmidium mucronatum* on *Rosa laxa hort*. Plant Pathol. **32** (1983): 67–72.

SIMPSON, W. R., and FENWICK, H. S.: Suppression of corn head smut with carboxin seed treatments. Plant Dis. Rep. **55** (1971): 501–503.

SISKEN, H. R., and NEWELL, J. E.: Determination of residues of Vitavax and its sulfoxide in seeds. J. Agric. Food Chem. **19** (1971): 738–741.

SMIRNOVA, K. F., ANDREEVA, E. I., PROCHENKO, T. S., and PETROVA, L. P.: Injury-bearing root and some results of a study of fungicides against its basic agents with *in vitro* experiments. Kihm. Primen. Pestits. Prep. (1977): 81–86.

SNEL, M., and EDGINGTON, L. V.: Fungitoxicity, uptake and transportation of two oxathiin systemic fungicides in bean. Phythopathology **58** (1968): 1068; **60** (1970): 1708–1716.

SPENCER, D. M.: Carnation rust caused by *Uromyces dianthii*. In: Annual Report of the Glasshouse Crops Research Institute. Res. Inst. P. (1979): 136. Plant Pathol. **28** (1979): 10–16.

STINGL, H., HÄRTEL, K., and HEUBACH, G.: HOE 2989, ein neues systematisches Fungizid, wirksam gegen verschiedene Basidiomyceten (Rost- und Brandpilze sowie *Rhizoctonia*). VII. Int. Congr. Plant Protect., Paris (1970): 205.

TEN HAKEN, P., and DUNN, C. L.: Structure-activity relationships in a group of carboxanilides systematically active against broad bean rust. *Uromyces fabae* and wheat rust (*Puccinia recondita*). Proc. 6th Brit. Insecticide-Fungicide Conf. (1971): 455–462.

TOLLENAAR, H., BERATTO, E. M., and NAREA, G. C.: Preliminary note on the control of smut of oats in Chile. Agr. Tec. (Santiago de Chile) **29** (1969): 32–33.

TYLER, L. J.: Treatment of winter wheat seed with systemic fungicides for control of loose smut. Plant Dis. Rep. **53** (1969): 733–736.

UDEOGALANYA, A. C. C., and CLIFFORD, B. C.: Control of barley brown rust, *Puccinia hordei* Orth. by benodanil and oxycarboxin in the field and the effects on yield. Crop Protect. **1** (1982): 299–308.

UMGELTER, H.: Experience with Plantvax for control of rust diseases of ornamentals and vegetable crops. Gesunde Pflanzen **3** (1969): 53–60.

VENKATA RAM, C. S.: Systemic control of *Exobasidium vexans* on tea with 1,4-oxathiin derivatives. Phytopathology **59** (1969): 125–128.

WALLACE, H. A. H.: Comparative seed treatment trials 1969 Can. Plant Dis. Surv. **49** (1969): 49–53.

WALLNOEFER, P. R., KOENIGER, M., SAFE, S., and HUTZINGER, O.: Metabolism of the systematic fungicide carboxin by *Rhizopus japonicus*. Int. J. Environ. Anal. Chem. **2** (1972): 37–43.

WELLS, HOMER, D.: Effectiveness of two 1,4-oxathiin derivatives for control of *Tolyposporium* smut of pearl millet. Plant Dis. Rep. **51** (1967): 468–469.

WHITE, G. A., and THORN, G. D.: Thiophenecarboxamide fungicides: structure-activity relationships with the succinate dehydrogenase complex from wild-type and carboxin-resistant mutant strains of *Ustilago maydis*. Pesticide Biochem. Physiol. **14** (1980): 26–40.

ZADOKS, J. C., KODDLE, A., and HOOGKAMER, W.: Effect of derivatives of 1,4-oxathiin on *Puccinia horiana* in *chrysanthemum morifolium*. Netherl. J. Plant Pathol. **75** (1969): 193–196.

ZIMMER, D. E.: Efficacy of 1,4-oxathiin fungicides for control of seeding safflower rust. Plant Dis. Rep. **51** (1967): 586–588.

Chapter 9

Mechanism of action of carboxin fungicides and related compounds

T. Schewe*) and H. Lyr**)

*) Humboldt University of Berlin, Institute of Biochemistry, Berlin, Germany
**) Biologische Bundesanstalt für Land- und Forstwirtschaft, Institut für Integrierten Pflanzenschutz, Kleinmachnow, Germany

Introduction

The interesting systemic features of carboxin and oxycarboxin and their selectivity for Basidiomycetes (Edgington and Barron 1967) were unique for fungicides at the time of their discovery. These compounds stimulated progress towards new systemic fungicides.

The primary site of attack of these fungicides is in the mitochondria. This observation was confirmed by several groups (Mathre 1968, 1970, 1971; Ragsdale and Sisler 1970; Lyr et al. 1971). Other effects such as inhibition of nucleic acid synthesis and of protein synthesis proved to be secondary events. Such a target was somewhat unexpected in the light of the generally accepted conclusion that the mitochondrial system is an ancient and conservative one. But studies with carboxins have revealed differences in the sensitivity of the mitochondria of higher plants and of fungi as well as differences even among the fungi. This seems to be different from the strobilurins, a new fungicide group, acting on site III of the respiratory chain in mitochondria (cf. chapter 19).

Biochemistry of the mechanism of action

Mathre (1970) was the first who recognized that the oxidation of succinate is the critical site of the fungicidal action of the carboxins. Studies on the mechanism of action in the succinate oxidase system have revealed that the carboxins act on Complex II (succinate-ubiquinone oxidoreductase system) of the mitochondrial electron transfer chain (White 1971; Lyr et al. 1972; Ulbrich and Mathre 1972; Schewe et al. 1973; Mowery et al. 1976; Grivennikova and Vinogradov 1985) (Fig. 9.1). This conclusion was also confirmed by genetical studies; carboxin-resistant mutants of *Aspergillus nidulans, Ustilago maydis* or *Ustilago hordii* exhibited diminished sensitivity of succinate-ubiquinone oxidoreductase or succinate oxidase activities towards carboxin and related compounds as well as altered enzymatic properties and stability of Complex II activity (Georgopoulos et al. 1972; Georgopoulos et al. 1975; Ben-Yephet et al. 1975; Gunatilleke et al. 1975; Georgopoulos and Ziogas 1977; White et al. 1978; Keon et al. 1991). It should be emphasized that the action of carboxins is not restricted to fungal mitochondria. A carboxin-sensitivity was also shown for the Complex II activities of the mitochondria of rat

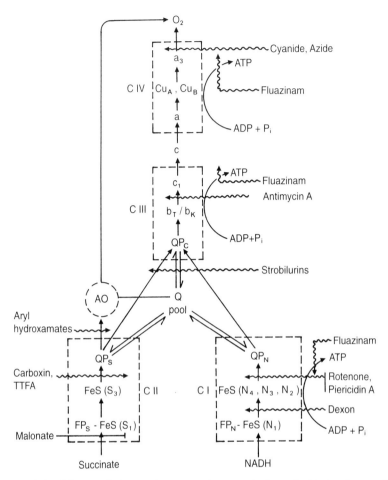

Fig. 9.1 Scheme of the respiratory pathway and sites of attack of some inhibitors, at site I (rotenone, piericidin A, dexon), site II (carboxins, TTFA), site III (antimycin A, strobilurines; compare chapter 19), site IV (cyanide, azide), including the possible relation to the "alternative pathway of respiration" (via AO, "alternative oxidase").

liver (MATHRE 1971; SCHEWE et al. 1973), bovine heart (SCHEWE et al. 1973; MOWERY et al. 1976) higher plants (MATHRE 1971; SCHEWE et al. 1974; DAY et al. 1978) as well as for respiring membranes of *Micrococcus denitrificans* (TUCKER and LILLICH 1974).

There are, however, differences in the concentration of carboxins needed for half-inhibition of the enzymatic activity. The relative tolerance of the succinate oxidase activity of higher plants may be one of the prerequisites for the systemic efficacy against fungal pathogens with low toxicity to the host organism.

More precise knowledge on the mechanism of action of carboxins arose from the contemporary progress in the field of the chemistry and enzymology of succinate dehydrogenase and its interaction with the other parts of the mitochondrial electron transfer system, especially from that of bovine heart. The use of mammalian mitochondrial systems as models for the target of the fungicidal action of carboxins appears to be possible or even advantageous for the following reasons;

i) These mitochondria show a sensitivity comparable to those of carboxin-sensitive fungi
ii) Many carboxin-sensitive fungi possess alternative pathways of succinate oxidation, which are absent in mammalian mitochondria, so that a serious complication in the studies with fungal mitochondria is alleviated
iii) Bovine heart mitochondria are the only source until now from which reconstitutively active succinate dehydrogenase could be extracted. More detailed studies on the mode of action of carboxins in the Complex II of the electron transfer chain in the mitochondria provides evidence that the succinate dehydrogenase *per se* cannot be the site of action. This conclusion based primarily on the observation that soluble succinate dehydrogenase prepared from either bovine heart mitochondria (SCHEWE et al. 1972; MOWERY et al. 1976; VINOGRADOV et al. 1980), or carboxin sensitive fungi as *Ustilago maydis* (ULRICH and MATHRE 1972), and *Trametes versicolor* (LYR et al. 1972) is not inhibited by carboxins at all.

Interaction with receptor structures

Only 50% of the succinate-phenazine methosulfate oxidoreductase activity (PMS reductase) is inhibited, i.e. that part which includes ubiquinone (coenzyme Q), whereas the other part of the PMS reductase activity being insensitive, is due to a direct interaction with the succinate dehydrogenase (MOWERY et al. 1976). The reduction of the Fe—S centres S_1 and S_3 of the succinate dehydrogenase by succinate is not inhibited by carboxins (ACKRELL et al. 1977), whereas real succinate dehydrogenase inhibitors such as malonate should do so. By contrast, EPR measurements revealed that carboxins inhibit the reoxidation of the two Fe—S centres S_1 and S_3 after having been reduced by succinate (ACKRELL et al. 1977). Assuming a linear electron flow through the chain histidyl-flavin$\rightarrow S_1 \rightarrow S_3 \rightarrow Q$ this behaviour implies that carboxins affect the electron transfer from the reduced form of S_3 to Q. An identical site of action is likely also true for the succinate oxidase inhibitor thenoyltrifluoroacetone (TTFA). The effects of carboxin and TTFA resemble each other in many respects, although differences have been reported (ULRICH and MATHRE 1972; SCHEWE et al. 1973; MOWERY et al. 1977). From binding experiments it had been suggested that both inhibitors compete for the same specific binding site, but these experiments were complicated by the obvious simultaneous presence of unspecific binding sites not giving rise to inhibition of the enzymic activity (COLES et al. 1978; RAMSEY et al. 1981). These authors demonstrated also the reversibility of the carboxin action at least for Complex II preparation from bovine heart. The site of action of carboxins is analogous to that of rotenone and piericidin A with respect to the fact that the latter inhibitors also interrupt the electron transfer between an Fe—S-cluster and ubiquinone.

It has been established that the succinate-ubiquinone oxidoreductase system is composed of four polypeptides having molecular masses of approximately 70 kDa, 30 kDa, 15 kDa and 7.5 kDa. The 70 kDa subunit contains the histidylflavin prosthetic group and an iron-sulfur moiety including the 2 Fe—2 S cluster S_1 which gives rise to a typical EPR signal on reduction of succinate dehydrogenase by succinate. The 30 kDa subunit is another Fe—S component containing centre S_3. It was formerly believed to be a HiPIP-type cluster (High-Potential Iron Protein) because it exhibits an EPR signal on the oxidized state as is the case with HiPIP's from bacteria (for review see BEINERT and ALBRACHT 1982). Recently it was shown that the oxidized species of centre S_3 is a 3 Fe cluster (ACKRELL *et al.* 1984) which is apparently converted to an EPR-silent 4 Fe—4 S cluster on reduction by succinate. Moreover, these authors discuss the possibility that the centre S_2 which appears in EPR spectroscopy only upon addition of sodium dithionite, may represent a superreduced state of this 4 Fe—4 S-cluster. The two large polypeptides represent the soluble, carboxin-insensitive succinate dehydrogenase which transfers electrons from succinate to artificial acceptors (phenazine methosulfate, ferricyanide, Wurster's blue) but not to ubiquinone, its physiological electron acceptor. Recombination of soluble succinate dehydrogenase with the two small polypeptides confers succinate-ubiquinone oxidoreductase activity (ACKRELL *et al.* 1980; VINOGRADOV *et al.* 1980). The 15 kDa polypeptide has the properties of a ubiquinone-binding protein (YU and YU 1981). It is assumed that the ubiquinone-binding proteins enable the ubiquinone to react with the proximate electron carriers of the respiratory chain and protect the ubisemiquinone radicals, which appear to be obligatory intermediates of electron transfer, from dismutation yielding oxidized and reduced ubiquinone.

Apparently the interaction of Fe—S clusters with ubiquinone is a critical step in the electron transfer. Unfortunately this electron transfer is not yet understood from the chemical point of view.

Ubiquinone occurs in excess over the other electron carrier of the respiratory chain and is thought to exert a "pool" function in the electron transfer chain, but only a small proportion of it is rapidly reduced by succinate or NADH. The behaviour may be rationalized in the light of the recently discovered Q-binding proteins (YU and YU 1981). Distinct Q-binding proteins have been proposed to function in the complex I, II and III of the electron transfer system (Fig. 9.1). Their interaction with Q may predispose the latter to specific electron transfer reactions. The obvious absence of Q-protein(s) in soluble succinate dehydrogenase preparations appears to be the reason for the lack of both Q reductase activity and carboxin sensitivity. This hypothesis was experimentally supported by the work of VINOGRADOV and coworkers, who isolated a protein fraction consisting of three small polypeptides (molar mass 13 kDa) from submitochondrial particles depleted of succinate dehydrogenase. Addition of this protein fraction to reconstitutively active succinate dehydrogenase resulted in a highly active, carboxin-sensitive succinate-Q oxidoreductase activity as measured with 2,6-dichlorophenolindophenol as acceptor (VINOGRADOV *et al.* 1980). The reconstitution also greatly increased the stability of the enzyme. Moreover, the 13 kDa polypeptide appears to mask the "low Km" ferricyanide oxidoreductase activity of soluble succinate dehydrogenase suggesting a location of this component proximal to the Fe—S centre S 3 (VINOGRADOV *et al.* 1990).

The involvement of protein(s) conferring Q reductase activity and carboxin-sensitivity in fungi is supported by the studies of GEORGOPOULOS *et al.* (1972, 1975), in which they showed that carboxin-resistance in *Ustilago maydis* is due to a single gene mutation which gives rise to a fairly labile succinate dehydrogenase activity which is resistant to low concentrations of carboxin. Such a lability is also typical for soluble succinate dehydrogenase preparations, where it is due to a destruction of the Fe—S centre S_3 by oxygen. Apparently the Q proteins protect this Fe—S cluster from aerobic inactivation which may be favoured by the complicated mechanism of redox change (ACKRELL et al. 1984). Obviously only the complex between succinate dehydrogenase and Q-protein(s) is able to interact with carboxins, which is also evident from studies of COLES et al. (1978) and RAMSAY et al. (1981). The latter authors showed that extraction of soluble dehydrogenase from complex II by perchlorate abolishes the specific binding of a radiolabelled carboxin derivative. In a photoaffinity labelling experiment using a labelled azido derivative of carboxin a sizable covalent binding to the small polypeptides of complex II was observed.

Moreover, carboxin has been shown to protect submitochondrial particles from alkali induced dissociation of "soluble succinate dehydrogenase", which contains the two large subunits but not the small polypeptides of Complex II (GRIVENNIKOVA and VINOGRADOV 1985). Carboxin does apparently not interfere with the interaction of QPs (see Fig. 9.1) with ubiquinone inasmuch as the inhibition by carboxin of the succinate-ubiquinone oxidoreductase activity of reconstituted Complex II has been shown to be non-competitive towards coenzyme Q_2 (TUSHURASHVILI et al. 1985).

All of the aforementioned data indicate that the complex of the Fe—S cluster S_3 and specific small polypeptide(s), called QP_S according to the proposal of YU and YU, constitute the carboxin receptor in which the Fe—S cluster must be reduced. The reduced form represents presumably a 4Fe—4S cluster (ACKRELL et al. 1984). Unfortunately its interaction with carboxin cannot be studied by EPR spectroscopy, since it is diamagnetic in this redox state. Strong evidence for the participation of the FeS—S_3 subunit of Complex II in the carboxin receptor has been provided by the recent work of KEON et al. (1991). These authors isolated and sequenced a gene which confers carboxin-resistance in a mutant of *Ustilago maydis*. The mutation was shown to be due to a single amino acid exchange (leucine for histidine) in the gene encoding the Fe—S cluster Ip, which is associated with the S_3 iron-redox centre (BROOMFIELD and HARGREAVES 1992).

It should be emphasized that the lack of drastic effects of carboxin on the EPR signal of oxidized S_3 (ACKRELL et al. 1977) does not exclude its interaction with reduced cluster S_3. It has been reported, however, that carboxin triggers the oxidative destruction of the cluster S_3 by dioxygen (ref. 13 in MOWERY et al. 1976). This effect may be rationalized by an interference with the stabilization of cluster S_3 afforded by QP_S so that cluster S_3 becomes susceptible to degradation by oxygen as it is in the case with soluble succinate dehydrogenase as well as with the particulate preparations from the carboxin-resistant mutants reported by GEORGOPOULOS et al. (1972, 1975).

Since carboxins appear to act on the complex between reduced S_3 and QPs, they should possess a binding affinity for at least one of these components. Some time ago we proposed a model for the interaction of carboxin with an Fe—S cluster (LYR et al. 1975; SCHEWE et al. 1979; SCHEWE and LYR 1987) which meets the following requirements:

i) the present knowledge on the spatial structure of 2Fe—2S or 4Fe—4S ferredoxins
ii) X-ray structural data of the carboxin molecule
iii) concordance with the structure-activity relationship of the inhibitory effects of a broad spectrum of carboxin congeners.

The model is in fair agreement with most of the experimental data from the modern carboxin research, but remains to be confirmed by appropriate approaches, such as X-ray crystal structure analyses. The model has been primarily designed to explain the affinity of carboxin and its congeners towards a Fe—S cluster; it does not, however, consider the interaction of carboxins with the polypeptide moieties of both Fe—S protein S_3 and QPs that appear to be obligatory components of the receptor. Moreover, it is not in line with the apparent stoichiometry of 0.5 moles of carboxin per mole of Complex II, which was estimated by both equilibrium dialysis and affinity labelling (for details see review by WHITE and GEORGOPOULOS 1992).

The elucidation of the mechanism of action of carboxins were substantially supported by synthesis of various chemical analogues, which demonstrated, that only a certain central configuration of the molecule, namely N-phenylbutenamide, or according to TEN HAKEN and DUNN (1971) "cis-crotonanilide" is required for activity. The 1,4-oxathiin ring could be replaced by different planar or nonplanar ring systems such as benzene, dihydropyran, pyridine, furan, pyrazole, thiophene and triazole. This means, that this part of the molecule does not show special structural requirements for the binding to the specific receptor. The common basic structure can be described as indicated in Figure 8.1t.

Since the methyl group can be replaced by halogens (POMMER et al. 1974) such as iodine in benodanil (LÖCHER et al. 1974) or chlorine in 2-chloropyridyl-3-(3'-tert-butyl)-carboxanilide (ZEEH et al. 1977), it follows that it is not involved in the specific interaction with the receptor region. Comparing all structures only the very general expression "carboxamide-fungicides" is adequate.

It is surprising that a replacement of the benzene ring of the aniline moiety by cyclohexane does not appreciably alter the spectrum of activity. Extensive structure-activity comparisons have been made by HARDISON (1971), MATHRE (1971), TEN HAKEN and DUNN (1971), WHITE and THORN (1975), MÜLLER et al. (1977), HUPPATZ et al. (1984), SNEL et al. (1970), WHITE and THORN (1980), WHITE et al. (1986), WHITE and GEORGOPOULOS (1986), WHITE (1988) and WHITE (1989). A considerable change in the spectrum of activity appeared only with compound F 427 from the Uniroyal Comp. in which the anilide ring is substituted in the o-position by a phenyl group (Fig. 8.1 u). According to EDGINGTON and BARRON (1967) this substitution extended the fungitoxic activity to some, but not all Deuteromycetes (*Aspergillus, Botrytis, Drechslera* and others) and to some Phycomycetes such as *Thamnidium elegans, Cunninghamella echinulata*, whereas Basidiomycetes as *Rhizoctonia solani* and *Polyporus giganteus* remained sensitive as with other oxathiins.

In summary it follows that for the interaction with the receptors the following parts of the molecule are involved:

i) the hydrophobic ring or a long-chain alkyl group at the amide nitrogen
ii) the cis-methyl (or halogen) group in 2-position of the oxathiin ring
iii) the vinylogous carbonyl grouping including both an electrophilic centre at C-2 and a nucleophilic one at the oxygen atom
iv) a weakly nucleophilic amide nitrogen.

The putative essential reactive sites of the carboxin molecule are shown in Fig. 9.2. This "pharmacophor" is compatible with most of the carboxin congeners hitherto described. An exception constitute some compounds with low but sizable inhibition of Complex II activity in which the double bond in α, β-position to the carboxyl group is absent or shifted (WHITE and GEORGOPOULOS 1992); there is however no evidence that these compounds act by the same mechanism of action like carboxin. The pharmacophor

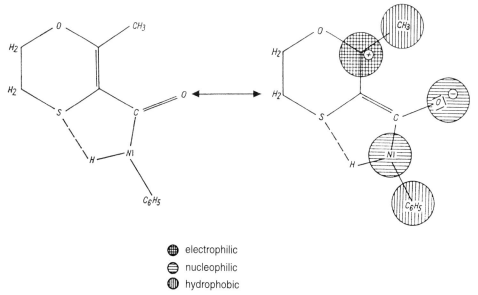

- ⊕ electrophilic
- ⊖ nucleophilic
- ⦿ hydrophobic

Fig. 9.2 Reactive groups of the carboxin molecule presumed to be involved in receptor binding. Cross-hatched: electrophilic centre; horizontally hatched: nucleophilic centres; perpendicularly hatched: hydrophobic interaction sites.

structure depicted in Fig. 9.2 was also the rationale for the receptor model mentioned above. The earlier receptor model for carboxin (SCHEWE et al. 1979) presupposes a multi-centre binding to reduced Fe—S centre S_3 and hydrophobic parts of the polypeptide moieties. One basic assumption is the interaction of the electrophilic carbonyl oxygen with an Fe atom of the Fe—S cluster. This assumption is strongly supported by the fact that thenoyltrifluoroacetone which acts at the same specific binding site (COLES et al. 1978) is known to form such complexes with inorganic ferric iron. The two essential non-polar groups of the carboxin molecule may serve as anchors for the fixation in hydrophobic regions of the receptor. These hydrophobic regions may be formed by QP_S or other small polypeptides of Complex II. The photoaffinity labelling of the small but not of the large polypeptides by a carboxin analogue possessing a azido group at the phenyl residue (WHITE et al. 1983) strongly supports such an assumption. Phospholipids may be excluded with regard to an involvement in the specific receptor binding, since the carboxin-sensitivity of soluble succinate dehydrogenase is conferred solely by the addition of the small polypeptide fraction (VINOGRADOV et al. 1980). On the other hand, a large proportion of unspecific binding of carboxin may involve binding to the phospholipids (COLES et al. 1978; WHITE et al. 1983). One may speculate that the formation of the specific complex with the receptor is preceded by an accumulation of the fungicide in the phospholipid bilayer.

The specificity of carboxins for some groups of fungi may be based on the specific steric assembly of the receptor site. The changed selectivity of F 427 (o-phenyl derivative) indicates that the hydrophobic orientation caused by hydrophobic amino acid sequences in the active centre might be different in Basidiomycetes and Ascomycetes.

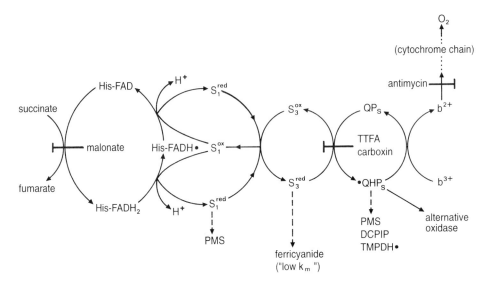

Fig. 9.3 Present knowledge on the electron transfer reactions through the succinate-ubiquinone oxidoreductase complex (Complex II) of the mitochondrial respiratory chain. S_1 and S_3 designate Fe—S clusters. PMS (phenazine methosulfate), DCPIP (2,6-dichlorophenolindophenol), TMPD (N,N,N′,N′-tetramethylphenylenediamine, reduced form of "Wurster's blue") and ferricyanide are artificial electron acceptors; full lines: physiological routes, broken lines: artificial routes; horizontal bars: blockage of electron transfer by inhibitors. Carboxin blocks the one-electron transfer from the reduced, EPR-silent form of Fe—S cluster S_3 to ubiquinone (Q) bound to the specific polypeptide QPs, so that the formation of ubisemiquinone radical (.QH) is prevented and therefore the electron flow through both the cytochrome chain and the alternative oxidase is interrupted.

Carboxin action and alternative pathway of respiration

The observation that many carboxin-sensitive fungi possess an alternative cyanide- and antimycin A − resistant respiratory pathway raises the question whether the two processes are related. Carboxins inhibit the oxidation of succinate via both cytochrome chain and alternative respiratory pathway (SHERALD and SISLER 1972; LYR and SCHEWE 1975). An indispensable role of the alternative respiratory pathway for the action of carboxins must be however excluded since mammalian mitochondria are highly carboxin-sensitive, but do not possess cyanide-insensitive respiration. The nature of the alternative oxidase has been the subject of extensive studies within the last two decades. A spectrally detectable component has not been observed. Therefore it is reasonable to assume that alternative respiration may be due to the autoxidation of an obligatory member of the respiratory chain. In an earlier paper we proposed that the Fe—S centre S_3 may function as "alternative oxidase" (LYR and SCHEWE 1975). Later studies of RICH and BONNER (1978) have indicated a more likely role of ubisemiquinone radicals. As mentioned above their sizable occurrence presupposes the presence of ubiquinone-binding proteins, which prevent spontaneous dismutation. It appears likely that a special type of Q proteins renders the ubisemiquinone radicals susceptible to the direct reaction with dioxygen forming oxidized ubiquinone and superoxide anion radical that is in turn converted by mitochondrial superoxide dismutase to hydrogen peroxide. Inhibitors of the main respiratory pathway such as antimycin A or cyanide trigger this process by enhancing the level of ubisemiquinone radicals, since their oxidation via the cytochrome chain is interrupted.

In this manner the special Q proteins may act as "ubisemiquinone oxidases". It is not yet established whether a particularly modified QP_S itself is the alternative oxidase or whether it is another Q protein. The selective inhibition of the alternative pathway by certain hydroxamates argues in favour of the latter possibility.

The putative "ubisemiquinone oxidase" may be identical with the 36 kDa protein that has been recently demonstrated in *Hansenula anomala* grown in the presence of both antimycin A and o-phenanthroline (MINAGAWA *et al.* 1990); this polypeptide was synthesized in an inactive form and constituted the alternative oxidase upon addition of ferrous iron. The observation that succinate is in many cases a much better substrate for the alternative pathway than NADH or NAD^+-dependent substrates may indicate a direct functional interaction of QP_S and "ubisemiquinone oxidase" without involvement of the Q pool. One has to take into consideration the occurrence of more than one alternative oxidase (AINSWORTH *et al.* 1980). Such a heterogeneity would explain the conflicting results in the literature (DAY *et al.* 1978; ZIOGAS and GEORGOPOULOS 1979; ZIOGAS and GEORGOPOULOS 1984) concerning the action of carboxin and thenoyltrifluoroacetone on the alternative pathway.

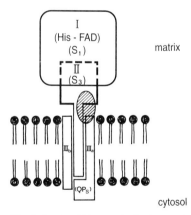

Fig. 9.4 Possible topological assembly of the polypeptides of Complex II in the inner mitochondrial membrane as indicated by photoaffinity labelling experiments and other approaches. The Roman numbers refer to the polypeptide subunits. The two large subunits are located at the matrix surface of the membrane, whereas the small polypeptides including QP_S are imbedded in the phospholipid bilayer. Alkali-treatment of the membrane causes dissociation of soluble succinate dehydrogenase consisting of subunits I and II, which only catalyses the electron transfer to artificial acceptors but not to ubiquinone. Carboxin (hatched ellipse) is intercalated between subunit II and QP_S like a clip. It interacts with the reduced form of Fe—S clusters S_3 residing on subunit II via its electrophilic and nucleophilic reaction centres (see Fig. 9.2) as well as with a hydrophobic region of subunit III a (QP_S) by means of its phenyl group. Owing to this "clip" effect of carboxin the mobility of the subunits II and III a is restricted which in turn causes both inhibition of electron transfer to Q and protection from alkali-induced dissociation of subunit I/II.

Possible mechanisms of side effects

It may be of interest, that carboxin can be oxidized to the inactive sulfoxide (Fig. 8.1 c) by mitochondria of sensitive and insensitive fungi as well as by riboflavins and flavin enzymes. Light strongly increases the conversion. Catalase has a protective effect. Only a small amount of the sulfoxide is further oxidized to the biologically active oxycarboxin (Fig. 8.1 b) (LYR et al. 1974). This implies that carboxin can act as oxygen radical scavenger. One may speculate, that this could be the background of the growth-stimulating effect described by VON SCHMELING and KULKA (1969) and the reason for a reduction of ozone injury (HOFSTRA et al. 1978). For further literature on the carboxins, see the articles by KUHN (1984) and by WHITE and GEORGOPOULOS (1992).

Acknowledgement: We thank Prof. Dr. H. D. SISLER for linguistic corrections.

References

ACKRELL, B. A. C., BALL, M. B., and KEARNEY, E. B.: Peptides from complex II active in reconstitution of succinate-ubiquinone reductase. J. Biol. Chem. **255** (1980): 2761–2769.
- KEARNEY, E. B., COLES, C. J., SINGER, T. P., BEINERT, H., WAN, Y. P., and FOLKERS, K.: Kinetics of the reoxidation of succinate dehydrogenase. Arch. Biochem. Biophys. **182** (1977): 107–117.
- - MIMS, W. B., PEISACH, J., and BEINERT, H.: Iron-sulfur cluster 3 of beef heart succinate-ubiquinone oxidoreductase is a 3-iron cluster. J. Biol. Chem. **259** (1984): 1015–1018.

AINSWORTH, P. J., BALL, A. J. S., and TUSTANOFF, E. R.: Cyanide-resistant respiration in yeast II. Characterization of a cyanide-insensitive NAD (P)H oxidoreductase. Arch. Biochem. Biophys. **202** (1980): 187–200.

BEINERT, H., and ALBRACHT, S. P. J.: New insights, ideas and unanswered questions concerning iron-sulfur clusters in mitochondria. Biochim. Biophys. Acta **683** (1982): 245–277.

BEN-YEPHET, Y., DINOOR, A., and HENIS, Y.: The physiological basis of carboxin sensitivity and tolerance in *Ustilago hordei*. Phytopathology **65** (1975): 936–942.

BROOMFIELD, P. L. E., and HARGREAVES, J.: A single amino-acid change in the iron-sulphur protein subunit of succinate dehydrogenase confers resistance to carboxin in *Ustilago maydis* Curr. Genet. **22** (1992): 117–121.

COLES, CH. J., SINGER, T. P., WHITE, G. A., and THORN, G. D.: Studies on the binding of carboxin analogs to succinate dehydrogenase. J. Biol. Chem. **253** (1978): 5573–5578.

DAY, D. A., ARRON, G. P., and LATIES, G. G.: The effect of carboxins on higher plant mitochondria. FEBS Lett. **85** (1978): 99–102.

EDGINGTON, L. V., and BARRON, G. L.: Fungitoxic spectrum of oxathiin compounds. Phytopathology **57** (1967): 1256.

GEORGOPOULOS, S. G., ALEXANDRI, E., and CHRYSAYI, M.: Genetic evidence for the action of oxathiin and thiazole derivatives on the succinic dehydrogenase system of *Ustilago maydis* mitochondria. J. Bacteriol. **110** (1972): 809–817.
- CHRYSAYI, M., and WHITE, G.: Carboxin resistance in the haploid, the heterozygous diploid, and the plant parasitic dicaryotic phase of *Ustilago maydis*. Pesticide Biochem. Physiol. **5** (1975): 543–551.
- and ZIOGAS, B. N.: A new class of carboxin-resistant mutants of *Ustilago maydis*. Netherl. J. Plant Pathol. **83** (1977): 235–242.

GRIVENNIKOVA, V. G., and VINOGRADOV, A. D.: Interaction of mitochondrial succinate: ubiquinone reductase with thenoyltrifluoracetone and carboxin. Biochimiya **50** (1985): 375–383.

GUNATILLEKE, I. A. U. N., ARST, H. N., and SCAZZOCCHIO, C.: Three genes determine the carboxin sensitivity of mitochondrial succinate oxidation in *Aspergillus nidulans*. Genet. Res., Camb. **26** (1975): 297–305.
HAKEN, P. TEN, and DUNN, C. L.: Structure-activity relationships in a group of carboxanilides systemically active against broad bean rust (*Uromyces fabae*) and wheat rust (*Puccinia recondita*). Proc. 6th Brit. Insecticide Fungicide Conf. (1971): 453–463.
HARDISON, J. R.: Relationships of molecular structure of 1,4-oxathiin fungicides to chemotherapeutic activity against rust and smut fungi in grasses. Phytopathology **61** (1971): 731–735.
HOFSTRA, G., LITTLEJOHNS, A., and WUKASCH, R. T.: The efficacy of the antioxidant ethylenediurea (EDU) compared to carboxin and benomyl in reducing yield losses from ozone in navy bean. Plant Dis. Rep. **62** (1978): 350–352.
HUPPATZ, J. L., PHILLIPS, J. N., and WITRZENS, B.: Structure-activity relationships in a series of fungicidal pyrazole carboxanilides. Agricult. Biol. Chem. **48** (1984): 45–50.
KEON, J. P. R., WHITE, G. A., and HARGREAVES, J. A.: Isolation, characterization and sequence of a gene conferring resistance to the systemic fungicide carboxin from the maize smut pathogen, *Ustilago maydis*. Curr. Genet. **19** (1991): 475–481.
KUHN, P. J.: Mode of action of carboxamides. In: TRINCI, A. P. J., and RYLEY, J. F. (Eds.): Mode of Action of Antifungal Agents. Brit. Mykol. Soc. (1984): 155–183.
LEROUX, P., and BERTHIER, G.: Resistance to carboxin and fenfuram in *Ustilago nuda* (Jens.) Rostr., the causal agent of barley loose smut. Crop Protection //7 (1988): 16–19.
LÖCHER, F., HAMPEL, M., and POMMER, E.-H.: Ergebnisse bei der Rostbekämpfung mit Benodil (BAS 31700 F). Meded. Fac. Landbouwwet. Rijksuniv. Gent **39** (1974): 1079–1089.
LYR, H., LUTHARDT, W., and RITTER, G.: Wirkungsweise von Oxathiin-Derivaten auf die Physiologie sensitiver und insensitiver Hefe-Arten. Z. Allg. Mikrobiol. **11** (1971): 373–385.
— RITTER, G., and CASPERSON, G.: Zum Wirkungsmechanismus des systematischen Fungizids Carboxin. Z. Allg. Mikrobiol. **12** (1972): 271–280.
— — and BANASIAK, L.: Detoxication of carboxin. Z. Allg. Mikrobiol. **14** (1974): 313–320.
— and SCHEWE, T.: On the mechanism of the cyanide-insensitive alternative pathway of respiration in fungi and higher plants and the nature of the alternative terminal oxidase. Acta Biol. Med. Germ. **34** (1975): 1631–1641.
— — MÜLLER, W., and ZANKE, D.: Zum Problem der Selektivität sowie der Struktur-Rezeptorbeziehungen von Carboxin und seinen Analogen. In: LYR, H., and POLTER, C. (Eds.): Systemic Fungicides. Int. Symp. Reinhardsbrunn 1974. Akademie-Verlag, Berlin 1975, pp. 153–166.
MATHRE, D. E.: Uptake and binding of oxathiin systemic fungicides by resistant and sensitive fungi. Phytopathology **58** (1968): 1464–1469.
— Mode of action of oxathiin systemic fungicides. I. Effect of carboxin and oxycarboxin on the general metabolism of several basidiomycetes. Phytopathology **60** (1970): 671–676.
— Mode of action of oxathiin systemic fungicides III. Effect on mitochondrial activities. Pesticide Biochem. Physiol. **2** (1971): 216–224.
MINAGAWA, N., SAKAJO, S., KOMIYAMA, T., and YOSHIMOTO, A.: Essential role of ferrous iron in cyanide-resistant respiration in *Hansenula anomala*. FEBS Lett. **267** (1990): 114–116.
MOWERY, P. C., ACKRELL, B. A. C., SINGER, T. P., WHITE, G. A., and THORN, G. D.: Carboxins, powerful selective inhibitors of succinate oxidation in animal tissues. Biochem. Biophys. Res. Commun. **71** (1976): 354–361.
— STEENKAMP, D. J., ACKRELL, B. A. C., SINGER, T. P., and WHITE, G. A.: Inhibition of mammalian succinate dehydrogenase by carboxins. Arch. Biochem. Biophys. **178** (1977): 495–506.
MÜLLER, W., SCHEWE, T., LYR, H., and ZANKE, D.: Wirkungsmechanismus der Atmungshemmung durch die Systemfungizide der Carboxingruppe. Wirkung von Oxathiinderivaten und -analoga auf nicht phosphorylierende submitochondriale Partikeln aus Rinderherz sowie *Trametes versicolor* und *Trichoderma viride*. Z. Allg. Mikrobiol. **17** (1977): 359–372.

POMMER, E.-H., GIRBENSOHN, B., KÖNIG, K.-H., OSIEKA, H., and ZEEH, B.: Development of new systemic fungicides with carboxanilide structure. Kemia-Kemi **1** (1974): 617–618.

RAGSDALE, N. N., and SISLER, H. D.: Metabolic effects related to fungitoxicity of carboxin. Phytopathology **60** (1970): 1422–1427.

RAMSAY, R. R., ACKRELL, B. A. C., COLES, C. J., SINGER, T. P., WHITE, G. A., and THORN, G. D.: Reaction site of carboxanilides and of thenoyltrifluoroacetone in complex II. Proc. Nation. Acad. Sci. USA **78** (1981): 825–828.

RICH, P. R., and BONNER, W. D.: The nature and location of cyanide and antimycin resistant respiration in higher plants. In: DEGN, H., et al. (Ed.): Functions of Alternative Terminal Oxidases. Vol. 49. Proc. 11th FEBS Meeting Copenhagen. Pergamon Press, Oxford 1978, pp. 149–158.

SCHEWE, T., HIEBSCH, CH., GARCIA PARRA, M., and RAPOPORT, S.: Effect of some respiratory inhibitors on the respiratory enzymes of the mitochondria from the cauliflower (*Brassica oleracea*). Acta Biol. Med. Germ. **32** (1974): 419–426.

– and LYR, H.: Mechanism of action of carboxin fungicides and related compounds. In: LYR, H., (Ed.): Modern Selective Fungicides. Gustav Fischer Verlag, Jena 1987, pp. 133–142.

– MÜLLER, W., LYR, H., and ZANKE, D.: Ein molekulares Rezeptormodell für Carboxin. In: LYR, H., and POLTER, C. (Eds.): Systematic Fungicides. Symposium 1977. Abh. Akad. Wiss. DDR 2N, Berlin, pp. 241–251.

– RAPOPORT, S., BÖHME, G., and KUNZ, W.: Zum Angriffspunkt des Systemfungicids Carboxin in der Atmungskette. Acta Biol. Med. Germ. **31** (1973): 73–86.

SCHMELING, B. V., and KULKA, M.: Regulation of plant growth. Can. Patent 828, 771 (1969); U.S. Patent 3,454,391 (1969).

SHERALD, J. L., and SISLER, H. D.: Selective inhibition of antimycin A-insensitive respiration in *Ustilago maydis* and *Ceratocystis ulmi*. Plant and Cell Physiol. **13** (1972): 1039–1052.

SNEL, M., SCHMELING, B. V., and EDGINGTON, L. V.: Fungitoxicity and structur-activity relationships of some oxathiin and thiazole derivatives. Phytopathology **60** (1970): 1164–1169.

TUCKER, A. N., and LILLICH, T. T.: Effect of the systemic fungicide carboxin on electron transport function in membranes of *Micrococcus denitrificans*. Antimicrob. Agents and Chemotherapy **6** (1974): 522.

TUSHURASHVILI, P. R., GAVRIKOVA, E. V., LEDENEV, A. N., and VINOGRADOV, A. D.: Studies on the succinate dehydrogenating system. Isolation and properties of the mitochondrial succinate-ubiquinone reductase. Biochim. Biophys. Acta **809** (1985): 145–159.

ULRICH, J. T., and MATHRE, D. E.: Mode of action of oxathiin systemic fungicides. V. Effect on electron transport of *Ustilago maydis* and *Saccharomyces cerevisiae*. J. Bacteriol. **110** (1972): 628–632.

VINOGRADOV, A., GAVRIKOV, V. G., and GAVRIKOVA, E. V.: Studies on the succinate dehydrogenating system II. Reconstitution of succinate-ubiquinone reductase from the soluble components. Biochim. Biophys. Acta **592** (1980): 13–27.

WHITE, G. A.: A potent effect of 1,4-oxathiin systemic fungicides on succinate oxidation by a particulate preparation from *Ustilago maydis*. Biochem. Biophys. Res. Commun. **44** (1971): 1212–1219.

– Furan carboxamide fungicides: structure-activity relationships with the succinate dehydrogenase complex in mitochondria from a wild-type strain and a carboxin-resistant mutant strain of *Ustilago maydis*. Pestic. Biochem. Physiol. **31** (1988): 129 ff.

– Substituted 2-methylbenzanilides and structurally related carboxamides: inhibition of complex II in mitochondria from a wild-type strain and a carboxin-resistant mutant strain of *Ustilago maydis*. Pestic. Biochem. Physiol. **34** (1989): 255 ff.

– and GEORGOPOULOS, S. G.: Thiophene carboxamide fungicides structure-activity relationships with the succinate dehydrogenase complex from wild-type and carboxin-resistant mutant strains of *Aspergillus nidulans*. Pestic. Biochem. Physiol. **25** (1986): 188 ff.

– – Target sites of carboxamides. In: KÖLLER, D. (Ed.): Target Sites of Fungicide Action. CRC Press Inc., Boca Raton Florida 1992, pp. 1–25.
– PHILIPPS, J. N., HUPPATZ, J. L., WITRZENS, B., and GRANT, S. J.: Pyrazole carboxanilide fungicides. I. Correlation of mitochondrial electron transport inhibition and antifungal activity. Pestic. Biochem. Physiol. **25** (1986): 163 ff.
– and THORN, G. D.: Structure-Activity relationships of carboxamide fungicides and the succinic dehydrogenase complex of *Crytococcus laurentii* and *Ustilago maydis*. Pesticide. Biochem. and Physiol. **5** (1975): 380–395.
– – Thiophene carboxamide fungicides: structure-activity relationships with the succinate dehydrogenase complex from wild-type and carboxin-resistant mutant strains of *Ustilago maydis*. Pestic. Biochem. Physiol. **14** (1980): 26–40.
– – and GEORGOPOULOS, S. G.: Oxathiin carboxamides highly active against carboxin-resistant succinic dehydrogenase complexes from carboxin-selected mutants of *Ustilago maydis* and *Aspergillus nidulans*. Pesticide Biochem. Physiol. **9** (1978): 165–182.
– – ACKRELL, B. A. C., KEARNEY, E. B., RAMSEY, R. R., and SINGER, T. P.: Site of action of carboxamides in mitochondrial complex II. In: MIYAMOTO, J., *et al.* (Ed.): IUPAC Pesticide Chemistry. Pergamon Press, Oxford 1983, pp. 141–146.
YU, CH., and YU, L.: Ubiquinone-binding proteins. Biochim. Biophys. Acta **639** (1981): 99–128.
ZEEH, B., LINHART, F., and POMMER, E.-H.: Nicotinic anilides. German Offenlegungsschrift 26 11 601 (1977): 1–17.
ZIOGAS, B. N., and GEORGOPOULOS, S. G.: The effect of carboxin and of thenoyltrifluoroacetone on cyanide-sensitive and cyanide-resistant respiration of *Ustilago maydis* mitochondria. Pesticide Biochem. Physiol. **11** (1979): 208–217.
– – Mitochondrial electron transport in a carboxin-resistant, antimycin A-sensitive mutant of *Ustilago maydis*. Pesticide Biochem. Physiol. **22** (1984): 24–31.

Chapter 10

Morpholine fungicides and related compounds

E.-H. POMMER

Limburgerhof, FRG

Introduction

In 1965, KÖNIG, POMMER and SANNE reported the fungicidal activity of N-substituted tetrahydro-1,4-oxazines or morpholine derivatives. They were able to demonstrate that morpholine derivatives with major N-positioned cyclo-aliphatic ot long-chain alkyl groups exhibited good to very good fungicidal activity, particularly against powdery mildew fungi. As a result of the systematic approach to this new class of compounds, the active compounds dodemorph (POMMER and KRADEL 1967a) and tridemorph (KRADEL et al. 1968) were produced and introduced into the market in 1965 and 1969 respectively. Whereas dodemorph found favour in the control of powdery mildew fungi in ornamental plant cultivation (KRADEL and POMMER 1967a, b), tridemorph gained importance in the first instance as a fungicide for the control of barley mildew, which had become a serious

Table 10.1 Development of morpholine and piperidine derivatives as fungicides for commercial use

Common name Code number	Tradename at the time of introduction Formulation	Year of introduction (by)	Main area of application
Dodemorph BAS 238 ... F	BASF-Mehltaumittel EC 400 g/l	1965 (BASF)	Powdery mildew on ornamentals
Tridemorph BAS 200 ... F 220 ... F	Calixin Ed 750 g/l	1969 (BASF)	Powdery mildew on cereals (barley); sigatoka disease (bananas)
Aldimorph	Falimorph EC 662 g/l	1980 (Fahlberg-List)	Powdery mildew on cereals (barley)
Fenpropimorph BAS 421 ... F Ro 14-3169	Corbel/Mistral EC 750 g/l	1980 (BASF/Maag)	Powdery mildew and rust on cereals (wheat and barley)
Fenpropidine Ro 12-3049	Patrol/Cetros EC 750 g/l	1986 (ICI/Maag)	Powdery mildew and rust on cereals
Trimorphamide VUAgT-866 BAS 463 ... F	Fademorf EC 200 g/l	1981 (Vysk. Ustav Agrochem. Technol.)	Powdery mildew on grapes and apples

Table 10.2 Morpholine and piperidine derivatives in commercial use as fungicides

Structure	Chemical name	Patent No.	LD_{50} rat p.o. mg/kg
Dodemorph — $(CH_2)_{11}$ CH–N (2,6-dimethylmorpholine ring)	4-cyclododecyl-2,6-dimethyl-morpholine	DE 1 198 125	1 800
Tridemorph — $CH_3-(CH_2)_{12}-N$ (2,6-dimethylmorpholine ring)	4-tridecyl-2,6-dimethyl-morpholine	DE 1 164 152	860
Aldimorph — $CH_3-(CH_2)_{11}-N$ (2,5/2,6-dimethylmorpholine ring)	N-n-dodecyl-2,5/2,6-dimethyl-morpholine	DD 140 412	3 500
Fenpropimorph — $(CH_3)_3C$–⟨phenyl⟩–$CH_2-CH(CH_3)-CH_2-N$ (2,6-dimethylmorpholine)	cis-4-3-(4-tert-butyl-phenyl)-2-methylpropyl-2,6-dimethyl-morpholine	DE 2 656 747 DE 2 752 096	3 515
Fenpropidine — $(CH_3)_3C$–⟨phenyl⟩–$CH_2-CH(CH_3)-CH_2-N$ (piperidine)	1-[3-(p-tert-butylphenyl)-methylpropyl]	DE 2 752 135	1 800
Trimorphamide — morpholine-$N-CH(CCl_3)-N(H)-CH=O$	N-2,2,2-tri-chloro-1-(4-morpholinyl)-ethyl-formamide	CS 193 659	2 820

problem in barley growing in Western Europe at the end of the 1950's. Over the past few years this compound has also been increasingly used for other crops, such as bananas and rubber. The active compound aldimorph, which belongs to the same class of substances as dodemorph and tridemorph, was introduced by JUMAR and LEHMANN in 1982 as a cereal mildew fungicide.

A substantial improvement in the field of morpholine derivatives with a fungicidal activity was achieved through the study of substances on the basic of arylalkylamines. The result of these studies was the active compound fenpropimorph; POMMER and HIMMELE (1979) and BOHNEN and PFIFFNER (1979) reported its fungicidal properties and possibilities of its use in the control of cereal mildew and rust fungi.

HIMMELE and POMMER (1980) demonstrated that the composition and structure of the 3-(p-tert.-butylphenyl)-2-methylpropyl group is the "optimum arylalkyl residue". Like fenpropimorph, cycloamines with this substituted on the nitrogen generally have more or less good activity against powdery mildew and rust fungi in cereals. BOHNEN *et al.* in 1986 presented a new fungicide which had a piperidine ring as the cycloamine component: fenpropidine. On account of its relationship with the morpholine fungicides and its comparable mode of action ("two-site" ergosterol inhibitor fungicide; see Chapter 11) and also its spectrum of fungicidal activity, fenpropidine is included in this chapter. Consistent with the term "morpholine derivative", fenpropidine is described as a piperidine derivative.

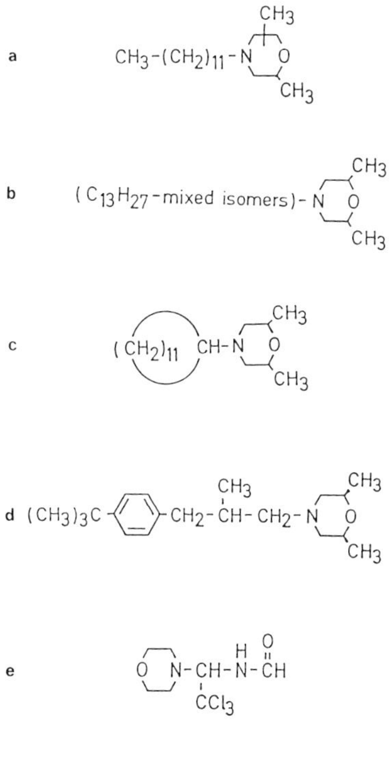

a $CH_3-(CH_2)_{11}-N\begin{smallmatrix}CH_3\\O\\CH_3\end{smallmatrix}$

b $(C_{13}H_{27}\text{-mixed isomers})-N\begin{smallmatrix}CH_3\\O\\CH_3\end{smallmatrix}$

c $(CH_2)_{11}\ CH-N\begin{smallmatrix}CH_3\\O\\CH_3\end{smallmatrix}$

d $(CH_3)_3C-\bigcirc-CH_2-\underset{CH_3}{CH}-CH_2-N\begin{smallmatrix}CH_3\\O\\CH_3\end{smallmatrix}$

e $O\ N-\underset{CCl_3}{CH}-\overset{H}{N}-\overset{O}{CH}$

f $(CH_3)_3C-\bigcirc-CH_2-\underset{CH_3}{CH}-CH_2-N\bigcirc$

Fig. 10.1 Structure formulae of morpholine fungicides and Fenpropidine. a) Aldimorph, b) Tridemorph, c) Dodemorph, d) Fenpropimorph, e) Trimorphamide, f) Fenpropidine.

The active compound trimorphamide, which is mainly active against powdery mildew fungi (DEMECKO and JARAS 1975), also belongs to the N-substituted tetrahydro-1,4-oxazines.

The morpholine and piperidine derivatives which have been introduced as commercial products are listet in Table 10.1.

The structural formulae (Fig. 10.1) and the chemical names of these six active compounds, together with their patent numbers and the LD_{50} values for rat are given in Table 10.2.

Uptake and transport in the plant

The fist investigations on the systemic properties of morpholine derivatives were carried out with dodemorph (KRADEL and POMMER 1967c) Soil treatment experiments made using emulsions containing 250 ppm active ingredients showed that barley plants, artificially inoculated with *Erysiphe graminis,* could be kept free of infection over a period of 16 days. Barley plants readily take up tridemorph *via* their roots and transport this acropetally with the transpiration stream into the leaves. It was possible to measure 7.2 ppm of this substance in leaf extracts of plants 4 h after their roots had been submerged in a nutrient solution containing 50 ppm tridemorph; after 48 h an equilibrium had been established.

In comparison with the root uptake, absorption of tridemorph by the leaf surface occurs more slowly. After the active substances has penetrated the leaf, however, further transport within the tissue occurs relatively rapidly (POMMER *et al.* 1969). Similr results were obtained by OTTO and POMMER (1973) in experiments with ^{14}C-labelled tridemorph. When 5 ml of a suspension which contained 0.15% active ingredient were pipetted into the leaf sheats of barley seedlings at the 1-leaf growth stage, acropetal transport of the active ingredient could be detected in an autoradiogram as soon as 8 h after application.

Due to the predominantly upward transport of the compound, a certain degree of impoverishment of the compound in the lower leaf parts was observed 54 h after application. When the labelled active compound was applied to a barley plant in the 1-node growth stage into a leaf sheath above this node, then within 2 days transport of the active compound into all leaves above the node could be detected, plant parts which developed later did not certain any radioactivity; basipetal transport does no occur. In experiments with cucumber plants the same authors found that labelled tridemorph applied to the leaves was distributed in the entire growing plant. Here, the tendency of the active compound to accumulate in the marginal zones of the leaves could prove to be disadvantageous if as a result the concentration of the active compound in the leaf centres were too greatly reduced. In bananas an equilibrium in concentration between treated and non-treated leaves occurs very slowly due to the low rate at which the active compound is absorbed through the leaf surfaces. In contrast to barley, in wheat tridemporph is taken up better by the morphological upper leaf surface than by the lower (POMMER and KRADEL 1971). Since the wheat upper leaf surface are considerably more densely covered with stomata than the lower leaf surface, the stomata may possibly be regarded as important gates of entry for tridemorph. The fact that the leaf surfaces are more densely covered with fine hairs may play a role in the lower uptake of the active compound, since the small air bubbles trapped between the hairs may greatly impede the wetting of the epidermical cells by tridemorph. DEYMANN (1981) observed that wheat plants take up lower quantities of tridemorph than barley plants. The degradation of the actice compound is accelerated during the first week following treatment. With fenpropimorph no statistically reliable differences have been observed between barley and wheat plants. After treatment with equal quantities of active compounds, the amount of fenpropimorph found in barley and wheat plants is greater than that of tridemorph. Furthermore, the author found that the distribution of the active compound varied depending upon the substance used and the growth stage of the plant. Treatment with tridemorph led in younger and older plants to an accumulation of the active substance in the leaf tips such that the lower parts of the leaves became impoverished (cf. OTTO and POMMER (1973). In fenpropimorph treatments it was found that plants at the heading growth stage showed prolonged retention of the active compound, although here, too, the active compound accumulated in the leaf tips. Uptake tridemorph and fenpropimorph is lower in older cereal plants than in younger ones, and their degradation occurs more slowly. No basipetal transport of fenpropimorph has been detected (POMMER and HIMMELE 1979; BOHNEN and PFIFFNER 1979; DEYMANN 1981).

Fenpropimorph is characterized by a pronounced vapour phase which can lead to secondary distribution of the active compound, as has been demonstrated in glasshouse experiments with *Erysiphe graminis* on infected wheat and barley plants (BOHNEN and PFIFFNER 1979; POMMER and HIMMELE 1979). When infected, untreated wheat and barley plants were placed together with plants treated with fenpropimorph, the infection could be stopped at a distance of up to one metre between the plants. According to DEYMANN (1981), the activity of fenpropimorph *via* the vapour phase does not play an important role in practice. JUMAR and LEHMANN (1982), working with *Hordeum distichon* and *Sinapis alba*, established that relatively high quantities of aldimorph were taken up by the plants both after leaf as well as root application. According to these authors, the non-metabolized compound shows only a low rate of translocation. However, the fungicidally active metabolites are more mobile and are responsible for the systemic activity. Experimental studies with ^{14}C-labelled aldimorph revealed comparable findings (DOBE et al. 1986).

BOHNEN et al. (1986) ascertained that fenpropidine is rapidly aborded by the roots and, to a minimal degree, by the rest of the plant. Acropetal translocation in the plant occurs in the xylem with the transpiration flow. No basipetal transport of fenpropidine was observed. KAESBOHRER and NAU (1992) reported a rapid uptake by the green parts of plants, followed by acropetal and translaminar distribution of the fungicide.

Despite of repeated inoculation with *Erysiphe cichoracearum*, cucumber plants were able to be kept completely free of infection over a period of 29 days following initial soil treatment with 250 mg trimorphamide (HUDECOVA et al. 1978). After 41 days the fungicide still provided 40% protection.

Influence of morpholine and piperidine derivatives on pathogenesis

This section deals with the influence of morpholine and piperidine derivatives on the pathogenesis of plant pathogenic fungi. Their mode of action will be described in Chapter 11.

SCHLÜTER and WELTZIEN (1971) investigated the effect of dodemorph and tridemorph on the pathogenesis of barley mildew by means of cytological and cytochemical methods. In these studies they observed only incomplete haustoria formation, i. e. the growth of the finger-shaped outgrowths was impeded. At the same time, the authors found an irreversible inhibition of sporulation. As an inhibitor or haustoria development, fenpropimorph interferes relatively late in the development of an infection; it also exhibits a directly damaging and germination-inhibiting effect on conidia of *Erysiphe graminis* on barley and wheat. In this respect fenpropimorph is more active than tridemorph (DEYMANN and WELTZIEN 1980).

BOHNEN and PFIFFNER (1979) were able to demonstrate with barley mildew that 30 – 36 hours after a curative application of fenpropimorph the young hyphae growing out from the periphery of existing infection centres grew in a hump-backed fashion away from the epidermis. The tips of these hyphae ballooned to 2 – 5 times their original size, finally collapsing and drying up. The production of conidia was interrupted. The mycelium which had grown prior to the treatment, and also the conidia which had already been formed, lost their turgor and dies. SAUR and LÖCHER (1984) established that fenpropimorph had a good effect in stopping *Erysiphe graminis* on wheat at low as well as at high temperatures.

ZOBRIST et al. (1982) investigated the morphological and microstructural effects of fenpropimorph on the germination of uredospores, and also the formation of appressoria and uredospores of oat crown rust. Following applications of 750 ppm fenpropimorph to the leaves, the germination of uredospores and the infection process proceeded in the same way as on untreated plants. However, the pathogen died after penetrating the leaf tissue before uredospores could be formed. After treatment with 300 ppm active compound, *Puccinia coronata* developed up to the formation of uredospores. However, the spores were for the most part collapsed or had developed only rudimentary cell walls.

Fluorescence microscopy investigations of the effect of fenpropimorph on *Ustilago maydis* and *Penicillium italicum* (KERKENAAR and BARUG 1984) indicated that this substance causes morphological changes in the sporidia of *Ustilago maydis* and in the germinating conidia of *Penicillium italicum*. The sporidia of *Ustilago maydis* appeared swollen, distorted, multicellular and sometimes branched; conidia of *Penicillium italicum* swelled in size and the extension of germ tubes was strongly inhibited. The mycelium of *Penicillium italicum* also exhibited a much enlarged hyphal diameter and relatively short distance between the septae.

In glasshouse trials, BOHNEN et al. (1966) demonstrated after preventive fenpropidine application of 50 mg/l to combat powdery mildew on barley that the germination of conidia was practically unaffected. However, the development of appressoria was moderately checked, and the formation of "secondary hyphae" was completely suppressed 24 to 32 hours after germination. After curative applications of 50 mg/l to combat powdery mildew on barley, it was observed that hyphae at the periphery of the infection site rose from the epidermis and that this was followed by swelling and, subsequently, collapse. Mycelium, conidia and basal cells which had developed before the application of fenpropidine lost their turgor, collapsed, and withered.

The behaviour of morpholine and piperidine derivatives in soil and plants

In 1973, OTTO and DRESCHER investigated the behaviour of tridemorph in the soil using tridemorph which had been radioactively labelled on the C-atoms in positions 2 and 6 of the morpholine ring. They were able to demonstrate that tridemorph is strongly adsorbed onto the soil particles (loamy sand and peaty soil), so that leaching into deeper soil layers and subsequent contamination of the surface water is prevented. In addition to this, they were able to establish that tridemorph is not persistent in the soil, and is biologically degraded. The initial step in this process is the formation of tridemorph-N-oxide; 2,6-dimethylmorpholine and CO_2 have been detected as further degradation products.

The persistence of aldimorph in the soil was established by using ^{14}C-labelled active compound (SIEBER et al. 1981). In laboratoy experiments, after the addition of 20.6 ppm active compound to two experimental soils, the half-life values of aldimorph were 18 and 26 days. Ivestigations carried out under field conditions also produced favourable results. Under these conditions it was possible to establish a half-life value of 10 days following irrigation. Aldimorph showed a great affinity for the soil. After irrigation with 200 ml precipitation, the active compound was retained in the upper 5 cm layer of soil.

Fenpropimorph does not persist in soil; it is degraded by oxidation of the tertiary butyl group. Furthermore, there is oxidation and opening of the dimethylmorphopline ring. The half-life was found to be 15 days in moderately humus loamy sand, and up to 93 days in a very humus loamy sand. Leaching the degradation of fenpropimorph in plants is similar to that in soil (Anon. 1991), leaching into the soil does not occur (Anon. 1982).

Investigations into the effect of fenpropimorph on the soil microflora revealed that the addition of 10 and 100 ppm active compound to a humus sandy soil and a loamy soil had no influence on the total number of viable cells, their respiration, various enzyme activities, or nitrification (POMMER and HIMMELE 1979).

The metabolism of aldimorph in treated barley plants starts relatively quickly. During this process, polar water-soluble metabolites are mainly found, which contain the intact morpholine ring (JUMAR and LEHMANN 1982). Experimental results from these authors indicate that primary hydroxylation of the terminal methyl group of the alkyl side-chain or of one of the methyl groups attached to the ring probably takes place, followed by conjugation with glucose. In addition to this, it is very probable that N-carboxymethyl-2,6-dimethylmorpholine is formed by beta-oxidation.

Fenpropidine degraded extensively in soil; this was proved by the production of $^{14}CO_2$. The compound itself and its metabolites had little or no tendency to leach (BOHNEN et al. 1986).

Extensive degradation in plants takes place relatively rapidly. In wheat the principal pathway involves hydroxylation on the piperidine ring and oxidation of tertiary butyl group. The half-life in wheat and barley plants is about 4 to 11 days (Anon. 1991).

Experiments carried out with radioactively labelled fenpropimorph in wheat revealed that the active compound is extensively degraded in the plant and metabolised largely to four identified degradation products; the half-life value was found to be approximately 5 days. In the grain, the main part of the radioactivity is built into starch, indicating that the degradation of the active compound is complete (Anon. 1984).

The use of morpholine, piperidine and fenpropidine derivatives in agriculture

When describing the use of morpholine derivatives in agriculture, it must be borne in mind that tridemorph, for example, is a fungicide that has been known for many years, and consequently there is a large amount of information available on the use of this compound. In contrast, aldimorph, fenpropidine and trimorphamide are relatively young products, so that to date there is only a limited amount of literature available. Fenpropimorph is predominantly used in cereal crops, so that experiments with this compound have been concentrated on its use in these crops. Finally, dodemorph has established itself as a fungicide for the control of powdery mildew fungi in ornamental plants; consequently, the literature available describing experiences with the use of this compound is not extensive.

Dodemorph

In commercial products (as EC formulations), the active compound dodemorph is not used in its free base form but as acetate. The two cis- and trans-stereoisomers of the 4-cyclodecyl-2,6-dimethylmorpholine are present in an approximately equal ratio (KRADEL and POMMER 1967a). Using barley and wheat powdery mildew, it could be demonstrated that the two stereoismers applied separately or as a cis-, trans-stereoisomeric mixture were equally effective (POMMER 1984). KRADEL et al. (1969) reported experiments conducted over a period of several years in which dodemorph (cyclomorph had been originally proposed as the common name) was compared with tridemorph for the control of barley mildew. Similar results were obtained by using application rates of 1.0–1.5 kg dodemorph/ha and 0.4–0.6 kg tridemorph/ha. Since dodemorph exhibits a good degree of activity against mildew, but also because it is sufficiently well tolerated by a number of crops, even after repeated application, this fungicide was developed for the use in ornamental plant cultivation (KRADEL and POMMER 1967a; KRADEL et al. 1969; KRADEL et al. 1970). A favoured area of application are roses (FROST and PATTISSEN 1971). In the control of *Podosphaera leucotricha* on apples severe fruit russetting developed in the sensitive varieties Cox's Orange Pippin and Golden Delicious; the activity against mildew was equal to that of dinocap. Good results against *Erysiphe cichoracearum* were achieved on cucumbers and pumkins.

Occassionally a yellowing of the leaf margins was observed when the compound was applied under glass (KRADEL et al. 1967).

In order to extend the range of activity and also to intensify its efficacy against mildew, the following mixtures with dodemorph have been developed into commercial products:

    dodemorph + dodine
    dodemorph + nitrothal-isopropyl

Aldimorph

The active compound aldimorph is formulated as an emulsifiable concentrate and its recommended for the control of *Erysiphe graminis* in spring and winter barley. SIEBER et al. (1981) recommended 0.8 kg a.i./ha as the application rate to be used. On account of its low phytotoxicity, this compound can also be applied to other sensitive crops, such as wheat. According to JUMAR and LEHMANN (1982), the active compound aldimorph is not a uniform substance, but consists chiefly of a mixture of 4-n-dodecyl-2,6-dimethylmorpholine (cis-, trans-mixture) and a smaller proportion of 4-n-dodecyl-2,5-dimethylmorpholine (also a cis-, trans-mixture), which act synergistically.

Tridemorph

The active compound tridemorph is not a uniform; it is a reaction mixture of $C_{11}-C_{14}$-4-alkyl-2,6-dimethylmorpholine (cis-, trans-mixture) homologues containing 60–70% 4-tridecyl-isomers and 26% 4-dodecyl isomers (Anonym 1984). Tridemorph was initially developed in particular for the control of barley and wheat mildew (POMMER and KRADEL 1967a; KRADEL et al. 1970; KRADEL and POMMER 1971). The rates of application lie between 375 and 560 g a.i./ha cereal. Tridemorph is distinguished by its plant tolerance in barley. Wheat exhibits varying reactions to tridemorph; SIDDIQUI and HAAHR (1971) established genetically based differences in the sensitivity of *Triticum aestivum* mutants. HAMPEL and LANG (1972) pointed out that with strong sunshine leaf burning occasionally occurs, which, however, does not cause any damage. In order to be able to investigate the tolerance of winter wheat varieties to tridemorph, BERGMANN et al. (1983) developed a test using 2,3,5-triphenyl tetrazolium chloride (TTC) to determine the phytotoxic effect. It was thus possible to detect a clear dependence on the concentration. In agreement with investigations on entire plants in the glasshouse, an increasing phytotoxic effect in the following order was observed: barley < wheat < cucumber. Tridemorph exhibits a wide range of activity against pathogens belonging to the classes Ascomycetes and Basidiomycetes as well as out of the group of the Deuteromycetes.

In addition to cereal mildew, positive results are available for the control of the following fungi (Tab. 10.3).

When tridemorph is used for the control of *Mycosphaerella* spp. causing leaf diseases in bananas, spray oil is added to the fungicide mixture in order to increase the penetration into the leaf tissue and to reduce the rat of vaporization of the active compound (HAMM 1983).

TRAN VAN CANH (1984) reported a new method for the direct control of *Rigidoporus lignosus*, the causal organism of white root disease in *Hevea* (rubber tree), with tridemorph. By applying 10 g a.i. tridemorph (formulated as EC) per tree every 6 months, *Hevea* is well protected against an infection by *Rigidoporus*. In trials over three years the drenching has proved to be very effective. MAPPES and HIEPKO (1984) reported good results achieved against *Rigidoporus lignosus*, *Phellinus noxius* and *Ganoderma* sp. with a tridemorph-bitumen formulation which had been applied to exposed infected roots.

A treatment of tea leaves with a fungicide emulsion containing 0.11% tridemorph inhibited sporulation of *Exobasidium vexans*; at a concentration of 0.375% it eradicated the fungal infection in situ. In the field, satisfactory disease control that was achieved with tridemorph at 225 and 420 g/ha under mild and moderate rain fall conditions broke down under severe conditions (VENKATA RAM 1974).

The pink disease of *Hevea*, caused by *Corticium salmonicolor*, could be effectively controlled under field conditions with a natural rubber latex formulation which contained 2% tridemorph (YEOH and TAN 1974).

LIM (1976) described a low volume procedure for the control of *Oidium secondary* leaf fall in *Hevea* with a formulation of tridemorph plus a refined white oil. Stem canker and black scurf of potato, caused by *Rhizoctonia solani*, was controlled by misting with 20 g tridemorph/l (CHAND et al. 1981).

The following combinations of tridemorph with other fungicides, to extend the range of activity on in connection with antiresistance strategy, are on the market:

tridemorph + maneb
tridemorph + carbendazim + maneb
tridemorph + epiconazole
tridemorph + propiconazole
tridemorph + triadimenol
tridemorph + anilazine
tridemorph + chlorothalonil + propiconazole

Table 10.3 In addition to cereal mildew, positive results in the control of the following fungi are available

Pathogen	Disease	Crop	Source
Erysiphe cichoracearum *Spaerotheca fuliginea*	Powdery mildew	Cucurbits	
Erysiphe cichoracearum	Powdery mildew	Tobacco	
Erysiphe polygoni	Powdery mildew	Legumes	MUNSHI (1983)
Erysiphe betae	Powdery mildew	Sugar beet	MEEUS (1977)
Mycosphaerella musicola	Yellow sigatoka	Banana	HAMPEL and POMMER (1971)
Mycosphaerella fijiensis var. *difformis*	Black sigatoka	Banana	MAPPES (1979)
Mycosphaerella fijiensis	Black leaf streak	Banana	MAPPES (1979)
Cladosporium musae	Leaf speckle	Banana	MAPPES (1979)
Corticium salmonicolor	Pink disease	Hevea (rubber)	YEOH and TAN (1974)
Ceratocystis fimbriata	Moldy rot	Hevea	
Oidium hevea	Powdery mildew	Hevea	LIM (1976)
Fomes lignosus (*Rigidoporus lignosus*)	White root disease	Hevea	TRAN VAN CANH (1984)
Fomes noxius (*Phellinus noxius*)	Stem rot	Hevea	TRAN VAN CANH (1984)
Ganoderma philippii	Rod root disease	Hevea	TRAN VAN CANH (1984)
Geotrichum candidum	Sour rot	Citrus	
Exobasidium vexans	Blister blight	Tea	VENKATA RAM (1974)
Oidium mangiferae	Powdery mildew	Mango	

Side effects of tridemorph

In laboratory experiments tridemorph emulsion was toxic to the pink mite (*Acaphylla theae*), purple mite (*Calacarus carinatus*) and scarlet mite (*Brevipalpus australis*) on tea leaves (CHANDRASEKRAN 1980).

Field applications of tridemorph had no effect on the aphid fungal pathogen *Entomophthora* on wheat (ZIMMERMANN and BASEDOW 1980).

ZIMMERMANN (1975) carried out in vitro experiments to study the effect of systemic fungicides on the germination of spored and mycelial growth of beneficial entomopathogenic Fungi imperfecti *Beauveria bassania, B. tenella, Metarrhizium anisopliae, Paecilomyces farinosus* and *P. fumosoroseus*. At a concentration of 0.22% tridemorph he observed an impairment of mycelial growth.

The antibacterial activity of tridemorph was detected against a wide spectrum of bacterial species (OROS and SULE 1983) *Corynebacterium* species were found to be the most sensitive (MIC values = 10-100 ppm), whereas *Agrobacterium* and *Erwinia* species were generally the most resistant (MIC values = >1,000 ppm). Electrolyte studies have indicated that irreversible membrane damage is involved in the sensitivity of bacteria to tridemorph.

Tridemorph does not negatively effect the symbiotic N-fixation by *Rhizobium trifolii* in root nodules of white clover (FISHER 1976). In a further study, FISHER (1981), established that in the presence of higher tridemorph concentrations in the soil, the formation of root nodules in white clover was reduced. However, after application of the recommended rates of tridemorph under field conditions no influence on the N-fixation is to be expected.

The treatment of winter wheat with a mixture of compounds, which contained, amongst other, 750 g tridemorph/ha, had no influence on the decomposition of the straw. In laboratory experiments to investigate the destruction of cellulose, SCHROEDER (1979) showed that tridemorph influences the rate of degradation.

Fenpropimorph

It was possible at a very early stage in the development of fenpropimorph to establish that the configuration characteristics of the 2,6-dimethylmorpholine component are important for the fungicidal activity. Of the two stereoisomeric forms, the cis-dimethylmorpholine compound exhibited considerably better activity agains *Puccinia recondita* and *Erysiphe graminis* on wheat than the corresponding trans-form (POMMER and HIMMELE 1979; HIMMELE and POMMER 1980). The marked activity of fenpropimorph against mildew and rust fungi in cereals, in addition to its good compatibility in wheat have been confirmed in field trials (BOHNEN and PFIFFNER 1979; HAMPEL et al. 1979; SAUR et al. 1979). Under laboratory and glasshouse conditions (BOHNEN et al. 1979) determined the range of activity indicated in Table 10.4.

Spectrum of activity

Strong fungicidal side effects against species of *Uromyces, Hemileia, Rhizoctonia, Pyricularia* and *Rhynchosporium* have been detected. To control these and other pathogens the concentrations given in Table 10.4 and 10.5 are necessary under laboratory conditions.

In seed dressing tests with cabbage seed naturally infected with *Alternaria brassica* and *Phoma lingam,* fenpropimorph showed at an application rate of 2.5 g a.i./kg seed an activity comparable to that of iprodione (MAUDE et al. 1984).

Both seed treatments were highly effective and produced healthy transplants and later healthy crops.

Scheffer *et al.* (1988) carried out trials to control Dutch elm disease (*Ophiostoma ulmi*) by injections of fenpropimorph phosphate and fenpropimorph sulphate at rates of only 7.5 or 10 g per tree. There was complete suppression of the disease within two years. The compound was transported throughout the tree and into the new annual ring.

Table 10.4 Main activities under laboratory, glasshouse or climate chamber conditions (Bohnen *et al.* 1979)

Pathogen	Host plant	ED_{75} values (MIC)
Foliar treatment		mg (a.i.)/l spray
Erysiphe graminis	barley	4
Erysiphe graminis	wheat	70
Puccinia triticina	wheat	300
Puccinia dispersa	rye	200
Puccinia graminis	wheat	250
Puccinia coronata	oat	75
Puccinia sorghi	maize	400
Puccinia striiformis	wheat	300
Uromyces spp.	bean	275
Seed treatment		mg (a.i.)/kg seed
Tilletia tritici	wheat	250
Ustilago avenae	oat	400
Pyrenophora graminea	barley	350
Rhizoctonia solani	cotton	400

Table 10.5 Fungicidal side effects under laboratory, glasshouse or climate chamber conditions (Bohnen *et al.* 1979)

Pathogen	Host plant	ED_{75} values (MIC) in mg a.i./l spray mixtures, kg seed or agar culture-medium
Foliar treatment		*in vivo*
Cercospora arachidicola	groundnut	400
Sphaerotheca pannosa	roses	200
Uncinula necator	vine	100
Septoria nodorum	wheat	~1 000
Agar culture medium		*in vitro*
Poria vaporaria		0.1
Lenzites trabea		0.2
Ceratocystis ulmi		0.2
Ustilago nuda		1
Septoria avenae		2
Ustilago maydis		3

Fenpropimorph, formulated as an emulsifiable concentrate, is currently primarily used for the control of cereal diseases caused by *Erysiphe graminis, Puccinia* species and *Rhynchosporium secalis*; the rate of application is 750 g a.i./ha (HAMPEL et al. 1979; ATKIN et al. 1981; GOEBEL 1983). Trials on a range of winter wheat and winter and spring barley cultivars, under conditions of differing disease intensity, have confirmed that fenpropimorph gives very good efficacy against powdery mildew with good control persisting, at some sites, for up to 45 days after a single application in the spring (ATKIN et al. 1981). In connection with the control of rust fungi, RATHMELL and SKIDMORE (1982) emphasize the eradicative and curative activity of fenpropimorph; they also mention the redistribution of the active compound via the vapour phase. A persistence of 4 – 5 weeks is stated.

The stop-effect of fenpropimorph is emphasized by various authors (BOHNEN and PFIFFNER 1979; POMMER and HIMMELE 1979; HOPP et al. 1984). GOEBEL (1983) could demonstrate that even if 5 – 10% of the leaf surface was covered with mildew, it was possible to use fenpropimorph without disadvantage to its efficacy. At low temperature and low rates of evaporation the wheat mildew in the crop stand has reportedly been stopped and killed even in cases where the infected leaf surface was considerably larger (FRAHM 1983). Comprehensive investigations to clarify environmental influences on the efficacy of fenpropimorph against *Erysiphe graminis* on wheat have been carried out by SAUR and LÖCHER (1984). They investigated the effect of varying day temperatures and precipitation after fungicide application (750 g a.i./ha in 400 l of water) with beginning to medium (3.8% of the leaf area infected) mildew attack. According to these investigations, fenpropimorph exhibits at low as well as at higher temperatures a good stop effect and persistence. The tests for resistance to wash-off indicated that a rapid penetration of fenpropimorph into the plant occured after application. If the crop was irrigated immediately after application of the compound, then no losses of active compound occurred when the plants were "dry" when they were treated; when they were dew- or rain-wet at the time of treatment then the losses were only negligible. Further experiments under field condition produced almost identical results.

To broaden the range of activity of fenpropimorph, or in connection with anti-resistance strategies, the following combinations have been developed into commercial products:

fenpropimorph + carbendazim
fenpropimorph + chlorothalonil
fenpropimorph + carbendazim + chlorothalonil
fenpropimorph + carbendazim + mancozeb
fenpropimorph + iprodione
fenpropimorph + prochloraz
fenpropimorph + epiconazole
fenpropimorph + propiconazole

Fenpropidine

Spectrum of activity
The main area of application of fenpropidine lies in the control of powdery mildew in cereals, with additional efficacy against rust fungi, and a secondary effect against *Rhynchosporium* in barley (Table 10.6). In agar plate tests, fenpropidine with low concentrations of active ingredient inhibited the growth of quite a number of fungi. (Table 10.7). In extensive field trials, fenpropidine was tested in different cereal-producing countries.

Table 10.6 Main activities and secondary effects under field conditions (BOHNEN et al. 1986)

Pathogen	Crop	g a.i./ha or g/kg seed	% Control
Main activity			
Erysiphe graminis	barley	500 – 750	>90
E. graminis	wheat	500 – 750	>90
Puccinia hordei	barley	750	>75
P. striiformis	wheat	750	>75
P. recondita	wheat	750	>75
P. graminis	wheat	750	>75
P. coronata	oats	750	>75
Secondary effects			
Rhyncosporium secalis	barley	750	<75
Hemileia vastatrix	coffee	1 000 – 2 000	>75
Helminthosporium gram.	seed	1	>90

Table 10.7 Fungicidal side effects under laboratory conditions (BOHNEN et al. 1986)

Pathogen	LC 50 (mg a.i.)	Pathogen	LC 50 (mg a.i.)
Poria placenta	0.2	Ceratocystis ulmi	0.2
Penicillium digitatum	0.5	Alternaria solani	0.6
Colletotrichum lindemuthianum	1.8	Lenzites trabea	2.5
Chaetomium globosum	3	Cladosporium cucumerinum	3
Ustilago maydis	5	Piricularia oryzae	5
Aspergillus niger	10	Gaemannomyces graminis	10
Monilia laxa	10	Pellicularia rolfsii (soil)	25
Pseudocercosporella herpotrichoides	20		

The compound is formulated as an emulsifiable concentrate containing 750 g active ingredient/litre: BOHNEN et al. (1986) reported that control of mildew in barley was good to excellent with 500 to 750 g a.i./ha at the beginning of infestation. The eradication effect with 750 g fenpropidine/ha was rapid and powerful. In wheat, the threat of early and/or persistent infestation required additional treatments; a definite eradicative activity was noted. The secondary effect against rust diseases observed in cereal mildew trials with 500 to 750 a.t./ha was good, but less than that achieved with fenpropimorph at 500 to 750 a.i./ha. Under field conditions, the secondary effects of fenpropidine against barley leaf blotch (*Rhynchosporium secalis*) and glume blotch (*Septoria nodorum*) are insufficient for control.

With an application rate of 500 to 750 a.i./ha, fenpropidine protects cereal crops against *Erysiphe graminis* independently of temperature for 4 to 5 weeks, and with 750 g a.i./ha against *Puccinia* species for 3 to 4 weeks.

In glasshouse and small field plot trials, fenpropidine, like fenpropimorph, shown a marked fungicidal efficacy against *Erysiphe graminis* on wheat and barley.

Coussin (1988) reported the effectiveness of fenpropidine against powdery mildew and brown rust on wheat. The control of powdery mildew was especially good (>90%) with application of 562 g a.i./ha.

In experiments in which agar disks covered with *Armillaria* species were drenched with fenpropidine, Turner and Fox (1988) observed a high inhibition of growth. A protectant assay using wood blocks confirmed the fungistatic effect of fenpropidine.

To widen the range of activity or in connection with antiresistance strategies, the following combinations have been developed into commercial products:

> fenpropidine + fenpropimorph
> fenpropidine + prochloraz
> fenpropidine + fenpropimorph + prochloraz
> fenpropidine + hexaconazole

Trimorphamide

The active compound trimorphamide is used either as a wettable powder or as an emulsifiable concentrate. Its main area of application is in the control of powdery mildew fungi; however, the substance also exhibits a pronounced activity against *Venturia inaequalis* on apples (Seidl and Lansky 1981).

Good results have been obtained in field tests against the following organisms:

– *Erysiphe graminis* on barley and wheat (Klimach 1981)
– *Erysiphe cichoracearum* on cucumbers (Demecko 1975)
– *Sphaerotheca mors-uva* on gooseberries
– *Sphaerotheca pannosa* on roses (Swiech 1981)
– *Podosphaera leucotricha* on apples (Demecko and Pandlova 1977)
– *Uncinula necator* on grapes (Gaher and Hudecova 1978)
– *Botrytis cinerea* on grapes (Vaselashkin et al. 1989)

Malenin and Todorova (1983) treated grape vines at 2-weekly intervals with 0.6% trimorphamide emulsions and observed a very good curatice effect as well as prophylactic activity against powdery mildew.

Resistance

Triazole derivatives have been widely use for a number of years to control cereal mildew. As a consequence of the increased selection pressure, mildew populations, e.g. in barley, have appeared which show a reduced sensitivity to this type of active compounds (Fletcher and Wolfe 1981; Hollomon 1982; Buchenauer 1984).

Morpholines, like triazoles and pyrimidines, belong to the ergosterol biosynthesis inhibitors; they interfere, however, with ergosterol biosynthesis in different ways (see

Chapter 11 and 13). HOLLOMON (1982) investigated the effects of tridemorph on barley powdery mildew. He compared these to those of cycloheximine (protein synthesis inhibitor), triadimenol (sterol biosynthesis inhibitor) and ethirimol (inhibitor of adenosine metabolism). Tridemorph, like triadimenol, stopped the mildew development after the first haustorium had been formed. No strong cross-sensitivity was observed between tridemorph and the other fungicides, although triadimenol and ethirimol showed some evidence of negative cross-insensitivity.

Successive sowings of glasshouse-grown barley plants were treated with 50 mg to 2,000 mg/l tridemorph and inoculated with an initially fungicide-sensitive isolate of *Erysiphe graminis* f.sp. *hordei*. The time required for symptoms to appear, compared with that for untreated plants, decreased with successive sowings. This was interpreted as evidence of the increase in the frequency of fungicide-tolerant propagules in the pathogen population. Effective mildew control was, however, obtained by the use of tridemorph in trial plots and field crops (WALMSLEY-WOODWAR et al. 1979a). The same authors (1979b) found considerable variation among the isolates of *Erysiphe graminis* f.sp. *hordei*, both in pathogenicity and in the level of tolerance to tridemorph. Tolerance was detected as the mean number of pustules on all treated material expressed as a percentage of the number on the untreated plants.

In model experiments with *Botrytis cinerea*. BISCHOFF (1983) isolated two strains which had different levels of resistance, different growth rates and different cross resistances. The author supposed that a series of genetic variations within the fungal cell was responsible for this. Cross-resistance to dodine and carboxin persisted, but not to drazoxolon. This would appear to indicate that tridemorph resistance depends on variations in the respiratory chain at site I.

Laboratory isolates of *Penicillium italicum* with varying levels of resistance to fenarimol showed cross-resistance to other fungicides which inhibit ergosterol biosynthesis (bitertanol, etaconazole, fenapanil and imazalil) but not to fenpropimorph (DE WAARD et al. 1982; DE WAARD and VAN NISTELROY 1982). In contrast, all isolates with a relatively high degree of resistance to ergosterol biosynthesis inhibitors exhibited increased sensitivity to fenpropimorph (negatively correlated cross resistance).

In glasshouse experiments over a two year period (POMMER and HIMMELE 1979) or of more than three years (BUTTERS 1984) it was not possible to induce an adaptive resistance to fenpropimorph. There are no indications to date that the sensitivity of powdery mildew populations alters following the applications of fenpropimorph.

LORENZ and POMMER (1984) reported a fenpropimorph monitoring programme designed to establish the fungicide sensitivity of wild populations, the variation in the range of these values and whether the sensitivity of powdery mildew populations to triazole fungicides is influenced by the use of fenpropimorph. In examples drawn from the different regions, the fenpropimorph values were found to be independent of treatment and lie in the range of values found in ten wild populations.

Investigating the effects of fenpropidine on DMI-resistant strains of *Erysiphe graminis* f.sp. *hordei* and *Rhynchosporium secalis*, GIRLING et al. (1988) observed no cross-resistance between the DMI-fungicide propiconazole and fenpropidine or tridemorph.

ROBERTSON et al. (1990) investigated the sensitivities of barley powdery mildew isolates, collected in Southern Scotland during 1988–1990, to tridemorph, fenpropimorph and fenpropidine. The pattern of sensitivity of the isolates to the fungicides differed. In general, the isolates were more sensitive to fenpropimorph and fenpropidine than to tridemorph. In tests with detached leaf segments, the pathogen showed a 12-fold range of sensitivity to tridemorph, a 10-fold range to fenpropimorph, and a 44-fold range to fenpropidine. There was no apparent connection between history of fungicide contact and sensitivity to the tested fungicides.

Single-colony isolates of *Erysiphe graminis* f. sp. *hordei* collected in Eastern Scotland and England in 1988 by BROWN *et al.* (1991) were classified as sensitive or resistant (isolates with markedly reduced sensitivity) to fenpropidine, fenpropimorph and tridemorph. The responses of the isolates to tridemorph were correlated only slightly with responses to fenpropidine and fenpropimorph. Compared with field application rates of these fungicides, the levels of resistance detected were low. There was a lack of cross-resistance between fenpropidine and fenpropimorph on the one hand, and tridemorph on the other. The authors hypothesise that in resistant isolates the mechanism of reduced activity may be associated with differences in the chemical structures and the modes of action of these three fungicides. The pattern of race-specific virulences of resistant and sensitive isolates indicated that resistance to fenpropidine and fenpropimorph evolved several times, and therefore suggests that the same gene(s) causes resistance to both chemicals.

In continuation of this programme, and including another which was begun independently in Switzerland in 1988; LORENZ *et al.* (1992) reported that in the period from 1984 to 1988 in Germany, and from 1982 to 1988 in Switzerland, there were virtually no changed in the sensitivity of wheat powdery mildew isolates to fenpropimorph. Signs of a clearly recognizable shift in sensitivitiy were observed in 1989 and in the year that followed. However, this development has stabilised, and the field performance of fenpropimorph at the recommended rates has remained unaffected. In order to avoid a shift in powdery mildew sensitivity to fenpropimorph, the authors recommend application of triazole/morpholine fungicides as an anti-resistance strategy.

References

Anonym: Corbel Getreidefungizid. BASF Aktiengesellschaft und Dr. Maag AG. Sept. 1982.
Anonym: Calixin-Fungizide for the control of Sigatoka and other diseases in bananas. BASF Aktiengesellschaft 1984.
Anonym: Agrochemical Handbook. 3rd edit. Royal Soc. of Chemistry, Cambridge 1991.
ATKIN, J. C., PARSONS, R. G., RIELEY, C. E., and WATERHOUSE, S.: Control of cereal diseases with fenpropimorph and fenpropimorph mixtures in the United Kingdom. Proc. Brit. Crop Prot. Conf. − Pests and Diseases (1981): 307−316.
BERGMANN, H., MÜLLER, R., and NEUHAUS, W.: Bestimmung der phytotoxischen Wirkung von Calixin auf Kulturpflanzen. Syst. Fungizide Antifungale Verbindungen. Abh. Akad. Wiss. DDR, Abt. Math., Naturwiss., Techn. 1 N (1982): 355−359.
BISCHOFF, G.: Tridemorph resistance in *Botrytis cinerea.* Syst. Fungiz. Antifungale Verbindungen. Abh. Akad. Wiss. DDR, Abt. Math., Naturwiss. Techn., 1 N (1982): 361−364.
BOHNEN, K., and PFIFFNER, A.: Fenpropemorph, ein neues, systematisches Fungizid zur Bekämpfung von echten Mehltau- und Rostkrankheiten im Getreidebau. Medelingen Fakulteit Landbouwwetenschappen Gent. **44/2** (1979): 487−497.
− − SIEGLE, H., and ZOBRIST, P.: Fenpropidin, a new systemic cereal mildew fungicide. Proc. Brit. Crop. Prot. Conf. − Pests and Dis. (1986): 27−32.
− SIEGLE, H., and LÖCHER, F.: Further experiences with a new morpholine fungicides for the control of powdery mildew and rust diseases on cereals. Proc. Brit. Crop Prot. Conf. − Pests and. Diseases (1979): 541−548.
BROWN, J. K. M., SLATER, S. E., and SEE, K. A.: Sensitivity of *Erysiphe* f. sp. *hordei* to morpholine and piperidine fungicides. Crop Prot. **10** (1991): 445−454.
BUCHENAUER, H.: Stand der Fungizidresistenz bei Getreidekrankheiten am Beispiel der Halmbruchkrankheit und des echten Mehltaus. Gesunde Pflanzen **36** (1984): 161−170.

BUTTERS, J., CLARK, J., and HOLLOMON, D. W.: Resistance to inhibitors of stereol biosynthesis in barley powdery mildew. Lecture held at XXXVI Intern. Symp. on Crop Protect., Gent, 8.5.1984.

CHAND, T., COPELAND, R. B., and LOGAN, C.: Fungicidal control of stem canker and black scurf of potato. Tests Agrochem. Cultiv. **2** (1981): 22–23.

CHANDRASEKARAN, R.: Acaricidal action of calixin against tea mites. Pesticides **14(5)** (1980): 16–17.

COUSSIN, G.: Qu est-ce que la fenpropidine?. Def. Veg. **42** (252) (1988): 15–18.

DEMECKO, J., and JARAS, A.: Biological properties of N-(1-formamido-2,2,2-trichlorethyl) morpholine. Dokl. Soobshch.-Mezhdunar. Kongr. Zashch. Rast., 8th, Vol. 3, 1 (1975): 178–183. (Orgkom. VIII Mezhdunar. Kongr. Zashch. Rast.: Moscow.)

DEMECKO, J., and PANDLOVA, M.: Effect of trimorfamide on *Podosphaera leucotricha* Salm. Agrochemia **17** (1977): 314–316.

DE WAARD, M. A., GROENEWEG, H., and VAN NISTELROOY, J. G. M.: Laboratory resistance to fungicides which inhibit ergosterol biosynthesis in *Penicillium italicum*. Netherland Journal of Planth Pathology **88** (1982): 99–112.

— and VAN NISTELROOY, J. G. M.: Toxicity of fenpropimorph to fenarimol-resistant isolates of *Penicillium italicum*. Netherland Journal of Plant Pathology **88** (1982): 231–236.

DEYMANN, A.: Untersuchungen zu Aufnahme und Transport der systemischen Fungizide Tridemorph und Fenpropimorph und ihrer Wirkungsweise gegen den echten Mehltau an Getreide *Erysiphe graminis* D. C. Marchal. Diss. Rheinische Friedrich-Wilhelm-Universität, Landwirtschaftliche Fakultät, Bonn-Poppelsdorf 1981.

— and WELTZIEN, H. C.: Wirkung von Tridemorph und Fenpropimorph auf den echten Mehltau an Getreide. *Erysiphe graminis* D. C. Marchal. Mededelingen Fakulteit Landbouwwetenschappen Gent. **45/2** (1980): 129–136.

DOBE, H., SIEBER, K., NEUMANN, ST., and JACOB, F.: Untersuchungen zur Aufnahme und Translokation von Aldimorph in ausgewählten Pflanzenspezies. Arch. Phytopathol. Pflanzenschutz, Berlin **22** (1986): 415–425.

FISHER, D.: Effects of some fungicides on *Rhizobium trifolii* and its symbionic relationship with white clover. Pesticide Science **7(1)** (1976): 10–18.

— Effects of some fungicides used against cereal pathogens on the growth of *Rhizobium trifolii* and its capacity to fix nitrogen in white clover. Annals Applied Biology **98(1)** (1981): 101–107.

FLETCHER, J. T., and WOLFE, M. S.: Insensitivity of *Erysiphe graminis* f. sp. *hordei* to triadimefon, triadimenol and other fungicides. Proc. Brit. Crop Prot. Conf. – Pest and Diseases (1981): 633–640.

FRAHM, J.: Lehren aus einem schwierigen Mehltaujahr. top agrar Nr. 10 (1983): 52–54.

FROST, A., and PATTISSEN, N.: Some results obtained in the United Kingdom using dodemorph for the control of powdery mildews in roses and other ornamentals. Proc. 6th Brit. Insect. and Fung. Conf. (1971): 349–354.

GAHER, S., and HUDEVOCA, D.: Practical use of the fungicidal activities of trimorphamide. Agrochemia. **18** (1978): 185–189.

GIRLING, I. J., HOLLOMON, S. J., KENDALL, S. J., LOEFFLER, R. S. T., and SENIOR, I. J.: Effects of fenpropidin on DMI-resistant strains of *Erysiphe graminis* f. sp. *hordei* and *Rhynchosporium secalis*. Proc. Brit. Crop. Prot. Conf.-Pests and Dis. (1988): 385–390.

GOEBEL, G.: Corbel-systemisches Fungizid für Getreide mit neuer Wirkungsart. Gesunde Pflanze **35** (1983): 243–247.

HAMM, R.: On the behaviour of Calixin in bananas. BASF Symposium, Central America (1983): 15–24.

HAMPEL, M., and LANG, H.: Erfahrungen bei der Bekämpfung des Getreidemehltaus (*Erysiphe graminis*) in den Jahren 1970 und 1971. BASF-Mitteilungen für den Landbau April 1972.

HAMPEL, M., LÖCHER, F., and SAUR, R.: Two year field trial results with fenpropimorph in cereals. Mededelingen Fakulteit Landbouwwetenschappen Gent. **44** (1979): 511−518.

— and POMMER, E.-H.: Results of trials with tridemorph on the control of *Mycosphaerella musicola* in bananas. Proc. 6th Brit. Insect. and Fung. Conf. (1971): 842−847.

HIMMELE, W., and POMMER, E.-H.: 3-phenylpropylamines, a new class of systemic fungicides. Angewandte Chemie (Intern. Ed. Engl.) **19** (1980): 184−189.

HOLLOMON, D. W.: The effects of tridemorph on barley powdery mildew: its mode of action and cross sensitivity relationships. Phytopathologische Zeitschrift **105** (1992): 279−287.

HOPP, H., SIEGLE, H., SAUR, R., and RISCH, H.: Wirkungsunabhängigkeit der Getreidemehltaufungizide von unterschiedlichen Temperaturen und verschieden starkem Ausgangsbefall unter besonderer Berücksichtigung von Fenpropimorph. Mitt. Biol. Bundesanstalt Land- u. Forstwirtsch. Berlin-Dahlem Heft 223−244. Deutsche Pflanzenschutz-Tagung Gießen (1984): 214.

HUDECOVA, D., KOVAC, J., and GAHER, S.: Biological activity of trimorfamide against *Erysiphe cichoracearum* on cucumbers. Agrichemia **18** (1978): 361−363.

JUMAR, A., and LEHMANN, H.: The special properties of fungicidal N-n-alkyldimethyl morpholines synthesized by an unorthodox route. Proc. 5th Intern. Congr. Pest. Chem., Kyoto 1982.

KAESBOHRER, M., and NAU, K.-L.: Fenpropidin − ein neuer Wirkstoff zur Krankheitsbekämpfung in Getreide. Mitt. Biol. Bundesanstalt Land- u. Forstwirtsch. Berlin-Dahlem **283** (1992): 432.

KERKENAAR, A., and BARUG, D.: Fluorescence microscope studies of *Ustilago maydis* and *Penicillium italicum* after treatment with imazalil or fenpropimorph. Pesticide Sci. **15** (2) 1984: 199−205.

KLIMACH, A.: Effectiveness of Fademorf EK-20 against *Erysiphe graminis* DC, in winter wheat and spring barley. Agrochemia **21** (1981): 51−53.

KÖNIG, K.-H., POMMER, E.-H., and SANNE, W.: N-substituierte Tetrahydro-1,4-oxazine, eine neue Klasse fungizider Verbindungen. Angewandte Chemie **77** (1965): 327−333.

KRADEL, J., EFFLAND, H., and POMMER, E.-H.: Mehrjährige Ergebnisse bei der Bekämpfung des Getreidemehltaus (*Erysiphe graminis*) an Sommergerste. Mededelingen Fakulteit Landbouwwetenschappen Gent. **33** (1968): 997−1004.

− − − Response of barley varieties to the control of powdery mildew with cyclomorph and tridemorph. Proc. 5th Brit. Insect. Fung. Conf. (1969): 16−19.

— and POMMER, E.-H.: Vierjährige Versuche mit Cyclododecyl-2,6-dimethyl-morpholinacetat, einem neuen Wirkstoff gegen echte Mehltaupilze. BASF Mitteilungen für den Landbau. Mai 1967 (1967a).

− − Erfahrungen mit einem neuen Wirkstoff gegen Echte Mehltaupilze. Gartenwelt. **67** (1967b): 185−186.

− − Some remarks and results on the control of powdery mildew in cereals. Proc. 4th Brit. Insect. Fung. Conf. (1967c): 170−175.

− − and WILL, H.: Langjährige Erfahrungen bei der Mehltaubekämpfung in Zierpflanzen mit Dodemorph. BASF Mitteilungen für den Landbau. Juni 1970.

LIM, T. M.: Low-volume spray of an oil-based systemic fungicide for controlling *Oidium* secondary leaf fall. Proc. Rubber Res. Inst. Malays. Plant Conf. (1976): 231−242.

LORENZ, G., and POMMER, E.-H.: Investigations into the sensitivity of wheat powdery mildew populations towards fenpropimorph. Brit. Crop Prot. Conf. − Pests and Diseases (1984): 489−494.

— SAUR, R., SCHELBERGER, K., FORSTER, B., KUENG, R., and ZOBRIST, P.: Long term monitoring results of wheat powdery mildew sensitivity towards fenpropimorph and strategies to avoid the development of resistance. Proc. Brit. Crop Prot. Conf. − Pests and Dis. (1992): 171−176.

MALENIN, I., and TODOROVA, M.: New fungicides for the control of powdery mildew. Lozar. Vinar. **32** (1983): 28−31.

MAPPES, D.: Results of new trials with tridemorph against *Mycosphaerella* spp. on bananas. 9th Intern. Congr. Plant Prot. and 71th. Ann. Meet. Am. Phytopath. Soc. Washington (1979): Ref. No. 606.
— and HIEPKO, G.: New possibilities for controlling root diseases of plantation crops. Mededelingen Fakulteit Landbouwwetenschappen Gent **49**(2a) (1984): 283–292.
MAUDE, R. B., HUMPHERSON-JONES, F. M., and SHURING, C. G.: Treatments to control *Phoma* and *Alternaria* infections of brassica seeds. Plant Pathology **33** (1984): 525–535.
MEEUS, P.: Fungicidal protection against powdery mildew in sugar beets. Driemaand. Publ., Belg. Inst. Verbetering Beit **45**(3) (1977): 131–141.
MUNSHI, G. D.: Efficacy of fungicides for the control of powdery mildew of pea. J. Res. (Punjab Agricult. Univ.) **20**(4) (1983): 485–488.
OROS, G., and SULE, S.: Antibacterial activity of calixin on pythopathogenic bacteria. Syst. Fungiz. Antifungale Verbindungen. Abh. Akad. Wiss. DDR, Abt. Math., Naturwiss., Tech. 1 N (1983): 417–420.
OTTO, S., and DRESCHER, N.: The behaviour of tridemorph in soil. Proc. 7th Brit. Insect. Fung. Conf. (1973): 57–64.
— and POMMER, E.-H.: Über das systemische Verhalten von Tridemorph (Untersuchungen zur Translokation von ^{14}C-markierten Tridemorph an Gerste, Gurken und Bananen). Mededelingen Fakulteit Landbouwwetenschappen Gent. **38** (1973): 1493–1506.
POMMER, E.-H.: Chemical structure-fungicidal activity relationships in substituted morpholines. Pest. Sci. **15** (1984): 285–295.
— and HIMMELE, W.: Die Bekämpfung wichtiger Getreidekrankheiten mit Fenpropemorph, einem neuen Morpholinderivat. Mededelingen Fakulteit Landbouwwetenschappen Gent. **44**/2 (1979): 499–510.
— and KRADEL, J.: Substituierte Dimethylmorpholin-Derivate als neue Fungizide zur Bekämpfung echter Mehltaupilze. Mededelingen Fakulteit Landbouwwetenschappen Gent **32** (1967a): 735–744.
— — Beiträge zur Wirkungsweise von Tridemorph gegen *Erysiphe graminis*. Mededelingen Fakulteit Landbouwwetenschappen Gent **36**/1 (1971): 120–125.
— OTTO, S., and KRADEL, J.: Some results concerning the systemic action of tridemorph. Proc. 5th Brit. Insect. Fung. Conf. (1969): 347–353.
RATHMELL, W. G., and SKIDMORE, A. M.: Recent advances in the chemical control of cereal rust diseases. Outlook on Agriculture **11** (1982): 37–43.
ROBERTSON, S., GILMOUR, J., NEWMAN, D., and LENNARD, J. H.: Sensitivity of barley powdery mildew isolates to morpholine fungicides. Proc. Brit. Crop. Prot. — Pests and Dis. (1990): 1159–1162.
SAUR, R., HOPP, H., SIEGEL, H., and RISCH, W.: Fenpropimorph, ein neues Fungizid zur Bekämpfung von Getreidekrankheiten — dreijährige Versuchsergebnisse aus dem Freiland. — Mitt. Biol. Bundesanstalt Land- u. Forstwirtsch. Berlin-Dahlem **191** (1979): 183.
— and LÖCHER, F.: Untersuchungen von Umwelteinflüssen auf die Anwendung von Fenpropimorph gegen *Erysiphe graminis* an Weizen. Mitt. Biol. Bundesanstalt Land- u. Forstwirtsch. Berlin-Dahlem Heft 223–244. Deutsche Pflanzenschutz-Tagung Gießen (1984): 213.
SCHEFFER, R. J., BRAKENDORFF, A. C., KERKENAAR, A., and ELGERSMA; D. M.: Control of Dutch elm disease by the sterol biosynthesis inhibitors fenpropimorph and fenpropidin. Neth. J. Pl. Path. **94** (1988): 161–173.
SCHLÜTER, K., and WELTZIEN, H. C.: Ein Beitrag zur Wirkungsweise systemischer Fungizide auf *Erysiphe graminis*. Mededelingen Fakulteit Landbouwwetenschappen Gent. **36** (1971): 1159–1164.
SCHROEDER, D.: Effect of agrochemical compounds on the decomposition of straw and cellulose in soil. Z. Pflanzenern. Bodenkunde **142**(4) (1979): 616–625.

SEIDL, V., and LANSKY, M.: Subsidiary effects of the antioidial fungicide Fademorph EC-20. Ved. Pr. Ovocnarske. **8** (1981): 117−128.

SIDDIQUI, K. A., and HAAHR, V.: Different reactions of wheat mutants to a systemic fungicide. Naturwiss. **58** (1971): 415−416.

SIEBER, K., LINK, V., and KÖNNIG, M.: Zum Rückstandsverhalten des Mehltaufungizids Falimorph. Nachrichtenbl. Pflanzensch. DDR (1981): 135−137.

SWIECH, T.: Effectiveness of Fademorf EK-20 against *Sphaerotheca pannosa* Lev. var. *rosae* Wor. and *Diplocarpon rosae* Wolf in roses. Agrochemia **21** (1981): 53−55.

TRAN VAN CANH: Lutte contre le *Fomes*: Nouvelle methode d'ètude. Caoutshouse et Plastiques Nr. 617/618 (1983).

TRAN VAN CANH: A new method of direct control of *Rigodoporus lignosus* causal agent of white rot disease of *Hevea*. Intern. Rubber Conf., 17.−19th Sept. 1984, Colombo.

TURNER, J. A., and FOX, R. T. V.: Prospects for the chemical control of *Armillaria* species. Proc. Brit. Crop Prot. Conf. − Pests and Dis. (1988): 235−240.

VASELSHKU, E. G., GAMOVA, O. V., and DOROKHOV, B. L.: Fungicides in relation to wine quality. Sadovod. Vinograd. Mold. **10** (1989): 24−26.

VENKATA RAM, C. S.: Calixin, a systemic fungicide effective against blister blight. Pesticides **8**(5) (1974): 21−25.

WALMSLEY-WOODWARD, D., LAWS, F. A., and WHITTINGTON, W. J.: Studies on the tolerance of *Erysiphe graminis* f. sp. *hordei* to systemic fungicides. Ann. Appl. Biol. **92**(2) (1979a): 199−202.

− − − The characteristics of isolates of *Erysiphe graminis* f. sp. *hordei* verying in response to tridemorph and ethirimol. Ann. Appl. Biol. **92**(2) (1979b): 211−219.

YEOH, C. S., and TAN, A. M.: Natural rubber latex formulation for controlling pink disease. Proc. Rubber Res. Inst. Malays. Plant Conf. (1974): 171−177.

ZIMMERMANN, G.: Effect of systemic fungicides on different entomopathogenic *Fungi imperfecti in vitro*. Nachrichtenbl. Deutsch. Pflanzenschutzdienstes **27**(8) (1975): 113−117.

− and BASEDOW, T.: Field tests on the effect of fungicides on the mortality of cereal aphids caused by *Entomophthoraceae* (*Zygomycetes*). Z. Pflanzenkrankh. Pflanzenschutz **87**(2) (1980): 65−72.

ZOBRIST, P., COLOMBO, V. E., and BOHNEN, K.: Action of fenpropimorph on exterior structures of *Puccinia coronata* on oat as revealed by scanning electron microscopy. Phytopath. Z. **105** (1982): 11−19.

Chapter 11

Mechanism of action of cyclic amine fungicides: morpholines and piperidines.

A. KERKENAAR

Plaggewagen 12
1261 KG Blaricum, The Netherlands

Introduction

Nowadays it has been generally accepted that cyclic amine fungicides like tridemorph, dodemorph, fenpropimorph, and amorolfine belonging to the morpholines and fenpropidin and piperalin belonging to the piperidines act primarily on sterol biosynthesis in fungi but, most of them, also in higher plants (KERKENAAR 1987, 1990; MERCER 1988, 1991).

Many different modes of action have been proposed before. The earliest work on the mode of action of tridemorph indicated a primary effect on the respiratory electron transport chain (BERGMANN 1979; BERGMANN et al. 1977; MÜLLER and SCHEWE 1975, 1976; RAPOPORT 1974). However, this has been questioned by several authors (FISHER 1974; GARG and MEHROTRA 1975; KERKENAAR and KAARS SIJPESTEIJN 1979; KATO et al. 1980). Primary effects have been proposed furthermore on cell membrane function in fungi (KAARS SIJPESTEIJN 1972) and higher plants (BUCHENAUER 1975), on protein synthesis (FISHER 1974) and lipid biosynthesis (POLTER and CASPERSON 1979).

The first indications that tridemorph might be an inhibitor of sterol biosynthesis came from the observation that the compound had many features in common with the sterol demethylation inhibitors. Similarities were found in the antimicrobial spectrum, the morphology, the alleviating effects of unsaturated lipophilic compounds and the alterations in the neutral lipid pattern. (KERKENAAR et al. 1979), together with the occurrence of cross-resistance between tridemorph and sterol C-14 demethylation inhibitors (BARUG and KERKENAAR 1979, 1984). Inhibition of sterol biosynthesis was already reported by LEROUX and GREDT (1978) but this effect was attributed to inhibition of protein synthesis.

This chapter considers the various cyclic amines in relation to growth, morphology and biochemical target sites in fungi and higher plants and their effects on organelles, cell-free systems and enzymes.

Effects on growth, morphology and biochemical activities in whole cells and organisms

A. Fungi

Growth

Cyclic amines act against many fungi, but are particularly effective against powdery mildews where they interfere with haustorial development (HOLLOMON 1982; SCHLÜTER and WELTZEIN 1971; DE WAARD et al. 1992). In non-obligate fungi in vitro cyclic amines have little or no effect on fungal growth until after one or more cell doublings have occurred. Spore germination is not inhibited by concentrations which are ultimately lethal or strongly inhibited to hyphal growth (KATO et al. 1980; KERKENAAR et al. 1979, 1984; KERKENAAR 1983; KERKENAAR and BARUG 1984). In Ustilago maydis only a moderate reduction of dry-weight increase is evident after $6-8$ h of exposure to tridemorph or fenpropimorph, whereas striking evidence for inhibition of sporidial multiplication is obtained after $2-4$ h (KERKENAAR et al. 1979; KERKENAAR 1983).

Amorolfine inhibits a large number of medically important fungi (POLAK 1987, 1988) and fenpropidin and piperalin inhibit both growth of sporidia of Ustilago maydis, piperaline beeing less effective than fenpropidin (SCHNEEGURT and HENRY 1992). LEROUX et al. (1988) described the effect of cyclic amines on growth of various strains of Pseudocercosporella herpotrichoides. ZIOGAS et al. (1991) reported the effects of fenpropimorph, fenpropidin, and tridemorph on Nectria haematococca var. cucurbitae and DEBIEU et al. (1992) on Fusarium species. Dilution of the terminal sterol, e.g. ergosterol, fucosterol or brassicasterol depending on the species, is probably required before the effect on growth becomes evident. The inhibition of dry-weight increase is accompanied by the first visible irregular deposition of cell wall material (KERKENAAR and BARUG 1984; KERKENAAR et al. 1986).

Morphology and cytology

Treatment of Ustilago maydis sporidia in liquid cultures with morpholine fungicides gives rise to production of large sporidia which are multicellular and frequently branched (KERKENAAR et al. 1979; KERKENAAR 1983). Each cell contains only one nucleus (KERKENAAR, unpublished results). Similar results are obtained using Candida albicans treated with amorolfine (POLAK et al. 1986).

Treatment of Botrytis cinerea, Botrytis allii and Penicillium italicum (KATO et al. 1980; KERKENAAR et al. 1979; KERKENAAR 1983) leads to short excessively branched germ tubes. Mycelium of Penicillium italicum, treated with fenpropimorph, shows much enlarged hyphal diameters and relatively short distances between septa (KERKENAAR and BARUG 1984).

Fenpropimorph causes an irregular deposition of β-1,3 and β-1,4 polysaccharides, probably chitin, in Ustilago maydis, Penicillium italicum and Ophiostoma ulmi (KERKENAAR and BARUG 1984; KERKENAAR et al. 1986). However, this phenomenon shows up much later than the accumulation of sterols different from the terminal sterol, ergosterol.

Respiration

Tridemorph does not inhibit respiration of whole cells of *Ustilago maydis, Saccharomyces cerevisiae, Torulopsis candida, Botrytis allii, Cladosporium cucumerinum* and *Botrytis cinerea* (KATO *et al.* 1980; KERKENAAR *et al.* 1979). Fenpropimorph does not inhibit respiration of *Ustilago maydis* KERKENAAR *et al.*, unpublished results). In previous publications it has been claimed that tridemorph inhibits fungal respiration (BERGMANN *et al.* 1975; BERGMANN 1979). However, KERKENAAR and KAARS SIJPESTEIJN (1979) demonstrated that these effects were probably obtained with the formulated product Calixin, at least in tests with whole cells. The synergistic effect of antimycin A or rotenone and tridemorph, observed by BERGMANN *et al.* (1979) and by KERKENAAR *et al.* (unpublished results), it at least partly due to its tenside character.

Protein, RNA and DNA synthesis

Tridemorph has little or no effect on protein or RNA synthesis in either *Botrytis cinerea* or *Ustilago maydis* at concentrations which strongly inhibit ergosterol biosynthesis (KATO *et al.* 1980; KERKENAAR *et al.* 1979). Moreover, tridemorph does not inhibit protein synthesis in a cell-free system prepared from powdery mildew conidia (HOLLOMON 1982). Inhibition of DNA synthesis by tridemorph has been reported within 2 h of treatment (KERKENAAR and KAARS SIJPESTEIJN 1979). ZIOGAS *et al.* (1990) reported the increase of sectoring in a heterozygous diploid strain of *Aspergillus nidulans* by fenpropimorph, at concentrations highly inhibitory to growth, due to chromosome nondisjunction.

Lipid synthesis

Most of the literature concerns the effects of cyclic amines of fungal sterol biosynthesis. Because of the complexity of the material each compound will be dealt with separately.

I Morpholines

Tridemorph

Treatment of *Botrytis cinerea* with tridemorph hardly inhibits incorporation of [^{14}C]acetate into the various lipid fractions (KATO *et al.* 1980). Analysis of the 4-desmethylsterol fraction reveals large differences between control and treated mycelium. The amounts of ergosterol and episterol were reduced and there was an accumulation of fecosterol, ergosta-8,22,24(28)-trien-3β-ol and ergosta-8,22-dien-3β-ol (cf. Fig. 11.1). Inhibition of $\Delta^8 \to \Delta^7$ isomerization causes the accumulation of these sterols retaining the C8(9) double bond.

Treatment of *Ustilago maydis* with tridemorph inhibits the incorporation of [^{14}C]acetate into desmethylsterols and increases the incorporation into C4-desmethylsterols and free fatty acids (KERKENAAR *et al.* 1979). Accumulation of sterols retaining the Δ^8 as well as the Δ^{14} double bonds leads to the concept of inhibition of sterol Δ^{14}

reductase (KERKENAAR *et al.* 1981). Because the Δ^8 sterols have been identified only by mass spectrometry, and misinterpretation due to the similarities in mass spectra of fecosterol and ignosterol is rather easy, it has been suggested that their identification was wrong. Moreover, evidence has been presented that inhibition of Δ^{14} reductase is more likely to be harmful to fungal growth than inhibition of the $\Delta^8 \rightarrow \Delta^7$ isomerase (KERKENAAR *et al.* 1981). The picture is further complicated by the observations that in *Saccharomyces cerevisiae* treatment with tridemorph leads to the accumulation of probably fecosterol (KATO 1983). The identity of fecosterol was confirmed by proton-NMR spectroscopy. The proton absorption of the olefinic hydrogen at C-15 is absent (KATO 1983). This result indicates that the $\Delta^8 \rightarrow \Delta^7$ isomerase is inhibited. However, MERCER *et al.* showed that inhibition of sterol biosynthesis by tridemorph in *Saccharomyces cerevisiae* results in the presence of about 40% of the total sterols retaining the Δ^8 as well as Δ^{14} double bonds and about 30% only the Δ^8 double bond (BALOCH *et al.*

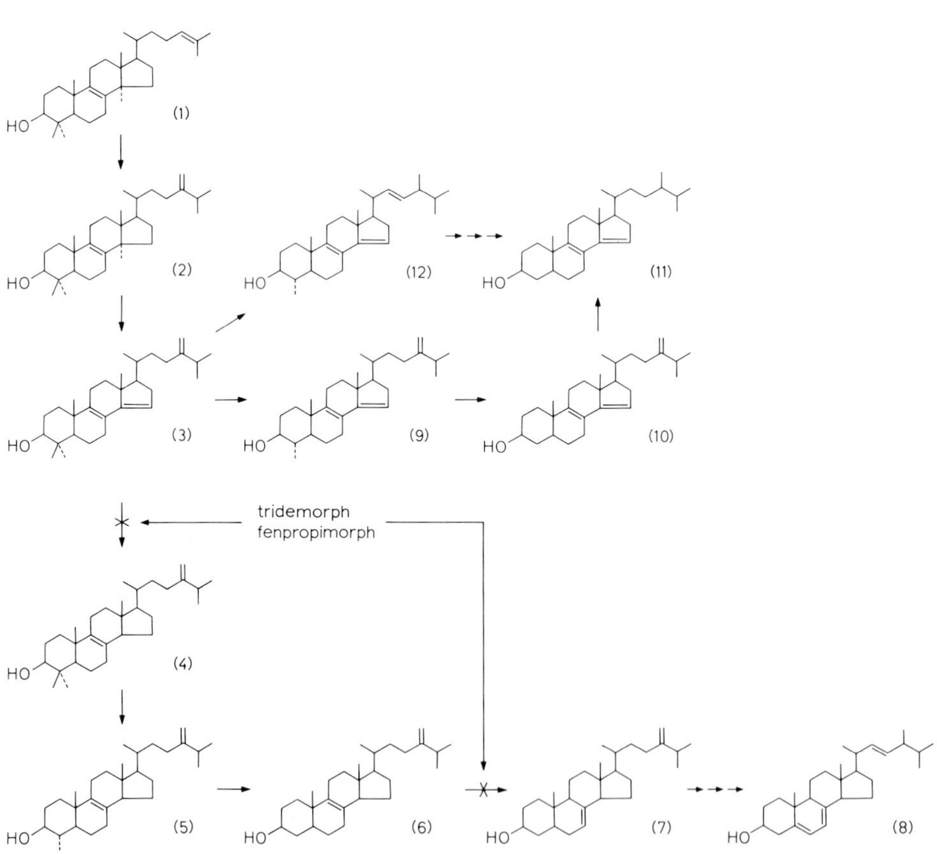

Fig. 11.1 Biosynthetic pathway of ergosterol indicating the presumed sites of inhibition by tridemorph and fenpropimorph. (1) Lanosterol, (2) 24-methylenedihydrolanosterol, (3) 4,4-dimethylergosta-8,14,24(28)-trien-3β-ol, (4) 4,4-dimethylergosta-8,24(28)-dien-3β-ol, (5) 4α-methylergosta-8,24(28)-dien-3β-ol, (6) fecosterol, (7) episterol, (8) ergosterol, (9) 4α-methylergosta-8,14,24(28)-trien-3β-ol, (10) ergosta-8,14,24(28)-trien-3β-ol, (11) ignosterol. (12) 4α-methylergosta-8,14,22-trien-3β-ol.

1984a, 1984b). These results indicate inhibition of the sterol Δ^{14} reductase as well as of the $\Delta^8 \to \Delta^7$ isomerase in *Saccharomyces cerevisiae*. On the other hand, in *Ustilago maydis* tridemorph only causes the accumulation of Δ^8 sterols; sterols retaining the Δ^8 as well as the Δ^{14} double bonds are absent (BALOCH *et al.* 1984a, 1984b). LEROUX and GREDT (1983) give similar results for *Ustilago maydis*. For *Saccharomyces cerevisiae*, *Botrytis cinerea* and *Penicillium expansum,* after tridemorph treatment, they also report the presence of only Δ^8 sterols. In budding fungi, like *Ustilago maydis* and *Saccharomyces cerevisiae,* fecosterol and ergosta-8-en-3β-ol accumulate, whereas in filamentous fungi, like *Botrytis cinerea* and *Penicillium expansum,* fecosterol and ergosta-8,22,24(28)-trien-3β-ol accumulate.

Tridemorph treatment of *Saprolegnia ferax* leads to inhibition of the terminal Δ^5 sterols: fucosterol, 24-methylenecholesterol, desmosterol and cholesterol. The following sterols accumulate: zymosterol, fecosterol and stigmasta-8,24(28)-dien-3β-ol (BERG *et al.* 1983; BERG 1983). Similar results are obtained using another four species from the order of the *Saprolegniales*: *Achlya americana, Dictyuchus monosporus, Apodachlya completa* and *Lagenidium callinectum* (BERG 1983). *Achlya radiosa,* treated with tridemorph, accumulates mainly stigmasta-8,24(28)-dien-3β-ol (KERKENAAR *et al.,* unpublished results).

Tridemorph treatment of *Pyricularia oryzae* causes inhibition of sterol biosynthesis and leads to accumulation of ergosta-5,8-dien-3β-ol (BERG *et al.* 1984; BERG 1984). In *Nectria haematococca* var. *cucurbitae* tridemorph treatment gives rise to accumulation of fecosterol (ZIOGAS *et al.* 1991) indicating the inhibition of $\Delta^8 \to \Delta^7$ isomerase.

Recently ADAMS *et al.* (1992) described the effect of tridemorph on mutants of *Neurospora crassa* with well-characterized defects in sterol biosynthesis. Erg-1 is deficient in the $\Delta^8 \to \Delta^7$ isomerase enzyme and proved resistant to tridemorph. Erg-3 is deficient in the Δ^{14} reductase and is sensitive to tridemorph.

These results indicate inhibition of the $\Delta^8 \to \Delta^7$ isomerase as the most frequently cited site of inhibition by tridemorph.

Fenpropimorph

Fenpropimorph treatment leads to the accumulation of ignosterol in *Ustilago maydis* and of ergosta-8,14,24(28)-trien-3β-ol in *Penicillium italicum* (KERKENAAR 1983; KERKENAAR *et al.* 1984). This suggests that fenpropimorph inhibits the $\Delta^{8,14}$ sterol reductase. However, when *Ustilago maydis* is treated with fenpropimorph at low oxygen pressure, Δ^8 sterols accumulate (KERKENAAR 1983). This suggests that under these conditions the $\Delta^8 \to \Delta^7$ isomerase can also be a target of fenpropimorph. Surprisingly treatment of the yeast-like and filamentous forms of the pleomorphic fungus *Ophiostoma ulmi* with fenpropimorph leads to different results with regard to sterol composition (KERKENAAR 1990). The yeast-like cells accumulated $\Delta^{8,14}$ sterols, whereas the filamentous cells accumulated Δ^8 sterols, indicating inhibition of the Δ^{14} reductase and $\Delta^8 \to \Delta^7$ isomerase, respectively.

In *Ustilago maydis* like in *Saccharomycytes cerevisiae* fenpropimorph has been reported to inhibit the $\Delta^8 \to \Delta^7$ isomerase, but not the sterol Δ^{14} reductase, due to the accumulation of only fecosterol and ergosta-8-en-3β-ol. The same applies to treatment by fenpropimorph of *Botrytis cinerea* and *Penicillium expansum,* where accumulation of fecosterol and probably ergosta-8,22,24(28)-tri-en-3β-ol takes place (LEROUX and GREDT

1983). However, BALOCH et al. (1984) have reported that fenpropimorph inhibits Δ^{14} reductase along with the $\Delta^8 \rightarrow \Delta^7$ isomerase in *Saccharomyces cerevisiae* and *Ustilago maydis* (BALOCH et al. 1984a, 1984b). Accumulation of 4,4-dimethyl-cholesta-8,14,24-trien-3β-ol, 4,4-dimethylcholesta-8,14-dien-3β-ol and ignosterol in *Saccharomyces cerevisiae* leads to the concept of inhibition of sterol Δ^{14} reduction. More than 60% in the fenpropimorph-treated cells retains the Δ^8 as well as the Δ^{14} double bonds; only 2% retains the Δ^8 double bond in the form of fecosterol. In *Ustilago maydis* about 30% of the sterols retains the Δ^8 as well as the Δ^{14} double bonds in the form of ignosterol, whereas about 52% of the desmethyl sterols is ergost-8-en-3β-ol and ergosta-8,22-dien-3β-ol (BALOCH et al. 1984a, 1984b).

In *Pyricularia oryzae* fenpropimorph treatment causes the accumulation of ignosterol indicating the inhibition of the sterol Δ^{14} reductase (BERG 1983; BERG et al. 1983).

Fenpropimorph causes the accumulation of Δ^8- and $\Delta^{8,14}$ sterols in *Botrytis cinerea, Pseudocercosporella herpotrichoides* and *Pyrenophora teres* (STEEL et al. 1989) indicating inhibition of $\Delta^8 \rightarrow \Delta^7$ isomerase and Δ^{14} reductase. In *Nectria haematococca* var. *cucurbitae* fenpropimorph accumulation has been reported of Δ^8 and $\Delta^{8,14}$ sterols as well as squalene, indicating inhibition of $\Delta^8 \rightarrow \Delta^7$ isomerase, Δ^{14} reductase, and squalene epoxidase respectively (ZIOGAS et al. 1991).

Finally, DEBIEU et al. (1992) reported accumulation of $\Delta^{8,14}$ sterols in fenpropimorph-sensitive strains of *Fusarium* and accumulation of $\Delta^{8,14}$ and Δ^8 sterols or only Δ^8 sterols in tolerant ones. In *Fusarium nivale*, highly sensitive to fenpropimorph, only Δ^8 sterols accumulate indicating the inhibition of $\Delta^8 \rightarrow \Delta^7$ isomerase. Squalene has been reported to accumulate to very high amounts in several strains, however, they did not detect a correlation between squalene accumulation and fenpropimorph sensitivity.

Fenpropimorph efficacy to mutants with the defects described under the head tridemorph was less to the *erg-1* mutant than for tridemorph but the *erg-3* proved surprisingly completely sensitive (ADAMS et al. 1992).

These results indicate inhibition of the sterol Δ^{14} reductase as the most frequently cited site of inhibition by fenpropimorph, but also $\Delta^8 \rightarrow \Delta^7$ isomerase inhibition has been frequently cited.

Dodemorph

Only one publication deals with the effect of dodemorph on sterol biosynthesis. This morpholine compound inhibits sterol biosynthesis, causing the accumulation of fecosterol and ergosta-8-en-3β-ol in *Ustilago maydis* and *Saccharomyces cerevisiae* and fecosterol and ergosta-8,22,24(28)-trien-3β-ol in *Botrytis cinerea* and *Penicillium expansum* (LEROUX and GREDT 1983). These results indicate the inhibition of $\Delta^8 \rightarrow \Delta^7$ isomerase.

Amorolfine

Amorolfine induces an increase of the total sterol content of cells of *Candida albicans*, due to the accumulation of sterols not present in control cells. The main sterol that accumulates is ignosterol indicating an inhibition of Δ^{14} reductase. However, also inhibition of the $\Delta^8 \rightarrow \Delta^7$ isomerase is assumed (POLAK et al. 1983) (cf. chapter 21).

Other fungicides containing a morpholine ring

Besides the above mentioned fungicides, the following fungicides contain a morpholine ring in their structure: trimorphamide and aldimorph. Data on the mode of action of these compounds are lacking. However, trimorphamide is probably a C-14-demethylation inhibitor (LEROUX, personal communication) and aldimorph, being closely related to tridemorph, is considered to act in the same way as tridemorph. It will be interesting to study the effect of trimorphamide in a mutant with an insensitive C-14-demethylation site.

II Piperidines

Fenpropidin

Accumulation of $\Delta^{8,14}$-sterols in fenpropidin treated cells has been reported by BALOCH et al. (1984a, 1984b). Fenpropidin treatment of *Nectria haematococca* leads to accumulation of ignosterol and 8,14,24(28)-ergosta-3β-ol as well as of fecosterol (ZIOGAS et al. 1991) and of *Ustilago maydis* to accumulation of ignosterol and 5,8,14,24(28)-ergosta-tetraen-3β-ol (SCHNEEGURT and HENRY 1992). These results strongly indicate Δ^{14} reductase as the most sensitive site of inhibition besides $\Delta^8 \rightarrow \Delta^7$ isomerase. Moreover, the accumulation of 5,8,14,24(28)- ergosta-tetraen-3β-ol indicate also inhibition of $\Delta^{24(28)}$ sterolreductase (SCHNEEGURT and HENRY 1992). They also report that at high concentrations of fenpropidin (100 µM/ml) squalene and 2,3-oxidosqualene accumulate indicating actions at earlier steps in sterol biosynthesis.

Fenpropidin showed the same activity in sterol mutants of *Neurospora crassa* described under the head of tridemorph (ADAMS et al. 1992). The activity on the *erg-3* is unexpected.

Piperaline

Piperaline causes the accumulation of Δ^5-sterols in *Ustilago maydis* (SCHNEEGURT and HENRY 1992). They suggest that the most sensitive site of inhibition is the $\Delta^8 \rightarrow \Delta^7$ isomerase. At high concentrations of piperaline (100 µM/ml) squalene and 2,3-oxidosqualene accumulate indicating action at earlier steps in sterol biosynthesis. Confirmation in a cell-free system is required.

B. Plants and tissue cultures

Growth and miscellaneous effects excluding lipid synthesis

Growth of bramble cells (*Rubus fruticosus*) is only slightly reduced after tridemorph (10 µg/ml) treatment (SCHMITT et al. 1981). At lower concentrations growth of cells is not significantly reduced (SCHMITT et al. 1981, 1982). Fenpropimorph is completely ineffective (BENVENISTE et al. 1984). Tridemorph and dodemorph induce a drastic leakage of betacyanin from discs of beet roots and an efflux of electrolytes from discs of bean leaves and sections of barley leaves (BUCHENAUER 1975). A pretreatment of barley seedlings by root drench with tridemorph and dodemorph results in an increased leakage of electrolytes from sections of primary leaves (BUCHENAUER 1975).

A mitose repressive effect of tridemorph on the meristematic cells of *Allium cepa* has been observed. This effect is dosage dependent. Tridemorph has been revealed as a strong C-mitotic agent (CORTÉS et al. 1982).

Tridemorph as Calixin has been reported to cause phytotyxicity in barley, wheat and cucumber, depending on the concentration (BERGMANN et al. 1982). Calixin causes chlorosis in wheat (*Triticum sevistivum*), varieties Indus 66, Nayab and C591 (SIDDIQUI and HAAHR 1971). Fenpropimorph inhibits growth of potato seedlings (ELENWA et al. 1983). Tridemorph and fenpropimorph treatment of maize seedlings lead to growth inhibition even at the lowest concentration (1 µg/ml). Higher concentrations resulted in stronger inhibition. Spikes and ears are formed after 2–3 months, but they are smaller than the same parts from control plants (BENVENISTA et al. 1984). Application of 250 µM of fenpropimorph retarded growth of shoot, primary leaf, and root of wheat and maize (KHALIL and MERCER 1991). At 5 µM fenpropimorph was not phytotoxic. Phytotoxicity of fenpropimorph in wheat is probably related to a different cellular target than sterol biosynthesis according to COSTET-CORIO and BENVENISTE (1988).

Structural analogues of tridemorph and fenpropimorph show relatively low toxicity to maize seedlings and bramble cells (TATON et al. 1987). A significant decrease in growth is observed at concentrations higher than 10 mg/l with bramble cells. In maize seedlings treated at 5 mg/l some growth inhibition was observed.

Growth as measured by dry-weight production in carrot, tobacco and soybean cultures is inhibited by tridemorph (HOSOKAWA et al. 1984).

Lipid synthesis

Tridemorph

Tridemorph (10 µg/ml) induces in bramble cell suspensions an increase of the total quantity of sterols. A strong accumulation of 9β, 19-cyclopropyl sterols and the disappearance of Δ^5 sterols, normally present in untreated cells, is observed. The major sterols accumulating in treated cells are cycloeucalenol and 24-methylenepollinastanol (SCHMITT et al. 1981). Tridemorph inhibits probably the cycloeucalenol-obtusifoliol isomerase in bramble cells (cf. Fig. 2).

Treatment with tridemorph modified drastically the sterol composition of maize seedlings (BLADOCHA and BENVENISTE 1983). Various 9β, 19-cyclopropyl sterols accumulate after treatment with 20 µg/ml of tridemorph. The major sterols of the treated plants are 24-methylenepollinastanol, 24-dehydrocycloeucalenol and cycloeucalenol. When the fungicide is applied at 5 µg/ml, the initial Δ^5 sterols of the leaves are replaced by both cyclopropyl sterols and Δ^8 sterols (cf. Fig. 2) indicating that besides the cycloeucalenol-obtusifoliol isomerase also the $\Delta^8 \rightarrow \Delta^7$ isomerase is inhibited (BLADOCHA and BENVENISTE 1983; BENVENISTE et al. 1984).

HOSOKAWA et al. (1984) reported the accumulation of sterols with a cyclopropane ring

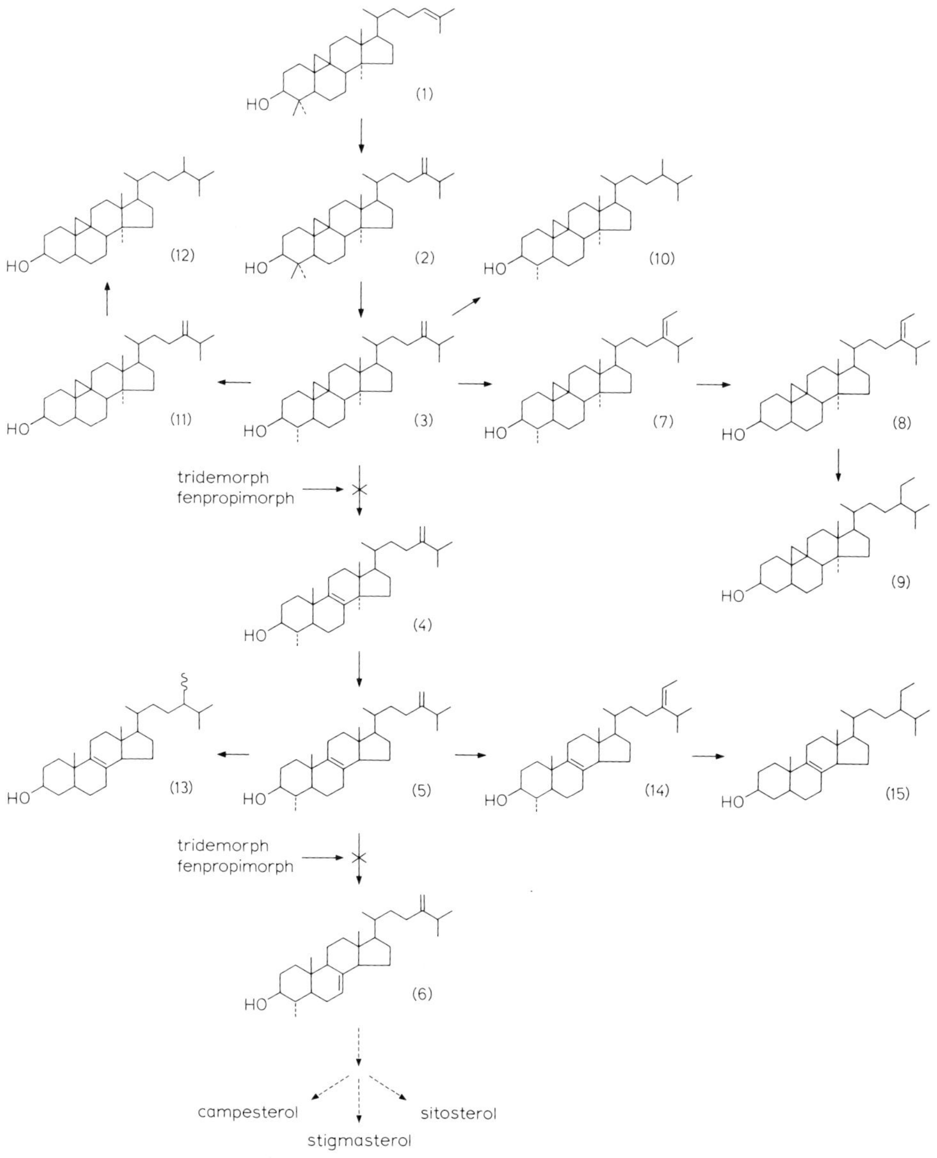

Mechanism of action of cyclic amine fungicides: morpholines and piperidines 193

Fig. 11-2 Biosynthetic pathways of campesterol, stigmasterol and sitosterol in higher plants indicating the presumed sites of inhibition by tridemorph and fenpropimorph. (1) Cycloartenol, (2) 24-methylenecycloartenol, (3) cycloeucanol, (4) obtusifoliol, (5) 4α-methyl-5α-ergosta-8,24(28)-dien-3β-ol, (6) 24-methylene lophenol, (7) cyclofontumienol, (8) 24-ethylidenepollinastanol, (9) (24ξ)-ethylpollinastanol, (10) 24(28)-dihydrocycloeucalenol, (11) 24-methylenepollinastanol, (12) (24R)-24-methylpollinastanol, (13) (24ξ)-24-methyl-5α-cholest-8-en-3β-ol, (14) 4α-methyl-5α-stigmasta-8,Z-24(28)-dien-3β-ol, (15) (24R)-24-ethyl-5α-cholest-8-en-3β-ol.

upon treatment of carrot, tobacco and soybean cultures; about 33% of total sterol in carrot, 58% in tobacco and 49% in soybean. An increase in the concentration of cyclopropyl sterols results from the accumulation of cycloeucalenol, 24-methylenecycloartanol, 24-methylenepollinastanol and 24-methylpollinastanol in tridemorph-treated cultures. Tridemorph appears to be an effective inhibitor of the cycloeucalenol-obtusifoliol isomerase in carrot, tobacco and soybean cells.

The other major tridemorph-induced modification related to the sterol composition is that tridemorph causes a reduction of sterols with a 10-carbon chain and an increase of sterols with a 9-carbon side chain. In addition, the level of 24-methylene sterols increases in the fungicide-treated cultures, especially in carrot and tobacco cultures. Tridemorph appears to be an effective inhibitor of the second alkylation at carbon 24 (HOSOKAWA et al. 1984).

A minor effect of tridemorph may be to prevent the removal of 14 α-methyl groups in soybean cultures. The other secondary effect of the fungicide is on the accumulation of sterols with $\Delta^{5,24(28)}$ double bonds, indicating the inhibition of the $\Delta^{24(28)}$ double bond reduction (HOSOKAWA et al. 1984). The latter observation in maize seedlings is also made by BLADOCHA and BENVENISTE (1985). In addition they reported that tridemorph affects indirectly the stereochemistry of the $\Delta^{24(25)}$ bond hydrogenation involved in the last step of the C-24 alkylation process.

Fenpropimorph

Fenpropimorph treatment of maize seedlings causes the accumulation of 9β, 19-cyclopropyl sterols in roots and leaves (BENVENISTE et al. 1984). When plants are treated with low concentrations (1 µg/ml) of fenpropimorph, Δ^8 sterols are detected in addition to to cyclopropyl sterols. Thus fenpropimorph, like tridemorph, acts on the cycloeucalenol-obtusifoliol isomerase and the $\Delta^8 \rightarrow \Delta^7$ isomerase (BENVENISTE et al. 1984).

Fenpropimorph treatment of maize seedlings results in the presence of 9β, 19-cyclopropyl sterols containing over 90% of 24-methyl and only 3% of 24-ethyl sterols in both roots and leaves. In both organs of control plants over 75% of the 24-ethyl sterols is present in the Δ^5 sterols (BENVENISTE et al. 1984). Similar effects have been reported for the activity of fenpropimorph on sterol composition of maize by MERCER et al. 1989. Also the effects on wheat (COSTET-CORIO and BENVENISTE 1988; KHALIL and MERCER 1991) and barley (MERCER et al. 1989) lead to accumulation of the same sterols indicating inhibition of 9β, 19-cycloeucalenol-obtusifoliol isomerase, $\Delta^8 \rightarrow \Delta^7$ isomerase, and Δ^{22} desaturase.

Female grasshoppers reared on wheat seedlings treated with fenpropimorph showed a drastic reduction in their cholesterol content and reproductive ability (COSTET et al. 1987).

N-benzyl decalin treatment of bramble cells leads to accumulation of Δ^8-sterols (RAHIER et al. 1985; TATON et al. 1987). Structure analogues having the azadecalin skeleton like N-(1,5,9-trimethyldecyl)-4α, 10-dimethyl-8-aza-trans-decal-3β-ol and n-(3-(4-tert-butyl phenyl-)2-methyl)-propyl-8-aza-4α,10-dimethyl-trans-decal-3β-ol cause mainly the accumulation of 9β, 19-cyclopropyl sterols in bramble cells and maize seedlings (TATON et al. 1987). Treatment of bramble cells and maize seedlings by the N-oxide derivatives of the N-benzyl decalin resulted in a drastic accumulation of $\Delta^{8,14}$ sterols.

Effects on organelles and enzymes of fungi and plants

The effects of tridemorph on three membrane fractions, prepared from roots of maize seedlings after treatment, are reported by BENVENISTE et al. (1984). One fraction consists of crude mitochondria, the other two are considered to be enriched in endoplasmic reticulum and plasmalemma, respectively. More than 98% of the total sterols in membranes from control plants are Δ^5 sterols, whereas in membranes from treated plants less than 7% is present in this form. On the other hand, cyclopropyl sterols which are not detected in control membranes, accounted to for more than 90% of the total sterols in membranes from treated plants. In addition, the sterol profile was almost identical in the three membrane fractions from treated plants and very similar to the sterol profile of the whole tissue (BENVENISTE et al. 1984). Similar results have been reported for the effect of fenpropimorph on the microsomal membrane fractions of *Saccharomyces cerevisiae* (STEEL et al. 1987). No sterols were found in the cytosolic fraction.

When microsomes from maize seedlings are incubated in the presence of cycloeucanol with various amounts of tridemorph, the fungicide strongly inhibits the cycloeucanol-obtusifoliol isomerase ($I_{50} = K_1 = 0.5-1$ µM; K_m for cycloeucalenol = 100 µM) (RAHIER et al. 1983; SCHMITT et al. 1982). The cycloeucalenol-obtusifoliol isomerase has a higher affinity for tridemorph than for its substrate cycloeucalenol. Fenpropimorph also strongly inhibits the cycloeucalenol-obtusifoliol isomerase in a cell-free extract from maize seedlings (BENVENISTE et al. 1984; RAHIER et al. 1983).

Tridemorph has been reported to inhibit both the NADH-oxidase and the succinate cytochrome C oxydoreductase system of non phosphorylating electron-transfer particles from bovine heart mitochondria (MÜLLER and SCHEWE 1976). Reinvestigations with tridemorph double distilled, and investigations with fenpropimorph using this system reveal that especially the NADH-oxidase is fairly sensitive, although sterol synthesis is inhibited at lower concentrations (LYR et al., unpublished results). The EC_{50} of tridemorph is 28 µM for succinate cytochrome oxydoreductase and 16 µM for NADH-oxidase. For fenpropimorph these values are 33 and 8 µM, respectively.

Tridemorph as Calixin inhibits the chitin synthetase of *Mucor rouxii in vitro* (LYR and SEYD 1978). At 10^{-4} M of tridemorph the incorporation of UDP-^{14}C-glucosamine in chitin is 59% of the control. The EC_{50} is reported to be 5×10^{-4} M. The inhibition is ascribed to the lytic activity of tridemorph, due to the alkyl chain and positive charge of the morpholine ring. Phospholipids are believed to be essential for the activity of the enzyme and lysis leads to inactivation of chitin synthetase.

BALOCH et al. (1984 b) has described the effects of morpholines on the sterol Δ^{14} reductase using microsomes of *Saccharomyces cerevisiae*. The yeast S_8 preparation from anaerobically grown yeast has been used in 0.1 M phosphate buffer, pH 6.2. Several concentrations of the fungicides are tested. The substrate is radiolabelled ignosterol and the product ergost-8-en-3β-ol. The radioactivity in ignosterol and ergost-8-en-3β-ol in the control incubation is used to determine the % conversion of substrate to product. Taking the conversion in the control as 100%, the degree of inhibition in the fungicide-containing samples is calculated. Determination of the I_{50} values from dose-response curves reveals that fenpropimorph is a much potent inhibitor of the Δ^{14} reductase than tridemorph (100-fold) and that fenpropidin with the same N-alkyl substituents, but a different N-heterocycle as fenpropimorph, has an almost identical I_{50} value as fenpropimorph. From these observations they conclude that the Δ^{14} reductase inhibitory activity of these three fungicides resides principally in the N-alkyl substituent rather than in the heterocyclic ring.

Confirmation in enzyme assays which have been developed in the mean time (BALOCH and MERCER 1987; MERCER 1987; RAHIER et al. 1985; 1986) has been obtained. Differences in efficacy of the parent morpholines in the various enzyme systems are detected. BALOCH and MERCER (1987) showed in the sterol $\Delta^8 \to \Delta^7$ isomerase enzyme assay that (i) the cis-2,6-dimethylmorpholine isomer of fenpropimorph is more inhibitory than the trans-2,6-dimethyl isomer or the morpholine analogue of fenpropidin, (ii) demethylated fenpropimorph is more inhibitory than the trans-2,6-dimethylmorpholine isomer of fenpropimorph, (iii) cis-3,5-dimethylfenpropidin, the piperidine analogue of cis-fenpropimorph is more inhibitory than fenpropidin and (iv) trans-3,5-dimethyl-fenpropidin is less inhibitory than fenpropidin. RAHIER et al. (1985) reported I_{50} values in the 9β,19-cycloeucalenol-obtusifoliol isomerase enzyme for tridemorph, fenpropimorph, and dodemorph of 0.4 µM, 0.4 µM, and 5 µM, respectively. The N-oxide of fenpropimorph proved also to be a good inhibitor of this enzyme. The S-enantiomer in the propyl moiety of fenpropimorph is an order of magnitude more effective than the R-enantiomer.

Whereas the morpholines, tridemorph and fenpropimorph, have been reported to inhibit the $\Delta^5 \to \Delta^4$-ketosteroid isomerase of various types of cholesterol oxidase (HESSELINK et al. 1989, 1990), the piperidines, fenpropimorph, are ineffective.

Tridemorph and fenpropimorph inhibit in microsomes of higher plants the $\Delta^{5,7}$ sterol-Δ^7 reductase (TATON and RAHIER 1991).

In addition to the effect of these established cyclic amines a high number of structural analogues have been tested in these enzyme systems (AKERS et al. 1991; RAHIER et al. 1985, 1986, 1990; RAMOS TOMBO and BELLUS 1991; TATON et al. 1987). Very potent enzyme inhibitors have been obtained but under field conditions none showed significant advantages (AKERS et al. 1991).

Discussion

Although various authors have demonstrated that cyclic amines strongly inhibit sterol biosynthesis, it is still questioned whether this inhibition is the sole basis of toxicity in various organisms. Inhibition of growth in fungi and higher plants may not only result from the effect of the drug on the sterol profile of membranes but also from side effects. NADH-oxidase and succinate cytochrome C reductase are sensitive to inhibition (LYR et al., unpublished results). In chloroplasts a repression of the netto assimilation in wheat leaves is observed which can not be explained by inhibition of sterol biosynthesis (LYR et al., unpublished results). COSTET-CORIO and BENVENISTE (1988) ascribe phytotoxicity in wheat to another, unknown target, than sterol biosynthesis. However, mutants resistant to cyclic amines affected in sterol biosynthesis have been frequently described in fungi and higher plant cells (ADAMS et al. 1992; ELLIS et al. 1991; FALANDRE, DE et al. 1987; SCHALLER et al. 1991).

Gram-positive bacteria like *Bacillus subtilus, Streptococcus lactis, Streptococcus faecalis* and *Mycobacterium phlei*, are sensitive to tridemorph (KERKENAAR et al. 1979). Tridemorph as Calixin is also active against the phytopathogenic bacteria *Corynebac-*

terium imichiganense, Corynebacterium fascians and other *Corynebacterium* spp. *in vitro* (OROS and SÜLE 1983). Also effects of Calixin have been reported on soil organisms and on the nitrogen cycle (BANERJEE and BANERJEE 1991) and on T-2 toxin production by *Fusarium sporotrichoides* (MOSS and FRANK 1985).

Since sterols are of limited occurrence in most bacteria (OURISSON and ROHMER 1982), it is difficult to believe that the antibacterial activity of tridemorph originates from interference with sterol biosynthesis.

The effects, altered morphology and inhibition of dry-weight increase, observed on fungal growth, are preceded by an irregular deposition of polysaccharides, probably chitin. These effects, however, appear much later than the observed effects on sterol biosynthesis. The irregular deposition of probably chitin is contradictory to the reported inhibition of chitin synthetase from *Mucor*. The latter inhibition is suggested to be a consequence of phospholipid breakdown, whereas the former effects are attributed to the altered sterol pattern. These and other effects based on membrane changes could account for a lethal action in filamentous fungi and a highly detrimental action in budding fungi, due to differences in cell wall composition (KERKENAAR and BARUG 1984).

In fungi two sites of action have been indicated for cyclic amines: the $\Delta^8 \rightarrow \Delta^7$ isomerase and the Δ^{14} reductase. Also in higher plants two sites have been described: the $\Delta^8 \rightarrow \Delta^7$ isomerase and the cycloeucalenol-obtusifoliol isomerase (*cf.* Fig. 11.3). In the

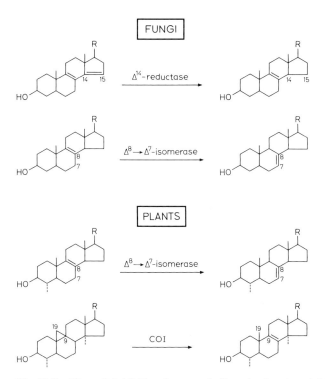

Fig. 11.3 Sites of inhibition by morpholines in fungi and higher plants: Δ^{14} reductase and $\Delta^8 \rightarrow \Delta^7$ isomerase in fungi, and $\Delta^8 \rightarrow \Delta^7$ isomerase and cycloeucalenol-obtusifoliol isomerase (COI) in higher plants.

case of tridemorph or fenpropimorph and fungi the inhibition site various from organism to organism and seems to be dependent on culture conditions. Oxygenation seems to play an important role as indicated by KERKENAAR (1990) in the way the sterols accumulate. In higher plants the inhibition site may vary from organism to organism and organ to organ. If the cycloeucalenol-obtusifoliol isomerase or the Δ^{14} reductase is completely inhibited, inhibition of the $\Delta^8 \rightarrow \Delta^7$ isomerase cannot be detected. The potential of cyclic amines to inhibit in principal these sites has been proven in enzyme assays. In spite of this, the real situation in intact organs and cells might be completely different. Metabolism of the compounds has to be considered. In higher plants N-oxidation of tertiary amines has been suggested (TATON et al. 1987). This mechanism has also been suggested for fungi and even correlated to the sensitivity of the various species (GASZTONIY and JOSEPOVITS 1984). This may also be in line with the explanation of the results of KERKENAAR (1983; 1990) on the effect of limited oxygen supply on sterol composition and with those of DEBIEU et al. (1992) with regard to differences in sensitivity in *Fusarium* species. Even the sensitivity of bacteria which do not synthesize sterols can be explained if the N-oxidation is a validate hypothesis. N-oxido-cyclic amines are potent biocides (KERKENAAR 1990).

At the moment evidence is available, obtained from *in vitro* as well as from *in vivo* experiments, that N-oxidation of the cyclic amines alters the affinity for certain enzymes e.g. Δ^{14} reductase.

The differences existing between higher plants and fungi could be explained partly in considering that the cycloeucalenol-obtusifoliol isomerase is present only in photosynthetic eukaryotes and had never been found in non-photosynthetic eukaryotes (BENVENISTE et al. 1984; BERG 1983; BERG et al. 1983). However, using tridemorph and fenpropimorph as inhibitors of sterol biosynthesis in the protozoan *Acanthamoeba polyphaga*, RAEDERSTORFF and ROHMER (1987) observed accumulation of 9β, 19-cyclopropyl- and Δ^8-sterols, showing inhibition patterns in protozoa similar to higher plants, 9β,19-cycloeucalenol-obtusifoliol- and $\Delta^8 \rightarrow \Delta^7$-isomerase, and the presence of the former mentioned enzyme in a non-photosynthetic organism. On the other hand, it is assumed that in higher plants the sterol Δ^{14} reductase is also active like in fungi (BOTTEMA and PARKS 1978) and mammals (PAIK et al. 1984). Surprisingly, this enzyme has not been reported to become inhibited in spite of the fact that SCHMITT et al. (1981) reported the accumulation of small amounts of $\Delta^{8,14}$ sterols in bramble cells. $\Delta^8 \rightarrow \Delta^7$ Isomerase in contrast has been reported to become inhibited. The enzymes Δ^{14} reductase, cycloeucalenol-obtusifoliol isomerase and $\Delta^8 \rightarrow \Delta^7$ isomerase show similarities. It has been suggested that $\Delta^8 \rightarrow \Delta^7$ isomerization reaction starts with a Δ^8 hydrogenation (BENVENISTE et al. 1984; BERG 1983; BERG et al. 1983). This hydrogenation is assumed to be an NADPH/H$^+$-dependent reduction of the Δ^8 double bond, similar to the NADPH/H$^+$-dependent reduction of the Δ^{14} double bond. Also the ring opening of the 9β, 19-cyclopropane ring could depend on such a mechanism. BENVENISTE et al. (1984) propose the formation of a carbonium ion high energy intermediate upon addition of a proton at C-14 for the reductase, C-8 for the cycloeucalenol-obtusifoliol isomerase and C-8 for the $\Delta^8 \rightarrow \Delta^7$ isomerase. These carbonium ions are very close to each other (less than 0.15 nm). It is known that N-alkyl morpholines (pK_a around 7−8) lead to the formation of ammonium (Morpholinium) ions at pH 7.4. They suggest that morpholinium ions, having electronic and broad structural similarities with the three carbonium ion high energy intermediates, would have a high affinity for the three enzymes and would be able to inhibit them. The likelihood of this hypothesis has been verified by these researchers and others (AKERS et al. 1991; HESSELINK et al. 1990; RAHIER et al. 1985, 1986; RAMOS TOMBO and BELLUS 1991; TATON and RAHIER 1987; TATON 1991). In this concept fits also the inhibition of the $\Delta^5 \to \Delta^4$ ketosteroid isomerase and the plant $\Delta^{5,7}$-sterol-Δ^7 isomerase. However, the hypothesis that cyclic amines act by virtue of being sterol mimics has to be viewed with some caution because these compounds may act in some cases as some kind of allosteric inhibitors (AKERS et al. 1991; HESSELINK et al. 1990).

The inhibition of the NADH-oxidase may fit in the concept of inhibition of NADPH-oxydoreductase and probably also the effect on the 2,3,5-triphenyltetrazoliniumchloride (TTC) reaction in the TTC test in plants as indicator of phytotoxic effects (LYR et al., unpublished results).

The differences in effects of tridemorph and fenpropimorph on $\Delta^8 \to \Delta^7$ isomeres and Δ^{14} reductase is believed to be due to the structural differences of these fungicides. It is hypothesized that the 2,4-dimethylmorpholine ring binds more efficiently to the sterol B ring than other rings, resulting in inhibition of the $\Delta^8 \to \Delta^7$ isomerase. The N-substituent of fenpropimorph gives a better fit and more efficient binding to the sterol moiety, consisting of the rings C and D plus side chain in the enzyme system than the N-substituent of tridemorph. The observations that trans-1,4-cyclohexyl isomers of fenpropimorph are much more potent inhibitors, and presumably sterol Δ^{14} reductase inhibitors, than the cis-1,4-isomers, are in agreement with this hypothesis. The trans-1,4-isomers have an essentially planar structure that matches the sterol rings C, D and side chain well (BALOCH et al. 1984).

Recently some researches indicated accumulation of squalene in fungi upon treatment with cyclic amines (ZIOGAS et al. 1991; SCHNEEGURT and HENRY 1992) and for this reason inhibition of squalene epoxidase. This seems rather a consequence of inhibition on later steps than inhibition of the enzyme itself. Others (DEBIEU et al. 1992) reported no correlation of squalene accumulation and fenpropimorph treatment. Also here the final prove lies in the inhibition of the enzyme.

In conclusion: the cyclic amines inhibit various types of isomerases and reductases involved in the biosynthesis of sterols in vitro and in vivo based on a general principle. Most sensitive are the 9β,19-cycloeucalenol-obtusifoliol isomerase, the $\Delta^8 \to \Delta^7$ isomerase, and Δ^{14} reductase. In vitro the $\Delta^5 \to \Delta^4$ ketosteroid isomerase and the $\Delta^{5,7}$-sterol-Δ^7 reductase are inhibited.

The final result on sterol biosynthesis in target organism will be the resultante of various parameters like growth conditions, oxygenation, and uptake as well as metabolism of the cyclic amines.

References

ADAMS, I., ELLIS, S. W., DE WAARD, M. A., and GRINDLE, M.: Molecular analysis of the mode of action of morpholine fungicides. In press.

AKERS, A., AMMERMANN, E., BUSCHMANN, E., GÖTZ, N., HIMMELE, W., LORENZ, G., POMMER, E.-H., RENTZEA, C., RÖHL, F., SIEGEL, H., ZIPPERER, B., SAUTER, H., and ZIPPLIES, M.: Chemistry and biology of novel amine fungicides: attempts to improve the antifungal activity of fenpropimorph. Pestic. Sci. 31 (1991): 521–538.

BALOCH, R. I., and MERCER, E. I.: Inhibition of sterol $\Delta^8 \to \Delta^7$ isomerase and Δ^{14} reductase by fenpropimorph, tridemorph and fenpropidin in cell-free enzyme systems from *Saccharomyces cerevisiae*. Phytochemistry **26** (1987): 663–668.
— — WIGGINS, T. E., and BALDWIN, B. C.: Where do morpholines inhibit sterol biosynthesis? Brit. Crop Protect. Conf. on Pests Dis. vol. 3 (1984a): 893–898.
— — — — Inhibition of ergosterol biosynthesis by tridemorph, fenpropimorph, and fenpropidin. Phytochemistry **23** (1984b): 2219–2226.
BANERJEE, A., and BANERJEE, A. K.: Effect of the fungicides tridemorph and vinclozolin on soil microorganisms and nitrogen metabolism. Folia Microbiol. **36** (1991): 567–571.
BARUG, D., and KERKENAAR, A.: Resistance in mutagenic induced mutants of *Ustilago maydis* to fungicides which inhibit sterol biosynthesis. Pestic. Sci. **15** (1984): 78–84.
— — Cross-resistance of UV-induced mutants of *Ustilago maydis* to various fungicides which interfere with ergosterol biosynthesis. Mede. Rijksfac. Landbouwwetensch. Gent **44** (1979): 421–427.
BENVENISTE, P., BLADOCHA, M., COSTET, M.-F., and EHRHARD, A.: Use of inhibitors of sterol biosynthesis to study plasmalemma structure and functions. In: GOUDIA, E. M. (Ed.): Membranes and compartmentalization in the regulation of plant functions. Ann. Proc. Phytochem. Soc. Europe, vol. **24** (1984): 283–300.
BERG, D.: Biochemical mode of action of fungicides. Studies on ergosterol biosynthesis inhibitors. Burdick and Jackson Symposium 188th ACS National Meeting, Philadelphia, Pennsylvania, USA, August 27, 1984, pp. 1–26.
— KRAEMER, W., REGEL, E., BUECHEL, K.-H., HOLMWOOD, G., PLEMPEL, M., and SCHEINPFLUG, H.: Mode of action of fungicides: studies on ergosterol biosynthesis inhibitors. Brit. Crop Protect. Conf. on Pests Dis. vol. 3 (1984): 887–892.
BERG, L. R.: The effect of triarimol and tridemorph on the sterol composition of five species of *Oomycetes*. Ph. D. Thesis, University of Maryland, College Park, MD 1983, pp. 1–43.
— PATTERSON, G. W., and LUSBY, W. R.: Effects of triarimol and tridemorph on sterol biosynthesis in *Saprolegnia ferax*. Lipids **18** (1983): 448–452.
— — Phylogenetic implications of sterol biosynthesis of sterol biosynthesis in the *Oomycetes*. Exp. Mycol. **10** (1986): 175–183.
BERGMANN, H.: Die Wirkung von Tridemorph auf *Torulopsis candida*. In: LYR, H., and POLTER, C. (Eds.:) Systemfungizide. Akademie-Verlag, Berlin 1975, pp. 215–224.
— Wirkung von Tridemorph auf *Torulopsis candida*. Z. Allg. Mikrobiol. **19** (1979): 155–162.
— LYR, H., KLUGE, E., and RITTER, G.: Untersuchungen zur Wirkungsweise von Tridemorph. In: LYR, H., and POLTER, C. (Eds.): Systemfungizide. Akademie-Verlag, Berlin, 1975 pp. 183–188.
— MÜLLER, R., and NEUHAUS, W.: Bestimmung der phytotoxischen Wirkung von Calixin auf Kulturpflanzen. In: LYR, H., and POLTER, C. (Eds.): Systemische Fungizide und antifungale Verbindungen. Akademie-Verlag, Berlin 1982, pp. 335–359.
BLADOCHA, M., and BENVENISTE, P.: Manipulation by tridemorph, a systemic fungicide, of the sterol composition of maize leaves and roots. Pl. Physiol., Lancaster **71** (1985): 756–762.
BOTTEMA, C. K., and PARKS, L. W.: Δ^{14}-sterol reductase in *Saccharomyces cerevisiae*. Biochim. Biophys. Acta **531** (1978): 301–307.
BUCHENAUER, H.: Various response of cell membranes of plant and fungal, cells to different systemic fungicides. Proc. Int. Plant Prot. Moscow **3** (1975): 94–110.
CORTÉS, F., ESCALZA, P., MORENO, J., and LOPEZ-CAMPOS, J. L.: Effects on the fungicide tridemorph on mitosis in *Allium cepa*. Cytobios. **34** (1982): 181–190.
COSTET, M. F., EL ACHOURI, M., CHARLET, M., LANOT, R., BENVENISTE, P., and HOFFMANN, J. A.: Ecdysteroid biosynthesis amd embryonic development are disturbed in insects (*Locusta migratoria*) reared on plant diet (*Triticum sativum*) with a selectively modified sterol profile. Proc. Natl. Acad. Sci. USA **84** (1987): 643–647.

COSTET-CORIO, M.-F., and BENVENISTE, P.: Sterol metabolism in wheat treated by N-substituted morpholines. Pestic. Sci. **22** (1988): 343–357.

DEBIEU, D., GALL, C., GREDT, M., BACH, J., MALOSSE, C., and LEROUX, P.: Ergosterol biosynthesis and its inhibition by fenpropimorph in *Fusarium* species. Phytochemistry **31** (1992): 1223–1233.

ELENWA, E. N., HOLGAT, M. E., and CLIFFORD, D. R.: Chemical control of gangrene in potatoes. Ann. Appl. Biol. Suppl. **102** (1983): 70–71.

ELLIS, S. W., ROSE, M. E., and GRINDLE, M.: Identification of a sterol mutant of *Neurospora crassa* deficient in $\Delta^{14,15}$-reductase activity. J. Gen. Microbiol. **137** (1991): 2627–2630.

FALANDRE, A. DE, BOUVIER-FOURCADE, I., SENG, J.-M., and LEROUX, P.: Induction and characterization of *Penicillium caseicolum* mutants resistant to ergosterol biosynthesis inhibitors. Appl. Environm. Microbiol. **53** (1987): 1500–1503.

FISHER, D. J.: A note on the mode of action of the systemic fungicide tridemorph. Pestic. Sci. **5** (1975): 219–224.

GARG, D. K., and MEHROTRA, R. S.: Effect of some fungicides on growth, respiration and enzyme activity of *Fusarium solani* f. sp. *pisi*. Z. Pflanzenschutz **10** (1975): 570–576.

GASZTONYI, M., and JOSEPOVITS, G.: Metabolism of some sterol inhibitors in fungi and higher plants, with special reference to the selectivity of fungicidal action. Pestic. Sci. **15** (1984): 48–55.

HESSELINK, P. G. M., KERKENAAR, A., and WITHOLT, B., Novel inhibitors of *Nocardia erythropolis* cholesterol oxidase: screening and characterization. Pestic. Biochem. Physiol. **33** (1989): 69–77.

– – – Inhibition of microbial cholesterol oxidases by dimethylmorpholines. J. Steroid Biochem. **35** (1990): 107–113.

HOLLOMON, D. W.: The effect of tridemorph on barley powdery mildew: its mode of action and cross sensitivity relationships. Phytopath. Z. **105** (1982): 279–287.

HOSOKAWA, G., PATTERSON, G. W., and LUSBY, W. R.: Effects of triarimol, tridemorph and triparanol on sterol biosynthesis in carrot, tobacco and soybean suspension cultures. Lipids **19** (1984): 449–456.

KAARS SIJPESTEIJN, A.: Effects on fungal pathogens. Chapter 7, In: MARSH, R. W. (Ed.): Systemic Fungicides. 2nd ed., Longman, London 1972, p. 153.

KATO, T.: Biosynthetic processes of ergosterol as the target of fungicides. In: Pesticide Chemistry: human welfare and the environment. Proc. 5th Intern. Congr. Pesticide Chemistry, Kyoto, Vol. 3. Pergamon Press, Oxford 1983, pp. 33–41.

– SHOAMI, M., and KAWASE, Y.: Comparison of tridemorph with buthiobate in antifungal mode of action. J. Pestic. Sci. **5** (1980): 69–79.

KERKENAAR, A.: Mode of action of tridemorph and related compounds. In: Pesticide Chemistry: human welfare and the environment. Proc. 5th Intern. Congr. Pesticide Chemistry, Kyoto, vol. 3. Pergamon Press, Oxford 1983, pp. 123–127.

– The mode of action of dimethylmorpholines. In: FROMTLING, R. A. (Ed.): International telesymposium on recent trends in the discovery, development and evaluation of antifungal agents. J. R. Prous S. A., Barcelona, Spain 1987, pp. 523–542.

– Inhibition of the sterol Δ^{14}-reductase and $\Delta^8 \rightarrow \Delta^8$-isomerase in fungi. Biochem. Soc. Trans. **18** (1990): 59–61.

– and BARUG, D.: Fluorescence microscopic studies of *Ustilago maydis* and *Penicillium italicum* after treatment with imazalil and fenpropimorph. Pestic. Sci. **15** (1984): 199–205.

– and KAARS SIJPESTEIJN, A.: On a difference in the antifungal activity of tridemorph and its formulated product Calixin. Pestic. Biochem. Physiol. **12** (1979): 124–129.

– VAN ROSSUM, J. M., VERSLUIS, G. G., and MARSMAN, J. W.: Effect of fenpropimorph and imazalil on sterol biosynthesis in *Penicillium italicum*. Pestic. Sci. **15** (1984): 177–187.

- Scheffer, R. J., Brakenhoff, A. C., Nielander, H. B., and Elgersma, D. M.: On the chemical control of *Ophiostoma ulmi* with sterol biosynthesis inhibitors *in vitro* and *in vivo*. 6th Internat. Congr. Pestic. Chem. (August 10–15, Ottawa) 1986; Abstract.
- Uchiyama, M., and Versluis, G. G.: Specific effects of tridemorph on sterol biosynthesis in *Ustilago maydis*. Pestic. Biochem. Physiol. **16** (1981): 97–104.

Khalil, I. A., and Mercer, E. I.: Accumulation of 9β,19-cyclopropylsterol in cereals treated with fenpropimorph. J. Agric. Food Chem. **39** (1991): 404–407.

Leroux, P., and Gredt, M.: Effets de quelques fongicides systémiques sur la biosynthèse de l'ergostérol chez *Botrytis cinerea* Pers., *Penicillium expansum* Link et *Ustilago maydis* (DC) Cda., Ann. Phytopathol. **10** (1978): 45–60.

- – Études sur les inhibiteurs de la biosynthèse des stérols fongiques: I. – Fungicides provoquant l'accumulation des desméthylstérols. Agronomie **3** (1983): 123–130.
- – and Boeda, P.: Resistance to inhibitors of sterol biosynthesis in field isolates or laboratory strains of the eyespot pathogen *Pseudocercosporella herpotrichoides*. Pestic. Sci. **23** (1988): 119–129.

Lyr, H., and Seyd, W.: Hemmung der Chitin-Synthetase von *Mucor rouxii in vitro* durch Fungizide und andere Wirkstoffe. Z. Allg. Mikrobiol. **18** (1978): 721–729.

Mercer, E. I.: The use of enzyme systems to assay the relative efficacy of ergosterol-biosynthesis-inhibiting fungicides. In: Lyr, H., and Polter, C. (Eds.): Systemic fungicides and antifungal compounds. Akademie-Verlag, Berlin 1987, pp. 115–120.
- The mode of action of morpholines. In: Berg, D., and Plempel, M. (Eds.): Sterol biosynthesis inhibitors: Pharmacological and agrochemical aspects. Ellis Horwood Ltd, Chichester, UK 1988, pp. 120–150.
- Morpholine antifungals and their mode of action. Biochem. Soc. Trans. **19** (1991): 788–793.
- Khalilk, I. A., and Wang, Z. X.: Effect of some sterolbiosynthesis-inhibiting fungicides on the biosynthesis of polyisoprenoid compounds in barley seedlings. Steroids **53** (1989): 393–412.

Moss, M. O., and Frank, J. M., Influence of the fungicide tridemorph on T-2 toxin production by *Fusarium sporotrichoides*. Trans. Br. Mycol. Soc. **84** (1985): 585–590.

Müller, W., and Schewe, T.: Das Systemfungizid Tridemorph auf Elektronentransportpartikel aus Rinderherz-Mitochondrien. In: Lyr, H., and Polter, C. (Eds.): Systemfungizide. Akademie-Verlag, Berlin 1982, pp. 417–420.

Oros, G., and Süle, S.: Antibacterial activity of Calixin on phytopathogenic bacteria. In: Lyr, H., and Polter, C. (Eds.): Systemische Fungizide und antifungale Verbindungen. Akademie-Verlag, Berlin 1982, pp. 417–420.

Ourisson, G., and Rohmer, M.: Prokaryotic polyterpenes: phylogenetic precursos of sterols. In: Bronner, F., and Kleinzeller, A. (Eds.): Current topics in membranes and transport, vol. 17: Membrane lipids of prokaryotes. Academic Press, New York 1982, pp. 153–182.

Paik, Y.-K., Trzaskos, J. M., Shafiee, A., and Gaylor, J. L.: Microsomal enzymes of cholesterol biosynthesis from lanosterol. Characterization, solubilization, and partial purification of NADPH-dependent $\Delta^{8,14}$-steroid 14-reductase. J. Biol. Chem. **259** (1984): 13413–13423.

Polak, A.: Antifungal activity *in vitro* of Ro 14-4767/002, a phenylpropyl-morpholine. Sabouraudia **22** (1983): 205–213.
- Morpholines in clinical use. In: Berg, D., and Plempel, M. (Eds.): Sterol biosynthesis inhibitors: Pharmacological and agrochemical aspects. Ellis Horwood Ltd, Chichester, UK 1988, pp. 430–448.
- and Dixon, D. M.: Antifungal activity of amorolfine (Ro 14-4767/002) *in vitro* and *in vivo*. In: Fromtling, R. A. (Ed.): International Telesymposium on recent trends in the discovery, development and evaluation of antifungal agents. J. R. Prous S. A., Barcelona, Spain 1987, pp. 555–573.

Polter, C., and Casperson, G.: Einfluß von Tridemorph auf Ultrastruktur und Lipidmetabolismus von *Botrytis cinerea.* In: LYR, H., and POLTER, C. (Eds.): Systemfungizide. Akademie-Verlag, Berlin 1979, pp. 225–241.

RAEDERSTORFF, D., and ROHMER, M.: Eur. J. Biochem. **164** (1987): 421–426.

RAHIER, A., BOUVIER, P., CATTEL, L., NARULA, A., and BENVENISTE, P.: Inhibition of 2,3-oxidosqualene-β-amyrin cyclase, (S)-adenosyl-L-methione-cycloartenol-C-24-methyltransferase and cycloeucalenol-obtusifoliol isomerase by rationally designed molecules containing a tertiary amine function. Biochem. Soc. Trans. **11** (1983): 537–543.

– NAVE, P., SCHMITT, P., and BENVENISTE, P.: Manipulation of sterol biosynthesis in higher plant cells and physiological consequences. In: BÖGER, P. (Ed.): Physiologische Schlüsselprozesse in Pflanze und Insekt. Universitäts-Verlag Konstanz GmbH Konstanz 1985, pp. 125–153.

– SCHMITT, P., HUSS, B., BENVENISTE, P., and POMMER, E.-H.: Chemical structure-activity relationships of the inhibition of sterol biosynthesis by N-substituted morpholines in higher plants. Pestic. Biochem. Physiol. **25** (1986): 112–124.

– TATON, M., and BENVENISTE, P.: Inhibition of sterol biosynthesis enzymes *in vitro* by analogues of high-energy carbocationic intermediates. Biochem. Soc. Trans. **18** (1990): 48–52.

– – SCHMITT, P., BENVENISTE, P., PLACE, P., and ANDING, C.: Inhibition of $\Delta^8 \to \Delta^7$-sterol isomerase and of cycloeucalenol-obtusifoliol isomerase by *N*-benzyl-8-aza-4α,10-dimethyl-*trans*-decal-3β-ol, an analogue of a carbocationic high energy intermediate. Phytochemistry **24** (1985): 1223–1232.

RAMOS TOMBO, G. M., and BELLUS, D.: Chirality and crop protection. Angew. Chem. Int. Ed. Engl. **30** (1991): 1193–1215.

RAPOPORT, S. M.: Zur Wirkung des Calixins auf Elektronentransportartikel aus Rinderherz-Mitochondrien. Symposium on Systemic Fungicides (Abstract), Reinhardsbrunn, May 1974, p. 29.

SCHALLER, H., MAILLOT-VERNIER, P., BENVENISTE, P., and BELIARD, G.: Sterol composition of tobacco calli selected for resistance to fenpropimorph. Phytochemistry **30** (1991): 2547–2554.

SCHEFFER, R. J., BRAKENHOFF, A. C., KERKENAAR, A., and ELGERSMA, D. M.: Control of Dutch elm disease by the sterol biosynthesis inhibitors fenpropimorph and fenpropidin. Neth. J. Pl. Path. **94** (1988): 161–173.

SCHLÜTER, K., and WELTZIEN, H. C.: Ein Beitrag zur Wirkungsweise systemischer Fungizide auf *Erysiphe graminis*. Meded. Rijksfac. Landbouwwetensch. Gent **36** (1971): 1159–1164.

SCHMITT, P., BENVENISTE, P., and LEROUX, P.: Accumulation of 9β,19-cyclopropyl sterols in suspension cultures of bramble cells cultured with tridemorph. Phytochemistry **20** (1981): 2153–2158.

– RAHIER, A., and BENVENISTE, P.: Inhibition of sterol biosynthesis in suspension cultures of bramble cells. Physiol. Vég. **20** (1982): 559–571.

SCHNEEGURT, M. A., and HENRY, M.: Effects of piperalin and fenpropimorph on sterol biosynthesis in *Ustilago maydis*. Pestic. Biochem. Physiol. **43** (1992): 45–62.

SIDDIQUI, K. A., and HAAHR, V.: Different reactions of wheat mutants to a systemic fungicide. Naturwissenschaften **58** (1971): 415–416.

STEEL, C. C., BALOCH, R. I., MERCER, E. I., and BALDWIN, B. C.: The intracellular location and physiological effects of abnormal sterols in fungi grown in the presence of morpholine and functionally related fungicides. Pestic. Biochem. Physiol. **33** (1989): 101–111.

TATON, M., BENVENISTE, P., and RAHIER, A.: Comparative study of the inhibition of sterol biosynthesis in *Rubus fruticosus* suspension cultures and *Zea mays* seedlings by N-(1,5,9-trimethyldecyl)-4α,10-dimethyl-8-aza-*trans*-decal-3β-ol and derivates. Phytochemistry **26** (1987): 385–392.

– and RAHIER, A.: Identification of $\Delta^{5,7}$-sterol-Δ^7-reductase in higher plant microsomes, Biochem. Biophys. Res. Commun. **181** (1991): 465–473.

WAARD, M. A. DE, BANGA, M., and ELLIS, S. W.: Characterization of the sensitivity of *Erysiphe graminis* f. sp. *tritici* to morpholines. Pestic. Sci. **34** (1992): 374–375.

ZIOGAS, B. N., VITORATOS, A. G., SIDERIS, E. G., and GEORGOPOULOS, S. G.: Effects of sterol biosynthesis inhibitors on mitosis. Pestic. Biochem. Physiol. **37** (1990): 254–265.

– OESTERHELT, G., MASNER, P., STEEL, C. C., and FURTER, R.: Fenpropimorph: a three site inhibitor ergosterol biosynthesis in *Nectria haematocca* var. *cucurbitae*. Pestic. Biochem. Physiol. **39** (1991): 74–83.

Chapter 12

DMI fungicides

K. H. Kuck*, H. Scheinpflug** and R. Pontzen*

* Institute for Plant Diseases, BAYER AG, Monheim, Germany
** formerly Institute for Plant Diseases, BAYER AG, Monheim, Germany

Introduction

In the second half of the 1960's, several independent research groups synthesised numerous fungicides from chemically heterogeneous classes. Subsequently these fungicides have been shown to have a common biochemical target in fungal metabolism, namely the biosynthesis of ergosterol.

In 1967, BASF presented the first morpholine derivative, dodemorph. Within the same year, several imidazole and pyrimidine derivatives were patented by various other companies. The imidazole clotrimazole, miconazole, econazole and isoconazole were patented by Bayer and Janssen and the pyrimidines triarimol, fenarimol and nuarimol were patented by Eli Lilly.

In 1968 the piperazine derivative triforine from Boehringer and the first triazole compound, fluotrimazole, from Bayer were patented.

With Sumitomo's presentation of buthiobate, a pyridine derivative, in 1970, at least one representative of the chemical groups now classified as ergosterol biosynthesis inhibiting fungicides (EBIs) had been synthesised. Although ergosterol is the main sterol of most fungi, there are a lot of exceptions. Several powdery mildew fungi and rust fungi contain no ergosterol. The major sterol of powdery mildew was identified as ergosta-5,24(28)-dienol, and rust fungi synthesise mainly Δ^7-sterols such as stigmast-7-enol, stigmasta-5,7-dienol and stigmasta-7,24(28)-dienol (LOEFFLER et al. 1984; JACKSON and FREAR 1968; KOELLER 1992). Therefore the term "sterol biosynthesis inhibitor" (SBI) is often used instead of EBI.

Biochemical investigations by the groups of SISLER (cited in SISLER et al. 1984), KATO et al. (1974), BUCHENAUER (1977) and others have shown that — with the exception of the morpholines and the more recently introduced compound fenpropidin — all above mentioned compounds have a common site of action within the fungal sterol biosynthesis pathway. That is an inhibition of demethylation at position 14 of lanosterol or 24-methylene dihydrolanosterol, which are precursors of sterols in fungi (chapter 13).

All compounds with this mechanism of action have been grouped together as demethylation inhibitors (DMIs) in order to distinguish them from the morpholines. Morpholines are sterol biosynthesis inhibitors but have been shown to inhibit later steps in fungal sterol biosynthesis (see chapter 11).

The imidazole and triazole derivatives — often designated as azoles — have proven to be the most important DMI fungicides when measured by the number of compounds reaching an advanced stage of development or registration.

The imidazole compounds have become an important tool in the medical field as antimycotics (see chapter 21) as well as in the field of agriculture as fungicides. The triazoles have developed to the most important and largest class of fungicides. A breakthrough in this field was the presentation of triadimefon in the early seventies. This compound combined a broad spectrum of fungicidal activity against important pathogens of cereals with excellent systemic properties and low application rates.

Spectrum of fungicidal activities

In spite of their chemical heterogeneity *in vitro* as well as *in vivo* the spectrum of fungicidal activity of most DMI fungicides has some similarities. *In vitro* studies have shown that most fungi from the Ascomycetes, Basidiomycetes, and Fungi Imperfecti are more or less sensitive to DMIs. Oomycetes, which are important in the field of agriculture, are generally not sensitive to DMIs.

Nearly all of these compounds are effective to some extent against powdery mildew and rust fungi, which are obligate parasites that cause yield losses in many economically important crops. A multitude of other leaf spot pathogens such as *Pyrenophora* spp., *Mycosphaerella* spp., *Venturia* spp. and *Septoria* spp. are also effectively controlled by several of these compounds.

For a long time DMIs have had only a side effect on two economically important pathogens, *Botrytis cinerea* and *Pseudocercosporella herpotrichoides*. Meanwhile, compounds, such as prochloraz and flusilazole, provide an acceptable level of control of eye spot caused by *Pseudocercosporella herpotrichoides,* and tebuconazole has a good efficacy against *Botrytis cinerea* on grapes.

Substantial progress has been made in the field of cereal seed treatment since the introduction of systemic and plant compatible products such as triadimenol and imazalil. These products have replaced the former mercury-based seed dressings. Nowadays, DMI fungicides are used to control a broad complex of seed- and soilborne pathogens such as *Ustilago* spp., *Tilletia* spp., *Typhula incarnata* and *Pyrenophora graminearum*. Additionally, early plant infections by airborne pathogens such as powdery mildew and rust can be controlled by some systemic DMIs.

Systemic properties

Variation in chemical structure and physico-chemical properties influence the compound's ability to enter the outer barriers of the plant and to be subsequently translocated. All DMIs are able to penetrate the plant cuticle and/or the seed coat to some extent. Having penetrated the plant surface, a compound may be further translocated either in the apoplast or symplast (EDGINGTON 1981) (see chapter 4).

Symplastic (or basipetal) transport, that is movement in the living part of the plant (phloem or cytoplasm), would allow for the control of root and stem diseases following foliar application. Thus far, DMIs have shown only minimal symplastic translocation within plants.

What is thought to be basipetal transport of DMIs is sometimes misidentified in those cases where the vapour pressure of a compound is high enough to allow translocation of biologically effective amounts of chemical via the gas phase. In dense plant stands this route of transport has been shown to improve fungicidal efficacy (SCHEINPFLUG and PAUL 1977), but may cause experimental errors in small plot experiments (JENKYN et al. 1983).

The use of systemic fungicides has considerably facilitated and improved the control of plant diseases. Pathogens which have already become established in the plant can be controlled by a curative treatment. In many cases this application provides control even when first infections are already established. The systemic uptake of chemicals into a plant not only prevents them from weathering, but may also result in a more complete control of fungal pathogens, since plant parts not covered during the initial spray treatment are protected through redistribution of the fungicide. Since all systemic DMI fungicides are translocated predominately in the apoplast, protection is confined to such cases where sufficient quantities of active ingredient have been applied to the basal parts of a leaf or shoot.

Side effects on plants

Distinct plant growth regulatory effects on monocotyledonous and — generally stronger effects — on dicytoledonous plants have been reported since the first use of DMI fungicides. Typical side effects on the plant which may be seen following foliar application are: shorter shoots and internodes, and/or smaller, darker green leaves, reduced transpiration, increased resistance against different kinds of stress (heat, chilling, air pollutants) and delayed senescence (BUCHENAUER and RÖHNER 1981; FLETCHER and HOFSTRA 1988). Non-specific signs of plant incompatibility such as necrosis or leaf drop can be found less often within this group.

Investigations on the biochemical mechanisms causing these plant growth regulator activities have revealed two possible targets: inhibition of gibberellin and/or plant sterol biosynthesis.

The anti-gibberellin effect of the DMI fungicides is widely accepted and is based on the observations that in DMI-treated plants the content of gibberellin or gibberellin-like activity is reduced and that the growth retardant action could be reversed, either fully or in part, by application of exogenous gibberellic acid (BUCHENAUER and RÖHNER 1981; SISLER et al. 1984). Further biochemical studies with nuarimol, triarimol and triadimenol revealed that these compounds, like structurally related plant growth regulators such as ancymidol and paclobutrazole, are inhibitors of gibberellin biosynthesis by blocking the oxidation of entkaurene to ent-kaurenoic acid mediated by a cytochrome P-450 (SISLER et al. 1984; BURDEN et al. 1987a; GRAEBE 1987).

However, with some DMI fungicides, especially at higher concentrations, the reversal of growth-retardant action by gibberellic acid was poor. Therefore, inhibition of plant sterol biosynthesis was discussed to explain the growth regulator activities of DMI fungicides. The inhibitory effect of the DMIs on phytosterol biosynthesis was found shortly after their introduction (BUCHENAUER 1977; SCHMITT and BENVENISTE 1979) and is caused by blocking the demethylation of obtusifoliol (BURDEN et al. 1987b; TATON et al. 1988). Meanwhile it was demonstrated with several DMI fungicides including their separated enantiomers, that inhibition of plant sterol biosynthesis is mainly responsible for growth retardation (KOELLER 1987; BURDEN et al. 1989).

Thus, plant growth regulator activity of DMI fungicides seems to be based on the inhibition of gibberellin and/or plant sterol biosynthesis whereby the relative contribution to both these processes varies for a given DMI. However, it cannot be ruled out that there are other putative cytochrome P-450 dependent steps in plant metabolism which are also affected by DMI fungicides, for example brassinolide biosynthesis (BURDEN et al. 1989).

Resistance to DMI fungicides

Shortly after the introduction and field use of the first systemic single site inhibitor fungicides (the benzimidazoles and hydroxypyrimidines), resistance of certain fungi had been found. Therefore, soon after the introduction of the first DMI fungicides, research on the risk of resistance was initiated. Since no naturally-occurring resistant organisms were obvious during the first years of agriculture use, laboratory studies were conducted using artificially induced resistant strains.

The essence of a multitude of publications reviewed by DEKKER (1985), BRENT and HOLLOMON (1988) and KOELLER and SCHEINPFLUG (1989) can be summarised in the following statements:

1) Most artificially induced laboratory strains initially showed a relatively low level of resistance and the resistance factors normally were comparatively small (around 10 – 100).
2) Resistant strains of phytopathogenic fungi were usually reduced in fitness and virulence and were not as competitive as the wild strains, although several exeptions have been found.
3) Genetic studies with several fungi such as *Aspergillus nidulans, Venturia inaequalis, Cladosporium cucumerinum* and *Erysiphe graminis* gave non-uniform results. Mostly, resistance against DMI fungicides was shown to be based on several genes (up to ten different loci were detected), but there is also evidence for only one resistance gene.
4) A lot of models have been developed to explain the resistance of different plant pathogens to DMI fungicides: reduced uptake of DMIs involving a constitutive energy-dependent efflux of the inhibitor from the treated mycelium (DE WAARD and VAN NISTELROY 1979, 1988), altered lipid or cell wall composition and reduced affinity of fungicide to the target site (KOELLER 1992). But up to now the resistance mechanism is not fully understood on the molecular level.
5) Strains with resistance to one DMI fungicide generally were cross-resistant to other DMI fungicides. However, the resistance factors often differs considerably between various DMIs.

Based on results and information available from the first use of DMI fungicides under field conditions up to that time, DEKKER (1982) gave the prognosis that with DMIs the risk of resistance should be relatively low.

With the increased use of DMIs at the beginning of the 1980s, field studies of powdery mildew in cucumbers and cereals became more frequent and there are increasing reports of resistance in field populations. SCHEPERS (1985) summarised his experience from several years of work with cucumber powdery mildew (*Sphaerotheca fuliginea*) from commercial greenhouses. According to this author, in 1981 the sensitivity of powdery mildew strains isolated from greenhouses was found to be lower than that of wild strains. In 1982 and 1983, a further decrease in sensitivity had been found. SCHEPERS demonstrated a clear correlation between reduced sensitivity and frequency of fungicide application. So far the level of resistance to DMIs found in SCHEPERS' isolates seems to be only moderate. In this study triforine was weak in performance and bitertanol, fenarimol and imazalil still provided an acceptable level of control when applied at shorter intervals.

The level of resistance found in *Sphaerotheca fuliginea* strains originating from the Mediterranean countries seems to be distinctly higher (HUGGENBERGER et al. 1984).

WOLFE and FLETCHER (1981) reported for the first time about an increased frequency of barley powdery mildew strains with a decreased sensitivity to triadimefon and triadimenol. Thereafter, other authors published similar findings (GILMORE 1984; LIMPERT & FISCHBECK 1983; DE WAARD et al. 1986; LIMPERT 1987). A few years later field isolates of grape powdery mildew were found that were resistant against DMIs although the resistance factors varied considerably. Often the highest levels of resistance affected those DMI fungicides which were the most used in the field (STEVA et al. 1988; STEVA & CLERJEAU 1990).

In some areas also *Rhynchosporium secalis* and *Venturia inaequalis* were shown to be less sensitive to DMI treatment (HUNTER et al. 1986; STANIS and JONES 1985).

Even though most of the DMI fungicides have an extremely broad spectrum of fungicidal activity, resistance problems have thus far only been found with a couple of plant pathogens. DMI fungicides are still fully effective against a multitude of other pathogens. From the examples discussed above, it may be concluded that in comparison to other groups of fungicides (such as the benzimidazoles and the phenylamides), resistance problems with DMIs should not appear suddenly with large populations having a high level of resistance. If resistance should develop, it would tend to develop stepwise within individual fungal species.

The most topical information on resistance development within the SBI fungicides is currently available from publications of the SBI working group of the Fungicide Resistance Action Committee (FRAC). This group publishes yearly reports on the current resistance situation and gives recommendations for the use of DMIs and Morpholines in order to prevent a further resistance development (Anonymous 1992, 1993).

Piperazines, pyridines and pyrimidines

Triforine (Fig. 12.1) was introduced in 1969 (SCHICKE and VEEN 1969) and remains the only piperazine derivative which is in agricultural use. Information on the broad spectrum of activity of triforine has been published by FUCHS and DRANDAREVSKI (1973). Today the compound is primarily applied as a foliar treatment against powdery mildew, rust, scab and several other pathogens such as *Colletotrichum* and *Monilinia* spp. Triforine was also introduced as a systemic seed treatment of barley for control of powdery mildew (ROHRBACH 1977). Unlike most other SBI fungicides, triforine has a distinct effect against red spider mites (*Panonychus ulmi*) (MANTINGER and VIGL 1975).

The systemic properties of triforine are well documented. Following leaf application, triforine penetrates the tissue (DRANDAREVSKI and MAYER 1974) and is then translocated over short distances, thus enabling a curative action against fungal pathogens such as

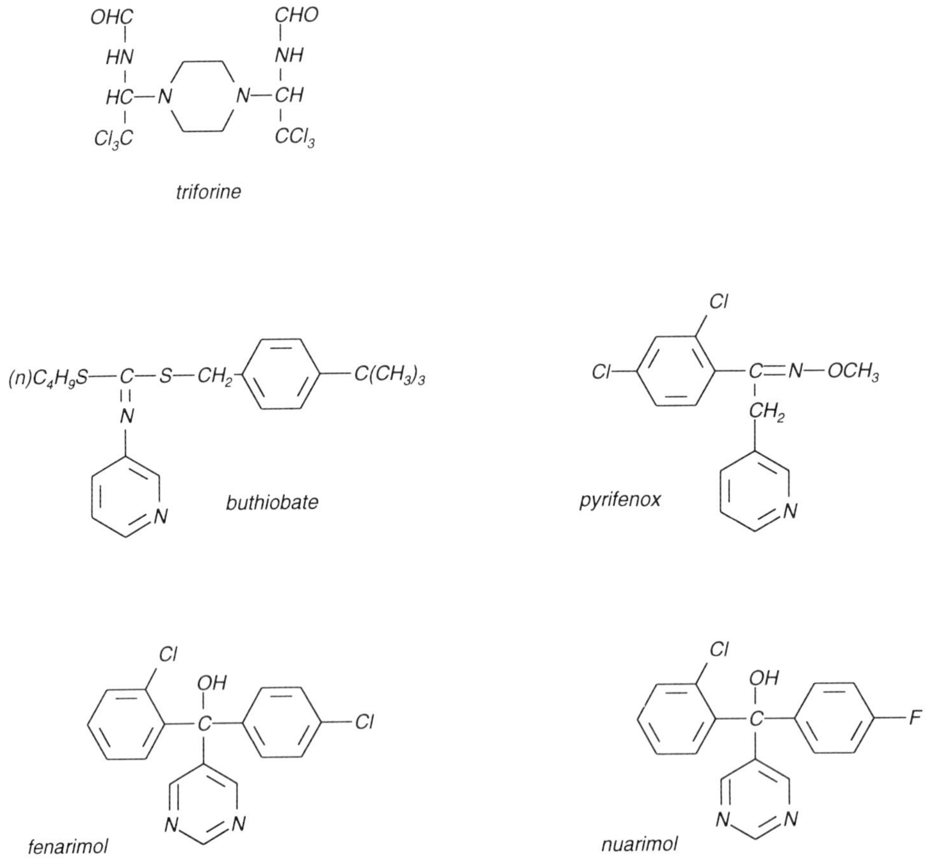

Fig. 12.1 Chemical structures of piperazin-, pyridine- and pyrimidine-fungicides.

Venturia inaequalis. When applied to the roots, triforine is readily taken up and translocated to stems and leaves. VON BRUCHHAUSEN and STIASNI (1973) reported that after soil drench application in sandy soil, barley shoots contained up to 7.5 ppm active ingredient two days after treatment. Soil with a high organic matter content decreased the rate of uptake (FUCHS et al. 1976a, b). An apparent lack of accumulation of triforine in the roots (EBENEBE et al. 1974) and a rather high turnover of the compound in plants (FUCHS et al. 1972) seem to account for a maximum concentration in the shoots 2 – 8 days after application of the chemical (FUCHS and OST 1976).

In contrast to most imidazole and triazole derivatives triforine does not cause typical growth regulator effects on treated plants. Papers dealing with plant incompatibility of triforine describe only non-specific symptoms in ornamentals (ATTABHANYA and HOLCOMB 1976) or European larch (BOUDIER 1981). Several authors report that triforine inhibits the germination of apple (CHURCH and WILLIAMS 1978) and blueberry (BRISTOW 1981) pollen.

Surprisingly, triforine has been shown to be very phytotoxic to certain lettuce cultivars. Inheritance studies by GLOBERSON and ELIASI (1979) and SMITH (1979) showed that resistance to leaf damage caused by triforine is recessive and is determined by a single gene.

Buthiobate (Fig. 12.1) was the first pyridine-based SBI fungicide that reached a commercial status. KATO et al. (1975) demonstrated through *in vitro* studies that buthiobate inhibited growth of several Ascomycetes and Fungi Imperfecti at low rates, but was ineffective against Basidiomycetes. In practice, this compound is mainly used in Japan against powdery mildew fungi.

Buthiobate's mobility in plants seens to be limited. OHKAWA et al. (1976) reported an absorption of the compound by leaves and roots but only minimal translocation into other parts of the plant. Using paper discs, KATO et al. (1975) were able to demonstrate a distinct vapour action of buthiobate against *Sphaerotheca fuliginea* on cucumbers.

Pyrifenox is a pyridine derivative (Fig. 12.1) from Maag Ltd. The compound was shown to be an effective systemic foliar fungicide in pomefruits, grapes, peanuts, stone fruits, and vegetables against a wide range of leaf spot pathogens. ZOBRIST et al. (1986) report that a rate of 5 g a.i./hl or at 80 g a.i./ha was sufficient to control scab (*Venturia inaequalis*) and powdery mildew (*Podosphaera leucotricha*) on apples. Other diseases such as powdery mildew on grapes (*Uncinula necator*), and several leaf spot diseases such as *Cercospora* spp., *Cercosporidium* spp. *Septoria* spp. and blossom blight (*Monilina* spp.) are also effectively controlled by this fungicide.

ZOBRIST et al. (1986) demonstrated a rapid penetration into treated leaves and mention that with seed and root application pyrifenox was shown to be translocated in the acropetal direction in the shoot.

The plant compatibility of pyrifenox is very good. According to MASNER and KERKENAAR (1988) DMI-typical symptoms such as stunting effects have not been observed.

A series of three substituted pyrimidin-5-ylmethanol fungicides (Fig. 12.1), triarimol, fenarimol and nuarimol, were discovered and developed by Eli Lilly since the late sixties. Triarimol (EL-273), the first of the three products presented to the public (BROWN et al. 1970), was an effective fungicide against scab and powdery mildew in fruits but was deleted from further development.

Fenarimol (Fig. 12.1) like triarimol has been mainly developed in fruits and vegetables. The compound shows activity against a broad spectrum of powdery mildew, scabs, rust and a multitude of leaf spot pathogens. According to BROWN and HALL (1981), fenarimol has been primarily used to provide protective and curative control of scab (*Venturia inaequalis*) and powdery mildew (*Podosphaera leucotricha*) on apples. On grapes, cucurbits, tomatoes, peppers and peach it has been used against other powdery mildew fungi. *In vitro*, fenarimol controls a wide range of Ascomycetes, Basidiomycetes and Fungi Imperfecti at low rates (BUCHENAUER 1979). As expected, there is a strong similarity in biological activity to the structurally related compound nuarimol.

Greenhouse studies from BROWN and HALL (1981) demonstrated the fenarimol is rapidly absorbed by leaf tissue. These authors demonstrated an apoplastic movement through the leaf, and a measureable vapour phase effect against powdery mildew using treated paper discs, separated or not, from the leaf by aluminium foil discs.

Root uptake of fenarimol has been reported (BUCHENAUER and RÖHNER 1982), but seems to be insufficient to provide disease control in crops grown under field conditions.

Another pyrimidin-5-ylmethanol derivative, ancymidol, is a potent growth regulant in higher plants (COOLBAUGH et al. 1982). Therefore, it is not surprising that a distinct growth retardant side effect of the fungicidal analogues fenarimol and nuarimol has sometimes been reported. BUCHENAUER (1977) reported an inhibition in shoot elongation of tomato and wheat plants, and

Table 12.1 Piperazine, Pyridine, and Pyrimidine Fungicides

1) Common name[1])	1) Inventing/Developing Company
2) Experimental No.	2) Patent No., Year[4])
3) Chemical name[2])	3) Year of presentation
4) Trade name(s) of products for spray application[3]) & combination partner(s)	4) Acute toxicity[5])
	5) Solubility (water)[6])
5) Trade name(s) of products for seed treatment[3])	6) Vapour pressure[7])
& combination partner(s)	7) Partition coefficient[8])
6) Use as foliar treatment (crop/pathogen) application rate (ai)	8) Literature
7) Use as seed treatment (crop/pathogen) application rate (ai)	
8) Other uses	

1) **Triforine**	1) Boehringer/Celamerck
2) CELA W 254	2) DOS 1 901 421, 1968
3) N,N'-[piperazine-1,4-diylbis[((2,2,2-trichloroethylene)bisformamide	3) 1969
	4) >16.000 mg/kg
4) Saprol®, Funginex®, Triforine 20®	5) 6 mg/l
	6) 27 µPa [25 °C]
5) Prodressan®	7) log $P_{(o/w)}$: 2.2
6) Stone fruit: *Monilinia* ssp., powdery mildew, rust, *Coccomyces hiemalis* 0.019 – 0.029%	8) Drandarewski & Schicke 1976
Apples: powdery mildew, scab, rust. 0.024%	Adlung & Drandarewski 1971
Berry fruits: powdery mildew, rust, *Monilinia* spp. 0.029%	
Grapes, Hops, Mango: powdery mildew, Vegetables: powdery mildew, rust *Colletotrichum* ssp. 190 – 380 g/ha	
various leaf spot diseases 0.019 – 0.029%	
Others: powdery mildew, rust, leaf spot diseases in tobacco, sugar beets, ornamentals	

1) **Buthiobate**	1) Sumitomo
2) S-1 358	2) DOS 21 19 174, 1970
3) Butyl [4-(1,1-dimethylethyl)phenyl]methyl-3-pyridinyl-carbonimidodithioate	3) 1975
	4) 2.700 – 4.900 mg/kg
4) Denmert®	5) 0.96 mg/l
	6) 6×10^{-7}
6) Vegetables, Fruits, Ornamentals: powdery mildew 0.007 – 0.02%	8) Kato et al. 1975

1) **Pyrifenox**	1) Hoffmann-La Roche/Maag
2) Ro 15-1 297, ACR 3 675, FD 4 060	2) EP 49 854, 1980
3) 2',4'-dichloro-2-(3-pryridyl) acetophenone O-methyloxime	3) 1986
	4) 2.900 mg/kg

Table 12.1 (Continued)

1) Common name[1]		1) Inventing/Developing Company
2) Experimental No.		2) Patent No., Year[4]
3) Chemical name[2]		3) Year of presentation
4) Trade name(s) of products for spray application[3]		4) Acute toxicity[5]
	& combination partner(s)	5) Solubility (water)[6]
5) Trade name(s) of products for seed treatment[3]		6) Vapour pressure[7]
	& combination partner(s)	7) Partition coefficient[8]
6) Use as foliar treatment (crop/pathogen)	application rate (ai)	8) Literature
7) Use as seed treatment (crop/pathogen)	application rate (ai)	
8) Other uses		

Pyrifenox (continued)

4) Dorado®, Corona®   5) 115 mg/l
   Rondo®   & captan   6) 1.9 mPa [25 °C]
5) Rondo® M, Furado®   & mancozeb   7) log $P_{(o/w)}$: 2.5
6) Pomefruits: *Podosphaera leucotricha,*   5–7.5 g/hl   8) Zobrist et al. 1986
   Venturia inaequalis
   Stonefruits: *Monilinia* spp.   0.005–0.01%
   Grapes: *Uncinula necator*   37.5–50 g/ha
   Peanuts: *Mycosphaerella* spp.   70–140 g/ha
   Others: powdery mildew and leaf spot diseases in pecans, sugar beets, vegetables
   Blackcurrants: *Sphaerotheca mors-uvae,*   80 g/ha   Hughes & Wilson 1988
   Pseudopeziza ribis

---

1) **Fenarimol**   1) Eli Lilly [now DowElanco]
2) EL 222   2) Fr 1.569940, 1967
3) (±)-2,4′-dichloro-α-(pyrimidin-5-yl)benzhydryl alcohol   3) 1975
4) Rubigan®, Bloc®   4) 2.500 mg/kg
   Rimidin plus®   & carbendazim   5) 13.7 mg/l [25 °C]
    & maneb   6) 13 µPa [25 °C]
   Splendor®   & carbendazim   7) log $P_{(o/w)}$: 3.69
    & oxycarboxin
6) Pome fruits: *Venturia* spp.,   0.0018–0.0036%   8) Kelley & Jones 1981
   Podosphaera leucotricha   Verheyden 1981
   Fruits: powdery mildew, *Monilinia* spp., rust   Beraud et al. 1980
   Sugar beets: *Cercospora beticola*   90–120 g/ha
   Soybeans: *Cercospora* spp., *Diaporthe* spp., *Glomerella* spp.
   Peanuts: *Mycosphaerella* spp.   90–120 g/ha
   Turf: *Sclerotinia homoeocarpa,*   60–6000 g/ha
   Rhizoctonia spp.
   Ustilago spp., *Fusarium* spp.
   Grapes: *Uncinula necator*   0.0018–0.0036%   Vergnes & Pistre 1982

Table 12.1 (Continued)

1) Common name[1])	1) Inventing/Developing Company
2) Experimental No.	2) Patent No., Year[4])
3) Chemical name[2])	3) Year of presentation
4) Trade name(s) of products for spray application[3])	4) Acute toxicity[5])
& combination partner(s)	5) Solubility (water)[6])
5) Trade name(s) of products for seed treatment[3])	6) Vapour pressure[7])
& combination partner(s) application rate (ai)	7) Partition coefficient[8])
6) Use as foliar treatment (crop/pathogen) application rate (ai)	8) Literature
7) Use as seed treatment (crop/pathogen) application rate (ai)	
8) Other uses	

1) **Nuarimol**		1) Eli Lilly [DowElanco]
2) EL 228		2) FR 1.569940, 1967
3) (±)-2-chloro-4′-fluoro-α-(pyrimidin-5-yl)benzhydryl alcohol		3) 1975
		4) 1.250 mg/kg (m) 2.500 mg/kg (f)
4) Trimidal®, Triminol®		5) 26 [pH 7, 25 °C]
		6) <2.7 µPa [25 °C]
5) Trimidal bejdse®, Gauntlet®, Murox® Elanco Beize®, Trimidal spezial®	& imazalil	7) log P$_{(o/w)}$: 3.18
6) Cereals: *Erysiphe graminis*, *Rhynchosporium secalis*	40 – 90 g/ha	CASANOVA et al. 1977 FRATE et al. 1979
Sugar beets: *Cercospora beticola*, *Erysiphe betae*		
Bananas: *Mycosphaerella* spp.	90 g/ha	
7) Cereals: *Ustilago* spp., *Tilletia* spp., *Pyrenophora* spp., powdery mildew *Fusarium* spp., *Rhynchosporium secalis*, *Leptosphaeria nodorum*	5 – 40 g/100 kg	LUZ & VIEIRA 1982 PIENING et al. 1983

[1]): approved or proposed common name; former proposals are given in brackets (); [2]): IUPAC nomenclature was used preferably; [3]): examples, not all inclusive; [4]): year of priority date; [5]): oral acute toxicity in rats (LD$_{50}$); [6]): standard conditions: temperature: 20 °C, pH 7, variations are given in brackets []; [7]): vapour pressure at 20 °C, variations are given in brackets []; [8]): Partion coefficient in octanol/water or equivalent determined by HPLC; RT = room temperature

Table 12.1 and its equivalents in the following sections provide an overview on some physico-chemical and biological properties of DMI fungicides. This information should be considered as a physico-chemical or biological profile and does **not** constitute a label or recommendation. Some (or for newly presented compounds all) of the uses of compounds may not yet or no longer be registered or not registered in a given country.

Data presented in these tables were obtained from technical data sheets from the respective companies or from a current literature review. The listed uses for each compound and the trade names are generally not all inclusive due to space limitations. General or specific literature citations are given beside their respective indication.

Abdel-Rahman (1977) noted a reduction of terminal shoot elongation and a reduction of fruit thinning in apples following application of fenarimol. A detailed analysis on the biochemical interrelationships of fungitoxicity and plant growth regulation within the pyrimidin-5-ylmethanols has been published by SISLER *et al.* (1984).

In contrast to fenarimol, which was developed for use in fruits and vegetables, **nuarimol** was developed mainly for use against cereal diseases. As a seed treatment in cereals, nuarimol is used in mixture with imazalil to control several seed- and soilborne pathogens such as *Ustilago* spp. and *Fusarium* (*Gerlachia*) *nivale*. In addition, control of the airborne pathogen *Erysiphe graminis* is obtained on young seedlings (CASANOVA *et al.* 1977). Growth retarding effects of nuarimol from seed treatments were mentioned by several authors (CASANOVA *et al.* 1977; DÖHLER and MERTZ 1979; BUCHENAUER 1977) who mostly regarded the effect as being transitory. As a foliar fungicide, nuarimol is used in cereals against powdery mildew and leaf blotch (*Rhynchosporium secalis*). Similar to fenarimol, nuarimol is further recommended against a broad range of fungal diseases in crops such as stone fruits, bananas, sugar beets and peanuts.

The systemic activity of nuarimol is similar to that of fenarimol. The somewhat higher hydrophilicity of nuarimol, as compared to fenarimol, indicates that the compound is more readily translocated in the apoplast of the plant. CARDOSO *et al.* (1979) and BUCHENAUER and RÖHNER (1982) describe a systemic uptake of the chemical into soybean hypocotyls and barley, respectively, after seed or soil treatment. This uptake is also obvious from the effect the fungicide has on powdery mildew of cereal seedlings grown from treated seed.

Imidazoles used in agriculture

Imazalil (Fig. 12.2), the first agriculture imidazole fungicide, is mainly used today as a seed treatment in cereals but has additional applications as a foliar treatment in vegetables and ornamentals and as a post harvest treatment in citrus fruits. Information

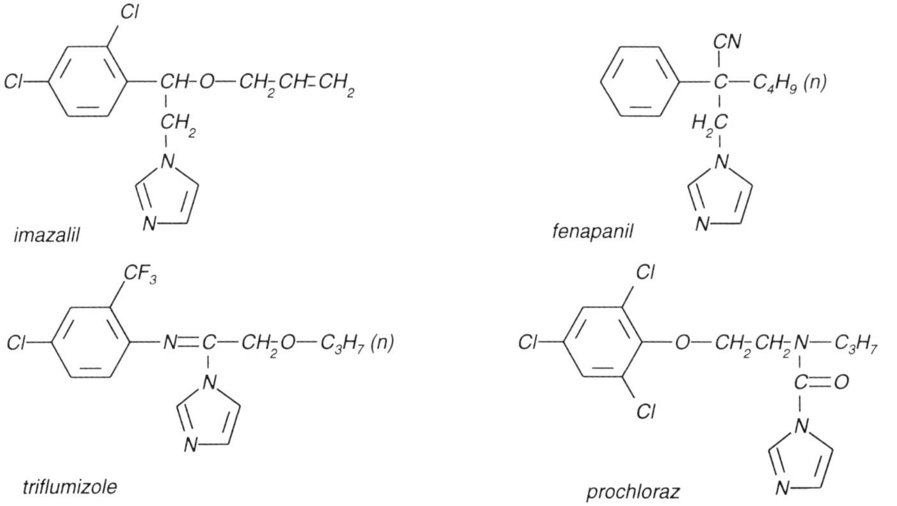

Fig. 12.2 Chemical structures of imidazole-fungicides.

on the broad *in vitro* activity of imazalil is available from studies of BUCHENAUER and RÖHNER (1979).

Because of imazalil's specific activity against seed- and soilborne pathogens, such as *Pyrenophora* (especially against *P. graminea*), *Fusarium* and *Septoria*, it is most often used in combination with other fungicides. Fungicides such as guazatine, triadimenol and nuarimol, which replaced mercury-containing seed dressing products, broaden the spectrum of activity of imazalil.

With the aid of tritium-labelled imazalil, VONK and DEKHUIJZEN (1979) demonstrated that imazalil is taken up by plants after seed treatment or root application. In 3 week-old barley plants grown from seed treated with 1.8 g imazalil/100 kg seed, the authors found about 6% of the labelled material to be present in the leaves. This resulted in an extractable concentration of only 0.07 ppm active ingredient in the leaves. REISDORF *et al.* (1983) confirmed these results and concluded that the excellent activity of imazalil against *Pyrenophora graminea* may not be derived from the relatively low concentrations of the product taken up into the plants, but mainly from its fungicidal activity at the seed surface. REISDORF *et al.* (1983) reported that high concentrations of imazalil caused distinct growth retardation of roots and shoots of barley seedlings. BUCHENAUER and RÖHNER (1979) observed similar symptoms at the rate of 25 – 50 g ai/100 kg seed. Physiological studies by these authors showed that imazalil caused both an inhibition of gibberellic acid biosynthesis and of C-4-desmethylsterol production in barley seedlings.

Fenapanil (Fig. 12.2) also known as phenapronil or fenapronil, is a broad spectrum imidazole fungicide from Rohm and Haas which has been discontinued in the meantime. Reports on the effectiveness of fenapanil deal mainly with its activity against scab and powdery mildew in pome fruits (SHABI *et al.* 1981; KELLY and JONES 1981), and as a cereal seed treatment against various seed- and soilborne diseases (HOFFMANN and WALDHER 1981; HANSEN 1981).

According to KELLEY and JONES (1982) about 50% of fenapanil, when applied to apple leaves, was taken up within 1 – 3 days. MARTIN and EDGINGTON (1981, 1982) who used barley and soybean plants for studies on the systemicity of fenapanil stated that it exhibited primarily apoplastic transport in barley. In soybeans fenapanil was found to be ambimobile with 13.5% of the material available for translocation transported basipetally.

Contrary to the usual growth regulating pattern of DMIs, fenapanil inhibited growth of roots and shoots of barley seedlings stronger than those of soybeans. Symptoms described in barley were stunting and chlorosis of the foliage with some necrosis at the leaf tips (MARTIN and EDGINGTON 1982).

Prochloraz (Fig. 12.2), presented in 1977 by the Boots Company, is a broad spectrum fungicide which has a pronounced activity against Ascomycetes and Fungi Imperfecti, and somewhat lesser activity against Basidiomycetes (BIRCHMORE *et al.* 1977). The fungicidal profile of prochloraz in cereals includes activity against *Pseudocercosporella herpotrichoides*. Fast growing (W-type) and slow growing (R-type) strains are equally well affected. This is unique among the azolyl fungicides that have reached the marked stage. In addition, good activity is seen against pathogens such as leaf blotch, net blotch, and *Septoria* spp. Prochloraz has shown no cross resistance to benzimidazole fungicides: This fact permits a wide use especially in cereal stands where resistance to carbendazim has been detected.

In France and other European countries, combination formulations with carbendazim are available for improved control of eyespot in wheat. This combination product also showed efficacy in oil seed rape against a complex of pathogens such as *Phoma lingam, Sclerotinia sclerotiorum,* and *Alternaria* spp..

Plant compatbility of the normally used EC formulations of prochloraz on various broadleaved crops is critical. A prochloraz-manganese complex, formulated as a wettable powder, is recommended for use in most dicotyledonous plants (BIRCHMORE *et al.* 1979).

The mobility of prochloraz in plants is confined to translaminar or locosystemic movement. According to de SAINT-BLANQUAT and MY (1982) prochloraz is readily absorbed from plant surfaces, but not translocated over longer distances. The studies of COOKE et al. (1979), who used strawberries as a test plant system, confirm this finding.

Triflumizole (Fig. 12.2), the most recent imidazole compound, is used mainly in fruits and vegetables against a wide range of pathogens such as powdery mildew, scab and *Monilinia* spp. The compound is described as a preventive and curative fungicide which has translaminar but not fully systemic activity. More recently the compound has been developed as a cereal seed treatment against smuts and bunts.

HASHIMOTO et al. (1990) studied the translocation of triflumizole in cucumber plants in more detail. As can be expected the symplastic transport was negligible. After treating the central region of a cucumber leaf with C-14-labelled active ingredient the apoplastic transport was sufficient to show an even distribution of the radiolabel in the upper leaf parts at day 7 after application.

Table 12.2 Imidazole Fungicides

1) Common name[1])		1) Inventing/Developing Company
2) Experimental No.		2) Patent No., Year[4])
3) Chemical name[2])		3) Year of presentation
4) Trade name(s) of products for spray application[3])		4) Acute toxicity[5])
	& combination partner(s)	5) Solubility (water)[6])
5) Trade name(s) of products for seed treatment[3])		6) Vapour pressure[7])
	& combination partner(s)	7) Partition coefficient[8])
6) Use as foliar treatment (crop/pathogen)	application rate (ai)	8) Literature
7) Use as seed treatment (crop/pathogen)	application rate (ai)	
8) Other uses		
1) **Imazalil**		1) Janssen
2) R 23 979 (free base), R 27 180 (sulphate salt)		2) BP 1 244 530, 1969
3) (±)-allyl 1-(2,4-dichlorophenyl)-2-imidazol-1-ylethyl ether		3) 1972
4) Fungaflor®, Fungazil®, Fecundal®, Deccozil®		4) 227–343 mg/kg
		5) 293 mg/l
5) Panoctin® universal	& guazatine	6) 9.3 µPa
	& fenfuram	
Vincit® LU	& flutriafol	7) log $P_{(o/w)}$: 3.8
	& thiabendazol	
Baytan® universal	& triadimenol	
	& fuberidazol	
Vitavax® 202	& carboxin & thiram	
6) Vegetables, Ornamentals: powdery mildew, various leaf and stem diseases	0.005–0.03%	8) REISSDORF et al. 1983
Bananas: *Mycosphaerella* spp.,	4–5 g/100 kg	MELIN et al. 1975
7) Cereals: *Pyrenophora* spp., *Fusarium* spp., *Septoria* spp.	4–5 g/100 kg	
Potatoes: *Phoma exigua, Polyscytalum pustulans.*	10–30 g/1 000 kg	CAYLEY et al. 1981
8) Fruits: post harvest diseases	0.015–0.5%	

Table 12.2 (Continued)

1) Common name[1])	1) Inventing/Developing Company
2) Experimental No.	2) Patent No., Year[4])
3) Chemical name[2])	3) Year of presentation
4) Trade name(s) of products for spray application[3])	4) Acute toxicity[5])
& combination partner(s)	5) Solubility (water)[6])
5) Trade name(s) of products for seed treatment[3])	6) Vapour pressure[7])
& combination partner(s)	7) Partition coefficient[8])
6) Use as foliar treatment (crop/pathogen) application rate (ai)	8) Literature
7) Use as seed treatment (crop/pathogen) application rate (ai)	
8) Other uses	

1) **Fenapanil** (Phenapronil)		1) Rohm & Haas
2) RH-2 161		2) DOS 2 604 047, 1976
3) (±)-2-(imidazol-1-ylmethyl)-2-phenylhexanenitrile		3) 1978
4) Sisthane®		4) 1 590 mg/kg
6) Pomefruits: powdery mildew, scab, rust	0.03 – 0.062%	7) Shabi et al. 1981
Others: powdery mildew, rust		
leaf spots on cereals, fruits, vegetables		
7) Cereals: smuts, bunts, *Pyrenophora graminea*	30 – 120 g/100 kg	Hoffmann & Waldher 1981

1) **Prochloraz**		1) Boots/FBC/Schering
2) BTS 40 542, SN 80 109		2) GB 1 469 772, 1973
3) 1-{N-propyl-N-[2-(2,4,6-trichlorophenoxy)ethyl]carbamoyl}imidazole		3) 1977
4) Sportak®, Octave® (Manganese salt)		4) 1 600 – 2 400 mg/kg
Sportak Alpha®	& carbendazim	5) 34 mg/l [25 °C]
Rival®, Sprint®	& fenpropimorph	6) 0.15 mPa [25 °C]
Sponsor®	& fenpropidin	7) log $P_{(o/w)}$: 4.38
Sportak® Delta, Tiptor®	& cyproconazole	
5) Prelude®, Abavit®	& carboxin	
6) Cereals: *Pseudocerosporella h.*, *Pyrenophora teres*, *Rhynchosporium secalis*, *Septoria* spp., *Erysiphe graminis*	400 – 450 g/ha	8) Thomas & Cornier 1983 Harris & Barnes 1981
Rapes: *Sclerotinia sclerotiorum*, *Phoma lingam*		Wakerley & Russel 1985
Coffee: *Colletotrichum coffeanum*	100 – 200 g/ha	
Others: various diseases in ornamentals, turf, sugar beet, stone fruit, rubber etc.		
7) Cereals: *Pyrenophora* spp., *Fusarium* spp., *Leptosphaeria nodorum*	20 – 50 g/100 kg	
Rice: *Gibberella fujikuroi*, *Helminthosporium oryzae*	12.5 g/hl	
8) Citrus, Avocados, Mangoes etc. (post harvest treatment): various moulds and rots	25 – 300 g/hl	Knights 1986
Mushrooms: *Verticillium fungicolam*, *Mycogone perniciosa*	3 – 15 g/10 m^2	V. Zaayen & V. Adrichem 1982

Table 12.2 (Continued)

1) Common name[1])		1) Inventing/Developing Company
2) Experimental No.		2) Patent No., Year[4])
3) Chemical name[2])		3) Year of presentation
4) Trade name(s) of products for spray application[3])		4) Acute toxicity[5])
	& combination partner(s)	5) Solubility (water)[6])
5) Trade name(s) of products for seed treatment[3])		6) Vapour pressure[7])
	& combination partner(s)	7) Partition coefficient[8])
6) Use as foliar treatment (crop/pathogen)	application rate (ai)	8) Literature
7) Use as seed treatment (crop/pathogen)	application rate (ai)	
8) Other uses		

1) **Triflumizole**		1) Nippon Soda/ Uniroyal
2) NF 114, A 815		2) DOS 2 814 041, 1978
3) (E)-4-chloro-α,α,α-trifluoro-N-(1-imidazol-1-yl-2-propoxy-ethylidene)-o-toluidine		3) 1982
		4) 1 057 mg/kg (m)
4) Trisosol®, Trifludol®, Trifmine®, Duotop®, Procure®		1 780 mg/kg (f)
		5) 12.5 mg/l
5) Trifumin®		6) 1.4 µPa [25 °C]
		7) log K_{ow}: 1.4
6) Apples: *Venturia inaequalis, Podosphera leucotricha, Gymnosporangium* spp.	0.015 – 0.03%	8) NAKATA et al. 1982 HICKEY (1983)
2 Stonefruits: *Monilinia* spp.	0.006 – 0.01%	
Grapes: *Uncinula necator*	140 – 280 g/ha	
Vegetables: powdery mildew, *Rhizoctonia* sp.	70 – 350 g/ha	
Ornamentals: powdery mildew, rust, *Rhizoctonia solani*	70 – 280 g/ha	
Cucurbits: powdery mildew		
7) Cereals: *Ustilago* spp., *Tilletia* spp.	30 g/100 kg	

[1]): approved or proposed common name; former proposals are given in brackets (); [2]): IUPAC nomenclature was used preferably; [3]): examples, not all inclusive; [4]): year of priority date; [5]): oral acute toxicity in rats (LD_{50}); [6]): standard conditions: temperature: 20 °C, pH 7, variations are given in brackets []; [7]): vapour pressure at 20 °C, variations are given in brackets []; [8]): Partition coefficient in octanol/water or equivalent determined by HPLC; RT = room temperature

Similar to other imidazole fungicides the in vitro activity of triflumizole is more pronounced against *Ascomycetes* and *Deuteromycetes* than against *Basidiomycetes* (HASHIMOTO et al. 1986).

Triazoles

Fluotrimazole (Fig. 12.3) was first mentioned in the literature by GREWE and BÜCHEL (1973). Following introduction and testing, it became the first triazole fungicide to be marketed. This compound is characterized by a very narrow usable spectrum of fungicidal activity, restricted primarily to powdery mildew fungi. *In vitro*, a broader spectrum has been demonstrated (BUCHENAUER

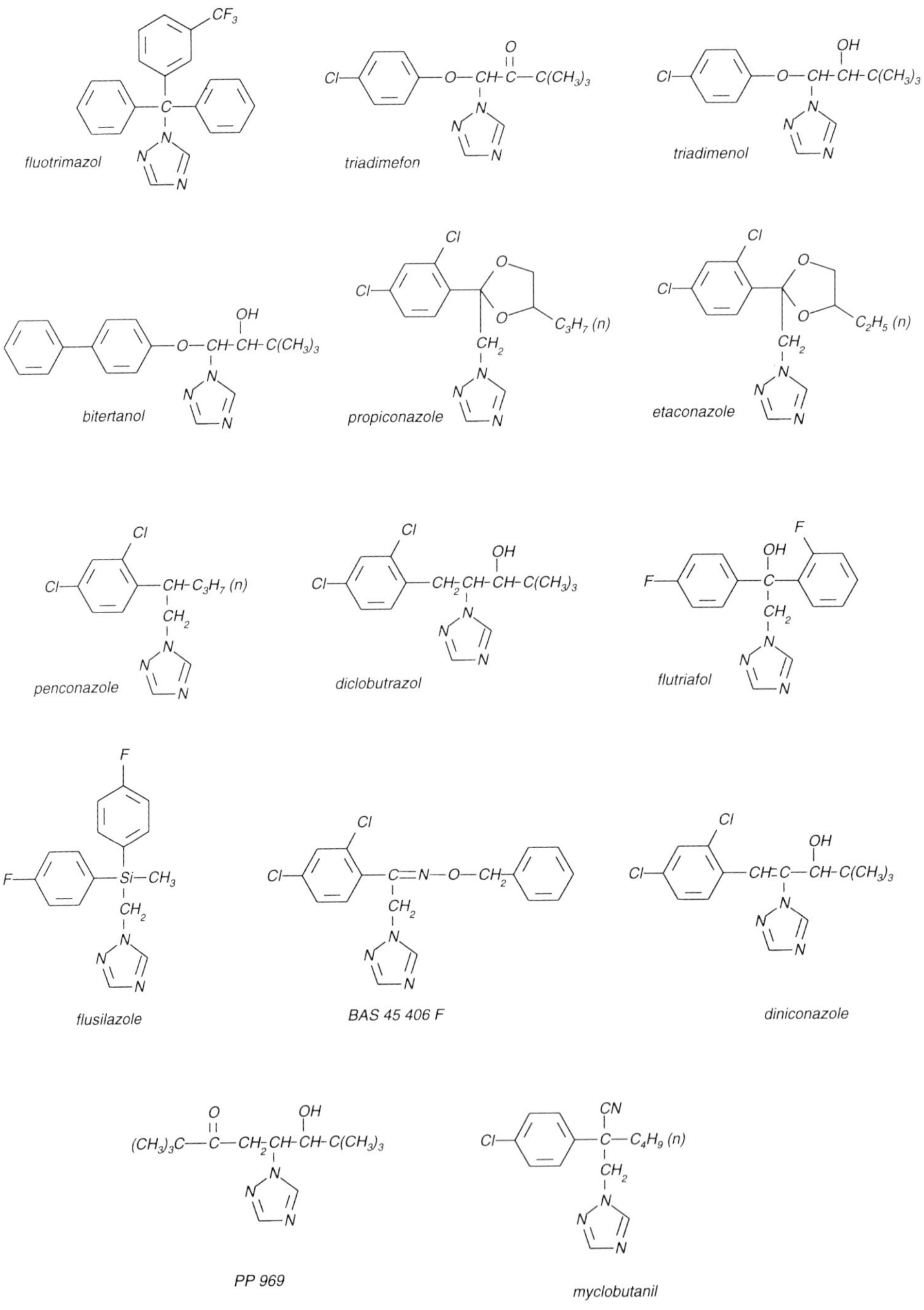

Fig. 12.3a Chemical structures of triazole-fungicides.

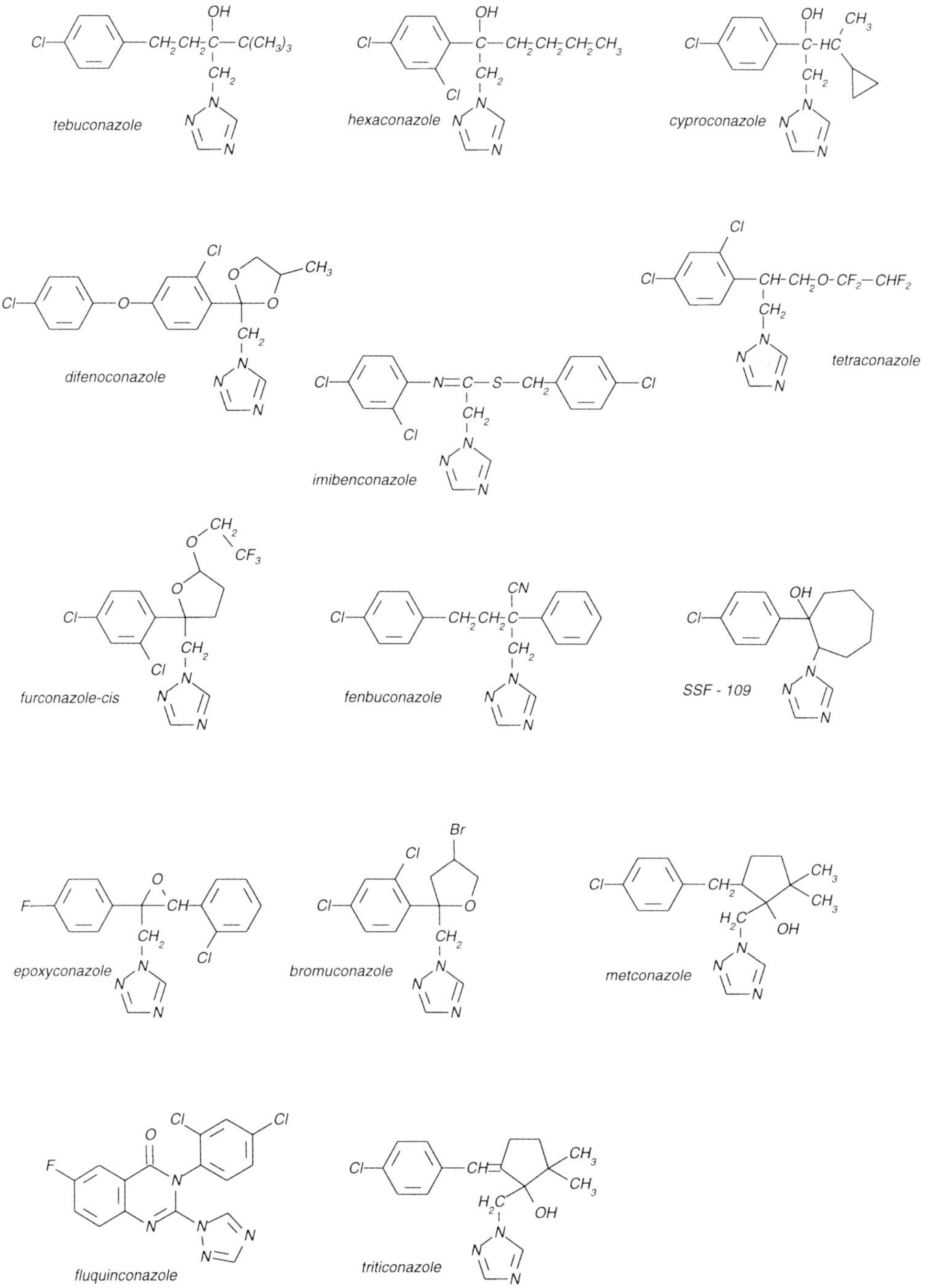

DMI fungicides 221

Fig. 12.3 b

1979a). The importance of fluotrimazole was limited due to the introduction of broad spectrum SBI fungicides such as triadimefon.

Information on the systemic properties of fluotrimazole was given by STEFFENS and WIENECKE (1981). Six days after soil treatment, only 0.3−0.4% of C-14 labelled active ingredient was taken up by wheat and tomato plants. Phytotoxic side effects of fluotrimazole have seldom been seen probably due to the limited uptake of this compound.

Triadimefon (Fig. 12.3), the first systemic broad spectrum triazole fungicide, was presented to the public a short time after fluotrimazole (GREWE and BÜCHEL 1973; KASPERS et al. 1975). In vitro testing of triadimefon demonstrated its effectiveness at 0.1 to 2 ppm against a broad spectrum of Ascomycetes, Basidiomycetes and Fungi Imperfecti (BUCHENAUER 1979a): Triadimefon soon became known under its trade name Bayleton®. The advantage of triadimefon over other fungicides was that it provided simultaneous protective and curative control of several economically important diseases in important crops such as cereals.

When triadimefon is applied to plants, it is quickly reduced to its corresponding secondary alcohol, triadimenol (Fig. 12.3). In wheat leaves, about 60% of the active ingredient which penetrates the epidermis is reduced to triadimenol within 2 days (KUCK 1987). This reduction is regarded as an activation since triadimenol is the more active of the two fungicides. The fungicidal activity of the precursor compound triadimefon is itself difficult to define since most fungi are also able to perform the reduction step (GASZTONYI and JOSEPOVITS 1979).

Triadimefon is a racemic mixture of two optical isomers (− and +) whose fungicidal activities are similar (KRÄMER et al. 1983).

The systemic properties of triadimefon contribute to a high degree to its curative efficacy. A multitude of investigations have shown that triadimefon (and/or its metabolite triadimenol) rapidly enters root, stem and leaf tissues and is translocated acropetally in the apoplast (FÜHR et al. 1978; BRANDES et al. 1978; KRAUS 1981; SANDERS et al. 1978; BUCHENAUER 1975). For example FÜHR et al. (1978) found that in barley up to 51% of the ^{14}C-labelled triadimefon applied to the basal half of leaves had been translocated to the apical half within 12 days. In these experiments basipetal translocation accounted for only 0.3%. KRAUS (1981) reported that only 40% of the applied triadimefon was removable with water from grape leaves 45 minutes after application.

A marked vapour phase activity of triadimefon was shown against powdery mildew on cucumber and barley during greenhouse studies by SCHEINPFLUG and PAUL (1977). Field experiments conformed that the gas phase may contribute to the performance of the fungicide (SCHEINPFLUG et al. 1978).

Several aspects of the plant growth regulatory side effects of triadimefon and triadimenol have been studied by FÖRSTER et al. (1980a, b, c) using barley plants. These authors reported that in plants grown from treated seed a transient retardation of roots, shoots and coleoptiles and a disturbed geotropism of the plants could be observed. Thirty days after sowing differences could no longer be found in the fresh and dry weights of shoots. BUCHENAUER and RÖHNER (1981) concluded from physiological studies that triadimefon and triadimenol interfere in gibberellin and sterol biosynthesis in barley seedlings by inhibiting oxidative demethylation reactions.

As most other SBI fungicides, triadimefon causes stronger plant growth regulatory side effects on dicotyledonous than on monocotyledonous plants. In red raspberries earlier ripening and shorter and thinner canes were reported by PEPIN et al. (1980). On apple trees reduction of leaf area, shortening of leaf petioles (ABDEL-RAHMAN 1977) and increase in fruit set (STRYDOM et al. 1981) have been reported.

More recently, FLETCHER and co-workers have investigated possible beneficial side effects of triadimefon. Treated wheat, peas and soybeans showed a reduced transpiration which prevented the leaves of water-stressed plants from wilting and increased yield (FLETCHER and NATH 1984). Beans proved to be more resistant to ozone, and cabbage and barley seedlings were shown to be more resistant to chilling after root application of triadimefon (FLETCHER and HOFSTRA 1985; FLETCHER 1985).

Triadimenol (Fig. 12.3) was initially introduced as the active ingredient of the systemic seed treatment Baytan®. Due to its broad fungicidal spectrum (BUCHENAUER 1979a) and its systemic uptake into the plant, triadimenol proved to be active against a wide range of seed-, soil- and airborne pathogens of cereal seedlings.

Smuts (*Ustilago* spp.), bunts (*Tilletia* spp. with the exception of *T. controversa*), *Pyrenophora* diseases and early powdery mildew and rust infections are simultaneously controlled by a Baytan seed treatment (FROHBERGER 1978): For complete control of *Pyrenophora graminea* and *Fusarium* (*Gerlachia*) *nivale,* imazalil and fuberidazol have been used as cofungicides.

Triadimenol has been further developed as a foliar fungicide. Triadimenol has shown advantages over triadimefon in curative and eradicative treatments. Due to its close chemical relationship to triadimefon the application rates and the fungicidal spectrum of triadimenol are essentially the same. Moreover, triadimenol (and triadimefon) were the first triazole fungicides which were succesfully used for the systemic control of coffee rust via soil application of a granule formulation.

Triadimenol resembles triadimefon in being a highly systemic fungicide. The uptake of ^{14}C-triadimenol into cereal seedlings after seed treatment has been studied by STEFFENS *et al.* (1982). According to these authors, up tp 7.7% of the applied radioactivity was taken up by the leaves of wheat seedlings within 51 days. The largest quantity of radioactivity was found in the first three leaves, with the primary leaf containing 3.9% and leaves two and three containing 1.4 and 0.5% of the applied radioactivity, respectively.

The reduction of triadimefon introduces a second chiral centre into its metabolite triadimenol. Thus triadimenol can exist in the two diastereomeric forms triadimenol A (= I or threo) and triadimenol B (= II or erythro). Each of these consists of two enantiomers. Thus, triadimenol A includes the enantiomeric forms A (−) (= 1S, 2R) and A (+) (= 1R, 2S), while triadimenol B contains the enantiomers B (−) (= 1S, 2S) and B (+) (= 1R, 2R).

The fungicidal properties of the diasteromeric forms, and more recently, of the enantiomers, as well as the reduction triadimefon by fungi and plants have been studied intensively. BUCHENAUER (1979b) reported that triadimenol A was more fungitoxic than triadimenol B and remarked that various fungi differed significantly in their capability to reduce triadimefon and in their capability of producing certain diastereomeric mixtures. According to KRÄMER *et al.* (1983) the fungicidal activity of individual enantiomers is in the order A (−) > B (+) > A (+), B (−). Efforts to explain the level of sensitivity to triadimefon of a given fungus with its capability to reduce triadimefon (GASZTONYI and JOSEPOVITS 1978, 1979) were recently re-examined on the enantiomeric level by DEAS *et al.* (1984a, b). These authors demonstrated that the amount and the pattern of reduction of tridimefon to triadimenol gives an explanation for the high sensitivity of some fungi to triadimefon. However, this study fails to explain the low sensitivity of some fungal species such as *Fusarium culmorum*. It was therefore concluded that a simple relationship between sensitivity to triadimefon and preferential metabolic production of the most active A (−) enantiomer does not exist.

Bitertanol (Fig. 12.3), although structurally related to triadimenol and triadimefon, is distinctly different from these compounds in regard to its usable spectrum of fungicidal activity. *In vitro* studies show the effects of bitertanol against the Fungi Imperfecti to be more pronounced than those of triadimefon (KRAUS 1979).

When applied as a protective or curative foliar fungicide, the main uses of bitertanol are for the control of scab (*Venturia* spp.) in pome fruits, leaf diseases (*Puccinia* spp., *Mycosphaerella* spp.) in peanuts, bananas, and rusts, powdery mildews and *Monilinia* diseases in vegetables, ornamentals and stone fruits.

As a seed treatment, bitertanol is effective against certain seed- and soilborne pathogens of cereals. Bitertanol controls dwarf bunt of wheat (*Tilletia controversa*), a pathogen which is difficult to control with other fungicides.

As with triadimenol, bitertanol is a mixture of four enantiomeric forms. According to BÜCHEL (1984) the 1S, 2R enantiomer was the most active form in experiments involving bean rust (*Uromyces phaseoli*).

Due to its high lipophilicity, bitertanol penetrates plant surfaces readily and is locally redistributed, but is hardly translocated over longer distances in the plant (SCHEINPFLUG and VAN DEN BOOM 1981; BRANDES et al. 1988), in accordance with its low vapour pressure, redistribution of the compound via a gas phase is minimal.

Because of the low mobility of bitertanol inside and outside the plant, growth regulator effects are seldom seen (KELLEY and JONES 1981; BRANDES et al. 1979).

Propiconazole (Fig. 12.3), originally discovered by Janssen, was presented in 1979 by Ciba-Geigy (URECH et al. 1979), and rapidly became one of the most succesful SBI fungicides of its time.

Chemically speaking, propiconazole is closely related to its ethyl analogue, etaconazole. Biologically, it is characterised as a very broad spectrum fungicide which is active at low rates. Propiconazole was initially developed for the control of a complex of cereal pathogens but also showed efficacy in a multitude of other crops such as peanuts, grapes and bananas.

In cereals, propiconazole is widely used in mixture with combination partners such as carbendazim, chlorothalonil, and fenpropidine in order to complete the control of *Septoria* spp. and other cereal pathogens (SMITH and SPEICH 1981).

Quantitative studies have not been published, but URECH and SPEICH (1981) described propiconazole as a highly systemic compound which was absorbed by leaves and stems of cereal plants within 24 hours of application and transported acropetally in the plant.

The vapour phase activity of propiconazole was estimated to be biologically similar to that of triadimefon (RATHMELL and SKIDMORE 1982).

Propiconazole is a typical SBI fungicide in regard to its plant growth regulator activities. BUCHENAUER et al. (1981), who investigated the plant growth regulatory effects of several triazoles, stated that when applied as a barley seed treatment at $25-50$ g ai/100 kg propiconazole caused a growth inhibition of leaves, roots and coleoptiles. BUCHENAUER et al. also described plant alterations which might be considered beneficial, such as a delayed senescence of chlorophyll and an increased tolerance of the seedlings towards drought, frost and salt stress.

The presence of two asymmetric carbon atoms in the molecules of propiconazole and etaconazole results in four stereoisomers. The fungicidal activity is mainly based on the 2S-isomers with little differences between 2S, 4S- and ''2S, 4R-isomers (EBERT et al. 1988).

Etaconazole (Fig. 12.3), the second triazole fungicide presented by Ciba-Geigy in 1979 (STAUB et al. 1979), had basically the same biological properties as the related compound popiconazole. Meanwhile, the sale of etaconazole was discontinued worldwide. According to URECH et al. (1979), this compound was especially suited for the control of diseases of deciduous fruit. Similar to penconazole, which was introduced more recently by Ciba-Geigy, etaconazole provides simultaneous control of scab and powdery mildew on apples and pears. Additional activity has been shown against a broad spectrum of diseases in peanuts, cucumber, peaches and other agriculturally important crops (STAUB et al. 1979).

The plant-growth-regulatory side effects of etaconazole on barley seedlings were reported by BUCHENAUER et al. (1981) to be more pronounced than those of propiconazole. In apples, several reports describe smaller, thicker, darker green leaves and retarded tree growth when compared to untreated check trees (KELLEY and JONES 1981; SZKOLNIK 1981).

Etaconazole is a fully systemic fungicide, which is transported acropetally in plants. In apple leaves KELLEY and JONES (1982) reported an uptake of approximately 90% of the compound within 12 hours. SZKOLNIK (1983) studied the pronounced vapour phase activity of etaconazole. When applied to the shading cloth in greenhouse, etaconazole controlled several genera of powdery mildew in crops such as apples, grapes, vegetables and roses.

The antifungal activity of the isomeric forms of etaconazole has been described (VOGEL et al. 1983; HEERES 1984). The activity of the individual isomers depended strongly on the fungi used as test organisms. For example, *in vitro* tests with *Botrytis cinerea* showed the order 2S, 4R (cis) > 2S, 4S (trans) > 2R, 4S (cis), 2R, 4S (trans). Against *Erysiphe graminis* on barley the 2R, 4S-isomer was the most active, followed by the 2S, 4R and 2R, 4R-forms, and the 2S, 4S-isomer was the weakest one in this test system.

Penconazole (Fig. 12.3), introduced by Ciba-Geigy in 1983 (EBERLE et al. 1983), is another triazole fungicide patented by Janssen and developed by Ciba-Geigy. According to the authors mentioned above, this compound has a broad *in vitro* spectrum of antifungal activity. It is especially active in controlling diseases such as powdery mildew on grapes and apples, scab on apples and pears, and black rot on grapes. In order to improve the protective activity of penconazole against scab and to prevent the selection of resistant strains mixtures with a residual fungicide such as captan are recommended.

Very little has been published on the systemicity of penconazole. EBERLE et al. 1983) mention that the compound penetrated rapidly into plant tissues and was translocated acropetally. These authors stated that in their trials symptoms of phytotoxicity such as reduction of leaf size could not be detected.

ICI presented **diclobutrazol** (Fig. 12.3) in 1979 (BENT and SKIDMORE 1979) as a foliar fungicide for the control of cereal diseases. Besides control of powdery mildew, *Rhynchosporium secalis* and *Typhula incarnata*, diclobutrazol provides especially good control of *Puccinia* spp. in wheat and barley. Good control of coffee leaf rust has also been reported (JAVED 1981).

According to BENT and SKIDMORE (1979) diclobutrazol is a systemic fungicide which shows translaminar movement and acropetal translocation. These authors reported some vapour activity against cereal powdery mildew in greenhouse trials. This activity was estimated to be weak under field conditions (RATHMELL and SKIDMORE 1982).

Diclobutrazol, like the chemically related compound triadimenol, contains two chiral centres. It should be noted that the common name diclobutrazol applies only to the 2RS, 3RS diastereomer which is over 100 times more active as a fungicide than the 2RS, 3SR diastereomer. Resolution of the 2RS, 3RS form showed that the 2R, 3R enantiomer was responsible for almost all of the fungicidal activity (BALDWIN and WIGGINS 1983).

The plant-growth-regulatory side effects of diclobutrazol are rather distinct. BENT and SKIDMORE

(1979) reported a reduction in shoot growth and a darkening of foliage in apples. In greenhouse tests using various plant species soil drenches tended to have greater plant growth regulating effects than foliar sprays. Dicotyledonous plants were affected more than monocotyledonous species.

BUCHENAUER et al. (1981) mentioned that diclobutrazol caused stronger growth regulating effects than etaconazole and propiconazole in barley.

Flutriafol (Fig. 12.3) is a highly systemic triazole fungicide from ICI, which was presented in 1983 (SKIDMORE et al. 1983). Flutriafol, when used alone as a foliar fungicide, controls many major cereal pathogens such as powdery mildew, rust, *Rhynchosporium secalis* and *Pyrenophora* spp. Mixtures with carbendazim are necessary for good control of eyespot and improved control of *Fusarium* spp. Mixtures with captafol have shown increased activity against *Septoria nodorum* (NORTHWOOD et al. 1983).

SKIDMORE et al. (1983) tested flutriafol as a cereal seed treatment at 7.5 g/100 kg seed and reported good control of smuts and bunts (*Ustilago* spp., *Tilletia* spp.). The addition of a co-fungicide, such as imazalil and thiabendazol, was needed to be fully effective against *Pyrenophora* and *Fusarium* spp. Doses higher than 10 g/100 kg seed caused slight delays in seedling emergence in wheat.

In combination with ethirimol and thiabendazol flutriafol is further marketed as a seed treatment in cereals under the trade name of Ferrax®. This treatment has shown good activity against powdery mildew strains with decreased sensitivity to inhibitors of C-14-demethylation of dihydrolanosterol, such as triadimenol or flutriafol (NORTHWOOD et al. 1984; NOON et al. 1988a).

The high mobility of flutriafol in plants is proved by the good systemic control of *Erysiphe graminis* and *Puccinia hordei* in winter barley after seed treatment and the experimental use as a soil granule or soil drench application for control of coffee rust caused by *Hemileia vastatrix*. SHEPHARD (1985) compared the translocation of propiconazole and flutriafol in barley leaves with the aid of a biotest and found the maximum of control at the leaf tip when flutriafol was applied as a droplet in the basal leaf part whereas the fungicidal inhibition by propiconazole, under the same conditions, was confined to the basal leaf parts.

Flusilazole (Fig. 12.3) was introduced by Du Pont 1984 and is the first agrochemical or pharmaceutical azole compound to contain silicon in the molecule (MOBERG et al. 1985).

The broad-spectrum fungicide has shown very good results against several major diseases of cereals such as powdery mildew, rust fungi, *Septoria* diseases, and *Pyrenophora* species. Unlike most other triazole fungicides the fungicidal spectrum of flusilazole includes *Pseudocercosporella* foot rot (FORT and MOBERG 1984) at the rate of 300 g/ha. When used in mixture with carbendazim not only the W-type but also the R-type of the eyespot pathogen can be controlled if BCM resistance is not present.

Excellent activity has also been shown against several important tree fruit diseases such as *Venturia* spp., *Podosphaera leucotricha* and *Monilinia* spp. In grapes powdery mildew caused by *Uncinula necator* and black rot caused by *Guignardia bidwellii* are controlled. From the multitude of other uses the high activity against Sigatoka diseases in bananas and against *Cercospora* leaf spot in sugar beet can perhaps be selected as examples.

In greenhouse tests, BRUHN et al. (1985) showed that flusilazole was able to penetrate plant tissue and to move within the plant's transpiration stream, allowing for preventative and curative activity. O'LEARY and JONES (1987) examined the uptake of flusilazole by apple leaves in more detail. At 10, 20 and 30 °C apple leaf disks absorbed 15, 32, and

54%, respectively, of the applied active ingredient within 24 hours. The systemic mobility of flusilazole is sufficient to allow a migration into the acropetal parts of treated wheat leaves but is not high enough to produce a marked accumulation at the leaf tip. DAVIS et al. (1985) reported that at test rates, flusilazole caused no phytotoxic effects on tree fruits.

BAS 454 06 F (Fig. 12.3) is another systemic triazole compound, which was presented by POMMER and ZEEH in 1983. The product was under investigation against a wide range of pathogens on crops which usually are treated with broad spectrum DMIs. In the meantime, the development of BAS 454 06 F has been discontinued.

Diniconazole (Fig. 12.3) belongs to a group of triazole derivatives containing potent, broad spectrum fungicides as well as effective plant growth regulators.

The compound is reported to be active against scab and powdery mildew in apples as well as in grapes against *Uncinula necator*. In cereals diniconazole is used, mostly in combination with iprodione and other partners, against rusts, powdery mildew and several leaf spot pathogens. TAKANO et al. (1983) showed control of *Ustilago* spp., *Tilletia* spp. and *Pyrenophora graminea* when diniconazole was used as a seed treatment. According to FUNAKI et al. (1983), the compound exhibited curative properties and was partially systemic.

Stereoisomeric forms exist within the structure of diniconazole, which express different biological properties. The R(−) isomer was the more potent fungicide, while the S(+) isomer was shown to be a better plant growth regulator (FUNAKI et al. 1983). KVIEN et al. (1987) studied the effect of the two isomers of diniconazole on peanut growth and development. The S(+) isomer which made up 19% of the diniconazole formulation tested reduced stem height by 33%, leaf area index by 16%, and total vegetative dry weight by 19%, but had no effect on average leaf size.

The 4-chlorophenyl analogue of diniconazole, S 3307, has been described as a plant growth regulator (IZUMI et al. 1984, 1985).

PP 969 (Fig. 12.3) was an experimental fungicide with uncommon systemic properties which has been deleted from further development. As PP 969 is a highly water soluble compound with a low partition coefficient (log P octanol/water = 1.95), this compound seems to be relatively ineffective in penetrating the cutinized surface of leaves but is highly mobile in woody plants when applied as a soil drench or stem injection (SHEPHARD and FRENCH 1983; SHEPHARD 1985). Single application as a soil drench or injection has provided control of leaf diseases of coffee, bananas and apples for up to 30 weeks.

Myclobutanil (Fig. 12.3) is an azole fungicide from Rohm and Haas, which has been introduced into the marked since 1986. This compound is structurally similar to the older imidazole fungicide fenapanil. Myclobutanil has been developed as a foliar fungicide against a broad spectrum of fungal diseases on (mostly) dicotyledonous plants such as rusts, powdery mildews, scab and various leaf spot diseases. So for example in apple powdery mildew and scab are controlled simultaneously whereas in vines *Uncinula necator* and *Guignardia bidwelli* are well controlled (ORPIN et al. 1986). QUINN et al. (1986) point out that myclobutanil is especially active on powdery mildews, rusts and fungi whose perfect stage are in the locoloascomycetidae (sensu Alexopoulos and Mims) such as *Helminthosporium* (*Pyrenophora* spp. *Cochliobolus* spp), *Septoria* spp., *Cercospora* spp. (*Mycosphaerella* perfect stage) and *Venturia inaequalis*.

Myclobutanil is also being used on monocotyledonous plants as a seed treatment against a complex of soil- and seed-borne seedling diseases. From the activity of myclobutanil against loose smuts, it can be assumed that it is at least partially systemic in plants.

According to QUINN et al. (1986) myclobutanil appeared to be intermediate in systemic translocation between triadimefon and propiconazole. The same authors describe the phytotoxicity and growth regulator effects to be relatively low for an EBI compound.

Tebuconazole (Fig. 12.3), presented to the public in 1986 (SCHEINPFLUG and KASPERS 1986; KUCK and BERG 1986), is a broad spectrum fungicide which can be used in a wide range of crops as a foliar fungicide as well as a seed treatment in cereals.

As a foliar fungicide in wheat and barley, tebuconazole at 125 – 312 g ai/ha controls rust species, *Septoria diseases*, powdery mildew and leaf spot pathogens such as *Rhynchosporium secalis* and *Pyrenophora* species. Mainly in ears, *Fusarium* infections are reduced significantly. Tebuconazole has an effect against W Type strains but not against R type strains of eyespot (REINECKE et al. 1986).

Two major grape diseases, grey mould, caused by *Botrytis cinerea*, and powdery mildew, caused by *Uncinula necator*, are evenly controlled by tebuconazole, partly in mixture with diclofluanid or tolylfluanid as a non-systemic partner (BRANDES and KASPERS 1989).

In oilseed rape, which is subject to attack by pathogens such as *Sclerotinia sclerotiorum, Leptosphaeria maculans* and *Alternaria* spp., tebuconazole controls these pathogens at 250 – 375 g ai/ha (KASPERS and SIEBERT 1989) as well as having a positive growth-regulating side effect. In peanuts, tebuconazole is developed as a fungicide against *Mycosphaerella* species, whereas in bananas Black and Yellow Sigatoka can be controlled (WYBOU 1989).

Used as a cereal seed treatment, tebuconazole effectively controls, at only 1 – 3 g ai/100 kg seed, several seed- und soilborne pathogens such as *Ustilago* spp., *Tilletia* spp. and some *Pyrenophora* and *Fusarium* pathogens.

Evaluations on the systemic properties of tebuconazole (KUCK and THIELERT 1987) have shown that the acropetal translocation of the compound holds an intermediate position between the locosystemic compound, bitertanol, and the highly mobile triazole, triadimenol. In cereal leaves a remarkably even distribution throughout the leaf can, therefore, be found in studies with radiolabelled active ingredient. As shown before with triadimenol (KUCK et al. 1982) tebuconazole also inhibits most effectively the early mycelial growth of rust fungi whereas the formation of fungal appressoria and haustoria was only weakly affected (HAENSSLER and KUCK 1987).

Tebuconazole is a mixture of two enantiomers, the S (−)- and the R (+)-isomer. The S (−) form has been shown to be clearly more fungitoxic (BERG et al. 1987) although the fungitoxicity of the R (+) isomer reaches the level of that of triadimenol with some fungi.

Hexaconazole (Fig. 12.3), the most recent triazole from ICI, was presented in 1986 (SHEPARD et al. 1986). On apples hexaconazole gave good control of scab (*Venturia inaequalis*) and powdery mildew (*Podosphaera leucotricha*) at 1 – 2 g / hl. On peanuts, the compound gives control of both early and late leaf spot (*Mycosphaerella* spp.). Against coffee rust (*Hemileia vastatrix*) only 30 g ai/ha were necessary to achieve control. In vines hexaconazole is active against powdery mildew at 0.6 to 1.5 g ai/hl and against black rot (*Guignardia bidwellii*) at the rate of 1.5 – 2 g ai/hl (HEANEY et al. 1986).

In wheat hexaconazole has demonstrated excellent control of rusts (*Puccinia* spp.) and *Septoria* spp. and good control of powdery mildew. In combination with carbendazim also eyespot and *Fusarium* ear blight can be controlled (WALLER et al. 1990).

Among the other crops where hexaconazole has shown good results the activity on bananas at 30 – 60 g ai/ha against Black Sigatoka, and on peaches at 2 – 5 g ai/hl against brown rot (*Monilinia fructicola*), scab (*Cladosporium carpophilum*) and powdery mildew (*Sphaerotheca pannosa*) have to be mentioned.

The systemic mobility of hexaconazole is sufficient to give curative and translaminar control after foliar treatment but also to give some systemic control after root drench application (SHEPHARD *et al.* 1986).

Cyproconazole (Fig. 12.3), chemically characterised by its cyclopropyl moiety, is a fully systemic broad spectrum triazole fungicide which was discovered by Sandoz.

In cereals at low rates of 40 – 100 g ai/ha excellent activity was observed against rusts and powdery mildews. Supplementary control of *Leptosphaeria nodorum* on wheat and of *Cochliobolus sativus* on barley have been reported by GISI *et al.* (1986a) whereas weaker effects were found against eye spot (*Pseudocercosporella*) on wheat and of *Pyrenophora* spp. on barley. Combinations with carbendazim are, therefore, offered.

Recommendations for use of cyproconazole in other crops include for example scab and powdery mildew in apple and pear, powdery mildew in grapes and cucumbers, *Monilinia* spp. in stone fruits, *Exobasidium vexans* in tea, and coffee rust (GISI *et al.* 1986b).

Cyproconazole, when applied to the soil, showed distinct systemic control of fungal pathogens. A rapid translocation in the transpiration stream, comparable to that of triadimenol, was indirectly proved (GISI *et al.* 1986b). At doses higher than the recommended field rates cyproconazole showed the same growth regulation symptoms as many other triazoles such as stunting, crinkling and foliage reduction (GISI *et al.* 1986b).

Cyproconazole is a mixture of four isomers in a ratio of $1:1:1:1$ with decreasing fungicidal activity in the sequence $B(-) \leq A(-) \ll A(+) < B(+)$ (GRABSKI and GISI 1990). In contrast to other triazole fungicides no single isomer was more active than the mixture of them, indicating some synergistic interactions between the enantiomers.

Difenoconazole (Fig. 12.3) is a new broad-spectrum triazole synthesised and developed by Ciba-Geigy. In cereals the product is especially suited for the control of the later season disease complex caused by fungi such as *Septoria* spp., rusts and sooty moulds (RUESS *et al.* 1988). In apples difenoconazole showed excellent protective and curative control of *Venturia inaequalis* at rates of 2.5 – 10 g a.i./hl combined with good activity against other apple pathogens such as powdery mildew, *Alternaria mali* and leaf rust. In grapes besides powdery mildew also black rot and red fire disease (*Pseudopezicula tracheiphila*) are controlled. In vegetable crops several leaf spot diseases and especially *Alternaria* species are covered by the spectrum of activity of difenoconazole. As a cereal seed treatment difenoconazole is used against *Ustilago* spp. and *Tilletia* spp. inclusive wheat dwarf bunt (*T. controversa*).

DAHMEN & STAUB (1992a, b) provided more details on uptake and translocation of difenoconazole. After soil drench treatments only small amounts of the compound are taken up by the root system and transported into the leaves of tomato and peanut plants. Together with other observations the authors judged these findings as an indication for a rather weak apoplastic transport in plants. In penetration studies with peanut and tomato leaves about 10 – 20% of the applied active ingredient penetrated within the first hour.

Imibenconazole (Fig. 12.3), the first triazole compound from Hokko, was presented in 1988 and is mainly developed in broad leaved crops. Laboratory and greenhouse studies showed a broad spectrum of activity against fungi of the Ascomycetes, Basidiomycetes and Deuteromycetes (OHYAMA et al. 1988). In field trials, doses from 2.5–7.5 g ai/hl controlled scab, powdery mildew and rust on apples and on Japanese pears. The activity of imibenconazole against powdery mildew and anthracnose (*Elsinoe ampelina*) of grapes was demonstrated at 3.8–15 g ai/hl. Other uses of imibenconazole as a foliar fungicide are in various vegetables and ornamentals as the low phytotoxicity gives no particular restrictions according to OHYAMA et al. (1988).

In addition to the foliar applications in dicotyledonous crops, control of wheat bunt (*Tilletia caries*) by seed treatment has been shown at 15–60 g ai/100 kg seed.

Tetraconazole (Fig. 12.3), presented in 1988, is the first azole fungicide introduced by the Montedison Group. *In vitro* the activity of the compound is less pronounced than that of reference triazoles such as penconazole and propiconazole. However, GARAVAGLIA et al. (1988) point out that in greenhouse tests the product was more effective, especially against obligate parasites, than most standard compounds. In field tests, the same authors found a strong activity against powdery mildew of wheat and barley and good control of brown rust (*Puccinia recondita*). In other crops several pathogens such as powdery mildew and scab in apples, *Cercospora beticola* in sugar beets, and powdery mildew in grapes are effectively controlled.

At rates substantially higher than those usually needed GARAVAGLIA et al. (1988) report only very weak growth retarding effects of tetraconazole on peas and beans which were fully reversible by application of gibberellic acid.

The fungitoxicity of the R(+) and the S(−) enantiomers of tetraconazole has been reported for a series of fungi. In all cases the R(+) enantiomer was more fungitoxic than the S(−) isomer (BIANCHI et al. 1991).

Furconazole-cis (Fig. 12.3) was presented in 1988 by Rhône-Poulenc as a broadly acting triazole fungicide but has since been deleted from further development. At use rates which varied mostly between 10 and 100 g a.i./ha, furconazole-cis showed good results in apples against powdery mildew and scab, in grapes against grape powdery mildew, in sugar beets against *Cercospora beticola* and powdery mildew, and in cereals against powdery mildew and rust (ZECH et al. 1988; GOUOT et al. 1988). Diseases of tropical crops where furconazole-cis performed well were Yellow Sigatoka of banana (*Mycosphaerella musae*) and coffee rust (*Hemileia vastatrix*).

Fenbuconazole (Fig. 12.3) is, like other azole fungicides from Rohm and Haas, fenapanil and myclobutanil, chemically characterised by its nitrile substituent. Presented in 1988, fenbuconazole has been developed initially for the foliar treatment of cereal crops and in stone fruits for the control of *Monilinia* spp. as primary target. In wheat and barley at the rate of 75 g a.i./ha especially *Septoria* diseases and rusts are well controlled (DRIANT et al. 1988). Good results are also reported for *Rhynchosporium secalis* whereas for the control of powdery mildew mixtures with a morpholine partner are needed.

Other crop uses where the product has shown good results include grapes (*Uncinula necator, Guignardia bidwelli*), pome fruit (*Venturia* spp., *Gymnosporangium* spp.), sugar beets (*Cercospora beticola*), peanuts (*Puccinia arachidis, Cercospora arachidicola, Cercosporidium personatum*) and bananas (*Mycosphaerella* spp.).

DRIANT et al. (1988) mention further that fenbuconazole is readily translocated upward in the plant whereas basipetal transport is low. The compound was more effective when applied as a preventative treatment but also showed eradicative activity.

Epoxyconazole (Fig. 12.3), presented to the public in 1990 under its code number BAS 480 F, is a new broad spectrum triazole with potential uses in cereals and a range of other crops. In cereals an pronounced activity against rusts, *Septoria* diseases and *Rhynchosporium secalis* has been pointed out by AMMERMANN et al. (1990). Among the broad range of other cereal pathogens which are well conrolled especially the activity against eyespot (*Pseudocercosporella herpotrichoides*) at the rate of 187 g/ha has to be mentioned (SAUR et al. 1990, 1991).

Other disease complexes where epoxyconazole showed promising results are *Cercospora beticola* in sugar beet, *Mycosphaerella* spp. and *Puccinia arachidis* in peanuts and *Alternaria brassicae* in oilseed rape.

AKERS et al. (1990) investigated the uptake and the systemic translocation of epoxyconazole in special studies. According to their findings the absorption of the molecule is strongly affected by the type of formulation and the addition of surfactant. When [^{14}C]-labelled active ingredient was applied to the basal parts of wheat leaves as a single droplet 4 days later the whole acropetal part of the leaf was very uniformly labelled. In cytological studies the same authors describe that at higher concentrations not only the mycelial growth of *Uromyces appendiculatus* was inhibited but that additionally the formation of functional haustoria was strongly reduced.

SSF-109 (Fig. 12.3) is another new triazole fungicide which was presented in 1990 to the public by Shionogi. According to MURABAYASHI et al. (1990, 1991) the compound offers protection against a broad spectrum of plant diseases and against grey mould (*Botrytis cinerea*) in particular. The authors report that *in vitro* SSF-109, which is a **cis**-isomer, was generally more effective than the corresponding **trans**-isomer. Against *Rosellinia necatrix*, however, the trans-isomer showed stronger fungicidal activity. In plant tests against cucumber grey mould, powdery mildew, *Sclerotinia* rot, sheat blight, oat crown rust and rice blast the differences between the two isomers were sometimes less pronounced. On the other hand the **trans**-isomer had a much stronger growth regulating effect on lettuce and rice seedlings.

Bromuconazole (Fig. 12.3), the second triazole fungicide presented by Rhône Poulenc in 1988 is used in a wide range of crops. In cereals bromuconazole provides good control of stem based infections of eyespot (*Pseudocercosporella herpotrichoides*) and of *Micronectriella nivalis* (*Fusarium nivale*) at rates of 300 g a.i./ha. Against foliar diseases such as *Puccinia* spp., *Septoria tritici*, *Erysiphe graminis*, *Pyrenophora* spp., and *Rhynchosporium secalis* rates close to 200 g a.i./ha gave good results (PEPIN et al. 1990). A special mention was made concerning the good efficacy of bromuconazole against *Fusarium roseum* on wheat ears.

In fruits and vegetables besides the pathogens which are usually controlled by DMI fungicides such as scab in apples and powdery mildew in apples, grapes and cucurbits, various species of *Alternaria* were well controlled in oilseed rape, potatoes and carrots with rates of 125 – 400 g a.i./ha.

PEPIN et al. (1990) report further that at temperatures ranging from 10 to 30 °C about 20 – 30% of the active ingredient was rapidly absorbed by wheat leaves. Following application of bromuconazole as a droplet in the lower third of barley leaves the compound was found to be evenly distributed within the apical part of the leaf after 24 hours.

Metconazole (Fig. 12.3) is a new triazole compound which is under development mainly for cereal diseases. The compound was first synthesised and patented by Kureha and is under development together with the Shell group of companies.

In vitro studies showed that the **cis** (1 **RS**, 5 **SR**) isomer is substantially more active than the corresponding **trans** isomer. Accordingly, the **cis**-isomer (code no. WL 136 184) seems to be under further development.

Against foliar diseases in cereals good control of *Septoria tritici*, *Leptosphaeria nodorum*, rusts, *Pyrenophora teres* and *Rhynchosporium secalis* was reported at rates of only 30–90 g active ingredient per hectare. Furthermore, a moderate effect against powdery mildew and *Fusarium* species has been demonstrated (SAMPSON *et al.* 1992). As a seed treatment on wheat and barley, metconazole provided control of smut and bunt fungi as well as of leaf stripe (*Pyrenophora graminae*) at fairly low rates of 2.5 to 7.5 g ai/100 kg seed. Systemic control of stripe rust (*Puccinia striiformis*) was also reported.

In broad-leaved crops metconazole showed activity against various pathogens such as scab in apples and *Cercospora beticola* in sugar beet.

Fluquinconazole (Fig. 12.3) is the second new triazole to be presented to the public in 1992. Chemically fluquinconazole is the first quinazolinone based triazole compound. RUSSEL *et al.* (1992) who described the spectrum of activity in the field gave special emphasis to the good results against *Venturia inaequalis* in apples at rates between 3.25 and 10 g ai/hl. Beside scab also powdery mildew of apples (*Podosphaera leucotricha*) was controlled at similar rates. In greenhouse tests with apple plants a systemic movement of the compound in the acropetal direction has been demonstrated. As to be expected, no evidence for phloem transfer was found.

In cereals rates of 125 to 375 g ai/ha of fluquinconazole controlled rusts and *Septoria* species whereas the effects against the leaf spot pathogens, *Rhynchosporium secalis* and *Pyrenophora teres*, and cereal powdery mildew were described to be only moderate. Like most other triazoles fluquinconazole did not control the R type population of eyespot (*Pseudocercosporella herpotrichoides*).

Triticonazole (Fig. 12.3) was presented by Rhône-Poulenc as a triazole fungicide with quite unusual properties. The compound is under development exclusively as a cereal seed treatment agent. According to MUGNIER *et al.* (1992) a seed treatment at the rate of 90–120 g a.i./100 kg seed with triticonazole is suited to control several winter cereal diseases such as rusts and *Septoria* diseases up until the ear emergence stage. Therefore foliar treatments could be avoided. Beside foliar pathogens also triazole sensitive strains of eyespot (*Pseudocercosporella herpotrichoides*) could be controlled by a seed treatment.

DMI fungicides   233

Table 12.3 Triazole Fungicides

1) Common name[1]	1) Inventing/Developing Company
2) Experimental No.	2) Patent No., Year[4]
3) Chemical name[2]	3) Year of presentation
4) Trade name(s) of products for spray application[3] & combination partner(s)	4) Acute toxicity[5]
	5) Solubility (water)[6]
5) Trade name(s) of products for seed treatment[3]	6) Vapour pressure[7]
& combination partner(s)	7) Partition coefficient[8]
6) Use as foliar treatment (crop/pathogen) application rate (ai)	8) Literature
7) Use as seed treatment (crop/pathogen) application rate (ai)	
8) Other uses	

1) **Fluotrimazol**	1) Bayer
2) BAY BUE 0 620	2) DOS 1 795 249, 1968
3) diphenyl-(3-fluoromethylphenyl)-1,2,4-triazole-1-yl-methane	3) 1973
4) Persulon®	4) >5 000 mg/kg
5)	5) 0.5 mg/l
	6) 36 µPa
6) Fruits, Ornamentals, Vegetables: powdery mildew	8) Jeffrey et al. 1975 Le Bon et al. 1978

1) **Triadimefon**		1) Bayer
2) Bay MEB 6 447		2) DOS 2 201 063, 1972
3) 1-(4-chlorophenoxy)-3,3-dimethyl-1-(1H-1,2,4-triazol-1-yl)-butanone		3) 1973
4) Bayleton®		4) ~1 000 mg/kg
		5) 64 mg/l
		6) 20 µPa
		7) log $P_{(o/w)}$: 3.11
6) Cereals: *Erysiphe graminis, Puccinia* spp., *Rhynchosporium secalis, Septoria tritici, Typhula incarnata*	125 – 250 g/ha	8) Frohberger 1975 Scheinpflug et al. 1978 Martin & Morris 1979
Apples: *Podosphaera leucotricha, Gymnosporangium* sp.	0.0025 – 0.01%	Kolbe 1982 a, b
Grapes: *Unicula necator*	0.0025 – 0.005%	
Coffee: *Hemileia vastatrix*	125 – 500 g/ha	Kaspers & Patel 1980
Vegetables, Ornamentals: powdery mildew, rust	0.0025 – 0.0125%	Noegel et al. 1977
Turf: various diseases		Sanders et al. 1978
Others: powdery mildew, rust, leaf spot diseases on mango, hops, sugar cane, small fruits etc.		

1) **Triadimenol**	1) Bayer
2) BAY KWG 0519	2) DOS 2 324 010, 1973
3) (1RS,2RS; 1RS,2SR)-1-(4-chlorophenoxy)-3,3-dimethyl-1-(1H-1,2,4-triazol-1-yl) butan-2-ol	3) 1977

Table 12.3 (Continued)

1) Common name[1]		1) Inventing/Developing Company
2) Experimental No.		2) Patent No., Year[4]
3) Chemical name[2]		3) Year of presentation
4) Trade name(s) of products for spray application[3]		4) Acute toxicity[5]
	& combination partner(s)	5) Solubility (water)[6]
5) Trade name(s) of products for seed treatment[3]		6) Vapour pressure[7]
	& combination partner(s)	7) Partition coefficient[8]
6) Use as foliar treatment (crop/pathogen)	application rate (ai)	8) Literature
7) Use as seed treatment (crop/pathogen)	application rate (ai)	
8) Other uses		

Triadimenol (continued)		
4) Bayfidan®		4) ~ 700 mg/kg
Colt®, Dorin®, Ondene®	& tridemorph	5) RS + SR: 62 mg/l
Matador®, Silvacur®	& tebuconazole	RR + SS: 32 mg/l
Vynoc®, Mac® 2	& sulphur	6) RS + SR: 0.041 µPa
		RR + SS: 0.24 µPa
5) Baytan® universal	& fuberidazole	7) RS + SR: log $P_{(o/w)}$: 3.08
	& imazalil	RR + SS: log $P_{(o/w)}$: 3.28
Baytan® spezial	& fuberidazole	
Domestin®	& bitertanol	
	& fuberidazole	
6) Cereals: powdery mildew, rust, *Rhynchosporium secalis*, *Septoria tritici*, *Typhula incarnata*	125 g/ha	
Apples: *Podosphaera leucotricha*, *Gymnosporangium* sp.	0.005 – 0.01 %	
Grapes: *Uncinula necator*, *Guignardia bildwellii*	0.0025 – 0.005 %	
Coffee: *Hemileia vastatrix*	125 – 250 g/ha	
Cucurbits: powdery mildew		
Bananas: *Mycosphaerella* spp.	100 – 150 g/ha	
7) Cereals: *Ustilago* spp., *Tilletia* spp. *Pyrenophora* spp., *Micronectriella nivales*, *Erysiphe graminis*, *Puccinia* spp., *L. nodorum* *Cochliobolus sativus*, *Septoria tritici*	15 – 37.5 g/dt	8) FROHBERGER 1978 TRAEGNER-BORN & VAN DEN BOOM 1978
8) Coffee (soil application): *Hemileia vastatrix*		

1) **Bitertanol**	1) Bayer
2) BAY KWG 0599	2) DOS 2 324 010, 1973
3) 1-((biphenyl)-4-yloxy)-3,3-dimethyl-1-(1H-1,2,4-triazol-1-yl)-butan-2-ol (20:80 ratio of (1RS,2RS) and (1RS,2SR) isomers)	3) 1978
4) Baycor®, Baymat®	4) > 5.000 mg/kg

Table 12.3 (Continued)

1) Common name[1]		1) Inventing/Developing Company
2) Experimental No.		2) Patent No., Year[4]
3) Chemical name[2]		3) Year of presentation
4) Trade name(s) of products for spray application[3]		4) Acute toxicity[5]
	& combination partner(s)	5) Solubility (water)[6]
5) Trade name(s) of products for seed treatment[3]		6) Vapour pressure[7]
	& combination partner(s)	7) Partition coefficient[8]
6) Use as foliar treatment (crop/pathogen)	application rate (ai)	8) Literature
7) Use as seed treatment (crop/pathogen)	application rate (ai)	
8) Other uses		

Bitertanol (continued)

5) Sibutol®		5) RS + SR: 2.9 mg/l
Sibutol® Combi	& fuberidazole	RR + SS: 1.6 mg/l
Domestin®	& triadimenol	6) RS + SR: 0.22 nPa
	& fuberidazole	RR + SS: 2.5 nPa
		7) RS + SR: log $P_{(o/w)}$: 4.1
		RR + SS: log $P_{(o/w)}$: 4.4
6) Pome fruits: *Venturia* spp.	0.005 – 0.025%	8) BRANDES et al. 1979
Gymnosporangium spp.		
Stonefruits: *Monilinia* sp., *Stigmina carpophila*	0.0125 – 0.0375%	
Bananas: *Mycosphaerella* spp.,		SCHEINPFLUG &
Guignarida musae		VAN DEN BOOM 1981
Peanuts: *Puccinia arachidis*, *Phoma arachidicola*		
Sugar beet: *Cerospora beticola*, *Erysiphe betae*	0.25 – 0.6 kg/ha	
7) Cereals: *Tilletia* sp., *Ustilago* spp., *Gerlachia nivale*, *Tilletia controversa*	10 – 75 g/1 000 kg	TRAEGNER-BORN & KASPERS 1981

1) **Propiconazole**		1) Janssen/Ciba-Geigy
2) CGA 64250		2) DOS 2 551 560, 1974
3) ± -1-[2-(2,4-dichlorophenyl)-4-propyl-1,3-dioxolan-2-ylmethyl]-1H-1,2,4-triazole		3) 1979
4) Tilt®, Radar®, Banner®, Desmel®		4) 1 517 mg/kg
Tilt® C	& carbendazim	5) 110 mg/l
Tilt® CT	& chlorothalonil	
Turbo® TR	& tridemorph	
Simbo®, Tilt® Top	& fenpropimorph	
Archer®	& fenpropidine	
5) Panogen®	& guazatine	6) 0.13 mPa
Benit®	& thiabendazol	7) log $P_{(o/w)}$: 3.72
6) Cereals: *Erysiphe graminis*, *Puccinia* spp., *Septoria tritici*, *Pyrenophora teres*, *Rhynchosporium secalis*	125 g/ha	8) URECH et al. 1979 URECH & SPEICH 1981
Peanuts: *Mycosphaerella arachidicola*, *M. berkeleyi*	100 – 150 g/ha	

Table 12.3 (Continued)

1) Common name[1])	1) Inventing/Developing Company
2) Experimental No.	2) Patent No., Year[4])
3) Chemical name[2])	3) Year of presentation
4) Trade name(s) of products for spray application[3])	4) Acute toxicity[5])
& combination partner(s)	5) Solubility (water)[6])
5) Trade name(s) of products for seed treatment[3])	6) Vapour pressure[7])
& combination partner(s) application rate (ai)	7) Partition coefficient[8])
6) Use as foliar treatment (crop/pathogen) application rate (ai)	8) Literature
7) Use as seed treatment (crop/pathogen) application rate (ai)	
8) Other uses	

Propiconazole (continued)

6) Grapes: *Uncinula necator, Guignardia bidwellii*   0.0025%   SCHWINN & URECH 1981
   Bananas: *Mycosphaerella* spp.   100 g/ha   MOURICHON & BEUGNON 1982
   Others: leaf spot diseases, rust, powdery mildew, scab in soybeans, etc.
7) *Tilletia caries, Ustilago* spp.   1–5 g/100 kg

---

1) **Etaconazole**   1) Janssen/Ciba-Geigy
2) CGA 64251   2) DOS 2551560, 1974
3) 1-[2-(2,4-dichlorophenyl)-4-ethyl-1,3-dioxolan-2-ylmethyl]-1H-1,2,4-triazole   3) 1979
   4) 1343
4) Vangard®, Sonax®   5) 80 mg/l
   6)
6) Pome fruits: *Venturia inaequalis*   0.002%   8) SZKOLNIK 1981
   Podosphaera leucotricha
   Gymnosporangium spp.
   Stonefruits: *Monilinia* spp.,   KELLEY & JONES 1981
   Coccomyces hiemalis
   Citrus: *Geotrichum candidum*,   GUTTER 1982
   Penicillium spp.   SCHACHNAI 1982

---

1) **Penconazole**   1) Janssen/Ciba-Geigy
2) CGA 71818   2) DOS 2735872, 1976
3) 1-(2,4-dichloro-β-propylphenethyl)-1H-1,2,4-triazole   3) 1983
4) Topas®, Topaz®, Omnex®   4) 2125 mg/kg
   Topas® C   & captan   5) 70 mg/l
   Topas® MZ   & mancozeb
   Preface®   & dinocap
5)   6) 0.21 mPa
   7)
6) Pomefruits: *Podosphaera leucotricha*,   0.0025%   8) EBERLE et al. 1983
   Venturia spp.   BOSSHARD et al. 1985
   Grapes: *Uncinula necator, Guignardia bidwellii*   0.0015–0.005%
   Others: powdery mildew, rust a. o.   0.0025–0.015%

Table 12.3 (Continued)

1) Common name[1]		1) Inventing/Developing Company
2) Experimental No.		2) Patent No., Year[4]
3) Chemical name[2]		3) Year of presentation
4) Trade name(s) of products for spray application[3]		4) Acute toxicity[5]
	& combination partner(s)	5) Solubility (water)[6]
5) Trade name(s) of products for seed treatment[3]		6) Vapour pressure[7]
	& combination partner(s)	7) Partition coefficient[8]
6) Use as foliar treatment (crop/pathogen)	application rate (ai)	8) Literature
7) Use as seed treatment (crop/pathogen)	application rate (ai)	
8) Other uses		

1) **Diclobutrazol**		1) ICI
2) PP 296		2) DOS 2 737 489, 1976
3) (2R,3R) and (2S,3S)-1-[2,4-dichlorophenyl]-4,4-dimethyl-2-(1,2,3-triazol-1yl) pentan-3-ol		3) 1979
		4) ~4.000 mg/kg
4) Vigil®		5) 9 mg/l
		6) 1.3 – 2.6 µPa
		7) log $P_{(o/w)}$: 3.8
6) Cereals: *Puccinia* spp., *Erysiphe graminis, Rhynchosporium secalis, Typhula incarnata*	125 g/ha	8) Bent & Skidmore 1979
Coffee: *Hemileia vastatrix*		Javed 1981

1) **Flutriafol**		1) ICI
2) PP 450		2) EP 0 015 756, 1979
3) RS-2,4'-difluoro-α-(1H-1,2,4-triazol-1-yl-methyl)-benzhydryl alcohol		3) 1983
4) Impact®		4) 1 140 mg/kg (m); 1 480 mg/kg (f)
Early Impact®, Impact® RM, Yellow®	& carbendazim	5) 130 mg/l
Cicéro®, Impact® TX, Halo®	& chlorothalonil	6) 0.4 µPa
Cyclone®	& iprodione	
Antares®	& fentin hydroxid	
5) Ferrax®	& ethirimol	7) log $P_{(o/w)}$: 2.32
	& thiabendazole	
Vincit®, Vincit® LU	& imazalil	
Vincit® F	& thiabendazole	
6) Cereals: *Puccinia* spp., *Erysiphe graminis, Septoria* spp., *Rhynchosporium secalis, Pyrenophora teres*, sooty moulds (ear)	95 – 125 g/ha	8) Skidmore et al. 1983 Dawson et al. 1984 Northwood et al. 1984
Sugar beet: *Cercospora beticola*	62.5 g/ha	
Cocoa: *Oncobasidium theobromae*	6.25 – 25 g/hl	
Oilseed rape: *Pseudocercosporella capsellae*	125 g/ha	Noon et al. 1988 b

Table 12.3 (Continued)

1) Common name[1])		1) Inventing/Developing Company
2) Experimental No.		2) Patent No., Year[4])
3) Chemical name[2])		3) Year of presentation
4) Trade name(s) of products for spray application[3])		4) Acute toxicity[5])
	& combination partner(s)	5) Solubility (water)[6])
5) Trade name(s) of products for seed treatment[3])		6) Vapour pressure[7])
	& combination partner(s)	7) Partition coefficient[8])
6) Use as foliar treatment (crop/pathogen)	application rate (ai)	8) Literature
7) Use as seed treatment (crop/pathogen)	application rate (ai)	
8) Other uses		

Flutriafol (continued)

7) Cereals: *Tilletia* spp., *Ustilago* spp., *Leptosphaeria nodorum*	5 – 15 g/100 kg	Noon et al. 1988a Godwin et al. 1984
8) Coffee (soil application): *Hemileia vastatrix*		

1) **Flusilazole**		1) DuPont
2) DPX H 6573		2) EP 0068813, 1981
3) bis (4-fluorophenyl) methyl (1H-1,2,4-triazol-1-ylmethyl)silane		3) 1983
4) Punch®, Nustar®, Olymp®, Benocap®, Capitan®		4) 674 mg/kg (f) 1.110 mg/kg (m)
Initial, Pluton	& fenpropimorph	
Cérélux®; Gral®	& tridemorph	5) 45 mg/l [pH 7.8]
Punch® C; Harvesan®	& carbendazim	6) 14.7 mPa [25 °C]
Triumph®	& chlorothalonil	7)
6) Apples: *Venturia inaequalis*, *Podosphaera leucotricha*, *Gymnosporangium* spp.	70 – 140 g/ha	8) Davies et al. 1985 Cagnieul & Labit 1985
Grapes: *Uncinula necator*, *Guignardia bidwellii*	35 – 140 g/ha	
Cereals: *Puccinia* spp., *Erysiphe graminis* *Pseudocercosporella herpotrichoides*, *Septoria* spp., *Pyrenophora* spp., *Rhynchosporium* spp.	100 – 300 g/ha	Moberg et al. 1985
Peanuts: *Cercospora* spp.	70 – 140 g/ha	
Bananas: *Mycosphaerella* spp.	100 g/ha	
Sugar beets: *Cercospora beticola*, *Erysiphe betae*		

1)		1) BASF
2) **BAS 45406 F**		2) DOS 3139370, 1981
3) 1-(2,4-dichlorophenyl)-2-(1H-1,2,4-triazol-1-yl)ethanon-O-(phenylmethyl)-oxime		3) 1983
		4) 4640 mg/kg
6) Cereals: *Erysiphe graminis*, *Leptosphaeria nodorum*	38 – 1000 g/ha	8) Pommer & Zeeh 1983

Table 12.3 (Continued)

1) Common name[1])	1) Inventing/Developing Company
2) Experimental No.	2) Patent No., Year[4])
3) Chemical name[2])	3) Year of presentation
4) Trade name(s) of products for spray application[3]) & combination partner(s)	4) Acute toxicity[5])
	5) Solubility (water)[6])
5) Trade name(s) of products for seed treatment[3]) & combination partner(s)	6) Vapour pressure[7])
	7) Partition coefficient[8])
6) Use as foliar treatment (crop/pathogen) application rate (ai)	8) Literature
7) Use as seed treatment (crop/pathogen) application rate (ai)	
8) Other uses	

BAS 45 406 F (continued)
6) Coffee: *Hemileia vastatrix*
   Apples: *Venturia inaequalis*
   Grapes: *Uncinula necator*
   Peanuts: *Cercospora* spp.

---

1) **Diniconazole**
2) S-3308, XE-779
3) (E)-(RS)-1-[(2,4-dichlorophenyl)-4,4-dimethyl-2-(1H-1,2,4-triazol-1-yl)pent-1-en-3-ol
4) Mixor®, Sumi® 8, Spottless®

   Sumidione®   & iprodione
   Sumistar®   & iprodione
           & carbendazim

5) Geriko®
   Geriko® double   & iprodione
6) Apples: *Venturia inaequalis,*   28–56 g/ha
        Podosphera leucotricha
        Gymnosporangium juniperi-virginianae
    Stonefruits: *Monilinia* spp.
    Vegetables: rust, powdery mildew
    Wheat: *Puccinia* spp., *Erysiphe*   28–60 g/ha
        graminis
    Grapes: *Uncinula necator*   24–84 g/ha
7) *Ustilago* spp., *Pyrenophora* spp.   7.5–15 g/100 kg

1) Sumitomo; Chevron
2) DOS 3010560, 1979
3) 1983
4) 639 mg/kg (m)
    474 mg/kg (f)
5) 40 mg/l [25 °C]
6) 2.93 mPa

8) Funaki et al. 1983

---

1)
2) **PP 969**
3) (5RS,6RS)-6-hydroxy-2,2-7,7-tetramethyl-5-(1H-1,2,4-triazole-1-yl)octan-3-one
4)
8) Granular or drench application in:
    Coffe: *Hemileia vastatrix*
    Bananas: *Mycosphaerella musicola*
    Apples: *Podosphaera leucotricha,*
        Venturia inaequalis

1) ICI
2) EP 27685, 1979
3) 1983
4)
5) 3.600 mg/l
8) SHEPHARD et al. 1983

Table 12.3 (Continued)

1) Common name[1]		1) Inventing/Developing Company
2) Experimental No.		2) Patent No., Year[4]
3) Chemical name[2]		3) Year of presentation
4) Trade name(s) of products for spray application[3]		4) Acute toxicity[5]
	& combination partner(s)	5) Solubility (water)[6]
5) Trade name(s) of products for seed treatment[3]		6) Vapour pressure[7]
	& combination partner(s)	7) Partition coefficient[8]
6) Use as foliar treatment (crop/pathogen)	application rate (ai)	8) Literature
7) Use as seed treatment (crop/pathogen)	application rate (ai)	
8) Other uses		

1) **Myclobutanil**		1) Rohm & Haas
2) RH-3 866		2) US 4 366 165, 1977
3) 2-(4-chlorophenyl)-2-(1H-1,2,4-triazol-1-1-ylmethyl)hexanenitrile		3)
4) Systhane®, Rally®, Nova®		4) 1.600 mg/kg (m)
Systhane® C	& captan	2.290 mg/kg (f)
Baktane®, Gana®	& mancozeb	5) 142 mg/l [25 °C]
5) Ravyl®		6) 0.213 mPa [25 °C]
		7)
6) Apples: *Venturia inaequalis*, *Podosphaera leucotricha*, *Gymnosporangium juniperi virginianae*	0.0045 – 0.0075%	8) Orpin *et al.* 1986 Quinn *et al.* 1986
Grapes: *Uncinula necator*, *Guignardia bidwellii*	30 – 70 g/ha	
Wheat: *Erysiphe graminis*, *Puccinia* spp., *Septoria* spp., *Fusarium* spp., *Pyrenophora tritici-repentis*	224 – 280 g/ha	
7) Cereals: *Ustilago* spp., *Tilletia* spp., *Pyrenophora* spp., *Fusarium* spp.	20 – 60 g/100 kg	

1) **Tebuconazole** (Terbuconazole)		1) Bayer
2) BAY HWG 1 608		2) EP 40 345, 1980
3) (RS)-1-(4-chlorophenyl) 4,4-dimethyl-3-(1H-1,2,4-triazol-1-ylmethyl) pentan-3-ol		3) 1986
4) Folicur®, Horizon®, Elite®, Lynx®		4) ~4 000 mg/kg
Matador®, Silvacur®	& triadimenol	5) 32 mg/l
Libero®	& carbendazim	6) 1.3 µPa
Aurore®	& tridemorph	7) log $P_{(o/w)}$: 3.7
Folicur® E	& dichlofluanid	
5) Raxil®		
6) Cereals: *Puccinia* spp., *Erysiphe graminis*, *Septoria* diseases, *Fusarium* spp. (ear), *Pyrenophora* spp., *Rhynchosporium secalis*	125 – 312 g/ha	8) Kaspers *et al.* 1987 Heatherington & Meredith 1988

Table 12.3 (Continued)

1) Common name[1])		1) Inventing/Developing Company
2) Experimental No.		2) Patent No., Year[4])
3) Chemical name[2])		3) Year of presentation
4) Trade name(s) of products for spray application[3])		4) Acute toxicity[5])
	& combination partner(s)	5) Solubility (water)[6]
5) Trade name(s) of products for seed treatment[3])		6) Vapour pressure[7])
	& combination partner(s)	7) Partition coefficient[8])
6) Use as foliar treatment (crop/pathogen)	application rate (ai)	8) Literature
7) Use as seed treatment (crop/pathogen)	application rate (ai)	
8) Other uses		

Tebuconazole (Terbuconazole) **(continued)**

6) Rape oilseed: *Pyrenopeziza* sp., *Sclerotinia* sp., *Alternaria* sp., *Botrytis cinerea*, *Leptosphaeria* sp.	250 – 375 g/ha	KASPERS & SIEBERT 1989
Grapes: *Uncinula necator*, *Botrytis cinerea*		BRANDES & KASPERS 1989
Peanuts: *Mycosphaerella* spp., *Puccinia arachidis*	125 – 250 g/ha	RUDOLPH & NOEGEL 1989
Banana: *Mycosphaerella* spp.	75 – 100 g/ha	WYBOU 1989
Apples: scab, powdery mildew	0.005 – 0.0075%	BRANDES et al. 1990
Coffe: *Hemileia vastatrix*	125 – 250 g/ha	
7) Cereals: *Ustilago* spp., *Tilletia* spp., *Pyrenophora* spp., *Fusarium* spp., *L. nodorum*	1 – 3 g/100 kg seed	WAINWRIGHT & LINKE 1987

1) **Hexaconazole**		1) ICI
2) PP 523		2) EP 0 015 756, 1979
3) (RS)-2-(2,4-dichlorophenyl)-1-(1H)-1,2,4-triazol-1-yl) hexane-2-ol		3) 1986
4) Anvil®, Planète® Aster		4) 2.189 mg/kg (m) 6.071 mg/kg (f)
Jupiter®	& fenpropidine	5) 18 mg/l
Sirius®	& chlorothalonil	6) 20 µPa
Planète® R	& carbendazim	
		7) log $P_{(o/w)}$: 3.9
6) Cereals: *Erysiphe graminis*, *Puccinia* spp., *Septoria* diseases, *Puccinia* spp.	250 g/ha	8) WALLER et al. 1990
Grapes: *Uncinula necator*, *Guignardia* sp., *Pseudopeziza tracheiphila*	1 – 2 g/hl	HEANEY et al. 1986
Bananas: *Mycosphaerella* spp.	100 – 150 g/ha	
Peanuts: *Mycosphaerella* spp.	20 – 50 g/ha	BROWN et al. 1988
Coffee: rust, *Cercospora coffeicola*, *Mycena citricolor*	50 – 100 g/ha	
Apples: powdery mildew, scab, *Gymnosporangium* sp., *Alternaria mali*	1 – 2 g/hl	
Vegetables: powdery mildews, rusts, *Septoria* spp., *Alternaria* spp.	30 – 40 g/ha	

Table 12.3 (Continued)

1) Common name[1])		1) Inventing/Developing Company
2) Experimental No.		2) Patent No., Year[47]
3) Chemical name[2])		3) Year of presentation
4) Trade name(s) of products for spray application[3])		4) Acute toxicity[5])
	& combination partner(s)	5) Solubility (water)[6])
5) Trade name(s) of products for seed treatment[3])		6) Vapour pressure[7])
	& combination partner(s)	7) Partition coefficient[8])
6) Use as foliar treatment (crop/pathogen)	application rate (ai)	8) Literature
7) Use as seed treatment (crop/pathogen)	application rate (ai)	
8) Other uses		

1) **Cyproconazole**		1) Sandoz
2) SAN 619 F		2) GB-A-2136423, 1982
3) (2RS,3RS;2RS,3SR)-2-(4-chlorophenyl)-3-cyclopropyl-1-(1H-1,2,4-triazol-1-yl) butan-2-ol		3) 1986
4) Alto®, Atemi®		4) 1020 mg/kg (m)
		1333 mg/kg (f)
Alto® Elite, Alto® Marathon	& chlorothalonil	5) 140 mg/l [22 °C]
Alto® Ambel	& carbendazim	6) 30.7 µPa
Alto® 'R	& carbendazim	7) log $P_{(o/w)}$: 2.9
	& pyrazophos	
Alto® Major	& tridemorph	
6) Cereals: *Erysiphe graminis*, *Puccinia* spp., *Septoria nodorum*, *Cochliobolus sativus* etc.	40 – 100 g/ha	8) Gisi et al. 1986a, b
Stonefruits: *Monilinia* spp.	1.0 – 1.2 g/hl	
Coffee: *Hemileia vastatrix*	50 g/ha	
Sugar beets: *Cercospora beticola*, *Ramularia beticola*, *Oromyces betae*	40 – 80 g/ha	
Grapes: *Uncinula necator*	10 g/ha	
Peanuts: *Cercospora* spp., *Sclerotium* white mould, rust	62 – 87 g/ha	
Banana: Sigatoka diseases	80 – 100 g/ha	
Apples: scab, powdery mildew	1.0 – 1.2 g/hl	
Others: various diseases in cucurbits, tea, banana etc.		

1) **Difenoconazole**		1) Ciba-Geigy
2) CGA 169374		2) EP 65485, 1981
3) cis,trans-3-chloro-4-[4-methyl-2-(1H-1,2,4-triazol-1-ylmethyl)-1,3-dioxolan-2-yl] phenyl 4-chlorophenyl ether		3) 1988
4) Score®, Geyser®, Dragon®		4) 1453 mg/kg
Arix®	& propiconazole	5) 3.3 mg/l
Eria®	& carbendazim	
5) Dividend®		6) 120 nPa
		7) log $P_{(o/w)}$: 4.3

Table 12.3 (Continued)

1) Common name[1])		1) Inventing/Developing Company
2) Experimental No.		2) Patent No., Year[4])
3) Chemical name[2])		3) Year of presentation
4) Trade name(s) of products for spray application[3])		4) Acute toxicity[5])
	& combination partner(s)	5) Solubility (water)[6])
5) Trade name(s) of products for seed treatment[3])		6) Vapour pressure[7])
	& combination partner(s)	7) Partition coefficient[8])
6) Use as foliar treatment (crop/pathogen)	application rate (ai)	8) Literature
7) Use as seed treatment (crop/pathogen)	application rate (ai)	
8) Other uses		

Difenoconazole (continued)

6) Grapes: *Uncinula necator, Pseudopeziza t., Guignardia bidwellii*	3 – 5 g/hl	8) RUESS et al. 1988
Apples: *Venturia inaequalis,*	1.5 – 2.5 g/hl	DAHMEN & STAUB
Podosphaera leucotricha, Alternaria mali	2.5 – 10 g/hl	1992 a, b
Sugar beet: *Cercospora beticola, Ramularia beticola, Erysiphe betae*	100 – 125 g/ha	KNAUF-BEITER et al. 1992
Peanuts: leaf spot, web blotch, scab, rust	125 g/ha	
Cereals: *Septoria* diseases, *Puccinia* spp., Sooty moulds	125 – 250 g/ha	
Others: *Alternaria* spp. in potatoes, onions etc.		
7) Cereals: *Ustilago* spp., *Tilletia* spp. (incl. *T. controversa*) etc.	15 – 60 g/100 kg	

1) **Imibenconazole** (timibenconazole)		1) Hokko
2) HF 6305 (HF-8 505)		2) DOS 3 238 306, 1981
3) 4-chlorobenzyl N-(2,4-dichlorophenyl)-2-(1H-1,2,4-triazol-1-yl) thioacetamidate		3) 1988
4) Manage®		4) >5.000 mg/kg
		5) 1.7 mg/l
		6) 85 nPa [25 °C]
6) Pome fruits: scab, powdery mildew	5 – 7.5 g/hl	8) OYAMA et al. 1988
Grapes: *Uncinula necator, Elsinoe*	7.5 – 15 g/hl	
7) Cereals: *Tilletia caries*	15 – 60 g/dt	

1) **Tetraconazole**		1) Montedison/Agrimont
2) M 14360		2) EP 0272 679, 1986
3) (±)-2-(2,4-dichlorophenyl)-3-(1H-1,2,4-triazol-1-yl)-propyl 1,1,2,2-tetrafluoroethyl ether		3) 1988
		4) 1 248 mg/kg (m)
4) Eminent®, Arpège®		1.031 mg/kg (f)
Eminent® Star	& chlorothalonil	5) 150 mg/l
		6) 1.6 mPa
		7) log $P_{(o/w)}$: 3.1

Table 12.3 (Continued)

1) Common name[1])		1) Inventing/Developing Company
2) Experimental No.		2) Patent No., Year[4])
3) Chemical name[2])		3) Year of presentation
4) Trade name(s) of products for spray application[3])		4) Acute toxicity[5])
	& combination partner(s)	5) Solubility (water)[6])
5) Trade name(s) of products for seed treatment[3])		6) Vapour pressure[7])
	& combination partner(s)	7) Partition coefficient[8])
6) Use as foliar treatment (crop/pathogen)	application rate (ai)	8) Literature
7) Use as seed treatment (crop/pathogen)	application rate (ai)	
8) Other uses		

Tetraconazole (continued)

5) Cereals: *Erysiphe graminis, Puccinia* spp., *Septoria* diseases, *Rhynchosporium secalis*	75 – 125 g/ha	8) GARAVAGLIA et al. 1988
Apples: *Podosphaera leucotricha*, scab	2 – 4 g/hl	
Sugar beet: *Cercospora beticola, Uromyces betae, Erysiphe betae, Ramularia betae*	100 g/ha	
Grapes: *Uncinula necator*	1.5 g/hl	
Others: rusts and powdery mildew on peach, ornamentals		

1) **Furconazole-cis**		1) Rhône-Poulenc
2) LS 840 606		2) EP 0 323 443, 1984
3) (2<u>RS</u>,5<u>RS</u>)-5-(2,4-dichlorophenyl)tetrahydro-5-(1<u>H</u>-1,2,4-triazol-1-yl methyl)-2-furyl 2,2,2-trifluoroethyl ether (IUPAC)		3) 1988
		4) 450 – 900 mg/kg
4) Belier®		5) 21 mg/l
		6) 14.5 µPa
6) Cereals: *Erysiphe graminis*		8) GOUOT et al. 1988
Apples: powdery mildew, scab	10 – 25 g/ha	ZECH et al. 1988
Grapes: *Uncinula necator*	20 g/ha	
Sugar beets: *Cercospora beticola*	100 g/ha	
Bananas: Yellow Sigatoka		
Coffee: rust	45 – 120 g/ha	

1) **Fenbuconazole** (fenethanil)		1) Rohm & Haas
2) RH 7592, RH 57592		2) US 880 990, 1986
3) (R,S)-4-(4-chlorophenyl)-2-phenyl-2-(1H-1,2,4-triazol-1-ylmethyl) butyronitrile		3) 1988
4) Indar®		4) >2.000 mg/kg (m)
Indar® Twin, Ténéré®	& fenpropidine	5) 3.8 mg/l [25 °C]
Indar® Mega, Troika®	& prochloraz	6) 4.9 µPa
	& carbendazim	7) log $P_{(o/w)}$: 3.23
6) Cereals: *Septoria* spp., *Puccinia* spp., *Rhynchosporium secalis*	65 – 75 g/ha	8) DRIANT et al. 1988
Apples: *Venturia inaequalis*	2 – 3 g/hl	

DMI fungicides   245

Table 12.3 (Continued)

1) Common name[1]	1) Inventing/Developing Company
2) Experimental No.	2) Patent No., Year[4]
3) Chemical name[2]	3) Year of presentation
4) Trade name(s) of products for spray application[3]	4) Acute toxicity[5]
& combination partner(s)	5) Solubility (water)[6]
5) Trade name(s) of products for seed treatment[3]	6) Vapour pressure[7]
& combination partner(s)	7) Partition coefficient[8]
6) Use as foliar treatment (crop/pathogen) application rate (ai)	8) Literature
7) Use as seed treatment (crop/pathogen) application rate (ai)	
8) Other uses	

Fenbuconazole (fenethanil) **(continued)**
6) Grapes: *Uncinula necator, Guignardia bidwellii, Botrytis cinerea*
   Sugar beet: *Cercospora beticola*   65 – 280 g/ha
   Stone fruit: *Monilinia* spp.   18 – 70 g/ha
7) Cereals: *Tilletia* spp., *Ustilago* spp., *Pyrenophora* spp., *Fusarium* spp.   15 – 60 g/100 kg

---

1)
2) **SSF 109**
3) (+/−)-<u>cis</u>-1-(4-chlorophenyl)-2-(1<u>H</u>-1,2,4-triazol-1-yl)cycloheptanol

6) Cucumber and other crops: *Botrytis cinerea*

1) Shionogi
2) EP 0 152 031, 1984
3) 1990
4) 153 – 214 mg/kg
5) 125 mg/l [25 °C]
7) log $P_{(o/w)}$: 3.03
8) Murabayashi et al. 1990

---

1) **Epoxyconazole**
2) BAS 480 F
3) (2<u>RS</u>,3<u>SR</u>)-1-[3-(2-chlorophenyl)-2-(4-fluorophenyl)oxiran-2-ylmethyl] 1<u>H</u>-1,2,4-triazole
4) Opus®
   Opus® Team   & fenpropimorph
   Tango®   & tridemorph
6) Cereals: *Puccinia* spp., *Erysiphe* spp.,   125 – 188 g/ha
   Septoria diseases, *Pyrenophora* spp.,
   Rhynchosporium secalis, eyespot
   Sugar beets: *Cercospora beticola*   125 g/ha

   Peanuts: *Mycosphaerella* spp., *Puccinia*   62 – 250 g/ha
   arachidis
   Oilseed rape: *Alternaria brassicae*   125 – 250 g/ha
   Others: various diseases in ornamentals, bananas, vegetables, stone fruits etc.

1) BASF
2) EP 0 094 564, 1982
3) 1990
4) >5.000 mg/kg
5) 6.63 mg/l
6) <10 µPa
7) log $P_{(o/w)}$: 3.44
8) Saur et al. 1991
   Saur et al. 1991

   Ammermann et al. 1990

Table 12.3 (Continued)

1) Common name[1])	1) Inventing/Developing Company
2) Experimental No.	2) Patent No., Year[4])
3) Chemical name[2])	3) Year of presentation
4) Trade name(s) of products for spray application[3])	4) Acute toxicity[5])
& combination partner(s)	5) Solubility (water)[6])
5) Trade name(s) of products for seed treatment[3])	6) Vapour pressure[7])
& combination partner(s) application rate (ai)	7) Partition coefficient[8])
6) Use as foliar treatment (crop/pathogen) application rate (ai)	8) Literature
7) Use as seed treatment (crop/pathogen) application rate (ai)	
8) Other uses	

1) **Bromuconazole**		1) Rhône-Poulenc
2) LS 860263		2) EP 0246982, 1986
3) 1-[(2RS,4RS;2RS,4SR)-4-bromo-2-(2,4-dichlorophenyl) tetrahydrofurfuryl]-1$\underline{H}$-1,2,4-triazole		3) 1990
4) Granit®, Edenor®		4) 365 mg/kg
Granit® TR, Edenor® TR	& tridemorph	5) 50 mg/l
		6) 4 µPa [25 °C]
6) Cereals: Eyespot, *Fusarium nivale*	300 g/ha	8) PEPIN *et al.* 1990
Puccinia spp., *Septoria* spp.,	200 g/ha	DURONI & GAULLIARD
Erysiphe graminis, *Pyrenophora* spp.,		1992
Rhynchosporium secalis		
Apples: Scab, powdery mildew	20 – 30 g/ha	
Grapes: *Uncinula necator*	20 g/ha	
Oilseed rape, potatoes, carrots: *Alternaria* spp.	125 – 400 g/ha	
Bananas: *Mycosphaerella* spp.	100 – 150 g/ha	
Stone fruit: *Monilinia* spp.	50 – 100 g/ha	
Others: various diseases in turf, coffee, peanuts, pecan etc.		

1) **Metconazole**		1) Kureha/Shell
2) WL 136184 (cis (1$\underline{RS}$,5$\underline{SR}$) isomer)		2) EP 0029222, 1987
3) (1$\underline{RS}$,5$\underline{RS}$;1$\underline{RS}$,5$\underline{SR}$)-5-(4-chlorobenzyl)-2,2-dimethyl-1-(1$\underline{H}$-1,2,4-triazol-1-ylmethyl)cyclopentanol		3) 1992
		4) 1459 mg/kg
4) Caramba®		5) 15 mg/l
6) Cereals: *Septoria* diseases, *Puccinia* spp., *Rhynchosporium secalis*, *Pyrenophora teres*, *Erysiphe graminis*	48 – 90 g/ha	8) SAMPSON *et al.* 1992
7) Cereals: *Tilletia caries*, *Ustilago* spp., *Pyrenophora* spp., *Puccinia striiformis*	0.5 – 7.5 g/100 kg	

Table 12.3 (Continued)

1) Common name[1])	1) Inventing/Developing Company
2) Experimental No.	2) Patent No., Year[4])
3) Chemical name[2])	3) Year of presentation
4) Trade name(s) of products for spray application[3])	4) Acute toxicity[5])
& combination partner(s)	5) Solubility (water)[6])
5) Trade name(s) of products for seed treatment[3])	6) Vapour pressure[7])
& combination partner(s) application rate (ai)	7) Partition coefficient[8])
6) Use as foliar treatment (crop/pathogen)	8) Literature
7) Use as seed treatment (crop/pathogen) application rate (ai)	
8) Other uses	

1) **Fluquinconazole**		1) Schering	
2) SN 597 265		2) US 4 824 469, 1984	
3) 3-(2,4-dichlorophenyl)-6-fluoro-2-(1$\underline{H}$-1,2,4-triazol-1-yl)quinazolin-4-(3$\underline{H}$)-one		3) 1992	
		4) 112 mg/kg	
		5) 1 mg/l	
4) Castellan®		6) 6.4 nPa	
		7) log $P_{(o/w)}$: 3.2	
6) Apple: *Venturia inaequalis*, *Podosphaera leucotricha*	5 – 10 g/hl	8) Russel *et al.* 1992	
Cereals: *Erysiphe graminis*, *Puccinia* spp., *Septoria* spp.	125 – 375 g/ha		
Grapes: *Uncinula necator*	2.5 – 5 g/hl		
Others: diseases in oilseed rape, sugar beet, peanuts, coffee, stone fruit etc.	125 – 500 g/ha or 5 – 10 g/hl		

1) **Triticonazole**		1) Rhône-Poulenc	
2) RP 727, RPA 4 000 727		2) EP 0 378 953, 1988	
3) (±)-($\underline{E}$)-5-(4-chlorobenzylidene)-2,2-dimethyl-1-(1H-1,2,4-triazol-1-ylmethyl)cyclopentanol		3) 1992	
5) Concept 727®			
Real®	& anthraquinone		
7) Cereals: *Ustilago nuda*, *Tilletia caries*, *Pyrenophora gramineum*, *Fusarium roseum*, *Septoria nodorum*	5 g/100 kg	8) Mugnier *et al.* 1992	
Erysiphe spp., *Puccinia* spp., *Septoria tritici*, *Rhynchosporium secalis*, *Pseudocercosporella herpotrichoides*	90 – 120 g/100 kg		

[1]): approved or proposed common name; former proposals are given in brackets (); [2]): IUPAC nomenclature was used preferably; [3]): examples, not all inclusive; [4]): year of priority date; [5]): oral acute toxicity in rats (LD$_{50}$); [6]): standard conditions: temperature: 20 °C, pH 7, variations are given in brackets []; [7]): vapour pressure at 20 °C, variations are given in brackets []; [8]): Partion coefficient in octanol/water or equivalent determined by HPLC; RT = room temperature

References

ABDEL-RAHMAN, M.: Morphological effects of fungicides on apple trees. Proc. Amer. Phytopathological Soc. **4** (1977): 213.

ADLUNG, K. G., and DRANDAREWSKI, C. A.: The evaluation of "CELA W524", a systemic fungicide, for the control of powdery mildew and apples scab. Proc. 6th Brit. Insectic. Fungic. Conf. (1971): 577–586.

AKERS, A., KÖHLE, H. H, and GOLD, R. E.: Uptake, transport and mode of action of BAS 480 F, a new triazole fungicide. Proc. 1990 Brighton Crop Prot. Conf. – Pests and Diseases – (1990): 837–845.

AMMERMANN, E., LOECHER, F., LORENZ, G., JANSSEN, B., KARBACH, S., and MEYER, N.: BAS 480 F – a new broad spectrum fungicide. Proc. 1990 Brighton Crop Prot. Conf. – Pests and Diseases – (1990): 407–414.

Anonymous: Sterolbiosynthesehemmer – Resistenzgefährdung und empfohlene Antiresistenz-Strategien. Gesunde Pflanzen **44** (1992): 361–365.

Anonymous: Combattre les résistances aux fongicides inhibiteurs de la biosynthèse des stérols. Phytoma – La Defense des végétaux **447** (1993): 12–18.

ATTABHANYA, A., and HOLCOMB, G. E.: Control of *Fusarium* wilt of mimosa with systemic fungicides. Plant Disease Rep. **60** (1976): 56–59.

BALDWIN, B. C., and WIGGINS, T. E.: Biochemical studies on the diclobutrazol series of systemic fungicides. Symp. Systemische Fungizide und Antifungale Verbindungen. Reinhardsbrunn 1983, Abstracts p. 4.

BENT, K. J., and SKIDMORE, A. M.: Diclobutrazol: a new systemic fungicide. Proc. 1979 Brit. Crop. Protect. Conf. (1979): 477–484.

BERAUD, J. M., GUEGUEN, F., LECA, J. L., TUSSAC, M., and BONQUET, G.: Qu'est-ce que le fenarimol? La Defense des Vegetaux **34** (1980): 17–24.

BERG, D., BORN, L., BÜCHEL, K. H., HOLMWOOD, G., and KAULEN, J.: HWG 1608 – Chemie und Biochemie eines neuen Azolfungizids. Pflanzenschutz-Nachrichten Bayer **40** (1987): 111–132.

BIANCHI, D., CESTI, P., SPEZIA, S., GARAVAGLIA, C., and MIRENNA, L.: Chemoenzymatic synthesis and biological activity of both enantiomeric forms of tetraconazole, a new antifungal triazole. J. Agric. Food Chem. **39** (1991): 197–201.

BIRCHMORE, R. J., BROOKES, R. F., COPPING, L. G., and WELLS, W. H.: BTS 40542 – a new broad spectrum fungicide. Proc. 1977 Brit. Crop. Protect. Conf. (1977): 593–598.

– WELLS, W. H., and COPPING, L. G.: A new group of fungicidally-active metal co-ordination compounds based on prochloraz. Proc. 1979 Brit. Crop. Protect. Conf. (1979): 583–601.

BON, Y. LE, BOURDIN, J., and BERTHIER, G.: Efficacy of various fungicides against rose mildew (*Spaerotheca pannosa* var. *rosae*). Phytiatrie-Phytopharmacie **27** (1978): 199–205.

BOSSHARD, E., SIEGFRIED, W., and SCHUEPP, H.: Experience with sterol synthesis inhibitor fungicides for scab control. Schweiz. Z. Obst Einbau **121** (1985): 166–173.

BOUDIER, B.: Essai de lutte contre le dessèchement des aiguilles du a Meria laricis sur melezes d'Europe (Larix decidua) en pepinieres. Revue Forestière Francaise **33** (1981): 394–399.

BRANDES, W., STEFFENS, W., FÜHR, F., and SCHEINPFLUG, H.: Further studies on translocation of [^{14}C] triadimefon in cucumber plants. Pflanzenschutz-Nachr. Bayer **31** (1978): 132–144.

– DEHNE, H. W., and KUCK, K. H.: On the protective and curative action of ®Baycor against the pathogen causing apple scab (*Venturia inaequalis*) and on the behaviour of the preparation on the apple leaf. Pflanzenschutz-Nachrichten Bayer **41** (1988): 285–299.

– and KASPERS, H.: Tebuconazole – ein neues *Botrytis*-Fungizid für den Weinbau. Pflanzenschutz-Nachrichten Bayer **42** (1989): 149–161.

– – and KRÄMER, W.: ®Baycor, a new foliar-applied fungicide of the biphenyloxytriazolylmethane group. Pflanzenschutz-Nachr. Bayer **32** (1979): 1–16.

- – SIEBERT, R., and DEHNE, H. W.: Applications of tebuconazole in fruit-growing. Pflanzenschutz-Nachrichten Bayer **43** (1990): 217–226.
- BRENT, K. J., and HOLLOMON, D. W.: Risk of resistance against sterol biosynthesis inhibitors in plant protection. In: BERG, D., and PLEMPEL, M. (Eds.): Sterol Biosynthesis Inhibitors – Pharmaceutical and Agrochemical Aspects. Ellis Horwood, Chichester, and VCH, Weinheim 1988, pp. 332–346.
- BRISTOW, P. R.: Effect of triforine on pollen germination and fruit set in highbush blueberry. Plant Dis. **65** (1981): 350–353.
- BROWN, I. F., and HALL, H. R.: Certain biological properties of fenarimol applicable to its field use. Proc. 1981 Brit. Crop. Protect. Conf. (1981): 573–578.
- – – and MILLER, J. R.: EL-273, a curative fungicide for the control of *Venturia inaequalis*. Phytopathology **60** (1970): 1013–1014.
- BROWN, M. C., SHEPHARD, M. C., and FRANK, J. A.: Hexaconazole – useful properties in the control of coffee, peanut and apple diseases. Proc. 1988 Brighton Crop Prot. Conf. – Pests and Diseases – (1988): 229–234.
- BRUCHHAUSEN, V. VON, and STIASNI, M.: Transport of the systemic fungicide CELA W 524 (Triforine) in barley plants. II. Uptake and metabolism. Pesticide Sci. **4** (1973): 767–773.
- BRUHN, J. A., FORT, T. M., and DENIS, S. J.: DPX-H 6573: a broad spectrum fungicide for the control of field crop diseases. National Meeting Amer. Phytopathological Soc., August 1985.
- BÜCHEL, K. H.: History of Azole Chemistry. Burdick and Jackson Symp.: 188th ACS National Meeting, Philadelphia 1984.
- BUCHENAUER, H.: Systemisch-fungizide Wirkung und Wirkungsmechanismen von Triadimefon (MEB 6447). Mitt. Biol. Bundesanst. Land- u. Forstw. Berlin-Dahlem **165** (1975): 154–155.
- – Mode of action and selectivity of fungicides which interfere with ergosterol biosynthesis. Proc. 1977 Brit. Crop Protect. Conf. (1977): 699–711.
- – Comparative studies on the antifungal activity of triadimefon, triadimenol, fenarimol, nuarimol, imazalil and fluotrimazole *in vitro*. Z. Pflanzenkrankh. Pflanzenschutz **86** (1979a): 341–354.
- – Conversion of triadimefon into two diastereomeres, triadimenol I and triadimenol II, by fungi and plants. IX. Intern. Congr. Plant Pathol., Washington USA (1979b): Abstr. no. 939.
- – KOHTS, T., and ROOS, H.: Zur Wirkungsweise von Propiconazol (Desmel®), CGA 64251 and Dichlobutrazol (Vigil®) in Pilzen und Gerstenkeimlingen. Mitt. Biol. Bundesanst. Land- u. Forstw. Berlin-Dahlem **202** (1981): 310–311.
- – and RÖHNER, E.: Zum Wirkungsmechanismus von Imazalil in Pilzen und Pflanzen. Proc. V. Intern. Symp. Systemfungizide Rheinhardsbrunn 1977, Akademie-Verlag, Berlin 1979, pp. 175–185.
- – – Effect of triadimefon on growth of various plant species as well as on Gibberellin content and sterol metabolism in shoots of barley seedlings. Pesticide Biochem. Physiol. **15** (1981): 58–70.
- – – Aufnahme, Translokation und Transformation von Triadimefon in Kulturpflanzen. Z. Pflanzenkrankh. Pflanzenschutz **89** (1982): 385–389.
- BURDEN, R. S., CARTER, G. A., CLARK, T., COOKE, D. T., CROKER, S. J., DEAS, A. H. B., HEDDEN, P., JAMES, C. S., and LENTON, J. R.: Comparative activity of the enantiomers of triadimenol and paclobutrazol as inhibitors of fungal growth and plant sterol and gibberellin biosynthesis. Pest. Sci. **21** (1987a): 253.
- – CLARK, T., and HOLLOWAY, P. J.: Effects of sterol biosynthesis-inhibiting fungicides and plant growth regulators on the sterol composition of barley plants. Pestic. Biochem. Physiol. **27** (1987b): 289–300.
- – COOKE, D. T., and CARTER, G. A.: Inhibition of sterol biosynthesis and growth in plants and fungi. Phytochemistry **28** (1989): 1791–1804.

CAGNIEUL, P., and LABIT, B.: DPX H6573: nouveau fongicide systémique à large spectre. In: SMITH, I. M. (Ed.): Fungicides for Cop Protection. Proc. Bordeaux Mixture Centenary Meeting, BCPC Publications Monograph No. 31 (1985): 237–240.

CARDOSO, J. E., HILDEBRANDT, A. C., and GRAU, L. R.: Growth inhibition of *Rhizoctonia solani* by five fungicides and their uptake by soybean seedlings. Fitopatol. Bras. **4** (1979): 11–15.

CASANOVA, A., DÖHLER, R., FARRANT, D. M., and RATHMELL, W. G.: The active of the fungicide nuarimol against diseases of barley and other cereals. Proc. Brit. Crop Protect. Conf. (1977): 1–7.

CAYLEY, G. R., HIDE, G. A., and TILLOTSON, Y.: The determination of imazalil on potatoes and its use in controlling potato storage diseases. Pesticide Sci. **12** (1981): 103–109.

CHURCH, R. M., and WILLIAMS, R. R.: Fungicide toxicity to apple pollen in the anthers. J. Horticult. Sci **53** (1978): 91–94.

COOLBAUGH, R. C., SWANSON, D. J., and WEST, C. A.: Comparative effects of ancymidol and its analogs on growth of peas and ent-kaurene oxidation in cell-free extracts of immature *Marah macrocarpus* endosperm. Plant. Physiol. **69** (1982): 707–711.

COOKE, B. K., PAPPAS, A. C., JORDAN, V. W. L., and WESTERN, N. M.: Translocation of benomyl, prochloraz and procymidone in relation to control of *Botrytis cinerea* in strawberries. Pesticide Sci. **10** (1979): 467–472.

DAHMEN, H., and STAUB, T.: Biological characterization of uptake, translocation, and dissipation of difenoconazole (CGA 169374) in wheat, peanut, and tomato plants. Plant Disease **76** (1992a): 523–526.

– – Protective, curative, and eradicant activity of difenoconazole against *Venturia inaequalis, Cercospora arachidicola,* and *Alternaria solani*. Plant Disease **76** (1992b): 774–777.

DAVIS, A. E., FORT, T. M., DENIS, S. J., and HENRY, M. J.: DPX-H 6573: a broad spectrum fungicide for the control of tree-fruit diseases. Nat. Meet. Amer. Phytopathol. Soc., August, 1985.

DAWSON, M., TORCHEUX, R., and HORRELLOU, A.: Qu'est-ce que le "Impact" Sopra®? La Defense des Végétaux **226** (1984): 77–85.

DEAS, A. H. B., CLARK, T., and CARTER, G. A.: The enantiomeric composition of triadimenol produced during metabolism of triadimefon by fungi. Part I: Influence of dose and time of incubation. Pesticide Sci. **15** (1984a): 63–70.

– – – The enantiomeric composition of triadimenol produced during metabolism of triadimefon by fungi. Part II: Differences between fungal species. Pesticide Sci. **15** (1984b): 71–77.

DEKKER, J.: Counter measures for avoiding fungicide resistance. In: DEKKER, J., and GEORGOPOULOS, S. G. (Eds.): Fungicide Resistance in Crop Protection. Pudoe, Wageningen 1982, pp. 177–186.

– The development of resistance to fungicides. Progr. Pesticide Biochem. and Toxicol., Vol. 4 (1985): 166–209.

DÖHLER, R., and MERTH, M. V.: Nuarimol – ein neues systemisches Fungizid zur Bekämpfung von Krankheiten in Gerste. Mitt. Biol. Bundesanst. Land- u. Forstw. Berlin-Dahlem **191** (1979): 185.

DRANDAREVSKI, C. A., and MAYER, E.: Eine Methode zur Untersuchung der Penetration von systemischen Fungiziden durch Blattkutikula und Epidermis. Meded. Fac. Landbouwwet. Rijksuniv. Gent **39** (1974): 1127–1143.

– and SCHICKE, P.: Formation and germination of spores of *Venturia inaequalis* and *Podosphaera leucotricha* following curative treatment with triforine. Z. Pflanzenkrankh. Pflanzenschutz **83** (1976): 385–396.

DRIANT, D., HEDE-HAUY, L., PERROT, A., QUINN, J. A., and SHABER, S. H.: RH 7592, a new triazole fungicide with high specific activity on cereals and other crops. Proc. 1988 Brighton Crop Prot. Conf. – Pests and Diseases – (1988): 33–40.

EBENEBE, C., BRUCHHAUSEN, V. VON, and GROSSMANN, F.: Dosage-response curve of wheat brown rust to triforine supplied via root treatment. Pesticide Sci. **5** (1974): 17–24.

EBERLE, J., RUESS, W., and URECH, P. A.: CGA 71818, a novel fungicide for the control of grape and pome fruit diseases. Proc. 10th Intern. Congr. Plant Protect., Vol. 1, Brighton (1983): 376–383.

EBERT, E., ECKHARD, W., JÄKEL, K., MOSER, P., SOZZI, D., and VOGEL, C.: Quantitative structure activity relationships of fungicidally active triazoles: analogs and stereoisomers of propiconazole and etaconazole. Z. Naturforsch. **44c** (1988): 85.

EDGINGTON, L. V.: Structural requirements of systemic fungicides. Ann. Rev. Phytopathol. **19** (1981): 107–124.

FLETCHER, R. A.: Plant growth regulating properties of sterol-inhibiting fungicides. In: PUROHIT, S. S. (Ed.): Hormonal Regulations of Plant Growth and Development. Vol. 2. Agro Botanical Publ. 1985.

– and HOFSTRA, G.: Triadimefon a plant multi-protectant. Plant and Cell Physiol. **26** (1985): 775–780.

– – Triazoles as potential plant protectants. In: BERG, D., and PLEMPEL, M. (Ed.): Sterol biosynthesis inhibitors. Ellis Horwood Ltd., Chichester, and VCH, Weinheim 1988, pp. 321–331.

– and NATH, V.: Triadimefon reduces transpiration and increases yield in water stressed plants. Physiolog. Plant **62** (1984): 422–426.

FÖRSTER, H., BUCHENAUER, H., and GROSSMANN, F.: Side effects of the systemic fungicides triadimefon and triadimenol on barley plants. I. Influence on growth and yield. Z. Pflanzenkrankh. Pflanzenschutz **87** (1980a): 473–492.

– – – Side effects of the systemic fungicides triadimefon and triadimenol on barley plants II. Cytokinin-like effects. Z. Pflanzenkrankh. Pflanzenschutz **87** (1980b): 640–653.

– – – Side-effects of the systemic funicides triadimefon and triadimenol on barley plants. III. Further effects of metabolism. Z. Pflanzenkrankh. Pflanzenschutz **87** (1980c): 717–730.

FORT, T. M., and MOBERG, W. K.: DPX-H 6573, a new broad spectrum fungicide candidate. Proc. 1984 Brit. Crop. Protect. Conf., Vol. 2 (1984): 413–419.

FRATE, C. A., LEACH, L. D., and HILLS, F. J.: Comparison of fungicide application methods for systemic control of sugar beet powdery mildew. Phytopathology **69** (1979): 1190–1194.

FROHBERGER, P. E.: New approaches to the control of cereal diseases with triadimefon (MEB 6447). VIII. Intern. Plant Protect. Congr. 1975, Moscow, Section III (1975): 247–258.

– Baytan®, ein neues systemisches Breitband-Fungizid mit besonderer Eignung für die Getreidebeizung. Pflanzenschutz-Nachr. Bayer **31** (1978): 11–24.

FUCHS, A., and DRANDAREVSKI, C. A.: Wirkungsbreite und Wirkungsgrad von Triforine *in vitro* und *in vivo*. Z. Pflanzenkrankh. Pflanzenschutz **80** (1973): 403–417.

– and OST, W.: Translocation, distribution and metabolism of triforine in plants. Arch. Environment. Contamin. Toxicol. **4** (1976): 30–43.

– VIETS-VERWEIJ, M., and VRIES, F. W. DE: Metabolic conversion in plants of the systemic fungicide triforine (N,N'-bis-[1-formamido-2,2,2-trichloroethyl]-piperazine; CELA W524). Phytopathol. Z. **75** (1972): 111–123.

– VRIES, F. W. DE, and AALBERS, M. J.: Uptake, distribution and metabolic fate of ^{3}H-triforine in plants. I. Short-term experiments. Pesticide Sci. **7** (1976a): 115–126.

– – – Uptake, distribution and metabolic fate of ^{3}H-triforine in plants. II. Long-term experiments. Pesticide Sci **7** (1976b): 127–134.

FÜHR, F., PAUL, V., STEFFENS, W., and SCHEINPFLUG, H.: Translokation von ^{14}C-Triadimefon nach Applikation auf Sommergerste und seine Wirkung gegen *Erysiphe graminis* var. *hordei*. Pflanzenschutz-Nachr. Bayer **31** (1978): 116–131.

FUNAKI, Y., ISHIGURI, Y., KATO, T., and TANAKA, S.: Structure-activity relationship of a new fungicide S-3308 and its derivatives. Proc. Intern. Congr. Pesticides Chem. 5th, 1982 (publ. 1983) Vol. 1: 309–314.

GARAVAGLIA, C., MIRENNA, L., PUPPIN, O., and SPAGNI, E.: M 14360, a new broad-spectrum and versatile antifungal triazole. Proc. 1988 Brighton Crop Prot. Conf. – Pests and Diseases – (1988): 49–56.

GASZTONYI, M., and JOSEPOVITS, G.: Translocation and metabolism of triadimefon in different plant species. Acta Phytopath. Acad. Scient. Hung. **13** (1978): 403–415.

– – The activation of triadimefon and its role in the selectivity of fungicide action. Pesticide Sci. **10** (1979): 57–65.

GILMOUR, J.: Comparison of some aspects of mildew fungicides use on spring barley in South-East Scotland in 1982 and 1983. Proc. Brit. Crop Protect. Conf. Brighton 1984, Vol. 1 (1984): 109–114.

GISI, U., SCHAUB, F., WIEDMER, H., and UMMEL, E.: SAN 619 F, a new triazole fungicide. Proc. 1986 Brit. Crop Protect. Conf. (1986): 33–40.

– RIMBACH, E., BINDER, A., ALTWEGG, P., and HUGELSHOFER, U.: Biological profile of SAN 619 F and related EBI-fungicides. Proc. Brit. Crop Protect. Conf. (1986): 857–864.

GLOBERSON, D., and ELIASI, R.: The response to Saprol® (systemic fungicide) in lettuce species and cultivars and its inheritance. Euphytica **28** (1979): 115–118.

GOUOT, J. M., GREINER, A., MERINDOL, B., ZECH, B., and GAULLIARD, J. M.: LS840606, un fongicide polyvalent nouveau. Proc. 2nd Intern. Conf. Plant Diseases. ANPP, Bordeaux (1988).

GRABSKI, L., and GISI, U.: Sensitivity and sterol profiles of plant pathogenic fungi treated with cyproconazole Tag. Ber. Akad. Landwirtsch. Wiss. DDR **291** (1990): 95–99.

GRAEBE, J. E.: Giberellin biosynthesis and control. Ann. Rev. Plant Physiol. **38** (1987): 419–465.

GREWE, F., and BÜCHEL, K. H.: Ein neues Mehltaufungizid aus der Klasse der Trityltriazole. Mitt. Biol. Bundesanst. Land- u. Forstw. Berlin-Dahlem **151** (1973): 208–209.

GUTTER, Y.: Comparative effectiveness of Sonax, thiabendazole and sodium orthophenylphenate in controlling green mould of citrus fruits. Z. Pflanzenkrankh. Pflanzenschutz **89** (1982): 332–336.

HAAS, E.: Spritzversuche gegen *Oidium* der Reben. Obstbau Weinbau **21** (1984): 82–83.

HAENSSLER, G., and KUCK, K. H.: Microscopic investigations on the pathogenesis of brown rust of wheat (*Puccinia recondita* f. sp. *tritici*). Pflanzenschutz-Nachrichten Bayer **40** (1987): 153–180.

HANSEN, K. E.: Experiments with cereal seed dressings: II. Field experiments. Tidsskrift for Plante avl. **85** (1981): 77–92.

HARRIS, R. G., and BARNES, G.: Prochloraz: the control of net blotch and *Septoria* in winter cereals. Proc. 1981 Brit. Crop Protect. Conf. (1981): 268–274.

HASHIMOTO, S., SANO, S., MURAKAMI, A., MIZUNO, M., NISHIKAWA, H., and YASUDA, Y.: Fungitoxic properties of triflumizole. Ann. Phytopathol. Soc. Japan **52** (1986): 599–609.

HASHIMOTO, Y., SUGIMOTO, S., SOEDA, Y., SANO, S., NAKATA, A., and HASHIMOTO, S.: Absorption, translocation and metabolism of the fungicide triflumizole in cucumber plants. J. Pesticide Sci. **15** (1990): 375–383.

HEANEY, S. P., ATGER, J. C., and ROQUES, J. F.: Hexaconazole: a novel fungicide for use against diseases on vines. Proc. 1986 Brighton Crop Prot. Conf. – Pests and Diseases – (1986): 363–370.

HEATHERINGTON, P. J., and MEREDITH, R. H.: Disease control with HWG 1608 on cereals and rape. Proc. 1988 Brighton Crop Prot. Conf. (1988): 953–958.

HEERES, J.: Structure-activity relationship in a group of azoles, with special reference to 1,3-Dioxolan-2-ylmethyl derivatives. Pesticide Sci. **15** (1984): 268–279.

HICKEY, K. D.: Efficacy of UBI-A815 experimental fungicide for preventative and after-infection control of apple scab. Phytopathology **73** (1983): 367.

HOFFMANN, J. A., and WALDHER, J. T.: Chemical seed treatment for controlling seedborne and soilborne common bunt of wheat. Plant Dis. **65** (1981): 256–259.

HUGGENBERGER, F., COLLINS, M. A., and SKYLAKAKIS, G.: Decreased sensitivity of *Sphaerotheca* fuliginea to fenarimol and other ergosterol-biosynthesis inhibitors. Crop Protect. **3** (1984): 137–149.

HUGHES, A., and WILSON, D.: Development of pyrifenox for control of foliar diseases on apples and blackcurrants in the UK. Proc. 1988 Brighton Crop Prot. Conf. – Pests and Diseases – (1988): 223–228.

HUNTER, T., JORDAN, V. M. L., and KENDALL, S. J.: Fungicide sensitivity changes in *Rhynchosporium secalis* in glasshouse experiments. Proc. 1986 Brit. Crop Prot. Conf. (1986): 523–530.

IZUMI, K., YAMAGUCHI, I., WADA, A., OSHIO, H., and TAKAHASHI, N.: Effects of a new plant growth retardant (E)-1-(4-Chlorophenyl)-4,4-dimethyl-2-(1,2,4,-triazol-1-yl)-1-penten-3-ol (S 3307) on growth and gibberellin content of rice plants. Plant and Cell Physiol. **25** (1984): 611–617.

– KAMIYA, Y., SAKURAI, A., OSHIO, H., and TAKANASHI, N.: Studies of side of action of a new plant growth retardant (E)-1-(4-Chlorophenyl)-4,4-dimethyl-2-(1,2,4-triazol-1-yl)-1-penten-3-ol (S 3307) and comparative effects of its stereo-isomers in a cell-free system from *Cucurbita maxima*. Plant and Cell Physiol. **26** (1985): 821–827.

JACKSON, L. L., and FREAR, D. S.: Lipids of rust fungi. II. Stigmast-7-enol and stigmasta-7,24(28)-dienol in flax rust uredospores. Phytochemistry **7** (1968): 651.

JAVED, Z. U. R.: Field trials with new and recommended fungicides for leaf rust control during 1980. Kenya Coffee **46** (1981): 239.

JEFFREY, R. A., ROWLEY, N. K., and SMAILES, A.: The effect of fluotrimazole on powdery mildew of cereals. Proc. 1975 Brit. Insect. Fung. Conf. (1975): 429–436.

JENKYN, J. F., DYKE, G. V., and TODD, A. D.: Effects of fungicide movement between plants in field experiments. Plant Pathol. **32** (1983): 311–324.

KASPERS, H., and PATEL, N. K.: Versuche zur Bekämpfung des Kafferostes (*Hemileia vastatrix* Berk. et Br.) in Kenya. Pflanzenschutz-Nachr. Bayer **33** (1980): 152–164.

KASPERS, H. et al.: 1,2,4-Triazoles derivatives, a new class of protective and systemic fungicides. VIII. Intern. Plant Prot. Congr. Moscow, Section III (1975): 398–401.

– and SIEBERT, R.: ®Folicur (Tebuconazole) – Possibilities for the use against rape diseases. Pflanzenschutz-Nachrichten Bayer **42** (1989): 131–148.

KATO, T., TANAKA, S., UEDA, M., and KAWASE, Y.: Effects of the fungicide, S-1358, on general metabolism and lipid biosynthesis in *Monilinia fructigena*. Agricult. Biol. Chem. **38** (1974): 2377–2384.

– – YAMAMOTO, S., KAWASE, Y., and UEDA, M.: Fungitoxic properties of a N-3-pyridylimidadithiocarbonate derivative. Ann. Phytopathol. Soc. Japan **41** (1975): 1–8.

KELLEY, R. D., and JONES, A. L.: Evaluation of two triazole fungicides for postinfection control of apple scab. Phytopathology **71** (1981): 737–742.

– – Volatility and systemic properties of etaconazole and fenapanil in apple. Canad. J. Plant Pathol. **4** (1982): 243–246.

KNAUF-BEITER, G., FLEISCHHACKER, C., and MITTERMEIER, L.: Difenoconazole, a new fungicide against *Cercospora beticola* on sugar beet. Proc. 1992 Brighton Crop Prot. Conf. – Pests and Diseases – (1992): 651–656.

KNIGHTS, I. K.: Developments in the use of prochloraz for tropical fruit disease control. Proc. 1986 Brit. Crop Prot. Conf. (1986): 331–338.

KOELLER, W.: Plant growth regulator activities of stereochemical isomers of triadimenol. Physiol. Plant **71** (1987): 309.

– Antifungal agents with target sites in sterol functions and biosynthesis. In: KOELLER, W. (Ed.): Target sites of fungicide action. CRC Press, London 1992, pp. 119–206.

– and SCHEINPFLUG, H.: Fungal resistance to sterol biosynthesis inhibitors: a new challenge. Plant Disease **71** (1987): 1066.

KOLBE, W.: Zehn Jahre Versuche mit ®Bayleton zur Mehltaubekämpfung im Getreidebau (1971 bis 1981). Pflanzenschutz-Nachr. Bayer **35** (1982a): 36–71.

- Versuche zur Bekämpfung des Obstbaumkrebses mit ®Bayleton. Pflanzenschutz-Nachr. Bayer **35** (1982b): 152–170.
KRÄMER, W., BÜCHEL, K. H., and DRABER, W.: Structure activity correlation in the azoles. In: MIYAMOTO, J., and KEARNEY, P. C. (Eds.): Pesticide Chemistry: Human Welfare and the Environment. Vol. 1. Pergamon Press, Oxford 1983, pp. 223–232.
KRAUS, P.: Studies on the mechanism of action of ®Baycor. Pflanzenschutz-Nachr. Bayer **32** (1979): 17–30.
- Untersuchungen zur Aufnahme und zur Verteilung von ®Bayleton in Weinreben. Pflanzenschutz-Nachr. Bayer **34** (1981): 197–212.
KUCK, K. H.: Studies on the uptake of ®Bayleton on wheat leaves. Pflanzenschutz-Nachr. Bayer **40** (1987): 1–28.
- SCHEINPFLUG, H., TIBURZY, R., and REISENER, H. J.: Fluorescence microscopy studies of the effect of ®Bayleton and ®Baytan on growth of stem rust in the wheat plant. Pflanzenschutz-Nachr. Bayer **35** (1982): 209–228.
- and THIELERT, W.: On the systemic properties of HWG 1608, the active ingredient of the fungicides ®Folicur and ®Raxil. Pflanzenschutz-Nachr. Bayer **40** (1987): 133–152.
KVIEN, C. S., CSINOS, A. S., ROSS, L. F., CONKERTON, E. J., and STYER, C.: Diniconazole's effect on peanut (*Arachis hypogaea* L.) growth and development. J. Plant Growth Regul. **6** (1987): 233–244.
LIMPERT, E.: Frequencies of virulence and fungicide resistance in the European barley mildew population in 1985. J. Phythopathology **119** (1987): 298–311.
- and FISCHBECK, G.: Regionale Unterschiede in der Empfindlichkeit des Gerstenmehltaus gegen Triadimenol und ihre Entwicklung im Zeitraum 1980 bis 1982. Phytomedizin **13** (1983): 19 (abstracts).
LOEFFLER, R. S. T., BUTTERS, J. A., and HOLLOMON, D. W.: The lipid contents of powdery mildews and their sensitivity to fungicides. Proc. 1984 Brit. Crop Prot. Conf. (1984): 911.
LUZ, W. C., and VIEIRA, J. C.: Seed treatment with systemic fungicides to control *Cochliobolus sativus* on barley. Plant Dis. **66** (1982): 135–136.
MANTINGER, H., and VIGL, J.: Ergebnisse des Mehltauversuchs. Obstbau Weinbau **12** (1975): 341–344.
MARTIN, R. A., and EDGINGTON, L. V.: Comparative systemic translocation of several xenobiotics and sucrose. Pesticide Biochem. Physiol. **16** (1981): 87–96.
- - Antifungal, phytotoxic and systemic activity of fenapanil and structural analogs. Pesticide Biochem. Physiol. **17** (1982): 1–9.
- and MORRRIS, D. B.: ®Bayleton als systemisches Fungizid zur Bekämpfung von Blattkrankheiten bei Sommer- und Wintergerste in Großbritannien. Pflanzenschutz-Nachr. Bayer **32** (1979): 31–82.
MASNER, P., and KERKENAAR, A.: Effect of the fungicide pyrifenox on sterol biosynthesis in *Ustilago maydis*. Pestic. Sci. **22** (1988): 61–69.
MELIN, P., PLAUD, G., DEZENAS DU MONTCEL, H., and LAVILLE, E.: Activé comparée de l'imazalil sur la cercosporiose du bananier au Cameroun. Fruits **30** (1975): 301–306.
MOBERG, W. K., BASARAB, G. S., CUOMO, J., and LIANG, P. H.: Biologically active organosilicon compounds: Fungicidal Silylmethyltriazoles. 190th National Meeting Am. Chemical Soc., Chicago 1985.
MOURICHON, X., and BENGNON, M.: Comparative effectiveness of ®Tilt (CGA 64259) on banana leaf spot in the Ivory Coast. Fruits **37** (1982): 595–597.
MUGNIER, J., CHAZALET, M., and AXIOTIS, S.: La molécule fongicide 727. Phytoma – La Défense des vegetaux **441** (1992): 29–32.
MURABAYASHI, A., MASUKO, M., SHIRANE, N., and HAYASHI, Y.: SSF-109, a novel triazole fungicide: Synthesis and biological activity. Proc. 1990 Brighton Crop Prot. Conf. (1990): 423–430.

– – NIIKAWA, M., SHIRANE, N., FURUTA, T., HAYASHI, Y., and MAKISUMI, Y.: Antifungal and plant growth inhibitory activities of stereo and optical isomers of 2-triazolylcycloalkanol derivatives. J. Pesticide Sci. **16** (1991): 419–427.

NOEGEL, K. A., et al.: ®Bayleton: A potential fungicide for ornamental crops. Hortsci. **12** (1977): 408.

NAKATA, A., IKURA, K., and WAKAI, A.: Structure-activity in imidazole ring chemicals and fungitoxic properties of NF-114. Abstr. 5th Int. Congr. Pestic. Chem. (1982): IIb-9.

NOON, R. A., GIBBARD, M., NORTHWOOD, P. J., and HEANEY, S. P.: "Ferrax" seed treatment – disease control and growth benefits. Proc. 1988 Brighton Crop Prot. Conf. – Pests and Diseases – (1988 a): 941–946.

– NORTHWOOD, P. J., BROWN, M. C., MONTURY, A., and CHARLET, C.: Flutriafol based formulations for control of oilseed rape diseases. Proc. 1988 Brighton Crop Prot. Conf. – Pests and Diseases – (1988 b): 947–952.

NORTHWOOD, P. J., HORELLOU, A., and HECKELE, K. H.: PP 450: Field experience with a new cereal fungicide. Proc. 10th Intern. Congr. Plant Protect. 1983, Vol. 3 (1983): 930.

– PAUL, J. A., and GIBBARD, M.: FF 4050 seed treatment – a new approach to control barley diseases. Proc. 1984 Brit. Crop. Protect. Conf. (1984): 47–52.

OHKAWA, H., SHIBAIKE, R., OKIHARA, V., MORIDAWA, M., and MIYAMOTO, J.: Degradation of the fungicide Denmert® (s-n-butyl-S′-p-tert-butylbenzyl N-3-pyridylthiocarbaonimidate, S-1358) by plants, soils and light. Agricult. Biol. Chem. **40** (1976): 943–951.

O'LEARY, A. L., and JONES, A. M.: Factors influencing the uptake of fenarimol and flusilazol by apple leaves. Phytopathology **77** (1987): 1564–1568.

ORPIN, C., BAUER, A., BIERI, R., FAUGERON, J. M., and SIDDDI, G.: Myclobutanil, a broadspectrum systemic fungicide for use in fruit, vines, and a wide range of other crops. Proc. 1986 Brit. Crop Protect. Conf. (1986): 55–62.

OHYAMA, H., WADA, T., ISHIKAWA, H., and CHIBA, K.: HF-6305, a new trizole fungicide. Proc. 1988 Brighton Crop Prot. Conf. – Pests and Diseases – (1988): 519–526.

PEPIN, H. S., MACPHERSON, E. A., and CLEMENTS, S. J.: Effect of triadimefon on the growth of Willamette red raspberry. Canad. J. Plant Sci. **60** (1980): 1203–1208.

PEPIN, R., GREINER, A., and ZECH, B.: LS860263 – a triazole fungicide with novel properties. Proc. Brighton Crop Prot. Conf. – Pests and Diseases – (1990): 439–446.

PIENING, L. J., DUCZEK, L. J., ATKINSON, T. J., and DAVIDSON, J. G. N.: Control of common root rot and loose smut and the phytotoxicity of seed treatment fungicides on Gateway barley. Canad. J. Plant Pathol. **5** (1983): 49–53.

POMMER, E. H., and ZEEH, B.: BAS 45406 F, a new triazole derivative for the control of phytopathogenic fungi. Proc. 4th Intern. Congr. Plant Pathol., Melbourne 1983.

QUINN, J. A., FUJIMOTO, T. T., EGAN, A. R., and SHABER, S. H.: The properties of RH 3866, a new triazole fungicide. Pestic. Sci. **17** (1986): 357–362.

RATHMELL, W. G., and SKIDMORE, A. M.: Recent advances in the chemical control of cereal rust diseases. Outlook on Agriculture **11** (1982): 37–43.

REINECKE, P., KASPERS, H., SCHEINPFLUG, H., and HOLMWOOD, G.: BAY HWG 1608, a new fungicide for foliar spray and seed-treatment use against a wide spectrum of fungal pathogens. Proc. 1986 Brit. Crop Protect. Conf. (1986): 41–46.

REISDORF, K., WURZER-FASSNACHT, U., and WALTHER, F. H.: Zur systemischen Wirkung von Imazalil. Z. Pflanzenkrankh. Pflanzenschutz **90** (1983): 641–649.

ROHRBACH, K. U.: Der Einsatz von Triforine als Beizmittel zu Sommergerste. Mitt. Biol. Bundesanst. Land- u. Forstw. **178** (1977): 147–148.

RUDOLPH, R. D., NOEGEL, K. A., and ROGERS, W. D.: Bekämpfung von Erdnußkrankheiten mit ®Folicur (Tebuconazole) in den USA. Pflanzenschutz-Nachrichten Bayer **42** (1989): 199–214.

RUESS, W., RIEBLI, P., HERZOG, J., SPEICH, J., and JAMES, J. R.: CGA 169374, a new systemic fungicide with a novel broad-spectrum activity against disease complexes in a wide range of crops. Proc. 1988 Brighton Crop Prot. Conf. − Pests and Diseases − (1988): 543−550.

RUSSEL, P. E., PERCIVAL, A., COLTMAN, P. M., and GREEN, D. E.: Fluquinconazole, a novel broad-spectrum fungicide for foliar application. Proc. 1992 Brighton Crop Prot. Conf. − Pests and Diseases − (1992): 411−418.

SAINT-BLANQUAT DE, A., and MY, J.: Qu'est-ce que le Prochloraz? La Defense des Vegetaux **221** (1983): 121−141.

SAMPSON, A. J., CAZENAVE, A., LAFFRANQUE, J. P., JONES, R. G., KUMAZAWA, S., and CHIDA, T.: Metconazole, an advance in disease control in cereals and other crops. Proc. 1992 Brighton Crop Prot. Conf. − Pests and Diseases − (1992): 419−426.

SANDERS, P. L., BURPEE, L. K., COLE, H. JR., and DUICH, J. M.: Uptake, translocation and efficacy of triadimefon in control of turfgrass pathogens. Phytopathology **68** (1978): 1482−1487.

SAUR, R., GOLD, R., and AMMERMANN, E.: BAS 480 F a new broad spectrum fungicide for control of cereal diseases. Meded. Fac. Landbouwwet. Rijksuniv. Gent **56** (1991): 479−489.

− LOECHER, F., and SCHELBERGER, K.: Experiences with BAS 480 F, a new triazole fungicide, for the control of cereal diseases in Western Europe. Proc. 1990 Brighton Crop Prot. Conf. − Pests and Diseases − (1990): 831−836.

SCHACHNAI, A.: Evaluation of the fungicides CGA 64251, Guazatine, Sodium o-Phenylphenate, and imazalil for the control of sour rot on lemmon fruits. Plant Dis. **66** (1982): 733−735.

SCHEINPFLUG, H., and VAN DEN BOOM, T.: Baycor, a new fungicide for tropical and subtropical crops. Pflanzenschutz-Nachr. Bayer **34** (1981): 8−28.

− and KASPERS, H.: Verbesserte Bekämpfung von Pflanzenkrankheiten durch ein neues Azolfungizid. Mitt. Biol. Bundesanst. Land- u. Forstwirtsch. Berlin-Dahlem **232** (1986): 193−194.

− and PAUL, V.: On the mode of action of triadimefon. Netherl. J. Plant-Pathol. **83** (1977): 105−111.

− − and KRAUS, P.: Studies on the mode of action of Bayleton® against cereal diseases. Pflanzenschutz-Nachr. Bayer **31** (1978): 110−115.

SCHEPERS, H. T. A. M.: Development and persistence of resistance to fungicides in *Sphaerotheca fuliginea* in cucumbers in the Netherlands. Doktorthesis Proefschrift van de Landvouwhogeschool te Wageningen (1985): 1−55.

SCHICKE, P., and VEEN, K. H.: A new systemic, CELA W 524 (N,N'-bis-[1-formamido-2,2,2-trichloroethyl)-piperazine) with action against powdery mildew, rust and apple scab. Proc. 5th Brit. Insect. Fung. Conf. (1969): 569−575.

SCHMITT, P., and BENVENISTE, P.: Effect of fenarimol on sterol biosynthesis in suspension cultures of bramble cells. Phytochemistry **18** (1979): 1659−1665.

SCHWINN, F., and URECH, P. A.: New approaches for chemical disease control in fruit and hops. Proc. 1981 Brit. Crop. Protect. Conf. (1981): 819−833.

SHABI, E., ELISHA, S., and ZELIG, Y.: Control of pear and apple diseases in Israel with sterolinhibiting fungicides. Plant Dis. **65** (1981): 992−994.

SHEPHARD, M. C.: Fungicide behaviour in the plant: systemicity. Proc. Bordeaux Mixture Centenary Meeting. Vo. 1 (BCPC Monograph No. 31) (1985): 99−196.

− FRENCH, P. N., and RATHMELL, W. G.: PP 969: A broad spectrum systemic fungicide for injection or soil application. Proc. 10th Intern. Congr. Plant Protect. (1983): 521.

− − Biochemical and cellular aspects of the antifungal action of ergosterol biosynthesis inhibition. In: TRINCI, A. P. J., and RYLEY, J. F. (Eds.): Mode of Action of Antifungal Agents. Cambridge Univ. Press, Cambridge 1984, pp. 257−282.

− NOON, R. A., WORTHINGTON, P. A., MCCLELLAN, W. D., and LEVER, B. G.: Hexaconazole: a novel triazole fungicide. Proc. 1986 Brit. Crop Prot. Conf. − Pests and Diseases − (1986): 19−26.

SISLER, H. D., RAGSDALE, N. N., and WATERFIELD, W. W.: Biochemical aspects of the fungitoxic and growth regulatory action of fenarimol and other pyrimidin-5-ylmethanols. Pesticide Sci. **15** (1984): 167–176.

SKIDMORE, A. M., FRENCH, P. N., and RATHMELL, W. G.: PP 450: a new broad-spectrum fungicide for cereals. Proc. 10th Intern. Congr. Plant Protect. 1983, Vol. 1 (1983): 368–375.

SMITH, J. M., and SPEICH, J.: Propiconazole: disease control in cereals in Western Europe. Proc. 1981 Brit. Crop Protect. Conf. (1981): 291–297.

SMITH, J. W. M.: Triforine sensitivity in lettuce (*Lactuca sativa*): A potentially useful genetic marker. Euphytica **28** (1979): 351–360.

STANIS, V., and JONES, A. L.: Reduced sensitivity to sterol-inhibiting fungicides in field isolates of *Venturia inaequalis*. Phytopathology **75** (1985): 1098–1101.

STAUB, T., SCHWINN, F., and URECH, P.: CGA-64251, a new broad spectrum fungicide. Phytopathology **69** (1979): 1046.

STEFFENS, W., FÜHR, F., KRAUS, P., and SCHEINPFLUG, H.: Uptake and distribution of ®Baytan in spring barley and spring wheat after seed treatment. Pflanzenschutz-Nachr. Bayer **35** (1982): 171–188.

– and WIENECKE, J.: Behaviour and fate of the ^{14}C-labelled mildew fungicide fluotrimazole in plant and soil. I. Radioactivity distribution in treated spring barley and in the soil of field lysimeters and uptake by untreated rotational crops. Z. Pflanzenkrankh. Pflanzenschutz **88** (1981): 343–354.

STEVA, H., CARTOLARO, P., CLERJEAU, M., LAFON, R., and GOMES DA SILVA, M. T.: Une resistance de l'oidium au Portugal? Phytoma **402** (1988): 49–50.

– and CLERJEAU, M.: Cross resistance to sterol biosynthesis inhibitor fungicides in strains of *Uncinula necator* isolated in France and Portugal. Med. Fac. Landbouww. Rijksuniv. Gent **55** (1990): 983–988.

STRYDOM, D. K., and HONEYBORNE, G. E.: Increase in fruit set of "Starking Delicious" apple with triadimefon. Hort Sci. **16** (1981): 51.

SZKOLNIK, M.: Physical modes of action of sterol-inhibiting fungicides against apple diseases. Plant Dis. **65** (1981): 981–985.

– Unique vapour activity by CGA-64251 (Vangard) in the control of powdery mildews roomwide in the greenhouse. Plant Dis. **67** (1983): 360–366.

TAKANO, H., OGURI, Y., and KATO, T.: Mode of action of (E)-1-(2,4-dichlorophenyl)-4,4-dimethyl-2-(1,3,4-triazol-lyl)-1-penten-3-ol (S-3308) in *Ustilago maydis*. J. Pesticide Sci. **8** (1983): 575–582.

TATON, M., ULLMANN, P., BENVENISTE, P., and RAHIER, A.: Interaction of triazole fungicides and plant growth regulators with microsomal cytochrome P-450-dependent obtusifoliol 14 α-methyl demethylase. Pestic. Biochem. Physiol. **30** (1988): 178.

THOMAS, J., and CORNIER, A.: Étude du prochloraz sur céréales. La Défense des Végétaux **221** (1983): 143–156.

TRÄGNER-BORN, J., and VAN DEN BOOM, T.: Über Ergebnisse von Freilandversuchen mit ®Baytan, einem neuen systemischen Getreidebeizmittel. Pflanzenschutz-Nachr. Bayer **31** (1978): 25–37.

– and KASPERS, H.: Zur Bekämpfung von Zwergsteinbrand an Winterweizen mit ®Sibutol. Pflanzenschutz-Nachr. Bayer **34** (1981): 1–7.

URECH, P. A., SCHWINN, F. J., SPEICH, J., and STAUB, T.: The control of airborne diseases of cereals with CGA 64250. Proc. 1979 Brit. Crop Prot. Conf. (1979): 508–515.

– and SPEICH, J.: Propriétés du CGA 64250 (Tilt®) et activité contre les maladies des cereales. Phytiatrie-Phytopharmacie **30** (1981): 21–26.

VERGNES, A., and PISTRE, R.: Sur les traitements contre l'Oidium. Progrès Agricole et Viticole **99** (1982): 528–530.

VERHEYDEN, C.: Curative control of apple scab (*Venturia inaequalis*). Meded. Fac. Landbouwwet. Rijksuniv. Gent **46** (1981): 955–960.

VOGEL, C., STAUB, T., RIST, G., and STURM, E.: The four isomers of etaconazole (CGA 64251) and their fungicidal activity. In: MIYAMOTO, S., and KEARNEY, P. C. (Eds.): Pesticide Chemistry: Human Welfare and the Environment. Vol. I, Pergamon Press, Oxford 1983, pp. 303–308.

VONK, J. W., and DEKHUIJZEN, H. M.: Transport and metabolism of ^{3}H-imazalil in barley and cucumber. Meded. Fac. Landbouwwet. Rijksuniv. Gent **44** (1979): 927–934.

WAARD DE, M. A., KIPP, E. M. C., HORN, N. M., and VAN NISTELROOY, J. G. M.: Variation in sensitivity to fungicides which inhibit ergosterol biosynthesis in wheat powdery mildew. Neth. J. Plant Pathol. **92** (1986): 21–23.

– and VAN NISTELROOY, J. G. M.: Mechanism of resistance to fenarimol in *Aspergillus nidulans*. Pesticide Biochem. Physiol. **10** (1979): 219–229.

– – An energie-dependent efflux mechanism for fenarimol in a wild-type strain and fenarimol-resistant mutants of *Aspergillus nidulans*. Pesticide Biochem. Physiol. **13** (1980): 255–266.

– – Accumulation of SBI fungicides in wild type and fenarimol-resistant isolates of *Penicillium italicum*. Pestic. Sci. **22** (1988): 22.

WAINWRIGHT, A., and LINKE, F.: Field experiments and observations in Great Britain on the activity of ®Folicur and ®Raxil against leaf infections and seed-borne diseases of cereals as well as of Folicur against rape diseases. Pflanzenschutz-Nachrichten Bayer **40** (1987): 181–212.

WAKERLEY, S. B., and Russel, P. E.: Prochloraz – a decade of development. Proc. Bordeaux Mixture Centenary Meeting, BCPC Monograph No. 31, Vol. 2 (1985): 257–260.

WALLER, C. D., ESCHENBRENNER, P., and GODWIN, J. R.: Hexaconazole a new flexible cereal fungicide. Proc. 1990 Brighton Crop Prot. Conf. – Pests and Diseases – (1990): 447–454.

WOLFE, M. S., and FLETCHER, J. T.: Insensitivity of *Erysiphe graminis* f. sp. *hordei* to triadimefon. Netherl. J. Plant Pathol. **87** (1981): 239.

WYBOU, A. P.: Field experiments with ®Folicur for the control of black Sigatoka (*Mycosphaerella fijiensis*) in banana growing at the Atlantic coast of Costa Rica. Pflanzenschutz-Nachrichten Bayer **42** (1989): 162–180.

ZAAYEN, A. VAN, and VAN ADRICHEM, J. C. J.: Prochloraz for control of fungal pathogens of cultivated mushrooms. Netherl. J. Plant Pathol. **88** (1982): 203–383.

ZECH, B., GOUOT, J. M., MERINDOL, B., and GREINER, A.: LS840606 – a new broad-spectrum fungicide. Proc. 1988 Brighton Crop Prot. Conf. – Pests and Diseases – (1988): 503–510.

ZOBRIST, P., BOHNEN, K., SIEGLE, H., and DORN, F.: Pyrifenox, a new pyridine fungicide against foliar and fruit diseases. Proc. 1986 Brit. Crop Protect. Conf. (1986): 47–53.

Chapter 13

DMI-fungicides — side effects on the plant and problems of resistance

H. BUCHENAUER

University Hohenheim, Institute of Phytomedizin, 70593 Stuttgart, FRG

13.1 Mode of action of N-heterocyclic plant growth retardants and side effects of structurally related fungicides in plants

During the 25 years fungicides with systemic properties have been discovered and developed with short intervals. The systemically active compounds differ from substances with non-specific mode of action by a pronounced selectivity. The selectivity of fungicides with systemic properties also implies a more specific mode of action; they often interfere with only one or few sites of fungal metabolism. Since systemic fungicides are taken up both by roots and foliage and translocated in the plants (predominantly in the xylem) they may induce numerous physiological effects in higher plants.

The carboxamides, benzimidazoles, hydroxy-pyrimidines and the acylalanines represent some examples of systemic fungicide classes. The sterol-biosynthesis inhibiting fungicides comprise a wide variety of different chemical groups. Regarding their mode of actions in fungi the sterol biosynthesis inhibitors (SBIs) used in agriculture may be divided into two groups: The inhibitors of $\Delta^8 \rightarrow \Delta^7$-isomerisation and Δ^{14}-reduction include the morpholine fungicides (e.g., tridemorph and fenpropimorph) and the 14α-demethylation inhibitors (DMIs) representing a most important group of modern systemic fungicides comprehend substituted pyrimidines, pyridines, piperazines, imidazoles and triazoles. Of the numerous physiological effects induced by many DMIs the plant growth regulating activity is the most obvious side effect and the research with DMIs resulted in chemical derivatives optimized as plant growth regulators.

Effects on growth

The pyrimidine derivatives ancymidol and flurprimidol are effective inhibitors of plant growth in monocotyledons and dicotyledons (TSCHABOLD et al. 1970; LEOPOLD and WRIGHT 1970; LEOPOLD 1971; COOLBAUGH and HAMILTON 1976; SHIVE and SISLER 1976; COOLBAUGH et al. 1982a; STERRETT and TWORKOSKI 1987). Both compounds are of commercial relevance as plant growth retardants especially in ornamentals and in turf grass.

The structurally related fungicides triarimol (SHIVE and SISLER 1976; COOLBAUGH and HAMILTON 1976; COOLBAUGH et al. 1982a), nuarimol (SHIVE and SISLER 1976; BUCHENAUER 1979; COOLBAUGH et al. 1982a; KONSTANTINIDOU-DOLTSINI et al. 1986a) and fenarimol (SHIVE and SISLER 1976; COOLBAUGH et al. 1982a) also exhibit plant growth-retarding activity.

In recent years, certain triazole-derivatives have achieved a high degree of interest: Paclobutrazol (LEVER et al. 1982; HEDDEN and GRAEBE 1985) and miconazole (IZUMI et al. 1984, 1985) are highly active plant growth retardants and may be applied in orchard trees, ornamental plants, bushes, grasses, cereals and rice (SWIETLIK and MILLER 1983; CURRY and WILLIAMS 1983; MCDANIEL 1983; MENHENETT et al. 1983). However, because of their relative high persistency, the use of these compounds is limited. Triapenthenol and BAS 111..W are less persistent (LÜRSSEN and REISLER 1987; JUNG et al. 1987; REED et al. 1989; REED and BUCHANAN 1990) and may be applied in oilseed rape (*Brassica napus*) and ornamental plants.

Inabenfide, a 4-substituted pyridine derivative, being used as antilodging compound in rice and the experimental imidazole-derivative HOE 074784 [1-(2,6-diethylphenyl)-imidazole-5-carboxamide] shows growth regulating activity in oilseed rape and cereals, especially rice (BUERSTELL et al. 1988).

Likewise, plant growth retarding effects of various triazole fungicides in growth chamber or greenhouse experiments have been reported. After seed treatment (0,25 g a.i./kg seed) of various triazole derivatives, retardation of elongation growth of primary leaves of barley, wheat, oats and rye seedlings increased in the following order: triadimenol < propiconazol < etaconazol < diclobutrazol < paclobutrazol. Primary leaves of barley and wheat seedlings were more affected by the triazole fungicides than those of oat and rye (BUCHENAUER et al. 1984). Root growth of cereal seedlings was less severely retarded than that of primary leaves by the triazole fungicides (BUCHENAUER and GROSSMANN 1977; FÖRSTER et al. 1980a; BUCHENAUER and RÖHNER 1981; BUCHENAUER 1984).

The norbornanodiazetine-derivative tetcyclasis (5-(4-chlorophenyl)3,4,5,9,10-pentaazatetra-cyclo-5,4-1,$0^{2,6}$,$0^{8,11}$,dodeca-3,9-diene) which is used as a dwarfing agent in the production of rice seedlings for transplanting, may be regarded as a nitrogen-containing heterocycle (JUNG et al. 1980).

Because of their asymmetric carbon atoms, a great number of the triazoles possess diastereomeric forms. The diastereomers or their enantiomers may determine their plant growth regulatory and fungicidal activity. For instance, triadimenol and paclobutrazol form, because of their chiral centres, two readily separable diastereomers.

While the diastereomers of triadimenol I (1 RS, 2 SR) exhibited a markedly higher fungitoxic activity than the diastereomeric form II (1 RS, 2 RS) with respect to their growth regulatory activity, both diastereomeric forms behaved contrarily (BUCHENAUER 1979; BURDEN et al. 1987). Triadimenol is now marketed predominantly as the 1 RS, 2 SR diastereomer (as in 80:20 Baytan) (BURDEN et al. 1987).

Paclobutrazol forms because of its two asymmetric carbon atoms the following two pairs of enantiomers: 2 RS, 3 RS and 2 RS, 3 SR. The + (−)-enantiomeric form (2 RS, 3 SR) showed high activity against powdery mildews and rusts of cereals and low plant regulatory activity in apple seedlings, whereas the (−) enantiomeric form (2 RS, 3 RS) displayed opposite effects (SUGAVANAM 1984; BURDEN et al. 1987). Paclobutrazol produced today consists almost exclusively of the 2 RS, 3 RS form.

The plant growth retardants reduce elongation growth of plants without changing the developmental patterns or being phytotoxic. This effect may be achieved primarily by reducing cell elongation and/or rate of cell division.

The shortening of shoots suggests that the activity of the meristematic tissue is affected by the growth retardants. While the activity of the apical meristems, which is responsible for the formation of new internodes and leaves and determine leaf arrangement, is only slightly affected, the growth retardants primarily affect subapical meristems determining internode elongation (DICKS 1980). Histological studies on stems of chrysanthemum showed that the number of mitotic figures and, thus, the cell division activity was reduced in the subapical meristems. Inhibition of cell elongation has been found in studies on shoot sections of sunflower, soybean and maize in the presence of low concentrations (e.g., 10^{-7} M) of tetcyclasis. The inhibition of longitudinal shoot and leaf growth suggests the retardants might interfere with specific steps of gibberellin biosynthesis (PHINNEY 1984).

Interaction between gibberellic acid and growth retardants and related compounds

Growth retardation (e.g., shoot height, leaf area) and other parameters induced by low concentrations of pyrimidine and triazole derivatives were fully reversed by application of gibberellic acid (GA), suggesting that growth retardation of plants by these compounds is attributed to interactions with GA-dependent processes (LEOPOLD and WRIGHT 1970; TSCHABOLD et al. 1970; LEOPOLD 1971; COOLBAUGH and HAMILTON 1976; SHIVE and SISLER 1976; BUCHENAUER and GROSSMANN 1977; BUCHENAUER and RÖHNER 1981; COOLBAUGH et al. 1982; WAMPLE and CULVER 1983; CURRY and WILLIAMS 1983; HEDDEN et al. 1989; GRZESIK et al. 1992).

Triazoles and GA_3 may significantly alter in vitro growth and differentiation. The plant growth regulators paclobutrazol and uniconazole reduced in vitro growth of moth bean callus, addition of gibberellic acid (GA_3) to the culture medium in combination with the triazoles restored callus growth to a level of the untreated control. Both retardants also reduced the percentage of cultures that formed roots as well as the mean number of roots per culture; GA_3 increased root formation and counteracted the inhibitory effect of the triazoles on rooting (SANKLA et al. 1991).

At higher concentrations of the inhibitors, however, growth retardation was only partly annulled by simultaneous GA application in bean (SHIVE and SISLER 1976), pea (COOLBAUGH et al. 1982) and barley and wheat seedlings (BUCHENAUER and GROSSMANN 1977; BUCHENAUER 1979; BUCHENAUER and RÖHNER 1981). The stunting effects at higher concentrations of the N-heterocyclic plant growth retardants were increasingly attributed to reduced rates of cell division (NITSCHE et al. 1985), suggesting that the growth inhibition may result from interactions of the retardants with metabolic sites other than those affecting GA biosynthesis.

Uptake and metabolism of triadimenol enantiomers in barley plants were studied following seed application by CLARK et al. (1986) and showed significant differences in metabolism of triadimenol enantiomers. Uptake and conjugation were greater for the 1S 2S enantiomer than for the 1R 2S, 1S 2R and 1R 2R enantiomers. A hexose conjugate was formed only from the 1R 2S enantiomer and the 1S 2S enantiomer formed an unidentified polar conjugate. While no racemisation or epimerisation of the 1S 2R, 1R 2R and 1S 2S enantiomers were detected, the 1R 2S enantiomer was partially converted to its 1R 2R epimer.

Interference with gibberellin synthesis in plants

Inhibition of cell elongation suggests that the compounds might interfere with gibberellin biosynthesis. Growth retardation of primary leaves of barley seedlings was attributed to a decreased cell extension: the number of cells per leaf of treated seedlings did not differ significantly from that of the control (FÖRSTER et al. 1980a).

It has been shown that ancymidol and triarimol reduced the extractable GA-like activity in bean plants (SHIVE and SISLER 1976). The triazole derivatives triadimenol, triadimefon, paclobutrazol, diclobutrazol, etaconazol and propiconazol and the pyrimidine derivative nuarimol and the imidazole compound imazalil diminished the activity of GA-like compounds in primary leaves of barley and wheat seedlings (BUCHENAUER and GROSSMANN 1977; BUCHENAUER and RÖHNER 1977, 1981; BUCHENAUER et al. 1984b; LENTON et al. 1987). Reduced levels of GAs have also been analyzed after treatment with uniconazole in rice (IZUMI et al. 1984), in BAS 111..W, LAB 150 978 and tetcyclasis-treated soybean seedlings (GROSSMANN et al. 1987), as well as in inabenfide treated rice (MIKI et al. 1990) and tetcyclasis-treated corn cockle (*Agrostemma githago*) (ZEEVAART 1985).

Soil treatment of oilseed rape seedlings with the growth retardant BAS 111..W caused 80% reduction in height 18 days after treatment and levels of all GAs were 20% or less than that of the control plants. Reduced shoot height was associated with lowered GA_1 concentration in the shoots. Foliar treatment at the same dosage reduced 18 days after treatment height by 50% and caused an 85% or greater reduction in the concentrations of GA_1 precursors GA_{20}, GA_{19} and GA_{44}. However, the levels of GA_1, GA_8 and GA_{29} were affected to a much smaller extent (HEDDEN et al. 1989).

STEFFENS et al. (1991, 1992) studied the gibberellin content of immature apple seeds from paclobutrazol-treated trees over three seasons. It had been observed that application of plant growth retardants such as paclobutrazol can increase flower bud formation. Floral induction in apple is thought to be closely associated with gibberellin (GA) metabolism and to be directly influenced by GAs that move from developing seeds to adjacent buds of bearing trees (HOAD 1984; LUCKWILL 1970; PHARIS and KING 1985). Paclobutrazol treatment can increase the levels of GA_4 and GA_7 and alter the ratio of GA_4 and GA_7 in immature apple seeds. GA_4 and GA_7 are the main GAs in apple seeds (HOAD 1978) and may be closely associated with apple flower bud initiation and development (PHARIS and KING 1985).

The induction of the synthesis of α-amylase in the GA-deficient dwarf mutant of rice, "Waito C", during germination was repressed in the presence of uniconazole. The inhibition of GA synthesis by uniconazole could be overcome by a subsequent treatment with GA_3, indicating that production of GA during germination is essential for the induction of α-amylase (MITSUNAGA and YAMAGUCHI 1993).

In general, GA levels were reduced by treatment of plants with N-heterocyclic growth retardants. However, the extent of the reduction was dependent on the concentrations and methods of application of the retardants, the time after treatment at which tissue was analysed, the nature of the GA and the plant species tested.

Studies of COOLBAUGH and HAMILTON (1976) and COOLBAUGH et al. (1978) indicated that ancymidol interfered with three reactions of the GA biosynthetic pathway by specific inhibition of ent-kaurene, ent-kaurenol and ent-kaurenal oxidation in microsomes of *Marah macrocarpus* and *Marah oreganus* (Fig. 13.1). These three cytochrome P-450-dependent oxidative reactions showed a similar degree of sensitivity to ancymidol (COOLBAUGH et al. 1978).

COOLBAUGH et al. (1982a) extented their studies by comparing several pyrimidine derivatives (including ancymidol as plant growth retardant and triarimol, fenarimol and nuarimol as fungicides) on their activity to retard growth of pea plants and to inhibit

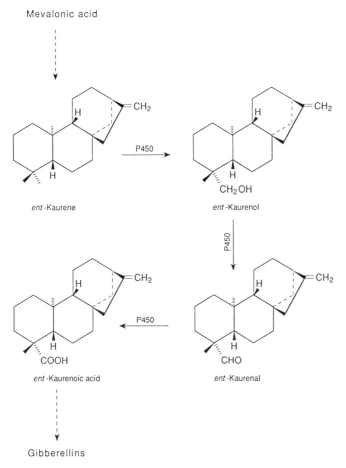

Fig. 13.1 The P 450-dependent synthesis of gibberellins from mevalonate.

ent-kaurene oxidation of microsome preparations of *M. macrocarpus*. They found a close correlation between the effectiveness of the substances as inhibitors of ent-kaurene oxidation and as plant growth retardants. The compounds tested could be grouped into two categories: Ancymidol, pyrimidol (EL-509), EL-93807, EL-75253 proved to be highly active on both processes, whereas the fungicides nuarimol, triarimol, fenarimol and EL-72303 showed a markedly reduced effectiveness both as inhibitors of elongation growth of pea plants and microsomal ent-kaurene oxidation in the cell free system.

All these plant growth inhibitors belonging to the pyrimidine, pyridine, triazole and imidazole – derivatives, have one common structural feature: the lone pair of electrones on the sp^2 hybridized nitrogen atom of the heterocycle. In each compound, this electron pair is located at the periphery of the molecule (RADEMACHER *et al.* 1987). The target enzymes of these compounds are monoxygenases containing cytochrome P-450 (HASSON and WEST 1976a, b). It is assumed that the lone electron pairs of the chemicals interact with the plant cytochrome P-450-dependent monoxygenases by binding to the 6th coordination position of protoheme iron of cytochrome P-450 by which oxygen is displaced from its binding site at the protoheme iron of cytochrome P-450. Consequently, cytochrome P-450-dependent monoxygenases are inactivated. Such membrane-bound microsomal enzymes catalyze many oxidative reactions in different metabolic pathways, and it seems that especially the interaction of the compounds containing the nitrogen heterocycle with those enzymes involved in the terpenoid metabolism results in the growth-retarding effects. The phytohormone groups of gibberellins, cytokinins and probably abscisic acid as well as the sterols are derived from the terpenoid pathway.

It has been demonstrated that these compounds effectively interfere with gibberellin biosynthesis by inhibition of ent-kaurene, ent-kaurenol and ant-kaurenal oxidations that are catalyzed by cytochrome P-450 oxidative enzymes (Fig. 13.1).

Comparative studies on the activity of the enantiomers of the triazole derivatives triadimenol (fungicide) and paclobutrazol (plant growth retardant) as inhibitors of fungal growth (using *Cladosporium cucumerinum* and *Sphaerotheca fuliginea*) as well as inhibitors of gibberellin biosynthesis (applying a cell-free *Cucurbita maxima* and *Malos pumila* ent-kaurene oxidase assay) revealed that chirality can have a decisive influence on the nature and degree of biological activity. The order of activity of triadimenol enantiomers with respect to fungitoxicity was: 1S, 2R > 1R, 2R > 1R, 2S > 1S, 2S and regarding inhibition of gibberellin biosynthesis was: 1R, 2S > 1S, 2S > 1R, 2R > 1S, 2R.

The relative fungitoxic activity of paclobutrazol enantiomers was: 2R, 3R > 2S, 3R ⩾ 2R, 3S ≈ 2S, 3S and activity of the enantiomers inhibiting gibberellin biosynthesis decreased as follows: 2S, 3S > 2R, 3S > 2R, 3R > 2S, 3R. These results indicate that the R configuration at the chiral carbon having the hydroxyl groups (carbon 2 of triadimenol, carbon 3 in paclobutrazol) is the prime determinant of fungitoxicity while enantiomers bearing the S configuration at this alcohol carbon are effective inhibitors of gibberellin biosynthesis (BURDEN *et al.* 1987; HEDDEN and GRAEBE 1985).

LUSTER and MILLER (1993) showed that the triazole plant growth retardant BAS 111..W binds to microsomal membranes isolated from endosperm of immature pumpkin (*Cucurbita maxima*) seeds and the spectral characteristics indicate binding of BAS 111..W to cytochrome P-450. Binding of the growth retardant to cytochrome was localized exclusively in the endoplasmic reticulum in endosperm cells of developing *C. maxima* seed. Binding competition experiments between ent-kaurene and BAS 111..W suggested that the substrate and the inhibitor compete for binding at the same site on ent-kaurene oxidase cytochrome P-450.

Interference with gibberellin synthesis in fungi

In comparative studies, the effects of the fungitoxic compounds triarimol, fenarimol and nuarimol and of the plant growth regulators ancymidol and EL-509 were tested on growth and gibberellin synthesis in *Gibberella fujikuroi* (*Fusarium moniliforme*). *G. fujikuroi* causes the "bakanae disease" of rice leading to an abnormal increase in longitudinal growth. While the three fungitoxic agents proved to be more effective in retardation of mycelium growth than ancymidol and EL-509, all five substances inhibited gibberellin production of the fungus in culture filtrates. The activity of GA-compounds closely paralleled that of growth retardation of *G. fujikuroi*. At subinhibitory concentrations of EL-509 and ancymidol, GA_3 synthesis in *G. fujikuroi* was stimulated. All compounds interfered in ent-kaurene oxidation of microsomal preparations from the mycelium of the fungus (COOLBAUGH et al. 1982b). It is assumed that in the oxidation reactions of ent-kaurene to ent-kaurenoic acid in the fungus G. fujikuroi, a cytochrome P-450 component of the microsomal mixed function oxygenases is involved (COOLBAUGH et al. 1978). However, the structure-activity relationships suggest that the enzymes from the higher plant and the fungus may differ in subtle properties.

The effects of triazole derivatives on growth, GA-like activity and sterol synthesis in G. fujikuroi were studied (KUTZNER and BUCHENAUER 1986). The compounds (each at 10^{-4}M) retarded the increase of mycelium dry weight of the fungus: etaconazol (95%), paclobutrazol (70%), diclobutrazol (60%), bitertanol (30%) and triadimenol (0%). All five compounds diminished the production of gibberellin-like substances of the fungus in the culture filtrates to a smaller extent than the dry weight increase. Furthermore, at 10^{-4}M the triazole derivatives interfered with ergosterol synthesis in G. fujikuroi. While ergosterol synthesis was inhibited, sterols containing C-4 and C-14 methyl groups accumulated.

The fungus Sphaceloma manihoticola causes in cassava plants the superelongation disease. Gibberellin A_4 was identified in the culture filtrate of the fungus (RADEMACHER and GRAEBE 1979). It may be assumed that the N-heterocyclic plant growth retardants interfere in gibberellin synthesis of S. manhoticola in a similar way as in G. fujikuroi.

Interference with sterol biosynthesis of plants

In general, cell elongation takes place in growth zones outside of the meristem and the regulation of this process seems to be closely linked to the gibberellin synthesis.

At higher concentrations of the plant growth retardants, an inhibition in cell division is often observed and studies indicated that gibberellins are not directly involved in this process.

In cell suspension cultures, the inhibition of cell division caused by tetyclasis could not be overcome by addition of GA_3 or kaurenoic acid (GROSSMANN 1988), suggesting that gibberellins are not directly involved in cell division. Similar to plant internodes, gibberellins induce in suspension cultures primarily an increase of cell expansion. However, the inhibiting effect of LAB 150 978 and ancymidol on cell division in rice (GROSSMANN 1988) and of paclobutrazol in celery cell suspensions (HAUGHAN et al. 1988) can be annulled by addition of cholesterol and/or stigmasterol. These results suggest that sterols might play a more important role in regulating cell division processes than gibberellins.

A double role of sterols in cell division of celery suspensions is proposed by HAUGHAN et al. (1988). While the major sterol components, the 4-desmethylsterols, are essential in the maintenance of membrane integrity, the minor 24-ethylsterol components might function as essential factors of cell proliferation (HAUGHAN et al. 1988).

Quantitative and qualitative changes in sterol pattern, depending on plant species and material analyzed, are generally detected especially at higher retardant concentrations. Decreased concentrations with campesterol, stigmasterol and sitosterol have been detected following seed treatment of paclobutrazol, triadimenol and nuarimol in shoots of barley seedlings following seed treatment (each $25 \text{ g} \times 100 \text{ kg seed}^{-1}$) (BUCHENAUER 1984), soil treatment with paclobutrazol and triapenthenol (each $5 \text{ mg} \times \text{pot}^{-1}$) in shoots and roots of barley (BURDEN et al. 1987a), soil treatment with triapenthenol ($0,1-1,6 \text{ mg} \times \text{pot}^{-1}$) in leaves and roots of rice (LÜRSSEN 1988), soil treatment with tetcyclasis (2 mg×pot^{-1}) or seed treatment in hydroponics in shoots of oats and plasma membrane preparations of shoots of oats (BURDEN et al. 1987b). On the other hand, higher levels of cholesterol as compared to the control were detected in shoots of oats. Soil treatment with tetcyclasis (about 5×10^{-6}) resulted in decreased concentrations of stigmasterol in plasma membrane preparations of winter rye shoots. In plasma membrane preparations of extracts of tetcyclasis-treated oat and rye shoots, the decreased stigmasterol contents were accompanied by an increased activity of the magnesium-dependent potassium-stimulated plasma membrane ATPase (COOKE et al. 1988). Seed treatment with BAS 111..W and tetcyclasis caused reduced concentrations of stigmasterol and sitosterol in maize (TATON et al. 1988).

Addition of paclobutrazol in hydroponic solution ($3,4 \times 10^{-5}$ M) resulted in lower sitosterol and compesterol concentrations in apple shoots and roots (WANG et al. 1988). Triadimefon and diclobutrazol applied as soil drenches inhibited sterol C-14-demethylation in maize and winter wheat seedlings. The inhibition was accompanied by an increase in the content of 4α, 14α-methyl sterols. They also affected second transmethylation reactions at C-24, this effect resulted in an increase of campesterol, a 24-methylsterol, and a decline in sitosterol and stigmasterol, the 24-ethylsterols (KHALIL et al. 1990; KHALIL and MERCER 1990a, b).

The separation of some chiral azole fungicides and plant growth regulators has provided useful tools for determining the relative importance of sterols and gibberellins to plant growth (KRÄMER et al. 1983; LÜRSSEN et al. 1987; SUGAVANUM 1984; BURDEN et al. 1987). The enantiomers of triadimenol showed a similar order of activity in retardation of shoot growth of etiolated wheat seedlings and inhibition of sterol synthesis: 1S, 2R > 1R, 2S > 1S, 2S > 1R, 2R (KÖLLER 1987). The 2S, 3S-enantiomer of the plant growth regulator paclobutrazol proved to be far more effective than the 2R, 3R-enantiomer in inhibiting pot-grown apple seedlings (SUGAVANUM 1984). These results corresponded to the effects of the enantiomers in a pumpkin endosperm ent-kaurene oxidase assay, indicating involvement of inhibition of gibberellin synthesis in growth retardation. On the other hand, the 2R, 3R-enantiomers of paclobutrazol proved to be potent inhibitors of both growth and sterol biosynthesis whereas the 2S, 3S- and 2S, 3R-forms were much less effective. These findings show that in cell cultures, growing mainly by cell division, inhibition of sterol biosynthesis appears to be the determinant of cell multiplication.

The enantiomers of the plant growth retardant triapenthenol displayed different activities in anti-gibberellin and anti-sterol effects. While the anti-gibberellin S,Z-enantiomer was responsible for the reduced elongation growth of internodes, the R,Z-form inhibited both fungal and phytosterol synthesis (LÜRSSEN 1987). In summary, the nature of plant growth responses depends on the activity of the azole compounds to interfere in gibberellin or sterol biosynthesis, the concentrations applied and the relative sensitivity of the plant tissues for growth (HAUGHAN et al. 1989).

For determining the role of sterols in growth retardation, the effect of the inhibitors on quantitative and qualitative changes in sterol components in meristematic tissues or their effects on isolated membrane preparations and cell compartments instead of whole plants should be studied. It may be concluded that only slight or moderate alterations in sterol components in meristematic tissue might result in growth-regulating effects.

In contrast, pronounced changes in sterol contents lead to phytotoxic effects. The R,Z-enantiomer of triapenthanol, inhibiting specifically sterol biosynthesis in fungi and plants, reduced root growth and inhibited new leaf formation and when applied at higher concentrations the compound induced necrosis (LÜRSSEN 1987), suggesting that a reduction in normal phytosterol levels and/or an accumulation of 14α-methylsterols may cause phytotoxic effects.

Treatment of barley and oat seedlings (at $0.1-10$ μM) with the experimental ICI 8-ketotriazole herbicide [4,4-dimethyl-1-(2-methoxyphenyl)-1-(1,2,4-triazol-1-yl)-1-penten-3-one] resulted in inhibition of the synthesis of phytosterols stigmasterol, sitosterol and campesterol and a cocomitant accumulation of 14α-methylsterols (obtusifoliol, dihydroobtusifoliol and 14α-methyl-Δ^8-ergostenol) in barley and (14α-methyl-Δ^8-ergostenol) in oat (VAN DEN BOSSCHE et al. 1991). At these concentrations, the 8-ketotriazole showed no ent-kaurene oxidase inhibition in cell-free preparations and no interference in sterol biosynthesis of *Cladosporium cucumerinum*. The herbicide [methyl-1-(2,2-dimethylindan-1-yl)-imidazole-5-carboxylate] R 69020 (CGA 201-029) blocked (at 0.1 μM) in roots of 7-day-old maize seedlings the phytosterols (stigmasterol, sitosterol and campesterol) whereas the 14α-methyl sterols (obtusifoliol and 14α-methylergosta-8-en-3β-ol) accumulated (VAN DEN BOSSCHE et al. 1991; STREIT et al. 1991). The compound was less effective in inhibition of ergosterol synthesis in *Candida albicans* and *Candida glabrata* and much higher concentrations (> 10 μM) were necessary to obtain 50% inhibition of cholesterol synthesis in human liver cells (VAN DEN BOSSCHE et al. 1991). These data indicate that the triazole herbicides are effective inhibitors of the plant P 450-dependent obtusifoliol 14α-demethylase.

Tobacco calli selected showing resistance to the 8-ketotriazole herbicide exhibited a new sterol profile with an exaggerated production of sterols and an esterification of the overproduced metabolites (MAILLOT-VERNIER et al. 1989, 1990). SCHALLER et al. (1992) hypothesized that resistance could be attributed to reduced binding of the inhibitor to cytochrome P450-dependent obtusifoliol 14α-demethylase due to a mutation affecting this enzyme.

The cytochrome P450-dependent 14α-demethylase of plants, fungi and mammalian cells exhibiting the same catalytic function, differ in their substrate affinity and sensitivity to the azole inhibitors. While the plant 14α-demethylase using primarily obtusifoliol, 24(28)-dihydro-obtusifoliol, 24(25)-dihydro-31-norlanosterol and 14α-methyl-24(28)-dihydrofecosterol as substrates, those of fungi and animal utilizing lanosterol or 24-methylenedihydrolanosterol (TATON and RAHIER 1991). In contrast to fungi and animals, 14α-demethylation in plants takes place only after the 4α-demethylation steps. The azole compounds coordinate with the P-450 iron and bind to a region of the aproprotein near the heme or substrate-binding domain (cf. chapter 24). Differences in the amino acid sequence in or near these domains may result in differential sensitivity to the azole compounds and substrate specifity and, finally, minor changes in chemical structures of the azole enantiomers may affect their binding apitude to the different types of cytochrome P-450-dependent demethylases.

Effects on other phytohormones and metabolic processes

In plants, cytochrome P-450-dependent monooxygenase are not exclusively involved in 14α-demethylation of obtusifoliol and oxidative demethylation at C19 of the gibberellin biosynthesis catalyzing ent-kaurene via ent-kaurenol and ent-kaurenal to ent-kaurenoic acid but also, e.g., in 4-hydroxylation of the monoterpene alkaloid precursor geraniol and in 4-hydroxylation of trans-cinnamic acid to form trans-p-coumaric acid, a precursor of flavonoids and lignin.

The interference of some of the azole compounds with gibberellin and sterol biosynthesis may explain some of the morphoregulating effects in plants. Furthermore, plant growth retardants also affect the endogenous levels of other phytohormones such as cytokinins, ethylene, abscisic acid and auxins as well as other physiological effects. The delayed senescence induced by the N-heterocyclic growth retardants has been attributed to changes in the plant hormone status, resulting in increased levels of cytokinins and polyamines and inhibition of formation of ethylene and abscisic acid.

The effects of the triazole type growth retardants uniconazole, LAB 150978, BAS 111..W and ancymidol as well as tetcyclasis in soybean seedlings (GROSSMANN et al. 1987), rice seedlings (IZUMI et al. 1988), sunflower (*Helianthus annuus*) cell suspensions (SAUERBREY et al. 1988), leaf disks of barley, soybean (GROSSMANN et al. 1987), and oilseed rape (GROSSMANN et al. 1988), mung beans (*Phaseolus aureus*) (HOFSTRA et al. 1989) and peas (*Pisum sativum*) (LAW and HAMILTON 1989) showed that the growth retardants hardly affected the levels of the auxin indole-3-acetic acid. The reduced auxin content in uniconazole-treated pea plants probably resulted from the low levels of gibberellins. Reduced ethylene production has been determined in mung bean seedlings treated with uniconazole (HOFSTRA et al. 1989), sunflower cell suspensions treated with LAB 150 978, BAS 111..W and tetcyclasis (SAUERBREY et al. 1987, 1988), leaf discs and seedlings of oilseed rape treated with tetcyclasis and BAS 111..W (GROSSMANN 1989). Uniconazole inhibited stress-induced ethylene production by heat and an auxin-like herbicide (triclopyr) (KRAUS et al. 1991) in wheat and soybean seedlings. Inhibition of ethylene production in heterotrophic sunflower cell suspensions by triazole-type growth retardants resulted in increased levels of 1-aminocyclopropane carboxylic acid (ACC); in the presence of the retardants, ACC was increasingly converted to N-malonyl-ACC (GROSSMANN 1992; GROSSMANN et al. 1991, 1993). It is assumed that in the conversion of ACC into ethylene by ethylene-forming enzymes, a cytochrome P-450-mediated monoxygenase is involved that is blocked by the retardants (KRAUS et al. 1992).

Inhibition of conversion of ACC into ethylene may result in enhanced reactions of methionine-derived S-adenosylmethionine into polyamines by providing aminopropyl groups derived from decarboxylated S-adenosylmethionine (ROBERTS et al. 1984; SLOCUM et al. 1984).

Increases in polyamine levels, e.g., spermidine and spermine, was induced by the pyrimidine-derivative ancymidol and the triazole compound paclobutrazol in roots of apple seedlings (WANG and FAUST 1986), in uniconazole-treated mung bean seedlings (HOFSTRA et al. 1989) and in sunflower cell suspensions treated with BAS 111..W (GROSSMANN 1993).

Polyamines are known to display various physiological effects in plants; they exhibit antisenescence properties, (SLOCUM et al. 1984), inhibit ethylene formation (BORS et al. 1989; FLORES et al. 1989) and act as antioxidants (BORS et al. 1989). For instance, injuries in tobacco leaves induced by ozone exposure were alleviated by exogeneously applied polyamines (BORS et al. 1989; LANGEBARTELS et al. 1991). The increased spermidine and spermine contents may appreciably contribute to the antisenescence effect induced by the N-heterocyclic plant growth regulators (GROSSMANN 1988; HOFSTRA et al. 1989). Moreover, polyamines may reduce ethylene production by inhibiting ACC synthase and conversion of ACC into ethylene. Thus, the increased polyamine contents in plants may intensify the inhibiting effects of N-heterocycles interfering in ethylene synthesis (FLORES et al. 1989; LI et al. 1992).

Cytokinins are considered as the major senescence-delaying phytohormones (NOODEN and LEOPOLD 1988). Increased cytokinin levels have also been observed in various plant tissues treated with N-heterocyclic plant growth retardants. Triadimefon and uniconazole have been reported to increase cytokinin levels in cucumbers (FLETCHER and ARNOLD

1986). The endogenous contents of cytokinins, such as trans-zeatin and its riboside significantly increased in uniconazole-treated rice seedlings (IZUMI et al. 1988), and soybean seedlings after treatment with LAB 150 978, BAS 111..W and tetcyclasis (GROSSMANN et al. 1987; GROSSMANN 1990). The increased cytokinin levels may contribute to the delayed senescence of treated plant tissues (BUCHENAUER and RÖHNER 1981).

A close correlation was found between delayed loss of total chlorophyll and increased levels of dihydrozeatin-9-glucoside and zeatin-9-glucoside in senescing cotyledons of pumpkin seedlings after soil treatment with BAS 111..W and in cotyledons of oilseed rape seedlings treated with this triazole growth retardant (GROSSMANN et al. 1991; GROSSMANN 1991). Stimulation of ethylene production by treatment with ACC or ethephon significantly reduced cytokinin levels (GROSSMANN et al. 1993). These results suggest a relationship between inhibition of ethylene synthesis by the growth retardants and the increased cytokinin levels. It is assumed that ethylene may decrease the level of cytokinins in plant tissues, possibly by enhancing their degradation (VAN STADEN et al. 1987; BOLLMARK and ELIASSON 1990).

Abscisic acid (ABA) is a sesquiterpenoid plant hormone. As a isoprenoid compound, ABA is biosynthetically related to gibberellins (GAs) and cytokinins, and these hormones are in some instances antagonistic to each other.

Studies indicated that the growth retardants may exert a dual role on endogenous ABA levels: ABA concentrations increased shortly after treatment in seedlings of various plant species and cell suspensions, e.g., in triadimefon-treated bean and barley seedlings (ASARE-BOAMAH et al. 1986), triapenthenol-treated oilseed rape (LÜRSSEN 1987), as well as in tetcyclasis, LAB 150978 and BAS 111..W-treated cell suspension cultures, detached leaves and intact seedlings of oilseed rape (HÄUSER et al. 1990). The increased levels of ABA may be attributed to the inhibition of the conversion of ABA into phaseic acid by the azole growth retardants. ZEEVAART et al. (1990) have shown that tetcyclasis inhibits in detached leaves of *Xanthium strumarium* the conversion of ABA into phaseic acid; phaseic acid does not exhibit biological activity. In the metabolic transformation of ABA into phaseic acid, a cytochrome P-450-dependent monoxygenase is involved (Fig. 13.2).

The increased ABA levels will affect stomatal closure which is accompanied by an reduced water consumption. Triapenthenol and several azole fungicides reduced water consumption in barley and oilseed rape plants (LÜRSSEN and REISER 1987; FÖRSTER 1980). On the other hand, when ABA contents were determined later appreciably lowered concentrations of this phytohormone have been analyzed, e.g., in leaves of apple seedlings treated with paclobutrazol (WANG et al. 1987) and in various tissues of soybean seedlings (HÄUSER et al. 1990; GROSSMANN 1990). The diminished ABA concentrations are probably caused by a stimulated catabolism and/or an inhibition of ABA biosynthesis. This biphasic response seems to depend on the concentrations of the retardants applied and the mode of treatment and the developmental stage of the plants.

The fungus *Cercospora rosicola* produces large amounts of abscisic acid (ABA) as a secondary metabolite. Biosynthesis of ABA is inhibited by a number of sterol synthesis inhibiting fungicides (e.g., etaconazole, fenarimol, flurprimidol, imazalil, nuarimol, triadimefon and triforine) (NORMAN et al. 1988) and plant growth regulators (e.g., ancymidol, paclobutrazol and uniconazole) and cytokinins (NORMAN et al. 1983, 1986).

Studies on the inhibition of ABA biosynthesis in *C. rosicola* by triarimol and other inhibitors of sterol synthesis showed that a cytochrome P-450 mixed function oxidase is probably involved in one or more steps of the ABA biosynthesis pathway (AL NIMRI and COOLBAUGH 1990) (Fig. 13.2).

N-heterocyclic growth retardants are causing various morphological and physiological effects in cultural plants. Triazoles were found to inhibit shoot more than root growth; consequently, an increased root to shoot ratio was observed usually in treated plants (BUCHENAUER and GROSSMANN 1977; FÖRSTER et al. 1980a; BUCHENAUER and RÖHNER 1981; BUCHENAUER et al. 1984; FLETCHER and NATH 1984; GROSSMANN 1987).

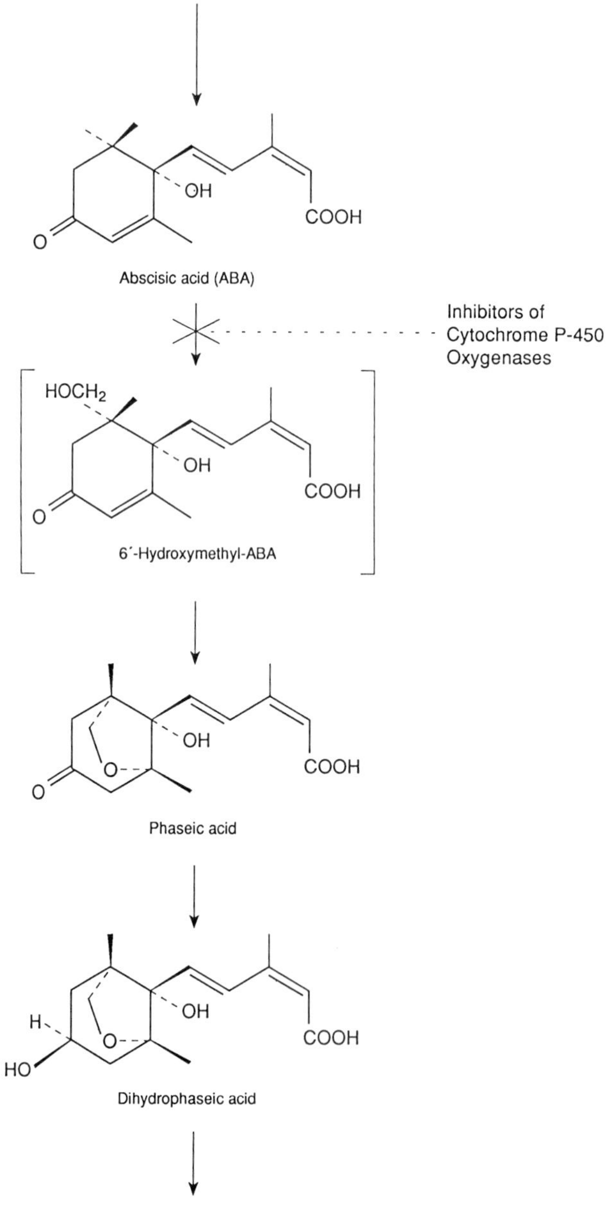

Fig. 13.2 Proposed interference of plant growth retardants in abscisic acid (ABA) metabolism by N-containing heterocyclic compounds.

Ancymidol caused lateral expansion of stem segments of oat (MONTAGUE 1975) and stems of bean seedlings (SHIVE and SISLER 1976). The diameter of epidermal cells in oat stem segments were increased (MONTAGUE 1975). Paclobutrazol-induced thickening and suppression of root elongation in pea was mainly due to increases in the diameter of cells and suppression of their length (WANG and LIN 1992). Similar morphological effects have been reported by ethylene, however, ethylene production was decreased by paclobutrazol (WANG and LIN 1992). Paclobutrazol induced in roots of apple and sweet orange seedlings comparable morphological and physiological alterations (increased contents of soluble sugar and starch) (WANG and STEFFENS 1985; VU and YELENOSKY 1992). At higher concentrations, pyrimidine and azole fungicides and growth retardants induced in bean and cereal seedlings ethylene-like responses (SHIVE and SISLER 1976; FÖRSTER et al. 1980a; KONSTANTINIDOU-DOLTSINI et al. 1986a), although ethylene synthesis was reduced by these chemicals.

Plants treated with N-heterocyclic growth retardants usually appear darker green. It has been shown that primary leaves of cereal seedlings were thicker and contained a higher water content than untreated leaves (FÖRSTER et al. 1980; KONSTANTINIDOU-DOLTSINI et al. 1986a).

The chlorophyll contents of primary leaves of barley and wheat were reduced after seed treatment with triadimefon, triadimenol and nuarimol when based on fresh weight but were increased when based on dry weight (FÖRSTER et al. 1980a; KONSTANTINIDOU-DOLTSINI et al. 1986a).

KANE and SMILEY (1983) reported that chlorophyll content of Kentucky blue grass was not affected by triadimefon and etaconazole although treated plants appeared greener than controls. Seedlings of maize treated with triadimefon and diclobutrazol as soil drenches appeared intensively green, the chlorophyll and carotenoid content per unit fresh or dry weight and per leaf were not significantly increased (KHALIL et al. 1990).

Increased chlorophyll contents in leaf tissues of plants treated with N-heterocyclic growth retardants have been reported by DAVIS et al. (1988), FLETCHER and HOFSTRA (1988) and GROSSMANN (1991). While xanthophyll, carotin and RNA contents were decreased in treated primary barley and wheat leaves, DNA and protein contents were unchanged. The concentrations of free amino acids were increased in barley and wheat seedlings (FÖRSTER et al. 1980b; KONSTANTINIDOU-DOLTSINI et al. 1986a). Also nitrate reductase activity was increased in nuarimol-treated wheat seedlings (KONSTANTINIDOU-DOLTSINI et al. 1986a) and in oilseed rape, triadimenol enhanced nitrate reductase and nitrate levels (SRIVASTAVA and FLETCHER 1992).

The effect of the triazoles and pyrimidines on endogenous levels of phytohormones may also account for the increase in nitrate reductase activity. Plant growth regulators auxins, gibberellins, cytokinins and ethylene are known to increase nitrate reductase activity in several systems (SCHMERDER and BORRISS 1986).

In senescing primary leaves treated with the triazole fungicides triadimefon and triadimenol, degradation of pigments and nucleic acids was delayed. The retarded diminution of the RNA contents could partly be attributed to higher synthetic processes maintained over a longer period of time in the leaf tissue of treated seedlings (FÖRSTER et al. 1980b).

Peroxidase activity was reduced in barley and wheat seedlings (FÖRSTER et al. 1980c; KONSTANTINIDOU-DOLTSINI et al. 1986b). Likewise, PAL and TAL-activities were lower in barley and wheat seedlings (BUCHENAUER 1979; KONSTANTINIDOU-DOLTSINI et al. 1981).

When wheat seedlings were grown from seeds (imbibed in uniconazole solutions) for 7 days in the dark, the leaves contained higher carotenoid levels than the controls and after exposure of the dark-grown seedlings to light the chlorophyll synthesis was more rapidly in leaves from uniconazole-treated seed (FLETCHER and HOFSTRA 1990). Addition of KCl to uniconazole enhances the effect of the triazole compound on chlorophyll synthesis.

The anti-senescing effects as well as the increased chlorophyll production may be attributed to the inhibition of ethylene synthesis and the increased cytokinin levels of triazole-treated plants (ABBAS et al. 1989; HOFSTRA et al. 1989; WANG and STEFFENS 1985; FLETCHER and ARNOLD 1986; IZUMI et al. 1986). Paclobutrazol applied to apple seedlings and rhododendron plants increased uptake of nutrients (e. g., N, P, K, Mg) and enhanced the efficient use of mineral nutrients (STEFFENS et al. 1985; GRZESIK et al. 1992). Foliar or drench applications of uniconazole increased in leaves of *Pyracantha* and *Photinia* N, P and Zn levels (FRYMIRE and COLE 1992).

N-heterocyclic growth retardants protect plants against various stresses including high and low temperatures, drought, air pollutants and heavy metals (BUCHENAUER et al. 1981; PENNYPACKER et al. 1982; DAVIS et al. 1988; FLETCHER and HOFSTRA 1988, 1990; SENARATNA et al. 1988; FLORES-NIMEDZ et al. 1993; SINGH et al. 1993). Triapenthenol protected tolerant and midtolerant soybean cultivars from injury caused by the herbicide metribuzin (VAVRINA and PHATAK 1988).

The antistress effects of the triazoles could be increased in combination with KCl (FLETCHER and HOFSTRA 1990). Under osmotic stress by polyethylene glycol (PEG), paclobutrazol diminished in primary roots of pea seedlings the increase of water potential, decreased the rate of water loss and increased catalase and peroxidase activity (WANG and LIN 1992). It is assumed that the higher levels of cytokinins, polyamines and carotenoids induced by the growth retardants might contribute to the protective action against environmental stresses.

Studies of RADICE and PESCI (1991) indicate that the two triazole fungicides (penconazole and propiconazole) exert early effects on plasma membrane activity, independently of their known activity on phytosterol and gibberellin biosynthesis: at low concentrations the triazoles inhibit specific mechanisms involved in proton transport (possibly H^+ pump) while at higher concentrations, they cause a disorganization of the plasmamembrane leading to a passive non specific release of intercellular solutes.

References

ABBAS, S., FLETCHER, R. A., MURR, D. P.: Alteration of ethylene synthesis in cucumber seedlings by triadimefon. Can. J. Bot. **67** (1989): 278–280.
AL-NIMRI, L. F., and COOLBAUGH, R. C.: Inhibition of abscisic acid biosynthesis in *Cercospora rosicola* by triarimol. J. Plant. Growth Regul. **9** (1990): 221–225.
ASARE-BOAMAH, H. K., HOFSTRA, G., FLETCHER, R. A., and DUMBROFF, E. B.: Triadimefon protects bean plants from water stress through its effects on abscisic acid. Plant Cell Physiol. **27** (1986): 383–390.
BOLLMARK, M., and ELIASSON, L.: Ethylene accelerates the breakdown of cytokinins and thereby stimulates rooting in Norway spruce hypocotyl cuttings. Physiol. Plant **80** (1990): 534–540.
BORS, W., LANGEBARTELS, C., MICHEL, C., SANDERMANN, H.: Polyamines as radical scavengers and protectants against ozone. Phytochemistry **28** (1989): 1589–1595.
BRITZ, S. J., and SAFTNER, R. A.: Inhibition of growth by ancymidol and tetcyclacis in the gibberellin-deficient dwarf-5 mutant of *Zea mays* L. and its prevention by exogenous gibberellin. J. Plant Growth Regul. **6** (1987): 215–219.

BUCHENAUER, H.: Untersuchungen zur Wirkungsweise und zum Verhalten verschiedener Fungizide in Pilzen und Kulturpflanzen. Habitationsschrift, Bonn 1979.
- Stand der Fungizidresistenz bei Getreidekrankheiten am Beispiel der Halmbruchkrankheit und des Echten Mehltaus. Gesunde Pflanzen **36** (1984): 132–142.
- and GROSSMANN, F.: Triadimefon: mode of action in fungi and plants. Netherl. J. Plant Path. **83** (1977): Suppl. 1, 93–103.
- and KEMPER, K.: Wirkungsweise von Propiconazol (CGA 64250) in verschiedenen Pilzarten. Meded. Fac. Landbouwwet. Rijksuniv. Gent **46** (1981): 909–921.
- KUTZNER, B., and KOHTS, T.: Wirkung verschiedener Triazol-Fungizide auf das Wachstum von Getreidekeimlingen und Tomatenpflanzen sowie auf die Gibberellingehalte und den Lipidstoffwechsel von Gerstenkeimlingen. J. Plant Dis. Prot. **91** (1984): 506–524.
- and RÖHNER, E.: Effect of triadimefon and triadimenol on growth of various plant species as well as on gibberellin content and sterol metabolism in shoots of barley seedlings. Pestic. Biochem. Physiol. **15** (1981): 58–70.
- - Einfluß von Nuarimol, Fenarimol, Triadimefon und Imazalil auf den Gibberellin- und Lipidstoffwechsel in jungen Gersten- und Weizenpflanzen. Mitt. Biol. Bundesanst. Land.- u. Forstwirtsch., Berlin-Dahlem, H. **178** (1977): 158.
BURDEN, R. S., CLARK, T., and HOLLOWAY, P. J.: Effects of sterol biosynthesis-inhibiting fungicides and plant growth regulators on the sterol composition of barley plants. Pestic. Biochem. Physiol. **27** (1987a): 289–300.
- COOKE, D. T., WHITE, P. J., and JAMES, C. S.: Effects of the growth retardant tetcyclasis on the sterol composition of oat (Avena sativa). Plant Growth Regul. **5** (1987b): 207–217.
BÜRSTELL, H. W., HACKER, E., and SCHMIERER, R.: HOE 074784 and analogues: a new synthetic group of highly active plant growth retardants in cereals on rape. In: COOKE, A. R. (Ed.): Proc. Plant Growth Regulator Soc. Amer. Plant growth regulator Society of America, Ithaca 1988 p. 185.
CLARK, T., VOGELER, K., and ISHIKAWA, I.: Comparative metabolism of the enantiomers of triadimenol in barley plants. Brit. Crop Prot. Conf. – Pests and Diseases 1986.
COOKE, D. T., BURDEN, R. S., CLARKSON, D. T., and JAMES, C. S.: Xenobiotic induced changes in membrane lipid composition: effects on plasma membrane ATPases. In: ATKIN, R. K., and CLIFFORD, D. R. (Eds.): Mechanisms and Regulations of Transport Processes. British Plant growth Regulator Group, Monograph No. 18 (1988).
COOLBAUGH, R. C., and HAMILTON, R.: Inhibition of entkaurene oxidation and growth by α-cyclopropyl-α-(p-methoxyphenyl)-5-pyrimidine methyl alcohol. Plant Physiol. **57** (1976): 245–248.
- HEIL, D. R., and WEST, C. A.: Comparative effects of substituted pyrimidines on growth and gibberellin biosynthesis in *Gibberella fujikuroi*. Plant Physiol. **69** (1982b): 712–716.
- HIRANO, S. S., and WEST, C. A.: Studies on the specificity and site of action of ancymidol, a plant growth regulator. Plant Physiol. **62** (1978): 571–576.
- SWANSON, D. J., and WEST, C. A.: Comparative effects of ancymidol and its analogs on growth of peas and ent-kaurene oxidation in cell-free extracts of immature *Marah macrocarpus* endosperm. Plant Physiol. **69** (1982a): 707–711.
CURRY, E. A., and WILLIAMS, M. W.: Promalin or GA_3 increase pedicel and fruit length and leaf size of "Delicious" apples treated with paclobutrazol. Hort Sci. **18** (1983): 214–215.
DAVIS, T. D., STEFFENS, G. L., and SANKHLA, N.: Triazole plant growth regulators. In: JANICK, J. (Ed.): Horticultural Reviews **10**. Timber Press, Portland, Oregon 1988, pp. 63–105.
DICKS, J. W.: Mode of action of growth retardants. In: CLIFFORD, D. R., and LENTON, R. (Eds.): Recent developments in the use of plant growth retardants. 1–14 British Plant Growth Regulator Group, Monograph 4 (1980).
FLETCHER, R. A., and ARNOLD, V.: Stimulation of cytokinin and chlorophyll synthesis in cucumber cotyledons by triadimefon. Physiol. Plant. **66** (1986): 197–201.

- and HORSTRA, G.: Triazoles as potential plant protectants. In: BERG, D., and PLEMPEL, E. (Eds.): Sterol biosynthesis inhibitors. Ellis Horwood Ltd., Cambridge 1988, pp. 321–331.
- – Improvement of uniconazole-induced protection in wheat seedlings. J. Plant Growth Regul. **9** (1990): 207–212.
- and NATH, V.: Triadimefon reduces transpiration and increases yield in water stressed plants. Physiol. Plant. **62** (1984): 422–426.
FLORES, H. E., PROTACIO, C. M., and SIGNS, M. W.: Primary and secondary metabolism of polyamines in plants. Rec. Adv. Phytochem. **23** (1989): 329–393.
FLORES-NIMEDZ, A. A., DÖRFFLING, K., and VERGARA, B. S.: Improvement of chilling resistance in rice by application of an abscisic acid analog in combination with the growth retardant tetcyclacis. J. Plant Growth Regul. **12** (1993): 27–34.
FÖRSTER, H., BUCHENAUER, H., and GROSSMANN, F.: Nebenwirkungen der systemischen Fungizide Triadimefon und Triadimenol auf Gerstenpflanzen. I. Beeinflussung von Wachstum und Ertrag. Z. Pfl. Krankh. Pflschutz **87** (1980a): 473–492.
- – – Nebenwirkungen der systemischen Fungizide Triadimefon und Triadimenol auf Gerstenpflanzen. II. Cytokininartige Effekte. Z. Pfl. Krankh. Pflanzenschutz **87** (1980b): 640–653.
- – – Nebenwirkungen der systemischen Fungizide Triadimefon und Triadimenol auf Gerstenpflanzen. III. Weitere Beeinflussungen des Stoffwechsels. Z. Pfl. Krankh. Pflanzenschutz **87** (1980c): 717–730.
FRYMIRE, R. M., and COLE, J. C.: Uniconazole effect on growth and chlorophyll content of pyracantha, photinia, and dwarf burford holly. J. Plant Growth Regul. **11** (1992): 143–148.
GROSSMANN, K.: Plant cell suspensions for screening and studying the mode of action of plant growth retardants. In: MARAMOROSH, K., and SATO, G. (Eds.): Advances in Cell Culture, Vol. 6. Academic Press, San Diego 1988, pp. 89–136.
- Screening of growth retardants and herbicides with heterotrophic plant cell suspension cultures. In: BÖGER, P., and SANDMANN, G. (Eds.): Target Assays for Modern Herbicides and Related Phytotoxic Compounds. Lewis Publ. Chelsea, Michigan, USA 1992.
- Plant growth retardants: their mode of action and benefit for physiological research. In: KARSSEN, C. M. VAN LOON, VREUGDENHILL, D. (Eds.): Progress in Plant Growth Regulation. Kluwer Publ., Dordrecht 1992, pp. 173–179.
- HÄUSER, CH., SAUERBREY, E., FRITSCH, H. J., SCHMIDT, O., and JUNG, J.: Plant growth retardants as inhibitors of ethylene production. J. Plant Physiol. **134** (1989): 538–543.
- KWIATKOWSKI, J., and HÄUSER, CH.: Phytohormonal changes in greening and senescing intact cotyledons of oilseed rape and pumpkin: influence of the growth retardant BAS 111..W. Physiol. Plant. **83** (1991): 544–550.
- – SIEBECKER, H., and JUNG, J.: Regulation of plant morphology by growth retardants: Effects on phytohormone levels in soybean seedlings determined by immunoassay. Plant Physiol. **84** (1987): 1018–1021.
- SAUERBREY, E., and JUNG, J.: Influence of growth retardants and ethylene-generating compounds on culture response of leaf explants from wheat (*Triticum aestivum* L.). J. Plant Physiol. **135** (1990): 725–731.
- SIEFERT, F., KWIATKOWSKI, J., SCHRAUDNER, M., LANGEBARTELS, C., and SANDERMANN, H., Jr.: Inhibition of ethylene production in sunflower cell suspensions by the plant growth retardant BAS 111..W: Possible relations to changes in polyamine and cytokinin contents. J. Plant Growth Regul. **12** (1993): 5–11.
GRZESIK, M., JOUSTRA, M. K., and MARCZYNSKI, S.: Effects of gibberellin A_3, paclobutrazol, chlormequat, and nutritional levels on the growth of rhododendron "Baden Baden". Gartenbauwissenschaft **57** (1992): 25–28.
HASSON, E. P., and WEST, C. A.: Properties of the system for the mixed function oxidation of kaurene and kaurene derivatives in microsomes of the immature seed of *Mara macrocarpus*: Cofactor requirements. Plant Physiol. **58** (1976a): 473–478.

- — Properties of the system for the mixed function oxidation of kaurene and kaurene derivatives in microsomes of the immature seed of *Mara macrocarpus*: electron transfer components. Plant Physiol. **58** (1976b): 479–484.
- HAUGHAN, P. A., BURDEN, R. S., LENTON, J. R., and GOAD, L. J.: Inhibition of celery cell growth and sterol biosynthesis by the enantiomers of paclobutrazol. Phytochemistry **28** (1989): 781–787.
- LENTON, J. R., and GOAD, L. J.: Sterol requirements and paclobutrazol inhibition of a celery cell culture. Phytochemistry **27** (1988): 2491–2500.
- HÄUSER, C., KWIATKOWSKI, J., RADEMACHER, W., and GROSSMANN, K.: Regulation of endogenous abscisic acid levels and transpiration in oilseed rape by plant growth retardants. J. Plant Physiol. **137** (1990): 201–207.
- HEDDEN, P., CROKER, S. J., RADEMACHER, W., and JUNG, J.: Effects of the triazole type plant growth retardant BAS 111..W on gibberellin levels in oilseed rape, *Brassica napus*. Physiol. Plant. **75** (1989): 445–451.
- and GRAEBE, J. E.: Inhibition of gibberellin biosynthesis by paclobutrazol in cell free homogenates in *Cucurbita maxima* endosperm and *Malus prunila* embryos. J. Plant Growth Regul. **4** (1985): 111–122.
- HOAD, G. V.: Hormonal regulation of fruit-bud formation in fruit trees. Acta Hortic. **149** (1984): 13–23.
- HOFSTRA, G., KRIEG, L. C., and FLETCHER, R. A.: Uniconazole reduces ethylene and 1-aminocyclopropane-1-carboxylic acid and increases spermine levels in mung bean seedlings. J. Plant Growth Regul. **8** (1989): 45–51.
- HUNTER, D. M., and PROCTOR, J. T. A.: Paclobutrazol affects growth and fruit composition of potted grapevines. Hort Science **27** (1992): 319–321.
- IZUMI, K., KAMIYA, Y., SAKURAI, A., OSHIO, H., and TAKAHASHI, N.: Studies of sites of action of a new plant growth retardants (E)-1-(4-chlorophenyl)-4,4-dimethyl-2-(1,2,4-triazol-1-yl)-1-penten-3-ol (S-3307) and comparative effects of its steroisomers in a cell free system from *Cucurbita maxima*. Plant Cell Physiol. **26** (1985): 821–827.
- NAKAGAWA, S., KOBAYASHI, M., OSHIO, H., SAKURAI, A., and TAKAHASHI, N.: Levels of IAA, cytokinins, ABA and ethylene in rice plants affected by a gibberellin biosynthesis inhibitor, uniconazole-P. Plant Cell Physiol. **29** (1988): 97–104.
- YAMAGUCHI, I., WADA, A., OSHIO, I., and TAKAHASHI, N.: Effects of a new plant growth retardants (E)-1-(4-chlorophenyl)-4,4-dimethyl-2-(1,2,4-triazol-1-yl)-penten-3-ol (S-3307) on the growth and gibberellin content of rice plants. Plant Cell Physiol. **25** (1984): 611–617.
- JUNG, J., LUIB, M., SAUTER, H., ZEEH, B., and RADEMACHER, W.: Growth regulation in crop plants with new types of triazole compounds. J. Agron. Crop Sci. **158** (1987): 324–332.
- KANE, R. T., and SMILEY, R. W.: Plant growth-regulating effects of systemic fungicides applied to Kentucky bluegrass. Agron. J. **75** (1983): 469–473.
- KHALIL, I. A., and MERCER, E. I.: Effect of diclobutrazol on the growth and sterol and photosynthetic pigment content of winter wheat. Pestic. Sci. **28** (1990): 271–281.
- — Effect of some sterol-biosynthesis-inhibiting fungicides on the biosynthesis of polyisoprenoid compounds in winter wheat seedling. Phytochemistry **29** (1990): 417–424.
- — and WANG, Z. X.: Effect of triazole fungicides on the growth, chloroplast pigments and sterol biosynthesis of maize (*Zea mays* L.). Plant Science **66** (1990): 21–28.
- KÖLLER, W.: Isomers of sterol synthesis inhibitors: fungicidal effects and plant growth regulator activities. Pestic. Sci. **18** (1987): 129–147.
- KONSTANTINIDOU-DOLTSINI, S.: Einfluß der systemischen Fungizide Nuarimol und Imazalil auf den Stoffwechsel von Weizenpflanzen. Diss., Univ. Stuttgart-Hohenheim 1981.
- BUCHENAUER, H., and GROSSMANN, F.: Einfluß der systemischen Fungizide Nuarimol und Imazalil auf Wachstum und Stoffwechsel von Weizenpflanzen. I. Wachstum. Z. Pfl. Krankh. Pflschutz **93** (1986a): 369–379.

– – – Einfluß der systemischen Fungizide Nuarimol und Imazalil auf Wachstum und Stoffwechsel von Weizenpflanzen. II. N-Metabolismus. Z. Pfl. Krankh. Pflschutz **93** (1986b): 614–623.

KRÄMER, W., BÜCHEL, K. H., and DRABER, W.: Structure-activity correlation in the azoles. In: MIYAMOTO, J., and KEARNEY, P. C. (Eds.): Pesticide Chemistry. Vol. I. Pergamon, Oxford 1983, pp. 223–232.

KRAUS, T. E., MURR, D. P., and FLETCHER, R. A.: Uniconazole inhibits stress-induced ethylene in wheat and soybean seedlings. J. Plant Growth Regul. **10** (1991): 229–234.

– – HOFSTRA, G., and FLETCHER, R. A.: Modulation of ethylene synthesis in acotyledonous soybean and wheat seedlings. J. Plant Growth Regul. **11** (1992): 47–53.

KUTZNER, B., and BUCHENAUER, H.: Effect of various triazole fungicides on growth and lipid metabolism of *Fusarium moniliforme* as well as on gibberellin contents in fungus filtrates. Z. Pfl. Krankh. Pflschutz **93** (1986): 597–607.

LANGEBARTELS, C., KERNER, K., LEONARDI, S., SCHRAUDNER, M., TROST, M., HELLER, W., and WANDERMANN, H.: Biochemical plant responses to ozone. I. Differential induction of polyamine and ethylene biosynthesis in tobacco. Plant Physiol. **95** (1991): 882–889.

LENTON, J. R., HEDDEN, P., and GALE, M. D.: Gibberellin insensitivity and depletion in wheat – consequences for development. In: HOAD, G. V., LENTON, J. R., JACKSON, M. B., and ATKIN, R. K. (Eds.): Hormone Action in Plant Development: A Critical Appraisal. Butterworths, London 1987, pp. 145–160.

LEOPOLD, A. C., and WRIGHT, W. L.: An apparent antagonist of gibberellin. Plant Physiol. **46** (1970): 19.

– Antagonism of some gibberellin actions by a substituted pyrimidine. Plant Physiol. **48** (1971): 537–540.

LEVER, B. G., SHEARING, S. J., and BATCH, J. J.: PP 333 – a new spectrum growth retardant: British Crop Protection Conference – Weeds 1982, Vol. 1., British Crop Protection Council, Croydon, England (1982): pp. 3–10.

LI, N., PARSONS, B. L., LIU, D., and MATTOO, A. K.: Accumulation of wound inducible ACC synthase transcript in tomato fruit is inhibited by salicylic acid and polyamines. Plant Mol. Biol. **18** (1992): 447–487.

LUCKWILL, L. C.: The control of growth and fruitfulness of apple trees. In: LUCKWILL, L. C., and CUTTING, C. V. (Eds.): Physiology of tree crops. Academic Press, London 1970, pp. 237–253.

LÜRSSEN, K.: Physiological effects of triazole PGRs in relation to their biochemical mode of action. Pestic. Sci. **21** (1987): 310–311.

– Triazole plant growth regulators: effects and mode of action. In: BERG, D., and PLEMPEL, M. (Eds.): Sterol biosynthesis inhibitors. Pharmaceutical and Agrochemical Aspects. VCH, Weinheim 1988, pp. 305–320.

– and REISER, W.: Triapenthenol – a new plant growth regulator. Pestic. Sci. **19** (1987): 153–164.

LUSTER, D. G., and MILLER, P. A.: Triazole plant growth regulator binding to native and detergent-solubilized plant microsomal cytochrome P450. Pestic. Biochem. Physiol. **46** (1993): 27–39.

MACKAY, C. E., HALL, J. CH., HOFSTRA, G., and FLETCHER, R. A.: Uniconazole induced changes in abscisic acid, total amino acids, and proline in *Phaseolus vulgaris*. Pestic. Biochem. Physiol. **37** (1990): 74–82.

MAILLOT-VERNIER, P., SCHALLER, H., BENVENISTE, P., and BELLIARD, G.: Biochemical characterization of a sterol mutant plant regenerated from a tobacco callus resistant to a triazole cytochrome P450-obtusifoliol-14-demethylase inhibitor. Biochem. Biophys. Res. Common. **165** (1989): 125–130.

– – – – In vitro selection of calli resistant to a triazole cytochrome P450-obtusifoliol-14-demethylase inhibitor from protoplasts of *Nicotiana tabacum* L. cv. *Xanthi*. Plant Physiol. **93** (1990): 1190–1195.

McDaniel, G. L.: Growth retardation activity of paclobutrazol on chrysanthemum. Hort Sci. **18** (1983): 199–200.

Menhenett, R., and Hanks, G. R.: Comparisons of a new triazole retardant PP 333 with ancymidol and other compounds on pot-grown tulips. Plant Growth Regul. **1** (1982/83): 173–181.

Miki, T., Kamiya, Y., Fukazawa, M., Ichikawa, T., and Sakurai, A.: Sites of inhibition by a plant-growth regulator, 4′-chloro-2′-(α-hydroxybenzyl)-isonicotinanilide (inabenfide) and its related compounds in the biosynthesis of gibberellins. Plant Cell Physiol. **31** (1990): 201–206.

Mitsunaga, S., and Yamaguchi, J.: Induction of α-amylase is repressed by uniconazole, an inhibitors of the biosynthesis of gibberellin, in a dwarf mutant of rice, Waito-c. Plant Cell Physiol. **34** (1993): 243–249.

Montague, M. J.: Inhibition of gibberellic acid-induced elongation in *Avena* stem segments by a substituted pyrimidine. Plant Physiol. **56** (1975): 167–170.

Nitsche, K., Grossmann, K., Sauerbrey, E., and Jung, J.: Influence of the growth retardant tetcyclasis on cell division and cell elongation in plants and cell cultures of sunflower, soybean, and maize. J. Plant Physiol. **118** (1985): 209–218.

Nooden, L. D., and Leopold, A. C. (Eds.): Senescence and Aging in Plants. Academic Press, San Diego 1988.

Norman, S. M., Bennett, R. D., Poling, S. M., Maier, V. P., and Nelson, M. D.: Paclobutrazol inhibits abscisic acid biosynthesis in *Cercospora rosicola*. Plant Physiol. **80** (1986): 122–125.

– Poling, S. M., Maier, V. P., and Orme, E. D.: Inhibition of abscisic acid biosynthesis in *Cercospora rosicola* by inhibitors of gibberellin biosynthesis and plant growth retardants. Plant Physiol. **71** (1983): 15–18.

– – – and Pon, D. L.: Abscisic acid biosynthesis in *Cercospora rosicola*: Sensitivity to inhibitors of sterol biosynthesis. Agric. Biol. Chem. **52** (1988): 1309–1310.

Pennypacker, B. W., Sanders, P. L., Gregory, L. V., Gilbride, E. P., and Cole, H. Jr.: Influence of triadimefon on the foliar growth and flowering of annual bluegrass. Can. J. Plant Pathol. **4** (1982): 259–262.

Pharis, R. P., and King, R. W.: Gibberellins and reproductive development in seed plants. Annu. Rev. Plant Physiol. **36** (1985): 517–568.

Phinney, B. O.: Gibberellin A_1, dwarfism and the control of shoot elongation in higher plants. In: Crozier, A., and Hillman, J. R. (Eds.): The biosynthesis and metabolism of plant hormones. 17–41. Soc. Exp. Biol. Sem. Ser. 23. Cambridge Univ. Press 1984.

Rademacher, W.: Gibberellins: metabolic pathways and inhibitors of biosynthesis. Target Sites of Herbicide Action (1989): 127–145.

– Fritsch, H., Graebe, J. E., Sauter, H., and Jung, J.: Tetcyclacis and triazole-type plant growth retardants: their influence on the biosynthesis of gibberellins and other metabolic processes. Pestic. Sci. **21** (1987): 241–252.

– and Graebe, J. E.: Gibberellin A_4 produced by *Sphaceloma manihoticola*, the cause of the superelongation disease of cassava (*Manihot esculenta*). Biochem. Biophys. Res. Commun. **91** (1979): 35–40.

– and Jung, J.: Comparative potency of various synthetic plant growth retardants on the elongation of rice seedlings. Z. Acker- und Pflanzenbau **150** (1981): 363–371.

– – Graebe, J. E., and Schwenen, L.: On the mode of action of tetcyclacis and triazole growth retardants. In: Menhenett, R., and Lawrence, D. K. (Eds.): Biochemical Aspects of Synthetic and Naturally Occurring Plant Growth Regulators. Monograph No. 11. British Plant Growth Regulator Group, Wantage, England 1984, pp. 1–11.

Radice, M., and Pesci, P.: Effect of triazole fungicides on the membrane permeability and on FC-induced H^+-extrusion in higher plants. Plant Sci. **74** (1991): 81–88.

REED, A. N., and BUCHANAN, D. A.: Translocation and metabolism of BAS 111 in apple seedlings. HortScience 25 (1990): 324–326.
− CURRY, E. A., and WILLIAMS, M. W.: Translocation of triazole growth retardants in plant tissue. J. Amer. Soc. Hortic. Sci. 114 (1989): 893–898.
ROBERTS, R. L., BOWERS, B., SLATER, M. L., and CABIB, E.: Chitin synthesis and localization in cell division cycle mutants of *Saccharomyces cerevisiae*. Mol. Cell. Biol. 3 (1983): 922–930.
SAUERBREY, E., GROSSMANN, K., and JUNG, J.: Influence of growth retardants on the internode elongation and ethylene production of sunflower plants. Physiol. Plant. 70 (1987): 8–12.
− − − Is ethylene involved in the regulation of growth of sunflower cell suspension cultures? J. Plant Physiol. 127 (1987): 471–479.
− − − Ethylene production by sunflower cell suspensions. Effect of plant growth retardants. Plant Physiol. 87 (1988): 510–513.
SCHALLER, H., MAILLOT-VERNIER, P., BELLIARD, G., and BENVENISTE, P.: Increased sterol biosynthesis in tobacco calli resistant to a triazole herbicide which inhibits demethylation of 14α-methyl sterols. Planta 187 (1992): 315–321.
SCHMERDER, B., and BORRIS, H.: Induction of nitrate reductase by cytokinin and ethylene in *Agrostemma githago* L. embryos. Planta 169 (1986): 589–593.
SCHOTT, P. E., KNITTEL, H., and KLAPPROTH, H.: Tetcyclasis: a new bioregulator for improving the development of young rice plants. In: ORY, R. L., and RITTIG, F. R. (Eds.): Bioregulators: Chemistry and Uses. American Chemical Society, Washington, D.C., 1984, pp. 45–63.
SANKLA, N., DAVIS, T. D., GEHLOT, H. S., UPADHYAYA, A., SANKLA, A., and SANKLA, D.: Growth and organogenesis in moth bean callus cultures as influenced by triazole growth regulators and gibberellic acid. J. Plant Growth Regul. 10 (1991): 41–45.
SENARATNA, T., MACKAY, C. E., MCKERSIE, B. D., et al.: Uniconazole-induced chilling tolerance in tomato and its relationship to antioxidant content. J. Plant Physiol. 133 (1988): 56–61.
SHIVE, J. B., and SISLER, H. D.: Effects of ancymidol (a growth retardant) and triarimol (a fungicide) on the growth, sterols and gibberellins of *Phaseolus vulgaris* (L.). Plant Physiol. 57 (1976): 640–644.
SINGH, V. P.: Uniconazole (S-3307) induced cadmium tolerance in wheat. J. Plant Growth Regul. 12 (1993): 1–3.
SLOCUM, R. D., KAUR-SAWHNEY, R., and GALSTON, A. W.: The physiology and biochemistry of polyamines in plants. Arch. Biochem. Biophys. 235 (1984): 283–303.
SRIVASTAVA, H. S., and FLETCHER, R. A.: Triadimenol increases nitrate levels and nitrate reductase activity in canola leaves. J. Exp. Bot. 43 254. (1992): 1267–1271.
STEFFENS, G. L.: Gibberellin biosynthesis inhibitors: Comparing growth-retarding effectiveness on apple. J. Plant Growth Regul. 7 (1988): 27–36.
− BYUN, and WANG, S. Y.: Controlling plant growth via the gibberellin biosynthesis system – I. Growth parameter alterations in apple seedlings. Physiol. Plant. 63 (1985): 163–168.
− LIN, J. T., STAFFORD, A. E., METZGER, J. D., and HAZEBROEK, J. P.: Gibberellin content of immature apple seeds from paclobutrazol-treated trees over three seasons. J. Plant Growth Regul. 11 (1992): 165–170.
− STAFFORD, A. E., and LIN, J. T.: Influence of an inhibitor of gibberellin biosynthesis, paclobutrazol, on apple seed gibberellin content. Physiol Plant. 83 (1991): 366–372.
STERRET, J. P.: XE-1019: Plant response, translocation, and metabolism. J. Plant Growth Regul. 7 (1988): 19–26.
− and TWORKOSKI, T. J.: Flurprimidol: plant response, translocation and metabolism. J. Am. Soc. Hortic. Sci. 112 (1987): 341–345.
− − Response of shade trees to root collar drenches of inhibitors flurprimidol and paclobutrazol. J. Plant Growth Regul. 5 (1987): 163–167.
STREIT, L., MOREAU, M., GAUDIN, J., EBERT, E., and VAN DEN BOSSCHE, H.: A novel imidazole carboxylic acid ester is a herbicide inhibiting 14α-methyl demethylation in plant sterol biosynthesis. Pestic. Biochem. Physiol. 40 (1991): 162–168.

SUGAVANAM, B.: Diastereoisomers and enantiomers of paclobutrazol: their preparation and biological activity. Pesticide Sci. **15** (1984): 296–302.

SWIETLIK, D., and MILLER, S.: The effect of paclobutrazol on growth and response to water stress of apple seedlings. J. Amer. Soc. Hort. Sci. **108** (1983): 1076–1080.

TATON, M., and RAHIER, A.: Properties and structural requirements for substrate specificity of cytochrome P-450-dependent obtusifoliol 14α-demethylase from maize (*Zea mays*) seedlings. Biochem. J. **277** (1991): 483–492.

– ULLMANN, P., BENVENISTE, P., and RAHIER, A.: Interaction of triazole fungicides and plant growth regulators with microsomal cytochrome P-450-dependent obtusifoliol-14α-methyl demethylase. Pest. Biochem. Physiol. **30** (1988): 178–189.

TSCHABOLD, E. E., TAYLOR, H. M., DAVENPORT, J. D., HACKLER, R. E., KRUNKALNS, E. V., and MEREDITH, W. S.: A new plant growth regulator. Plant Physiol. **46** (1970): 19.

VAN DEN BOSSCHE, H., MARICHAL, P., WILLEMSENS, G., and JANSSEN, P. A. J.: Effects of inhibitors on the P450-dependent metabolism of endogenous compounds in fungi, protozoa, plants and vertebrates. In: ARINC, E. *et al.* (Eds.): Molecular Aspects of Monooxygenases and Bioactivation of Toxic Compounds. Plenum Press, New York 1991, pp. 345–363.

VAN STADEN, J., FEATONBY-SMITH, B. C., MAYAK, S., SPIEGELSTEIN, H., and HALEVY, A. H.: Cytokinins in cut carnation flowers. II. Relationship between endogenous ethylene and cytokinin levels in the petals. Plant Growth Regul. **5** (1987): 75–86.

VAVRINA, C. S., and PHATAK, S. C.: Efficacy of triapenthenol as a safener against metribuzin injury in soybean (*Glycine max*) cultivars. J. Plant Growth Regul. **7** (1988): 67–75.

VU, J. C. V., and YELENOSKY, G.: Growth and photosynthesis of sweet orange plants treated with paclobutrazol. J. Plant Growth Regul. **11** (1992): 85–89.

WAMPLE, R. L., and CULVER, E. B.: The influence of paclobutrazol, a new growth regulator on sunflowers. J. Amer. Soc. Hort. Sci. **108** (1983): 122–125.

WANG, L. H., and LIN, C. H.: The effect of paclobutrazol on physiological and biochemical changes in the primary roots of pea. J. Experim. Bot. **43** (1992): No. 255, 1367–1372.

WANG, S. Y., and FAUST, M.: Effect of growth retardants on root formation and polyamine content in apple seedlings. J. Amer. Soc. Hort. Sci. **111** (1986): 912–917.

– and STEFFENS, G. L.: Effect of paclobutrazol on water stress-induced ethylene biosynthesis and polyamine accumulation in apple seedlings leaves. Phytochemistry **24** (1985): 2185–2190.

– – Effect of paclobutrazol on accumulation of organic acids and total phenols in apple wood. J. Plant Growth Regul. **6** (1987): 209–213.

WANG, S. Y., SUN, T., WHITAKER, B. D., and FAUST, M.: Effect of paclobutrazol on membrane lipids in apple seedlings. Physiol. Plant. **73** (1988): 560–564.

ZEEVAART, J. A. D.: Inhibition of stem growth and gibberellin production in *Agrostemma githago* L. by the growth retardant tetcyclasis. Planta **166** (1985): 276–279.

– Gage, D. A., and CREELMAN, R. A.: Recent studies of the metabolism of abscisic acid. In: PHARIS, R. P., and ROOD, S. B. (Eds.): Plant Growth Substances 1988. Springer-Verlag, Berlin. 1990, pp. 233–240.

13.2. Resistance of fungi to sterol demethylation inhibitors

With respect to their resistance risk, the fungicides may be classified according to GEORGOPOULOS (1985) into the following groups:
- low risk fungicides: development of resistance is considered as very unlikely; this is true for multisite inhibitors, e.g., sulphur, copper, dithiocarbamates, chlorothalonil and others
- moderate risk fungicides: resistance development is gradual and loss of activity is transitory; this group includes the following fungicides: dodine, guazatine, 2-aminopyrimidines, fentins, phosphorothiolates and ergosterol biosynthesis inhibitors
- high risk fungicides: the probability of emergence of resistance is high and often long lasting in the fungal population; fungicides with these properties are, e.g., benzimidazoles, acylalanines, carboxamides, aromatic hydrocarbons and dicarboximides

The systemic fungicides characterized by high selectivity and specificity, belonging to the groups of moderate and high risk fungicides, differ in their resistance risk. Resistance development to high risk fungicides is determined by a major gene and mutations result in high resistance levels. If a new high risk fungicide is introduced in practice and repeatedly applied on a large scale, serious problems may arise, especially when the fitness of the resistant population is not inferior to that of the wild type population.

Within the group of moderate risk fungicides, in general, resistance is controlled by several genes. Mutations of a single gene usually result in a low degree of resistance and resistance levels may increase by additional mutations of other genes (multistep mutations). Changes in resistance of the population are gradual and resistance problems to this fungicide group are less serious (GEORGOPOULOS 1985).

Fungal strains resistant to sterol demethylation inhibitors (DMIs) can easily be obtained in laboratory experiments by selection of untreated or mutagen-treated (either following irradiation with ultra violet light or treatment with N-methyl-N'-nitro-N-nitrosoguanidine) spores on agar media containing lethal concentrations of DMIs: *Cladosporium cucumerinum* (SHERALD et al. 1973; FUCHS, et al. 1977; BUCHENAUER and RÖHNER 1977), *Aspergillus nidulans* (DE WAARD and SISLER 1976; DE WAARD and GIESKES 1977), *Ustilago maydis* (BARUG and KERKENAAR 1979; WELLMANN and SCHAUZ 1992, 1993), *Botrytis cinerea* (LEROUX and GREDT 1984), *Pseudocercosporella herpotrichoides* var. *herpotrichoides* (KLEIN 1991). Irradiation of spore suspensions of a prochloraz-resistant strains of *P. herpotrichoides* var. *herpotrichoides*, exhibiting a middle degree of resistance, and selection on prochloraz-containing nutrient medium resulted in mutants with a higher degree of prochloraz resistance. Higher levels of resistance have also been obtained by sequential selection of fen-mutants of *Nectria haematococca* var. *cucurbitae* (KALAMARAKIS et al. 1991) and *Penicillium italicum* (DE WAARD and VAN NISTELROOY 1988). Generally, the resistant mutants are cross resistant to most of the DMIs (SHERLAND et al. 1973; DE WAARD and SISLER 1976; BUCHENAUER and RÖHNER 1977; BARUG and KERKENAAR 1979; KLEIN 1991). However, some mutants in different fungal species did not exhibit cross resistance. For instance, imazalil-resistant strains of *C. cucumerinum* (FUCHS et al. 1977; VAN TUYL 1977), *Phialophora cinerescens* (VAN TUYL 1977) and *A. nidulans* (DE WAARD and GIESKES 1977) proved to be sensitive to fenarimol. Similar results have been reported in a number of mutants of *U. maydis* (BARUG and KERKENAAR 1979, 1984).

Resistant mutants of *N. haematocca* var. *cucurbitae* — isolates to triadimenol exhibiting mutations at the tri-1 locus showed a high degree of resistance to triadimenol and were cross resistant to other triazole fungicides (e. g., triadimefon, bitertanol and propiconazol) but not to imidazole derivatives (e.g., imazalil, prochloraz) (KALAMARAKIS *et al.* 1991).

Studies of FUCHS and DE VRIES (1984) revealed that the diastereomers of some triazole fungicides displayed a differential toxicity against wild-type and resistant strains of *C. cucumerinum*. While the activity ratio of the diastereomers of bitertanol, diclobutrazol and etaconazol was almost 1 for the wild-type strain, this ratio increased as the degree of resistance of the mutants increased. On the other hand, the activity ratio of the diastereomers of triadimenol towards the wild was 1:36 and the ratio became progressively smaller with increasig resistance of the fungal strains.

Two isolates of *C. cucumerinum* widely differing in their sensitivity towards DMIs (the resistant isolate had a greater saturated/unsaturated fatty acid ratio) were tested regarding their sensitivity to triadimenol enantiomers (CARTER *et al.* 1989). The sensitive isolate of *C. cucumerinum* responded most sensitive to the 1S, 2R enantiomer of triadimenol, followed by the 1R, 2R; 1R, 2S and 1S, 2S enantiomers. These differences in mycelium growth showed good correlation with the inhibition of ergosterol synthesis and the accumulation of C-14α-methyl sterols, whereas even at the highest concentration tested none of the enantiomers showed any pronounced effects on growth or sterol composition of the resistant strain (CARTER *et al.* 1989).

After irradation of spores of a wild-type isolate of *Saccharomyces lipolytica* with UV-light and selection on triadimenol containing nutrient medium, a triadimenol resistant strain was obtained containing ergosterol as the main sterol component. However, triadimenol no longer interfered in C-14 demethylation. Cross resistance studies revealed that the l-enantiomer of triapenthenol showed positively correlated cross resistance to the triadimenol resistant strain while the d-enantiomer of triapenthenol proved to be more sensitive to the resistant strain than to the wild-type strain of *S. lipolytica* (KRÄMER *et al.* 1987).

Mutations in fungi for resistance to DMIs are sometimes associated to pleiotropic effects, which are defined as a single gene mutation affecting besides resistance to a toxicant also other characteristics of the cell (GEORGOPOULOS 1977). Resistance to DMIs in fungi was often associated with reduced pathogenicity and fitness. The degree of resistance of triforine and triarimol-resistant strains of *C. cucumerinum* appeared to be negatively correlated with their pathogenicity (FUCHS and VIETS-VERVEIJ 1975; FUCHS *et al.* 1977). Reduced fitness has also been observed with mutants of *A. nidulans*. The degree of resistance was found to be inversely proportional to germ tube elongation and mycelium growth. Mutant strains of *C. cucumerinum* (SHERALD and SISLER 1975; FUCHS and DRANDAREVSKY 1976; BUCHENAUER and RÖHNER 1977) and *A. nidulans* (DE WAARD and SISLER 1976; DE WAARD and GIESKES 1977) showed a diminished capacity in spore production and the spores produced were often not viable (SHERALD and SISLER 1975; FUCHS and DRANDAREVSKY 1976). Prochloraz-resistant laboratory mutants of *P. herpotrichoides* var. *herpotrichoides* often differed from the wild-type strain on fungicide-free nutrient medium in mycelium growth rate and spore production and conidiophores were partly deformed and their number compared to the wild type was often significantly reduced. The resistant strains showed in laboratory studies a lower competitive ability than the wild type strain when spore mixtures were used (KLEIN 1991). However, resistance to DMIs was not always accompanied by reduced pathogenicity. Mutants of *N. haematococca* var. *cucurbitae* selected on fenarimol, nine resistant genes have been recognized. The fen-mutants were inferior in conidia production, spore germination, germ tube elongations and showed a somewhat higher sensitivity to elevated temperatures (KALAMARAKIS *et al.* 1991). Mutants of *P. expansum* resistant to imazalil (VAN TUYL 1977) and *P. italicum* resistant to fenarimol (DE WAARD and VAN NISTELROOY

1984) were not inferior in their virulence compared to the wild-type strains. Triadimenol-resistant strains of *N. haematococca* var. *cucurbitae* (resistance was due to a mutation at the tri-1 locus) showed no adverse effects on growth, sporulation and virulence for squash seedlings (KALAMARAKIS *et al.* 1991).

Amongst imazalil-resistant mutants of *A. nidulans,* some strains showed a higher sensitivity to unrelated fungitoxic agents, e. g., acriflavine, cycloheximide, chloramphenicol and neomycin as well as to lower temperatures (VAN TUYL 1977). Furthermore, several imazalil and fenarimol resistant strains of *A. nidulans* were more sensitive to 8-azaguanide, p-fluorophenylalanine, D-serine and thiourea (DE WAARD and VAN NISTELROOY 1979) and to primaricin (DE WAARD and SISLER 1976). Triadimenol-resistant mutants of *B. cinerea* showed a higher sensitivity to cycloheximide than the wild-type strain (LEROUX and GREDT 1978 b). Some of the *U. maydis* (KERKENAAR and BARUG 1984) and *P. italicum* mutants (DE WAARD and NISTELROOY 1984) exhibited an increased sensitivity to fenpropimorph. Prochloraz-resistant strains of *P. herpotrichoides* var. *herpotrichoides*, especially those with a higher degree of resistance, were more sensitive to fenpropimorph (KLEIN 1991). On the other hand, triadimefon-resistant *U. avenae* strains exhibited a high degree of cross resistance to the inhibitor of chitin synthesis, nikkomycin (KRÄMER *et al.* 1987).

VAN TUYL (1977) studied the genetics of resistance in *A. nidulans* to imazalil. Eight loci allocated to six different linkage groups were identified in an analysis of 21 imazalil resistant mutants. Two other genes for resistance to cycloheximide impart also resistance to imazalil. Mutations at particular loci occurred more frequently than at other sites. The large number of at least 10 different genes involved in imazalil resistance indicates that the loci conferring resistance are distributed over the genome of *A. nidulans.* Allelic mutations in a single locus often resulted in different degrees of resistance, indicating that mutations in a single gene may differ. Generally, the levels of resistance to imazalil were low (VAN TUYL 1977). These findings imply that several alterations at certain sites in the fungal cell may occur.

Studies of BUTTERS *et al.* (1986) indicated that in recombination experiments with DMI-sensitive and DMI-resistant *Erysiphe graminis* f. sp. *hordei*-isolates resistance was under multiple gene control. Also results with *Venturia inaequalis*-isolates indicated that more than one gene controlled sensitivity to fusilazole and myclobutanil (SHOLBERG and HAAG 1993). Polygenic resistance to DMI-fungicides has also been found in *N. haematococca* var. *cucurbitae* (KALAMARAKIS *et al.* 1991).

Investigations with heterozygous imazalil-resistant strains of *A. nidulans* showed that the degree of resistance of such strains was intermediate between sensitive diploid and haploid strains. By combining different single gene mutations for imazalil resistance the recombinant strain showed a higher level of resistance, indicating additive interaction of the single gene mutations (VAN TUYL 1977). In mutants of *N. haematococca* var. *cucurbitae* (resistance was due to fen-mutations), exhibiting small degrees of resistance to DMIs, levels of resistance were increased by recombinations (KALAMARAKIS *et al.* 1991).

The decreased fitness or pathogenicity in resistant fungal strains to DMIs which might be regarded also as pleiotropic effect may be considered as a mechanism of resistance (DE WAARD and VAN NISTELROOY 1979, 1980). Resistance in *A. nidulans* to fenarimol was attributed to differences in uptake of the toxicant by the wild-type and resistant strains. The wild-type strain rapidly accumulated fenarimol during the 10 min of incubation followed by a slow release of the compound and with extended incubation period an equilibrium was reached. It was suggested that the amount of fenarimol accumulating during the initial phase is sufficient for saturation and, thus, for inhibition of the target site. While the influx proved to be a passive process, efflux which is inducible was regarded as an energy-dependent process.

On the other hand, the resistant strains of *A. nidulans* took up the fungicide only in a constant low level during incubation. It was assumed that the low concentrations of the toxicants in the fungal cells from the beginning of exposure is the result of a high energy-dependent efflux activity. Because of the high efflux rate of fenarimol in resistant strains, inhibitory concentrations are not reached at the target site. Efflux may be inhibited by chemicals that interfere in plasma membrane ATPase activity (e.g., Na VO4), uncouple oxidative phosphorylation (carbonyl cyanide m-chlorophenylhydrazone; CCCP), inhibit respiration (N,N-dicyclohexylcarbodiimide; DCCD), inhibit nonspecifically cell metabolism (e.g., cationic agents, $CuSO_4$, cetylpyridinium bromide; CPB) or by low temperature. (DE WAARD and VAN NISTELROOY 1979, 1980, 1982). It was suggested that the decreased pathogenicity or fitness of the resistant strains that are characterized by a high energy-dependent efflux activity may be attributed to higher constitutive membrane transport activities that will have impaired effects on other cell processes.

Resistance to DMIs based on an increased energy dependent efflux has also been found in *Penicillium italicum* (DE WAARD and VAN NISTELROOY 1984; DE WAARD and VAN NISTELROOY 1988), *Monilia fructicola* (NEY 1988) and *N. haematococca* var. *cucurbitae* (KALAMARAKIS et al. 1991).

These results indicate that this resistance mechanism may be of general significance in resistant DMI mutants obtained in laboratory studies. It is of interest whether the energy dependent efflux mechanism is involved in resistance to DMIs of phytopathogenic fungi occuring in practice. It may be assumed that mutants with decreased fitness or virulence will not readily develop in the field (DE WAARD and FUCHS 1982).

Another resistance mechanism to DMIs has been proposed for *Saccharomyces cerevisiae, Candida albicans* and *Ustilago maydis* concerning target mutations at the cytochrome P-450 14α-demethylase. The mutations leading to alterations of conformation of the enzyme may result in reduced or complete loss of affinity of DMIs to the cytochrome P-450 14α-demethylase. (AOYAMA et al. 1987; HITCHCOCK et al. 1987; WALSH et al. 1982; KRÄMER et al. 1987). The defective enzyme, however, is also unable to catalyze the C-14 demethylation and this results in markedly reduced vitality of the mutants.

Sensitivity of sterol 14α-demethylase activity in cell-free extracts of *Penicillium italicum* of the wild-type isolate and isolates with increasing degrees of resistance to imazalil was studies by GUAN and DE WAARD (1993). However, the results indicated that affinity of sterol 14α-demethylase of wild-type and DMI-resistant isolates to imazalil did not correlate with resistance.

A third resistance mechanism based on a single gene mutation at sterol $C5-6$ desaturation of the ergosterol biosynthesis pathway has been reported by WATSON et al. (1988, 1989). Analysis of nine ketoconazole-resistant strains of *S. cerevisiae* showed that resistance of all strains was due to a single gene mutation in the ergosterol biosynthesis pathway, however not at the target steps of lanosterol 14α-demethylation but at $C5-6$ desaturation. In the presence of azole treatment (e.g., fluconazole), the mutants accumulated 14-methyl fecosterol, resulting from the inhibition of 14α-demethylation and the mutation block at $C 5-6$ desaturation. 14-methyl fecosterol provides a functional sterol for growth, this compound, accumulating in the DMI-resistant strains, may either substitute the depleted ergosterol or stabilize the membrane and protect it against disruptive effects of sterol precursors. On the other hand, in the wild-type strain, possessing a functional $C 5-6$ desaturase, 14-methyl-3,6-diol is accumulated in the presence of azoles besides lanosterol, both sterols accumulated exhibit detrimental effects on membranes. It is concluded that the accumulation of 14-methyl fecosterol is the basis of azole resistance.

A fourth mechanisms has been described in *Ustilago avenae*-sporidia by rapid removal of the fungicide from the target through accumulation in the vacuole. In DMI-resistant sporidia, a pronounced accumulation of triadimenol in lipid-rich vacuoles has been detected, by which the cytoplasmic triazole concentration was simultaneously reduced (HIPPE 1987).

Structural and biochemical modifications in the cell wall have been proposed as a further mechanism of resistance by WELLMANN and SCHAUZ (1993). Resistance to triadimefon in mutants of *Ustilago maydis* was only detected in intact sporidia. The permeability of the plasma membrane of protoplasts from sensitive and resistant strains was affected by the triazole fungicide to the same extent. It was suggested that the cell wall might play an important role as a factor of resistance mechanism to triadimefon. This assumption was supported by the finding that the enzymatic cell wall lysis of the resistant sporidia required more time than that of the sensitive ones and it was suggested that the altered cell wall might represent a "penetration barrier" to the fungicide and that an enhanced binding of the compound to the cell wall may occur.

Detoxification of DMIs by fungi as an additional resistance mechanism has been described for resistant strains of *C. cucumerinum* (FUCHS and DE VRIES 1984) and *P. italicum* (GUAN et al. 1992).

Finally, an inducible mechanism of resistance has been assumed by SMITH and KÖLLER (1990) in mutants of *U. avenae*. Treatment with DMIs induced at the beginning similar effects both in sensitive and resistant sporidia: accumulation of sterol precursors, ergosterol depletion, growth inhibition and morphological alterations. Contrary to the sensitive strain, these effects were only transient in the resistant sporidia; after several hours of exposure of fungicide treatment, growth and ergosterol biosynthesis normalized. This recovery from the DMI action was dependent on a de novo protein synthesis and probably on the activation of "resistance genes" induced either by the inhibitor itself or by the initially accumulated sterol precursors.

The findings that fungal strains resistant to DMIs obtained in laboratory experiments are inferior to wild-type strains concerning their fitness and virulence might be of great significance with respect to development of resistance in practice, since in absence of selection pressure, generally, the competitive ability of resistant strains is lower compared to the wild type strains. Furthermore, it was concluded that development of resistance to DMIs in practice would be retarded since the majority of resistant strains might be controlled by recommended fungicide dosages (FUCHS and DRANDAREVSKY 1976; BROWN and HALL 1979). However, there are indications that laboratory strains of *P. expansum* and *P. italicum* resistant to imazalil and fenarimol, respectively, (VAN TUYL 1977; DE WAARD and VAN NISTELROOY 1984) as well as *P. herpotrichoides* var. *herpotrichoides* mutants resistant to prochloraz (KLEIN 1991) did not differ appreciably from the wild-type strains in their pathogenicity.

Decreased sensitivity of powdery mildew of cucumber (*Sphaerotheca fuliginea*) as well as of barley (*Erysiphe graminis* f. sp. *hordei*) and wheat (*Erysiphe graminis* f.sp. *tritici*) powdery mildew to DMIs in greenhouse and in the field has been described. In the Netherlands, for control of cucumber powdery mildew in the greenhouses triforine as the first DMI has been applied from 1972 to 1977. After introduction of imazalil and fenarimol in 1977 and 1981, respectively, the use of DMIs increased. Monitoring experiments in 1981 revealed that the sensitivity of *S. fuliginea* was already lower than that of control isolates and during 1982 and 1983, the number of isolates with decreased sensitivity to DMIs further increased. Under high disease pressure, triforine failed to control cucumber powdery mildew, the efficacy of imazalil was limited and normal application rates of fenarimol still controlled the resistant isolates (SCHEPERS 1983, 1984, 1985b). Resistance of *S. fuliginea* to DMIs had also been reported by HUGGENBERGER (1984). Positively correlated cross resistance of *S. fuliginea* isolates to other DMIs (e.g., bitertanol, buthiobate) was found (SCHEPERS 1983; HUGGENBERGER 1984). Resistant greenhouse isolates of *S. fuliginea* showed no reduced competitive ability compared to the reference isolates (SCHEPERS 1985a).

With the continuing use of DMIs to control powdery mildew of barley and wheat in various western European countries, the population of powdery mildews became increasingly less sensitive during 1981 and 1984 (FLETCHER and WOLFE 1981; WOLFE et al. 1982, 1983, 1984; LIMPERT and FISCHBECK 1983; BENNETT and VAN KINTS 1983; BUTTERS et al. 1984; BUCHENAUER 1983; BUCHENAUER et al. 1984a; BUCHENAUER and HELLWALD 1985; DE WAARD et al. 1984).

Isolates of powdery mildew of barley and wheat with decreased sensitivity to triadimefon showed cross resistance to other DMIs, such as triadimenol, propiconazole, diclobutrazol, flusilazol, prochloraz and nuarimol. The lack of cross resistance to the morpholine derivatives (tridemorph and fenpropimorph) agreed with the different site of interference of these chemicals in the sterol pathway (KATO et al. 1980; KERKENAAR et al. 1981). No cross sensitivity to pyrazophos was detected (BUCHENAUER et al. 1984; BUCHENAUER and HELLWALD 1985). The decrease in sensitivity of barley powdery mildew isolates to DMIs was associated with an increased sensitivity to ethirimol (HOLLOMON 1982; BUCHENAUER and HELLWALD 1985).

The portion of isolates with higher levels of resistance decreased as the selection pressure decreased during winter and spring and the powdery mildew population tended to an intermediate sensitivity (WOLFE et al. 1984).

The development of resistance of powdery mildew of cucumber and barley against DMIs resembles observations in build-up of resistance in cucumber powdery mildew to dimethirimol and in barley powdery mildew to ethirimol during 1970 and 1975.

It is assumed that resistance in powdery mildew fungi is not determined by one major gene but rather by different genes, similar to imazalil resistance in *A. nidulans* (VAN TUYL 1977). This may possible be the cause for low development of high levels of resistance in powdery mildew fungi.

Field-trial data of KENDALL et al. (1993) revealed that performance of control of *Rhynchosporium secalis* in barley by triadimenol and less extent by propiconazole had declined since its introduction. The resistant isolates displayed a similar degree of pathogenicity as the sensitive isolates. Laboratory tests with a large number of isolates revealed that triadimenol selected for a bimodal population distribution while propiconazole produced a gradual shift of an unimodal population towards less sensitive strains. The degree of resistance to the triazole compound tebuconazole was lower than to triadimenol or propiconazole. This may probably reflect differences in the mode of action (BERG et al. 1987). No cross resistance to the imidazole derivative prochloraz was observed.

FEHRMANN (1993) reported on long term monitoring on prochloraz sensitivity of *P. herpotrichoides*. After 6 years of treatment of wheat (one or two applications per year with prochloraz at the same field site) a shift towards lower sensitivity was observed and the less sensitive strains to prochloraz proved to be less pathogenic.

Antagonism and synergism

According to the primary mode of action of DMIs, it should be expected that exogeneously applied ergosterol would annul the toxicity of these compounds. However, addition of ergosterol to the culture medium only partially alleviated growth inhibition of some fungi by the fungicides but did not restore growth (SHERALD et al. 1973; RAGSDALE and SISLER 1972, 1973; LEROUX et al. 1976; BUCHENAUER 1979) and in various cases ergosterol did not exhibit any antagonistic activity.

Partial reversal of toxicity of DMIs in some fungi has been described (e.g., *C. cucumerinum, B. cinerea, U. maydis, U. avenae, A. nidulans, C. albicans*) by different lipophilic compounds, e.g., squalene, farnesol, progesterone, testosterone, β-carotene and vitamin A, phospholipids (e.g., phosphatidyl inosit, phosphatidylglycerol, phosphatidylserin, phosphatidylcholin phosphatidylethanolamin), acylglycerides (α-mono-olein, 1,2-diolein, triolein) and free fatty acids (e.g., oleic acid, linoleic acid) (SHERALD et al. 1973; LEROUX and GREDT 1978; KERKENAAR et al. 1979; BUCHENAUER 1979; DE WAARD and VAN NISTELROOY 1982; YAMAGUCHI 1977, 1978; KURODA et al. 1978). α-Tocopherol antagonized the antifungal activity of prochloraz (KLEIN 1991). Toxicity of a number of DMIs was reduced by non-ionic detergents (e.g., TWEEN 20 and 40) (SHERALD et al. 1973, BUCHENAUER 1979; DE WAARD and VAN NISTELROOY 1982), calcium and magnesium chlorides as well as by fungicides (e.g., carboxin and dialkyldithiocarbamates) (DE WAARD and VAN NISTELROOY 1982).

Apparently, different factors operate in alleviation of the toxicity of DMIs. Antagonism may be due to complex formation between the fungicide and the antagonist (e.g., non-ionic detergents) in the medium; other possible explanations of antagonism are effes on fungal membranes that result in reduced uptake or a decreased affinity of the fungicide to the membrane-bound target enzyme. The variable results with certain antagonists may be due to the different activity of these compounds in the different fungi tested.

Numerous different compounds displayed synergistic activity to the action of DMIs; these included hydrochloric acid, sodium hydroxide, cationic and anionic surfactants, sodium orthovanadate, respiratory inhibitors (e.g., oligomycin, amphotericin, nystatin, diclohexylcarbodiimide) (DE WAARD and VAN NISTELROOY 1982, 1984 a, b; KLEIN 1991). It is assumed that an increased solubility of the toxicants by some compounds (e.g., HCl, NaOH) in the medium might be a possible mechanism of synergism. Other synergistic mechanisms may include increased uptake of the fungicides which may be attributed either to membrane alterations (e.g., cationic and anionic agents) or to inhibition of energy-dependent efflux mechanism of the fungicide (e.g., respiratory inhibitors, sodium orthovanadate) (DE WAARD and VAN NISTELROOY 1984 a, b). Synergism between the following DMIs and other antifungals has been reported:

- Fenarimol and chlorothalonil or phthalimides (DE WAARD 1984 a, b)
- Propiconazol and pyrazophos or sulphur (ZEUN and BUCHENAUER 1991; KOHTS and BUCHENAUER 1984)
- Prochloraz and mancozeb (KLEIN 1991)

Applications of DMIs in mixture with fungicides exerting an unrelated mode of action and showing synergistic interactions in practice may extent the antifungal spectrum and counteract the development of resistance.

References

AOYAMA, Y., YOSHIDA, Y., NISHINO, T., KATSUKI, H., MAITRA, U. S., MOHAN, V. P., and SPRINTON, D. B.: Isolation and characterization of an altered cytochrome P-450 from a yeast mutant defective in lanosterol 14α-demethylation. J. Biol. Chem. **262** (1987): 1460–1466.

BARUG, D., and KERKENAAR, A.: Cross resistance of UV-induced mutants of *Ustilago maydis* to various fungicides which interfere with ergosterol-biosynthesis. Meded. Fac. Landbouwwet. Rijksuniv. Gent **41** (1979): 421–427.

BENNETT, F. G. A., and VAN KINTS, T. M. C.: Powdery mildew of wheat. Ann. Rep. Pl. Breed. Inst. (1983): 85–87.

BERG, D., BORN, L., BÜCHEL, K.-H., HOLMWOOD, G., and KAULEN, J.: HWG 1608 – Chemie und Biochemie eines neuen Azolfungizides. Bayer Pfl. Schutz Nachr. **40** (1987): 111–132.

BROWN, I. F., and HALL, H. R.: Induced and natural tolerance of fenarimol (EL-222) in *Cladosporium cucumerinum* and *Venturia inaequalis*. Phytopathology **69** (1979): 914.

BUCHENAUER, H.: Untersuchungen zur Wirkungsweise und zum Verhalten verschiedener Fungizide in Pilzen und Kulturpflanzen. Habitationsschrift, Bonn 1979.

– Wirkungsweise moderner Fungizide in Pilzen und Kulturpflanzen. Ber. Deutsch. Bot. Ges. **96** (1983): 427–457.

– BUDDE, K., HELLWALD, K. H., TAUBE, E., and KIRCHNER, R.: Decreased sensitivity of barley powdery mildew isolates to triazole and related fungicides. Proc. Brit. Crop Protec. Conf. – Pests and Diseases, Vol. **2** (1984a): 483–488.

– and HELLWALD, K.-H.: Resistance of *Erysiphe graminis* on barley and wheat to sterol C-14-demethylation inhibitors. EPPO Bull. **15** (1985): 459–466.

– and RÖHNER, E.: Einfluß von Nuarimol, Fenarimol, Triadimefon und Imazalil auf den Gibberellin- und Lipidstoffwechsel in jungen Gersten- und Weizenpflanzen. Mitt. Biol. Bundesanst. Land- u. Forstwirtsch., Berlin-Dahlem, H. **178** (1977): 158.

BUTTERS, J., CLARK, J., and HOLLOMON, D. W.: Resistance to inhibitors of sterol biosynthesis in barley powdery mildew. Meded. Fac. Landbouww. Rijksuniv. Gent **49** (1984): 143–151.

– – – Recombination as a means of predicting fungicide resistance in barley powdery mildew. Proc. 1986 Br. Crop Prot. Conf. **2** (1986): 561–565.

CARTER, G. A., KENDALL, S. J., BURDEN, R. S., JAMES, C. S., and CLARK, T.: The lipid composition of two isolates of *Cladosporium cucumerinum* do not explain their differences in sensitivity to fungicides which inhibit sterol biosynthesis. Pestic. Sci. **26** (1989): 181–192.

DE WAARD, M. A., and FUCHS, A.: Resistance to ergosterol-biosynthesis inhibitors II. Genetic and physiological aspects. In: DEKKER, J., and GEORGOPOULOS, S. G. (Eds.): Fungicide resistance in crop protection. Pudoc, Wageningen 1982, pp. 87–100.

– and GIESKESS, A.: Characterization of fenarimol-resistant mutants of *Aspergillus nidulans*. Netherl. J. Plant Path. **83** (1977): 177–188.

– KIPP, E. C. M., and VAN NISTELROOY, J. G. M.: Varietie in gevoeligheid van tarwemeeldauw voor fungiciden die die ersterolbiosynthese remmen. Gewasbescherning **15**, 8 (Abstr.) 1984.

– and SISLER, H. D.: Resistance to fenarimol in *Aspergillus nidulans*. Meded. Fac. Landbouwwet. Rijksuniv. Gent **41** (1976): 571–578.

– and VAN NISTELROOY, J. G. M.: Mechanism of resistance to fenarimol in *Aspergillus nidulans*. Pesticide Biochem. Physiol. **10** (1979): 219–229.

– – An energy dependent efflux mechanism for fenarimol in a wild-type strain and fenarimol resistant mutant of *Aspergillus nidulans*. Pesticide Biochem. Physiol. **13** (1980): 255–266.

– – Antagonistic and synergistic activities of various chemicals on the toxicity of fenarimol to *Aspergillus nidulans*. Pesticide Sci. **13** (1982): 279–286.

– – Differential accumulation of fenarimol by a wild-type isolate and fenarimol-resistant isolates of *Penicillium italicum*. Netherl. J. Plant Pathol. **90** (1984a): 143–153.

– – Effects of phthalimide fungicides on the accumulation of fenarimol by *Aspergillus nidulans*. Pesticide Sci. **15** (1984b): 56–62.

– – Stepwise development of laboratory resistance to DMI fungicides in *Penicillium italicum*. Pestic. Sci. **22** (1988): 371–382.

– – Stepwise development of laboratory resistance to DMI-fungicides in *Penicillium italicum*. Neth. J. Plant Pathol. **96** (1990): 321–329.

FEHRMANN, H.: Recent results from long-term monitoring on prochloraz sensitivity in field populations of *Pseudocercosporella herpotrichoides*. In: Exploring the depths of eyespot (1993): 179–183.

FLETCHER, J. T., and WOLFE, M. S.: Insensitivity of *Erysiphe graminis* f. sp. *hordei* to triadimefon, triadimenol and other fungicides. Proc. Brit. Crop. Prot. Conf. – Pests and Diseases Vol. 2 (1981): 633–640.

FUCHS, A., and DE VRIES, F. W.: Diastereomer-selective resistance in *Cladosporium cucumerinum* to triazole-type fungicides. Pestic. Sci. **15** (1984): 90–96.

– DE RUIG, S. P., VAN TUYL, J. M., and DE VRIES, F. W.: Resistance to triforine: a nonexistant problem? Netherl. J. Plant Pathol. **83** (1977): 189–250.

– and VIETS-VERWEIJ, M.: Permanent and transient resistance to triarimol and triforine in some phytopathogenic fungi. Meded. Fac. Landbouwwet. Rijksuniv. Gent **40** (1975): 699–706.

GEORGOPOULOS, S. G.: Development of fungal resistance to fungicides. In: SIEGEL, M. R., and SISLER, H. R. (Eds.): Antifungal Compounds. Vol. 2. Marcel Dekker Inc., New York 1977, pp. 439–495.

GEORGOPOULOS, S. G.: The genetic basis of classification of fungicides according to resistance risk. EPPO Bulletin **15** (1985): 513–517.

GUAN, J., and DE WAARD, M. A.: Inhibition of sterol 14α-demethylase activity in *Penicillium italicum* does not correlate with resistance to the DMI fungicide imazalil. Pestic. Biochem. Physiol. **46** (1993): 1–6.

– VAN LEEMPUT, L., POSTHUMUS, M. A., and DE WAARD, M. A.: Metabolism of imazalil by wild-type and DMI-resistant isolates of *Penicillium italicum*. Neth. J. Plant Pathol. **98** (1992): 161–168.

HIPPE, S.: Combined application of low temperature preparation and electron microscopic autoradiography for the localization of systemic fungicides. Histochemistry **87** (1987): 309–315.

HITCHCOCK, C. A., BARRETT-BEE, K. J., and RUSSELL, N. J.: The lipid composition and permeability to azole- and polyene-resistant mutant of *Candida albicans*. J. Med. Vet. Mycol. **25** (1987): 29–33.

HOLLOMON, D. W.: The effects of tridemorph on barley powdery mildew: its mode of action and cross sensitivity relationships. Phytopath. Z. **105** (1982): 279–287.

HUGGENBERGER, F., COLLINS, M. A., and SKYLAKAKIS, G.: Decreased sensitivity of *Sphaerotheca fuliginea* to fenarimol and other ergosterol biosynthesis inhibitors. Crop Protec. **3** (1984): 137–149.

KALAMARAKIS, A. E., DE WAARD, M. A., ZIOGAS, B. N., and GEORGOPOULOS, S. G.: Resistance to fenarimol in *Nectria haematococca* var. *cucurbitae*. Pestic Biochem. Physiol. **40** (1991): 212–220.

KALAMARAKIS, A. E., DEMOPOULOS, V. P., ZIOGAS, B. N., and GEORGOPOULOS, S. G.: A highly mutable major gene for triadimenol resistance in *Nectria haematococca* var. *cucurbitae*. Neth. J. Plant Pathol. **95** (1989): 109–120 (Suppl. 1).

KATO, T., SHOAMI, M., and KAWASE, Y.: Comparison of tridemorph with buthiobate in antifungal mode of action. J. Pesticide Sci. **5** (1990): 69–79.

KERKENAAR, A., and BARUG, D.: Fluorescence microscope studies of *Ustilago maydis* and *Penicillium italicum* after treatment with imazalil or fenpropimorph. Pesticide Sci. **15** (1984): 199–205.

– and KAARS SIJPESTEIJN, A.: On the antifungal mode of action of tridemorph. Pesticide Biochem. Physiol. **12** (1979): 195–204.

KLEIN, U.: Zur Resistenz bei *Pseudocercosporella herpotrichoides* (Fron) Deighton var. *herpotrichoides* gegenüber Prochloraz. Dissertation, 1991.

KOHTS, T., and BUCHENAUER, H.: Some aspects of the mode of action of pyrazophos. Br. Crop Prot. Conf. – Pests and Dis. **3** (1984): 917–922.

KRÄMER, W., BERG, D., and KRÖLLER, D. W.: Chemical synthesis and fungicidal resistance. In: FORD, M. G., HOLLOMON, D. W., KHAMBAY, B. P. S., and SAWICKI, R. M. (Eds.): Combating resistance to xenobiotics. Ellis Horwood, Chichester 1987, 291—305.

KURODA, S., UNO, J., and ARAI, T.: Target substances of some antifungal agents in the cell membrane. Antimicrob. Agents Chemother. **13** (1978): 454—459.

LEROUX, P., and GREDT, M.: Effet de l'imazalil sur la biosynthèse de l'ergostérol chez *Penicillium expansum* Link. Comptes Rendus Académie des Sci. Paris Ser. D, **286** (1978a): 427—429.

— — Effets de quelques fongicides systemiques sur la biosynthèse de l'ergostérol chez *Botrytis cinerea* Pers., *Penicillium expansum* Link et *Ustilago maydis* (DC.) (da.) Ann. Phytopath. **10** (1978b): 45—60.

— — Resistance to fungicides which inhibit ergosterol biosynthesis in laboratory strains of *Botrytis cinerea* and *Ustilago maydis*. Pesticide Sci. **15** (1984): 85—89.

LIMPERT, E., and FISCHBECK, G.: Regionale Unterschiede in der Empfindlichkeit des Gerstenmehltaus gegen Triadimenol und ihre Entwicklung im Zeitraum 1980 bis 1982. Phytomedizin **13** (1983): 19.

NEY, C.: Untersuchungen zur Resistenz von *Monilia fructicola* (Wint.) Honey gegenüber Ergosterol-Biosynthesehemmern. Diss., Universität Basel, 1988.

RAGSDALE, N. N., and SISLER, H. D.: Inhibition of ergosterol biosynthesis in *Ustilago maydis* by the fungicide triarimol. Biochem. Biophys. Res. Commun. **46** (1972): 2048—2053.

— — Mode of action of triarimol in *Ustilago maydis*. Pesticide Biochem. Physiol. **3** (1973): 20—29.

SCHEPERS, H. T. A. M.: Decreased sensitivity of *Sphaerotheca fuliginea* to fungicides which inhibit ergosterol biosynthesis. Netherl. J. Plant Path. **89** (1983): 185—187.

— Persistence of resistance to fungicides in *Sphaerotheca fuliginea*. Netherl. J. Plant Path. **90** (1984): 165—171.

— Fitness of isolates of *Spherotheca fuliginea* resistant or sensitive to fungicides which inhibit ergosterol biosynthesis. Netherl. J. Plant Path. **91** (1985a): 65—76.

— Changes over a three-year period in the sensitivity to ergosterol biosynthesis inhibitors of *Sphaerotheca fuliginea* in the Netherlands. Netherl. J. Plant Pathol. **89** (1985b): 105—118.

SHERALD, J. L., RAGSDALE, N. N., and SISLER, H. D.: Similarities between the systemic fungicides triforine and triarimol. Pestic. Sci. **4** (1973): 719—727.

— and SISLER, H. D.: Antifungal mode of action of triforine. Pesticide Biochem. Physiol. **5** (1975): 477—488.

SHOLBERG, P. L., and HAAG, P. D.: Sensitivity of *Venturia inaequalis* isolates from British Columbia to flusilazole and myclobutanil. Can. J. Pl. Pathol. **15** (1993: 102—106.

SMITH, F. D., and KÖLLER, W.: The expression of resistance of *Ustilago avenae* to the sterol demethylation inhibitor triadimenol is an induced response. Phytopathology **80** (1990): 584—590.

VAN TUYL, J. M.: Genetics of fungal resistance to systemic fungicides. Meded. Landbouwhogeschool Wageningen 77—2 (1977): 1—136.

WALSH, R. C., and SISLER, H. D.: A mutant of *Ustilago maydis* deficient in sterol C-14 demethylation: Characteristics and sensitivity to inhibitors of ergosterol biosynthesis. Pestic. Biochem. Physiol. **18** (1982): 122—131.

WATSON, P. F., ROSE, M. E., ELLIS, S. W., ENGLAND, H., and KELLY, S. L.: Defective sterol C5—6 desaturation and azole resistance: a new hypothesis for the mode of action of azole antifungals. Biochem. Biophys. Res. Comm. **164** (1989): 1170—1175.

— — and KELLY, S. L.: Isolation and analysis of ketoconazole resistant-mutants of *Saccharomyces cerevisiae*. J. Med. Vet. Mycol. **26** (1988): 153—162.

WELLMANN, H., and SCHAUZ, K.: DMI-resistance in *Ustilago maydis*. I. Characterization and genetic analysis of triadimefon-resistant laboratory mutants. Pestic. Biochem. Physiol. **43** (1992): 171—181.

– – DMI-resistance in *Ustilago maydis*. II. Effect of triadimefon on regenerating protoplasts and analysis of fungicide uptake. Pestic. Biochem. Physiol. **46** (1993): 55–64.

WOLFE, M. S., MINCHIN, P. N., and SLATER, S. E.: Powdery mildew of barley. Ann. Rep. Pl. Breed, Inst. (1982): 92–96.

– – – Dynamics of triazole sensitivity in barley mildew, nationally and locally. Proc. Brit. Crop. Prot. Conf. – Pests and Diseases **2** (1984): 465–470.

– SLATER, S. E., and MINCHIN, P. N.: Fungicide insensitivity and host pathogenicity in barley mildew. Proc. 10th Intern. Congr. Pl. Protect. **2** (1983): 645.

YAMAGUCHI, H.: Antagonistic action of lipid components of membranes from *Candida albicans* and various other lipids on two imidazole antimycotics, clotrimazole and miconazole. Antimicrob. Agents Chemother. **12** (1977): 16–25.

– Protection by unsaturated lecithin against the imidazole antimycotics, clotrimazole and miconazole. Antimicrob. Agents Chemother. **13** (1978): 423–426.

ZEUN, R., and BUCHENAUER, H.: Synergistic interactions of the fungicide mixture pyrazophos-propiconazole against barley powdery mildew. Z. Pfl. Krankh. PflSchutz **98** (1991): 526–538.

Chapter 14

Benzimidazole and related fungicides

C. J. DELP

Agricultural Consultant (formerly Du Pont Co.) Sarasota, FL, USA

Introduction

A new era in fungicide use began in the late 1960's with the introduction of the benzimidazole fungicides. The benzimidazoles, and the thiophanates which are transformed to benzimidazoles, are effective at relatively low doses for the inhibition of a broad range of fungi. Even more important is their systemic action in the host plant, that provides control of pathogens after infection. The enthusiastic acceptance of benzimidazoles by plant pathologists throughout the world has resulted in countless reports on new and improved disease control. A specific site of action contributes to their high selective activity, and unfortunately also to the risk of resistance problems. Since the benzimidazoles have been used widely as the most effective control measures of so many destructive pathogens, resistance can be a major problem. Reports of new benzimidazole resistance problems and in-depth studies have pioneered this field of fungicide research. This chapter includes a selection and interpretation of some of the literature on the subject — not a review of the more than 10,000 references. Among the many general articles on this subject are MARSH (1972), ERWIN (1973), SIEGEL and SISLER (1977), NENE and THAPLIYAL (1979), and VYAS (1984 and 1988). The genetics of fungal sensitivity and mode of action of benzimidazoles are covered in chapter 15.

Chemicals

Although the thiophanates are not benzimidazoles until transformed, they will be referred to as benzimidazoles in this chapter. Since the introduction of benomyl (DELP and KLOPPING 1968), other benzimidazole compounds with similar properties have been developed for practical plant disease control. Carbendazim, the degradation product of benomyl (CLEMONS and SISLER 1969) and thiophanate-methyl (VONK and SIJPESTEIJN 1971), appears to be the chemical active in fungi for these fungicides. The names and structures of the six most important benzimidazole fungicides are listed in table 14.1.

Table 14.1 Major benzimidazole fungicides

Common Name	Chemical Name Patent co./no./date	Manufacturer Trade Name®	Structure
Benomyl	methyl 1-(butylcarbamoyl)-2-benzimidazole-carbamate Du Pont USP 3,541,213 1966	Du Pont Benlate® Chinoin Fundazol®	
Carbendazim	methyl benzimidazole-2-yl carbamate (MBC) Du Pont USP 3,852,460 1967	Du Pont Delsene® BASF Bavistin® Hoechst Derosal®	
Fuberidazole	2-(2'-furyl)-1H-benzimidazole Bayer DAS 1,209,799 1964	Bayer Neo-Voronit®	
Thiabendazole	2-(4'-thiazolyl)-benzimidazole (TBZ) Merck USP 3,017,415 1960	Merck Mertect®	
Thiophanate-methyl	dimethyl 4,4'-o-phenylene-bis (3-thioallophanate) Nippon Soda DOS 1,806,123 1967	Nippon Soda Cercobin-M® Topsin-M®	

Spectrum of bio-activity

These related fungicides have similar patterns of selective action (BOLLEN and FUCHS 1970; EDGINGTON *et al.* 1971), but their relative toxicity and effectiveness for disease control differs. Most Ascomycetes, some of the Basidiomycetes and Deuteromycetes, and none of the Phycomycetes are sensitive. Sensitivity in the cabbage club root pathogen, *Plasmodiophora brassica*, appears to be an exception. Their selective toxicity provides safety to animals, bees, and plants with a corresponding high degree of activity against certain fungi, insects, mites and worms.

The benzimidazoles have been developed for different uses because of differences in activity and marketing strategies. Thiabendazole, first sold primarily as an anthelmintic, was then developed on crops for postharvest fruit treatments, and later a special formulation was developed for treatment of elm trees infected with *Ceratocystis ulmi*. Fuberidazole has been used primarily in Europe for cereal seed treatment. Benomyl, carbendazim, and thiophanates have been developed for the uses listed in table 14.2. The effects on systemic pathogens like *Verticillium*, *Fusarium* and *Ceratocystis* were promising under controlled conditions, but distribution in soil and host is frequently insufficient for practical results.

An important non-plant pathogenic fungus controlled by benzimidazoles is *Pithomyces chartarum,* the causal agent of facial eczema in sheep and cattle (CAMPBELL and SINCLAIR 1968).

Table 14.2 Major practical applications

Some of the diseases controlled by benzimidazole fungicides:

Crops	Pathogens
Almond (*Prunus*)	– *Monilinia* spp.
Asparagus	– *Phoma asparagi*
Avocado (*Persea*)	– *Cercospora perseae, Colletotrichum gloeosporioides, Fusarium* spp., and *Verticillium albo-atrum*
Banana (*Musa*)	– *Ceratocystis paradoxa, Colletotrichum musae, Fusarium* spp., *Mycosphaerella musicola, M. fijiensis* var. *difformis, Nigrospora, Penicillium* spp.
Beans (*Phaseolus*)	– *Botrytis cinerea, Colletotrichum* spp., *Sclerotinia sclerotiorum* and *Cercospora* spp.
Blueberry (*Vaccinium*)	– *Botrytis cinerea, Gloeosporium* sp., and *Monilinia vacciniicarombosi*
Bushberry (*Rubus*)	– *Botrytis cinerea, Penicillium* spp., and *Sphaerotheca humuli*
Carrot (*Daucus*)	– *Cercospora carota*
Cassava (*Manihot*)	– *Cercospora* spp.
Celery (*Apium*)	– *Cercospora apii,* and *Septoria apii*
Chestnut (*Castanea*)	– *Colletotrichum castanea*
Citrus	– *Botrytis cinerea, Colletotrichum gloesporiodes, Diplodia natalensis, Elsinoe fawcetti, Guignardia citricarpa, Mycosphaerella citri, Penicillium* spp., *Phomopsis citri,* and *Oidium tingitaninum*
Clove (*Eugenia*)	– *Cylindrocladium quinqueseptatum*
Coffee	– *Cercospora coffeicola, Colletotrichum coffeanum,* and *Pellicularia koleroga*
Cole Crops, Canola or Rape (*Brassica*)	– *Fusarium oxysporum, Mycosphaerella brassicola, Phoma lingam,* and *Sclerotinia sclerotiorum*
Cucurbits & Melons (*Cucumis*)	– *Cladosporium cucumerinum, Colletotrichum gossypii, Colletotrichum lagenarium, Erysiphe cichoracearum, Fusarium* spp., *Mycosphaerella citrullina, Oidium* spp., *Rhizoctonia solani,* and *Sclerotinia sclerotiorum*
Currant (*Ribes*)	– *Drepanopeziza* sp., *Sphaerotheca humuli,* and *Cronartium* sp.
Eggplant (*Solanum*)	– *Colletotrichum capsici, Colletotrichum gloeosporioides,* and *Corynespora melongenae*
Garlic (*Allium*)	– *Penicillium* sp., and *Sclerotinia sclerotiorum*
Grape (*Vitis*)	– *Botrytis cinerea, Gloeosporium ampelophagum, Guignardia bidwellii, Melanconium fuligenum, Pseudopezizza* sp., and *Uncinula necator*
Lettuce (*Lactuca*)	– *Botrytis cinerea,* and *Sclerotinia sclerotiorum*
Mango (*Mangifera*)	– *Colletotrichum gloeosporiodes,* and *Oidium*
Mulberry (*Morus*)	– *Phyllactinia* sp.

Table 14.2 (continued)

Some of the diseases controlled by benzimidazole fungicides:
Crops	Pathogens
Mushroom	– *Dactylium* sp., *Mycogone perniciosa*, and *Verticillium malthousi*
Olive (*Olea*)	– *Cycloconium oleaginum*
Onion (*Allium*)	– *Botrytis allii*, *Colletotrichum*, *gloeosporiodes*, and *Fusarium oxysporum*
Ornamentals and Trees	– *Ascochyta*, *Botrytis cinerea*, *Ceratocystis* spp., *Cerospora* spp., *Colletotrichum* spp., *Corynespora* spp., *Cylindrocladium*, *Didymellina* spp., *Diplocarpon rosae*, *Entomosporium* sp., *Fabrae maculata*, *Fusarium* spp., *Gnomonia leptostyla*, *Oidium* spp., *Ovulinia azaleae*, *Penicillium* spp., *Phomopsis* spp., *Phyllostictina* spp., *Ramularia* spp., *Rhizoctonia* spp., *Sclerotinia* spp., *Sphaerotheca pannosa*, and *Thielaviopsis* spp.
Papaya (*Carica*)	– *Oidium caricae*
Peanut (*Arachis*)	– *Ascochyta* spp., *Aspergillus* spp., *Cercospora arachidicola*, *Cercosporidium personatum*, and *Mycosphaerella arachidicola*
Pea (*Pisum*)	– *Mycosphaerella pinoides*, and *Oidium* spp.
Pecan (*Carya*)	– *Cercospora fusca*, *Cladosporium effusum*, *Cristulariella pyramidalis*, *Leptothyrium caryae*, *Microsphaera alni* and *Mycosphaerella caryigena*
Pepper (*Capsicum*)	– *Botrytis cinerea*, *Fusarium piperi*, and *Fusarium solani*
Persimmon (*Diospyros*)	– *Cercospora kaki*, and *Phyllactinia kaki*
Pineapple (*Ananas*)	– *Fusarium moniliforme*, *Rhizoctonia* sp., and *Thielaviopsis paradoxa*
Plantain (*Plantago*)	– *Colletotrichum musae*, *Fusarium* spp., and *Penicilium* spp.
Pome Fruit (*Malus* and *Pyrus*)	– *Botrytis cinerea*, *Cladosporium* spp., *Gloeodes pomigena*, *Gymnosporangium* spp., *Marssonina mali*, *Microthyriella rubi*, *Mycosphaerella pomi*, *Penicillium* spp., *Phyllactinia pyri*, *Physalospora* spp., *Podosphaera leucotricha*, *Rosellinia necatrix*, *Valsa ceratosperma*, and *Venturia* spp.
Potato (*Solanum*)	– *Fusarium solani*, *Oospora* sp., *Phoma* sp., and *Rhizoctonia solani*
Rice (*Oryza*)	– *Acrocylindrium oryzae*, *Cercospora oryzae*, *Gibberella fujikuroi*, *Piricularia oryzae*, *Thanatephorus cucumeris*, and *Rhizoctonia solani* (stem rot)
Rubber (*Hevea*)	– *Ceratocystis fimbriata*, *Microcyclus ulei*, and *Mycosphaerella* spp.
Soybean (*Glycine*)	– *Cercospora kikuchii*, *C. sojina*, *Colletotrichum truncatum*, *Diaporthe* spp., *Phomopsis sojae*, and *Septoria glycinea*
Stone Fruit (*Prunus*)	– *Cladosporium carpophilum*, *Coccomyces hiemalis*, *Cytospora leucostoma*, *Fusarium* spp., *Monilinia* spp., *Oidium* spp., *Phomospsis persicae*, and *Sphaerotheca pannosa*
Strawberry (*Fragaria*)	– *Botrytis cinerea*, *Cercospora* sp., *Fusarium oxysporum*, *Mycosphaerella fragariae*, *Odium* spp., *Sphaerotheca humali*, and *Verticillium* spp.
Sugar Beet (*Beta*)	– *Cercospora beticola*
Sugar Cane (*Saccharum*)	– *Ceratocystis paradoxa*, and *Cercospora* spp.

Table 14.2 (continued)

Some of the diseases controlled by benzimidazole fungicides:

Crops	Pathogens
Sweet Potato and Yam (*Ipomoea* and *Dioscores*)	– *Ceratocystis fimbriata*, and *Monilochaetes infuscans*
Tea (*Thea*)	– *Gloeosporium theae-sinensis*, *Elsinoe leucospila*, and *Rosellinia necatrix*
Tobacco (*Nicotiana*)	– *Ascochyta nicotianae*, *Cereospora nicotianae*, *Helicobasidium mompa*, and *Pellicularia filamentosa*
Tomato (*Lycopersicon*)	– *Botrytis cinerea*, *Cerospora* spp., *Cladosporium* sp., *Colletotrichum phomoides*, *Corynespora melongenae*, *Fusarium oxysporum*, *Odium* spp., *Phoma destricitiva*, *Sclerotinia sclerotiorum*, and *Septoria lycopersici*
Turf (*Poa*, *Agrostis*, etc.)	– *Fusarium* spp., *Gloeotinia temulenta*, *Pellicularia filamentosa*, and *Sclerotinia homeocarpa*
Wheat (*Triticum*)	– *Erysiphe graminis*, *Fusarium* spp., *Pseudocercosporella herpotrichoides*, *Rhynchosporium* sp., *Septoria avenae*, and *Ustilago tritici*

Non-target effects and undeveloped uses

The target organisms discussed above are so responsive to benzimidazole treatment that it is natural for some sensitive, non-target organisms to also be affected. VYAS in 1988 reviewed much of the literature in this area. Some mycorrhizal fungi are sensitive to benzimidazoles mixed into soil (FITTER and NICHOLS 1988; CAREY 1992). Of course, contact with the benzimidazole is most critical to any non-target effects. For example, mites in a treated apple tree or earthworms feeding on treated foliage under an apple tree may be exposed to active rates of a benzimidazole during the season of treatment for apple disease control. But earthworms and microorganisms below the soil surface would contact very little benzimidazole, and most of the chemical present is so tightly bound to the soil surface that it is not available for reaction.

Population suppression effects are significant for the spider mites *Pananychus ulmi* and *Tetranychus urticae* and their predators *Amblyseius fallacis* (NAKASHIMA and CROFT 1974) and *Agistemus fleschneri* (CHILDERS and ENNS 1975). This mite ovacide activity, reported in 1968 by DELP and KLOPPING, has not been developed commercially because these and other non-target mites develop resistance to benzimidazoles rapidly and population suppression effects may be only temporary.

Benzimidazoles have also been shown to control some insects and diseases of cultured insects and bees. Examples include reductions in the population of aphids (BINNS 1970; DELORME 1976; ENGELHARD and POE 1971; BAILISS et al. 1978) and cabbage maggots *Hylemya brassicae* (REYES and STEVENSON 1975). HARVEY and GAUDET (1977) found that 75 ppm of benomyl in the artificial diet of spruce budworm reduced the microsporidian levels and also reduced worm growth and fertility. The protozoan-microsporida *Nosema*, a parasite which constitutes a major problem in laboratory culturing of many insects, is controlled by adding benomyl to the artificial diet of alfalfa weevils *Hyper postica* (HSIAO 1973). Bees of the genera *Aphis, Megachile* and *Nomia* may be affected by the chalkbrood disease caused by the fungus, *Ascosphaera* spp. Benomyl treatments can reduce chalkbrood and at the same time are safe to bees (MOELLER and WILLIAMS 1976).

Thiabendazole and thiaphanate are commercial anthelmintics, and the other benzimidazoles have varying degrees of animal worm activity. Their effects on nematodes and earthworms cover a broad number of genera if sufficient chemical is contacted. There are reports of decreased root penetration by nematodes, elimination of the ability of the dagger nematode, *Xiphinema*, to transfer tobacco ring spot virus, and reduction in populations of the rice nematode, *Aphelenchoides,* which causes white tip disease.

The non-target effects on earthworms have been the subject of considerable concern. Under some conditions where earthworms feed on benzimidazole treated residues, populations are temporarily reduced. For example, population decline was seen in earthworms which fed on apple foliage on the orchard floor in the UK (STRINGER and LYONS 1974). Excessive use where the presence of earthworms is critical should be avoided. On the other hand, some airport managers have used high rates to reduce the danger of earthworms on runway (TOMLIN *et al.* 1981).

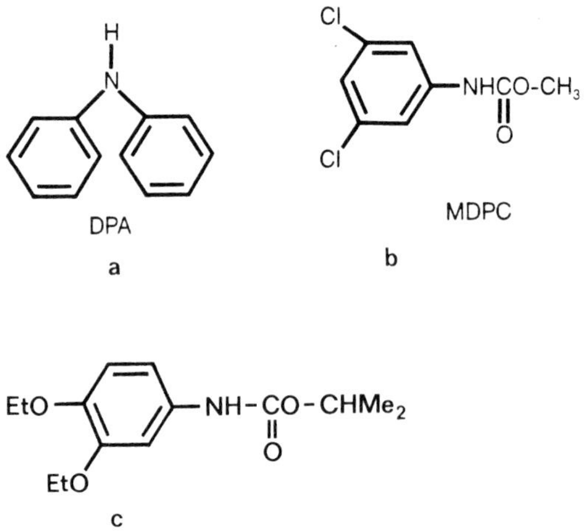

Fig. 14.1 Chemical structures of compounds exhibiting a negative cross resistance. a) diphenylamine, b) N-(3,5 dichlorophenyl)-carbamate, c) diethofencarb

Behaviour on and in plants

The benzimidazoles have characteristics on and in plants that distinguished them from traditional fungicides in the 1960's. They are effective protectants, and also may penetrate the host plant to inhibit post-infection or may move in the apoplast to untreated portions of the plant for preventive or curative effects. In practice, most disease control is accomplished by direct contact or protectant action. They are tightly bound to plant surfaces and degrade slowly, thus they have desirable residual activity. Excessive deposits serve for surface redistribution. Penetration into plants varies greatly among compounds and also varies with adjuvants applied with the compound, the kind of plant, and the location or maturity of the plant. For example, UPHAM and DELP (1973) found 20 times more active compound in herbaceous plants when applying benomyl as compared to applying carbendazim. Thiophanate and benomyl penetrate apple cuticle more rapidly than carbendazim while thiabendazole has the slowest penetration rate (SOLEL and EDGINGTON 1973).

Translaminar movement will stop the infection process into untreated leaf surfaces opposite the site of application in plants where penetration is sufficient. Although there is some evidence of detectible movement out of a treated leaf, the quantitaties are too small for practical diseases control. Movement in the leaf is in the transpiration water, and carbendazim accumulates at the margins and tip.

Uptake by roots is more efficient and results in more complete distribution throughout a plant than from foliar treatment. Areas of low transpiration, for example some fruit, tend to accumulate less benzimidazole from systemic movement.

It is remarkable that these chemicals which are highly toxic to a broad number of pathogenic fungi are so safe even when taken systematically into host plant. The use of benzimidazoles at recommended rates for plant disease control has resulted in very few undesirable plant responses. Massive doses, especially when applied on seed and soil treatments, may delay emergence or cause stunting or chlorosis of some plants. When used on certain apple varieties, benomyl (and to a lesser extent the other benzimidazoles) may amplify the naturally occurring fruit-finish problems of russeting or opalescence. Growers (mostly of ornamentals in subtropical areas) claimed over 500 million US dollars in crop injury from Benlate® DF treatments in 1991, but the causes of the alleged plant damage have not been determined from this unusual situation.

Most plants have no injurious responses, but there are other plant responses which are noteworthy. Under some conditions, treatments have resulted in antisenescent, cytokinin and yield boosting effects, and masking of virus symptoms and ozone injury. The cytokinin-like properties of benzimidazoles (SKENE 1972; THOMAS 1974) may contribute to the observations of delayed senescence, cereals tillering, and symptom masking. Although the yield boosting effects on soybeans and peanuts appear to be directly related to the control of diseases, this is not the case with cereals. Both increased tillering and antisenescence have been correlated with increased yields of cereal treated with a benzimidazole in the absence of traditional disease symptoms (PEAT and SHIPP 1981; FEHRMANN et al. 1978; TRIPATHI et al. 1982). Some positive effects of benzimidazols may be due to their ozone protecting activity, perhaps by their weak cytokinin like effects. Soil drench systemic applications of benomyl or thiophanate (PELLISSIER et al. 1972; MOYER et al. 1974) as well as foliar sprays (MANNING et al. 1974) can reduce ozone injury on some crop foliage.

Among the plant responses attributed to carbendazim treatment are the masking of the mosaic and yellows symptoms in tobacco inoculated with tobacco mosaic virus and in lettuce infected with beet western yellows virus (TOMLINSON et al. 1976).

Behaviour in soil

Mixing relatively high rates of benzimidazoles in the soil of the root zone may result in practical control of soil-borne, root rot, vascular and foliage pathogens. This is possible with container-grown plants, incorporation into plant beds, and in-furrow applications with small, high-value crops with confined root systems like strawberries and some vegetables and ornamentals. Under other field conditions, disease control with soil applications has been disappointing because the compounds are so tightly absorbed to soil colloids and organic matter that they are virtually immobile and in some cases unavailable to affect sensitive organisms (PEEPLES 1974). Penetration into soil can be increased by the addition of some surfactants (PITBLADO and EDINGTON 1972). Carbendazim is relatively stable in soil, but studies with uneconomically high (100 ppm) rates of application may lead to unrealistic conclusions about the sensitivity of non-target soil organisms (BAUDE et al. 1974).

The temporary effects on sensitive organisms such as *Fusarium, Penicillium,* and *Trichoderma* appear to be compensated for by the increased growth of other less sensitive organisms. This compensation can maintain a balanced soil biological activity as measured by CO_2 release.

Resistance

Benzimidazole resistance represented the beginning of serious fungicide resistance problems. This happened in the 1970's because benzimidazoles were used widely, alone, and intensively for crop protection, because they are specific-site inhibitors, and also because most fungi contain resistant strains in their natural populations. Most resistant fungal populations are fit for survival, and benzimidazoles may be ineffective for several seasons after resistant strains dominate large crop production areas. There are some exceptions where these fungicides can be used after a period of abstinence. For example, powdery mildew of cucurbits in New York state was the first to be reported as resistant (SCHROEDER and PROVVIDENTI 1969), but other powdery mildew populations have developed resistance more slowly, and in some locations, rapidly revert to sensitive populations when the selective pressure from benzimidazoles is removed.

Because benomyl provided excellent control of sugar beet, peanut and celery *Cercospora* leaf spots; of apple and pear *Venturia* scab and *Botrytis* of grapes and other crops, it was frequently used exclusively. Under such conditions, resistance developed within 2 to 4 seasons. On the other hand, where mixtures of benomyl and unrelated fungicides have been used from the beginning, resistance has not become a problem or was delayed for several seasons (DELP 1980).

The basic benzimidazole manufacturers launched a cooperative study of potential resistance in *Pseudocercosporella herpotrichoides,* the cereal eyespot pathogen (DELP 1984). Organized as the Benzimidazole Working Group of industry's Fungicide Resistance Action Committee (FRAC), this group brought together key European investigators and coordinated studies and recommendations in France, Germany and the UK. The impact of resistance was often difficult to interpret since treatment of fields with resistant populations might fail to control eyespot but still result in yield benefits. High risk situations were identified, and alternative programs, which were generally more expensive, were recommended. Although many of the major uses of benzimidazoles have been lost, there are situations where strategies have been used successfully to delay resistance problems. In general strategies involve reduced exposure to selective fungicide pressure by use of fewer treatments, lower rates and mixtures with unrelated companion fungicides. Where mixtures were used from the beginning, some *Cercospora* and *Botrytis* populations retained sensitivity. Also, controlled use of a benzimidazole in the banana Sigatoka *Mycosphaerella* has prolonged effectiveness and even made reentry possible in areas where resistance was emerging. Twenty years after resistance problems became serious, these fungicides are still effective for the control of numerous pathogens in many areas of the world.

Cross-resistance, multiple-resistance and negative cross-resistance

Practical resistance problems for one benzimidazole are also problems for the others, because most fungal strains resistant to one benzimidazole also have reduced sensitivity to the other benzimidazoles (cross-resistance) (chapter 2). This precludes switching or combining among benzimidazoles to avoid or solve resistance problems. Although there is no evidence of cross-resistance to non-benzimidazole fungicides, it is possible for a benzimidazole-resistant strain to also develop resistance to an unrelated fungicide (multiple-resistance). Multiple-resistance is common in some *Penicillium pathogens* in citrus fruit packing houses as reported by Eckert (Dekker and Georgopoulos 1982), and in *Botrytis* (Elad et al. 1992).

There are reports of strains that are resistant to one benzimidazole and more sensitive to another (negative cross-resistance), but this phenomenon, as discussed by Davidse, has been of no practical value. On the other hand, there is a group of carbamate compounds active only against strains of fungi highly resistant to benzimidazoles. This negative cross-resistance was reported for N-phenyl carbamate herbicides and benzimidazole fungicides (Leroux and Gredt 1979). Methyl N-(3,5 dichloro-phenyl)-carbamate (MDPC) and diethofencarb (Fig. 14.1 b, c) represent a class of chemicals which are effective for plant disease control and can be combined with benzimidazoles to overcome resistance (Kato et al. 1984). Unfortunately strains resistant to both benzimidazole and MDPC developed rapidly (Faretra et al. 1989).

The concept may have been in practical use before it was understood that diphenylamine (DPA) (Fig. 14.1a) in combination with a benzimidazole controls both benzimidazole-sensitive and resistant *Penicillium expansum*, the cause of post-harvest decay of apples. Rosenberger and Meyer (1985) showed that DPA combined with a benzimidazole is more effective for the control of some resistant isolates than either used alone. DPA, used as a post-harvest treatment to control the physiological disorder known as storage scald, is effective for the control of strains highly resistant to benzimidazoles. Some strains of *P. expansum* are resistant to both DPA and benzimidazoles, but this combination may be the first commercial application of negative cross resistance for the solution of a fungicide resistance problem.

Residue, environmental and toxicology issues

Residue deposits are most important in relation to environmental and toxicology issues. From analysis of treated produce, residue tolerances are established well below the levels which have a toxic effect on test animals (Fao 1983). Residue analysis of crops treated for disease control taken throughout the world have been used to establish tolerances (the allowable and safe residues of benzimidazole on raw agricultural commodities). Typical residue tolerances are in Table 14.3. Analytical methods for benzimidazole can be found in Zweig and Sherma 1982.

Carbendazim is moderately stable to photodegradation (Fleeker and Lacy 1977). In plants and

Table 14.3 Typical* residue tolerances

Commodity	parts per million (mg/kg)
Meat, milk and eggs	0.1
Nuts including peanut and soybean	0.2
Sugarbeet roots (tops 15 ppm)	0.2
Cereal grains	0.5
Cucurbits, banana and avocado	1
Beans and most crops in the Netherlands	2
Celery & mango	3
Rice, strawberry & tomato	5
Pome fruits & berries	7
Citrus, grape and mushroom	10
Stone fruits	15
Pineapple	35

* tolerances selected from different products and countries

soil there is relatively slow decomposition to 2-aminobenzimidazole. HELWEG (1977), demonstrated biodegradation in soil. RHODES and LONG (1974), showed that carbendazim and 1-aminobenzimidazole are immobile in soil and do not leach or move significantly from the site of application. Adsorption and stability in soil is directly influenced by pH and organic matter. Thiabendazole appears to be even more strongly absorbed than carbendazim.

The conversion of benomyl and thiophanates to carbendazim varies greatly under different conditions. Some laboratory evidence for rapid conversion of benomyl in dilute solution may be misleading because under field use (BAUDE et al. 1973) a major portion of benomyl stays intact on plants. The residual benzimidazole on foliage 28 days after treatment of cucumber, banana, orange and grape is more than 60% benomyl. CHIBA and VERES (1981), confirmed these results on apple foliage. This probably has little influence on the contact or protectant activity because benomyl and carbendazim have similar protectant efficiency. But the presence of benomyl could have a pronounced effect on systemic and post-infection action since benomyl treatments accumulate higher systemic concentrations of active ingredient in the host than carbendazim treatments (UPHAM and DELP 1973). On the other hand, the slow transformation of thiophanates to carbendazim (VONK and SIJPESTEIJN 1971) has different practical implications since carbendazim is so much more active than intact thiophanates. It is possible to lose some effectiveness because of a transformation delay. There is little or no hydrolysis to carbendazim from pH 1 to 7 (25 °C to 27 °C), and the half-life at pH 9 is about 17 hours.

Compatibility of benzimidazole formulations with most other agricultural chemicals in tank mixtures has been good except with highly alkaline pesticides such as Bordeaux mixture or lime sulfur. Of course, it is necessary to determine the compatibility of each mixture by reference to the product labels and by small-scale, "jar" tests.

Toxicology studies for benzimidazoles show acute effects at values of greater than 7,500 mg/kg in rats. Reproductive, teratogenic and mutagenic studies also assure that these products are safe when used according to the label. Thiabendazole and thiophanates are administered to animals as anthelminthics. The United States Environmental Protection Agency (EPA) has conducted several special reviews of benomyl based on potential teratogenic and sperm effects. Minor modifications on the use of protective equipment have resulted from these reviews.

Conclusions

Benzimidazoles represent a group of highly effective, broad-spectrum, systemic fungicides which are widely used for efficient plant disease control. With the exception of the unusual situation in 1991 where growers claimed massive crop damage from Benlate® DF treatments, there have been few reports of phytotoxicity. Their mild cytokinin effects on some plants tend to retain chlorophyll and in some cases increase yields, and delay maturity, but otherwise there are few significant plant responses. Effects on non-target organisms are minimal because of selective toxicity and strong adsorption to plants and soil. Progressive limitations on use have occurred because of resistance in major pathogens where these products were used intensively and exclusively. With increased awareness of effective ways to cope with resistance, users should be less likely to experience resistance problems, and these fungicides will be among the favored diseases control agents for many years to come.

References

BAILISS, K. W., PARTIS, G. A., HODGSON, C. J., and STONE, E. V.: Some effects of benomyl and carbendazim on *Aphis fabae* and *Acyrthosphon pisum* on field bean (*Vicia faba*). Ann. appl. Biol. **89** (1978): 443–449.

BAUDE, F. J., GARDINER, J. A., and HAN, J. C.-Y. jr.: Characterization of residues on plants following foliar spray applications of benomyl. J. Agricult. Food Chem. **21** (1973): 1084–1090.

– PEASE, H. L., and HOLT, J.: Fate of benomyl on field soil and turf. J. Agricult. Food Chem. **22** (1974): 413–418.

BINNS, E. S.: Aphicidal activity of benomyl. Glasshouse Crops Res. Inst. Ann. Rep. 1969 (1970): 113.

BOLLEN, G. J., and FUCHS, A.: On the specificity of the *in vitro* and *in vivo* antifungal activity of benomyl. Netherl. J. Plant Pathol. **76** (1970): 299–312.

CAMPBELL, A. G., and SINCLAIR, D. P.: Control of facial eczema in lambs by use of fungicides. Proc. Ruakura Farmers' Conf. Week. N. Z. (1968): 3–12.

CAREY, P. D., FITTER, A. H., and WATKINSON, A. R.: A field study using the fungicide benomyl to investigate the effect of mycorrhizal fungi on plant fitness. Oecologia **90** (1992): 550–555.

CHIBA, M., and VERES, D. F.: Fate of benomyl and its degradation compound methyl 2-benzimidazolecarbamate on apple foliage. J. Agricult. Food Chem. **29** (1981): 588–590.

CHILDERS, C. C., and ENNS, W. R.: Field evaluation of early season fungicide substitutions on Tetranychid mites and the predators Neoseinlus follacis and Agistemus fleschneri in two Missouri apple orchards. J. Econ. Entomol. **68** (1975): 719–724.

CLEMONS, G. P., and SISLER, H. D.: Formation of a fungitoxic derivative from Benlate. Phytopathology **59** (1969): 705–706.

DEKKER, J., and GEORGOPOULOS, S. G.: Fungicide Resistance in Crop Protection. Centre for agricultural publishing and documentation, Wageningen, Netherl. 1982.

DELORME, R.: Evaluation en laboratoire de la toxicité pour *Diaeretiella* rapae (Hym. Aphidaidae) des pesticides utilises en traitement des parties aeriennes des plantes. Entomophaya **21** (1976): 19–29.

DELP, C. J.: Coping with resistance to plant disease control agents. Plant Disease **64** (1980): 651–657.

– Industry's response to fungicide resistance. Crop Protect. **3** (1984): 3–8.

- and KLOPPING, H. L.: Performance attributes of a new fungicide and mite ovicide candidate. Plant Disease Rep. **52** (1968): 95–99.
EDGINGTON, L. V., KHEW, K. L., and BARRON, G. L.: Fungitoxic spectrum of benzimidazole compounds. Phytopathology **61** (1971): 42–44.
ELAD, Y., YUNIS, H., and KATAN, T.: Multiple fungicide resistance to benzimidazoles, dicarboximides and diethofencarb in field isolates of *Botrytis cinerea* in Israel. Plant Pathol. **41** (1992): 41–46.
ENGELHARD, A. W., and POE, S. L.: Combinations of fungicides and insecticides for control of disease, insects and mites on chrysanthemums. Proc. Florida State Horticult. Soc. **84** (1971): 435–441.
ERWIN, D. C.: Systemic fungicides: disease control, translocation, and mode of action. Ann. Rev. Phytopathol. **11** (1973): 389–422.
FAO Plant Production and Protection Paper 56. Pesticide Residues in Food. Rep. joint meeting on pesticide residues held in Geneva, December 5–14, Rome **4.4** (1983): 12–20.
FARETRA, F., POLLASTRO, K. S., and DITONNO, A. P.: New natural variants of *Botryotinia fuckeliana* (*Botrytis cinerea*) coupling benzimidazole-resistance to insensitivity toward the N-phenylcarbamate diethofencarb. Phytopathol. Mediterr. **28** (1989): 98–104.
FEHRMANN, H., REINECKE, P., and WEIHOFEN, U.: Yield increase in winter wheat by unknown effects of MBC-fungicides and captafol. Phytopath. Z. **93** (1978): 359–362.
FITTER, A. H., and NICHOLS, R.: The use of benomyl to control infection by vesicular-arbuscular mycorrhizal fungi. New Phytol. **110** (2) (1988): 201–206.
FLEEKER, J. R., and LACY, H. M.: Photolysis of methyl 2-benzimidazolecarbamate. J. Agricult. Food. Chem. **25** (1977): 51–55.
HARVEY, G. T., and GAUDET, P. M.: The effects of benomyl on the incidence of microsporidia and the developmental performance of eastern spruce budworm (Lepidoptera: Tortricidae). Can. Entomol. **109** (1977): 987–993.
HELWEG, A.: Degradation and absorption of carbendazim and 2-aminobenzimidazole in soil. Pesticide Sci. **8** (1977): 71–78.
HSIAO, T. H., and HSIAO, C.: Benomyl: A novel drug for controlling a microsporidian disease of the alfalfa weevil. J. Inventebrate Pathol. **22** (1973): 303–304.
KATO, T., SUZUKI, D., TAKAHASHI, J., and KAMOSHITA, K.: Negatively correlated cross-resistance between benzimidazole fungicides and methyl N-(3,4-dichlorophenyl)-carbamate. J. Pesticide Sci. **9** (1984): 485–495.
LEROUX, P., and GREDT, M.: Phenomenes de resistance croisee negative chez *Botrytis cinerea* Pers. entre les fongicides benzimidazoles et des herbicides carbamates. Phytiatr. Phytopharm. **28** (1979): 79–86.
MANNING, W. J., FEDER, W. A., and VARDARO, P. M.: Suppression of oxidant injury by benomyl. J. Environment. Quality **3** (1974): 1–3.
MARSH, R. W. (Ed.): Systemic Fungicides. Longman, London 1972.
MOELLER, F. E., and WILLIAMS, P. H.: Chalkbrood research at Madison, Wisconsin. Am. Bee J. **116** (1976): 484–486.
MOYER, J., COLE, H., and LA CASSE, N. L.: Reduction of ozone injury on poa annua by benomyl and thiophanate. Plant Disease Rep. **58** (1974): 41–44.
NAKASHIMA, M. J., and CROFT, B. A.: Toxicity of benomyl to the life stages of *Amblyseius follacis*. J. Econ. Entomol. **67** (1974): 675–677.
NENE, Y. L., and THAPLIYAL, P. N.: Fungicides in Plant Disease Control. Oxford and IBH, New Delhi 1979.
PEAT, W. E., and SHIPP, D. M.: The effects of benomyl on the growth and development of wheat. EPPO Bull. **11** (1981): 287–293.
PEEPLES, J. L.: Microbial activity in benomyl-treated soils. Phytopathology **64** (1974): 857–860.
PELLISSIER, M., LACASSE, N. L., and COLE, H.: Effectiveness of benzimidazole, benomyl

and thiabendazole in reducing ozone injury to pinto beans. Phytopathology **62** (1972): 580–582.
PITBLADO, R. E., and EDGINGTON, L. V.: Movement of benomyl in field soils as influenced by acid surfactants. Phytopathology **62** (1972): 513–516.
REYES, A. A., and STEVENSEN, A. B.: Toxicity of benomyl to the cabbage maggot *Hyhemya brassicae* in greenhouse tests. Can. Entomologist **107** (1975): 685–687.
RHODES, R. C., and LONG, J. D.: Run-off and mobility studies on benomyl in soils and turf. Bull. Environment. Contamin. Toxicol. **12** (1974): 385–393.
ROSENBERGER, D. A., and MEYER, F. W.: Negatively correlated cross-resistance to diphenylamine in benomyl-resistant *Penicillium* expansum. Phytopathol. **75** (1985): 74–79.
SCHROEDER, W. T., and PROVVIDENTI, R.: Resistance to benomyl in powdery mildew of cucurbits. Plant Disease Resp. **53** (1969): 271–275.
SIEGEL, M. R., and SISLER, H. D.: Antifungal Compounds, Vol. 2, M. Dekker Inc., New York 1977.
SKENE, K. G. M.: Cytokinin-like properties of the systemic fungicide benomyl. J. Horticult. Sci. **47** (1972): 1979–1982.
STRINGER, A., and LYONS, C. H.: The effect of benomyl and thiophanate-methyl on earthworm populations in apple orchards. Pesticide Sci. **5** (1974): 189–196.
THOMAS, T. H.: Investigations into the cytokinin-like properties of benzimidazole-derived fungicides. Ann. appl. Biol. **76** (1974): 237–241.
TOMLIN, A. D., TOLMAN, J. H., and THORN, G. D.: Suppression of earthworm (*Lumbricus terrestris*) populations around an airport by soil applications of the fungicide benomyl. Protect. Ecol. **2** (1981): 319–323.
TOMLINSON, J. A., FAITHFULL, E. M., and WARD, C. M.: Chemical suppression of the symptoms of two virus diseases. Ann. appl. Biol. **84** (1976): 31–41.
TRIPATHI, R. K., KOMMAL, K., SCHLÖSSER, E., and HESS, W. M.: Effect of fungicides on the physiology of plants. Pesticide Sci. **13** (1982): 395–400.
UPHAM, P. M., and DELP, C. J.: Role of benomyl in the systemic control of fungi and mites on herbaceous plants. Phytopathol. **63** (1973): 814–820.
VONK, J. W., and SIJPESTEIJN, A. K.: Methyl benzimidazol-2-ylcarbamate, the fungitoxic principle of thiophanate-methyl. Pesticide Sci. **2** (1971): 160–164.
VYAS, S. C.: Systemic Fungicides, Chap. 8. Tata McGraw-Hill Publishing Co., New Delhi 1984.
– Nontarget Effects of Agricultural Fungicides. CRC Press Inc., Boca Raton 1988.
ZWEIG, G., and SHERMA, J.: Analytical methods for pesticides and plant growth regulators. Vol. 12, Chap. 2, Academic Press, New York 1982.

Chapter 15

Biochemical and molecular aspects of the mechanisms of action of benzimidazoles, *N*-phenylcarbamates and *N*-phenylformamidoximes and the mechanisms of resistance to these compounds in fungi

L. C. DAVIDSE[1]) and H. ISHII[2])

[1]) Enkhuizen, The Netherlands
[2]) Division of Plant Protection, Fruit Tree Research Station, MAFF, Japan

Soon after the introduction of benzimidazoles as fungicides and anthelminthic drugs in agriculture and veterinary medicine (chapter 14) their novelty and effectiveness but also the development of resistance in target fungi initiated intensive research to elucidate their mode of action in detail and the mechanisms involved in resistance. Results of the earlier genetic, biochemical and cytological studies which provided evidence for an effect of benzimidazoles on microtubule assembly, have been intensively reviewed (BURLAND and GULL 1984; CORBETT et al. 1984; DAVIDSE 1982, 1986, 1987; DAVIDSE and DE WAARD 1984; DEKKER 1985; LANGCAKE et al. 1983). Knowledge on the mechanism of action of the benzimidazoles rapidly increased when these compounds were recognized as effective tools in studying the genetics of tubulin and the organization and function of microtubules (MORRIS 1986). The genetics of benzimidazole resistance in plant pathogenic fungi, reviewed by ISHII (1992) showed that in most cases resistance was single gene based, confirming earlier results obtained with *Neurospora crassa* and *Aspergillus nidulans* (BORCK and BRAYMER 1974; VAN TUYL 1977). Cloning and characterization of β-tubulin genes from laboratory-induced resistant mutants of *N. crassa* and *A. nidulans* and transformation studies finally proved that resistance was based on single point mutations in the β-tubulin gene and gave more insight in the molecular aspects of the interaction between the tubulin molecules and the benzimidazoles (FUJIMURA et al. 1992 a; b; JUNG and OAKLEY 1990; JUNG et al. 1992; ORBACH et al. 1986). Characterization of the β-tubulin genes of laboratory-induced resistant and field-resistant strains of a number of plant pathogenic fungi identified a number of similar mutations at identical positions, thereby providing strong evidence that development of resistance to benzimidazole fungicides in field populations of target fungi in general is also based on single point mutations in their β-tubulin genes (COOLEY and CATEN 1989, 1993; KOENRAADT et al. 1992; MARTIN et al. 1992; YARDEN and KATEN 1993; WHEELER et al. 1994). Mutations conferring field resistance may represent only a small subset of the mutations recovered in laboratory experiments as has been demonstrated in *Botrytis cinerea* (YARDEN and KATAN 1993), *Rhynchosporium secalis* (WHEELER et al. 1994), *Venturia* and *Penicillia* species (KOENRAADT et al. 1992) which would allow the use of rapid diagnostic tests using DNA probes to characterize benzimidazole resistance in field strains of target fungi (HOLLOMON 1990; KOENRAADT and JONES 1992; MARTIN et al. 1992; WHEELER et al. 1994).

In this chapter the major developments in research that have led to the elucidation of the mechanisms of action of the benzimidazoles and the *N*-phenylcarbamates and the molecular basis of resistance will be discussed.

Mechanism of action and selectivity of benzimidazoles

Initial studies on the mechanism of action of the benzimidazoles focused on DNA and RNA synthesis. The observed inhibition of DNA synthesis, that in some fungi becomes evident after an initial lag period, appeared to be a secondary effect and blockage of nuclear division was primarily responsible for this effect (CLEMONS and SISLER 1971; DAVIDSE 1973; HAMMERSCHLAG and SISLER 1973).

In arresting nuclear division of fungi the benzimidazoles resemble the secondary plant metabolite colchicine. This compound inhibits mitosis and meiosis in animal and plant cells by interference with spindle formation (DUSTIN 1984). Microtubule assembly is prevented by the binding of colchicine to tubulin, the major component of microtubules. Tubulin is a heterodimer composed of the subunits α-tubulin and β-tubulin, both with molecular weights of circa 50,000 daltons (CLEVELAND and SULLIVAN 1985).

Initial binding experiments in which cell-free extracts of *A. nidulans* and ^{14}C-carbendazim were used, indicated that binding of the benzimidazoles to a cellular protein, with characteristics typical for tubulin, was involved in their mechanism of action (DAVIDSE and FLACH 1977). Binding was competitively inhibited by colchicine and nocodazole. Nocodazole, which is structurally related to carbendazim (Fig. 15.1) and was discovered in a screening programme for human antitumoral drugs, is highly inhibitory to mycelial growth of *A. nidulans*. Nocodazole is an effective inhibitor of mitosis in mammalian cells in culture (DE BRABANDER *et al.* 1975, 1976) and disrupts microtubules in a similar way as colchicine and competitively inhibits ^{3}H-colchicine binding to mammalian tubulin (HOEBEKE *et al.* 1976).

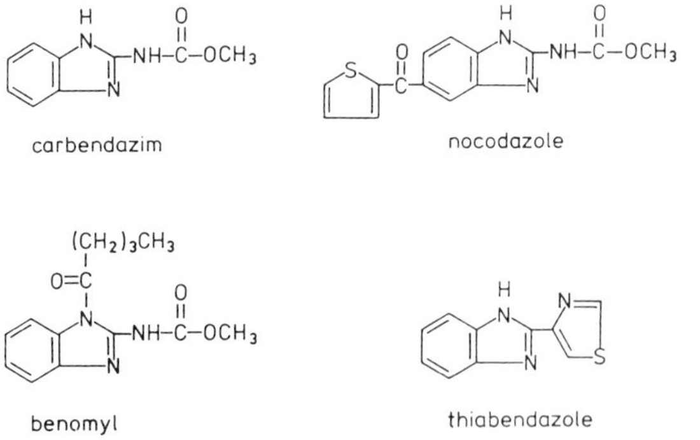

Fig. 15.1 Chemical structures of antifungal benzimidazoles.

Colchicine and nocodazole are effective inhibitors of *in vitro* microtubule assembly of tubulin from mammalian sources, whereas carbendazim only slightly affects this process (FRIEDMAN and PLATZER 1978; HOEBEKE *et al.* 1976). A structure-activity study (LACEY and WATSON 1985) indicated that the size or some colinear physico-chemical characteristic of the substituent in the 5 (or 6) position has a profound effect on the ability of benzimidazoles to inhibit *in vitro* mammalian tubulin. Assembly of yeast tubulin is sensitive to both nocodazole and carbendazim, but almost insensitive to colchicine. (Tab. 15.1; KILMARTIN 1981). Similar results were obtained with tubulin from myxamoebae of *Physarum polycephalum* (QUINLAN *et al.* 1981). Apparently 5 (or 6) substitution of the benzimidazole nucleus is essential for inhibition of mammalian microtubule assembly but not for that of fungal microtubule assembly.

Table 15.1 Effect of carbendazim, nocodazole and colchicine on *in vitro* assembly of yeast and pig brain tubulin[a]

Inhibitor	EC_{50}-value for inhibition of tubulin assembly (µM)		EC_{50}-value against growth of yeast (µM)
	pig brain	yeast	
carbendazim	>1,300	4	30
nocodazole	7	1	4
colchicine	3	2,000	>1,000

[a] KILMARTIN 1981

Interference of benzimidazoles with microtubule assembly in fungi *in vivo* has convincingly been demonstrated by HOWARD and AIST (1977, 1980), who studies the various effects of carbendazim on hyphal tip cells of *Fusarium acuminatum* by light and electron microscopy. A variety of effects could be ascribed to disappearance of microtubules. Effects included displacement of mitochondria from hyphal apices, disappearance of Spitzenkörpers which are presumed to function in hyphal linear elongation, reduction of linear growth rate, and metaphase arrest of all mitosis. D_2O, a known stabilizer of microtubules antagonized the action of carbendazim, additionally proving that carbendazim destabilizes microtubules. Ultrastructural aspects of carbendazim inhibition of mitosis in *A. nidulans* have been studied by KÜNKEL and HADRICH (1977) and KÜNKEL (1980). In the presence of carbendazim spindle formation did not take place, although the spindle pole bodies duplicated and in some cases even quadrupled indicating that both processes are independently regulated.

The experimental data and observations made in the initial studies on the mechanism of action of benzimidazoles strongly supported the idea that their biological activity is due to interference with microtubule assembly. These studies also indicated the approach to be followed in molecular-genetic research assuming that selectivity of benzimidazoles would be based on their differential binding to tubulin and that resistance of target fungi might be due to mutational changes in the tubulin genes.

Molecular basis of resistance to and selectivity of benzimidazoles

Studies on the mechanism of resistance to benzimidazole fungicides closely paralleled those on the mode of action. As will become evident the availability of genetically characterized resistant mutants of *A. nidulans* has been of utmost importance to elucidate the molecular basis of resistance to benzimidazoles. In *A. nidulans* three loci *benA*, *benB* and *benC* are involved in resistance (VAN TUYL 1977). Allelic *benA* mutants carrying the mutations *benA15* and *benA16* together with a wild-type strain were used in studies to determine binding affinity of carbendazim in cell-free extracts. Previous work had ruled out that resistance to carbendazim was due to a reduced uptake or increased metabolic conversion (DAVIDSE 1976). *BenA15* governs resistance to both carbendazim, thiabendazole and nocodazole, whereas *benA16* governs supersensitivity to carbendazim and nocodazole, but resistance to thiabendazole (Tab. 15.2). The strain carrying the lat-

Table 15.2 Sensitivity to benzimidazoles and the dissociation constants of the carbendazim-tubulin complex in strains of *Aspergillus nidulans*

Strain[a])	allele[b])	mutation	EC$_{50}$-value against growth (μM)[a]			Dissociation constant (μM)[a])
			carbendazim	nocodazole	thiabendazole	
003	wild-type	–	4.5	0.50	50	2.2
186	benA16	^{165}Val[c])	1.5	0.23	195	0.6
R	benA15	^{6}Leu[d])	95	20	800	27

[a]) Designation and data of DAVIDSE and FLACH 1977; [b]) Designation of SHEIR-NEISS et al. 1978; [c]) JUNG and OAKLEY 1990; [d]) JUNG et al. 1992

ter mutation had been isolated among several others on thiabendazole-containing medium (VAN TUYL et al. 1974). The affinity of carbendazim in cell-free extracts derived from the strain carrying the *benA15* mutation was lower and that in those derived from the strain carrying the *benA16* mutation was higher than in extracts of the wild-type strain (Tab. 15.2; DAVIDSE and FLACH 1977). Binding affinities of thiabendazole in cell-free extracts of *A. nidulans* could not be determined with ^{14}C-thiabendazole in the routinely used binding assay because of high aspecific binding of thiabendazole to components of the crude extracts and the rather low specific binding (DAVIDSE and FLACH 1978). Thiabendazole, however, inhibited ^{14}C-carbendazim-binding in cell free extracts derived from strains carrying the *benA16* mutation significantly less than those of the wild-type strain. Molecular analysis of these strains identified the *benA15* mutation as an A – T change in the 6th codon of the coding sequence of the β-tubulin gene (JUNG et al. 1992, Tab. 15.2). The mutation results in a histidine to leucine change at this amino acid position. In *benA16* two nucleotides, both in codon 165 are different from the wild type, changing the codon from GCC to GTT substituting the amino acid alanine in wild-type tubulin to valine (JUNG and OAKLEY 1990, Tab. 15.2). Hence it can be concluded that histidine at amino acid position 6 and alanine at 165 of β-tubulin have a role in binding benzimidazoles.

Binding studies with wild-type and resistant mutants of other benzimidazole sensitive species such as *B. cinerea* (GROVES et al. 1988), *Gibberella fujikuroi* (ISHII and TAKEDA 1989), *Fusarium oxysporum* f.sp. *lycopersici* (GESSLER et al. 1980), *Penicillium expansum*, *Penicillium brevicompactum* and *Penicillium corymbiferum* (DAVIDSE and FLACH 1977), *R. secalis* (KENDALL et al. 1994), *Venturia nashicola* (ISHII and DAVIDSE 1986) and species with natural resistance to benzimidazoles such as *Alternaria brassicae* and *Pythium irregulare* (DAVIDSE and FLACH 1977) showed that binding of carbendazim in cell-free extracts of resistant strains was low as compared with that in extracts of sensitive strains. The affinity of the target site to benzimidazoles apparently solely determines whether a benzimidazole has antifungal activity or not (Tab. 15.3).

Knowledge on the interaction of benzimidazoles with tubulin inspired ORBACH et al. (1986) to develop a fungal transformation system with the cloned β-tubulin gene from a benomyl-resistant strain of *N. crassa* as a dominant selectable marker. Their research did not only lead to an efficient fungal transformation system for fungi (FINCHAM 1989) but also initiated the molecular characterization of mutations in β-tubulins from several benzimidazole-resistant strains of *A. nidulans* (JUNG and OAKLEY 1990; JUNG et al. 1992) and *N. crassa* (FUJIMURA et al. 1992 a, b) and

Table 15.3 Sensitivity to benzimidazoles and the dissociation constants of the carbendazim- and thiabendazole-tubulin complex in strains of *Penicillium expansum*[a])

Strain	EC_{50} value against growth (µM)		Dissociation constant (µM)	
	carbendazim	thiabendazole	carbendazim	thiabendazole
S	0.4	7	0.9	34
SS	0.07	85	0.2	68
R	>2,500	225	10	N.D.

[a]) DAVIDSE and FLACH 1978

a number of plant pathogenic fungi (COOLEY et al. 1991; COOLEY and CATEN 1993; KOENRAADT et al. 1992; MARTIN et al. 1992; YARDEN and KATAN 1993; WHEELER et al. 1994). The results are summarized in Table 15.4. A comparison of the codon changes in the β-tubulin genes of laboratory-induced benzimidazole-resistant mutants with those of resistant field strains of plant pathogenic fungi showed that changes in the latter are restricted to codon 198 and 200. Apparently these changes have no influence on the competitive ability of the strains. Changes at other positions may in-

Table 15.4 Deduced amino acid substitutions in β-tubulins of laboratory mutants and field strains with resistance to benzimidazoles

Codon	Substitution	Organism
6	His (H) to Leu (L)	*A. nidulans*[a])
	His (H) to Tyr (Y)	*A. nidulans*[a]), *S. nodorum*[b])
50	Tyr (Y) to Asn (N)	*A. nidulans*[c])
	Tyr (Y) to Ser (S)	*A. nidulans*[c])
134	Gln (Q) to Lys (K)	*A. nidulans*[c])
165	Ala (A) to Val (V)	*A. nidulans*[d])
167	Phe (F) to Tyr (Y)	*A. crassa*[e])
198	Glu (E) to Ala (A)	*B. cinerea*[f,g]), *N. crassa*[b]), *P. aurantiogriseum*[j]), *P. expansum*[j]), *P. puberulum*[j]), *V. inaequalis*[j]), *V. pirina*[j])
	Glu (E) to Asp (D)	*A. nidulans*[a])
	Glu (E) to Gln (Q)	*A. nidulans*[a])
	Glu (E) to Gly (G)	*N. crassa*[b]), *R. secalis*[j]), *V. inaequalis*[j])
	Glu (E) to Lys (K)	*A. nidulans*[a]), *B. cinerea*[g]), *M. fructicola*[j]), *N. crassa*[h]), *P. aurantiogriseum*[j]), *P. digitatum*[j]), *P. italicum*[j]), *P. puberulum*[j]), *P. viridicatum*[j]), *R. secalis*[j]), *S. homoeocarpa*[j]), *V. inaequalis*[j]), *V. pirina*[j])
	Glu (E) to Val (V)	*P. expansum*[l]
200	Phe (F) to Tyr (Y)	*A. nidulans*[a]), *B. cinerea*[g]), *P. aurantiogriseum*[j]), *P. italicum*[j]), *V. inaequalis*[j]), *V. pirina*[l]
241	Arg (R) to His (H)	*S. saccharomyces*[k]
257	Met (M) to Leu (L)	*A. nidulans*[c])

[a]) JUNG et al. 1992; [b]) COOLEY and CATEN 1993; [c]) B. R. OAKLEY (*personal communication* to KOENRAADT et al. 1992); [d]) JUNG and OAKLEY 1990; [e]) ORBACH et al. 1986; [f]) MARTIN et al. 1992; [g]) YARDEN and KATAN 1993; [h]) FUJIMURA et al. 1992b; [i]) KOENRAADT et al. 1992; [j]) WHEELER et al. 1994; [k]) THOMAS et al. 1985.
Underlined names denote benzimidazole-resistant field strains.

fluence this ability, for instance by altering microtubule stability resulting in a change in temperature sensitivity, as has been observed in *A. nidulans* strains carrying resp. *benA17* (substitution at codon 50 from tyrosine to asparagine), *benA31* (substitution at codon 50 from tyrosine to serine), *benA33* (substitution at codon 134 from glutamine to lysine) or *benA11* (substitution at codon 257 from methionine to leucine) (OAKLEY and MORRIS 1981; pers. comm. of B. R. OAKLEY to KOENRAADT et al. 1992) and/or in altering microtubule architecture (GAMBINO et al. 1984). Site-directed mutagenesis of the β-tubulin gene at position 198 (FUJIMURA et al. 1990, 1992 a, b) unambiguously proved that substitution of glutaminic acid at position 198 in β-tubulin by either alanine, glycine or lysine results in resistance to benzimidazoles in *N. crassa*.

Assuming that the substitutions listed in table 15.4. affect the binding of benzimidazoles to tubulin, as has been demonstrated for those at codon 6 and 165 of β-tubulin of *A. nidulans,* amino acid regions of β-tubulin can be identified that are presumed to be involved in binding. If variation of the amino acid sequence at these regions of β-tubulins of various organisms correlates with the sensitivity of the organisms to benzimidazoles, it is likely that these regions form the binding site for benzimidazoles. The data in table 15.5, presenting the results of such an analysis, indeed indicate that variation exist among amino acid sequences in these regions. Since small gaps or insertions cause deviations in amino acid numbers among β-tubulins of various organisms, the numbering for β-tubulin of *A. nidulans* is used in this table and the numbering of other sequences is adapted. The most obvious conclusion from this comparison is that the regions 132–136, 163–169 and 196–202 are highly conserved in benzimidazole-sensitive filamentous fungi, *Saccharomyces cerevisiae*, *Schizosaccharomyces pombe* and the thiabendazole-sensitive protozoan *Pneumocystis carinii,* whereas the β-tubulins from these organism show slight variation in the other regions. *Candida albicans*, however, although having identical sequences in these regions as *S. cerevisiae* and *S. pombe* is highly resistant to benomyl (SMITH et al. 1988). It indicates that the other regions of the β-tubulin as well are important in determining binding affinity. *C. albicans* is the only organism of the list in table 15.5 having the acidic amino acid aspartic acid at position 48, which may suggest that the polar amino acids asparagine or serine present at this position in the β-tubulin of the other organisms play a role in binding of the benzimidazoles.

Both the two β-tubulin genes of the resistant mutant of *Trichoderma viride* have characteristics of benzimidazole-resistant β-tubulin genes (tyrosine at amino acid position 6 in *tub2* and lysine at position 198 of *tub1*) (GOLDMAN et al. 1993). Transformation of a wild-type strain with the *tub2* gene resulted in benzimidazole-resistant transformants. It proves that amino acid 6 plays a role in determining benzimidazole-resistance of resistant mutants of sensitive fungal species. None of the transformants, however, were as resistant as the donor strain. Surprisingly, transformation of a wild-type *T. viride* strain with the *tub1* gene did not result in resistant transformants. This may indicate that the presence of both genes is necessary for the expression of a high level of resistance. A definite conclusion has to await the sequencing of both homologous genes of the wild-type strain.

The fact that a valine for alanine substitution at position 165 in β-tubulin of *A. nidulans* decreases sensitivity to thiabendazole and increases sensitivity to carbendazim and nocodazole and that in the thiabendazole-sensitive nematode *Haemonchus contortus* the amino acid serine is present at this position whereas in benzimidazole resistant organisms either cysteine, leucine, methionine or asparagine is present at this position, suggest that this region is important in determining the selectivity of benzimidazoles. A search made by JUNG and OAKLEY (1990) revealed that 19 metazoan β-tubulins (11 mammal, 6 bird and 2 insect) have asparagine at position 165, 8 plant and 2 green algae β-tubulins have leucine or methionine at position 165, and 7 protozoan β-tubulins have either aspartic acid, cysteine, glutaminic acid, or methionine at this position. As postulated by JUNG and OAKLEY (1990) alanine at this position might be involved in the interaction with the R2 group of the antimicrotubule benzimidazoles. Site-directed mutagenesis at this position of β-tubulins from *A. nidulans* or *N. crassa* may provide final evidence for the hypothesis that alanine at position 165 is important in determining selectivity of benzimidazoles.

It is interesting to note that β-tubulin of the benzimidazole-resistant Oomycete *Achlya klebsiana* has the amino acid tyrosine at position 167. Substitution of the rather conservative phenylalanine at this position by tyrosine causes resistance in *N. crassa* (ORBACH et al. 1986). The β-tubulin of *A. klebsiana* at region 196–202 also resembles to a greater extent the β-tubulin genes of benzimidazole-resistant algae and plants than those of most fungi indicating the role of that region in determining benzimidazole sensitivity. The resemblance may also indicate that Oomycetes are more closely related to higher plants than to other groups of fungi.

Site-directed mutagenesis could also be used to demonstrate the importance of amino acid position 200 in determining selectivity to benzimidazoles. Whereas β-tubulins of benzimidazole-sensitive fungi have phenylalanine at this position and resistant mutants have tyrosine, β-tubulins of insensitive organisms have either methionine of tyrosine at this position, which suggest a role of phenylalanine at position 200 in binding of at least carbendazim and thiabendazole. The tyrosine for phenylalanine substitution at this position also confers resistance to nocodazole in *A. nidulans* (JUNG et al. 1992). This indicates that the presence of tyrosine at amino acid position 200 in mammalian β-tubulin at least partially accounts for some degree of resistance in insects, avians and mammals to benzimidazoles. Evidence for this may be found in the observation that although assembly in mammalian tubulin is sensitive to nocodazole, it is less affected than that of *S. cerevisiae* (Tab. 15.1; KILMARTIN 1981). Moreover, the dissociation constant of the nocodazole-rat brain tubulin complex is circa 6 μM (HOEBEKE et al. 1976) which is circa three times higher than that of the carbendazim-*A. nidulans* tubulin complex (Tab. 15.2). Since growth of the wild-type strain of *A. nidulans* is circa nine times more sensitive to nocodazole than to carbendazim, the affinity of rat brain tubulin for nocodazole apparently is relatively low as compared with that of *A. nidulans* tubulin. It can, therefore, be concluded that phenylalanine at position 200 in β-tubulin indeed plays an important role in binding of benzimidazoles.

The acidic amino acid glutaminic acid is highly conserved at position 198. The β-tubulin gene of species sensitive to carbendazim and thiabendazole as well as this gene of species that are insensitive, have glutaminic acid at this position. It suggests that although substitutions at position 198 lead to resistance, as has been demonstrated by site directed mutagenesis (FUJIMURA et al. 1992b), glutaminic acid at this position is not solely responsible for sensitivity. This amino acid might interact with some common feature of biological active benzimidazoles whereas other positions determine selectivity.

Studies on quantitative structure-activity relationships have demonstrated the role of the hydrophobicity and/or size of 5(6) substituents of the benzimidazole nucleus on the ability of benzimidazole carbamates to inhibit the assembly of sheep brain tubulin (LACEY and WATSON 1985). Apparently the interaction between benzimidazole com-

Table 15.5 Comparison of predicted amino acid sequences of β-tubulins in regions presumed to be involved in binding of benzimidazoles

Organism and sensitivity to carbendazim and thiabendazole	Codon 4...8	48...52	132...136	163....169	196.....202	239...243	255...259
	IVHLQ	NVYFN	GFQIT	MMATFSV	SDETFCI	CLRFP	VNMVP
High sensitivity:							
Aspergillus nidulans (*benA*)[a]	.....	.....	.....	.....	.....	.....	.....
Aspergillus nidulans (*tubC*)[a]	.....	.....	A..V.	.....	L	S....	.....
Acremonium coenophialum[b]	.....	S....	.....	.....	.....	.....	.....
Botrytis cinerea[c,d]	- - - -	- - aa	.....	.....	.....	.....	.....
Colletotrichum gloeosporioides[e]	..I..	S...A	.....	.....	.....	GF...	.L...
Colletotrichum graminicola (*tub1*)[f]	..I..	S...T	.....	.....	.....	GF...	..L..
Colletotrichum graminicola (*tub2*)[f]	.....	S....	.....	.....	.....	.....	.....
Epichloe typhina[g]	.....	S....	.....	.....	.....	.....	.....
Erysiphe graminis[h]	.....	.....	.....	.....	.....	S....	.....
Neurospora crassa[i]	.....	.....	.....	.....	.....	.....	.....
Septoria nodorum[j]	.....	.....	.....	.....	.....	.....	.....
Resistant mutant:							
Trichoderma viride R (*tub1*)[k]	.....	.....	.....	.....	.K....	S....	.....
Trichoderma viride R (*tub2*)[k]	.YI..	.....	.....	.....	.....	.....	.....
Intermediate sensitivity:							
Saccharomyces cerevisiae[l]	.I.IS	.....	.....	.....	.....	S..Y.	.L...
Schizosaccharomyces pombe[m]	.LI..	.....	..L..	.....	.....	S..Y.	.....
High sensitivity to thiabendazole:							
Pneumocystis carinii[n]	.....	.....	.....	I.SS..	T.....	.....	.....
Haemonchus contortus (*tubA*)[o]	...V.	...Y.	...L.	I.SS..	T.....	.....	.....
Haemonchus contortus (*tubB*)[o]	...V.	...Y.	...L.	I.SS..	T.....	.....	.....
Haemonchus contortus (*tubC*)[o]	...V.	...Y.	...L.	I..S..	T.....	.....	.....
Low sensitivity:							
Schizophyllum commune[p]	.....	S..Y.	.....	.C....	.....	S....	..L..
Dictyostelium discoideum[q]	..QI.	...Y.	...V.	.C....	A..VM.	.....	.LI..
Physarum polycephalum[r]	...I.	...Y.	...A.	.C....	A..VM.	.....	.....

312   L. C. Davidse and H. Ishii

Resistant:

Candida albicans[a]	.I..S	D....	.....	I.C.Y..	A..VM.L	S..Y.	.L..
Achlya klebsiana[c]	L..I.	..Y..	.....	..L...	A..CMVL	L....	LI..
Chlamydomonas reinhardtii[u]	..I..	.....	..VC	..L...	A..CMVL	S....	LI..
Arabidopsis thaliana[v]	.L.V.	..Y..	..VC	..L...	A..CMVL		LI..
Zea mays[w]	.L.I.	..Y..	..VC	..M...	A..CMVL		LI..
Drosophila melanogaster[x]	..I..	..Y..	..L..	I.N...	T...Y..		
Gallus gallus domestica[y]	..I..	..Y..	..L..	I.N...	T...Y..		
Homo sapiens[z]	..I..	S..Y.	..L..	I.N...	T...Y..		

[a] May et al. 1987; [b] Sequence data submitted to EMBL/GenBank/DDBJ Nucleotide Sequence Data Libraries by C. L. Schardl (1990) under the accession number X56847; [c] Martin et al. 1992; [d] Yarden and Katan 1993; [e] Buhr and Dickman 1993; [f] Panaccione and Hanau 1990; [g] Byrd et al. 1990; [h] Sheerwood and Sommerville 1990; [i] Orbach et al. 1986; [j] Cooley and Caten 1993; [k] Goldman et al. 1993; [l] Neff et al. 1983; [m] Hiraoka et al. 1984; [n] Dyer et al. 1992; [o] Geary et al. 1992; [p] Russo et al. 1992; [q] Trivinos-Lagos et al. 1993; [r] Werensklold et al. 1988; [s] Smith et al. 1988; [t] Cameron et al. 1990; [u] Youngblom et al. 1984; [v] Oppenheimer et al. 1988; [w] Hussey et al. 1990; [x] Rudolph et al. 1987; [y] Sullivan et al. 1985; [z] Lee et al. 1983; [aa] not determined.

pounds and tubulin has hydrophobic characteristics, in addition to the hydrophilic ones postulated by JUNG et al. (1993) for the interaction with histidine at aminoacid position 6. The hydrophobic interaction may take place at position 200 since substitution of the polar amino acid tyrosine for the neutral hydrophobic amino acid phenylalanine at this position confers resistance to benzimidazoles. The fact that substitution of the acidic amino acid glutaminic acid at position 198 by the basic one lysine causes high levels of resistance (FUJIMURA et al. 1990; KOENRAADT et al. 1992; YARDEN and KATAN 1993) leads FUJIMURA et al. (1990) to speculate that glutaminic acid at position 198 might interact with the imidazole moiety of the benzimidazoles. Substitution at this position, however, by the neutral amino acid alanine (KOENRAADT et al. 1992) and the acidic one apartic acid (JUNG et al. 1992) causes even greater levels of resistance to benomyl than a lysine for glutaminic acid substitution. Lysine at position 198 in β-tubulin of *A. nidulans*, however, gave a higher level of resistance to thiabendazole than asparagine, indicating that the interaction is highly complex.

Summarizing it can be concluded that these studies provide strong evidence for the involvement of at least the amino acid sequences 4–8, 48–52, 163–169 and 196–202 of the β-tubulin gene in the interaction with antimicrotubule benzimidazoles.

Negatively correlated cross-resistance between benzimidazoles, *N*-phenylcarbamates and *N*-phenylformamidoximes

A number of benzimidazole-resistant field strains of several fungal species show a higher sensitivity to *N*-phenylcarbamates. These compounds, being developed as herbicides, interfere with cellular and nuclear division of plant cells by affecting microtubule functioning (CORBETT et al. 1984). *N*-phenylcarbamates affect fungal mitosis as well (GULL and TRINCI 1973; WHITE et al. 1981).

This phenomenon of negatively correlated cross resistance of benzimidazole resistant strains to the *N*-phenylcarbamates barban and chlorpropham has been first noticed by LEROUX et al. (1979a, b) among a number of benzimidazole resistant strains of *B. cinerea* and *P. expansum*. Later on this phenomenon has also been found among resistant strains of *Pseudocercosporella herpotrichoides* (LEROUX and CAVALIER 1983a, b) and *Venturia nashicola* (ISHII et al. 1984). Only highly resistant strains of the latter fungus showed an increased sensitivity to *N*-phenylcarbamates, whereas intermediately and weakly-resistant strains did not show this phenomenon.

These observations initiated a search among *N*-phenyl compounds for compounds that did not show phytotoxicity but were still inhibitory to fungi (KATO et al. 1984). One such compound, methyl N-(3,5-dichlorophenyl)carbamate (MDPC, Fig. 15.2) inhibited growth of benzimidazole-resistant field isolates of *B. cinerea, Cercospora beticola, Fusarium nivale* and *Mycosphaerella melonis* on nutrient medium and controlled disease incited by these strains on their various hosts. Growth of benzimidazole sensitive isolates was not inhibited, nor could disease control be achieved with MDPC when benzimidazole-sensitive isolates were involved. MDPC induced similar morphological changes in germ tubes of benzimidazole resistant strains of *B. cinerea* as did carbendazim in that of sensitive strains. Furthermore MDPC arrested mitosis in a similar manner

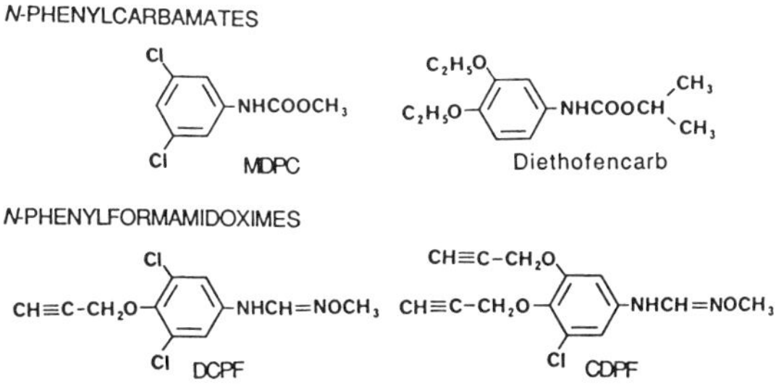

Fig. 15.2 Chemical structures of antifungal N-phenylcarbamates and N-phenylformamidoximes.

(SUZUKI et al. 1984) as carbendazim. Additional compounds that were found to be promising, are the N-phenylcarbamate isopropyl 3,4-diethoxyphenyl carbamate (NAKAMURA et al. 1986), which has been developed commercially as diethofencarb and the N-phenylformamidoximes such as N-(3,5-dichloro-4-propynyloxyphenyl)-N'-methoxyformamidine (DCPF) and N-(3-chloro-4,5-dipropynyloxyphenyl)-N'-methoxyformamidine (CDPF) (NAKATA et al. 1987). All of these compounds have a higher antifungal activity against benzimidazole highly-resistant strains than against wild-type isolates. Commercial exploitation of this phenomenon, however, is limited because strains with low and intermediate levels of resistance to benzimidazoles are sensitive to the N-phenylcarbamates and the N-phenylformamidoximes. Moreover, strains with a high level of resistance to both benzimidazoles and the N-phenyl compounds appear when populations of target fungi are exposed to mixtures of both types of fungicides (ISHII et al. 1992).

The observations that binding of DCPF and diethofencarb was higher in cell-free extracts of benzimidazole resistant mutants of, resp. *B. cinerea, V. nashicola* (ISHII and TAKEDA 1989) and *N. crassa* (FUJIMURA et al. 1992c) strongly supported the idea that the N-phenylformamidoximes and the N-phenylcarbamates like the benzimidazoles bind to tubulin. Analyses of the nucleotide sequences of β-tubulins from benzimidazole-resistant field strains of *Penicillia* and *Venturia* species revealed that an alanine or glycine substitution for glutaminic acid at position 198 conferred increased sensitivity to MDPC and diethofencarb, whereas a lysine or valine substitution for glutaminic acid at this position conferred increased sensitivity to MDPC only (KOENRAADT et al. 1992). The tyrosine for phenylalanine substitution at position 200 conferring benzimidazole resistance did not change the response of the strains to MDPC or diethofencarb. Similarly a mutation to benzimidazole resistance substituting respectively alanine for glutaminic acid in *B. cinerea* (YARDEN and KATAN 1993) and glycine for glutaminic acid in *R. secalis* (WHEELER et al. 1994) at position 198 increased sensitivity to diethofencarb, whereas a lysine for glutaminic acid substitution at this position did not change the response to diethofencarb. The latter phenomenon was also observed in benzimidazole-resistant strains of *Monilinia fructicola* and *Sclerotinia homoeocarpa* (KOENRAADT et al. 1992) and laboratory mutants of *N. crassa* (FUJIMURA et al. 1992).

Enhancement of benzimidazole activity by diphenylamine

The temperature dependent effect of diphenylamine (DPA) on benzimidazole activity against benzimidazole resistant strains of *P. expansum* is another intriguing phenomenon that has been observed in crop protection practice (ROSENBERGER and MEYER 1985). At 2 C (but not at ambient temperature) DPA proved to have an inhibitory effect on blue mold decay of apples caused by benomyl resistant strains. In combination with benomyl this effect was even more enhanced and resulted in acceptable decay control. An inhibitory effect of DPA was not noticed with benzimidazole sensitive strains. Growth in vitro of benzimidazole resistant strains was more sensitive to DPA than that of benzimidazole sensitive strains although exceptions occurred. Actual colony growth rates of resistant strains on unamended media were generally lower for benomyl resistant strains than for sensitive ones. The highly-resistant isolates had the lowest growth rate. Growth rates were enhanced by subinhibitory levels of both DPA and benomyl.

These observations are compatible with the idea that hyperstability of *P. expansum* microtubules is causing resistance to benzimidazoles and that DPA has a destabilizing effect on the mutant microtubules. At lower temperature the latter effect apparently becomes more pronounced. When benzimidazoles are additionally present reassembly is inhibited, as a consequence of which microtubules do not form.

Concluding remarks

Elucidation of the mechanism of action of a fungicide is a long process and requires the dedicated efforts of interested biologists from several disciplines, as is illustrated in this review for the benzimidazoles. The rapid development of resistance to benzimidazoles clearly has triggered the earlier research on the mode of action of the benzimidazoles and mechanisms of resistance. Initially these compounds were thought to resolve a number of problems in crop protection, but the development of resistant strains in target fungi was a serious drawback that obviously needed to be investigated.

When the target site of the benzimidazoles was identified as tubulin and resistance could be ascribed to amino acid substitutions in β-tubulin, cell biologists became excited about the value of these compounds as tools to study microtubule structure and functioning (OAKLEY 1985). This, indisputably, has led to a thorough understanding of the mode of action of the benzimidazoles, details of which are still being investigated. Although this knowledge has not solved resistance problems in crop protection, it greatly facilitates the characterization of resistant strains. Since mutations in β-tubulin leading to resistance in field populations of target fungi are restricted to codons 198 and 200, specific amplifications of an appropriate β-tubulin sequence using the polymerase chain reaction (PCR) and analysis of the amplified DNA sequence by allele-specific oligonucleotide probes is an elegant and rapid method to characterize strains collected from the field. This method can even be directly applied to lesions eliminating the need for laborious isolation procedures of the pathogen (KOENRAADT and JONES 1992). Direct characterization of either sensitive- or resistant β-tubulin using specific primers in a nested primer system may lead to an even more efficient system to characterize field populations (MARTIN *et al.* 1992; WHEELER *et al.* 1994).

It is beyond doubt that the benzimidazoles have contributed to a better understanding of the structure and functioning of microtubules. Our increasing knowledge in this area will help us to understand the fundamental aspects of cell structure and functioning. It immediately evokes the question whether this knowledge may lead to the discovery of new compounds, that can be used in plant disease control, veterinary medicine or cancer therapy. Since site-directed mutagenesis of β-tubulin of fungi is a reality now, structural modelling of β-tubulin sequences assumed to be involved in binding of benzimidazoles based on comparison of near-homologous sequences with known structures is possible. And since large number of benzimidazoles, N-phenylcarbamates and N-phenylformamidoximes are available to be used in structure-activity studies with characterized strains and conditions for binding studies have been established, it is possible to investigate in detail the molecular interactions involved in the mechanism of action of this group of inhibitors of microtubule assembly. It is predictable that progress in this type of studies will get momentum as soon as X-ray diffraction studies have elucidated the structure of tubulin. It is a challenge for academia and industry to explore this research area in a concerted action, which may lead to the specific design of new agents that interfere with microtubule functioning and that are candidates for commercialization in various application areas.

References

BORCK, K., and BRAYMER, H. D.: The genetic analysis of resistance to benomyl in *Neurospora crassa*. J. Gen. Microbiol. **85** (1974): 51–56.

BORISY, G. G., CLEVELAND, D. W., and MURPHY, D. G.: Molecular Biology of the Cytoskeleton. Cold Spring Harbor Press, New York 1984, 512 pp.

BUHR, T. L., and DICKMAN, M. B.: Isolation and characterization of a beta-tubulin gene from *Colletotrichum gloesporioides* f.sp. *aeschynomene*. Gene **124** (1993): 121–125.

BURLAND, T. G., and GULL, K.: Molecular and cellular aspects of the interaction of benzimidazole fungicides with tubulin and microtubules. In: TRINCI, A. P. J., and RYLEY, J. F. (Eds.): Mode of action of antifungal agents. Brit. Mycol. Soc. Symposia Series 8, Cambridge Univ. Press, Cambridge 1984, pp. 299–320.

BYRD, A. D., SCHARDL, C. L., SONGLIN, P. J., MOGEN, K. L., and SIEGEL, M. R.: The beta-tubulin gene of *Epichloe typhina* from perennial ryegrass (*Lolium perenne*). Curr. Genet. **18** (1990): 347–354.

CAMERON, L. E., HUTSUL, J.-A., THORLACIUS, L., and LEJOHN, H. B.: Cloning and analysis of beta-tubulin gene from a prototist. J. Biol. Chem. **265** (1990): 15245–15252.

CLEMONS, G. P., and SISLER, H. D.: Localization of the site of action of a fungitoxic benomyl derivative. Pestic. Biochem. Physiol. **1** (1971): 32–43.

CLEVELAND, D. W., and SULLIVAN, K. F.: Molecular biology and genetics of tubulin. Ann. Rev. Biochem. **54** (1985): 331–365.

COOLEY, R. N., VAN GORCOM, R. F. M., VAN DEN HONDEL, C. A. M. J. J., and CATEN, C.: Isolation of a benomyl-resistant allele of the β-tubulin gene from *Septoria nodorum* and its use as a dominant selectable marker. J. Gen. Microbiol. **137** (1991): 2085–2091.

– and CATEN, C. E.: Molecular analysis of the *Septoria nodorum* β-tubulin gene and characterization of a benomyl-resistance mutation. Mol. Gen. Genet. **237** (1993): 58–64.

CORBETT, J. R., WRIGHT, K., and BAILLIE, A. C.: The biochemical mode of action of pesticides. Academic Press, New York 1984, 382 pp.

DAVIDSE, L. C.: Antimitotic activity of methyl benzimidazol-2-ylcarbamate (MBC) in *Aspergillus nidulans*. Pestic. Biochem. Physiol. **3** (1973): 317–325.

– Metabolic conversion of methyl-benzimidazol-2-ylcarbamate (MBC) in *Aspergillus nidulans*. Pestic. Biochem. Physiol. **6** (1976): 538–546.

– Benzimidazole compounds; Selectivity and Resistance. In: DEKKER, J., and GEORGOPOULOS, S. G. (Eds.): Fungicide Resistance in Crop Protection. Pudoc, Wageningen 1982, pp. 60–70.

- Benzimidazole fungicides: Mechanism of action and biological impact. Ann. Rev. Phytopathol. **24** (1986): 43–65.
- Biochemical aspects of benzimidazole fungicides – action and resistance. In: Lyr, H. (Ed.): Modern Selective Fungicides – Properties, Applications, Mechanisms of Action. Longman Group UK Ltd., London, and VEB Gustav Fischer Verlag, Jena 1987, pp. 275–282.
- and de Waard, M. A.: Systemic Fungicides. In: Ingram, D. S., and Williams, P. H. (Eds.): Advances in Plant Pathology, Vol. 2. Academic Press, London 1984, pp. 191–257.
- and Flach, W.: Differential binding of methyl benzimidazol-2-yl carbamate to fungal tubulin as a mechanism of resistance to this antimitotic agent in mutant strains of *Aspergillus nidulans*. J. Cell Biol. **72** (1977): 174–193.
- – Interaction of thiabendazole with fungal tubulin. Biochem. Biophys. Acta **543** (1978): 82–90.

de Brabander, M., van de Veire, R., Aerts, F., Geuens, G., Borgers, M., Desplenter, L., and de Crée, J.: Oncodazole (R 17934): a new anticancer drug interfering with microtubules. Effects on neoplastic cells cultured in vitro and in vivo. In: Borgers, M., and de Brabander, M. (Eds.): Microtubules and Microtubule Inhibitors. North Holland/Elsevier, Amsterdam – New York 1975, pp. 509–521.
- – – Borgers, M., and Janssen, P. A. J.: The effects of methyl [5-(2-thienylcarbonyl)-1H-benzimidazol-2-yl]-carbamate (R 17934; NSC 238159), a new synthetic antitumoral drug interfering with microtubules, on mammalian cells cultured in vitro. Cancer Res. **36** (1976): 1011–1018.

Dekker, J.: The development of resistance to fungicides. In: Hutson, D. H., and Roberts, T. R. (Eds.): Progress in Pesticide Biochemistry and Toxicology, Vol. 4. Wiley and Sons Ltd., New York 1985, pp. 165–218.

Dustin, P.: Microtubules. 2nd edit. Springer-Verlag, Berlin–Heidelberg 1984, 482 pp.

Dyer, M., Volpe, F., Delves, C., Somia, N., Burns, S., and Scaife, J.: Cloning and sequence of a beta-tubulin cDNA from *Pneumocystis carinii*: possible implications for drug therapy. Mol. Microbiol. **6** (1992): 991–1001.

Fincham, J. R. S.: Transformation in fungi. Microbiol. Rev. **53** (1989): 148–170.

Friedman, P. A., and Platzer, E. G.: Interaction of anthelmintic benzimidazoles and benzimidazole derivatives with bovine brain tubulin. Biochim. Biophys. Acta **544** (1978): 605–614.

Fujimura, M., Oeda, K., Inoue, H., and Kato, T.: Mechanism of action of N-phenylcarbamates in benzimidazole-resistant *Neurospora* strains. In: Green, M. B., Lebaron, H. M., and Moberg, W. K. (Eds.): Managing Resistance to Agrochemicals. ACS Ser. 421, ACS Press, Washington. DC, 1990, pp. 224–236.
- – – – A single amino-acid substitution in the beta-tubulin gene of *Neurospora* confers both carbendazim resistance and diethofencarb sensitivity. Curr. Genet. **21** (1992a): 399–404.
- Kamakura, T., Inoue, H., Ino, S., and Yamaguchi, I.: Sensitivity of *Neurospora crassa* to benzimidazoles and N-phenylcarbamates: effect of amino acid substitutions at position 198 in β-tubulin. Pestic. Biochem. Physiol. **44** (1992b): 165–173.
- – and Yamaguchi, I.: Action mechanism of diethofencarb to a benzimidazole-resistant mutant in *Neurospora crassa*. J. Pestic. Sci. **17** (1992c): 237–242.

Gambino, J., Bergen, L. C., and Morris, N. R.: Effects of mitotic and tubulin mutations on microtubule architecture in actively growing protoplasts of *Aspergillus nidulans*. J. Cell Biol. **99** (1984): 830–838.

Geary, T, S., Nulf, S. C., Faureau, M. A., Tang, L., Prichard, R. K., Hatzenbuhler, N. T., Shea, M. H., Alexander, S. J., and Klein, R. D.: Three beta-tubulin cDNAs from the parasitic nematode, *Haemonchus contortus*. Mol. Biochem. Parasitol. **50** (1992): 295–306.

Gessler, C., Sozzi, D., and Kern, H.: Benzimidazol-fungicide: Wirkungsweise und Probleme. Ber. Schweizer. Bot. Ges. **90** (1980): 45–54.

GOLDMAN, G. H., TEMMERMAN, W., JACOBS, D., CONTRERAS, S., VAN MONTAGU, M., and HERRERA-ESTRELLA, A.: A nucleotide substitution in one of the β-tubulin genes of *Trichoderma viride* confers resistance to the antimitotic drug methyl benzimidazole-2-yl-carbamate. Mol. Gen. Genet. **240** (1993): 73–80.

GROVES, J. D., and FOX, R. T. V.: Modes of action of carbendazim and ethyl N-(3,5-dichlorophenyl) carbamate on field isolates of *Botrytis cinerea*. Brighton Crop Protection Conference – Pests and Diseases (1988): 397–402.

GULL, K., and TRINCI, A. P. J.: Griseofulvin inhibits fungal mitosis. Nature **244** (1973): 2920–2930.

HAMMERSCHLAG, R. S., and SISLER, H. D.: Benomyl and methyl-2-benzimidazole carbamate (MBC): biochemical, cytological and chemical aspects of toxicity to *Ustilago maydis* and *Saccharomyces cerevisiae*. Pestic. Biochem. Physiol. **3** (1973): 42–54.

HIRAOKA, Y., TODA, T., and YANAGIDA, M.: The *NDA3* gene of fission yeast encodes β-tubulin: a cold-sensitive *nda3* mutation reversibly blocks spindle formation and chromosome movement in mitosis. Cell **39** (1984): 349–358.

HOEBEKE, J., VAN NYEN, C., and DE BRABANDER, M.: Interaction of oncodazole (R 17934), a new antitumoral drug, with rat brain tubulin. Biochem. Biophys. Res. Comm. **69** (1976): 319–324.

HOLLOMON, D. W.: Molecular approaches to understanding the mechanisms of fungicide resistance. Brighton Crop Protection Conference – Pests and Diseases (1990): 881–888.

HOWARD, R. J., and AIST, J. R.: Effects of MBC on hyphal tip organization, growth, and mitosis of *Fusarium acuminatum*, and their antagonism by D_2O. Protoplasma **92** (1977): 195–210.

– – Cytoplasmic microtubules and fungal morphogenesis: Ultrastructural effects of methyl benzimidazol-2-yl carbamate determined by freeze-substitution of hyphal tip cells. J. Cell Biol. **87** (1980): 55–64.

HUSSEY, P. J., HAAS, N., HUSNPERGER, J., LARKIN, J., SNUSTAD, D. P., and SILFLOW, C. D.: The beta-tubulin gene family in *Zea mays*: two differentially expressed beta-tubulin genes. Plant Mol. Biol. **15** (1990): 957–972.

ISHII, H.: Target sites of tubulin-binding fungicides. In: KÖLLER, W. (Ed.): Target Sites of Fungicide Action. CBC Press Inc., Boca Raton 1992, pp. 43–52.

– and DAVIDSE, L. C.: Decreased binding of carbendazim to cellular protein from *Venturia nashicola* and its involvement in benzimidazole resistance. Brighton Crop Protection Conference – Pests and Diseases (1986): 567–573.

– and TAKEDA, H.: Differential binding of a N-phenylformamidoxime compound in cell free extracts of benzimidazole-resistant and – sensitive isolates of *Venturia nashicola, Botrytis cinera* and *Gibberella fujikuroi*. Neth. J. Pl. Path. **95** (1989) Supplement 1: 99–108.

– TOMIKAWA, A., VAN RAAK, M., and INOUE, I.: Limitations in the exploitation of N-phenylcarbamates, and N-phenylformamidoximes to control benzimidazole-resistant *Venturia nashicola* on Japanese pear. Plant Pathol. **41** (1992): 543–553.

– YANASE, H., and DEKKER, J.: Resistance of *Venturia nashicola* to benzimidazole fungicides. Meded. Fac. Landbouwwet. Rijksuniv. Gent **49** (1984): 163–172.

JUNG, M. K., and OAKLEY, B. R.: Identification of an amino acid substitution in the *benA*, β-tubulin gene of *Aspergillus nidulans* that confers thiabendazole resistance and benomyl supersensitivity. Cell Motil. Cytoskeleton **17** (1990): 87–94.

– WILDER, I. B., and OAKLEY, B. R.: Amino acid alterations in the *benA* (Beta-tubulin) gene of *Aspergillus nidulans* that confer benomyl resistance. Cell Motil. Cytoskeleton **22** (1992): 170–174.

KATO, T., SUZUKI, K., TAKAHASHI, J., and KAMOSHITA, K.: Negatively correlated cross-resistance between benzimidazole fungicides and methyl N-(3,5-dichlorophenyl)carbamate. J. Pestic. Sci. **9** (1984): 489–495.

KENDALL, S., HOLLOMON, D. W., ISHII, H., and HEANY, S. P.: Characterization of benzimidazole-resistant strains of *Rhynchosporium secalis*. Pest. Sci. **40** (1994): 175–181.

KILMARTIN, J. V.: Purification of yeast tubulin by self-assembly in vitro. Biochemistry **20** (1981). 3629–3633.

KOENRAADT, H., and JONES, A. L.: The use of allele-specific oligonucleotide probes to characterize resistance to benomyl in field strains of *Venturia inaequalis*. Phytopathology **82** (1992): 1354–1358.

— SOMERVILLE, S. C., and JONES, A. L.: Characterization of mutations in the beta-tubulin gene of benomyl-resistant field strains of *Venturia inaequalis* and other plant pathogenic fungi. Phytopathology **82** (1992): 1348–1354.

KÜNKEL, W.: Antimitotische Aktivität von Methylbenzimidazol-2-yl carbamat (MBC). I. Licht-, elektronenmikroskopische und physiologische Untersuchungen an keimenden Konidien von *Aspergillus nidulans*. Z. Allgem. Mikrobiol. **20** (1980): 113–120.

— and HADRICH, H.: Ultrastrukturelle Untersuchungen zur antimitotischen Aktivität von Methylbenzimidazol-2-yl carbamat (MBC) und seinem Einfluß auf die Replikation des Kern-assoziierten Organells ("centriolar plaque", "MTOC", "KCE") bei *Aspergillus nidulans*. Protoplasma **92** (1977): 311–323.

LACEY, E., and WATSON, T. R.: Structure-activity relationships of benzimidazole carbamates as inhibitors of mammalian tubulin, *in vitro*. Biochem. Pharmac. **34** (1985): 1073–1077.

LANGCAKE, P., KUHN, P. J., and WADE, M.: The mode of action of systemic fungicides. In: HUTSON, D. H., and ROBERTS, T. R. (Eds.): Progress in Pesticide Biochem. Toxicology. Vol. 3. John Wiley and Sons Ltd., New York 1983, pp. 1–109.

LEE, M. G. M.-S., LEWIS, S. A., WILDE, C. D., and COWAN, N. J.: Evolutionary history of a multigene family: an expressed human β-tubulin gene and three processed pseudogenes. Cell **33** (1983): 477–487.

LEROUX, P., and CAVELLIER, N.: Charactéristiques des souches de *Pseudocercosporella herpotrichoides* (Agent du piétin-verse des cerealis) résistantes auz fongicides benzimidazoles et thiophanates. La Défense des Végétaux **222** (1983 a): 231–238.

— — Phénomènes de résistance du piétin-verse auz benzimidazoles et aux thiophanates. Phytoma-Defense des cultures **353** (1983 b): 40–47.

— and GREDT, M.: Effets du barbane, du chlorbufame, du chlorprophame et du prophame sur diverses souches de *Botrytis cinerea* Pers. et de *Penicillium expansum* Link sensibles ou résistantes au carbendazim et au thiabendazole. Comptes Rendues Academie des Sciences Paris, Serie D **289** (1979 a): 691–693.

— — Phénomènes de résistance croisée négative chez *Botrytis cinerea* Pers. entre les fongicides benzimidazoles et des herbicides carbamates. Phytiatrie-Phytopharmacie **28** (1979 b): 79–86.

MARTIN, L. A., FOX, R. T. V., BALDWIN, B. C., and CONNERTON, I. F.: Use of polymerase chain reaction for the diagnosis of MBC resistance in *Botrytis cinerea*. British Crop Protection Conference — Pest and Diseases (1992): 207–214.

MAY, G. S., TSANG, L.-S., SMITH, H., FIDEL, S., and MORRIS, N. R.: *Aspergillus nidulans* β-tubulin genes are unusually divergent. Gene **55** (1987): 231–243.

MORRIS, N. R.: The molecular genetics of microtubule proteins in fungi. Exp. Mycology **10** (1986): 77–82.

NAKAMURA, S., KATO, T., NOGUCHI, H., TAKAHASHI, J., and KAMOSHITA, K.: Biological activity and mode of action of a resistance breaker, diethofencarb. Abstracts 6th International Congress of Pesticide Chemistry (1986): 3E-01.

NAKATA, A., SANO, S., HASHIMOTO, S., HAYAKAWA, K., NISHIKAWA, H., and YASUDA, Y.: Negatively correlated cross-resistance to *N*-phenylformamidoximes in benzimidazole-resistant phytopathogenic fungi. Ann. Phytopath. Soc. Japan **53** (1987): 659–662.

NEFF, N. F., THOMAS, J. H., GRISAFI, P., and BOTSTEIN, D.: Isolation of the β-tubulin gene from yeast and demonstration of its essential function in vivo. Cell **33** (1983): 211–219.

OAKLEY, B. R.: Microtubule mutants. Can. J. Biochem. Cell Biol. **63** (1985): 479–488.

- and MORRIS, N. R.: A β-tubulin mutation in *Aspergillus nidulans* that blocks microtubule function without blocking assembly. Cell **24** (1981): 837–845.
OPPENHEIMER, D. G., HAAS, N., SILFLOW, C. D., and SNUSTAD, D. P.: The β-tubulin gene family of *Arabidopsis thaliana*: preferential accumulation of the β1 transcript in roots. Gene **63** (1988): 87–102.
ORBACH, M. J., PORRO, E. B., and YANOVSKY, C.: Cloning and characterization of the gene for β-tubulin from a benomyl-resistant mutant of *Neurospora crassa* and its use as a dominant selectable marker. Mol. Cell. Biol. **6** (1986): 2452–2461.
PANACCIONE, D. G., and HANAU, R. M.: Characterization of two divergent beta-tubulin genes from *Colletotrichum graminicola*. Gene **86** (1990): 163–170.
QUINLAW, R. A., ROOBOL, A., POGSON, C. I., and GULL, K.: A correlation between in vivo and in vitro effects of the microtubule inhibitors colchicine, parbendazole, and nocodazole on myxamoebae of *Physarum polycephalum*. J. Gen. Microbiol. **122** (1981): 1–6.
ROSENBERGER, D. A., and MEYER, F. W.: Negatively correlated cross-resistance to diphenylamine in benomyl-resistant *Penicillium expansum*. Phytopathology **75** (1985): 74–79.
RUDOLPH, J. E., KIMBLE, M., HOYLE, H. D., SUBLER, M. A., and RAFF, E. C.: Three *Drosophila* beta-tubulin sequences: a developmentally regulated isoform ($\beta3$), the testes-specific isoform ($\beta2$), and an assembly-defective mutation of the testis-specific isoform ($\beta2t^8$) reveal both an ancient divergence in metazoan isotypes and structural constraints for beta-tubulin function. Mol. Cell. Biol. **7** (1987): 2231–2242.
RUSSO, P., JUUTI, J. T., and RAUDASKOSKI, M.: Cloning, sequence and expression of a beta-tubulin-encoding gene in the homobasidiomycete *Schizophyllum commune*. Gene **119** (1992): 175–182.
SHEIR-NEISS, G., LAI, M. H., and MORRIS, N. R.: Identification of a gene for β-tubulin in *Aspergillus nidulans*. Cell **15** (1978): 639–649.
SHERWOOD, J. E., and SOMERVILLE, S. C.: Sequence of the *Erysiphe graminis* f.sp. *hordei* gene encoding β-tubulin. Nucleic Acids Res. **18** (1990): 1052.
SMITH, H. A., ALLAUDEEN, H. S., WHITMAN, M. H., KOLTIN, Y., and GORMAN, A.: Isolation and characterization of a β-tubulin gene from *Candida albicans*. Gene **63** (1988): 53–63.
SULLIVAN, K. F., LAU, J. T. Y., and CLEVELAND, D. W.: Apparent gene conversion between β-tubulin genes yields multiple regulatory pathways for a single β-tubulin polypeptide isotype. Mol. Cell. Biol. **5** (1985): 2454–2465.
SUZUKI, K., KATO, T., TAKAHASHI, J., and KAMOSHILA, K.: Mode of action of methyl N-(3,5-dichlorophenyl)carbamate in the benzimidazole-resistant isolate of *Botrytis cinerea*. J. Pestic. Sci. **9** (1984): 497–501.
THOMAS, J. H., NEFF, N. F., and BOTSTEIN, D.: Isolation and characterization of mutations in the β-tubulin gene of *Saccharomyces cerevisiae*. Genetics **112** (1985): 715–734.
TRIVINOS-LAGOS, L., OHMACHI, T., ALBRIGHTSON, C., BURNS, R. G., ENNIS, H. L., and CRISHOLM, R. L.: The highly divergent alpha- and beta-tubulins from *Dictyostelium discoideum* are encoded by single genes. J. Cell Sci. **105** (1993): 903–911.
VAN TUYL, J. M.: Genetics of fungal resistance to systemic fungicides. Meded. Landbouwhogeschool **77-2** (1977): 1–127.
- DAVIDSE, L. C., and DEKKER, J.: Lack of cross-resistance to benomyl and thiabendazole in some strains of *Aspergillus nidulans*. Neth. J. Plant Pathol. **80** (1974): 165–168.
WERENSKIOLD, A. K., POETSCH, B., and HAUGLI, F.: Cloning and expression of a beta-tubulin gene of *Physarum polycephalum*. Eur. J. Biochem. **174** (1988): 491–495.
WHEELER, I., KENDALL, S., BUTTERS, J., and HOLLOMON, D.: Rapid detection of benzimidazole resistance in *Rhynchosporium secalis* using allelespecific oligonucleotide probes. In: HEANY, S. P., HOLLOMON, D., PARRY, D. A., SLAWSON, D., and SMITH, M. (Eds.): Proceedings BSPP Fungicide Resistance Conference. BSPC, Farnham UK, in press.
WHITE, E., SCANDELLA, D., and KATZ, E. R.: Inhibition by CIPC of mitosis and development in

Dictyostelium discoideum and the isolation of CIPC-resistant mutants. Dev. genet. **2** (1981): 99–111.

YARDEN, O., and KATAN, T.: Mutations leading to substitutions at amino acids 198 and 200 of beta-tubulin that correlate with benomyl-resistance phenotypes of field strains of *Botrytis cinerea*. Phytopathology **83** (1993): 1478–1483.

YOUNGBLOM, J., SCHLOSS, J. A., and SILFLOW, C. D.: The two β-tubulin genes of *Clamydomonas reinhardii* code for identical proteins. Mol. Cell. Biol. **4** (1984): 2686–2696.

CHAPTER 16

Oomycetes fungicides

16.1 Phenylamides and other fungicides against *Oomycetes*

F. SCHWINN* and T. STAUB**

*Institute of Microbiology, Swiss Fed. School of Technology, Zürich, Switzerland
**Ciba-Geigy Ltd., Crop Protection Division,
Basle, Switzerland

Introduction

In modern taxonomy it is generally accepted that fungal organisms having motile stages (= zoospores) in their life cycle, are phylogenetically different from the true fungi and consequently areto be placed in the kingdom protista (GRIFFITH et al. 1992; KENDRICK 1992). One of the divisions of protista is that of the *Oomycetes* (taxonomically correct: *Oomycota*), which comprises some 70 genera with more than 800 species, living partly in aquatic, partly in terrestrial biotopes as saprophytes or plant parasites. Despite their ecological diversity, they are a well defined unit of high physiological and biochemical uniformity, well separated from all other taxa. Their classification is shown in table 16.1.

Table 16.1 Taxonomy of the *Oomycetes*

Subdivision	**Mastigomycotina**		
Class:	**Oomycetes**		
Order:	*Peronosporales*		
Family:	*Pythiaceae*	*Peronosporaceae*	*Albuginaceae*
	(non-obligate)	(obligate-biotrophic)	(obligate)
		= Downy Mildews	
Genus:	*Pythium*	*Bremia*	*Albugo*
	Phytophthora	*Peronospora*	
		Peronosclerospora	
		Plasmopara	
		Pseudoperonospora	
		Sclerophthora	
		Sclerospora	

Among the *Oomycetes*, the *Peronosporales* comprise most of the plant parasitic genera and species, attacking plants from the seedling to the mature stage and causing diseases on leaves, fruits, stems, crowns, and roots on a wide range of annual and perennial crops in temperate and tropical climates. Some of them have an extremely narrow host spectrum, others may attack more than one hundred different plants species. Their wide distribution and their high potential of causing heavy epidemics within very short periods of favorable weather resulting in great economic losses makes them a devastating group of plant pathogens. (SCHWINN 1992). This holds particularly true for foliar pathogens such as late blight of potatoes, caused by *Phytophthora infestans* (INGRAM and WILLIAMS 1991) and the downy mildews (SPENCER 1981; LEBEDA and SCHWINN 1994). Their control had high priority since the beginning of modern plant protection (SCHWINN and URECH 1986).

The first effective foliar fungicides for practical use against late blight and downy mildews were copper compounds, particularly copper sulfate in combination with limestone, the famous Bordeaux mixture (introduced 1885) and cuprous oxide (introduced 1932). They were followed by the ethylene bis-dithiocarbamates (introduced from 1931–1962), the phthalimides captan, folpet and captafol (1949–1965), the triphenyl tin compounds (1954) and chlorothalonil, in (1963) (STAUB and HUBELE 1981). Despite being non-selective biocides these fungicides can be used in crop protection because they do not penetrate into the plant tissue and thus do not affect it. Apart from the organo tin compounds which have some curative effect, they are purely residual and protective fungicides, i.e. they protect only those plant parts against diseases which were treated. Since they stay on the plant surface, they are exposed to rainfall and weathering, a fact which requires repeated applications. The level of inherent fungitoxicity, their residual behavior and exposure to wash-off result in fairly high application rates in the order of 1 to 2.5 kg of active ingredient/ha.

Until the mid 1970's for the control of soil-borne root- or crown-infecting *Oomycetes* the main fungicides were non-specific biocidal soil sterilants like vapam or methylbromide particularly in horticultural crops. Their use is limited to plant-free periods due to their lack of selectivity between pathogens and crop plant. In addition, some of the foliar fungicides described above are used as soil drenches or root dips, such as captan, zineb, or mancozeb.

One group of *Oomycetes* which until the introduction of modern, systemic compounds was not amenable to chemical control, are the systemic downy mildews. They occur on a large range of host plants (Tab. 16.2) with a wide geographical distribution, and may cause considerable losses in staple crops like sorghum, maize and millet as well as in cash crops like tobacco, hops, brassicas and sunflower.

The control of diseases caused by *Oomycetes* always has been a major element in chemical plant protection. The economic significance of these pathogens is illustrated by the fact that in 1991 about 20% of the worldwide expenses for chemical disease control were devoted to their control as shown in figure 16.1. It also illustrates the relative

Table 16.2 Major systemic diseases, caused by *Oomycetes, Peronosporales*

Name	Host plant	Distribution
Peronosclerospora maydis	maize	Asia, Australia
Peronosclerospora sorghi	sorghum	Asia, Africa, America
Peronospora parasitica	brassicas	Europe, Africa, North America
Peronospora tabacina	tobacco	North and Central America, Australia, Europe
Plasmopara halstedii	sunflower	Europe, America, Africa, Asia
Pseudoperonopora cubensis	cucumber	Worldwide in humid and warm zones
Pseudoperonospora humuli	hops	Europe, North America
Sclerophthora macrospora	cereals, maize, rice, sugarcane	Asia, America, South Europe
Sclerospora graminicola	pearl millet	Asia, Africa, North America

economic importance of major diseases in this class, based on recent expenditures for their control. The dominating role of downy mildew of grapes and vegetables as well as late blight in potatoes is evident.

A

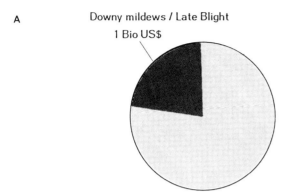

Downy mildews / Late Blight
1 Bio US$

Fig. 16.1 A Relative importance of diseases caused by *Oomycetes* (= Downy mildews) in the world fungicide market. Total: 4.7 Bio US $ (1991, industry level)

B

DOWNY MILDEW MARKET
(incl. Late Blight)

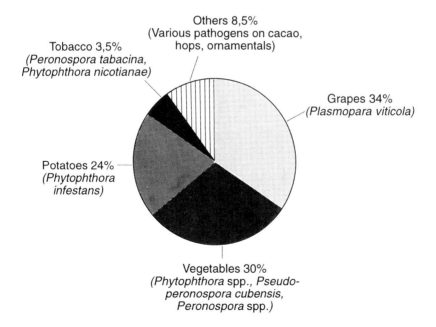

Source: Ciba-Geigy

Fig. 16.1 B relative importance of major crops in the Downy Mildew fungicides market.

Chemistry

During the past 20 years representatives of seven new classes of fungicides controlling diseases caused by *Oomycetes* were introduced:

- the carbamates,
- the isoxazoles,
- the cyanoacetamide oximes,
- the ethyl phosphonates,
- the phenylamides,
- the 2,6 dinitroanilides,
- the morpholines

Recently, the morpholines and the 2,6 dinitroanilides compounds belonging to the novel class of strobilurine analogues, i.e. the metoxyacrylates and the oxim+ether+esters reached advanced stages of development.

For the sake of completeness, we also list here the class of the benzamides with zarilamide (ICI A001) as the main representative (HEANEY et al. 1988). Even though this compound was recently dropped from further development, it is worth mentioning because it has a novel mode of action: it inhibits mitosis by interfering with the microtubule cytoskeleton (YOUNG 1991).

Finally, we mention andoprim (4-methoxy-N-(4,6-dimethyl pyrimidin-2-yl) aniline) which inhibits selectively *Phytophthora* species. Based on preliminary results, it interferes with active cellular transport systems (GRUNWALDT et al. 1990). For details see chapter 19.

Older fungicides with specific activity against *Oomycetes*, but little commercial impact, such as fenaminosulf, etridiazole and chloroneb, are not discussed here. They were reviewed by DAVIDSE and DE WAARD (1984) and are covered by chapter 5 of this book.

The chemical structure of typical representatives of the above classes, are listed in table 16.3. Whereas in groups 1, 2, 3, 4, 6 and 7 there is one commercial compound per group, there are two in the strobilurin analogues and five in the phenylamides. The first sales products in this latter group were the acylalanines furalaxyl and metalaxyl, (Tab. 16.3) followed by the acylamino butyrolactone ofurace, the acylalanine benalaxyl and the acylamino-oxazolidinone oxadixyl (STAUB and HUBELE 1981). Cyprofuram, launched by Schering, was withdrawn from the market after a short time. The history of discovery of the acylalanines was described by STAUB and HUBELE (1981), structure-activity relations were studied by HUBELE et al. (1983) and GOZZO et al. (1985), methods of preparation of enantiomers and their biological activity were investigated by MOSER and VOGEL (1978), SCHWINN and STAUB (1982) and HUBELE et al. (1983). An interesting aspect is the high water solubility of many of the specific *Oomycetes* fungicides, in comparison to that of fungicides with a broader spectrum (Tab. 16.4), as discussed by BRUIN and EDGINGTON (1983). More research is needed in order to find out the role of this factor in fungitoxicity against *Oomycetes*.

Major trade names and formulations as well as first literature quotations are summarized in table 16.5. It is remarkable that after a productive period from 1974 to 1983 several new *Oomycetes* fungicides have been described since 1988. This reflects the ongoing intensity and success of industrial research in this sector.

In general, wettable powder formulations are prevailing. Metalaxyl is sold in the broadest range of commercial formulations, making it the most versatile product in terms of ways of applications. The numerous mixtures of *Oomycetes* fungicides with conventional products are not mentioned here.

Table 16.3 Chemistry of recent fungicides against Oomycetes

Chemical class Common name*, originating company, code number Chemical structure (*after Worthing and Hance, 1991)	First report Chemical name* Patent number*
1. Carbamates 1.1. Propamocarb (Schering. SN 66752)	PIEROH et al. (1978) propyl 3-(dimethylamino) propyl carbamate (DEP 1 567 169; 1 643 040)
1.2. Prothiocarb (Schering. SN 41 703)**	BASTIAANSEN et al. (1974) S-ethyl (3-dimethylamonipropyl) thiocarbamate (DEP 1 567 169)
2. Isoxazoles 2.1. Hymexazol (Sankyo, F-319, SF-6 505)	TAKAHI et al. (1974 a, b) 5-methyl- 1,2-oxazol-3-ol (JPP 518 249; 532 202)
3. Cyanoacetamide-oximes 3.1. Cymoxanil (Du Pont, DPX 3 217)	SERRES and CARRARO (1976) 1-(2-cyano-2-methoxyiminoacetyl)-3-ethylurea (USP 3 957 847)
4. Ethyl phosphonates 4.1. Fosetyl (Rhone-Poulenc, LS 74-783)	BERTRAND et al. (1977), WILLIAMS et al. (1977), ethylhydrogen phosphonate, aluminium salt (FP 2 254 276)
5. Phenylamides 5.1. Acylalanines Furalaxyl (Ciba-Geigy, CGA 38 140)	SCHWINN et al. (1977 a, b) methyl N-(2-furoyl)-N(2,6-xylyl)-DL-alaninate (BEP 827 419; GBP 1 448 810)

Table 16.3 (Continued)

Chemical class Common name*, originating company, code number Chemical structure (*after Worthing and Hance, 1991)	First report Chemical name* Patent number*
Metalaxyl (Ciba-Geigy, CGA 48988)	URECH et al. (1977) methyl N-(2-methoxyacetyl)-N-(2,6-xylyl)-DL-alaninate (BEP 827671; GBP 1500581)
Benalaxyl (Montedison, M 9834)	BERGAMASCHI et al. (1981) methyl N-phenylacetyl-N-2,6-xylyl-DL-alaninate (BEP 873908; DEP 2903612; ITPA 19896-1978)
5.2. Acylamino-Butyrolactones Ofurace (Chevron, Ortho 20615)	LUKENS et al. (1978) (+/−)-α-2-chloro-N-2,6-xylylacetamido-γ-butyrolactone
5.3. Acylamino-Oxazolidinones Oxadixyl (Sandoz, SAN 371 F)	GISI et al. (1983) 2-methoxy-N-(2-oxo-1,3-oxazolidin-3-yl) acet-2′,6′-xylidide (BE 884661; GBP 2058-059)
Cyprofuram (Schering SN 78314)** 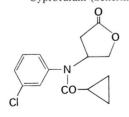	BAUMERT and BUSCHHAUS (1982) (+/−)-α-[N-3-chlorophenyl)cyclopropane-carboxamido] γ-butyrolactone (GBP 1603730)

Table 16.3 (Continued)

Chemical class Common name*, originating company, code number Chemical structure (*after Worthing and Hance, 1991)	First report Chemical name* Patent number*
6. Morpholines 6.1. Dimethomorph (American Cyanamid, CL 336379, Cl 183776, CME 151)	ALBERT et al. (1988) (E, Z) 4-[3-(4-chlorophenyl)-3-(3,4-dimethoxyphenyl) acryloyl] morpholine (EUP 120321)
7. 2,6-Dinitroanilines 7.1. Fluazinam (Ishihara Sangyo Kaisha, B-1216; IKF-1216)	ANEMA et al. (1992) 3-chloro-N-(3-chloro-5-trifluoromethyl-2-pyridyl)-α,α,α-2,6 dinitro-p-toluidine (USP 4331670)
8. Strobilurine analogues 8.1. BAS 490 F, Oximetherester	AMMERMANN et al. (1992) methyl-(E)-methoximino [α-(o-tolyloxyl)-0-tolyl] acetate (EUP 253213)

Table 16.3 (Continued)

Chemical class Common name*, originating company, code number Chemical structure (*after Worthing and Hance, 1991)	First report Chemical name* Patent number*
8.2. ICI A 5504, Methoxyacrylates	GODWIN et al. (1992) methyl (E)-2-{2-[6-(2-cyanophenoxy)pyrimidin-4-yloxy]phenyl}-3-methoxyacrylate
9. Benzamides Zarilamide (ICIA0001, PP001) ᐃ	HEANEY et al. (1988) (CRS)-4-Chloro-N-[cyano (ethoxy) methyl] benzamide
10. Anilino-Pyrimidines Andoprim	ZOLLFRANK and LYR (1988) (4-methoxy-N-[4,6 dimethyl pyrimidin-2-yl] aniline)

** = withdrawn from the market
ᐃ = no commercial development

Biological performance

The biological activity of the new *Oomycetes* fungicides has been described in several reviews (SCHWINN 1979; SCHWINN and URECH 1986; DAVIDSE and DE WAARD 1984; GRIFFITH et al. 1992). Therefore, it is discussed here only briefly, mainly under the aspect of practical use. An overview of the remarkably varying spectra of activity of fungicide groups of this chapter is given in table 16.6. Hymexazol on the one end of the scale controls only the *Pythiaceae Aphanomyces* spp., *Pythium* spp. and some *Phytophthora* spp. (KATO et al. 1990), but no *Peronosporaceae*. The phenylamides on the other end are active against all pathogens in the order of *Peronosporales*. Within the group of phenyl-

Table 16.4 Correlation between *Oomycete* selectivity and hydrophilicity of fungicides

Fungicide	Selectivity to *Oomycetes*	Walter solubility ($\mu g\ ml^{-1}$)
Residual		
Ethylbisdithiocarbamates	n[a]	0–10
Dialkyldithiocarbamates:		
thiram, ferbam	n	30–130
Fenaminosulf	o[a]	20,000
Systemic		
Chloroneb	n	8
Etridiazole	n	25
Cymoxanil	o	1,000
Hymexazol	o	85,000
Fosetyl-Al	o	120,000
Propamocarb	o	700,000[b]
Phenylamides: Metalaxyl	o	7,400
Ofurace	o	140
Benalaxyl	o	37
Oxadixyl	o	3,400
Dimethomorph	o	18
Fluazinam	n	?
BAS 490F	n	2
ICIA 5 504	n	10

a: not selective to *Oomycetes*; o: activity confined to *Oomycetes*.
Formulated at 70% a.i. in water as a true solution.

Adapted from Bruin and Edington (1983)

Table 16.5 Recent chemicals for control of *Oomycetes* (in order of their commercial introduction)

Common name	Trade name (S) Formulation (S)	First report
Hymexazol	Tachigaren; EC, SD, D	Takahi et al. (1974a, b)
Cymoxanil	Curzate; WP	Serres et al. (1976)
Fosetyl Al	Aliette; WP	Bertrand et al. (1977)
		Williams et al. (1977)
Furalaxyl	Fongarid; WP, G	Schwinn et al. (1977a, b)
Metalaxyl	Ridomil; WP, G	Urech et al. (1977)
	Acylon; WP	
	Apron; SD	
Propamocarb	Previcur N; SCW	Pieroh et al. (1978)
Ofurace	Patafol; Caltan; WP	Lukens et al. (1978)
Benalaxyl	Galben; WP, G	Bergamaschi et al. (1981)
Oxadixyl	Sandofan; WP	Gisi et al. (1983)
Fluazinam	Shirlan; EC,	Anema et al. (1992)
	Frowncide; EC	
Dimethomorph	Acrobat MZ;	Albert et al. (1988)
	Forum; WP, EC	
	Development Products	
BAS 490F	?; WP, SC	Ammermann et al. (1992)
ICIA 5 504	?	Godwin et al. (1992)

D = dust; EC = emulsifiable concentrate; G = granules; SCW = soluble concentrate on water basis; SD = seed treatment; WP = wettable powder

Table 16.6 Spectrum of activity of recent *Oomycetes* Fungicides

Common name	Pathogens on root/stem	Foliar pathogens	Activity against pathogens other than Oomycetes
Hymexazol	*Aphanomyces* *Pythium*	–	*Corticium sasaki* *Fusarium* spp.
Propamocarb	*Aphanomyces* *Pythium,* *Phytophthora*	*Bremia, Peronospora* *Phytophthora* *Pseudoperonospora*	–
Cymoxanil	–	*Plasmopara* *Peronospora* *Phytophthora* *Pseudoperonospora*	
Fosetyl	*Phytophthora*	*Bremia* *Plasmopora* *Pseudoperonospora*	*Phomopsis viticola* *Guignardia bidwelli* *Pseudopezzia tracheiphila*
Phenylamides (spectrum varies between individual compounds)	*Peronosclerospora* *Phytophthora* *Pythium* *Sclerospora* *Sclerophthora*	*Albugo* *Bremia* *Peronospora* *Peronosclerospora* *Phytophthora* *Plasmopara* *Pseudoperonospora* *Sclerospora* *Sclerophthora*	–
Fluazinam		*Phytophthora* *Pseudoperonospora*	*Botrytis cinerea*
Dimethomorph	*Phytophthora*	*Bremia* *Peronospora* *Phytophthora* *Plasmopara* *Pseudoperonospora*	
BASF 490F/ ICIA 5 504		*Phytophthora* *Plasmopara*	Foliar and glume diseases of cereals and rice, Powdery Mildews, *Venturia inaequalis, Aternaria* spp.

amides, however, there are strong differences in the relative activity of the individual molecules, with metalaxyl being the most active representative with the broadest use spectrum. Fluazinam, BAS490F and ICIA5044 shown usable activity only against a few genera within the *Peronosporaceae* and dimethomorph is not active against *Pythium* spp.

Whilst the activity of propamocarb, cymoxanil, the phenylamides and dimethomorph is confined to the *Oomycetes*, fosetyl and particularly fluazinam and the strobilurin analogues have a much wider spectrum of activity (Tab. 16.6). It is remarkably broad in the case of BAS490F and ICIA5504, covering a range of *Oomycetes, Ascomycetes, Basidiomycetes* and *Deuteromycetes*. (AMMERMANN et al. 1992; GODWIN et al. 1992).

Based on spectrum of biological activity, crop tolerance and systemicity, the spectrum of target crops can be summarized as shown in table 16.7. Hymexazol is used on just two crops: rice and sugarbeets, mainly as a seed dressing application or as a drench to the rice

Table 16.7 Main practical usage of modern *Oomycetes* fungicides (in order of their market introduction)

Compound	Main usage against	Main crops	Application method
Hymexazol	diseases of roots and stems in seedling stage	rice, sugarbeet	drench, seed dressing, dust
Cymoxanil	foliar diseases	grape vine, potato	spray
Fosetyl	foliar, stem and root diseases	grape vine, avocado, pineapple, citrus, ornamentals	spray, drench, dip, injection
Furalaxyl	diseases of root and stems	ornamentals	drench
Metalaxyl, Ofurace, Benalaxyl, Oxadixyl	foliar, stem and root diseases	grape vine, potato avocado, pineapple citrus, tobacco hops, maize, sorghum, millet	spray, drench, dip, granules, seed dressing
Propamocarb	diseases of roots and foliage	ornamentals, vegetables, potato	drench, spray
Fluazinam	foliar diseases	potato, cucumber	spray
Dimethomorph	foliar and root diseases	grape vine, potato vegetables, hops, tobacco	spray
Compounds under development			
BASF 490F	foliar and root diseases	grape vine, vegetables	spray
ICIA 5 504	foliar diseases	grape vine, potato	spray

seedling box. Besides its fungicidal effect, it has a direct growth-promoting activity in rice seedlings (OTA 1975) by stimulation of lateral root and root hair development.

Furalaxyl is a special product for use on ornamentals, a use segment which requires an extremely good crop tolerance against large numbers of cultivars of high economic value. Propamocarb shows a similar use profile with additional uses in vegetables and potatoes.

Cymoxanil has made strong inroads into grape vines and potatoes as a mixture partner for foliar use. In contrast, it has no useful activity against soil-borne pathogens. For use against foliar diseases, it has to be mixed with protective fungicides because of its rapid loss of activity. In combinations at low rates (0,1 kg ai/ha; 10–15 g ai/hl), it improves strongly the performance of traditional products such as Mancozeb or Folpet. This effect is mainly based on its curative action.

Fosetyl has a very broad use spectrum, reaching from foliar application against *Plasmopara viticola* in grapes to trunk injections in avocado trees against root rot caused by *Phytophthora cinnamomi* or dip treatment of pineapple tops against *Phytophthora nicotianae*. Its strong basipetal systemicity in green and woody tissue (cf. p. 335; Tab. 16.9) is an outstanding feature of this compound which makes it an extraordinarily flexible tool in the control of a range of diseases on many crops. Its main characteristics are:

- active against foliar downy mildews (not late blight) and root-infecting *Oomycetes*,
- both acropetal and basipetal translocation allows protection of new growth and protection of roots by foliar application,
- weak curative activity.

The phenylamides have the most complete range of activity. Basically they control all diseases caused by pathogens of the order *Peronosporales*, including the *Albuginaceae* (Tab. 16.6). However, there are strong differences in performance against various target pathogens between the individual chemicals of this class. Metalaxyl has the most complete spectrum of activity with remarkable variations of effectiveness between the various genera of fungi (KATO et al. 1990). Of all the new compounds discussed here the phenylamides are those with the highes inherent fungitoxic activity showing *in vitro* ED_{50} values of 0,01 to 3 ppm for the inhibition of mycelial growth (BRUIN 1980; SCHWINN and STAUB 1982). Dimethomorph inhibits mycelial growth (*P. megasperma*) with an ED50 of <1 µM (KUHN et al. 1990).

So far, metalaxyl is the most active, versatile and broadly used compound of this class. Its main biological features have been described by several authors (for references see SCHWINN and URECH 1986). They can be summarized as follows:

- high inherent fungitoxicity,
- protective and curative activity against all *Peronosporales*,
- rapid uptake, high acropetal systemicity, leading to protection of new growth,
- good persistence in plant tissue allowing extended spray intervals,
- control of systemic seed- and soil-borne diseases,
- weak on senescent plant tissue.

In conclusion, the fungicides against *Oomycetes* presented here, above all cymoxanil, fosetyl and the phenylamides have contributed to substantial progress in the practical control of this important group of pathogens. This holds particularly true for root and crown diseases which are well controlled by foliar application of fosetyl and by soil applicating of the phenylamides. In addition, the phenylamides control systemic diseases (Tab. 16.2) by seed dressing application which so far were not well controlled by chemicals. The most recent products (No 6–9 in table 16.3) enlarge the arsenal of weapons against Oomycete diseases. Whereas their spectrum of activity within this class is rather narrow and their inherent fungitoxicity is not higher than that of the commercial products, their novel modes of action are most valuable features in the context of resistance management. It is worth mentioning that apart from the speciality products propamocarb and furalaxyl, all fungicides of table 16.3 are used in practice in admixtures with protectants as illustrated in table 16.8. This concept leads to a broader spectrum of activity and, in addition, reduces the risk of resistance development.

Table 16.8 Mixture concepts for recent *Oomycetes* fungicides

Fungicide	Mixture partner	Rationale
Hymexazol	metalaxyl	enhancing activity
Cymoxanil	protectants*, oxadixyl	improving performance, broadening spectrum
Fosetyl	protectants*	improving performance, broadening spectrum
Phenylamides	protectants*, cymoxanil	anti-resistance strategy, broadening spectrum, improved end-of-season performance
Dimethomorph	protectants*	anti-resistance strategy, broading spectrum

For fluazinam, BAS 490F and ICIA 5 504 no mixture concepts are known yet.

*protectants = dithiocarbamates, chlorothalonil, dithianon, copper and others.

Uptake and transport in plants

From a practical point of view, perhaps the most important feature of the mode of action of the new *Oomycetes* fungicides is their systemicity. The varying degree of apoplastic and symplastic transport of these compounds offers characteristic possibilities for each group to protect plant parts away from the point of application. The range of systemicity is illustrated by cymoxanil with only locally systemic activity, by the phenylamides and dimethomorph with excellent apoplastic transport and by fosetyl with both excellent symplastic and good apoplastic transport (Tab. 16.9). First indications for the strobilurine analogues suggest a local systemicity that leads to a useful curative action.

Table 16.9 Systemicity of *Oomycetes* fungicides

Chemical	Characteristics of translocation		
	local (penetration)	apoplastic	symplastic
Propamocarb	+ +	+ +	−
Hymexazol	+ +	+	−
Cymoxanil	+ + +	+	−
Fosetyl	+ + +	+ +	+ + +
Phenylamides	+ + +	+ + +	+
Fluazinam	+	−	−
Dimethomorph	+ +	+ + +	−
BAS 490F	+ +	+	−
ICIA 5 504	+ +	+	−

+ + + = rapid, major factor for performance
+ + = intermediate factor, contributing to performance
+ = weak, slow transport
− = no transport in effective quantities

With cymoxanil, hymexazole and propamocarb the lack of useful long distance transport appears to be based on different limiting factors. Cymoxanil and hymexazole are taken up rapidly by plant roots (KLUGE 1978). All three compounds are metabolized rapidly inside the plants which prevents them from showing lasting systemic effects. The short duration of the local protection by cymoxanil (SERRES and CARRARO 1976) and its rapid metabolism to glycine (BELASCO et al. 1981) are clear evidence for the limiting role of *in vivo* stability in the duration of protection. With hymexazole two major glycosidic metabolites are formed which exhibit different biological activities. The O-glycoside is as fungitoxic as hymexazole itself while the N-glycoside is not fungitoxic but has plant growth promoting properties (KAMIMURA et al. 1974).

The phenylamides studied are taken up easily by roots, green stems and leaves and transported apoplastically with the transpiration stream. For metalaxyl it could be shown that limited basipetal transport does occur in tomato (STAUB et al. 1978) and avocado (ZAKI et al. 1981). On potatoes the sink effect of tubers appears to lead to an accumulation of small but sufficient quantities of metalaxyl (0.02 − 0.04 ppm) for protection against tuber rot by *Phytophthora infestans* (BRUIN et al. 1982; STEWART and McCLAMONT 1982) independent of foliar disease control. Transport in the plant tissue is sometimes confounded with distribution by vapour phase. The latter contributes to an even distribution of metalaxyl throughout leaves.

Based on performance data the symplastic transport of fosetyl is much faster than that of phenylamides (MUNNECKE 1982). However, while the parent compounds of the latter group are translocated apoplastically, the symplastic activity of fosetyl appears to be based on the transport of its metabolite H_3PO_3 (FENN and COFFEY 1984). Precise quantitative data on metabolism and transport of fosetyl and H_3PO_3 is lacking, but it seems that the distribution follows a source-sink relation (GROUSSOL et al. 1986). This can be exploited for a optimal timing of application. For a complete review of the phosphonates as antifungal compounds see GUEST and GRANT 1991.

Mode of action in target fungi and in host-pathogen interaction

1. Physiological level

An interesting aspect on the mode of action of modern *Oomycetes* fungicides is their specific inhibition of certain stages in the biology of their target fungi. *In vitro* most of them are more inhibitory to mycelial growth and sporulation than to spores and spore on germination (Tab. 16.10). Exceptions are propamocarb with weak activity against

Table 16.10 Mode of action of recent fungicides against *Oomycetes*

Compound	Sensitive stages of target fungi	Biochemical target area	Type of inhibition	Reversal of fungitoxicity by	References
Propamocarb	sporulation, mycelium	cell membrane	fungistatic	steroles L-methionine	PAPAVIZAS et al. (1978); BURDEN et al. (1988)
Hymexazole	mycelium, sporulation	RNA synthesis	?	?	NAKANISHI et al. (1983)
Cymoxanil	mycelium, zoospore germination	primary target unknown	fungistatic	amino acids	DESPREAUX et al. (1981); ZIOGAS and DAVIDSE (1987)
Fosetyl	mycelium, sporulation	amino acid metabolism, protein composition	fungistatic	H_3PO_4	FENN and COFFEY (1984); GRIFFITH et al. (1992)
Phenylamides	mycelium, sporulation	ribosomal RNA synthesis	fungistatic	?	KERKENAAR and KAARS SIJPESTEIJN (1981); DAVIDSE, next chapter
Dimethomorph	sporulation, mycelium	cell wall biogenesis	fungicidal/ fungistatic	yeast extract	KUHN et al. (1991)
Fluazinam	zoospore release and germination	uncoupling of phosphorylation	?	glutathione	GUO et al. (1991)
BAS 490F ICIA 5504	spore germination, sporulation, mycelium	mitochondrial respiration (cytochrome bc_1 complex)	?	?	BECKER et al. (1981); BRANDT et al. (1988, 1991); MANSFIELD and WIGGINS (1990)

mycelial growth (REICH et al. 1992) and strong effects an sporulation, fluazinam, interfering mainly with zoospore release and germination (ANEMA et al. 1992) and the strobilurine analogues BAS490F and ICIA5044, affecting mainly sporangial and zoospore germination (AMMERMANN et al. 1992; GODWIN et al. 1992).

In general, the action of the *Oomycetes* fungicides is rather fungistatic than fungicidal. The effects of dimethomorph on morphology and ultrastructure of *P. infestans* were studied by KUHN et al. 1991). The compound causes extensive proliferation and aberrant deposition of cell-wall material in hyphae, abnormal wall ingrowths, accumulation of vacuoles and membrane debris.

The biological site of action of phenylamides in the infection cycle has been studied for several airborne and soilborne *Oomycetes*. In *Plasmopara viticola* metalaxyl did not inhibit spore germination and initial penetration of the fungus into grape leaves. It inhibited strongly any further fungal growth inside the leaf after the formation of the first haustorium (STAUB et al. 1980). This is in strong contrast to the site of action of protective multi-site fungicides which only affect spores and germtubes on the leaf surface but do not affect fungal growth inside the leaves (Fig. 16.2). In mixtures of phenylamides with residual compounds the complementary modes of biological action often lead to synergistic effects the field. However, the exposure time of the pathogen to the systemic compounds is much longer than to the residual ones.

With soilborne *Oomycetes* the sites of action of metalaxyl appear to be analogous. In the infection cycle of *Phytophthora nicotianae* on tobacco seedlings no inhibition of zoospore release from sporangia nor of their migration to and encystment on root tips was found. Inhibition was observed only after penetration of root tissue had occurred. In addition, sporangia formation from infected root tissue and from chlamydospores was also strongly inhibited (STAUB and YOUNG 1980).

Another aspect of the mode of action of systemic *Oomycetes* fungicides is the question of indirect action via activation of latent host plant resistance mechanisms. For metalaxyl, propamocarb (REICH et al. 1992) and especially for fosetyl an indirect effect via the activation of the host's defense reaction has been claimed.

Based on lacking *in vitro* activity of fosetyl and the type of response of fosetyl-treated sensitive grape and tomato plants, respectively, which is similar to that of naturally resistant ones including the accumulation of phytoalexins, RAYNAL et al. (1980) and BOMPEIX et al. (1980, 1981) suggested these indirect effects to be the primary ones. However, resistant type responses and phytoalexin accumulation were also found on metalaxyl treated lettuce (CRUTE 1979), potatoes (BRUCK et al. 1980) and soybean (WARD et al. 1980), and some isolates also show a discrepancy between *in vivo* and *in vitro* sensitivity (STAUB et al. 1979). The latter ist also a general characteristic of prothiocarb (KAARS SIJPESTEIJN et al. 1974). For metalaxyl and prothiocarb, therefore, effects on host responses have also been implicated, but as secondary ones. The suppression of both phytoalexin production and protection on treated plants by inhibitors of the shikimic pathway (e. g. the herbicide glyphosate) provided additional evidence for some sort of involvement of host responses in the protection effects by fosetyl (FETTOUCHE et al. 1981) and metalaxyl (WARD 1984). Work by BOWER and COFFEY (1985) with fosetyl resistant mutants of *Phytophthora capsici* showed that *in vitro* resistance to this fungicide is paralleled by *in vivo* resistance on green pepper plants. Similar observations with metalaxyl are common. They suggest that the hypothezed triggered host resistance effects are not primary events in the mode of action of fosetyl. It appears rather that fosetyl or its degradation product H_3PO_4 acts directly on the target fungi leading to growth inhibition. This allows the defense reactions of normal susceptible plants to be triggered. Therefore, we suggest that as with many other systemic fungicides, the protection by fosetyl is the result of combined actions by the fungicide on the fungus (primary) and the plant on the fungus (secondary). For a detailed review see GUEST and GRANT 1991.

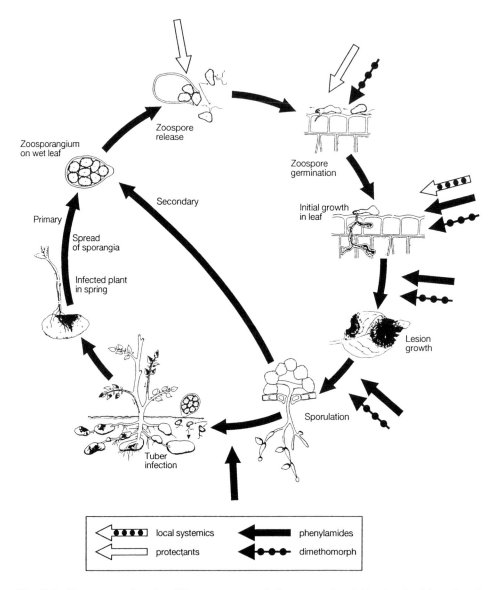

Fig. 16.2 Target steps for the different groups of *Oomycetes* fungicides in the life cycle of *Phytophthora infestans*.

2. Biochemical level

The biochemical mode of action of the fungicides treated in this chapter is summarized in table 16.10. Phenylamides interfere with RNA synthesis of their target fungi. Interestingly, RNA synthesis is also implicated as the biochemical mode of action of hymexazole in *Pythium aphanidermatum* (NAKANISHI and SISLER 1983) and of cymoxanil in a sensitive *Botrytis cinerea* mutant (DESPREAUX et al. 1981). *Botrytis* is not ordinarily a target for cymoxanil and it remains to be seen whether the mechanism is the same in sensitive *Oomycetes*.

The biochemical mode of action of propamocarb is still not known other than that its action *in vitro* can be reversed by the addition of sterols to the medium. However, for certain *Saprolegniales* which are sensitive to prothiocarb only, fungitoxicity is attributed to the degradation product ethylmercaptan and it can be reversed by the addition of l-methionine (KERKENAAR and KAARS SIJPESTEIJN 1977). The biochemical mode of action of Dimethomorph is not yet clearly known. KUHN et al. (1991) based on ultrastructural studies suggest that it acts on biochemical processes in cell wall biogenesis. Fluazinam has been shown to be an extraordinarily active uncoupler of oxidative phosphorylation in rat liver mitochondria. In the presence of glutathione it is rapidly metabolized (GUO et al. 1991). The strobilurine analogues BAS490F and ICIA5044 inhibit mitochondrial respiration by blocking electron transfer at the cytochrom bc_1 complex. More precisely, the binding site is ubihydroquinone- cytochrome-c oxydoreductase. (BRANDT et al. 1988; MANSFIELD and WIGGINS 1990) (cf. chapter 19).

Status of resistance and experience with counter-measures

Of the fungicides discussed in this chapter only the phenylamides have met with resistance problems in the field. The other *Oomycetes* fungicides have so far not produced field resistance. This is not necessarily due to their inherent lower proneness on the biochemical level, as was shown by the easy induction of resistant mutants to fosetyl in *Ph. capsici* (BOWER and COFFEY 1985). They may rather have escaped resistance problems because of their lower fungitoxic efficacy. This in turn made it necessary, especially for the foliar compounds, to use them from the start of introduction in mixtures with residual compounds. Without a mixture partner, neither cymoxanil nor fosetyl appear to achieve reliable protection against foliar *Oomycetes* under field conditions.

Of the newer *Oomycetes* fungicides, dimethomorph has been investigated concerning its resistance risk. For this fungicide adaptation, mutation and selection experiments failed to produce isolates with substantially decreased sensitivity (BISSBORT 1992). On the basis of these studies, the resistance risk associated with dimethomorph is estimated lower than that of phenylamides. In spite of these results however, its use is recommended only in mixture with residual compounds (ALBERT et al. 1988; BISSBORT 1992).

Field resistance to phenylamides has developed shortly after their market introduction in cases of intensive, continuous and exclusive use of the straight product, mainly in *Ph. infestans* and *Pl. viticola*. Meanwhile, it has also shown up in certain cases in other target pathogens. History, parameters and consequences of this development were reviewed by SCHWINN (1983), DAVIDSE and DE WAARD (1984) and MORTON and URECH (1988).

An important reaction to the development of resistance against this and other fungicide groups by industry was the formation of the Fungicide Resistance Action Committee (FRAC) (DELP 1984; WADE and DELP 1990). It consists of working groups for the major classes of resistance prone modern fungicides; e.g. for the phenylamides. Their main goal is to design and establish use strategies which help to safeguard the availability and effectiveness of the endangered fungicides.

For the phenylamides the working group recommended the use of pre-pack mixtures with residual compounds and a limited number of applications per season as the basic anti-resistance strategy (URECH 1985; URECH and STAUB 1985). Soil application against foliar pathogens is not recommended. In areas with substantial proportions of resistance in the pathogen population, the number of applications should not exceed two.

Still today experimental evidence on the effects of such anti-resistance strategies and tactics is very sparse and further research in this direction is badly needed. However, observational evidence suggests that the FRAC anti-resistance strategy works in slowing down the build-up of restistance and in preventing crop losses even in cases where resistance is detactable in the fungus population. In addition, model test with *Ph. infestans* populations, starting with a known proportion of phenylamide resistant spores, confirmed the slow-down of the selection process when mixtures with mancozeb were used (STAUB and SOZZI 1983; LEVY et al. 1991). For this purpose triple mixtures with curzate as a second systemic mixture component are also recommended (SAMOUCHA et al. 1987). Often mixtures of fungicides with different modes of action exhibit a synergistic activity against both the sensitive and the resistant part of the target population (GISI et al. 1985). Prepack-mixtures are also favored over the use of single products in alternation for more practical reasons: they enforce automatically the implementation of the anti-resistance strategy whereas the recommendations for alternations might not be followed. They also provide a safeguard against crop losses where resistance levels would lead to complete failures of the single product.

Where cautious anti-resistance strategies along the lines of the FRAC recommendations are followed, a stabilization and in some cases a decrease of resistance could be observed (DAVIDSE et al. 1989; STAUB 1991). Even in areas with detectable resistance present, the phenylamides continue to contribute significantly and uniquely to disease control programs when they are applied in mixture not more than two to three times early in the growing season. This is true even in the Netherlands and in Ireland, where they were withdrawn in 1981 and reintroduced a few years later, when strong epidemics could not be controlled by conventional fungicide programs (DAVIDSE et al. 1989; DOWLEY 1994). A similar stabilization occurred in the eastern part of Germany, when phenylamide use was brought in line with the FRAC recommendations in 1989 (PLUSCHKELL and OESER 1990). The scientific basis of this stable resistance situation in spite of continued use of phenylamides is not known; it suggests certain fitness deficits in the resistant populations. Factors such as increased sensitivity to high temperatures for *Pl. viticola* (PIGANEAU and CLERJEAU 1985) or decreased survival in tubers for *Ph. infestans* (WALKER and COOKE 1990; STAUB 1994) may play a role.

Detailed studies have been done on cross-resistance patterns in phenylamides. For all practical purposes cross-resistance exists among all phenylamides even though the resistance factors can vary in some strains (CLERJEAU et al. 1985; DIRIWÄCHTER et al. 1987; DAVIDSE in chapter 16.2). No cross-resistance exists between phenylamides and the other new *Oomycetes* fungicides or residual products (LEROUX and GREDT 1981; KENDALL and CARTER 1984; BOWER and COFFEY 1985). The only conflicting report indicating cross-resistance between phenylamides and unrelated fungicides is by COHEN and SAMOUCHA (1984). This group also found that phenylamide resistant isolates of *Ph. infestans* were more aggressive on a number of potato cultivars than phenylamide sensitive ones (KADISH et al. 1990). These two exceptional observations are not made outside the Middle East region and might be explained by the suggestion of FRY et al. (1992), who found by fingerprinting that the resistant isolates in Israel were from a "new", and possibly recently introduced population, that is inherently more aggressive than the "old" native population. The more aggressive population would by definition also be more difficult to control with unrelated fungicides. Fitness according to this suggestion is part of the genetic background of the new population and not genetically linked to resistance.

References

ALBERT, G., CURTZE, J., and DRANDAREVSKI, Ch. A.: Dimethomorph (DME 151) a novel curative fungicide. Proc. Brit. Crop Prot. Conf. − Pests and Diseases **1** (1988): 17−24.

AMMERMANN, E., LORENZ, G., SCHELBERGER, K., WENDEROTH, B., SAUTER, H., and RENTZEA, C.: BAS490F − A broad spectrum fungicide with a new mode of action. Proc. Brit. Crop Prot. Conf. − Pests and Diseases **1** (1992): 403−410.

ANEMA, B. P., BOUWMANN, J. J., KOMYOJI, T. and SUZUKI, K.: Fluazinam: a novel fungicide for use against *Phytophthora infestans* in potatoes. Proc. Brit. Crop Prot. Conf. − Pests and Diseases **2** (1992): 663−668.

BASTIAANSEN, M. G., PIEROH, E. A., and AELBERS, E.: Prothiocarb, a new fungicide to control *Phytophthora fragariae* in strawberries and *Pythium ultimum* in flower bulbs. Meded. Fac. Landbouwwet. Rijksuniv. Gent **39** (1974): 1019−1025.

BAUMERT, D., and BUSCHHAUS, H.: Cyprofuram, a new fungicide for the control of Phycomycetes. Meded. Fac. Landbouwwet. Rijksuniv. Gent **47** (1982): 979−983.

BECKER, W. F., JAGOW, G. VON, ANKE, T., and STEGLICH, W.: Oudemansin, strobilurin A, strobilurin B and myxothiazole: new inhibitors of the bc_1 segment of the respiratory chain with an E-β-methoxyacrylate system as common structural element. FEBS Letters **132** (1981): 329−333.

BELASCO, I. J., HAN, J. C. Y., CHRZANOWSKI, R. L., and BAUDE, F. J.: Metabolism of (14C) cymoxanil in grapes, potatoes and tomatoes. Pesticide Sci. **12**, 4 (1981): 355−364.

BERGAMASCHI, P., BORSARI, T., GARAVAGLIA, C., and MIRENNA, L.: Methyl N-phenylacetyl-N-2,6-xylyl-dl-alaninate (M 98834), a new systemic fungicide controlling downy mildew and other diseases caused by *Peronosporales*. Proc. Brit. Crop Prot. Conf. (1981): 11−18.

BERTRAND, A., DUCRET, J., DEBOURGE, J.-C., and HORRIERE, D.: Etude des propriétés d'une nouvelle famille de fongicides. Les monoéthylphosphites métalliques. Charactéristiques physicochimiques et propriétés biologiques. Phytiatr. Phytopharm. **26** (1977): 3−18.

BISSBORT, S.: Beurteilung einer möglichen Resistenzentwicklung gegenüber Dimethomorph, dargestellt an *Plasmopara viticola* (Berk. & Curtis) Berl. & de Toni und *Phytophthora cactorum* (Leb. & Cohn) Schroet. Dissertation. Wissenschaftlicher Verlag, Giessen 1992.

BOMPEIX, G., FETTOUCHE, F., and SAURDRESSAN, P.: Mode d'action du phoséthyl Al. Phytiatr. Phytopharm. **30** (1981): 257−272.

− RAVISE, A., RAYNAL, G., FETTOUCHE, F., and DURAND, M. C.: Modalités de l'obtention des nécroses bloquantes sur feuilles détachées de tomates par l'action du tris-O-éthyl phosphonate d'aluminium (phoséthyl d'aluminium), hypothése sur son mode d'action *in vivo*. Ann. Phytopathol. **12** (1980): 337−351.

BOWER, L. A., and COFFEY, M. D.: Development of laboratory tolerance to phosphorous acid, fosetyl-Al and metalaxyl in *Phytophthora capsici*. Can. J. Plant Pathol. **1** (1985): 1−6.

BRANDT, U., SCHÄGGER, H., and JAGOW, G. VON: Characterisation of binding of the methoxyacrylate inhibitors mitochondrial cytochrome c reductase. Europ. J. Biochem. **173** (1988): 499−506.

− and JAGOW, G. VON: Analysis of inhibitor binding to the mitochondrial cytochrome c reductase by fluorescence quench titration. Europ. J. Biochem. **195** (1991): 163−170.

BRUIN, G. C. A.: Resistance in *Peronosporales* to acylalanine-type fungicides. Ph. D. thesis Univ. Guelph, Ontario, Canada 1980, 110 pp.
— and EDGINGTON, L. V.: The chemical control of diseases caused by zoosporic fungi. In: BUCZACKI, S. T. (Ed.): Zoosporic Plant Pathogens. A Modern Perspective. Academic Press, London 1983, pp. 193–233.
— — and RIPLEY, B. D.: Bioactivity of the fungicide metalaxyl in potato tubers after foliar sprays. Canad. J. Pl. Path. **4** (1982): 353–356.
BRUCK, R. I., FREY, W. E., and APPLE, A. E.: Efect of metalaxyl, an acylalanine fungicide, on developmental stages of *Phytophthora infestans*. Phytopathology **70** (1980): 597–601.
BURDEN, R. S., CARTER, G. A., JAMES, C. S., CLARK, T., HOLLOWAY, P. I.: Selective effect of propamocarb and prothiocarb on the fatty acid composition of some Oomycetes. Brit. Crop Port. Conf. — Pests and Diseases **1** (1988): 403–408.
CLERJEAU, M., IRHIR, H., MOREAU, C., PIGANEAU, B., STAUB, T., and DIRIWAECHTER, G.: Etude de la résistance croisée au métalaxyl et au cyprofurame chez *Plasmopara viticola*: Evidence de plusieurs mécanismes de résistance independants. Proc. Centenary Meeting Bordeaux **2** (1985): 303–306.
— MOREAU, C., PIGANEAU, B., BOMPEIX, G., and MALFATTI, P.: Effectiveness of Fosetyl-Al against strains of *Plasmopara viticola* and *Phytophthora infestans* that have developed resistance to anilide fungicides. Proc. Brit. Crop Prot. Conf. — Pests and Diseases **2** (1984): 497–502.
COHEN, Y., and SAMOUCHA, Y.: Cross-resistance to four systemic fungicides in metalaxyl-resistant strains of *Phytophthora infestans* and *Pseudoperonospora cubensis*. Plant Dis. **68** (1984): 137–139.
CRUTE, I. R.: Lettuce mildew — Destroyer of quality. Agric. Res. Counc. (U. K.) Res. Rev. **5** (1979): 9–12.
DAVIDSE, L. C., and DE WAARD, M. A.: Systemic fungicides, In: INGRAM, D. S., and WILLIAMS, P. M. (Eds.): Advances in Plant Pathology **2** (1984): 191–257.
— HENKEN, J., VAN DALEN, A., JESPERS, A. B. K., and MANTEL, B. C.: Nine years of practical experience with phenylamide resistance in *Phytophthora infestans* in the Netherlands. Neth. J. Plant Pathol. (Suppl. 1) **95** (1989): 197–213.
DELP, C. J.: Industry's response to fungicide resistance. Crop Protection **3**, 1 (1984): 3–8.
DESPREAUX, D., FRITZ, R., and LEROUX, P.: Mode d'action biochimique du cymoxanil. Phytiatr. Phytopharm. **30**. No. 4 (1981): 245–255.
DIRIWAECHTER, G., SOZZI, D., NEY, C., and STAUB, T.: Cross resistance in *Phythophtora infestans* and *Plasmopara viticola* against different phenylamides and unrelated fungicides. Crop Protection. **6** (1987): 250–255.
DOWLEY, L. J.: Practical aspects of phenylamide resistance management. In: HEANEY, S. et al. (Eds.): Fungicide resistance. BCPC Monograph No 60 1994, pp. 147–153.
FENN, M. E., and COFFEY, M. D.: Studies on the *in vitro* and *in vivo* antifungal activity of fosetyl-Al and phosphorous acid. Phytopathology **74** (1984): 606–611.
FETTOUCHE, F., RAVISE, A., and BOMPEIX, G.: Suppression de la résistance induite — phoséthyl-Al — chez la tomate à *Phytophthora capsici* avec deux inhibiteurs — glyphosate et acide aminoxyacétique. Agronomie **1**, No. 9 (1981): 826.
FRY, W. E., GOODWIN, S. B., MATUSZAK, J. M., SPIELMAN, L. J., and MILGROOM, M. G.: Population genetics and intercontinental migrations of *Phytophthora infestans*. Ann. Rev. Phytopathology **30** (1992): 107–130.
GISI, U., BINDER, H., and RIMBACH, E.: Synergistic interactions of fungicides with different modes of action. Trans. Br. mycol. Soc. **85** (2), (1985): 299–306.
— HARR, J., SANDMEIER, R., and WIEDMER, H.: A new systemic oxazolidinone fungicide (San 371) against diseases caused by *Peronosporales*: Meded. Fac. Landbouwwet. Rijksuniv. Gent **48** (1983): 541–549.
GODWIN, J. R., ANTHONY, V. M., CLOUGH, J. M., and GODREY, C. R. A.: ICIA5504: a novel, broad

spectrum, systemic β-methoxyacrylate fungicide. Proc. Brit. Crop Prot. Conf. − Pests and Diseases **1** (1992): 435−442.

Gozzo, F., Garlaschelli, L., Boschi, P. M., Zagni, A., Oveereem, J. C., and de Vries, L.: Recent progress in the field of N-acylalanines as systemic fungicides. Pestic. Sci. **16** (1985): 277−286.

Griffith, J., Davis, A. J., and Grant, B. R.: Target sites of fungicides to control Oomycetes. In: Köller, W. (Ed.): Target sites of fungicide action. CRC Press, Boca Raton 1992, pp. 69−100.

Groussol, J., Delrot, S., Caruhel, P., and Bonneman, J.-L.: Design of an improved exudation method for phloem sap collection and its use for the study of phloem mobility of pesticides. Physiologie Végétale **24** (1986): 123−133.

Grunwaldt, G., Lyr, H., Wollgiehn, R., and Klepel, M.: Andoprim: a novel fungicide with an unconventional mode of action and special selectivity for *Phytophthora species*. Pestic. Sci. **30** (1990): 323−325.

Guest, D., and Grant, B.: The complex action of phosphonates as antifungal agents. Biol. Rev. **66** (1991): 159−187.

Guo, Z., Miyoshi, H., Komyoji, T., Taga, T., and Fujita, T.: Uncoupling activity of a newly developed fungicide, fluazinam [3-chloro-N-(3-chloro-2,6-dinitro-4-trifluoro methylphenyl)-5-trifluromethyl-2-pyridinamine]. Biochem. Biophy. Acta **1056** (1991): 89−92.

Heaney, S. P., Shephard, M. C., Crowley, P. J., and Shearing, S. J.: ICIA0001: a novel benzamide fungicide. Crop Prot. Conf.-Pests and Diseases (1988): 551−558.

Hubele, A., Kunz, W., Eckhardt, W., and Sturm, H.: The fungicidal activity of acylanilines. In: Miyamoto, J., and Kearney, P. C. (Eds.): IUPAC Pesticide Chemistry. Human Welfare and the Environment. Pergamon Press, Oxford-New York 1983, pp. 233−242.

Ingram, D. S., and Williams, P. H.: *Phytophthora infestans*, the cause of late blight of potato. Advances in Plant Pathology **7** (1991): 273 pp.

Kaars Sijpesteijn, A., Kerkenaar, A., and Overeem, J. C.: Observations on selectivity and mode of action of Prothiocarb. Meded. Fac. Landbouwwetensch. **39** (1974): 1027−1034.

Kadish, D., Grinberger, M., and Cohen, Y.: Fitness of metalaxyl-sensitive and metalaxyl-resistant isolates of *Phytopthora infestans* on susceptible and resistant potato cultivars. Phytopathology **80** (1990): 200−205.

Kamimura, S., Nishikawa, M., Saeki, H., and Takahi, Y.: Absorption and metabolism of 3-hydroxy-5-methylisoxazole in plants and the biological activites of its metabolites. Phytopathology **64**, 10 (1974): 1273−1281.

Kato, S., Coe, R., New, Laura, and Dick, W.: Sensitivities of various Oomycetes to hymexazol and metalaxyl. J. Gen. Microbiol. **136** (1990): 2127−2134.

Kendall, S. J., and Carter, G. A.: Resistance of isolates of *Phytophthora infestans* to fungicides. Proc. Brit. Crop Prot. Conf. − Pests and Diseases **2** (1984): 503−508.

Kendrick, B.: The fifth Kingdom. Mycologue Publications. Focus Information Group Inc., Newburyport, USA. 406 pp.

Kerkenaar, A., and Kaars Sijpesteijn, A.: On the mode of action of prothiocarb. Netherland J. Pl. Path. **83** Suppl. 1 (1977): 145−152.

− − Antifungal activity of metalaxyl and furalaxyl. Pestic. Biochem. Physiol. **15** (1981): 71−78.

Kluge, E.: Vergleichende Untersuchungen über die Wirksamkeit von Systemfungiziden gegen Oomyzeten. Arch. Phytopathol. Pflanzenschutz **14**, No. 2 (1978): 115−122.

Kuhn, P. J., Albert, G., Loyness, L. E., Lee, S. A., and Lindsley, M. C.: Studies on the antifungal activity *in vitro* of dimethomorph, a novel fungicide active against downy mildews and *Phytophthora*. 9th International Reinhardsbrunn Symposium, Modern fungicides and antifungal compounds. 1990, pp. 229−238.

− Pitt, D., Lee, Susan A., Wakley, G., and Sheppard, A. N.: Effects of dimethomorph on the morphology and ultrastructure of *Phytophthora*. Mycol. Res. **95** (1991): 333−340.

Lebeda, A., and Schwinn, F. J.: The downy mildews − an overview of recent research progress. Z. Pfl.kr. Pfl.schutz **101** (1994): 225−254.

LEROUX, P., and GREDT, M.: Phénoménes de résistance aux fongicides anti-mildious: Quelques résultats de laboratoire. Phytiatr. Phytopharm. 30, No. 4 (1981): 273–282.

LEVY, Y., COHEN, Y., and BENDERLY, M.: Disease development and buildup of resistance to oxadixyl in potato crops inoculated with *Phytophthora infestans* as affected by oxadixyl and oxadixyl mixtures: experimental and simulation studies. J. Phytopathol. 132 (1991): 219–229.

LUKENS, R. J., CHAM, D. C. K., and ETTER, G.: Ortho 20615, a new systemic for the control of plant diseases caused by Oomycetes. Phytopath. News 12 (1978): 142.

MANSFIELD, R. W., and WIGGINS, T. E.: Photoaffintity labelling of the β-methoxyacrylate binding site in bovine heart mitochondrial cytochome bc_1 complex, Biochem. Biophys. Acta 1015 (1990): 109–115.

MORTON, H. V., and URECH, P. A.: History of the development of resistance to phenylamide fungicides. In: DELP, C. S. (Ed.): Fungicide Resistance in North America. APS Press, St. Paul 1988, pp. 59–60.

MOSER, H., and VOGEL, C.: Preparation and biological activity of the enantiomers of CGA 48988, a new systemic fungicide. 4th Intern. IUPAC Congr. Zürich (1978): Abstracts II–310.

MUNNECKE, D. E.: Apparant movement of Aliette and Ridomil in *Persea indica* and its effect on root rot. Phytopathology 74 (1982): 970.

NAKANISHI, T., and SISLER, H. D.: Mode of action of Hymexazol in *Pythium aphanidermatum*. J. Pestic. Sci. 8 (1983): 173–181.

OTA, Y.: Plant growth promoting activities of 3-hydroxy-5-methylisoxazole. Jap. Agric. Res. Quarterly 9, No. 1 (1975): 1–7.

PAPAVIZAS, G. C., O'NEILL, N. R., and LEVIS, J. A. Fungistatic activity of propyl-N-(γ-dimethylaninopropyl) carbamate on *Pythium pp.* and its reversal by sterols. Phytopathology 68 (1978): 1667–1671.

PIEROH, E. A., KRASS, W., and HEMMEN, C.: Propamocarb, ein neues Fungizid zur Abwehr von Oomyzeten im Zierpflanzen- und Gemüsebau. Meded. Fac. Landbouwwet. Rijksuniv. Gent 43 (1978): 933–942.

PIGANEAU, B., and CLERJEAU, M.: Influence differentielle de la température sur la sporulation et la germination des sporocystes de souches de *Plasmopara viticola* sensibles et resistantes aux phenylamides. In: Fungicides for Crop Protection, BCPC Monogr. 31 (1985): 327–330.

PLUSCHKELL, H. J., and OESER, J.: Weitere Untersuchungsergebnisse zur Metalaxylresistenz bei *Phytophthora infestans* an Kartoffeln. Nachrichtenbl. Pflanzenschutz 44, 7 (1990): 158–159.

RAYNAL, G., RAVISE, A., and BOMPEIX, G.: Action du tris-O-éthylphosphonate d'aluminium (phoséthyl d'aluminium) sur la pathogénie de *Plasmopara viticola* et sur la stimulation des réactions de défense de la vigne. Ann. Phytopathol. 12 (1980): 163–175.

REICH, B., BUCHENAUER, H., BUSCHHAUS, H., und WENZ, M.: Wirkungsweise von Propamocarb gegenüber *Phytophthora infestans*. Mitt. Biol. Bundesanstalt Land- u. Forstwirtsch. 266 (1992): 423.

SAMOUCHA, Y., and GISI, U.: Use of two- and three-way mixtures to prevent buildup of resistance to phenylamide fungicides in *Phytophthora* and *Plasmopara*. Phytopathology 77 (1987): 1405–1409.

SCHWINN, F. J.: Control of Phycomycetes: a changing scene. Proc. Brit. Crop Prot. Conf. 10th 3 (1979): 791–802.

– New developments in chemical control of *Phytophthora*. In: ERWIN, D. C., BARTNICKI-GARCIA, S., and TSAO, P. H. (Eds.): *Phytophthora*: its Biology, Taxonomy, Ecology and Pathology. Amer. Phytopath. Soc., St. Paul, USA 1983, pp. 327–334.

– Significance of fungal pathogens in crop production. Pesticide Outlook 3 (1992): 18–25.

– and STAUB, T.: Biological properties of metalaxyl. In: LYR, H., and POLTER, C. (Eds.): Systemic Fungicides and Antifungal Compounds. Akademie-Verlag, Berlin 1982, pp. 123–133.

– – and URECH, P. A.: A new type of fungicide against diseases caused by Oomycetes. Meded. Fac. Landbouwwet. Rijksuniv. Gent 42 (1977a): 1181–1188.

- - - Die Bekämpfung falscher Mehltaukrankheiten mit einem Wirkstoff aus der Gruppe der Acylalanine. Mitt. Biol. Bundesanstalt Land- u. Forstwirtsch. Berlin-Dahlem **178** (1977b): 145-146.
- and URECH, P. A.: Progress in chemical control of diseases caused by Oomycetes. ACS Symposium Series **304** (1986): 89-106.
- SERRES, J. M., and CARRARO, G. A.: DPX-3217, a new fungicide for the control of grape downy mildew, potato blight and other *Peronosporales*. Meded. Fac. Landbouwwet. Rijksuniv. Gent **42** (1976): 645-650.
- SPENCER, D. M.: The downy mildews. Academic Press, London 1981, 636 pp.
- STAUB, T.: Fungicides resistance: Practical experience with antiresistance strategies and the role of integrated use. Annu. Rev. Phytopathol. (1991): 421-442.
- — Early experiences with phenylamide resistance and lessons for continued successful use. In: HEANEY, S. et al. (Eds.): Fungicide Resistance. BCPC Monograph No. 60, 1994, pp. 131-138.
- — DAHMEN, H., and SCHWINN, F. J.: Biological characterization of uptake and translocation of fungicidal acylalanines in grape and tomato plants. Z. Pflanzenkr. Pflanzenschutz **85** (1978): 162-168.
- — — Effects of Ridomil on the development of *Plasmopara viticola* and *Phytophthora infestans* on their host plants. Z. Pflanzenkr. Pflanzenschutz **87** (1980): 83-91.
- — — URECH, P., and SCHWINN, F.: Failure to select for *in vivo* resistance in *Phytophthora infestans* to acylalanine fungicides. Plant Dis. Rep. **63** (1979): 385-389.
- — and HUBELE, A.: Recent advances in the chemical control of Oomycetes. In: WEGLER, R. (Ed.): Chemie der Pflanzenschutz- und Schädlingsbekämpfungsmittel. Vol. 6. Springer-Verlag, Heidelberg 1981, pp. 389-422.
- — and SOZZI, D.: Recent practical experiences with fungicide resistance. 10. Internat. Congr. Plant Path. Brighton, U. K. **2** (1983): 591-598.
- — and YOUNG, T. R.: Fungitoxicity of metalaxyl against *Phytophthora parasitica* var. *nicotianae*. Phytopathology **70** (1980): 797-801.
- STEWART, H. E., and McCLAMONT, D. C.: The effect of metalaxyl and mancozeb on the susceptibility of potato tubers to late blight. Ann. Appl. Biol. **100**, suppl. (1982): 54-55.
- TAKAHI, Y., NAKANISHI, T., and KAMINURA, S.: Characteristics of hymexazol as a soil fungicide. Am. Phytopath. Soc. Japan **40** (1974a): 362-367.
- — — TOMITA, K., and KAMINURA, S.: Effects of 3-hydroxy isoxazoles as soil fungicides in relation to their chemical structure. Am. Phytopath. Soc. Japan **40** (1974b): 354-361.
- URECH, P. A.: Management of fungicide resistance in practice. Bulletin EPPO **15** (1985): 571-575.
- — SCHWINN, F. J., and STAUB, T.: CGA 48988, a novel fungicide for the control of late blight, downy mildews and related soil-borne diseases. Proc. Brit. Crop Prot. Conf., 9th **2** (1977): 623-631.
- — and STAUB, T.: The resistance strategy for acylalanine fungicides. EPPO Bull. **15** (1985): 539-543.
- — and DELP, C. J.: The fungicide resistance action committee. ACS Symposium Series (1990): 320-333.
- WALKER, A. S. L., and COOKE, L. R.: The survival of *Phytophthora infestans* in potato tubers — The influence of phenylamide resistance. Proc. Brighton Crop Prot. Conf. Pests and Diseases (1990): 1109-1114.
- WARD, E. W. B.: Suppression of metalaxyl activity by glyphosate: Evidence that host defence mechanisms contribute to metalaxyl inhibition of *Phytophthora megasperma* f. sp. *glycinae* in soybeans. Physiol. Pl. Pathol. **25** (1984): 381-386.
- — LAZORVITS, G., STOESSEL, P., BARRIE, S. D., and UNWIN, C. H.: Glyceollin production associated with control of *Phytophthora* rot of soybeans by the systemic fungicide metalaxyl. Phytopathology **70** (1980): 738-740.
- WILLIAMS, D. J., BEACH, B. G. W., HORRIERE, D., and MARECHAL, G.: LS 74-783, a new systemic

fungicide with activity against Phycomycete diseases. Proc. Brit. Crop Prot. Conf., 9th **2** (1977), 565–573.

WORTHING, C. R., and HANCE, R. J.: The pesticide manual. A world compendium. 9th Ed. Brit. Crop. Prot. Council 1991, 1141 p.

YOUNG, D. H.: Effects of zarilamide on microtubules and nuclear division in *Phytophthora capsici* and tobaco suspension-cultured cells. Pesticide Biochem. Physiol. **40** (1991): 149–161.

ZAKI, A. I., ZENTMYER, G. A., and LE BARON, H. M.: Systemic translocation of C-labeled metalaxyl in tomato, avocado, and *Persea indica*. Phytopathology **71** (1981): 509–514.

ZIOGAS, B. N., and DAVIDSE, L. D.: Studies on the mechanism of action of cymoxanil in *Phytophthora infestans*. Pestic. Biochem. Physiol. **29** (1987): 89–96.

ZOLLFRANK, G., and LYR, H.: Properties of Andoprim – a new *Phytophthora* fungicide. Pesticide Newsletters (1988): 40–41.

16.2 Phenylamide fungicides – Biochemical action and resistance

L. C. Davidse

Enkhuizen, The Netherlands

The group of phenylamide fungicides includes a number of structurally related chemicals with high activity against fungi belonging to the Peronosporales and a limited number of other fungi (Fuller and Gisi 1985). The group can be subdivided in the acylalanines metalaxyl, furaxyl and benalaxyl, the butyrolactones, ofurace and cyprofuram, and the oxazolidinones, oxadixyl. The phenylamide fungicides are structurally related to the chloroacetanilide herbicides, which provided the lead compounds in screening programs for fungicidal phenylamides. The selective activity of the phenylamides against almost exclusively Peronosporales makes these compounds excellent tools to study unique features of these fungi. This together with the ability of sensitive fungi to develop a high level of resistance to the phenylamides stimulated research on their mode of action and the mechanisms of resistance.

The antifungal activity of the phenylamides has been most intensively studied with metalaxyl. Details of the mechanism of action of metalaxyl will be discussed in the next section. A mechanism of resistance to metalaxyl that has been found in resistant strains of *Phytophthora megasperma* f. sp. *medicaginis* and *Phytophthora infestans* will be described in the following section. Subsequently the antifungal mode of action of various phenylamides including related herbicidal compounds is compared with that of metalaxyl.

Mode of action of metalaxyl

Among the biosynthetic processes studied in various fungi incorporation of radiolabelled uridine into RNA proved to be the most sensitive one to metalaxyl (Arp and Buchenauer 1981; Davidse et al. 1981a, 1983b; Fisher and Hayes 1982; Kerkenaar 1981; Wollgiehn et al. 1984). Incorporation of precursors into DNA, proteins and lipids is effected less rapidly or to a lesser extent. Respiration is not inhibited by metalaxyl. Since metalaxyl does not inhibit the uptake of uridine nor its conversion into UTP, inhibition of RNA synthesis must be responsible for the observed effects on uridine incorporation. However, even at concentrations that are fully inhibitory to growth, a complete inhibition of uridine incorporation does not occur. Depending on the fungal species used in the experiments incorporation is reduced by metalaxyl to 20–60% of the control value. It indicates that only part of the cellular RNA synthesis is sensitive to metalaxyl.

In eukaryotes RNA is synthesized by three different RNA polymerases each mediating the synthesis of a distinct product. RNA polymerase I (or A) synthesizes r(ibosomal) RNA, that accounts for the majority of the cellular RNA. RNA polymerase II (or B) produces m(essenger) RNA and RNA polymerases III (or C) gives rise to t(ransfer) RNA and the 5S RNA of the ribosomes. Each enzyme has its own characteristics that can be used to purify the individual enzymes. Sensitivity to α-amanitin is a useful property to characterize the polymerases. RNA polymerase II is highly sensitive to the toxin whereas RNA polymerase I is insensitive. Polymerase III of different organisms varies in sensitivity to α-amanitin but in general it is much less sensitive than polymerase II.

A similar differential sensitivity of the RNA polymerases to metalaxyl would explain the partial inhibition of RNA synthesis. Therefore, the effects of metalaxyl on the synthesis of the different classes of RNA were studied. In mycelium of *P. megasperma* f. sp. *medicaginis* synthesis of poly(A)-containing RNA, that represents most of the mRNA appeared to be less affected by metalaxyl than that of total RNA, the majority of which is rRNA (DAVIDSE et al. 1983b). rRNA appeared to be selectively inhibited (DAVIDSE and FLEUREN, unpublished results). Similar differential effects were observed in *P. nicotianae* (WOLLGIEHN et al. 1984). Inhibition of rRNA synthesis, therefore, can be considered to be the primary mode of action of metalaxyl.

Inhibition of rRNA synthesis would ultimately lead to inhibition of fungal growth because turnover of rRNA will deprive the cell of its ribosomes causing protein synthesis to decrease. It explains the inability of metalaxyl to inhibit germination of encysted zoospores of various fungi. Apparently the reproductive structures possess enough ribosomes to support germ tube formation. Several studies have shown that metalaxyl exerts its inhibiting effect against fungi on plant surfaces after penetration (BRUCK et al. 1980; HICKEY and COFFEY 1980; STAUB and YOUNG 1980; STAUB et al. 1980). Neither direct nor indirect germination of sporangia, nor the mobility, encystment and germination of zoospores is affected. Inhibition becomes apparent not until after formation of the primary haustorium. Spores apparently carry enough ribosomes along to sustain mycelial growth through critical stages of the life cycle of a fungus.

Inhibition of rRNA synthesis will also lead to accumulation of its precursors because rRNA is the major end product the nucleoside triphosphates are synthesized for. Nucleoside triphosphates are known to stimulate $\beta(1-3)$ glucan synthetase from various fungi (SZANISLO et al. 1985) and are involved in regulation of cell wall synthesis. Therefore, the observed thickening of cell walls of metalaxyl-treated hyphae (GROHMANN and HOFFMANN 1982; MÜLLER and LYR 1983) might be caused by elevated levels of nucleoside triphosphates enhancing cell wall synthesis.

The exact way metalaxyl interferes with rRNA synthesis is not known yet. In *in vitro* experiments metalaxyl does not inhibit the activity of partial purified polymerase I from *P. megasperma* (DAVIDSE et al. 1983b) and *P. nicotiana* (WOLLGIEHN et al. 1984). Endogenous RNA polymerase activity of nuclei isolated from these fungi, however, is sensitive to metalaxyl, indicating that metalaxyl only interferes with the intact polymerase-template complex. Circa 40% of the endogenous RNA polymerase activity of isolated nuclei of *P. megasperma* is metalaxyl sensitive and circa 30% α-amanitin sensitive (DAVIDSE et al. 1983b). The effects of metalaxyl and α-amanitin are additive indicating interference of metalaxyl with an RNA polymerase activity different from polymerase II. Synthesis of mRNA being highly sensitive to α-amanitin is apparently not affected by metalaxyl. It confirms the lack of any effect of metalaxyl on the synthesis of poly(A)-containing RNA in the intact organism.

Mechanism of resistance to metalaxyl

The mechanism of resistance to metalaxyl has been studied in strains of *P. megasperma* f. sp. *medicaginis* in which metalaxyl resistance was induced by nitrosoguanidine mutagenesis (DAVIDSE 1981) and in resistant strains of *P. infestans* obtained from potato fields where metalaxyl failed to control late blight (DAVIDSE et al. 1981 b, 1983 a). In the resistant strains uridine incorporation into RNA appeared to be completely insensitive to metalaxyl at concentrations which maximally inhibited that of the sensitive strains (DAVIDSE et al. 1984; DAVIDSE et al., unpublished results). Likewise endogenous nuclear RNA polymerase activity of resistant strains of both species was significantly less sensitive to metalaxyl than that of sensitive strains. Apparently, in both species a change in the target site of metalaxyl is responsible for resistance. The fact that both chemically induced resistance and resistance after natural selection have a similar basis proves once more the validity of laboratory studies aimed at evaluating the potential of a fungus to develop resistance to a fungicide.

Mode of action of other phenylamides in comparison with that of metalaxyl

The phenylamide fungicides originate from a screening program in which the antifungal activity displayed by a number of phenylamide herbicides has been optimized with the concurrent elimination of herbicidal activity (HUBELE et al. 1983). This development has been possible because herbicidal and antifungal activity require different structural features of the phenylamides. The presence of the chloroacetyl group seems important for herbicidal activity but not for antifungal activity. On the other hand, high antifungal activity requires the presence of an alanine methylester moiety, or an equivalent structure, which does not seem to be essential for herbicidal activity. Chirality dependency of herbicidal activity is also different from that of antifungal activity. R-enantiomers display considerable higher antifungal activity than S-enantiomers. In contrast herbicidal activity is higher with the S-enantiomers (MOSER et al. 1982). In view of this the primary mechanism of action of the phenylamides in fungi will likely be different from that in plants (DAVIDSE 1984).

The herbicidal phenylamides have been included in a number of studies on the antifungal mode of action of the phenylamides (DAVIDSE et al. 1984). Such comparative studies may yield valuable information about structure-activity relationships and about the relation, if any, between the antifungal and the herbicidal activity of the phenylamides.

Tables 16.11 and 12 summarize the results of a study in which the activities of four phenylamide fungicides and two antifungal phenylamide herbicides on growth, uridine uptake and uridine incorporation of a wild-type strain and a metalaxyl resistant strain of P. megasperma f. sp. medicaginis were compared. The existence of cross resistance of the resistant strain to the various phenylamides is evident when the respective EC_{50}-values against growth of the two strains are compared (Tab. 16.11). Endogenous nuclear RNA polymerase activity of the resistant strain is hardly affected by any of the compounds wheres that of the sensitive strain is sensitive to all phenylamides except propachlor (Tab. 16.12). The activity of propachlor might be too low in order to be detected in this assay. Interference with rRNA synthesis evidently is a common basis of the antifungal activity of phenylamides and resistance appears to be due to a change at the target site. Results obtained with wild-type and metalaxyl-resistant strains of *P. infestans* in similar studies support this idea. In this species the endogenous RNA-polymerase activity of isolated nuclei from resistant strains was also significantly less sensitive to inhibition by the phenylamides than that of wild-type strains (DAVIDSE *et al.* unpublished).

Table 16.11 Effects of phenylamides on radial growth on agar, uptake of uridine and incorporation of uridine into RNA of a wild-type (S) and a metalaxyl-resistant (R) strain of *Phytopthora megasperma* f. sp. *medicaginis*

Phenylamide	EC_{50} (μM) Radial growth[2]		Resistance factor (EC_{50} R)/ (EC_{50} S)	EC_{50} (μM) uridine uptake[3]		EC_{50} (μM) uridine incorporation[3]	
	S	R		S	R	S	R
metalaxyl	0.03	1,900	63,300	>1,000	>1,000	0.06	>1,000
benalaxyl	0.25	250	1,000	140	100	0.06	60
cyprofuram	10	900	90	1,000	>1,000	1.6	70
oxadixyl	1.2	1,800	1,500	>1,000	>1,000	1.1	>1,000
metolachlor	13	450	35	1,000	700	5.6	220
propachlor	75	200	3	420	420	130	130

[1]) Concentration of phenylamide at which the process indicated is inhibited for 50%.
[2]) Determined by measuring colony diameter on a synthetic agar medium (ERWIN and KATZNELSON 1961), amended with the different phenylamides at various concentrations.
[3]) 10 ml liquid cultures containing 20–30 mg dry weight of mycelium in a synthetic liquid medium (ERWIN and KATZNELSON 1961) were pulse labelled for 15 min with 0.5 µCi [³H] uridine with or without (control cultures) preincubation with the different phenylamides at various concentrations for 45 min. Uptake of [³H] uridine and its incorporation into RNA were determined using standard procedures (DAVIDSE *et al.* 1983 b).

Table 16.12 Effects of phenylamides at concentrations of 10 µg/ml on endogenous RNA polymerase activity of nuclei isolated from a wild-type (S) and a metalaxyl-resistant (R) strain of *Phytophthora megasperma* f. sp. *medicaginis*

Phenylamides	Activities as a percentage of control[1]	
	S	R
metalaxyl	58	91
benalaxyl	51	97
cyprofuram	70	96
oxadixyl	63	92
metolachlor	72	92
propachlor	93	91

[1]) Isolated nuclei were incubated in a reaction mixture containing in a total of 100 µl: 50 mM Tris-HCl (pH 7.9), 1 mM dithiothreitol, 1 mM ATP, GTP and CTP, 0.01 mM UTP, 2 µCi [5,6-³H] UTP, 4 mM MnCl₂, 50 mM ammonium sulphate, 10 µl of nuclear suspensions and the different phenylamides or solvent (control). Reaction mixtures were incubated at 25 °C and after 30 min acid-precipitable radioactivity was determined using standard procedures (DAVIDSE *et al.* 1983 b).

As is evident from table 16.11 the resistance level of the resistant strain is distinct for various phenylamides. Propachlor, benalaxyl and metolachlor display the highest inhibitory effect on growth of the resistant strain. It might either be due to a residual specific effect on RNA synthesis of the resistant strain or to a second mechanism of action, the effect of which becomes evident at higher concentrations. The herbicides and in particular benalaxyl also inhibit the uptake of uridine by both the sensitive and the resistant strain as a consequence of which uridine incorporation is also affected. Metalaxyl and oxadixyl do not inhibit uridine uptake; cyprofuram shows some effect but it is certainly less pronounced than that of benalaxyl and the two phenylamide herbicides. Apparently the latter three compounds have in addition to their effect on RNA synthesis a second mechanism of action that contributes to the inhibition of growth of the sensitive strain at higher phenylamide concentrations and that is solely responsible for growth inhibition of the resistant strain. Since cyprofuram is much less inhibitory to uridine uptake than benalaxyl but inhibits incorporation of the resistant strain to the same extent, this compound still specifically interferes with RNA synthesis. Apparently the change at the target site of the phenylamides that has led to almost complete resistance to metalaxyl does not completely prevent the interaction of cyprofuram with RNA synthesis at this site.

Similar results were obtained with a wild-type and a metalaxyl-resistant strain of *P. infestans* (DAVIDSE et al., unpublished results). Benalaxyl inhibited uridine uptake of both *P. infestans* strains and cyprofuram displayed significant activity on uridine incorporation of the resistant strain, whereas it did not inhibit uridine uptake. Uridine uptake by *P. infestans* was also not affected by metalaxyl and oxadixyl. Both compounds were also inactive at high concentrations on uridine incorporation of the resistant strain but were highly active on this process in the sensitive strain.

Inhibition of uridine uptake as displayed by benalaxyl and the phenylamide herbicides probably will not lead to a reduction of fungal growth because uridine is synthesized by the fungus itself. Inhibition of this process evidently is just one feature of the second mechanism of action of these phenylamides and indicates a general disturbance of metabolism as a consequence of which growth is affected. Suggestions about the nature of the second inhibitory mechanism of action can only be speculative. Interference with membrane functioning may be involved. Metolachlor and to a lesser degree cyprofuram causes lysis of protoplasts of *P. megasperma* f. sp. *medicaginis* (FISHER and HAYES 1985), whereas metalaxyl is almost inactive. Benalaxyl lyses zoospores of *Plasmopara viticola* at concentrations as low as 10 µg/ml (Gozzo et al. 1984). These observations and the effect on uridine uptake support the idea that interference with membrane functioning is a second mechanism of action of propachlor, metolachlor and benalaxyl.

The second mechanism of action of benalaxyl and the activity of cyprofuram to interfere specifically with RNA synthesis of the resistant strain could theoretically lead to a better performance of these compounds in disease control as compared with the other phenylamides, when phenylamide resistant strains are involved. Benalaxyl and cyprofuram, however, could not prevent damping-off of lucerne seedlings caused by a metalaxyl-resistant strain of *P. megasperma* f. sp. *medicaginis* in a laboratory test, even at concentrations that were almost phytotoxic (DAVIDSE, unpublished results). In greenhouse experiments both fungicides were unable to control disease development incited by metalaxyl-resistant strains of *P. infestans* and *Pseudoperonospora cubensis* on potato and cucumber plants, respectively (COHEN and SAMOUCHA 1984; KATAN 1982). In a detached-leaf assay and a leaf disc assay, however, cyprofuram inhibited sporulation of resistant strains of *P. infestans* at lower concentrations than metalaxyl but still at levels at least 10-fold higher than required for inhibition of sporulation of sensitive strains. Similarly the level of phenylamide resistance of *Plasmopara viticola* to cyprofuram was lower than that to metalaxyl (LEROUX and CLERJEAU 1985).

Field observations on late blight development in experimental plots treated with cyprofuram and benalaxyl, however, did not reveal any disease-controlling effect of these compounds when phenylamide-resistant strains were present at the experimental site. These data indicate that the second mechanism of action of benalaxyl and the ability of cyprofuram to inhibit RNA polymerase activity of phenylamide-resistant strains are only of limited value, if any, under practical conditions.

Knowledge of how the phenylamide target site actually has been changed in phenylamide resistant strains would be valuable in interpreting the interaction of cyprofuram with the altered site. It may help in redesigning phenylamide structures or even in the design of completely new structures that show increased affinity to the altered site. In view of this but also from a more fundamental point of view the site of interaction of the phenylamides in the synthesis of rRNA should be identified further.

Concluding remarks

Studies on the mode of action of the phenylamides identified these compounds as highly specific inhibitors of rRNA synthesis. This property makes the phenylamides excellent tools to study details of this complex process. Research along this line will lead to characterization of the component of the RNA polymerase-template complex that interacts with the phenylamides and its role in rRNA synthesis. When this component, presumably a protein, is characterized isolation of its structural gene would be feasible. Once this has been accomplished an extensive research area will be opened. A search for similar components in other fungi that are naturally resistant to the phenylamides would be possible, indicating how rRNA synthesis in fungi belonging to different taxonomic classes has evolved. From a practical point of view it may even lead to new fungicides that are specifically designed to interact with these components.

References

ARP, U., and BUCHENAUER, H.: Untersuchungen zum Wirkungsmechanismus von RE 20615 und metalaxyl in *Phytophthora cactorum* und zur Resistenzentwicklung des Pilzes gegenüber diesen Fungiziden. Mitt. Biol. Bundesanst. (1981): 236–237.

BRUCK, R. I., FRY, W. E., and APPLE, A. E.: Effect of metalaxyl, an acylalanine fungicide on development stages of *Phytophthora infestans*. Phytopathology **70** (1980): 597–601.

COHEN, Y., and SAMOUCHA, Y.: Cross resistance to four systemic fungicides in metalaxyl-resistant strans of *Phytophthora infestans* and *Pseudoperonospora cubensis*. Plant Disease **68** (1984): 137–139.

DAVIDSE, L. C.: Resistance to acylalanine fungicides in *Phytophthora megasperma* f. sp. *medicaginis*. Netherl. J. Plant Pathol. **87** (1981 b): 11–24.

- Antifungal activity of acylalanine fungicides and related chloroacetanilide herbicides. In: TRINCI, A. P. J., and RYLEY, J. F. (Eds.): Mode of Action of Antifungal Agents. Brit. Mycolog. Soc. Symp. Ser. 8, Cambridge Univ. Press, Cambridge 1984, pp. 239–255.
- DANIAL, D. L., and VAN WESTEN, C. J.: Resistance to metalaxyl in *Phytophthora infestans* in The Netherlands. Netherl. J. Plant Pathol. **89** (1983a): 1–20.
- GERRITSMA, O. C. M., and HOFMAN, A. E.: Mode d'action du metalaxyl. Phytiatrie-Phytopharmacie **30** (1981a): 235–244.
- — and VELTHUIS, G. C. M.: A differential basis of antifungal activity of acylalanine fungicides and structurally related chloroacetanilide herbicides in *Phytophthora megasperma* f. sp. *medicaginis*. Pesticide Biochem. Physiol. **21** (1984): 301–308.
- HOFMAN, A. E., and VELTHUIS, G. C. M.: Specific interference of metalaxyl with endogenous RNA polymerase activity in isolated nuclei from *Phytophthora megasperma* f. sp. *medicaginis*. Experiment. Mycol. **7** (1983b): 344–361.
- LOOYEN, D., TURKENSTEEN, L. J., and VAN DER WAL, D.: Occurrence of metalaxyl-resistant strains of *Phytophthora infestans* in Durch potato fields. Netherl. J. Plant Pathol. **87** (1981b): 65–68.

ERWIN, D. C., and KATZNELSON, H.: Studies on the nutrition of *Phytophthora cryptogea*. Canad. J. Microbiol. **7** (1961): 15–25.

FISHER, D. J., and HAYES, A. L.: Mode of action of the systemic fungicides furalaxyl, metalaxyl and ofurace. Pesticide Sci. **13** (1982): 330–390.
- — A comparison of the biochemical and physiological effects of the systemic fungicide cyprofuram with those of the related compounds metalaxyl and metolachlor. Crop Protect. **4** (1985): 501–510.

FULLER, M. S., and GISI, U.: Comparative studies of the *in vitro* activity of the fungicides oxadixyl and metalaxyl. Mycologia **77** (1985): 424–432.

GOZZO, F., GARAVAGLIA, C., and ZAGNI, A.: Structure-activity relationships and mode of action of acylalanines and related structures. Proc. of the 1984 Brit. Crop Protect. Conf. Pest and Diseases (1984): 923–928.

GROHMANN, U., and HOFFMANN, G. M.: Licht- und elektronenoptische Untersuchungen zur Wirkung von metalaxyl bei *Pythium*- und *Phytophthora*-Arten. Z. Pflanzenkrankh. u. Pflanzenschutz **89** (1982): 435–446.

HICKEY, E. L., and COFFEY, M. D.: The effects of Ridomil on *Peronospora pisi* parasitising *Pisum sativum*: an ultrastructural investigation. Physiol. Plant Pathol. **17** (1980): 199–204.

HUBELE, A., KUNZ, W., ECKHARDT, W., and STURM, E.: The fungicidal activity of acylalanines. In: DOYLE, P., and FUJITA, T. (Eds.): Pesticide Chemistry, Human Welfare and the Environment, Vol. 1, Synthesis and Structure-activity Relationships. Pergamon Press, Oxford 1983, pp. 233–242.

KATAN, T.: Cross-resistance of metalaxyl-resistant *Pseudoperonospora cubensis* to other acylalanine fungicides. Can. J. Plant Pathol. **4** (1982): 387–388.

KERKENAAR, A.: On the antifungal mode of action of metalaxyl, an inhibitor of nucleic acid synthesis in *Pythium splendens*. Pestic. Biochem. Physiol. **16** (1981): 1–13.

LEROUX, P., and CLERJEAU, M.: Resistance of *Botrytis cinerea* (Pers.) and *Plasmopora viticola* (Berk. & Curt.) Berl. and de Toni to fungicides in French vineyards. Crop Protect. **4** (1985): 137–160.

MOSER, H., RIKS, G., and SANTER, H.: Der Einfluß von Atropisomerie und chiralem Zentrum auf die biologische Aktivität des metolachor. Z. f. Naturforsch. **87b** (1982): 451–462.

MÜLLER, H. M., and LYR, H.: Morphologische und zytologische Veränderungen bei *Phytophthora infestans* (Mont.) de Bary und *Phytophthora cactorum* (Leb. et Cohn) Schroet. unter dem Einfluß von metalaxyl. In: LYR, H., and POLTER, C. (Eds.): Systemische Fungizide und Antifungale Verbindungen 1982 1N. Akademie-Verlag, Berlin 1983, pp. 403–410.

STAUB, T. H., DAHMEN, H., and SCHWINN, F. J.: Effects of Ridomil on the development of *Plasmopara viticola* and *Phytophthora infestans* on their host plants. Z. Pflanzenkrankh. u. Pflanzenschutz **87** (1980): 83–91.
— and YOUNG, T. R.: Fungitoxicity of metalaxyl against *Phytophthora parasitica* var. *nicotianae*. Phytopathology **70** (1980): 797–801.
SZANISZLO, P. J., KANG, M. S., and CABIB, E.: Stimulation of $\beta(1 \to 3)$ glucan synthetase of various fungi by nucleoside triphosphates: generalized regulatory mechanism for cell wall biosynthesis. J. Bacteriol. **161** (1985): 1188–1194.
WOLLGIEHN, R., BRÄUTIGAM, E., SCHEUHMANN, B., and ERGE, D.: Wirkung von Metalaxyl auf die Synthese von RNA, DNA und Protein in *Phytophthora nicotiana*. Z. Allgemeine Mikrobiol. **24** (1984): 269–279.

Chapter 17

2-Aminopyrimidine fungicides

D. W. HOLLOMON* and H.-H. SCHMIDT**)

*) Department of Agricultural Sciences, University of Bristol, Institute of Arable Crops Research, Long Ashton Research Station, Long Ashton, Bristol BS18 9AF, UK.
**) Federal Biological Research Centre for Agriculture and Forestry, Kleinmachnow, Germany.

Introduction

Many important crop plants are attacked by powdery mildews, and potential markets for fungicides which control these diseases are large. Because they are easily handled in the greenhouse, mildews are invariably included by the agrochemical industry in fungicide screening programmes. Some 30 years ago, routine screening at Jealott's Hill Research Station identified certain phosphorylated analogues of the insecticide diazinon (Fig. 17.1 a) as having protectant activity against powdery mildews. Additional structure-activity studies revealed a series of 2-amino-4-hydroxy pyrimidines (SNELL et al. 1966) which were not only specific against powdery mildews, but also systemic. Dimethirimol (Fig.17.1 b) was the first hydroxypyrimidine to be discovered and develoed (ELIAS et al. 1968; GEOGHEGAN and DE GRAAFF 1969). A second related fungicide, ethirimol, (Fig. 17.1 c) soon followed (BEBBINGTON et al. 1969), because of its better activity, especially against barley powdery mildew. Advances in formulation and seed treatment technology enabled ethirimol to be marketed as a seed treatment, with considerable agronomic benefits to farmers. Its widespread use helped confirm that fungicides could be used economically to control cereal leaf diseases, and that chemicals offered a viable alternative to breeding for disease resistance. Dimethirimol and ethirimol were effective largely against powdery mildews of herbaceous plants; bupirimate (Fig. 17.1 d) was developed later to control these diseases on woody plants and ornamentals (FINNEY et al.

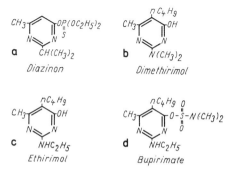

Fig. 17.1 Chemical structures for pyrimidine insecticides and fungicides. a) Diazinon, b) Dimethirimol, c) Ethirimol, d) Bupirimate.

1975). Since the introduction of bupirimate, there appears to have been little further interest in the chemistry of 2-aminopyrimidines. The chemical, physical and toxicological properties of commercial 2-aminopyrimidine fungicides are summarised in table 17.1.

Table 17.1 Some chemical, physical and toxicological properties of ethirimol, dimethirimol and bupirimate (derived from WORTHING, C. R., and WALKER, S. B., 1983)

	Ethirimol	Dimethirimol	Bupirimate
code name	PP 149	PP 675	PP 588
trade marks	Milgo, Milgo E, Milstem, Milcurb Super	Milcurb	Nimrod
melting point	159–160 °C (with a phase change at about 140 °C)	102 °C	50–51 °C
vapour pressure	267 µPa at 25 °C	1.46 mPa at 30 °C	67 µPa at 20 °C
solubility at room temperature	200 mg/l water, sparingly soluble in acetone, slightly soluble in diaceton, alcohol, chloroform, trichlorethylene and acqueous solutions of strong acids and bases	1.2 g/l water, 45 g/l acetone, 1.2 kg/l chloroform, 65 g/l ethanol, 360 g/l xylene	22 mg/l water, soluble in most organic solvents except paraffin hydrocarbons
acute oral LD_{50} for rats	6,340 mg/kg	2,350 mg/kg	approx. 4,000 (female rat)
no effect level in 2-year feeding tests	200 mg/kg diet	300 mg/kg diet	100 mg/kg diet

Practical application

Ethirimol

Although used primarily to control powdery mildews of cereals (*Erysiphe graminis*), ethirimol has also shown good effects against cucumber mildew (*E. cichoracearum* and *Sphaerotheca fuliginea*, BENT 1970; HARTMAN and SIEGEL 1973). Ethirimol also reduced damage caused by *E. betae*, and increased sugar content of beet, but its performance was inferior to that of sulphur and benomyl (BYFORD 1978). Yield increases in cereals following either protective or eradicative treatments to control mildew have been obtained in many European countries (Tab. 17.2). Differences between trials may reflect variation in disease levels, and so responses frequently depend on the cultivars used, and their susceptibility to mildew (CLIFFORD et al. 1971; KRÜGER 1969; PETKOVA 1970). WOLFE 1969 obtained greatest control with spring sown barley and wheat; autumn sown crops were less responsive. Indeed, yield increases in wheat are often difficult to achieve (BAUERS 1972; HANSEN 1978), and may require two applications during each season, preferably as mixtures with broader spectrum fungicides. In fact, only recently has ethirimol been introduced commercially for mildew control in wheat in France.

Yield increases reflect either increased numbers of ears, numbers of grain per ear, or increases in grain size (D'Arbigny and DAWSON 1973; BAUERS 1972; JENKYN 1974, 1978),

Table 17.2 Some examples of yield responses after application of ethirimol to control powdery mildew on barley and winter wheat

Country	Years	Number of tests	Dose rate (ai) seed dressing (g/100 kg seed)	spray (g/ha)	Yield response (%)	Remarks	Reference
Spring barley							
Austria	1973	many	360	—	18 (10–24)		1
	1973	many	420	—	18 (10–31)		1
Belgium	1970–1974		—	280 (1 application)	10–20		2
Czechoslovakia	1971	many	438	—	9		3
	1971		—	700	8		3
Denmark	1969–1976	72	240	480	7		4
Germany	1969–1971	15	600	—	111		5
	1971	6	360	—	6	mean results with 10 cultivars	6
	1976	5	—	280	6 (4–8)		7
UK	1969–1970		1,200	—	4 (3–6)	Mildew susceptible cultivar 'Golden Promise'	8
	1971		1,200	—	26	Severe mildew attack in 1971 only	8
	1971	5	—	900	16	Response dependent on cultivar used with yield losses on some cultivars	9
	1969–1971	28	573	—	8 (3–9)	Yield loss recorded in 1975	10
	1968–1974		400	—	12 (10–17)		11
	1970–1975		400	—	13 (7–30)		12
Winter barley							
UK	1969–1971	18	700	—	6		3
Winter wheat							
Czechoslovakia	1970	1	—	1,600 (2 applications)	26	Mildew susceptible cultivar 'Consul'	14
Germany (only in the former GDR)	1975–1981	29	—	280 (2 applications)	3	Only 3 tests showed significant yield increases	15
Poland	1971–1975	223	—	750 (2 applications)	10		16

1 = ZWATZ 1974; 2 = MEEUS et al. 1975; 3 = BENADA 1972; 4 = HANSEN 1978; 5 = BAUERS 1972; 6 = BAUMER and ULONSKA 1971; 7 = Official registration tests; 8 = CHANNON and BOYD 1973; 9 = GILMOUR 1971; 10 = JENKYN and MOFFAT 1975; 11 = SHEPHARD et al. 1975; 12 = WOLFE 1975; 13 = HALL 1971; 14 = BENADA 1971; 15 = MULLER 1984; 16 = JACZEWSKA 1983

and effects on these components are influenced by application time (BETHUNE 1973; CARVER and GRIFFITHS 1981). Slight yield reductions on some mildew resistant spring barley cultivars were recorded by CHANNON and BOYD (1973) and GILMOUR (1971) after ethirimol seed dressing at above normal recommended rates. Significant reductions in root growth associated with a decrease in chromosome volume in the spring barley cultivar Julia, were observed after treatment with ethirimol (16 ppm) in a culture solution (BENNETT 1971). There was also a small reduction in chiasma frequency of pollen mother cells of cultivars Julia and Sultan, but yield losses due to infertility appeared unlikely. However, deleterious effects of ethirimol on plants were not demonstrated if recommended dose rates were used.

Although primarily a prophylactic measure against early infection, seed treatment has proved especially economical against barley powdery mildew (BEBBINGTON et al. 1971; HALL 1971; KÜTHE 1972). Combinations with triazole fungicides like flutriafol (NORTHWOOD et al. 1984) and triadimenol have also been developed as seed treatments, and NOON et al. (1988) demonstrated with such combinations high levels of mildew control 14 to 16 weeks after drilling. Addition of thiabendazole and imazalil controls diseases caused by *Fusarium* and *Helminthosporium* species. Soil moisture levels must be adequate (COLLIER et al. 1979), otherwise insufficient ethirimol may be taken up by barley plants, so that seed dressings are not always effective under dry weather conditions (KING 1977).

Spray applications enable farmers to respond with more flexibility to mildew attacks in different cereal species. Treatment of barley in the early phases of a mildew epidemic (3 – 5% mildew infection on the first three leaves) are recommended (CARVER and GRIFFITHS 1981; NEUHAUS and REICH 1975), whereas in wheat flag leaves and ears need to be protected. MÜLLER (1984), who examined the efficacy of "Milgo E" in winter wheat, found good curative action against mildew when it was applied within the latent period of infection (up to 5 days after inoculation of cultivar Alcedo). Milgo E also showed stability against rainfall and in the former GDR it has proved suitable for aerial application (spray rate: 50 litres/ha).

To enlarge the spectrum of activity tank mixtures of ethirimol with other fungicides may be used, and in France mixtures with captafol gave additional control of rust fungi, *Rhynchosporium* (LESCAR 1977), and *Septoria* (MILLOU and D'ARBIGNY 1974). In tests of the IOBC/WPRS-working group the spray formulation, Milgo E, was proved to be harmless to useful parasites (Aphidius, Coccygomimus, Leptomastix, Phygadeuon, Trichogramma), predatory insects (Anthocoris, Bembidion, Chrysoperla) and to the fungus *Verticillium lecanii*). Only slightly harmful effects were found against species of the genera Cales and Encarsia. So ethirimol can be regarded as harmless to all beneficial organisms relevant in cereals, and recommended for use in integrated control programmes (HASSAN et al. 1988).

Dimethirimol

Dimethirimol is limited mainly to treatment against *E. cichoracearum* and *S. fuliginea* on cucurbits. There are also records of positive results on ornamentals like cineraria, sweet pea (GEOGHEGAN 1969) and chrysanthemum (KOBACHIDZE and TOSKINA 1971). Dimethirimol can be applied either as a soil drench to the stem base of plants (e.g. 0.25 g a.i. per cucumber plant) or as a spray treatment. In glasshouses soil drenching is preferred. As with ethirimol uptake by plant roots is positively correlated with soil moisture, and dimethirimol is loosely adsorbed onto soil particles and released slowly (GEOGHEGAN 1969). Its availability to plant roots is best in light alkaline soils.

Soil drenches have kept cucumber plants free from mildew for more than 6 weeks (BROOKS 1970; KOBACHIDZE and TOSKINA 1971). In some crops, especially irrigated melons (BROOKS 1970), a granule formulation was very effective. Dimethirimol may also have some curative action against *E. pisi*, the cause of powdery mildew of peas (GORSKA-POCZOPKO 1971).

Bupirimate

Bupirimate, the sulphamate ester of ethirimol, moves in woody plants (SHEPHARD 1981; TEAL and CAVELL 1975), and its spectrum of activity includes powdery mildews of various fruit trees (Tab. 17.3). The effective spray rate depends on plant, mildew species and method of application, but ranges mostly between 50 to 500 ppm bupirimate. FINNEY et al. (1975) reported that bupirimate, at a concentration of 75 to 150 mg/l suppressed sporulation of *Podosphaera leucotricha* more effectively than binapacryl (500 mg/l),

Table 17.3 Spectrum of activity of bupirimate

host plants	mildew species	Reference*)
woody plants		
apple	*Podosphaera leucotricha*	2, 3, 4
apricot	*Podosphaera tridactyla*	2
mango	*Oidium mangifera*	2
vine	*Uncinula necator*	2
peach	*Sphaerotheca pannosa*	2
rose	*Sphaerotheca pannosa*	2, 3
currant	*Sphaerotheca mors-uvae*	1, 2
gooseberry	*Sphaerotheca mors-uvae*	5
Herbaceous plants		
hop	*Sphaerotheca humuli*	3
strawberry	*Sphaerotheca humuli*	2
cucurbit	*Sphaerotheca fuliginea* (only *cucumber*)	2, 3
	Erysiphe cichoracearum	2
comfrey	*Erysiphe cichoracearum*	3
peas	*Erysiphe pisi*	2
sugar beet	*Erysiphe betae*	2
chrysanth	*Oidium chrysanthemum*	2, 3
gerbera	*Oidium* sp.	3
pepper	*Leveillula taurica*	2

*) 1 = BIELENIN et al. 1985
2 = FINNEY 1975
3 = unpublished data from registration tests in the former GDR 1978–1981
4 = KUNDERT 1977
5 = O'RIODAIN and KENNEDY 1981

dinocap (250–300 mg/l), sulphur (3,200 mg/l) or triforine (250 mg/l), but showed activity similar to thiophanate-methyl (500 mg/l), benomyl (250 mg/l) and ditalimfos (375 mg/l). In potted apple plants under glass, both HUNTER et al. (1981) and JAHN et al. (1986) found that bupirimate had significant curative effects against *P. leucotricha*, if applied within 4 days after inoculation. Applications 12 days before inoculation still gave 90 per cent disease control (JAHN et al. 1986). One reason for the remarkable protective and eradicative efficacy of bupirimate may be its good vapour phase activity (FINNEY et al. 1975; PETSIKOL-PANYOTAROU 1980; TEAL and CAVELL 1975). Simulated rainfall (30 mm) did not alter the effectiveness of the EC-formulation of bupirimate (Nimrod 25 EC) against *P. leucotricha* on potted apple plants (RATHKE and JAHN 1984), and the fungicide remained equally effective over the temperature range from 15 to 40 °C (RATHKE 1984).

Phytotoxicity caused by bupirimate is not normally a serious problem if recommended application rates are used (KUNDERT 1977). However, MANTINGER (cited by KUNDERT) recorded some detrimental effects (chlorosis, violet colour change and abscission of leaves) with the cultivars 'Morgenduft' and 'Jonathan', but this did not seem to affect yield. Investigating different cultivar x fungicide interactions (including bupirimate) with apples JEGER et al. (1983) found that spraying (irrespective of fungicide) was most effective on cultivars with a high level of mildew susceptibility (e. g. 'Golden Delicious'). Contrary to experience with ethirimol and dimethirimol, bupirimate did not effectively control foliar mildew infections when applied to young apple and cucumber plants as a soil drench (FINNEY et al. 1975).

The most effective way to apply bupirimate in orchards is by high volume spray (COOKE et al. 1977). In the former GDR an application of 375 ml bupirimate/ha by helicopter (spray volume 100 l/ha) only protects slight or moderately susceptible apple cultivars, when disease pressure is low. Surfactants and stickers may improve the fungicidal effectiveness of bupirimate (SHABI 1975). Equally, the effects of surfactants against overwintering mildew in apple buds were improved by addition of bupirimate (BENT et al. 1977), but it was necessary to reduce the amount of the surfactant PP222 (a nonylphenolethoxylate) in this mixture, to avoid phytotoxicity: In numerous experiments bupirimate has been proved to be non-toxic to phytoselid mites, of the genera Amblyseius, Metaseiulus, Typlodromus (EASTERBROOK 1984; KARG 1990; KARG et al. 1987), predatory insects from the families of Chrysopidae, Coccinellidae, Syrphidae, some hymenopterous insects (EATERBROOK et al. 1979; FISCHER-COLBRIE 1986; KARG et al. 1987) and fungal pathogens of aphids (HALL 1981). This, and its specificity as a powerful antisporulant, make bupirimate suitable for integrated pests management programmes in both orchards and glasshouse crops (BUTT et al. 1983). Nowadays bupirimate is recommended without restrictions for such purposes in pome fruits in Austria, Denmark, France, Germany, Great Britain, Italy, The Netherlands and Switzerland (SCHÄFERMEYER and DICKLER 1991).

2-Aminopyrimidines generally have little effect on the soil fauna, including earthworms. Ethirimol seed dressings influenced neither the phylloplane microflora of barley leaves (DICKINSON 1973) nor *Sporobolomyces* spp. and *Cladosporium* spp. on flag leaves of winter wheat (JENKYN and PREW 1973 a, b). However, other pathogenic fungi may colonize healthy leaves protected by treatment with mildew specific fungicides, and this may limit yield responses (JENKYN and MOFFATT 1975). Barley cultivars susceptible to brown rust were more severely infected by *Puccinia hordei* after treatments with ethirimol, than were untreated mildewed plants (LITTLE and DOODSON 1971). In some instances, spray treatments seemed especially to favour rust development (JENKYN 1974), but not in others (BETHUNE 1973). ROUND and WHEELER (1978) examined the competition for space on the leaf surface of barley seedlings (cultivar 'Zephyr') and showed that an ethirimol soil drench applied at the time of rust inoculation, but six days after inoculation with *E. graminis*, increased significantly both the numbers of rust pustules and their size.

In one of three trials carried out by KAMPE (1974), infection of spring barley with *R. secalis* following ethirimol seed treatment (350 g/dt) was twice that observed in untreated plots, and no yield responses were recorded. In other trials, ethirimol seems to have stimulated Rhynchosporium infection in winter barley, but not rye (BEER and BIELKA 1986), but this has not been a general observation in trials carried out at various Western European locations. Ethirimol seed treatments also increased the severity of leaf stripe (*Helminthosporium sativum*) of winter barley (cultivar 'Dura') following artificial inoculation in the greenhouse (SAUR and SCHÖNBECK 1975). On agar, ethirimol (1 and 10 ppm) stimulated growth of *H. sativum*, and on barley leaves stimulated germination of conidia, penetration and colonization. Sporulation on treated plants occurred five days earlier than on untreated ones. Helminthosporal, a toxin of *H. sativum*, inhibited root growth, reduced IAA content, but increased activity of IAA oxidase, peroxidase, and ion efflux, in ethirimol treated barley plants (SAUR 1976a, b; SAUR and SCHÖNBECK 1976). These changes are all likely to influence disease development, and it seems that there is some interaction with the host, perhaps involving damage to cell membranes.

In vitro effects

In vitro 2-aminopyrimidines inhibited spore germination, formation of appressoria and haustoria and, to a lesser extent, hyphal growth (BENT 1970; HOLLOMON 1979a). However, the concentration needed to affect germination was well above those normally found in plants (CALDERBANK 1971). Sporulation may also be reduced. Dimethirimol (50 ppm) failed to inhibit germination of conidia of *Botrytis fabae*, *Venturia inaequalis*, uredospores of *P. recondita*, but marginally (26%) inhibited germination of *Phytophthora infestans* zoosporangia. 2-Aminopyrimidines also had little effect on the growth of *B. cinerea* (GRINDLE 1981). Sporulation of *V. inaequalis* was reduced by bupirimate (RATHKE and JAHN 1989), whilst ethirimol showed reasonable activity against *Phomopsis viticola* in an amended agar growth assay (FARETRA et al. 1987). Susceptibility of different mildew species to 2-aminopyrimidines varied considerably with the minimum inhibitory concentration ranging from 0.005 ppm to 10 ppm for *E. communis* on *Brassica juncea* (GORTER and NEL 1974). A similar range of variation was encountered in laboratory tests against *E. graminis* f. sp. *hordei*, reflecting a broad spectrum of genetic variability in response to ethirimol (HOLLOMON 1981). The intrinsic toxicity of ethirimol compared very favourably with that of eleven other mildew fungicides (KLUGE and LYR 1983). Unlike ethirimol, bupirimate has excellent vapour phase activity, and a deposit of 25 µg inhibited growth of *S. fuliginea* some 5 mm away; 20 mm away inhibition of growth was about 50 per cent (PETSIKOL-PANAOTAROU 1980).

Uptake, metabolism and degradation

Availability to plants of 2-aminopyrimidines in soils depends very much on the organic matter present, and on acidity. Once adsorbed onto soil particles these fungicides are released slowly into soil moisture so providing a continuous supply of fungicide for uptake by the crop. The strong adsorption of ethirimol to acid peat, however, prevents its use in these soils (COLLIER et al. 1979; GRAHAM-BRYCE and COUTTS 1971). Uptake from spray formulations has received less attention, but foliar penetration of ethirimol into winter wheat from suspension concentrates was considerably better with aliphatic alcohol surfactants present, rather than nonylphenol or sorbitan products (HOLLOWAY et al. 1992). Within herbaceous plants 2-aminopyrimidines move in the transpiration stream, accumulating around leaf margins. Some may even be exuded on the leaf surface. In woody plants 2-aminopyrimidines are much less mobile (SHEPHARD 1973). Bupirimate remains within veins when applied as a root drench (TEAL and CAVELL 1975), but is sufficiently volatile to be redistributed as vapour across leaf surfaces (FINNEY et al. 1975).

Degradation of 2-aminopyrimidines in both cucumber and barley leaves is rapid, with a half-life of no more than 4 days (CALDERBANK 1971). Polar metabolites are generated principally through N-dealkylation, conjugation to form glucosides, and hydroxylation of the 5 n-butyl group (CAVELL et al. 1971). Some of these metabolites are also active against powdery mildews. Bupirimate is readily hydrolysed to ethirimol in acid solution on the leaf surface.

Isolated haustoria of *E. pisi* accumulated ethirimol to at least 60-fold the external concentration, which was in excess of its aqueous solubility (MANNERS and GAY 1980). Lipophilicity may play some role in this entry, but accumulation was similar regardless of whether ethirimol was protonated or not. It is not clear how relevant this uptake into haustoria is, since accumulation of a series of 2-aminopyrimidines was not related to their fungitoxicity (HOLLOMON 1984). Furthermore, the main effect on powdery mildew fungi is seen at appressoria formation, which occurs before haustoria form.

2-Aminopyrimidines readily polymerize on exposure of concentrated solutions to UV light in the laboratory (WELLS et al. 1979). Similar polymers have not been detected after exposure to sunlight on leaf surfaces, although other photochemical degradation products of ethirimol have been identified (CAVELL 1979).

Mechanism of action

Powdery mildews enter their hosts through the cuticle. An essential first step involves formation of an appressorium, and this step in development is inhibited by 2-aminopyrimidine fungicides. The critical biochemical events occur well before the appressoria are first seen, for only when applied within the first eight hours after inoculation did ethirimol prevent appressoria formation (HOLLOMON 1977). Exchange of cations and fluorescent dye between *E. graminis* and its host occurred soon after germination began (KUNOH et al. 1982), so fungicides may enter germinating conidia in the same way. Other steps in mildew development may be inhibited, although these may be of limited practical significance. Electron microscope studies have emphasised differences between ethirimol and DMI fungicides, and although haustorial membranes may eventually be disrupted (SMOLKA and WOLFE 1986), effects on nuclear function and ribosome loss occur first (HELLER et al. 1990).

Some indication of the likely mode of action of 2-aminopyrimidines followed attempts to reduce their toxicity with various metabolites (see HOLLOMON 1992 for refs). In experiments involving several powdery mildews, the purine base adenine, and its ribonucleoside, adenosine, were always good reversal agents. Kinetin (6-furfuryl adenine) and isopentenyl adenine both prevented appressoria formation on barley, and mildew strains resistant to ethirimol were cross-resistant to these purines. 2-Aminopyrimidines seemed, therefore, to interfere with purine metabolism, although after the purine ring formed, since *E. graminis* appears to be a purine auxotroph, obtaining purines it needs from its host (HOLLOMON 1979a).

Further studies showed that germinating *E. graminis* conidia incorporated adenine and adenosine into nucleic acids, but in the presence of ethirimol formation of inosine and adenosine nucleotides was inhibited and nucleic acid synthesis halted (HOLLOMON and Chamberlain 1981). Several enzymes involved in the reutilization of purines were examined in cell-free extracts from *E. graminis* conidia, but of these only adenosine deaminase (ADA-ase) was inhibited to any significant extent by ethirimol. ADA-ase catalyses the largely irreversible hydrolytic deamination of adenosine to inosine. Plants apparently lack ADA-ase, and instead deaminate 5'AMP at the nucleotide rather than nucleoside level. ADA-ase is present in many fungi, but only the enzyme from powdery mildew fungi was sensitive to ethirimol (HOLLOMON 1979b). This may account for the extreme specificity of 2-aminopyrimidines towards powdery mildews. Structure-activity studies provided further evidence that ADA-ase was a site of action of these pyrimidine fungicides, for analogues that were poor inhibitors of this enzyme, were also poor fungicides. Exceptions occurred, but these analogues may have been activated by conversion to ethirimol within barley during bioassay (HOLLOMON and CHAMBERLAIN 1981).

Inhibition of ADA-ase by 2-aminopyrimidine fungicides was pH dependent and non-competitive. Indeed, ethirimol hardly resembles adenosine in shape, and it does not bind tightly to ADA-ase. The 5 *n*-butyl group not only provides adequate lipophilicity for membrane permeability, but is essential for correct binding to the enzyme. Loss of both fungicide activity and ADA-ase inhibition follows hydroxylation of the butyl group, suggesting that it is buried within a hydrophobic region of ADA-ase, with the 2-substituted amino group towards the enzyme surface (HOLLOMON and CHAMBERLAIN 1981). In both mildew infected plants and *E. graminis* conidia, ADA-ase exists as a single, large molecular weight (MW 300,000) protein, with kinetic properties similar to those of ADA-ase from other organisms. In many of these organisms the catalytic subunit (MW 36,000) of ADA-ase is associated with a large accessory protein whose function is unknown (KELLEY *et al.* 1977). It is not known if mildew ADA-ase has a similar structure. Why inhibition of ADA-ase should have such drastic consequences on events leading to appressoria formation is not clear. *E. graminis* does not synthesise purines *de novo* so, apart from some limited reutilization of the purines already in conidia, infection requires a continuous supply of additional purines from the host. ADA-ase may play some part in linking the purine metabolism of the host with that of the powdery mildew fungus (BUTTERS *et al.* 1985). Reutilization of purines involves a matrix of reactions (Fig. 17.2), in which ADA-ase would channel purines derived from the larger 5'AMP pool to 5'GMP. However, neither inosine nor any subsequent metabolite involved in the pathway to 5-GMP synthesis reversed the toxicity of ethirimol (HOLLOMON 1979a), so clearly inhibition of ADA-ase must have other effects. ADA-ase is one of several enzymes that control adenosine levels and is likely, therefore, to alter the activity of enzymes such as S-adenosyl homocysteine hydrolase, which is regulated by adenosine and is involved in C-1 metabolism. Finally, ADA-ase ensures a "sink" for 5'AMP breakdown, so by increasing the energy charge (ATKINSON 1968) within conidia, many enzymes would be activated.

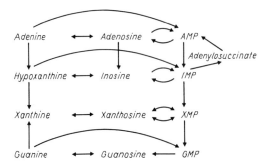

Fig. 17.2 Salvage pathway in purine metabolism.

Resistance situation

Use of dimethirimol in 1970 to control cucumber powdery mildew (*S. fuliginea*) soon encountered difficulties in a number of glasshouses throughout N.W. Europe. Control failures were associated with a decline in sensitivity of the mildew (BENT et al. 1971) and withdrawal of dimethirimol was recommended. As a result, sensitivity has now increased, but attempts to reintroduce dimethirimol have only been partially successful as resistant strains can still be isolated from diseased cucumbers (SCHEPERS 1984; OHTSUKA et al. 1991). Control of the same disease on field grown cucurbits has not, so far, encountered problems. A more gradual decrease in sensitivity to ethirimol followed its introduction as a seed treatment to control barley powdery mildew. Although continued use may not have led to any further decline in sensitivity, recommendation for use for ethirimol on autumn sown barley was withdrawn in 1973, in an attempt to reduce carry over of resistant strains which might otherwise infect spring sown crops. This may have reduced the frequency of resistant strains (SHEPHARD et al. 1975), which generally are less fit than wild-type strains (HOLLOMON 1978). The success of this policy in maintaining the effectiveness of ethirimol was difficult to judge, since broad spectrum triazole fungicides, introduced in 1978, soon replaced ethirimol. Population changes were further complicated by an association between ethirimol resistant strains and their virulence on barley cultivars with the Mla_{12} host-plant resistance gene (WOLFE and DINOOR 1973). No genetic link could be established (HOLLOMON 1981), suggesting that ethirimol resistance probably arose by chance in virulent strains on these Mla_{12} cultivars (Hassan, Sultan), which were widely grown at the time. These strains were then selected in the largely asexually reproducing mildew population.

The failure to detect discrete phenotypic groups in field populations suggests that several genetic factors control ethirimol resistance, and this was supported by analysis of progeny from crosses using standard bioassay procedures (HOLLOMON 1981). More recent studies have suggested that ethirimol resistance was controlled by a single gene, Eth1 (BROWN et al. 1992). In these experiments, seedling assays produced considerable variation well outside the range that could be handled by a conventional probit or regression analysis, and **P**rincipal **C**omponent **A**nalysis (PCA) was used instead. Further examination of these data suggests that the progeny groups identified by PCA may be artefacts, and result from the transformation of these data to generate the principal components.

So far, the performance of bupirimate has not been eroded in this way by the appearance of resistant strains. When the triazole seed treatment, triadimenol, was first introduced, it was far superior to ethirimol for mildew control on many barley varieties (Martin et al. 1981). Since 1981, sensitivity of barley mildew to triadimenol has changed significantly (Fletcher and Wolfe 1981; Wolfe et al. 1983; Butters et al. 1984; Heaney et al. 1984), and by 1986 control was unacceptable, especially on mildew-susceptible varieties. Triazole-resistant mildew strains often showed wild-type sensitivity to ethirimol (Butters et al. 1984), and in some situations ethirimol provided better mildew control than triadimenol (Northwood et al. 1984; Hollomon et al. 1985).

Evidence of negatively correlated cross-resistance between triadimenol and ethirimol (Butters et al. 1984) may simply reflect changes in field use of the two fungicides. In fact, a survey in 1984 revealed no significant correlations between ethirimol sensitivity and triadimenol sensitivity, but the mildew sampled did not include any populations that were either ethirimol resistant or triadimenol sensitive (Heaney et al. 1984). In field experiments, intensive selection can be applied to achieve populations with a wider range of sensitivities, especially to ethirimol. Selection for ethirimol resistance indeed increased sensitivity to triadimenol, but the reverse could not be shown, possibly because mildew at that site was already sensitive to ethirimol, and could not be shifted further (Hollomon et al. 1985). In a separate experiment use of each fungicide tended to induce greater sensitivity towards the other (Hunter et al. 1984), and mixtures of triadimenol and ethirimol limited development of resistance to each fungicide, gave better disease control, and increased yields. Mixed seed treatments containing ethirimol and triazoles like flutriafol or triadimenol (and thiabendazole to control Fusarium diseases) are now available, and provide an opportunity to assess how readily strains resistant to both these fungicides will be related. After six years use of the flutriafol: ethirimol (FERRAX) mixture as a seed treatment, no decrease in ethirimol sensitivity could be detected in field populations throughout the UK (Stott et al. 1990).

In three-year experiments Pons and Eule (1990), applying ethirimol with triadimefon or triadimenol either in mixtures or alternately, found no differences between these two strategies.

Future prospects

The introduction of effective broad spectrum systemic fungicides in the late seventies greatly reduced the demand for 2-aminopyrimidines. But just as resistance to both dimethirimol and ethirimol had earlier prompted concern, so evidence of erosion in the level of mildew control achieved by fungicides that inhibit the C-14 demethylation step in sterol biosynthesis, has re-emphasised the need to diversify fungicide use. Little attention has been given to 2-aminopyrimidines for several years, although a recent report of a new pyrimidine fungicide mepanipyrim (Maeno et al. 1990) suggests that further examination of pyrimidine chemistry might be rewarding. Mepanipyrim is not specific for control of powdery mildews, but instead offers broad spectrum activity against several important plant pathogens, including *Botrytis*, *Venturia* and *Molinia*. There must be considerable scope for new ADA-ase inhibitors since the metabolic steps in the purine salvage pathway are common throughout eukaryotes, yet clearly specificity exists in the enzyme target. Although inhibition of ADA-ase in humans is associated with autoimmune disease, sufficient selectivity probably exists in the target enzyme to provide scope for fungal inhibitors with satisfactory toxicological properties. In powdery mildews shifts in fungicide sensitivity are generally gradual (BRENT 1982), and overlap often remains between treated and untreated populations. This offers considerable scope for strategies using fungicides with different modes of action, either in mixtures, or as alternating programmes. By variation of application rates with fungicides differing in time and area, together with use of varietal resistance (WOLFE 1984, 1985), mildew populations may be stabilized, and loss of any one fungicide group through resistance prevented. Management of mildew populations in this way has generated renewed interest in ethirimol, and so far the mixed seed treatment with flutriafol has not heralded a return to the resistance problems of almost twenty years ago. Sensible use of ethirimol in strategies with azole and morpholine fungicides should maintain the usefulness of 2- aminopyrimidines for many years ahead.

References

ATKINSON, D. E.: The energy charge of the adenylate pool as a regulatory parameter: interaction with feedback modifiers. Biochemistry **7** (1968): 4030–4034.
BAUERS, C.: Erfahrungen bei der Mehltaubekämpfung im Getreide. Mitt. DLG, Frankfurt (M.) **87** (1972): 347–350.
BAUMER, M., and ULONSKA, E.: Chemische Mehltaubekämpfung. Bayer. landwirtsch. Jahrb., München **48**, Sonderheft 4 (1971): 107–113.
BEBBINGTON, R. M., BROOKS, D. H., GEOGHEGAN, M. J., and SNELL, B. K.: Ethirimol: a new systemic fungicide for the control of cereal powdery mildews. Chem. Ind. (1969): 1512.
– CHAMIER, O. D., and ALBRECHT, J.: Erfahrungen mit Milstem-Saatgutbehandlungen zur Bekämpfung von *Erysiphe graminis* DC in der Bundesrepublik. Mitt. Biol. Bundesanstalt, Berlin-Dahlem **146** (1971): 264.
BEER, W. W., and BIELKA, F.: Efficiency of fungicidal active ingredients against *Rhynchosporium*. Nachrichtenbl. Pflanzenschutz **40** (1986): 38–40.
BENADA, J.: Pouziti Milstemu proti padli na psenici. Agrochemia, Bratislava **11** (1971): 89–91.
– Pouziti systemovych fungicidu proti *Erysiphe graminis* DC. na jarnim jeemeni. Ochrana rostlin, Praha **45** (1972): 83–88.
BENNETT, M. D.: Effects of ethirimol on cytological characters in barley. Nature **230** (1971): 406.
BENT, K. J.: Fungitoxic action of dimethirimol and ethirimol. Ann. appl. Biol. **66** (1970): 103–113.
– COLE, A. M., TURNER, J. A. W., and WOOLNER, M.: Resistance of cucumber powdery mildew to dimethirimol. Proc. 6th Brit. Insect. Fungic. Conf. (1971): 274–282.
– SCOTT, P. D., and TURNER, J. A. W.: Control of apple powdery mildew by dormant season sprays – prospects for practical use. Proc. Brit. Crop Protect. Conf. – Pests and Diseases (1977): 331–339.
BETHUNE, J. C.: Trois annees d'essais de lutte chemique contre l'oidium sur orge de printemps. In: La lutte contre maladiés des ceréales. Report of papers, Versailles 28th February (1973): 311–322.
BIELENIN, A., CIMANOWSKI, J., BYSTYRZIENSKA, K., and PUCHALA, Z.: Control of black currant diseases. Fruit Sci. Rep. **12** (1985): 41–50.
BRENT, K. J.: Case study 4: Powdery mildews of barley and cucumber. In: DEKKER, J., and GEORGOPOULOS, S. G. (Eds.): Fungicide resistance on crop protection. Centre for Agricultural Publishing and Documentation, Wageningen 1982, pp. 219–230.
BROOKS, D. H.: The control of powdery mildew diseases with pyrimidine fungicides. In: VII. Intern. Congr. Plant Protect., Paris (1970): 247.
BUTT, D. J., JEGER, M. J., and SWAIT, A. A. J.: Supervised control of apple orchard diseases in England. Proc. 10th Intern. Congr. Plant Protect., Brighton (1983): 1005.

BUTTERS, J. A., BURRELL, M. M., and HOLLOMON, D. W.: Purine metabolism in barley powdery mildew and its host. Physiol. Pl. Path. **27** (1985): 65–74.
– CLARK, J., and HOLLOMON, D. W.: Resistance to inhibitors of sterol biosynthesis in barley powdery mildew. Meded. Fac. Landbouww. Rijksuniv. Gent. **49/2a** (1984): 143–151.
BYFORD, W. J.: Field experiments on sugar-beet powdery mildew, *Erysiphe betae*. Ann. appl. Biol. **88** (1978): 377–382.
CALDERBANK, A.: Metabolism and mode of action of dimethirimol and ethirimol. Acta Phytopathol. Acad. Sci. Hungaricae, Budapest **6** (1971): 355–363.
CARVER, T. L. W., and GRIFFITHS, E.: Relationship between powdery mildew infection, green leaf area and grain yield of barley. Ann. appl. Biol. **99** (1981): 255–266.
CAVELL, B. D.: Methods used in the study of the photochemical degradation of pesticides. Pesticide Sci. **10** (1979): 177–180.
– HEMMINGWAY, R. J., and TEAL, G.: Some aspects of the metabolism and translocation of the pyrimidine fungicides. Proc. 6th Brit. Insect. Fungic. Conf. (1971): 431–437.
CHANNON, A. G., and BOYD, A. G.: The effect of some fungicides on mildew of spring barley in the South-West of Scotland. Proc. 7th Brit. Insect. Fungic. Conf. (1973): 21–28.
CLIFFORD, B. C., JONES, I. T., and HAYES, J. D.: Interaction between systemic fungicides and barley genotypes: their implications in the control of mildew. Proc. 6th Brit. Insect. Fungic. Conf. (1971): 287–294.
COLLIER, G. F., GRAHAM-BRYCE, I. J., KNIGHT, A. G., and COUTTS, J.: Direct observation of the distribution of radiolabelled ethirimol in soil by resin impregnation and autoradiography. Pesticide Sci. **10** (1979): 50–56.
COOKE, B. K., HERRINGTON, P. J., JONES, K. G., and MORGAN, N. G.: Progress toward economical and precise top fruit spraying. Proc. Brit. Crop Protect. Conf. – Pests and Diseases (1977): 323–329.
D'ARBIGNY, P., and DAWSON, M. G.: Trois annees d'essai de lutte contre l'oidium des orges avec l'ethirimol. In: La lutte contre les maladies des cereales. Report of papers, Versailles 28th February (1973): 236–249.
DICKINSON, C. H.: Effects of ethirimol and zineb on phylloplane microflora of barley. Trans. Brit. Mycol. Soc. **60** (1973): 423–431.
EASTERBROOK, M. A: Chemical and integrated control of apple rust mite. Proc. Brit. Crop Protect. Conf. – Pests and Diseases (1984): 1107–1111.
– SOUTER, E. F., SOLOMON, M. G., and CRANHAM, J. E.: Integrated pest management in English apple orchards. Proc. Brit. Crop Protect. Conf. – Pests and Diseases (1979): 61–67.
ELIAS, R. S., SHEPHARD, M. C., SNELL, B. K, and STUBBS, J.: 5-n-butyl-2-dimethylamino-4-hydroxy-6-methylpyrimidine: a systemic fungicide. Nature **219** (1968): 1160.
FARETRA, F., MANTEGAZZA, G., ANTONACCI, E., and POLLASTRO, S.: Effectiveness *in vitro* of ergosterol biosynthesis inhibitors and some traditional compounds against *Phomopsis viticola* sacc. Diffesa delle Planta **10** (1987): 397–404.
FINNEY, J. R.: PP 588 (Bupirimate): A new fungicide for the specific control of powdery mildews on apples and other crops. Proc. 8th Intern. Congr. Plant Prothect., Moscow, Sect. III, Part I (1975): 235–236.
– FARRELL, G. M., and BENT, K. J.: Bupirimate – a new fungicide for the control of powdery mildews on apples and on other crops. Proc. 8th Brit. Insect. Fungic. Conf. (1975): 667–673.
FISCHER-COLBRIE, P.: Für den österreichischen Obstbau genehmigte Wirkstoffe und ihre Nebenwirkungen auf Nützlinge. Pflanzenschutz – offizielle Veröffentlichung der Bundesanstalt für Pflanzenschutz – Agrarwelt, Nr. 180 Pflanzenschutz Nr. 12 (1986): 4–9.
FLETCHER, J. T., and WOLFE, M. S.: Insensitivity of *Erysiphe graminis* f. sp. *hordei* to triadimefon, triadimenol and other fungicides. Proc. Brit. Crop Protect. Conf. – Pests and Diseases (1981): 633–640.
GEOGHEGAN, M. J. A.: Pyrimidine fungicide. Proc. 5th Brit. Insect. Fungic. Conf. (1969): 333–339.

- and DE GRAAFF, A.: Systemische fungiciden bestrijding van meeldauw. Netherl. J. Plant Path. **75** (1969): 277−278.
GILMOUR, J.: Fungicidal control of mildew on spring barley in South-East Scotland. Proc. 6th Brit. Insect. Fungic. Conf. (1971): 63−74.
- Performance of barley mildew fungicides in South-East Scotland. Proc. 10th Intern. Congr. Plant Protect., Brighton (1983): 643.
GORSKA-POCZOPKO, J.: Preliminary investigations on PP 675 activity against pea mildew. Acta Mycol., Warszawa **7** (1971): 159−167.
GORTER, G. J. M. A., and NEL, D. D.: Systemic fungicidal effects on powdery mildews of metabolic inhibitors and related compounds in laboratory tests. I. Pyrimidine and purine analogues. Phytophylactica **6** (1974): 209−212.
GRAHAM-BRYCE, I. J., and COUTTS, J.: Interactions of pyrimidine fungicides with soil and their influence on uptake by plants. Proc. 6th Brit. Insect. Conf. (1971): 419−426.
GRINDLE, M.: Variations among field isolates of *Botrytis cinerea* in their sensitivity to antifungal compounds. Pesticide Sci. **12** (1981): 305−312.
HALL, D. W.: Control of mildew in barley with ethirimol. Proc. 6th Insect. Fungic. Conf. (1971): 26−32.
HALL, R. A.: Laboratory studies on the effects of fungicide, acaricides and insecticides on the entomopathogenic fungus, *Verticillium lecanii*. Ent. expt. et appl. **29** (1981): 39−48.
HANSEN, K. E.: Forsog med kemisk bekaempelse af meldug (*Erysiphe graminis*) pa korn 1969−1976. Tidsskrift for Planteavl, Kobenhavn **82** (1978): 289−306.
HARTMAN, J. K., and SIEGEL, M. K.: Cucumber (*Cucumis sativus*), cantaloupe (*Cucumis melo*), powdery mildew (*Erysiphe cichoracearum*). Fungic. Nematic. Tests, Bradenton **29** (1973): 58.
HASSAN, S. A., BIGLER, F., BOGENSCHÜTZ, H., BOLLER, E., BRUN, J., CHIVERTON, P., EDWARDS, P., MANSOUR, F., NATON, E., OOMEN, P. A., OVERMEER, W. P. J., POLGAR, L., RIECKMANN, W., SAMSØE-PETERSEN, L., STÄUBLI, A., STERK, G., TAVARES, K., TUSET, J. J., VIGGIANI, G., and VIVAS, A. G.: Results of the fourth joint pesticide testing programme carried out by the IOBC/WPRS-working group "Pesticides and Beneficial Organisms". J. Appl. Ent. **105** (1988): 321−329.
HEANEY, S. P., HUMPHREYS, G. J., HUTT, R., MONTEIL, P., and JEGERINS, P. M. F. E.: Sensitivity of barley powdery mildew to fungicides in the UK Proc. Brit. Crop Protect. Conf. − Pests and Diseases (1984): 459−464.
HELLER, A., GROSSMANN, F., FRENZELL, B., and HIPPE, P.: A cytological study of the development of Erysiphe graminis in its host barley, as influenced by two fungicides ethirimol and propiconazole. Can. J. Bot. **68** (1990): 2618−2625.
HOLLOMON, D. W.: Laboratory evaluation of ethirimol activity. In: MCFARLANE, N. R. (Ed.): Crop Protect. Agents. Academic Press, London 1977, pp. 505−515.
- Evidence that ethirimol may interfere with adenine metabolism during primary infection of barley powdery mildew. Pesticide Biochem. Physiol. **10** (1979a): 181−189.
- Specificity of ethirimol in relation to inhibition of the enzyme adenosine deaminase. Proc. Brit. Crop Protect. Conf. − Pests and Diseases (1979b): 251−256.
- Genetic control of ethirimol resistance in a natural population of *Erysiphe graminis* f. sp. *hordei*. Phytopathology **71** (1981): 536−540.
- Target sites of hydroxypyrimidine fungicides. In: KÖLLER, W.: Target sites of fungicide action. CRC Press, Boca Raton 1992, pp. 31−41.
- and CHAMBERLAIN, K.: Hydroxypyrimidine fungicides inhibit adenosine deaminase in barley powdery mildew. Pesticide Biochem. Physiol. **16** (1981): 158−169.
LOCKE, T., and PROVEN, M.: Sensitivity of barley powdery mildew to ethirimol in relation to field performance. E.P.P.O. Bull. (1985): 467−471.
HOLLOWAY, P. J., WONG, W. W.-C., PARTRIDGE, H. J., SEAMAN, D., and PERRY, R. B.: Effects of some nonionic polyoxyethylene surfactants on uptake of ethirimol and diclobutrazol from suspension formulations applied to wheat leaves. Pestic. Sci. **34** (1992): 109−118.

HUNTER, L. D., BLAKE, P. S., and SOUTER, R. D.: Protectant and post-inoculation activities of fungicides for apple mildew control. Proc. Brit. Crop Protect. Conf. – Pests and Diseases (1984): 537–544.

HUNTER, T., BRENT, K. J., and CARTER, G. A.: Effects of fungicide regimes on sensitivity and control of barley powdery mildew. Proc. Brit. Crop Protect. Conf. – Pests and Diseases (1984): 471–476.

JACZEWSKA, A.: Evaluation of powdery mildew damage on wheat and attempts to chemical control. Zeszyty problemowe postepow nauk rolniczych, Warszawa **275** (1983): 51–57.

JAHN, M., BURTH, U., and RATHKE, S.: Ein Beitrag zur differenzierten Beurteilung von Fungiziden am Beispiel von *Podosphaera leucotricha* (Ell. et Ev.) Salm and *Venturia inaequalis* (Cooke) Aderh. Arch. Phytopathol. Pflanzenschutz, Berlin **22** (1986): 205–212.

JEGER, M. J., BUTT, D. J., and SWAIT, A. A. J.: Combining fungicide and partial resistance to control apple powdery mildew. Proc. 10th Intern. Congr. Plant Protect., Brighton (1983): 1004.

JENKYN, J. F.: Effects of mildew on the growth and yield of spring barley: 1969–1972. Ann. appl. Biol. **78** (1974): 281–288.

– Effects of chemical treatments for mildew control at different times on the growth and yield of spring barley. Ann. appl. Biol. **88** (1978): 369–376.

– and MOFFATT, J. R.: The effect of ethirimol seed dressings on yield of spring barley grown with different amounts of nitrogen fertilizer 1969–1971. Plant Pathol. **24** (1975): 16–21.

– and PREW, R. D.: Activity of six fungicides against cereal foliage and root diseases. Ann. appl. Biol. **75** (1973a): 241–245.

– – The effect of fungicides on incidence of *Sporobolomyces* spp. and *Cladosporium* spp. on flag leaves of winter wheat. Ann. appl. Biol. (1973b): 253–256.

KAMPE, W.: Verstärkter Befall durch *Rhynchosporium secalis* Dav. nach chemischer Bekämpfung von *Erysiphe graminis* D.C. bei Sommergerste (vorläufige Mitteilung). Nachr.-Bl. Dt. Pflanzenschutzdienst, Braunschweig **26** (1974): 148–150.

KARG, W.: Biologie der Raubmilben und ihre Bedeutung im integrierten Pflanzenschutz. Nachr.-Bl. Pflanzenschutz **44** (1990): 207–209.

– GOTTWALD, R., and FREIER, B.: Die Selektivität von Pflanzenschutzmitteln und ihre Bedeutung. Nachr.-Bl. Pflanzenschutz DDR **41** (1987): 218–223.

KELLEY, W. N., DADDONA, P. E., and VAN DER WEYDEN, M. B.: Characterization of human adenosine deaminase. In: Purine and Pyrimidine Metabolism. CIBA Foundation Symposium **NS48** (1977): 277–293.

KING, J. E.: Surveys of foliar diseases of spring barley in England and Wales, 1972–1975. Plant Pathol. **26** (1977): 21–29.

KLUGE, E., and LYR, H.: Vergleichende Untersuchungen über die Wirkung von Systemfungiziden gegen *Erysiphe graminis* DC. In: LYR, H., and POLTER, C. (Eds.): Systemische Fungizide und antifungale Verbindungen. Abhandl. Akad. Wiss. DDR 1982, Nr. IN, Akademie-Verlag, Berlin 1983, pp. 379–382.

KOBACHIDZE, N. I., and TOSKINA, V. A.: Effectivnost' rjada sistemnych fungicidov v bor'be s nekotorymi mucnistorosjannymi zabolevanijami. Bjulleten' vsesojuznogo naucnoissledovatel'skogo Instituta sascity rastenij, Leningrad **22** (1971): 31–32.

KRÜGER, W.: Auftreten und Bekämpfung des „Echten Mehltaus" *Erysiphe graminis*. Jahresber. Biol. Bundesanstalt, Braunschweig (1969): A97.

KUNDERT, J.: Apfelmehltaubekämpfung mit neuen organischen Fungiziden. Schweiz. Z. Obst- u. Weinbau, Wädenswil **113** (1977): 3–8.

KUNOH, H., YAMAMORI, K, and ISHIZAKI, H.: Cytological studies of the early stages of powdery mildew in barley and wheat. VIII. Autofluorescence at penetration sites of *Erysiphe graminis hordei* on living barley coleoptiles. Physiol. Plant Path. **21** (1982): 373–379.

KÜTHE, K: Erfahrungen bei der Getreidemehltaubekämpfung (*Erysiphe graminis* DC) in Mittelhessen 1971. Gesunde Pflanzen, Frankfurt (M.) **24** (1972): 72–76.

LESCAR, L.: Current practice in integrated cereal pest and disease control in North-Western-Europe (excluding Gr. Britain). Proc. Brit. Crop Protect. Conf. − Pests and Diseases (1977): 763−772.

LITTLE, R., and DODDSON, J. K.: The comparison of yields of some spring barley varieties in the presence of mildew and when treated with a fungicide. Proc. 6th Brit. Insect. Fungic. Conf. (1971): 91−97.

MAENO, S., MIURA, I., MASUDA, K., and NAGATA, T.: Mepanipyrin (KIF-3535), a new pyrimidine fungicide. Proc. Brighton Crop Prot. Conf. − Pests and Diseases (1990): 415−422.

MANNERS, J. M., and GAY, J.: Fluxes and accumulation of ethirimol in haustoria of *Erysiphe pisi* and protoplasts of *Pisum sativum*. Ann. appl. Biol. **96** (1980): 283−293.

MARTIN, T., MORRIS, D. B., and CHIPPER, M. E.: Triadimenol seed treatment on spring barley: results of a 60 site evaluation in the United Kingdom. Proc. Brit. Crop Protect. Conf. − Pests and Diseases (1981): 299−306.

MEEUS, P., MADDENS, K., and DOHET, J.: Cinq annees de lutte contre l'oidium de l'orge de printemps en Belgique. Rev. Agric., Bruxelles **28** (1975): 285−306.

MILLOU, J., and D'ARBIGNY, P.: Utilisation d'associations d'ethirimol et de captafol contre les maladies foliares. Meded. Fac. Landbouww. Rijksuniv. Gent **39** (1974): 1051−1078.

MÜLLER, P.: Untersuchungen zur Biologie und Bekämpfung des Weizenmehltaus (*Erysiphe graminis* DC f. sp. *tritici* Marchal) als Grundlage für die Modellierung des Epidemieverlaufes sowie die gezielte Anwendung von Fungiziden. Diss., Akad. Landwirtschaftswiss., Berlin 1984, 123 p.

NEUHAUS, W., and REICH, R.: Bedeutung des Getreidemehltaus beim Anbau von Intensivgetreide und Möglichkeiten zur Bekämpfung. Nachr. Bl. Pflanzenschutz DDR, Berlin, **29** (1975): 161−164.

NOON, R. A., GIBBARD, M., NORTHWOOD, P. J., and HEANEY, S. P.: Ferrax seed treatment − disease control and growth benefits. Proc. Brighton Crop Protect. Conf. Pests and Diseases (1988): 941−946.

NORTHWOOD, P. N., PAUL, J. A., GIBBARD, M., and NOON, R. A.: FF 4050 seed treatment − a new approach to control barley diseases. Proc. Brit. Crop Protect. Conf. − Pests and Diseases (1984): 47−52.

OHTSUKA, N., SOU, K, AMANO, T., NAKAZAWA, Y., YAMAO, A: Sensitivity of cucumber powdery mildew (*Sphaerotheca fuliginea*) to several fungicides. J. Pestic. Sci. **16** (1991): 271−273.

O'RIORDAIN, F., and KENNEDY, D.: A comparison of eleven fungicides for the control of powdery mildew (*Sphaerotheca morusvae*) of gooseberry. Proc. Brit. Crop Protect. Conf. − Pests and Diseases (1981): 555−561.

PETKOVA, M.: Vazmoznosti za chimicna borba s brasnjankata po psenicata pri uslovijata na severozapadna balgarija. Rast. Nauki. Sofija **16** (1979): 129−134.

PETSIKOL-PANAYOTAROU, N.: Action par vapeurs de quelques fongicides systemiques vis-a-vis de l'oidium des courges *Sphaerotheca fuliginea*. Meded. Fac. Landbouww. Rijksuniv. Gent **45** (1980): 199−206.

PONS, I., and EUL, A.: Dynamik der Fungizidresistenz in Gurkenmehltau-Populationen bei unterschiedlichen Bekämpfungsstrategien. Mitt. Biol. Bundesanstalt Land- u. Forstwirtschaft, Heft **266** (1990): 110.

RATHKE, S.: Einfluß der Temperatur auf die Wirkung ausgewählter Obstbaufungizide. Nachr. Bl. Pflanzenschutz DDR **38** (1984): 241−260.

− and JAHN, M.: Influence of rain on the effectiveness of fungicides against apple powdery mildew and apple scab. In: LYR, H., and POLTER, C. (Eds.): Systemic Fungicides and Antifungal Compounds. Tagungsber. Akad. Landwirtschaftswiss. DDR Nr. **222**, Berlin (1984): 253−258.

− − Effects of fungicides on sporulation of *Podosphaera leucotricha* Salm. and *Venturia inaequalis* (Cooke) Wint. Arch. Phytopathol. Pflanzenschutz **25** (1989): 459−464.

ROUND, P. A., and WHEELER, B. E. J.: Interactions of *Puccinia hordei* and *Erysiphe graminis* on seedling barley. Ann. appl. Biol. **89** (1978): 21−35.

SAUR, R.: Untersuchungen über den Einfluß systemischer Fungizide auf den Befall der Gerste mit *Helminthosporium sativum* P., K. & B. unter besonderer Berücksichtigung von Ethirimol. Diss., Univ. Bonn 1976a, p. 93.
— Untersuchungen über den Einfluß von Ethirimol auf die Pathogenese einer Helminthosporiose (*H. sativum*) an Gerste. Phytopathol. Z., Berlin (West) und Hamburg **87** (1976b): 304–313.
— and SCHÖNBECK, F.: Untersuchungen über den Einfluß systemischer Fungizide auf den Befall der Gerste mit *Helminthosporium sativum*. Z. Pflanzenkrankh. Pflanzenschutz **82** (1975): 173–175.
— — Einfluß von Ethirimol auf die Empfindlichkeit von Gerste gegenüber einem Toxin von *Helminthosporium sativum*. Meded. Fac. Landbouww. Rijksuniv. Gent **41** (1976): 511–516.
SCHÄFERMEYER, S., and DICKLER, E.: Vergleichende Untersuchungen zu Richtlinien für die integrierte Kernobstproduktion in Europa. Mitt. Biol. Bundesanstalt Land- u. Forstwirtschaft, Heft **271** (1991): 110 pp.
SCHEPERS, H. T. A. M.: Persistence of resistance to fungicides in *Sphaerotheca fuliginea*. Netherl. J. Plant Path. **90** (1984): 165–171.
SHABI, E.: Control of apple mildew by bupirimate and the influence of added spreader-sticker on its performance. Proc. 8th Brit. Insect. Fungic. Conf. (1975): 711–714.
SHEPHARD, M. C.: Barriers to uptake and translocation in herbaceous and woody plants. Proc. 7th Brit. Insect. Fungic. Conf. (1973): 841–850.
— Factors which influence the biological performance of pesticides. Proc. Brit. Crop Protect. Conf. — Pests and Diseases (1981): 711–721.
— BENT, K. J., WOOLNER, M., and COLE, A. M.: Sensitivity to ethirimol of powdery mildew from UK barley crops. Proc. 8th Brit. Insectic. Fungic. Conf. (1975): 59–66.
SMOLKA, S., and WOLFE, G.: Cytological studies on the mode of action of systemic fungicides on the host pathogen complex barley-powdery mildew (*Erysiphe graminis* f. sp. *hordei* Marchal). Pestic. Sci. **17** (1986): 249–255.
SNELL, B. K., ELIAS, R. S., and FREEMAN, P. F. H.: Pyrimidine derivatives and the use there of as fungicides. GBP 1,182,584 (1966).
STOTT, I. P. H., NOON, R. A., and HEANEY, S. P.: Flutriafol, ethirimol and thiabendazole seed treatment — an update on field performance and resistance monitoring. Proc. Brighton Crop Prot. Conf. — Pests and Diseases (1990): 1169–1174.
TEAL, G., and CAVELL, B. D.: Degradation of bupirimate fungicide on apples and in water. Proc. 8th Brit. Insect. Fungic. Conf (1975): 25–30.
WELLS, C. H., POLLARD, S. J., and SEN, D.: Phytochemistry of some systemic pyrimidine fungicides. Pesticide Sci. **10** (1979): 171–176.
WOLFE, M. S.: Pathological and physiological aspects of cereal mildew control using ethirimol. Proc. 5th Brit. Insect. Fungic. Conf. (1969): 8–15.
— Pathogen response to fungicide use. Proc. 8th Brit. Insect. Fungic. Conf. (1975): 813–892.
— Trying to understand and control powdery mildew. Plant Path. **33** (1984): 451–466.
— Integration of host resistance and fungicide use. EPPO Bull. **15** (1985): 563–570.
— and DINOOR, A.: The problems of fungicide tolerance in the field. Proc. 7th Brit. Insect. Fungic. Conf. (1973): 11–19.
— SLATER, S. E., and MINCHIN, P. N.: Fungicide intensitivity and host pathogenicity in barley mildew. Proc. 10th Intern. Congr. Plant Protect., Brighton (1983): 645.
WORTHING, C. R., and WALKER, S. B.: The pesticide manual — a world compendium. 7th edit. The Brit. Crop Protect. Council, London (1983).
ZWATZ, B.: Getreidemehltau: Erfahrungen 1973. Der Pflanzenarzt, Wien **27** (1974): 5–6.

Chapter 18

Organophosphorus fungicides

P. Braun and B. Schreiber

Hoechst AG, Frankfurt (Main), FRG

Introduction

The development of organophosphorus compounds started in Germany in the 1930s, and many of these compounds, being pesticides with insecticidal properties, are today used in practice. Their toxicity can be attributed to the fact that they provoke lethal neurophysiological damage by inhibiting the enzyme acetylcholin-esterase.

The fungicidal secondary action of the insecticide phorate (Thimet®) was described in 1958 by Erwin and Reynolds, followed by a large number of publications including the most recent ones (Roy 1990; Gupta and Roy 1988 and 1991). Since the sixties, a number of fungicidal commercial products have been made available from amongst this substance class and their activity extends mainly to mildews such as *Sphaerotheca fuliginea*, *Erysiphe* spp. and *Podosphaera leucotricha*, but also *Pyricularia oryzae*, *Rhizoctonia solani*, *Sclerotium rolfsii* and others. These organophosphorus fungicides frequently have an insecticidal or acaricidal secondary action.

The most important compounds can be classified into the following chemical groups (Table 18.1).

The spectra of action of the compounds differ greatly, which allows them to be used in a targeted fashion against various pathogens in a number of crops.

Phosphorothiolates are particularly active against blast and other rice diseases, while the phosphoric acid amides, which are no longer commercially available, and the phosphorothionates are mainly active against powdery mildews. Such selectivities can be based on the one hand on a series of biochemical targets, but, on the other hand, also on differences in penetration behavior for substances caused by different lipophilicity or hydrophilicity. For example, the uptake of phosphorothionates by powdery mildews proceeds particularly effectively through the gelatinous outer layer of the cell walls (McKeen et al. 1966; McKeen and Smith 1967; Belal 1971).

Kitazin P® (IBP) has replaced the analogous, older diethyl compound EBP (S-benzyl O,O-diethylphosphorothioate; Kitazin®) due to its advantageous greater solubility in water (Craig and Peberdy 1983a). Inezin® and Cerezin® (S-4-chlorophenyl-c-hexyl-methyl-phosphorothiolate), which is not mentioned above, were not competitive from the marketing point of view. Novel compounds from this group have not gained in importance in recent years. This leaves us with the following important structures (Fig. 18.1).

Table 18.1 Organophosphorus fungicides

Chemical group[1])	Common name	Trade mark	Inventor
Phosphoric acid esters	ESBP	Inezin®, Inejin®	SCHRADER (Bayer AG) 1957; SCHEINPFLUG and SCHRADER (Bayer AG) 1965
Phosphorothiolates	EBP	Kitazin®	SCHEINPFLUG, JUNG and SCHRADER (Bayer AG) 1963; KADO et al. (Ihara Noyaku Chem. Company) 1964
	IBP (Iprobenfos)	Kitazin P®	(Kumiai Chemical Industry Co.) 1966
	EDDP (Edifenphos)	Hinosan®	SCHRADER, MANNES and SCHEINPFLUG (Bayer AG) 1965
Phosphoric ester amides	Triamiphos	Wepsyn®	KOOPMANNS et al. (Philips Duphar) 1957
	Ditalimfos	Plondrel®, Laptran®, Frutogard®, Leucon®, Millie®	TOLKMITH and SENKBEIL (Dow Chem. Company) 1965
Phosphorothionates	Pyrazophos	Afugan®, Curamil®, Siganex®	SCHERER and MILDENBERGER (Hoechst AG) 1965
	Tolclofosmethyl	Rizolex®	(Sumitomo Chemical Company) 1979

[1]) According to C. FEST and K.-J. Schmidt (1973)

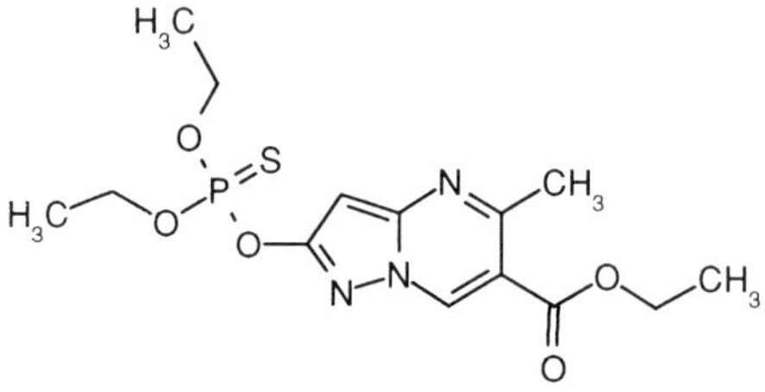

Pyrazophos, Afugan^R

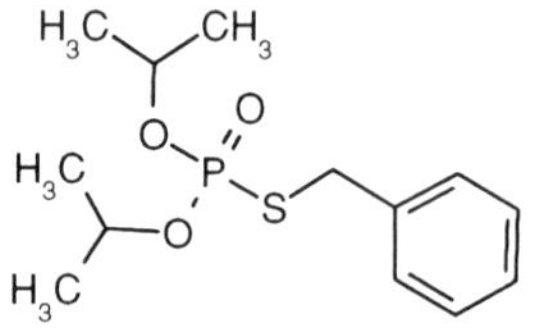

Iprobenfos, Kitazin P^R

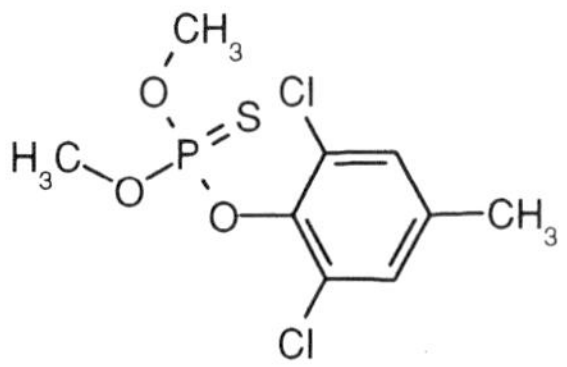

Tolclofos-methyl, Rizolex^R

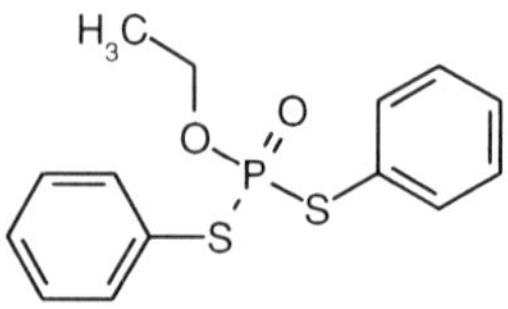

Edifenphos, Hinosan^R

Fig. 18.1 Organophosphorus fungicides.

Practical application

The most important field of application of the phosphorothiolates IBP and EDDP is rice growing, where *Pyricularia oryzae* (rice blast) and *Pellicularia* (*Corticium*) *sasakii* (sheath blight) are the most important pathogens. Both compounds can be used protectively and curatively, but curative properties are mainly shown by Edifenphos, which additionally has a good residual efficacy (YOSHINO 1988). Edifenphos is therefore particularly suitable for foliar application (OU 1980), where its action also extends to stem rot, *Helminthosporium* leaf spot and leaf scald.

Iprobenfos, which is a highly systemically acting compound, is particularly suitable for submerged application in the paddy fields. It is mainly used in the form of granule formulations (YOSHINO 1988; INOUE 1990). The substance is taken up via the roots and leaves and translocated in the xylem. The active substance is released slowly due to its poor solubility in water. This means that the efficacy of an application is retained over a period of 3–4 weeks (KOZAKA 1969; YOSHINAGA 1969). An additional advantage is a shortening of the stalks, which reduces the tendency to lodge without affecting length, or number, of ears.

In addition to *Pyricularia* and *Pellicularia*, which are the main fields of application, iprobenfos and edifenphos are also active against *Cochliobolus miyabeanus* and *Leptosphaeria salvanii*, and to some extent also against *Sphaerotheca fuliginea* and *Erysiphe gram.* f. sp. *hordei* (UESUGI 1970; UMEDA 1973; DE WAARD 1974).

Compounds from the group of the organophosphates which were available for controlling the powdery mildews are the phosphoric acid amides, and the phosphorothionate pyrazophos is still available. Apart from *Erysiphe* and *Sphaerotheca*, it also controls *Pyrenophora*, *Rhynchosporium* and *Colletotrichum* (SCHREIBER et al. 1984), and there is also some fungitoxic action against *Pyricularia oryzae* in rice (DE WAARD 1974). Pyrazophos has mainly protective, but also curative properties (SCHREIBER et al. 1984; QVARNSTROEM 1989). Depending on the plant species and the stage of development, it rapidly penetrates the tissue, uptake being promoted by the diethylthionophosphate moiety of the molecule (DE WAARD 1974). The active substance can be taken up by all plant organs, but in particular by shoot axis and leaves (SCHMIDT and BURTH 1977). Translocation in the plant preferentially takes place acropetally and translaminarly, and is essentially restricted to the vascular system and closely adjacent tissue regions (BELAL 1971; KOECHER and LOETZSCH 1976). Pyrazophos is currently commercially available against cereal pathogens, exclusively in the form of mixtures, for example with propiconazole (ZEUN and BUCHENAUER 1991) or flusilazole (GARNIER 1990).

In contrast, the spectrum of action of the related compound tolclofos-methyl is entirely different. Tolclofos-methyl was introduced against soil-borne diseases, especially *Rhizoctonia solani* (OHTSUKI and FUJINAMI 1982; FROST 1987; SMILEY 1990), but also acts against *Sclerotium rolfsii* (GANGWAR 1989). The compound shows good curative properties, but its systemic action is poor (CSINOS 1985) (Table 18.2).

Organophosphorus fungicides are mainly used protectively. The efficacy of the curative action is generally only sufficient 2–4 days after infection. Eradicative effects have not been described. In most annual crops the first application should coincide with the appearance of the symptoms. In top fruits, protective applications are necessary.

Most organophosphorus fungicides not only show a fungicidal action, but also insecticidal effects. In rice, for example, iprobenfos and edifenphos also act against rice cater-

Table 18.2 Organophosphorus fungicides – main field of practical application

Common name	Crop	Disease, Plant Pathogen	Formulation	Dosage (a.i./ha)	Application method	Remarks
Iprobenfos	Rice	*Pyricularia oryzae*	EC 48	400–600 g	2–4 foliar sprays	Side effects on rice leaf and plant hoppers
			Dust 2%	600–800 g	spread on leaves applicator	
			Dust 3%	900–1,200 g		
		Pyricularia oryzae *Leptosphaeria salvanii* *Rhizoctonia solani*	Granules 17%	5,100–8,500 g	spread on water (1–2 times)	
Edifenphos	Rice	*Pyricularia oryzae* *Cochliobolus miyabeanus*	EC 30, EC 50	300–500 g	2–4 foliar sprays	Side effects on rice leaf and plant hoppers and caterpillars
		Leptosphaeria salvanii *Rhizoctonia solani* *Micronectriella nivalis*	Dust 1.5% DL Dust 2.5%	450–600 g 750–1,000 g	spread on leaves (dust applicator) 2–4 times	
Pyrazophos	Apples, Cucumbers, Ornamentals, Hops, Mulberry, Vineyards	Powdery mildew	EC 30	0.01–0.03% and 0.03–0.05% a.i.	7–10 days spray interval	Russeting effects on some apple varieties, phytotoxic effects on some ornamental varieties. Good insecticidal effect on leaf miners.
	Banana	*Mycosphaerella*	special formulation mix with banana oil and water	300 g	aeral appl.	
	Cereals	*Erysiphe graminis* *Pyrenophora teres*	EC comb. with triazoles	250 g	1–2 applications at appearance of first symptoms	Side effect on *Rhynchosporium secalis*

Table 18.2 (Continued)

Common name	Crop	Disease, Plant Pathogen	Formulation	Dosage (a.i./ha)	Application method	Remarks
Tolclofos-methyl	Cotton, Potatoes, Ornamentals	*Rhizoctonia* spp. *Sclerotium rolfsii*	WP 10, WP 50, WP 75	200–400 g a.i./ 100 kg seeds 200 kg a.i./tons of tuber 1,000 ppm a.i. 10 kg	tuber spray soil incorporation	
	Sugar beet, Potato, Wheat & barley, Lettuce	*Rhizoctonia solani* *Rhizoctonia solani* *Typhula* sp. *Rhizoctonia solani*	WP 50	500–1,000 g 5,000–10,000 g 500–667 g 500 g	foliar or injection into soil dipping into solution before planting foliar, within 2 times foliar or injection into soil	
	Tomato Cucumber Eggplant	damping-off caused by *Rhizoctonia solani*		1,000 g 0.5% of seed weight	injection into soil seed dressing	
	Tomato Cucumber Eggplant Sugar beet	damping-off caused by *Rhizoctonia solani*	Dust 5%	1–2 g/m² in nursery bed 25–50 g/300 l soil for nursery bed	soil incorporation soil incorporation	before sowing in nursery
	Wheat & barley	*Typhula* sp.		1,500 foliar		

pillars and against leaf and plant hoppers. In the case of pyrazophos, too, the insecticidal component against aphids, leaf miners and some beetles in cereals, cucumbers and ornamentals is mentioned frequently (LEDIEU and HELYER 1983; PARELLA 1983; LINDQUIST et al. 1984; HELYER and LEDIEU 1985; HEIMBACH 1988; FRAMPTON 1988; SOTHERTON and MOREBY 1988; YATOM et al. 1988). Effects on honey bees are also described (CHOI et al. 1989; FERGUSON 1987). However, the insecticidal action and, in particular, the residual action of the fungicides remains less pronounced than that of customary insecticides.

Mode of action and resistance situation

a. Iprobenfos (IBP) and edifenphos (EDDP)

The action of phosphorothiolates takes place in late stages of the infection process. It is not so much spore germination and appressoria formation which are inhibited, but rather a powerful suppression of mycelial growth and sporulation is revealed (OU 1980). This emphasizes the curative activity of the compounds. Iprobenfos acts in particular on hyphal tips and in this way suppresses apical growth. Extensive vacuolization also takes place (ISHIZAKI et al. 1986). These effects occur even when concentrations are low.

The first physiological and biochemical investigations with phosphorothiolates, which initially also included Kitazin® (EBP), were published by KAKIKI et al. (1969) and MAEDA et al. (1970). In their work with *Pyricularia oryzae*, they demonstrated inhibition of chitin synthesis as one of the targets of the compound class. As yet, there are no definitive answers as to how this finding is to be classified when assessing the complete mode of action of the substances. The importance of this target for toxification of the fungal cell depends apparently on the fungal species. Inhibition of chitin synthase activity is regarded as a decisive factor by BRILLINGER (1979) with regard to *Coprinus cinereus*, CRAIG and PEBERDY (1983a) with regard to *Aspergillus nidulans* and BINKS et al. (1990) with regard to *Fusarium graminearum*. In contrast, DE WAARD (1974) also demonstrates an inhibitory action of iprobenfos and edifenphos in the case of the chitin-free oomycete *Pythium ultimum*.

The increase of the leakage of ^{32}P-orthophosphate from mycelial cells of *Pyricularia oryzae* under the influence of IBP and EDDP was observed by DE WAARD (1972 and 1974). He regards the adverse effect on the cell membrane permeability as the primary effect and reduced chitin synthesis as the consequence. In the work with *Mucor rouxii* with edifenphos, LYR and SEYD (1978) come to the same conclusion. However, IBP has no direct effect on the membrane permeability of *Aspergillus nidulans* (CRAIG and PEBERDY 1983a).

AKATSUKA et al. (1977), KODAMA et al. (1979), KODAMA et al. (1980); KODAMA and AKATSUKA (1983) and AKATSUKA and KODAMA (1984) explain the inhibition of transmethylation reactions of the phosphatidylcholine biosynthesis as the primary antifungal action of IBP and EDDP. This applies mainly to the methylation of phosphatidylethanolamine in the Greenberg's pathway of phospholipid biosynthesis (KODAMA et al. 1979; KODAMA and AKATSUKA 1983), but also to the methyl transfer from methionine into choline in the Kennedy's pathway (YOSHIDA et al. 1984), at least in the case of edifenphos (WIEBE 1990). However, there exist contradictory reports, for example by CRAIG and PEBERDY (1983b), who find that IBP has no influence on the phosphatidylcholine content of *Aspergillus nidulans.* Yet, the manipulation of phospholipid biosynthesis seems to be the main effect in the product-relevant fungus *Pyricularia oryzae.* Phospholipids in cell membranes play important roles in physiological functions. Imbalances in their biosynthesis cause damage to the membrane structure and hence result in changes in ion fluxes, permeabilities and transport processes (YOSHIDA and NOSE 1990; YOSHIDA et al. 1990). Irregular membrane-protein interactions and enzyme activities, e.g. chitin synthase, could be indirect results of changes in the lipid environment (BURDEN 1990; WIEBE 1990). The initially observed effects of IBP on the cell wall biosynthesis of *Pyricularia oryzae,* too, are probably a secondary effect caused by the factor mentioned above (BUCHENAUER 1990). It is possible that, depending on the fungal species, the main action of the phosphorothiolates is to be seen somewhere between influencing chitin synthase and seriously engaging in phospholipid biosynthesis.

In *Pyricularia oryzae* three main metabolic reactions with phosphorothiolates were found, the cleavages of P—S and S—C bonds and a hydroxylation of the benzene ring at the m-position in the case of iprobenfos or p-position in the case of edifenphos (UESUGI and TOMIZAWA 1971; TOMIZAWA and UESUGI 1972). The P—S and S—C cleavages of iprobenfos were mediated by inducible types of mixed function oxygenases (mfo) (KODAMA et al. 1982) whereas hydroxylation of the benzene ring seems to be mediated by constitutive enzymes (UESUGI and KATAGIRI 1983). Wild-type strains of *Pyricularia oryzae* cleave both P—S and S—C bonds of IBP and EDDP, moderately resistant strains mainly show S—C cleavage and virtually no P—S cleavage. Resistant strains lack degradation completely (UESUGI and KATAGIRI 1983; KUROGOCHI et al. 1985).

Pefurazoate or propiconazole, sterol demethylation inhibitors (DMIs) via inhibition of the cytochrome P450 (cyt P450)-mediated C14-demethylation, antagonize the action of phosphorothiolates in *Pyricularia oryzae* and vice versa. This is not observed in a resistant strain (SUGIURA et al., in preparation). Besides diverse substrate specificities of the cyt P450s, their concentration required for P—S cleavage might be apparently largely bound by DMIs and then is no longer available for the metabolization of IBP or EDDP. This suggests activation of the phosphorothiolates by P—S cleavage. Since metabolites which contain P are not fungi-toxic (UESUGI 1983), it has been suggested that unstable S-containing metabolites are the active principle (UESUGI and TAKENAKA 1992).

A direct inhibition of the exoenzyme cutinase, an esterase enzyme from *Fusarium solani* f. sp. *pisi* by, inter alia, edifenphos and iprobenfos, is reported by KOELLER et al. (1982), and, in the case of cutinase from *Colletotrichum gloeosporioides* by, inter alia, edifenphos, by DICKMAN et al. (1983), even at subtoxic concentrations. However, this antipenetrant action is insufficient against *Pyricularia oryzae* and only plays a marginal role (SISLER 1986).

Under field conditions, *Pyricularia oryzae* became slowly resistant to phosphorothiolates. IBP-resistant field strains are firstly described by KATAGIRI et al. (1980). The development of resistance is equally possible in various strains of *Pyricularia oryzae* (SAITO and KATO 1990). The isolates show crossresistance to other phosphorothiolates and also against isoprothiolane (a non-organophosphorus rice blast fungicide) and a negatively correlated crossresistance to phosphoramidates (UESUGI et al. 1974), possibly due to decreased detoxification of these chemicals. Wild-type strains were not inhibited by this group of compounds. Most field isolates have a moderate degree of resistance. They detoxify this fungicide more rapidly than sensitive isolates by preferential cleavage of the S—C bond.

Since a sufficient number of alternative rice blast fungicides with different mechanisms of action is available, fungicide resistance in the control of blast diseases is at present not a serious problem (INOUE 1990).

b. Tolclofos-methyl

In vitro experiments on the effect of tolclofos-methyl (TM) on *Sclerotium rolfsii* and *Rhizoctonia solani* show marked inhibition of spore germination and hyphal growth (GANGWAR and DASGUPTA 1989; CSINOS 1985) and suggest an action in early stages of the infection process. In addition, the formation of sclerotia is suppressed (HARI et al. 1989), and, in experiments using *Ustilago maydis*, the multiplication of sporidia is inhibited (NAKAMURA and KATO 1984). Sporidia turn into multicelled structures by the insertion of transverse walls; DNA synthesis and protein synthesis are markedly reduced. The center of action is probably cytokinesis. The inhibition of DNA synthesis and protein synthesis are possibly a consequence (NAKAMURA and KATO 1984). SOMASHEKAR et al. (1987) describe mutagenic and spindle-inhibiting properties and regard TM as a mitotic poison.

Like cytochalasin A, which is known to inhibit motile functions of the cells, TM causes substantial leakage in *Botrytis cinerea*. Flagellar movement in zoospores of *Phytophthora capsici* is strongly inhibited by TM in the same manner as under the influence of cytochalasin. It seems that these substances bind to actin filaments of fungal cells and in this way interfere with coordinated movements in cytoplasmic streaming, flagella movement, cytokinesis etc. (KATO 1983).

The effects also suggest a relationship with aromatic hydrocarbon fungicides (AHFs). LYR and EDLICH (1987) report stimulation of NADH-dependent GSSG-glutathione reductase and glucose oxidase, a flavin enzyme, for some AHFs and for TM. NADPH cytochrome c reductase is strongly inhibited by TM.

This results in the formation of various free radicals, which ultimately cause peroxidative degradation of phospholipids. This analogous action is also suggested by the positive cross-resistance between TM, the phenylpyrrole fenpiclonil and the dicarboximide iprodione, which can be found in some strains of *Botrytis cinerea*, *Penicillium expansum* and *Pseudocercosporella herpotrichoides* (LEROUX 1991). On the other hand ORTH et al. (1992) cannot find any lipid peroxidation and NADPH cytochrome P450 reductase inhibition in *Ustilago maydis*.

ANITHA et al. (1989) have evidence for in vitro tolerance of *Macrophomina phaseolina* against TM with positive cross-tolerance to carbendazime, thiophanate-methyl, mancozeb and zineb. It represents a quite unusual combination for cross-tolerance. The reasons for this are absolutely obscure.

Laboratory-produced TM-resistant *Rhizoctonia solani* isolates show the same pathogenicity as the wild type, but markedly reduced growth rate (VAN BRUGGEN and ARNESON 1984).

c. Pyrazophos

After pyrazophos has been taken up into the plant or the fungus, a metabolization mediated by mfo and enzymatic hydrolysis via the phosphate-analogous compound O-pyrazophos (PO) takes place to give 2-hydroxy-5-methyl-6-ethoxycarbonylpyrazolo (1,5-α)-pyrimidine (PP) and further water-soluble metabolites which have not been identified (GORBACH et al. 1975; DE WAARD 1974; LYR et al. 1992). Besides pyrazophos itself, no other substances can be detected intracellularly (DE WAARD 1974; LYR et al. 1992). Accordingly, the metabolites would have to be leaked extremely rapidly (DE WAARD 1974) or they do not penetrate very well because of their lower lipophilicity.

In many cases, the metabolite PP is distinguished as the actual fungitoxic structure, for example in *Ustilago maydis* (GEORGOPOULOS et al. 1975), in *Pyricularia oryzae* (DE WAARD and SCHEEPENS 1973; DE WAARD 1974; DE WAARD 1980) and *Helminthosporium* spp. (KOHTS 1985). In *Erysiphe graminis*, however, KOHTS (1985) finds that the action of pyrazophos is superior compared with PO and PP.

The absence of metabolites in the fungal cells caused LYR et al. (1992) to assume pyrazophos as the main antifungal compound under practical conditions. They detected a pyrazophos metabolizing activity in the cell wall fraction of *Mucor mucedo* and postulate a "Pyrazophos oxidase" on the outer side of the plasmalemma which seems not to be of mfo-type. Whether this membrane-bound enzyme is the target for pyrazophos and how the intracellular biochemical effects observed tie in remains to be elucidated.

A large number of publications deal with the pyrazophos-caused inhibition of various development stages of the fungi during the infection process. For example, DE WAARD (1971) demonstrates a severe suppression of spore germination and appressoria formation rate in powdery mildew of barley and cucumber, KOHTS and BUCHENAUER (1984a) on powdery mildew of barley. In *Helminthosporium*, the formation of conidia is also markedly reduced. SCHLUETER and WELTZIEN (1971 and 1977) characterized pyrazophos in its action against powdery mildew of barley as an inhibitor of infection pegs. Appressoria are still formed, but they remain without secondary hyphal initials. No haustoria are formed. Low pyrazophos concentrations may result in the formation of haustoria and secondary hyphae, but their growth soon comes to a halt. Pyrazophos does not induce encapsulation of the haustoria by the plants as is typically the case under the influence of ergosterol biosynthesis inhibitors (EBIs) (SMOLKA and WOLF 1986). The pronounced sporulation-inhibiting action of pyrazophos is particularly important against *Podosphaera leucotricha* (SZTEJNBERG et al. 1975; BLAKE et al. 1982; CIMANOWSKI 1987; RATHKE and JAHN 1989).

Damage on hyphae and conidia which can be observed under the optical microscope includes coagulation and shrinking of the cytoplasm in addition to bulging hyphal tips (SCHREIBER et al. 1984). However, direct destruction of the membranes, accumulation of lipids and the formation of lamellar bodies can be seen under the electron microscope (MUELLER et al. 1992).

An inhibition of NAD^+-dependent dehydrogenases from *Pyricularia oryzae* has been described for the metabolite PP (DE WAARD and SCHEEPENS 1973; DE WAARD and DEKKER 1973). Important enzymes from this group such as glycerinaldehyde phosphate dehydrogenase (GAPDH) needed for glycolysis and malate dehydrogenase (MDH) needed for the citrate cycle are markedly inhibited. The FAD-dependant succinate dehydrogenase remains unaffected. Pyrazophos and the metabolite PO have no inhibitory action (DE WAARD and SCHEEPENS 1973; DE WAARD 1974). With a view to these results, it can be assumed that PP exerts a general dehydrogenase-inhibitory action which is bound to cause substantial adverse effects during glycolysis and the citrate cycle, if it penetrates into the cell in significant amounts.

Long incubation times with pyrazophos of up to 4 hours substantially suppress oxygen uptake, as described for *Sphaerotheca*, *Pyricularia* and *Helminthosporium* (DE WAARD and SCHEEPENS 1973; DE WAARD 1974; KOHTS 1985). However, in shortterm experiments, pyrazophos shows only minor effects (DE WAARD 1974; DE WAARD 1975). The action of PP in equal concentrations is superior to that of pyrazophos and commences earlier. Moreover, a rising pH causes a loss of activity, which can be attributed to the dissociation of PP at pH > 5.7 (DE WAARD 1975).

The constant decrease in the intensity of respiration reflects the growing inhibition of glycolysis and the citrate cycle via dehydrogenase inhibition. The latter not only causes grave concentration shifts of the reactants involved, but simultaneously lowers the production of NADH. Since this co-substrate is the most important electron donor of the respiratory electron transport, this necessarily results in the respiration inhibition observed. A direct consequence of this inhibition is a reduced ATP concentration.

The effect of pyrazophos on nucleic acid biosynthesis was studied in *Pyricularia* by incorporation studies with ^{14}C-thymdidine (DNA) and ^{14}C-uridine (RNA). The protein biosynthesis was monitored by the incorporation of ^{14}C-phenylalanine. The effects of pyrazophos were weak; in contrast, these energy-intensive biosyntheses are greatly inhibited by PP (DE WAARD and DEKKER 1973; DE WAARD 1974; DE WAARD 1975).

The DNA and RNA polymerases, which build the nucleic acids, require, inter alia, the simultaneous presence of all 4 nucleoside triphosphates ATP, GTP, CTP and TTP/UTP in sufficiently great concentration for their activity. If only one of these pool quantities is reduced, this results in an inhibition of the biosynthesis rate. It is feasible that the PP-triggered respiration inhibition also has an effect on DNA synthesis and RNA synthesis via reduced ATP concentrations, which, in turn, must necessarily result in an inhibition of protein biosynthesis.

In contrast to edifenphos and iprobenfos, no significant increase of cell membrane permeability can be detected under the influence of pyrazophos or PP (phosphate-efflux with Pyricularia – DE WAARD and DEKKER 1973; DE WAARD 1974; DE WAARD 1975; conductivity with *Helminthosporium* – KOHTS 1985). The leakage effect, which is only slightly pronounced, is surprising in view of the above-mentioned membrane damage caused by pyrazophos. An effect on the main fraction of the plasma membranes, the phospholipids, has been detected (KOHTS 1985). The treated mycelia show not only a generally reduced incorporation of ^{14}C-acetate into the total lipid fraction, but also particularly reduced phospholipid contents combined with a weak inhibition of phosphatidyl-ethanolamine-N-methyl-transferase. This suggests parallels with the action of edifenphos and iprobenfos, which cause marked inhibition of this reaction. The concentration of free fatty acids increases substantially under the effect of pyrazophos (KOHTS 1985), which can be attributed to reduced acyltransferase activity and/or increased phospholipase activity. In addition, the destruction of membranes by lipid peroxidation is also discussed (MUELLER et al. 1992). Pyrazophos may cause a reduced total content of sterols, but the ratio of sterols to each other remains unaltered. This allows the conclusion that no inhibition of ergosterol biosynthesis takes place (ZEUN 1990).

Besides a direct interaction of pyrazophos with membrane phospholipids which is possible (KOHTS 1985), drastic changes in the membrane properties are caused in particular by the above described effects, since even minor shifts in the phospholipid pattern already have adverse effects on the physiological membrane functions and prevent an orderly procedure of transport and biosynthesis processes. In addition, the copious accumulation of free fatty acids can have a toxic effect on processes in the cell (RIETH 1977; SUMRELL et al. 1978). As demonstrated by microscopic symptoms, these defects also have a long-term action on the biosynthesis of cell wall material. However, pyrazophos does not inhibit synthesis of the chitin fraction directly (DE WAARD 1974).

In subtoxic concentrations, pyrazophos reduces the pigmentation of *Helminthosporium*. Larger dosage rates can result in the complete absence of mycelial pigmentation. In this case, the fungus excretes reddish pigments into the medium. In the case of PO and PP, this seems to be a minor phenomenon (KOHTS 1985). These findings suggest that pyrazophos engages in the pentaketide metabolism of melanine biosynthesis, as has already been depicted for tricyclazole (WOLOSHUK et al. 1980). The inhibition of melanine biosynthesis is combined with a loss of penetrability of the fungus into the host tissue (WOLOSHUK et al. 1983). However, the degree of effectiveness of tricyclazol remains higher, and the exact target site in the melanine reaction pathway remains undefined.

Due to the complex mode of action of pyrazophos, the development of a stable resistance is highly unlikely. Although resistant mutants can be produced in a laboratory by suitable treatment of fungal mycelia (GEORGOPOULOS et al. 1975), these mutants are generally inferior to the wild types with regard to vitality. This is also true for *Sphaerotheca* strains with reduced sensitivity from greenhouse crops (DEKKER and GIELINK 1979; SCHEPERS 1985; O'BRIEN et al. 1989). Resistant laboratory strains of *Pyricularia* are not capable of transforming pyrazophos into PP. Accordingly, the resistance seems to be based on a lack in the ability to metabolize (DEKKER 1976). Moreover, these strains show crossresistance with iprobenfos and edifenphos (DE WAARD and NISTELROOY 1980). There is no crossresistance of pyrazophos with EBIs (BUCHENAUER 1984; KOHTS and BUCHENAUER 1984b).

Synergistic effects with triazole fungicides were described for pyrazophos/propiconazole against powdery mildew of barley and wheat (ZEUN and BUCHENAUER 1991; ZEUN et al. 1992) and can also be expected for pyrazophos/flusilazole (GARNIER 1990).

References

AKATSUKA, T., KODAMA, O., and YAMADA, H.: A novel mode of action of Kitazin P® in *Pyricularia oryzae*. Agric. Biol. Chem. **41** (1977): 2111–2112.
– – A mode of action of anti-blast substances. J. Pestic. Sci. **9** (1984): 375–381.
ANITHA, R., REDDY, M. S., and RAO, K. C.: Acquired fungicide tolerance in *Macrophomina phaseolina* to thiophanate-methyl and tolclofos-methyl and their cross-tolerance. Ind. Phytopath. **42** (1989): 128–131.
BELAL, M. H.: Studien zur chemischen Bekämpfung des Echten Mehltaus. Thesis, Göttingen 1971.
BINKS, P. R., ROBSON, G. D., GOOSEY, M. W., HUMPHREYS, A., and TRINCI, A. P. J.: Chitin synthesis in *Fusarium graminearum* and its inhibition by edifenphos (Hinosan®). J. Gen. Microbiol. **137** (1990): 615–620.

BLAKE, P. S., HUNTER, L. D., and SOUTER, R. D.: Glasshouse tests of fungicides for apple powdery mildew control. (II) Eradicant and antisporulant activity. J. Horticult. Sci. **57** (1982): 407–412.

BRILLINGER, G. U.: Metabolic products of microorganisms 181. Chitin synthase from fungi a test model for substances with insecticidal properties. Arch. Microbiol. **121** (1979): 71–74.

BRUGGEN, A. H. C. VAN, and ARNESON, P. A.: Resistance in *Rhizoctonia solani* to tolclofos-methyl. Neth. J. Plant Pathol. **90** (1984): 95–106.

BUCHENAUER, H.: Situation of fungicide resistance in cereal diseases demonstrated with eye-spot disease and powdery mildew. Ges. Pflanzen **36** (1984): 161–170.

— Physiological reactions in the inhibition of plant pathogenic fungi. In: Chemistry of Plant Protection. Vol. 6. Springer-Verlag, Berlin 1990, pp. 217–292.

BURDEN, R. S., COOKE, D. T., and HARGREAVES, J. A.: Review — Mechanism of action of herbicidal and fungicidal compounds on cell membranes. Pestic. Sci. **30** (1990): 125–140.

CHOI, S. Y., KIM, Y. S., LEE, M. L., OH, H. W., and JEONG, B. K.: Studies on the acute and chronic toxicities of pesticides to honey bees, *Apis mellifera*. Korean J. Apiculture **4** (1988): 85–95.

CIMANOWSKI, J.: Antisporulant activity of some systemic fungicides against apple powdery mildew. Tag.-Ber. Akad. Landwirtsch.-Wiss. DDR, Berlin **253** (1987): 289–293.

CRAIG, G. D., and PEBERDY, J. F.: The mode of action of S-benzyl O,O-di-isopropyl phosphorothioate and of dicloran on *Aspergillus nidulans*. Pestic. Sci. **14** (1983a): 17–24.

— — The effect of S-benzyl O,O-di-isopropyl phosphorothioate (IBP) and dicloran on the total lipid, sterol and phospholipids in *Aspergillus nidulans*. FEMS Microbiol. Lett. **18** (1983b): 11–14.

CSINOS, A. S.: Activity of tolclofos-methyl (Rizolex®) on *Sclerotium rolfsii* and *Rhizoctonia solani* in peanut. Peanut Science **12** (1985): 32–35.

DE WAARD, M. A.: Effects of systemic and non-systemic compounds on the in vitro germination of powdery mildew conidia. Med. Fac. Landbouww. Rijksuniv. Gent **36** (1971): 113–119.

— On the mode of action of the organophosphorus fungicide Hinosan®. Neth. J. Plant Pathol. **78** (1972): 186–188.

— Mechanisms of action of the organophosphorus fungicide pyrazophos. Med. Landbouwhogesch. Wageningen **74** (1974): 1–97.

— Mode of action of the organophosphorus fungicide pyrazophos. Tag.-Bericht Int. Symp. Systemfungizide (1975): 197–202.

— Fungitoxic mechanisms of organophosphorus compounds. In: MINKS, A. K., and GRUYS, P. (Eds.): Integr. Control Insect Pests. Neth. Cent. Agric. Publ., Wageningen 1980, pp. 221–222.

— and DEKKER, J.: Mechanism of action of the organophosphorus fungicides pyrazophos, Hinosan®, and Kitazin®. Int. Congr. Plant Protect. Abstr. Pap. **2** (1973): 0462.

— and SCHEEPENS, P. C.: Mode of action of the organophosphorus fungicide pyrazophos. Proc. 25th Int. Symp. Crop Protect., Gent 1973.

— and VAN NISTELROOY, J. G. M.: Mechanism of resistance to pyrazophos in *Pyricularia oryzae*. Neth. J. Plant Pathol. **86** (1980): 251–258.

DEKKER, J.: Acquired resistance to fungicides. Ann. Rev. Phytopathol. **14** (1976): 405–428.

— and GIELINK, A. J.: Decreased sensitivity to pyrazophos of cucumber and gherkin powdery mildew. Neth. J. Plant Pathol. **85** (1979): 137–142.

DICKMANN, M. B., PATIL, S. S., and KOLATTUKUDY, P. E.: Effects of organophosphorus pesticides on cutinase activity and infection of papayas by *Colletotrichum gloeosporioides*. Phytopathology **73** (1983): 1209–1214.

ERWIN, D. C., and REYNOLDS, H. T.: The effect of seed treatment of cotton with Thimet, a systemic insecticide, on *Rhizoctonia* and *Pythium* seedling diseases. Plant Dis. Reptr. **42** (1958): 174–176.

FERGUSON, F.: Long term effects of systemic pesticides on honey bees. The Austral. Beekeeper **89** (1987): 49–54.

FEST, C., and SCHMIDT, K. J.: The chemistry of organophosphorus pesticides. Reactivity — Synthesis — Mode of action — Toxicology. Springer-Verlag, Berlin 1973.

FRAMPTON, G. K.: Effects of the foliar fungicide pyrazophos on cereal collembola. BCPC Mono. **40** (1988): 319–326.

FROST, N. M.: Rizolex® – a new fungicide for control of *Rhizoctonia solani* in potatoes. Proc. 4th Dan. Plant Prot. Conf. 1987: 287–294 (Dan).

GANGWAR, S. K., and DASGUPTA, B.: Rizolex® 50WP – a new fungicide to control the foot rot of betelvine (*Piper betle* L.) caused by *Sclerotium rolfsii* Sacc.. Proc. Ind. Acad. Sci. (Plant Sci.) **99** (1989): 265–269.

GARNIER, F.: Stark® – Fongicides pour les orges. Phytoma **418** (1990): 57–58.

GEORGOPOULOS, S. G., GEERLIGS, J. W. G., and DEKKER, J.: Sensitivity of *Ustilago maydis* to pyrazophos and one of its conversion products, and failure to induce resistance with UV-treatment. Neth. J. Plant Pathol. **81** (1975): 38–41.

GORBACH, S., THIER, W., KELLNER, H. M., SCHULZE, E. F., KUENZLER, K., and FISCHER, H.: Environmental impact of pyrazophos. 1. Contribution. Degradation in plants and in the rat. In: COULSTON, F., and KORTE, F. (Eds.): Environmental quality and safety Suppl. Vol. 3. Thieme, Stuttgart 1975, pp. 840–844.

GUPTA, R. L., and ROY, N. K.: Synthesis and fungicidal activity of O,O-diaryl S-isopropyl phosphorothioates. Proc. Indian natn. Sci. Acad. **B54** (1988): 287–290.

– – Synthesis and anti rice blast activity of S-alkyl S,S-diaryl phosphorotrithioates. Indian J. Chem. **30 B** (1991): 320–323.

HARI, B. V. S. C., CHIRANJEEVI, V., SITARAMAIAH, K., and SUBRAHMANYAM, K.: In vitro screening of fungicides against groundnut isolate of *Sclerotium rolfsii* Sacc. by soil vial technique. Pesticides **23** (1989): 47–49.

HEIMBACH, U.: Side effects of some fungicides on insects. Nachrichtenbl. Dt. Pflanzenschutzd. **40** (1988): 180–183.

HELYER, N. L., and LEDIEU, M. S.: Phytotoxicity trial of pyrazophos on chrysanthemum cultivars. Ann. appl. Biol. **106** (Suppl.) (1985): 116–117.

INOUE, S.: Trends in the chemical control of rice disease in Japan. Pestic. Outlook **1** (1990): 31–37.

ISHIZAKI, H., KOBAYASHI, I., and KUNOH, H.: Cytological and physiological studies of the effects of IBP on *Pyricularia oryzae*. (II) Effects on apical cells of single hyphae. Pestic. Sci. **17** (1986): 517–525.

KAKIKI, K., MAEDA, T., ABE, H., and MISATO, T.: Studies on mode of action of organophosphorus fungicide Kitazin®. (I) Effect on respiration, protein synthesis, nucleic-acid synthesis, cell wall synthesis and leakage of intra-cellular substances from mycelia of *Pyricularia oryzae*. Nippon Nogeikagaku Kaishi **43** (1969): 37–44 (Jpn).

KATAGIRI, M., UESUGI, Y., and UMEHARA, Y.: Development of resistance to organophosphorus fungicides in *Pyricularia oryzae* in the field. J. Pestic. Sci. **5** (1980): 417–421.

KATO, T.: Mode of antifungal action of a new fungicide, tolclofosmethyl. In: MIYAMOTO, J., and KEARNEY, P. C. (Eds.): Pesticide Chemistry: Human welfare and the environment. Vol. 3. Pergamon Press, Oxford 1983, pp. 153–157.

KODAMA, O., TAKASE, K., AKATSUKA, T., and UESUGI, Y.: Degradation of S-benzyl O,O-diisopropyl phosphorothiolate by mixed function oxidase of *Pyricularia oryzae*. J. Pestic. Sci. **7** (1982): 517–522.

– YAMADA, H., and AKATSUKA, T.: KITAZIN P®, inhibitor of phosphatidylcholine biosynthesis in *Pyricularia oryzae*. Agric. Biol. Chem. **43** (1979): 1719–1725.

– YAMASHITA, K., and AKATSUKA, T.: Edifenphos, inhibitor of phosphalidylcholine biosynthesis in *Pyricularia oryzae*. Agric. Biol. Chem. **44** (1980): 1015–1021.

– and AKATSUKA, T.: Kitazin P® and edifenphos, possible inhibitors of phosphalidylcholine biosynthesis. In: MIYAMOTO, J., and KEARNEY, P. C. (Eds.): Pesticide Chemistry: Human welfare and the environment. Vol. 3. Pergamon Press, Oxford 1983, pp. 135–140.

KOECHER, H., and LOETZSCH, K.: Zur systemischen Wirkung von Afugan® (Pyrazophos). Med. Fac. Landbouww. Rijksuniv. Gent **41** (1976): 635–644.

KOELLER, W., ALLAN, C. R., and KOLATTUKUDY, P. E.: Protection of *Pisum sativum* from *Fusarium solani* f. sp. *pisi* by inhibition of cutinase with organophosphorus pesticides. Phytopathology **72** (1982): 1425–1430.

KOHTS, T.: Untersuchungen zu einigen neuen Aspekten der Wirkungsweise von Pyrazophos (Afugan®) in *Erysiphe graminis* und *Helminthosporium*-Arten. Thesis, Bonn 1985.

— and BUCHENAUER, H.: Beiträge zur Wirkung von Pyrazophos (Afugan®) gegenüber *Erysiphe graminis* und *Helminthosporium teres*. Mittlg. Biolog. Bundesanstalt Berlin **223** (1984a): 231.

— — Some aspects of the mode of action of pyrazophos. Proc. Brit. Crop Prot. Conf. Pests Dis. **3** (1984b): 917–922.

KOZAKA, T.: Chemical control of rice blast in Japan. Rev. Plant Protect. Res. **2** (1969): 53.

KUROGOCHI, S., KATAGIRI, M., TAKASE, I., and UESUGI, Y.: Metabolism of edifenphos by strains of *Pyricularia oryzae* with varied sensitivity to phosphorothiolate fungicides. J. Pestic. Sci. **10** (1985): 41–46.

LEDIEU, M. S., and HELYER, N. L.: Pyrazophos: a fungicide with insecticidal properties including activity against chrysanthemum leaf miner (*Phytomyza syngenesiae*) (Agromyzidae). Ann. appl. Biol. **102** (1983): 275–279.

LEROUX, P.: Mise en évidence d'une similitude d'action fongicide entre le fenpiclonil, l'iprodione et le tolclofos-methyl. Agronomie **11** (1991): 115–117.

LINDQUIST, R. K., CASEY, M. L., HELYER, N., and SCOPES, N. E. A.: Leafminers on greenhouse chrysanthemum control of *Chromatomyia syngenesiae* and *Liriomyza trifolii*. J. Agric. Entomol. **1** (1984): 256–263.

LYR, H., and SEYD, W.: Chitin synthetase inhibition in *Mucor rouxii* under the in vitro effects of fungicides and other active substances. Z. Allg. Mikrobiol. **18** (1978): 721–729.

— and EDLICH, W.: On the mechanism of action of aromatic hydrocarbon fungicides. Tag.-Ber. Akad. Landwirtsch.-Wiss. DDR, Berlin **253** (1987): 69–75.

— POLTER, C., and BRAUN, P.: The mechanism of action of pyrazophos – a reinvestigation. Proc. 10th Int. Symp. Reinhardsbrunn "Modern fungicides and antifungal compounds" 1992. Ulmer Verl., Stuttgart (1993): 141–149.

MAEDA, T. H., ABE, H., KAKIKI, K., and MISATO, T.: Studies on the mode of action of organophosphorus fungicide, Kitazin®. (II) Accumulation of an aminosugar derivative on Kitazin®-treated mycelia of Pyricularia oryzae. Agric. Biol. Chem. **34** (1970): 700–709.

MCKEEN, W. E., MITCHELL, N., JARVIE, W., and SMITH, R.: Electronmicroscopy studies of conidial walls of *Sphaerotheca macularis*, *Penicillium levetium* and *Aspergillus niger*. Can. J. Microbiol. **12** (1966): 427–428.

— and SMITH, R.: The Erysiphe cichoracearum conidium. Can. J. Bot. **45** (1967): 1489–1496.

MUELLER, H. M., LYR, H., and BRAUN, P.: Ultrastructural changes in fungi under the influence of pyrazophos. Proc. 10th Int. Symp. Reinhardsbrunn "Modern fungicides and antifungal compounds" 1992. Ulmer Verl., Stuttgart (1993): 419–430.

NAKAMURA, S., and KATO, T.: Mode of action of tolclofos-methyl in *Ustilago maydis*. J. Pestic. Sci. **9** (1984): 725–730.

O'BRIEN, R. G., HUTTON, D. G., and VAWDREY, L. L.: Fungicide resistance and its management. Queensland Agric. J. **115** (1989): 13–15.

OHTSUKI, S., and FUJINAMI, A.: Rizolex® (tolclofos-methyl). Jpn. Pestic. Inf. **41** (1982): 21–25.

ORTH, A. B., SFARRA, A., PELL, E. J., and TIEN, M.: An investigation into the role of lipid peroxidation in the mode of action of aromatic hydrocarbon and dicarboximide fungicides. Pestic. Biochem. Physiol. **44** (1992): 91–100.

OU, S. H.: A look at worldwide rice blast disease control. Plant Dis. **64** (1980): 439–445.

PARELLA, M. P.: Evaluations of selected insecticides for control of permethrin-resistant *Liriomyza trifolii* (Diptera: Agromyzidae) on chrysanthemum. J. Economic. Entomol. **76** (1983): 1460–1464.

QVARNSTROEM, K.: Control of powdery mildew (*Erysiphe cichoracearum*) on cucumber plants. Växtskyddsnotiser **53** (1989): 54–57 (Swe).

Rathke, S., and Jahn, M.: Wirkung von Fungiziden auf die Sporulation von *Podosphaera leucotricha* (Ell. et Ev.) Salm. und *Venturia inaequalis* (Cooke) Wint.. Arch. Phytopathol. Pflanzenschutz **25** (1989): 459–464.

Rieth, H.: Die unterschiedliche Bedeutung der Fettsäuren in ihrer fungistatischen Wirkung. Fette-Seifen-Anstrichmittel **79** (1977): 120–121.

Roy, N. K.: Chloroalkyl phosphonates and phosphorothioates – a new group of fungicides. Proc. Indian natn. Sci. Acad. **B56** (1990): 305–310.

Saito, H., and Kato, S.: Distribution of IBP- and KSM-resistant strains in Aichi prefecture. Res. Bull. Aichi Agric. Res. Ctr. **22** (1990): 79–84.

Schepers, H. T. A. M.: Fitness of isolates of *Sphaerotheca fuliginea* resistant or sensitive to fungicides which inhibit ergosterol biosynthesis. Neth. J. Plant Pathol. **91** (1985): 65–76.

Schlueter, K., and Weltzien, H. C.: Ein Beitrag zur Wirkungsweise systemischer Fungizide auf *Erysiphe graminis*. Med. Fac. Landbouww. Rijksuniv. Gent **36** (1971): 1159–1164.

– – Additional studies on the mode of action of systemic fungicides on *Erysiphe graminis*. Nachrichtenbl. Dt. Pflanzenschutzd. **29** (1977): 17–20.

Schmidt, H. H., and Burth, U.: Mode of action, secondary effects and examples of the use of systemic fungicides for controlling mycoses in fruit and vegetables. Arch. Phytopathol. Pflanzensch. **13** (1977): 241–262.

Schreiber, B., Kötter, U., and Wagner, H.-J.: Afugan® – an alternative for the control of cereal diseases. Mittlg. Biol. Bundesanst. **223** (1984): 221–222.

Sisler, H. D.: Control of fungal diseases by compounds acting as antipenetrants. Crop Protect. **5** (1986): 306–313.

Smiley, R. W., Uddin, W., Ott, S., and Rhinhart, K. E. L.: Influence of flutolanil and tolclofos-methyl on root and culm diseases of winter wheat. Plant Dis. **74** (1990): 788–791.

Smolka, S., and Wolf, G.: Cytological studies on the mode of action of systemic fungicides on the host pathogen complex barley – powdery mildew (*Erysiphe graminis* f. sp. *hordei* Marchal). Pestic. Sci. **17** (1986): 249–255.

Somashekar, R. K., Venkatasubbaiah, P., and Gowda, M. T. G.: Cytological effects of a fungicide Rizolex® in *Allium cepa* L. Pesticides **21** (1987): 21–23.

Sotherton, N. W., and Moreby, S. J.: The effects of foliar fungicides on beneficial arthropods in wheat fields. Entomophaga **33** (1988): 87–99.

Sugiura, H., Hayashi, K., Tanaka, T., Takenaka, M., and Uesugi, Y.: Mutual antagonism between sterol demethylation inhibitors and phosphorothiolate fungicides on *Pyricularia oryzae* and the implications for their mode of action. J. Pestic. Sci. **39** (1993): 193–198.

Sumrell, G., Mod, R. R., and Mague, F. C.: Antimicrobial activity of some fatty acid derivates. J. Amer. Oil Chem. Soc. **55** (1978): 395–397.

Sztejnberg, A., Byrde, R. J. W., and Woodcock, D.: Antisporulant action of fungicides against *Podosphaera leucotricha* on apple seedlings. Pestic. Sci. **6** (1975): 107–111.

Tomizawa, G., and Uesugi, Y.: Metabolism of S-benzyl O,O-diisopropyl phosphorothioate (Kitazin P®) by mycelial cells of *Pyricularia oryzae*. Agric. Biol. Chem. **36** (1972): 294–300.

Uesugi, Y.: Development of organophosphorus fungicides. Jpn. Pestic. Inf. **2** (1970): 11–14.

– Pest Resistance to Pesticide. Plenum Press (1983): 481–503.

– Katagiri, M., and Noda, O.: Joint action of chemicals with phosphorothiolate fungicides on rice blast fungus. Ann. Phytopath. Soc. Jpn. **40** (1974): 252–260.

– – Metabolism of a phosphorothiolate fungicide IBP by strains of *Pyricularia oryzae* with varied sensitivity. In: Miyamoto, J., and Kearney, P. C. (Eds.): Pesticide Chemistry: Human welfare and the environment. Vol. 3 Pergamon Press, Oxford 1983, pp. 165–170.

– and Takenaka, M.: The mechanism of action of phosphorothiolate fungicides. Proc. 10th Int. Symp. Reinhardsbrunn "Modern fungicides and antifungal compounds" 1992. Ulmer Verl., Stuttgart (1993): 159–164.

- and TOMIZAWA, C.: Metabolism of S-benzyl O-ethyl phenylphosphonothioate (Inezin®) by mycelial cells of *Pyricularia oryzae*. Agr. Biol. Chem. **36** (1972): 313–317.
UMEDA, Y.: Hinosan, a fungicide for control of rice blast. Jpn. Pestic. Inf. **18** (1973): 25–34.
WIEBE, M. G., ROBSON, G. D., and TRINCI, A. P. J.: Edifenphos (Hinosan®) reduces hyphal extension, hyphal growth unit length and phosphalidylcholine content of *Fusarium graminearum* A3/5, but has no effect on specific growth rate. J. Gen. Microbiol. **136** (1990): 979–984.
WOLOSHUK, C. P., SISLER, H. D., CHRYSAYI TOKOUSBALIDES, M., and DUTKY, S. R.: Melanin biosynthesis in *Pyricularia oryzae*: Site of tricyclazole inhibition and pathogenicity of melanin-deficient mutants. Pestic. Biochem. Physiol. **14** (1980): 256–264.
- SISLER, H. D., and VIGIL, E. L.: Action of the antipenetrant tricyclazole on appressoria of *Pyricularia oryzae*. Physiol. Plant Pathol. **22** (1983): 245–259.
YATHOM, S., TAL, S., and CHEN, M.: The effects of insecticides on different stages of the leafminer *Liriomyza trifolii* (Burgess). Hassadeh **68** (1988): 784–787.
YOSHIDA, M., MORIYA, S., and UESUGI, Y.: Observation of transmethylation from methionine into choline in the intact mycelia of *Pyricularia oryzae* by ^{13}C NMR under the influence of fungicides. J. Pestic. Sci. **9** (1984): 703–708.
- KAWASAKI, A., YUKIMOTO, M., and NOSE, K.: Detection of the effects of fungicides on the cell membrane by proton nuclear magnetic resonance spectroscopy. Pest. Biochem. Physiol. **38** (1990): 172–177.
- and NOSE, K.: Changes in the relaxation times of water protons in fungal cells indicate fungicide effects on membrane water permeability. Pest. Biochem. Physiol. **38** (1990): 162–171.
- UESUGI, Y., NOSE, K., and LALITHAKUMARI, D.: Inhibition of phospholipid N-methylation by dibutyl N-methyl-N-phenyl-phosphoramidate in *Pyricularia oryzae* in relation to negative cross-resistance with phosphorothiolates. J. Pestic. Sci. **12** (1987): 513–515.
YOSHINAGA, E.: A systemic fungicide for rice blast control. Proc. 3rd Brit. Insect. Fungic. Conf., Brighton 1969, pp. 593–599.
YOSHINO, R.: Present status of occurrence and control of blast disease in Japan. Jpn. Pestic. Inf. **52** (1988): 3–8.
ZEUN, R.: Zur Wirkung synergistischer Fungizid-Kombinationen gegenüber *Erysiphe graminis*, *Septoria nodorum* sowie *Pyrenophora teres*. Thesis, Hannover 1990.
- and BUCHENAUER, H.: Synergistic effects of pyrazophos and propiconazole against Pyrenophora teres. J. Plant Dis. Protect. **98** (1991): 661–668.
- SACHSE, B., and BUCHENAUER, H.: Studies on the nature of the synergistic effects of pyrazophos and propiconazole in *Pyrenophora teres*. J. Plant Dis. Protect. **99** (1992): 273–285.

Chapter 19

Miscellaneous fungicides

Maya Gasztonyi*) and H. Lyr**)

*) Plant Protection Institute, Hungarian Academy of Sciences, Budapest, Hungary
**) Institut für Integrierten Pflanzenschutz der Biologischen Bundesanstalt Kleinmachnow, Germany

Introduction

Some important or newly developed compounds do not fit into the fungicide groups described in the other chapters, or are not yet as well known as other fungicides that an own chapter can be devoted to them. Therefore this chapter contains some smaller groups of fungicidal compounds or even single compounds, which are important enough to be mentioned in this book. Very often they have an aberrant mechanism of action from other groups of fungicides.

Some compounds described in the 1st edition of this book have now been omitted, because of minor importance nowadays, inspite of interesting features. These are: fenitropan, dinocap, binapacryl, nitrophthal-isopropyl, quinomethionate, thioquinox, fenaminosulf, sec-butylamine, drazoxolon, pyroxychlor. Their structures and descriptions can be found in the first edition (Lyr 1987).

Wherever possible we combined the compounds described here to groups based on their chemical structures. Some groups have been described very recently and may enlarge in future to new main fungicidal groups, but at present it is still premature to bring them in an own chapter.

Many of the new compounds have an interesting and new mechanism of action which demonstrates that still new targets do exist in fungal cells, which can be used for selective attacks on this group of organisms.

19.1 Guanidines

M. Gasztonyi

This group of compounds includes guanidine derivatives with a hydrophobic alkyl-chain and one or more basic nitrogen atoms. The family of the guanidines is represented here by three fungicides developed in more than ten years intervals, i.e. dodine, guazatine and iminoctadine (Fig. 19.1 a, b, c). Each of them were marketed as acetates, the number of acetate anions being dependent on the number of basic nitrogens in the molecule. There are some contradictions hardly to be avoided in treating of this group. Namely, the common name guazatin was first used in 1968 for the compound 1,1′-imino-di(octa-

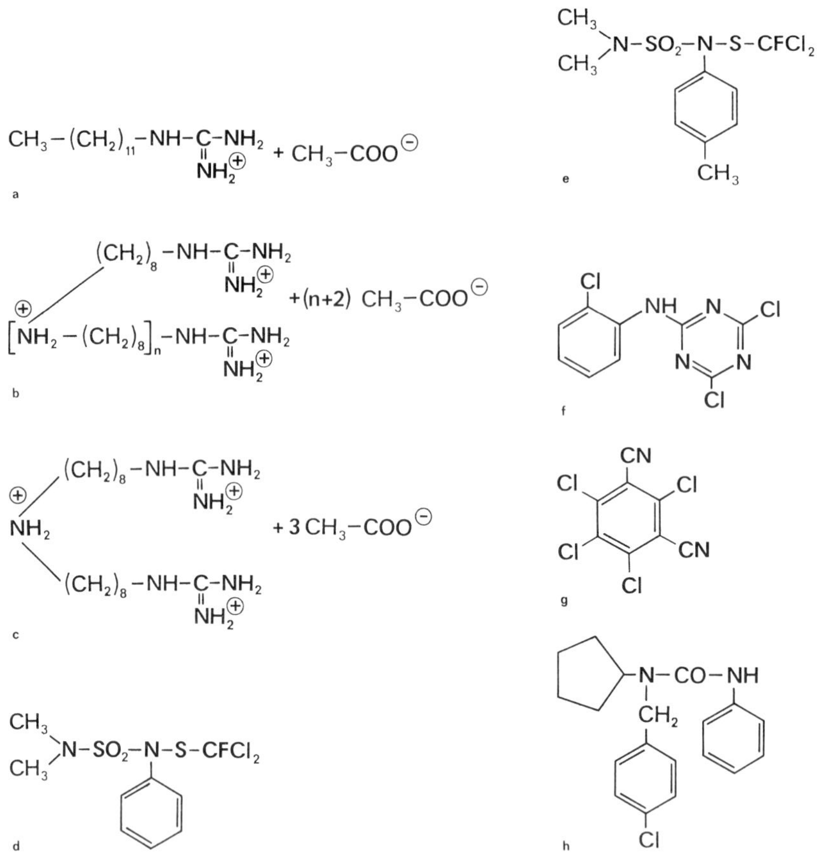

Fig. 19.1 Structure formulae of fungicides belonging to guanidines (a–c), sulphamides (d–e) and miscellaneous chemical groups (f–h). a) Dodine, b) Guazatine, c) Iminoctadine, d) Dichlofluanide, e) Tolylfluanide, f) Anilazine, g) Chlorothalonile, h) Pencycuron.

methylene)-diguanidine (WORTHING 1979), which was newly developed in the 70's under the name iminoctadine (MASUI et al. 1986), while guazatin since 1973 is the common name of a mixture of closely related compounds (HARTLEY and KIDD 1987).

The guanidines are protective fungicides on the whole, although also curative action was established in some cases (BYRDE 1969). Their spectra of activity and fields of application differ from one compound to another. Guanidines belong to the multi-site action fungicides, but some dominant characters of their mode of action may be established. The alteration of permeability of cell membranes (CORBETT 1974) and the interaction with mitochondrial membranes (SOLEL and SIEGEL 1984) may be considered as primary actions. In case of the simplest representative of the family (dodine) the direct inclusion of the lipophylic alkyl-chain into the membrane was assumed, while the guanidine residue remains in the aqueous phase (CORBETT 1974). Experiments with the compounds with more complex structure (e. g. iminoctadine) emphasize the possibility of the inhibition of lipid synthesis as a cause of alteration of membrane permeability (MASUI et al. 1986).

Fungal strains resistant to guanidines may be selected after continuous application for several years, as it was found in case of dodine (SZKOLNIK and GILPATRICK 1969). Cross-resistance to guazatine was also proved in dodine-resistant isolates. Negative cross-resistance between guanidines and fenarimol was demonstrated in laboratory isolates of several fungal species (DE WAARD and VAN NISTELROOY 1983).

Dodine

1-dodecylguanidinium acetate (Fig. 19.1 a) was introduced in 1956 by the American Cyanamid Co. as a foliar protective fungicide (USP 2867562). Its other name is doguadine. The w. p., liquid or dust formulations of dodine are known with the following trade names: "Carpene", "Curitan", "Cyprex", "Efuzin", "Melprex", "Syllit", "Venturol", "Vondodine".

The pure compound exixts as colourless crystals, with m. p. 136 °C, v. p. 0.01 mPa at 20 °C. Its solubility in water at 25 °C come to 0.063%. It is soluble in hot water and alcohol, readily soluble in mineral acids, but insoluble in most organic solvents. The acute oral LD_{50} for mae rats is $1-2$ g/kg.

In contrast to its introduction long ago, dodine is still on the market and used in USA (HICKEY 1991) and in Europe in orchards. The formulations can be used for protective control of scab (*Venturia* spp.), leaf spot (*Mycosphaerella* spp.), blossom brown rot and leaf blight in fruit cultivation. Foliar application is recommended especially against *Venturia* spp. and cherry leaf spot against which dodine has some eradicant action (WORTHING 1979). As a scab fungicide dodine has marked residual effect. Beside this, an excellent curative effect has been reported 30 hours after infection (BYRDE 1969). Local systemic effect has been demonstrated, but there is only a minor translocation to new growth (1 mg/kg) in seedling apple trees sprayed with ^{14}C-dodine (2 kg a. i./ha).

Both, the fungitoxicity and phytotoxicity are strongly connected with the length of the alkyl-chain. Although in a series of n-alkyl guanidine acetate homologues maximum phytotoxicity was found at C_{10} (BROWN and SISLER 1960), damage caused by dodine on apples, pears, currants and gooseberries was also reported. Green-skinned varieties of apples are particularly susceptible. Therefore in Europe the application of dodine on apples is recommended usually only before flowering (GREWE 1965). Dodine has also surface active properties. It is incompatible with anionic wetting agents, with lime and chlorobenzilate (BYRDE 1969).

The free base of dodine is moderately strong, so that the acetate is largely ionised at physiological pH-values, but a varying degree of hydrolysis will occur depending on pH, causing twenty times higher fungitoxic effect at pH 7.8 than at pH 5.1 (HASSALL 1982). Accumulation of large quantities is required for adverse effects on germination of spores. ED_{50} values from 2.0 to 2.5 mg/g conidial weight were obtained (MILLER 1969). It appears to be rather firmly attached to immobile anionic cell constituents such as phosphate and carboxyl groups (HASSALL 1982).

In plants dodine is converted to creatine via the action of a methyltransferase and a simultaneous oxidative cleavage of the dodecyl moiety. As intermediates guanidine as well as that substituted with short alkyl-groups, have been found (VONK 1983).

Dodine is one of the few multi-site action fungicides, against which fungal resistance developed in the field (SZKOLNIK and GILPATRICK 1969). Therefore its combined or alternated use with other fungicides is recommended, in particular on apple trees. In case of laboratory-induced resistance the existence of at least three unlinked genes has been demonstrated, each conferring a different resistance level (GEORGOPOULOS 1969, 1982).

Among the side effects of dodine may be mentioned the bactericidal action e.g. against *Xanthomonas campestris* (HICKEY 1991), as well as its effect at high concentration against most soil fungi except *Beauveria bassiana*. The latter property of dodine might be used in isolation techniques of certain fungi (BEILHARZ et al. 1982).

Guazatine,

acetates of a mixture of the reaction products from polyamines and carbamonitrile (HARTLEY and KIDD 1987). The components of the mixture contain one or more octamethylene-chains separated by imino groups and holding two guanidine radicals at the chain-ends (Fig. 19.1b). It was developed by KenoGard and introduced in 1973 in the praxis. The same common name was used formerly for the triacetate of bis(8-guanidino-octyl)-amine (WORTHING 1979) introduced in 1968 by Evans Medical Ltd. and Murphy Chemical Ltd. Physical form of the mixture is brown solid, melting at ca. 60 °C, stable in neutral or acidic aqueous media. It is soluble in water and in alcohol. Its formulations include liquids and powders. Trade names are "Kenopel", "Panoctine", "Panolil" and "Radam". Guazatine is of low oral and dermal toxicity to rats. Extensive toxicological studies have shown it to be neither mutagenic nor carcinogenic.

Guazatine salts are contact fungicides. In contrast to the structurally related dodine, the main field of guazatine application are seed dressings of cereals at $0.6-0.8$ g a. i./kg seed and post-harvest dips of pineapple, citrus and potato (WORTHING 1979; SCHACHNAI and BARASH 1982). It is now registered and marketed in most European countries against *Fusarium* spp., *Septoria nodorum* and *Tilletia caries* (CAMERON et al. 1986). It is also used by foliar application against *Piricularia oryzae* in rice. The seed treatment with guazatine has also bird repellent effect. Another advantageous side effect is its antifeeding activity against the larvae of the insect *Pseudoplusia includens* (MATOLCSY et al. 1988). Mixed formulations of guazatine with imazalil and/or fenfuram are also available. Other fungicides, primarily those belonging to DMIs may be tank-mixed with guazatine to increase the spectrum of its activity (KIDD et al. 1988).

Guazatine like dodine causes alterations in cellular permeability as well as in the acetate and glucose oxidation. Since experiments in vitro showed some enzymes involved in the oxidative processes to be not inhibited directly, the primary mechanism of action should involve a rapid effect on cellular permeability. The apparent loss in oxidative capacity might be explained by the inhibition of the uptake of certain substrates or by the loss of potassium from the cell (SOLEL and SIEGEL 1984).

The fate of guazatine in soil is defined primarily by its cationic character. Due to the rapid and strong adsorption to a soil inorganic fraction, it becomes resistant to microbial degradation and persistent for long time in soil. For the same reason, however, the soil residues are unavailable to the plants, thus their part in human uptake is negligible. Residues on directly sprayed crops would be considerably reduced through processing. The biological degradation of guazatine is initiated by deamidination steps (SATO and MAKI 1989).

Iminoctadine,

triacetate of 1,1'-imino-di(octamethylene)-diguanidine, DF 125, (Befran) is manufactured by Dainippon Inc. and Chemical Inc. (KIDD et al. 1988), (Fig. 19.1 c). Official tests have been carried out since 1975 in Japan and since 1984 in Europe. It is a crystalline solid with m.p. 143–144 °C, soluble in water and alcohols. Liquide and w.p. formulations are available with the trade name "Befrane". Iminoctadine is of low mammalian toxicity, no mutagenic activity has been detected. Its mode of action is the inhibition of lipid biosynthesis (MASUI et al. 1986), wich results in alteration of the membrane pemeability, similarly to that observed in case of the other guanidine-type fungicides.

Iminoctadine is a contact fungicide with broad spectrum of activity. It is effective against apple cancer (*Valsa ceratosperma*), dormant stage blossom blight (*Monilia mali*) and snow blight (*Fusarium nivale, Sclerotinia borealis*) of winter wheat. Citrus storage diseases have also been effectively controlled (MASUI et al. 1986). Trials in West Europe with small grain cereals demonstrated the effectivity of topically sprayed Befrane against diseases caused by *Septoria nodorum, S. tritici, Rynchosporium secalis, Pyrenophora teres, Puccinia recondita, Alternaria* spp. and *Fusarium* spp.

Interestingly, the first registration of iminoctadine in Europe is directed for its use against scab on apples and pears, similarly to dodine (KIDD et al. 1988). Extension of the use is to be expected, however, when the compounds belonging to guanidines will find their optimal fields of application among themselves.

Sulphamides

The family of sulphamides includes the N-substituted diamides of sulphuric acid. One of their biologically important substituents is the dichlorofluoromethylthio-group, which constitutes an essential part also of the molecules of another type of fungicides, the phthalimides. It is to be mentioned, that presence of one fluorine atom in the trihalogenated methylthio-group highly enhances the activity as compared with the trichloroderivatives, while the same difference among the phthalimides has not a great influence on the fungitoxicity (GREWE 1965). The fungicides discussed here among the sulphamides are dichlofluanid and tolylfluanid (Fig. 19.1 d, e). Some closely related compounds belonging to the alkylsulphonamides also are potent fungicides, they did not find, however, application in the agriculture (MATOLCSY et al. 1988).

The similarity between sulphamides and phthalimides manifests itself also in their biological properties. Both groups are protective fungicides with multi-site action (LUKENS and SISLER 1958; SIEGEL 1971). The mechanism of action could be similar, but a certain selectivity of sulphamides can be stated. The fungicidal activity is focussed on *Venturia* spp. and *Botrytis cinerea*, but they have a secondary action against powdery mildews and spider mites (GREWE 1968). Resistance to sulphamides or cross-resistance with other fungicides was not observed (LYR and CASPERSON 1982).

The N—S bond in the molecule is exposed to chemical, photochemical or microbial cleavage, resulting in degradation of residues.

Dichlofluanid,

N-dichlorofluoromethylthio-N′,N′-dimethyl-N-phenylsulphamide, DCF was introduced in 1965 by Bayer AG (Bayer 47 531), (Fig. 19.1 d) and protected by DAS 1 193 498. It is marketed in w. p. and dust formulations as "Euparen" and "Elvaron". The a. i. is a colourless powder with m. p. 105 °C. It is soluble in organic solvents and practically insoluble in water. The acute oral LD_{50} for rats is 2 500 mg/kg, but the toxicity to fishes is high (HARTLEY and KIDD 1987).

Dichlofluanid has a relative broad spectrum of activity and is active already at low doses in sensitive fungi. A higher activity against *Botrytis cinerea*, *Plasmopara viticola*, *Venturia inaequalis*, *Phytophthora infestans* and *Alternaria solani* in comparison to TMTD, zineb, captan and folpet could be stated. Important for its action seems to be an additional activity over the vapour phase. Latter property may play a role in comprising its effectivity to powdery mildew (GREWE 1968). Its practical application was stimulated by its protective and curative effect at lower doses than those of captan in controlling apple and pear scab. Additionally favoured its practical application the good effects against *Gloeosporium* fruit decay as preharvest spray (KASPERS 1968). A side effect against *Podosphaera leucotricha* in apples could be stated which allowed to lower the concentration of Morestan (chinomethionate) in mixtures. At concentrations of 0.1% side effects against spider mites in apples and hops were observed which allowed to reduce acaricidal sprays. A good effect against *Plasmopara viticola* and *Botrytis cinerea* in vineyards and against *Pseudoperonospora humuli* and partly *Sphaerotheca humuli* as well as against *B. cinerea* in hops recommended its practical application (KOLBE and CASPERS 1968). The field of application was enlarged also in the last decade, by utilizing its good effect against *Botrytis* in strawberry, stimulated by the lack of toxicity to bees.

Under conditions of agricultural practice only small amounts of the applied dichlofluanid remain on the crops until harvest, partly due to decomposition (KUBIAK and EICHHORN 1990) by hydrolysis and by light.

Besides the application in crop protection its properties allow use as an antifouling agent or as an additive for paints for wood preservation especially against blue staining fungi (cf. chapter 22 and 23).

Tolylfluanid

N-dichlorofluoromethylthio-N′,N′-dimethyl-N-p-tolylsulphamide (Fig. 19.1 e) is a homologue of the fungicide dichlofluanid belonging to the sulphamides. It was developed by Bayer AG and marketed with trade name "Euparen M". Tolylfluanid is a colourless crystalline solid with m. p. 95−97 °C (HARTLEY and KIDD 1987).

It is a foliar fungicide with protective action very similar to that of dichlofluanid, giving a good control of scab on apples and pears, *Botrytis* on grapes, ornamentals, etc. Applied on apples, it reduces also the infection of *Podosphaera leucotricha*, replacing or at least reducing the use of specific fungicides against powdery mildew (KOLBE 1972).

It has a secondary activity against spider mites on fruit. Very low toxicity to mammals and no toxicity to honey bees have been demonstrated.

Hydrolytic degradation in plants and in water produces dimethylaminosulphotoluidine (BRENNECKE 1988). Extensive investigation on the fate of tolylfluanid residues in stored fruits has been carried out recently by ROUCHAUD et al. (1991).

Anilazine

2,4-dichloro-6-(2-chloroanilino)-1,3,5-triazine (Fig. 19.1f) was chosen by WOLF et al. (1955) among 80 analogues of s-triazines for development as an agricultural fungicide. It was introduced between 1966–1968 by Bayer AG (Brit. 1120338), under the name "Dyrene". The trade names of w. p. formulations are "Botrysan", "Direx", "Direz", "Dyrene", "Kemate", "Tryasin", "Zinochlor". The pure compound exists as colourless crystals, m. p. 159–160 °C; v. p. is very low. It is stable in neutral and weakly acid media, practically insoluble in water, soluble in most organic solvents. Its acute oral LD_{50} for female rats is 2700 mg/kg, but toxicity to fishes is high.

Anilazine (Dyrene) is a broad-spectrum protective leaf-fungicide. It was first commercially used on turf grasses to control *Helminthosporium* blights, *Fusarium* snow mold and *Rhizoctonia* brown patch (WOLF et al. 1955). Its application was later extended to the protection of cereals, coffee, vegetables and other crops against diseases caused by *Alternaria*, *Helminthosporium* and *Cercospora* spp. At present in USA, the main field of its application is the control of fungal diseases on vegetables (JOHNSTON 1991). In Europe it is used preferably against *Botrytis cinerea* on ornamental plants (PERKOW 1983). On tomato plants it is highly effective against early blight (*Alternaria solani*) and grey leaf spot (*B. cinerea*), but less effective against late blight (*Phytophthora infestans*). Blooming plants should not be treated.

Anilazine is incompatible with oils and alkaline materials, but it is preferably applied in combinations with other fungicides, e. g. benomyl, chlorothalonil, triadimenol, tridemorph (KIDD, HARTLEY et al. 1988). The recently observed synergistic effect of Zn^{2+} and Cu^{2+} ions applied in combinations with anilazine (GOSS and MARSHALL 1985), makes it possible to reduce the concentration of a. i. Results of experiments (MALCOM and BLUETT 1986) indicated, that anilazine was superior to captafol for control of *Septoria nodorum* when used in mixture with triadimefon or triadimenol. These findings may promote and extend the application of anilazine.

Anilazine is quickly and strongly adsorbed by fungal spores. ED_{50} value of spore germination in *Neurospora sitophila* is 1.53 mg/g. The ultimate site of action is not known, however the distribution of electrons in the triazine ring encourages nucleophilic substitution reactions (LUKENS 1969, 1971). Cellular amino and sulfhydryl groups are particularly reactive with anilazine (CORBETT 1974), so that it seems likely that it causes inhibition of a variety of cell processes by non-specific interactions with vital cell components. The fungicide character of this type of s-triazines depends on the phenylamino substituent, while the alkylamino derivatives possess herbicidal activity.

Non-essential thiol and amino groups of fungi can destroy the fungicide. One or two chlorine atoms of the triazine ring can take part in this reaction. The degradation in soil is very quick. Its photochemical degradation on silica gel is rather intensive, with a half-life time of 2 days (HULPKE et al. 1983).

Occurrence of resistance in *Sclerotinia homoeocarpa* on turf grasses to anilazine has been reported from the USA (NICHOLSON et al. 1971), but further development of resistance has not been confirmed and anilazine is still used without loss of effectivity (SANDERS 1991).

Chlorothalonile

Tetrachloroisophthalonitril, (TPN), (Fig. 19.1 g) was introduced in 1963 by Diamond Alkali Co. (Diamond Shamrock Corp.). Trade names are "Daconil", "Bombardier", "Bravo", "Clortocaffaro", "Exotherm Termil", "Faber", "Notar", "Repulse". The pure a.i. exists as colourless crystals with m.p. 250−251 °C and relatively high vapour pressure. It is a stable and non corrosive compound. The acute LD_{50} for albino rats is extremely high (more than 10 g/kg) (WORTHING 1979). Because of its simple chemical structure, good efficiency, low toxicity and valuable other properties it is still in practical use on a broad scale.

Chlorothalonile is a contact fungicide with broad spectrum of activity. It can be used for the protection of many vegetables and agricultural crops between doses of 0.6−2.5 kg a.i./ha. In Canada it is preferably used to control late blight and early blight in potatoes (HILL 1991). In other countries it is used against *Septoria nodorum* in winter wheat (O'REILLY et al. 1986), as well as against *Alternaria* spp., *Cercospora* spp. and downy mildew diseases in vegetables and fruits (JOHNSTON 1991; HICKEY 1991). Preventive application for several grapevine diseases may have a good effect, although when applied after bloom chlorothalonile can produce berry skin russeting (GOFFINET and PEARSON 1991). Beside of its fungicidal properties it is used as algicide or as a preservative in paints and adhesives.

The use of chlorothalonile is promoted by introducing mixtures with other fungicides. Combined formulation of chlorothalonile and nuarimol provided good control of scab and powdery mildew on apple (HUGGENBERGER et al. 1986). Combinations containing other EBI's, metalaxyl, cymoxanil, vinclozolin, folpet or maneb are marketed in several countries (KIDD, HARTLEY et al. 1966).

Microbial degradation in soil goes through dechlorination with a half-life time of 3 weeks. Repeated application of chlorothalonile reduces the number of soil bacteria and fungi involved in its degradation, which leads to the longer presence of fungicide residues in soil (KATAYAMA et al. 1991). Although chlorothalonile is extremely toxic to fish in acute lethal exposures, faunal impact in the ponds were generally of a smaller magnitude than were predicted by bioassay results. Factors such as dilution, adsorption and degradation are thought to have attended the initial pond concentrations of chlorothalonile, thereby reducing their toxicity (ERNST et al. 1991).

The mechanism of action was thoroughly investigated by VINCENT and SISLER (1968), TURNER and BATTERSHELL (1970) and TILLMAN et al. (1973). Chlorothalonile reacts in vitro with glutathione, coenzyme A, 2-mercaptoethanol and other compounds forming several S-derivatives (VINCENT and SISLER 1968). By this the SH-content in cells is significantly reduced which results in a lethal inhibition of a number of thiol dependent reactions (enzymes) in fungal cells. The glutathione level is an important regulator of normal cell metabolism which is readily affected by chlorothalonile. In addition a direct interaction with SH-groups of enzymes can occur, leading to irreversible enzyme inactivations. Because of this unspecific attack within the cell a certain selectivity is surprising.

Pencycuron

1-(4-chlorobenzyl)-1-cyclopentyl-3-phenylurea, Bay NTN 19701 (Fig. 19.1 h) was introduced by Bayer AG and Nitokuno, and has been registrated from the late eighties in many European and Asian countries with trade name "Monceren". The pure compound exists as colourless crystals with m. p. 130 °C. Dust, w. p. and liquid formulations are available (KIDD et al. 1988). The toxicities towards mammals, birds, fish and earthworm are quite low, with no indication of teratogenicity or mutagenicity.

Pencycuron is a non-systemic fungicide with protective action. It has an extremely narrow spectrum of activity, controlling diseases caused by *Rhizoctonia solani* and *Pellicularia* spp. in potatoes, rice and ornamentals. The most important field of its application is potato tuber treatment. The selectivity manifests itself even within fungal species, only a part of anastomosis groups of *R. solani* being sensitive to pencycuron (UEYAMA et al. 1990).

The fungitoxic action appears in inhibition of hyphal growth, and the activity is dependent on temperature. The mode of action is unclear at present. Some data indicate a possible influence on the chitin synthesis in target fungi, but no full correlation between inhibition of the incorporation of ^{14}C-glucosamine and the fungal sensitivity has been found (UEYAMA et al. 1990). Thus chitin synthesis inhibition itself seems not to be the primary site of action. On the other hand, distortions of germ tubes similar to those produced by the known antimicrotubular fungicides were observed on sensitive fungal species. Moreover, some phenotypes resistant to carbendazim showed an increased sensitivity to pencycuron (LEROUX and GREDT 1990). These observations are not yet enough, pencycuron to be put among the fungicides with antimicrotubular action like benzimidazoles. However, the recently increasing number of various compounds acting on fungal microtubules (e. g. zarilamide, rhizoxine) raises the question, whether fungal mitotic apparate may become a more common target of future fungicides, also chemically different from those known before.

References

BEILHARZ, V. C., PARBERY, D. G., and SWART, H. J.: Dodine: A selective agent for certain soil fungi. Trans. Brit. Mycol. Soc. **79** (1982): 507–511.

BRENNECKE, R.: Methode zur gaschromatographischen Bestimmung von Rückständen der Fungizide Euparen und Euparen M in Pflanzenmaterial und Getränken. Pflanzenschutz Nachrichten Bayer **41** (1988): 136–172.

BROWN, J. F. and SISLER, H. D.: Mechanisms of fungitoxic action of N-dodecyl-guanidine acetate. Phytopathology **50** (1960): 830–839.

BYRDE, R. J. W.: Non aromatic Organics. In: TORGESON, D. C. (Ed.): Fungicides. Vol. 2. Academic Press, New York – London 1969, pp. 531–578.

CAMERON, D. G., HYTTEN-CAVALLIUS, I., JORDOW, E., KLOMP, A. O., and WIMSCHNEIDER, W.: Guazatine – its use as a foliar fungicide in cereals and oilseed rape. Proc. 1986 Brit. Crop. Prot. Conf. – Pests and Diseases Vol. 3 (1986): 1201–1207.

CORBETT, J.R.: The Biochemical Mode of Action of Pesticides. Academic Press, New York – London 1974.

DE WAARD, M. A., and VAN NISTELROOY, J. G. M.: Negatively correlated cross-resistance to dodine in fenarimol-resistant isolates of various fungi. Netherl. J. Plant Pathol. **89** (1983): 67–73.

ERNST, W., DOE, K., JONAH, P., YOUNG, T., JULIEN, G., and HENNIGAR, P.: The toxicity of chlorothalonil to aquatic fauna and the impact of its operational use on the pond ecosystem. Arch. Environ. Contam. Toxicol. **21** (1991): 1–9.

GEORGOPOULOS, S. G.: The problem of fungicide resistance. Bio Science **19** (1969): 971–973.

– Genetical and biochemical background of fungicide resistance. In: DEKKER, J., and GEORGOPOULOS, S. G. (Eds.): Fungicide Resistance in Crop Protcection. Pudoc, Wageningen 1982, pp. 46–62.

GOFFINET, M. C., and PEARSON, R. C.: Anatomy of russeting induced in Concord grape berries by the fungicide chlorothalonil. Am. J. Enol. Vitic. **42** (1991): 281–289.

GOSS, V., and MARSHALL, W. D.: Synergistic antifungal interactions of zinc and copper with anilazine. Pesticide Sci. **16** (1985): 163–171.

GREWE, F.: Rückblick auf 25 Jahre Fungizidforschung. Pflanzenschutznachrichten Bayer **18**, Sonderheft (1965): 45–74.

– Euparen (Dichlofluanid), ein neues polyvalentes Fungizid mit besonderer Wirkung gegen Grauschimmel (*Botrytis cinerea* Pers.). Pflanzenschutz Nachrichten Bayer **21** (1968): 147–170.

HARTLEY, D., and KIDD, H. (Eds.): The Agrochemicals Handbook. 2nd Ed., Nottingham 1987.

HASSALL, K. A.: The Chemistry of Pesticides. MacMillan Press Ltd., London–Basingstoke 1982.

HICKEY, K. D.: Fungicide Benefits Assessment. – Fruits and Nuts, – East. National Agricultural Impact Assessment Program. 1991.

HILL, T. L., and STRATTON, G. W.: Interactive effects of the fungicide chlorothalonil and the herbicide metribuzin towards the fungal pathogen *Alternaria solani*. Bull. Environ. Contam. Toxicol. **47** (1991): 97–103.

HUGGENBERGER, F., FARRANT, D. M. and BACCI, L.: Combined formulations of nuarimol with captan, chlorothalonil or mancozeb for the control of powdery mildew and scab on apples. Proc. 1986 British Crop. Prot. Conf. – Pests and Diseases, Vol. 1. (1986): 299–306.

HULPKE, H., STEIGH, R., and WILMES, R.: Ligh-induced transformation of pesticides on silica gel as a model system for photodegradation on soil. In: MATSUNAKA, S., HUTSON, D. H., and MURPHY, S.D. (Eds.): Pesticide Chemistry. Human Welfare and the Environment. Vol. 3. Pergamon Press, Oxford 1983, p. 323–332.

JOHNSTON, S. A.: Fungicide Benefits Assessment. – Vegetables, – East. National Agricultural Impact Assessment Program. 1991.

KASPERS, H.: Über Anwendungsmöglichkeiten von Euparen zur Bekämpfung von Kelch- und Lagerfäulen. Pflanzenschutz Nachrichten Bayer **21** (1968): 243–256.

KATAYAMA, A., ISEMURA, H., and KUWATSUKA, S.: Suppression of chlorothalonil dissipation in soil by repeated application. J. Pestic. Sci. **16** (1991): 233–238.

KIDD, H., HARTKEY, D., KENNEDY, J. M., and JAMES, D. R.: European Directory of Agrochemical Products. Vol. 1. Fungicides. 3rd Ed., Nottingham 1988.

KOLBE, W.: Untersuchungen über Euparen M als Fungizid im Kernobstbau unter besonderer Berücksichtigung der Spinnmilbenwirkung und Sortenverträglichkeit. Pflanzenschutz Nachrichten Bayer **25** (1972): 123–162.

– Untersuchungen über die *Botrytis*-Bekämpfung mit Euparen im Erdbeerbau (1969–1981). Pflanzenschutz Nachrichten Bayer **35** (1982): 104–119.

– and KASPERS, H.: Die Bekämpfung pilzlicher Krankheiten im Hopfenbau mit organischen Fungiziden unter Berücksichtigung der akariziden Nebenwirkung. Pflanzenschutz Nachrichten Bayer **21** (1968): 178–303.

KUBIAK, R., and EICHHORN, K. W.: Lysimeter studies under practical field conditions. – Residues in vines and wine after repeated application of ^{14}C-dichlofluanid. 7th Internat. Congr. Pestic. Chem., Hamburg 1990. Abstr. Vol. 3. (1990): 321.

LEROUX, P., and GREDT, M.: Cellular microtubules: targets for the fungicides pencycuron and zarilamide. 7th Intern. Congr. Pestic. Chem., Hamburg 1990. Abstr. Vol. 1. (1990): 289.

LUKENS, R. J.: Heterocyclic nitrogen compounds. In: TORGESON, D. C. (Ed.): Fungicides. Vol. 2. Academic Press, New York – London 1969, pp. 396 – 446.
— Chemistry of Fungicidal Action. Chapman and Hall, London 1971.
— and SISLER, H. D.: Chemical reactions involved in the fungitoxicity of captan. Phytopathology **48** (1958): 235 – 244.
MALCOM, A. J., and BLUETT, D. J.: The performance of anilazine in mixture with triazole fungicides against late diseases of winter wheat in the United Kingdoom. Proc. 1986 Brit. Crop Prot. Conf. — Pests and Diseases Vol. 1. (1986): 79 – 85.
MASUI, M., JOSHIOCA, N., HARRIS, R. J., ROOS, H., and CORNIER, A.: DF 125 — a new broad spectrum fungicide. Proc. 1986 Brit. Crop Prot. Conf. — Pests and Diseases Vol. 1. (1986): 63 – 70.
MATOLCSY, GY., NADASY, M., and ADRISKA, V.: Pesticide Chemistry. Akademiai Kiado, Budapest 1988, p. 451.
MILLER, P.: Mechanisms for reaching the size of action. In: TORGESON, D. C. (Ed.): Fungicides. Vol. 2. Academic Press, New York – London 1969, pp. 2 – 60.
NICHOLSON, J. F., MEYER, W. A., SINCLAIR, J. B. and BULLER, J. D.: Turf isolates of *Sclerotinia homeocarpa* tolerant to dyrene. Phytopathol. Z. **72** (1971): 169 – 172.
O'REILLY, P., DOWNES, M. J., and BANNON, E.: Earl detection and control of *Septoria nodorum* in winter wheat. Proc. 1986 Brit. Crop Prot. Conf. — Pests and Diseases Vol. 3. (1986): 1035 – 1039.
PERKOW, W.: Wirksubstanzen der Pflanzenschutz- und Schlädlingsbekämpfungsmittel. Paul Parey, Berlin – Hamburg 1983.
ROUCHAUD, J., GUSTIN, F., CREEMERS, P., GOFFRINGS, G., and HERREGODS, M.: Fate of the fungicide tolylfluanid in the pear cold stored in controlled or non-controlled atmosphere. Bull. Environ. Contam. Toxicol. **46** (1991): 499 – 506.
SANDERS, P. L.: Fungicide Resistance in the United States. National Agricultural Pesticide Impact Assessment Program. 1991.
SATO, K., and MAKI, S.: Fate and behavior of the fungicide guazatine in the environment. J. Pestic. Sci. **14** (1989): 383 – 393.
SCHACHNAI, A., and BARASH, I.: Evaluation of the fungicides CGA 64251, guazatine, sodium o-phenylphenate and imazalil for control of sour rot of lemon fruit. Plant Dis. **66** (1982): 733 – 735.
SIEGEL, M.: Reactions of the fungicide folpet (N'-trichloromethylthiophthalimide) with a non-thiol protein. Pestic. Biochem. Physiol. **1** (1971): 234 – 240.
SOLEL, Z., and SIEGEL, M. R.: Effect of the fungicide guazatine and dodine on growth and metabolism of *Ustilago maydis*. Z. Pflanzenkrankh. Pflanzenschutz **91** (1984): 273 – 285.
SZKOLNIK, M., and GILPATRICK, J. D.: Apparent resistance of *Venturia inaequalis* to dodine in New York apple orchards. Plant Dis. Rep. **53** (1969): 861 – 865.
UEYAMA, I., ARAKI, Y., KUROGOCHI, S., YONEYAMA, K., and YAMAGUCHI, I.: Mode of action of phenylurea fungicide pencycuron in *Rhizoctonia solani*. 7th Intern. Congr. Pestic. Chem., Hamburg 1990. Abstr. Vol. 1. (1990): 290.
VONK, J. W.: Metabolism of Fungicides in Plants. In: HUTSON, D. H., and ROBERTS, T. R. (Eds.): Pesticide Biochemistry and Toxicology. Vol. 3. John Wiley and Sons, Chichester – New York 1983, pp. 111 – 162.
WOLF, C. N., SCHULDT. P. H., and BALDWIN, M. M.: s-triazine derivatives — a new class of fungicides. Science **121** (1955): 61 – 62.
WORTHING, C. R.: The Pesticide Manual. 6th Edit. Brit. Crop Protect. Council., Croydon 1979.
— The Pesticide Manual. 7th Edit. Brit. Crop Protect. Council., Croydon 1984.

19.2 Triphenyl-tin-Derivatives

H. LYR

Although organic-tin-compounds are a rather old group of fungicides, many have been used up to recent years, partly as unselective biocides. These were mainly alkyl derivatives, such as tributyl-tin-oxide (TBTO) (see chapter 22). For plant disease control only triphenyl-tin derivatives have been used, because of their much higher selectivity. Main compounds are triphenyl-tin-acetate (TPTA), and triphenylhydroxide (TPTH), or of minor importance TPT-Chloride. TPTA and TPTCl hydrolyze more or less rapidly to the corresponding hydroxide (TPTH) (Fig. 19.2). TPTA is 2 times more water soluble than TPTH (4.3 ppm at 20° C), and is for this reason less toxic and easier to formulate as WP. In contrast to other organic tins, triphenyl-tin compounds are rather selective in their action against fungal species for unknown reasons.

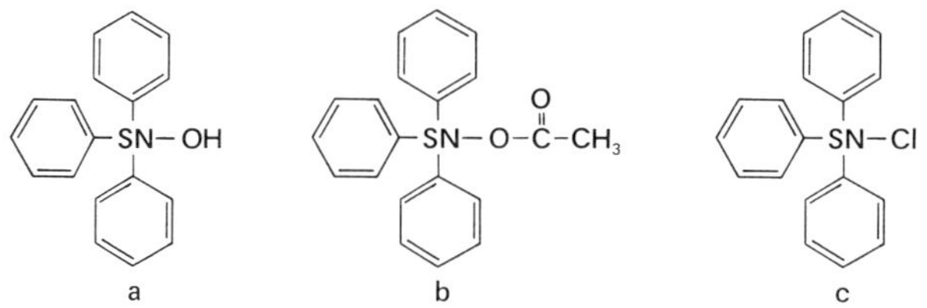

Fig. 19.2 Chemical structures of fentin fungicides. a) fentin hydroxide, b) fentin acetate, c) fentin-chloride.

Their main fields of application are: control of *Phytophthora infestans* and *Alternaria* in potatoes; *Ramularia* spp. in celery; *Cercospora beticola* in sugar beets; *Pyricularia oryzae* in rice, (as side effect they control algae and snails in paddy rice); *Pseudoperonospora* and *Alternaria* in garlic and onions; *Pseudoperonospora* and *Phytophthora* in hops.

The products have been registered in more than 26 countries and are difficult to replace because of fungal resistance or other problems involving competing compounds. They are also used in cotton, cacao, pepper, tobacco, groundnuts, beans, eggplants, tomatoes, strawberries, poplars, date palms, manioc and pecan nuts.

In spite of their high acute toxicity (LD 50 female rats 140 – 298 mg TPTA/kg), several favourable properties kept them on the market.

These include rapid degradation and adsorption in the upper layers of the soil, and therefore low risk of water contamination, low risk for resistance, low amounts needed for fungal control, low volatility, degradation to non toxic compounds by sunlight.

Because of phytotoxicity in younger leaves (WOOD 1984), triphenyltins are often formulated in mixtures with maneb (Brestan 10, 20, 60) or mancoceb, which decrease the phytotoxicity like safeners for unknown reasons, and additionally enlarge the spectrum of activity.

In sugar beets some mixtures with triazoles are in use. TPTH and TPTA are incompatible with organic solvents (EC-formulations) due to phytotoxicity.

TPTA and TPTH are toxic for fish, but not for bees. Antifeeding or repellent effects have been observed in some insects (*Leptinotarsa decemlineata, Spodoptera littoralis, Heliothis, Ascotis* and others (in soybeans) (WORTHING 1979).

Spore germination of *Cercospora beticola* is inhibited at 1 ppm. This may be also the mechanism for the protection of potato tubers against *Phytophthora infestans* blight by late applications before harvest. The formulations have a good protective effect up to 14 days and a good rain stability.

In Greece a loss of activity of TPTA/H for control of *Cercospora beticola* in sugar beets was observed by CHRYSAI-TOKOUSBALIDES in 1980 after long term use of these compounds. There exists no cross resistance to benzimidazoles. BMC (carbendazim) resistant strains were controlled by TPTH. In mixtures BMC resistance develops selectively.

Although many organic tins have been intensively investigated in toxicological and biochemical respects, most publications describe actions of alkyl-tin compounds mainly for *in vitro* tests with isolated mitochondria. Regarding the mechanism of action of TPTH or TPTA the following general effects have been observed:
- inhibition of oxidative phosphorylation in mitochondria (VAN DER KERK 1971; SELWYN 1976),
- swelling of mitochondria (STOCKDALE et al. 1970; ALDRIDGE et al. 1981),
- lytic actions at higher concentrations (STOCKDALE et al. 1970),
- increase of oxygen consumption in potato leaves (BAUMANN 1958),
- inhibition of glutathion-transferase activity of *Phytophthora infestans* (POLTER unpubl.),
- induction of lipid peroxidation (SEIDEL 1991 unpubl.).

An interaction of TPT-compounds with hydrophobic parts of membranes does occur (SELWYN et al. 1970). In isolated mitochondria an uncoupling of oxidative phosphorylation similar, but not identical with that of oligomycin, was observed. However, these compounds can be considered as ,,energy transfer inhibitors. This can be connected to the swelling of the organelles. ATP is hydrolysed, therefore ATP-ases may be directly, or more probably indirectly impaired. HAUER et al. (1981) observed that the action of TPTCl in isolated mitochondria and intact cells of *Rhodotorula gracilis* is different. Oxygen consumption and the ATP level decreased only in isolated mitochondria, but not in intact cells. They assume, that a plasma membrane ATP-ase (P-type) is the target responsible for H+/K+ exchange and Ca-transport. Whether a direct interaction with an ATP-ase does occur, or only an interaction with hydrophobic phospholipids in the surrounding, which impairs the enzyme, remains an open question. No direct interaction of TPTH with thiol groups seems to occur.

Differences in phytotoxicity in various plants under various conditions probably depend on the cuticula penetrability of the compounds (PAQUET and WILKIN 1971; BOCK 1981).

SEIDEL et al. (1993) found in potato cell cultures a correlation of growth inhibition and induction of a lipid peroxidation by TPTA and TPTH. Maneb and zineb counteracted both effects like typical safeners. This was also observed with some other compounds (SEIDEL 1993).

Anilino-Pyrimidines

This new class of fungicides (KRAUSE et al. 1984) is represented by three selected compounds. These are "Andoprim" (Fig. 19.3a), "Mepanipyrim" (Fig. 19.3b), and "Pyrimethanil" (Fig. 19.3c). Their mechanism of action is different from that of other groups of fungicides. All compounds are hydrophobic and have a low toxicity for mammals ($LC_{50} > 5000$ mg/kg oral application in rats).

Andoprim

This compound was developed by VEB Fahlberg-List Magdeburg (former GDR). Its main and most interesting activity is directed toward *Phytophthora* species and other

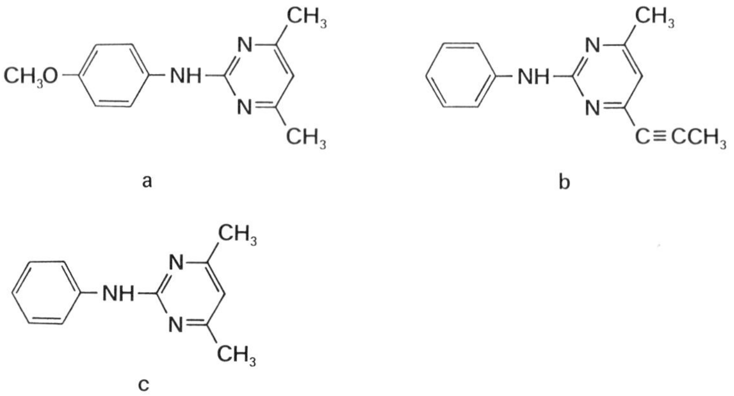

Fig. 19.3 Chemical structures of anilinopyrimidine fungicides. a) Andoprim, b) Menapyrim, c) Pyrimethanil.

Table 19.1 Spectrum of activity of andoprim in a malt agar disk test. Inhibition of radial growth of the mycelia is expressed as % of controls. (Fungicide concentrations in mg/l)

Fungi	Concentration		
	1	3	10
Phytophthora infestans	8	48	100
Ph. cactorum	–	57	80
Ph. cinnamoni	–	90	100
Ph. citricola	50	100	100
Ph. palmivora	40	88	100
Ph. parasitica	45	97	100
Ph. megasperma	25	58	80
Pythium irregulare	–	–	33
Rhizoctonia solani	1	6	16
Pyricularia oryzae	0	0	9

Peronosporales among the *Oomycetes*. A selected spectrum of activity (ZOLLFRANK and LYR 1988) is shown in Table 19.1.

Andoprim exhibits a relatively high and rather selective activity for *Phytophthora* species (Tab. 19.1) and also for *Peronosporales*, such as *Plasmopara, Pseudoperonospora* a. o. *Phytophthora infestans* was completely controlled on tomatos in the greenhouse by 200 ppm. Within the genus *Phytophthora* some differences in the sensitivity to this compound do exist. *Pythium* was surprisingly insensitive.

Andoprim shares this property with dimethomorph and ICI A 0001, in contrast to the activity of phenylamides.

Other fungi are not sensitive, but at higher doses some activity was detected in host/parasite combinations for *Pellicularia sasakii* and a few other fungal species. *Cercospora beticola*, *Sclerotinia* and *Penicillium digitatum* are not inhibited.

Andoprim acts fungistatically, and mainly as a protectant with a weak curative effect. Sometimes it has phytotoxic side effects.

Andoprim inhibits spore germination of *Phytophthora infestand* to 50% at 1 mg/l, metalaxyl for comparison only at > 20 mg/l. No resistant or tolerant *Phytophthora* strains could be selected.

The mechanism of action appears to be different to that of other fungicides. No effects on respiration, lipid-, DNA-, RNA- or protein biosynthesis were detected. Only an inhibition of uptake of amino acids and precursors of DNA and RNA could be measured. The causes of this effect remain to be elucidated (GRUNWALD et al. 1990). In *Phythophtora infestans* a significant reduction of the glutathione level was observed (ELLNER 1991).

For unknown reasons andoprim failed to adequately control *Phytophthora infestans* on potato in the field and therefore did not reach the market.

Mepanipyrim

This compound (KIF 3535) (Fig. 19.3 b) was developed by Kumai Chem. Industry Co., Japan. The properties are described by MASUDA and NAGATA (1990). The structure is similar to that of andoprim, but the pyrimidine ring bears a propionyl group. This changes the spectrum of activity dramatically (Tab. 19.2).

Mepanipyrim has a high activity in greenhouse tests against grey mould (*Botrytis cinerea*) and apple scab (*Venturia inaequalis*). It controls benzimidazole and/or dicarboximide resistant strains of grey mould in cucumber plants and has a good effect in strawberries and aubergines. In apples, 200 mg a.i./l achieved excellent control of leaf and fruit scab. Good efficacy was also described for the control of *Alternaria alternata* in pear and apple, of brown rot (*Monilinia* spp.) and of black spot of roses (*Diplocarpon rosae*) at rates of 167–500 mg a.i./l.

According to MIURA et al. 1993 menapyrim inhibits the uptake of various substrates at 10–100 ppm in *Botrytis cinerea*. At 1 ppm the secretion of host-cell wall lytic enzymes, such as cutinase, pectinase and cellulase, is impaired leading to their intracellular accumulation. This is assumed to inhibit the infection process by the parasite and by this menapyrim acts as antifungal agent. The mechanism of action of menapyrim and pyrimethanil seems to be identical and unique among other fungicides.

Table 19.2 Spectrum of activity of mepanipyrim in an *in vitro* test (LD_{50} values) according to MASUDA and NAGATA 1990. (× means poor, + good, ++ excellent activity in glass house tests)

Fungi	LC_{50} concentration (mg/l)	Effect *in vivo*
Phytophthora infestans	>300	×
Pythium debaryanum	>300	×
Cochliobolus miyabeanus	<0.1	
Erysiphe graminis	–	+
Sclerotinia sclerotiorum	7	+
Venturia inaequalis	<0.1	++
Alternaria alternata	<0.1	+
Botrytis cinerea	<0.1	++
Cercospora beticola	>300	×
Cladosporium fulvum	>300	×
Fusarium oxysporum	>300	×
Monilinia fructicola	–	++
Penicillium digitatum	0.7	×
Pyricularia oryzae	>300	×
Rhizoctonia solani	>300	×

Pyrimethanil

Pyrimethanil (N-(4,6-dimethylpyrimidin-2-yl) aniline) (Fig. 19.3c) is a new fungicide within the group of anilinopyrimidines. It is being developed by Schering AG as a foliar fungicide to control grey mould (*Botrytis cinerea*) on vine, fruits and vegetables. It is highly active against all strains of this fungus, and has no cross resistance relationship to other known botryticides. Pyrimethanil is of low toxicity (rat oral LC_{50} about 5000 mg/kg). The solubility in water is 0.121 g/l. The log P value is 2.48 (octanol/water). It will be marketed under various trade names including "Scala".

Pyrimethanil is currently available as a flowable formulation (SC 300 and 400 g/l) and a waterdispersable granulates (WDG. 80%). Biological properties: The fungicide exhibits protectant and curative control, but no significant systemic activity by leaf to leaf transfer. However, pyrimethanil shows a high level of systemicity when applied to the root system. A vapour phase activity has also been observed, by which unsprayed tomato plants were protected by sprayed plants standing nearby. The compound also controls strains resistant to dicarboximides, benzimidazole related fungicides, and diethofencarb. Strains resistant to pyrimethanil have thus far not been isolated or selected. The spectrum of activity for some fungi is shown in table 19.3.

Table 19.3 Spectrum of activity of pyrimethanil on malt agar, measured as inhibition of radial growth (in %) related to controls (LYR unpubl. results).

Fungi	Concentrations in mg/l			
	0.1	1	10	30
Septoria nodorum	7	70	99	100
Rhizoctonia solani	3	6	40	73
Ustilago maydis	–	5	9	25
Sclerotinia sclerotiorum	24	100	100	100
Botrytis cinerea	20	90	100	100
Drechslera teres	61	82	88	89
Alternaria solani	11	76	86	88
Pullularia pullulans	–	0	4	17
Ascochyta fabae	0	47	93	93
Phytophthora megasperma	–	0	50	72

Pyrimethanil has no effect on wine fermentation. It has excellent efficacy against *Botrytis cinerea* in vine, strawberries, tomatoes, onions and flower bulbs, as well as in peas, beans, cucumbers, aubergines and in ornamentals, such as *Erica, Rhododendron, Begonia, Cyclamen*.

It is a valuable partner in modern *Botrytis* antiresistance management programs (NEUMANN, WINTER and PITTIS 1992). The mechanism of action seems to be different from that of other known fungicides. It inhibits germ tube extension and decreases the number of host epidermal cells killed at individual penetration sites.

It has been shown neither to inhibit respiration nor to act as an uncoupler of oxidative phosphorylation. It does not cause lipid peroxidation or affect the integrity or osmotic stability of fungal cells. There is not effect on the uptake of lysine, uridine, thymidine, acetate, glucoseamine, or glucose, and on the biosynthesis of proteins, RNA, DNA and chitin. Preliminary tests showed that ergosterol biosynthesis is not inhibited. It is not an inhibitor of adenosine deaminase.

Pyrimethanil inhibits the secretion of extracellular proteins, including fungal hydrolases associated with pathogenesis (MILLING et al. 1994). Own experiments revealed that the formation of the toxic enzyme glucoseoxidase is suppressed (in prepar). This may explain its observed effect on the ability of *Botrytis* ssp. to kill host cells during the early stages of infection. The exact mechanism of action needs further investigations.

Cyprodinil

(CGA 219417) (N-(4-cyclopropyl-6-methyl-pyrimidin-lyl)aniline) is a new member of the anilinopyrimidine fungicides. Structurally related to menapyrim, the cyclopropyl side chain instead of the propionyl group enlarges the spectrum of activity. Besides *Botrytis* and *Venturia*, *Pseudocercosporella*, *Erisyphe*, *Septoria* and *Rhynchosporium* can be controlled. The mechanism of action is an inhibition of aminoacid synthesis (KÜHL and RAUM 1994).

Phenylpyrroles

This new group of fungicides is derived from the antifungal compound pyrrolnitrin produced by *Pseudomonas pyrocinia* and was developed by CIBA-GEIGY for commercial use through various substitution changes of the natural compound (Fig. 19.4).

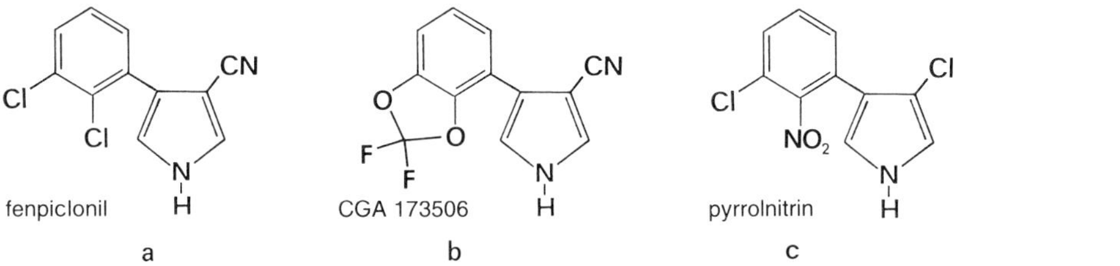

Fig. 19.4 Chemical structures of phenylpyrrol fungicides. Fenpiclonil (BERET), Fludioxonil = CGA 173506 (SAPHIRE), Pyrrolnitril (natural compound).

Fenpiclonil

Two highly active and partly selective compounds have reached the market. One of the first to be developed was fenpiclonil (Fig. 19.4a) (trade name "Beret"), which is used mainly as a seed dressing agent because of its light sensitivity (NEVILL et al. 1988; NYFELER et al. 1990). The high activity of fenpiclonil makes it valuable for the control of some important *Fusarium* species which are hard to control, if benzimidazole resistance problems have evolved (KOCH et al. 1992). The compound has a low water solubility (2 ppm) and a very low acute oral toxicity for rats (> 5000 mg/kg).

There is no fungal cross resistance between Beret and most other fungicides. Its combinations are well tolerated by plants, and they do not delay emergence or development of seedlings, as DMIs can do by inhibition of gibberellic acid biosynthesis. Beret is used in cotton, oilseed rape, peas, and potato tubers in various combinations. Table 19.4 shows activities for some fungal species in an *in vitro* test.

Table 19.4 Inhibition of fungal radial growth by fenpiclonil (% of controls). Concentrations in malt agar dishes in mg/l (LYR unpubl.)

Fungi	Concentrations (mg/l)				
	100	30	10	1	0.1
Alternari solani	98	89	77	45	30
Fusarium sulfureum	100	99	98	70	18
Fusarium oxysporum	60	58	25	4	0
Gerlachia nivalis	100	100	98	75	14
Drechslera teres	100	100	99	77	46
Septoria nodorum	92	91	82	41	21
Rhizoctonia solani	99	94	87	39	9
Pythium ultimum	80	78	43	10	0
Phytophthora nicotianae	50	49	44	8	0
Ustilago maydis	59	50	41	16	0
Sclerotinia sclerotiorum	100	100	100	100	90

For seed dressing in wheat, rye and barley, fenpiclonil is usually combined with other compounds, such as difenoconazole, imazalil and metsulfovax (carboxanilide) in order to control a broad spectrum of pathogens including *Gerlachia nivalis, Tilletia caries, Tilletia controversa, Urocystis occulata, Drechslera teres* a.o. (BRANDL and HAGMEYER 1990). Fenpiclonil protects stored potato tubers against *Fusarium sulfureum*, which is the most destructive pathogen during storage. Fenpiclonil, applied at 20–25 g/t directly after harvesting, has excellent efficacy including MBC resistant strains (STACHEWITCZ et al. 1990; LEADBEATER and KIRK 1992). Fenpiclonil, applied at 40–50 g/t in spring before planting, gave the same or better efficacy then pencycuron, validamycin or tolclofos-methyl for control of *Rhizoctonia solani, Helminthosporium solani, Polyscytalum pustulans* and *Colletotrichum coccodes*. In combinations with imazalil side effects for *Phoma exigua* var *foveata* were observed (STACHEWITCZ et al. 1990).

Fludioxonil

Further progress was made with the development of compound fludioxonil (CGA 173506) ("Saphire") (Fig. 19.4b). Substitutions in the phenyl ring make it more light stable and it can be used as a foliar fungicide against pathogenic genera, such as *Botrytis* and *Monilinia*. It is active at low rates for seed treatment against *Fusarium, Septoria, Tilletia* and *Helminthosporium*, or in rice against *Gibberella*. No cross resistance between fludioxonil and benzimidazoles or guanidines was found (GEHMANN et al. 1990). As a seed dressing agent at 5 g a.i./100 kg seeds it controls *Gerlachia nivalis, Fusarium culmorum, F. graminearum, Tilletia caries* and seed born *Septoria nodorum*. At 25 g a.i./100 kg seeds it is effective against *Gibberella fujikuroi* in rice (LEADBEATER et al. 1990; KOCH et al. 1992). However, FARETRA and POLLASTRO (1992) obtained resistant mutants of *Botrytis cinerea (Botryotinia fuckeliana)* in the laboratory with various degrees of resistance to CGA 173506. The R-strains were cross resistant to dicarboxamides and most of them were hypersensitive to high osmotic pressure. This confirms the results of LEROUX (1991), who observed strong cross resistance also in *Penicillium expansum*, and a moderate cross resistance in *Pseudocercosporella herpotrichoides*. These facts point to a related machanism of resistance to the dicarboximides. HILBER et al.

(1992) stated, that in spite of an easy selection of R-strains in the laboratory, R-strains have so far not yet been isolated from the field. Cross (or double) resistant strains of *Botrytis* for CGA 173506 and vinclozolin showed decreased fitness and pathogenicity. The R-genes for both compounds do not segregate independently, which means, that they are either identical or closely linked. This was studied in detail by HILBER (1992). Therefore it will be crucial to develop a reliable antiresistance strategy for this novel compound for the efficient control of *Botrytis*.

JESPERS and DE WAARD (1992) used *Fusarium sulfureum*, a very sensitive species, for investigations of the mechanism of action phenylpyrroles.

Fenpiclonil is readily taken up by the fungus, and at low concentrations does not affect oxygen consumption, nuclear division, DNA-, RNA-, protein-, chitin-, ergosterol- and phospholipid synthesis. However, accumulation of amino acids, such as glycine, phenylalanine, and aspartic acid, or of 2-desoxyglucose is instantaneously affected at concentration of 4.2 – 42 µM. The fungicide inhibits accumulation and incorporation of glucose and mannose into hyphal wall glycans. The transport associated phosphorylation of glucose seems to be the site of action of fenpiclonil (JESPERS 1994). This suggests that membrane bound transport processes are impaired. The ED_{50} value for the strain used, is about 10 µM. The background of this effect needs further investigations.

Strobilurins

This is a new class of fungicides (also entitled β-methoxyacrylates (MOA), and includes compounds which are currently being developed by BASF AG, ICI-Agrochemical and others (AMMERMANN et al. 1992; GODWIN et al. 1992).

Their origin is a natural antifungal compound, strobilurin A, exudated by the Basidiomycete *Strobilurus tenacellus*, which grows in the forest on dead cones of *Pinus sylvestris* (ANKE et al. 1977). Strobilurins are chemically closely related to oudemansins (Fig. 19.5) produced by the fungus *Oudemansiella mucida* (ANKE et al. 1979).

The natural compounds are decomposed by light, and therefore not active enough as plant protection agents. Chemical derivatisation resulted in new insights regarding the structure/activity relationships, and some active and more stable compounds were synthesised. Some of them (strobilurin E) also exhibit cytostatic and antiviral effects (ANKE and STEGLICH 1989). Similar approaches have been performed by BEAUTEMENT et al. (1991) and by CLOUGH et al. (1992 and 1993).
Biological data:
Strobilurins are selective, but broad spectrum fungicides. Examples of their activity *in vitro* are given in table 19.5 (according to AMMERMANN et al. 1992).

The spectrum of activity shows quite unexpected differences even within the same genera (i. e. *Phytophthora*, *Sclerotinia*), and is different from that of other fungicides.

Glass house tests indicated a strong curative effect against *Venturia inaequalis*, but systemic movement by root application in rice could not be observed with BAS 490 F. The main fields of practical applications could be apple scab, powdery and downy mildew in vine, rice blast and sheath blight, several important pathogens in cereals, such as mildew, rusts, *Septoria nodorum, Pyrenophora teres* a. o., *Cercospora beticola* and *Erysiphe betae* in sugar beets, *Ph. infestans* and *Alternaria solani* in potatoes. Similar effects were obtained with ICI 5504 (GODWIN et al. 1992) (Fig. 19.5). Here even a systemic effect is described. Both compounds have a low acute oral toxicity in rats (> 5 000 mg/kg).

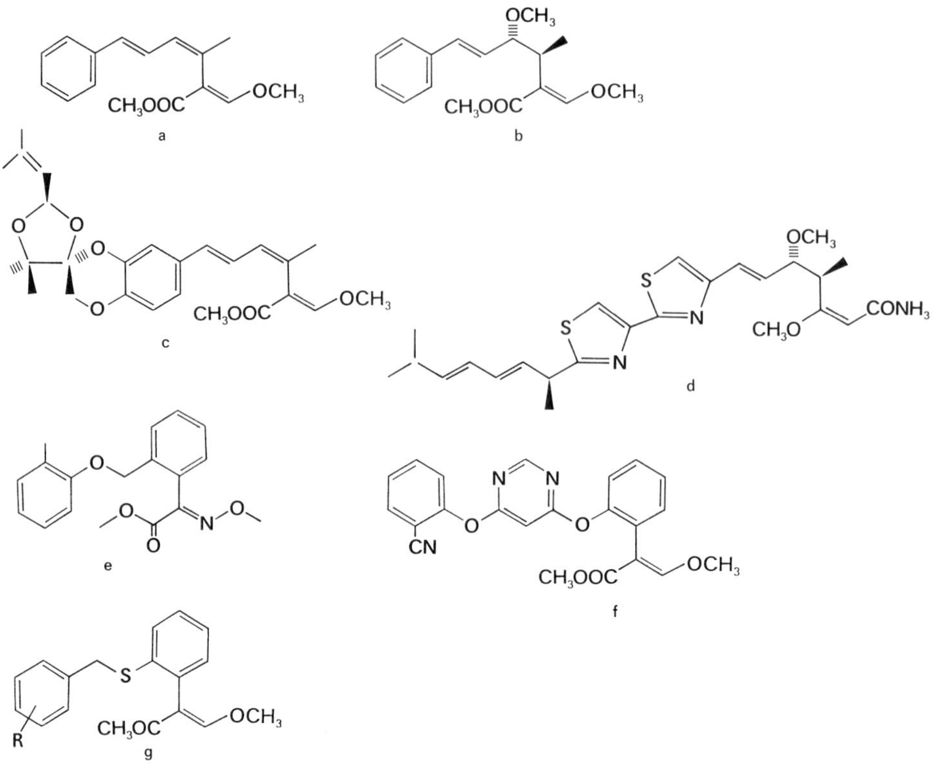

Fig. 19.5 Chemical structures of strobilurines (β-methoxyacrylates). a) Strobilurin A, b) Oudemansin A, c) Strobilurin E, d) Myxothiazol, e) BAS 490 F, f) ICI A 5504, g) compounds from SCHERING AG.

Table 19.5 *In vitro* activity of BAS 490 F against various fungi on agar plates (LC 50 values a.i. in mg/l)

Alternari solani	<1	*Penicillium digitatum*	<1
Aspergillus niger	>500	*Penicillium expansum*	<1
Botrytis cinerea	>500	*Phaeosphaeria nodorum*	<1
Chaetomium globosum	<50	*Phytophthora cactorum*	>500
Cochliobolus sativus	<50	*Phytophthora infestans*	<1
Coniophora puteana	<1	*Pythium ultimum*	>500
Corticium rolfsii	<1	*Rhizopus stolonifer*	<10
Fusarium culmorum	<50	*Sclerotinia fructigena*	>500
Leptosphaeria maculans	<10	*Sclerotinia sclerotiorum*	<1
Leptosphaeria salvanii	<1	*Trichoderma viride*	<10
Macrophomina phaseoli	>500	*Venturia inaequalis*	<1
Mucor circinelloides	>500	*Verticillium dahliae*	<10
Nectria galligena	>500	*Serpula himantoides*	<1

The mechanism of action of strobilurins is well investigated, because of their general biochemical interest. They inhibit a new target within the mitochondrial respiratory chain at site III (BECKER *et al.* 1981).

Several fungicides are already known to be inhibitors of the respiratory chain in fungi. These include carboxins, dexone, rotenone, piercidin A, antimycin A. Their sites of inhibition are indicated in Fig. 9.1. The binding sites are only partly identified.

Complex III (cytochrome bc complex) is very similar and essential in most organisms. Bacteria, such as *Paracoccus denitrificans* or the autotrophic *Rhodobacter* have only 3 peptides in the bc-complex (cytochrome b, cytochrome c_1, and an iron-sulfur-protein). Fungi have in addition 6 other subunits consisting of small peptides without catalytic properties and of as yet unknown function (perhaps necessary for orientation of the complex in the membrane). Mammals have 11 subunits (v. JAGOW 1993). The use of these new fungicides has helped to elucidate in more detail the electron flow from ubiquinone to cytochrome c via the bc-complex. Its high efficiency in energy conservation can now be better explained, because two routes of electrons are possible (Fig. 19.6). One branch represents a linear chain (upper part in Fig. 19.6), the other branch allows a cyclic electron flow, which makes feed back reactions possible. The consequence of the oxidoreduction shuttle of ubiquinone is a chemiosmotic potential at the mitochondrial membrane, as was postulated by MITCHELL in 1975. This serves as the driving force for the ATP synthesis (Fig. 19.7). Dissipation of this potential (perhaps by fluazinone or fentin-compounds) has also fungicidal effects, because the ATP synthesis is strongly impaired. Strobilurins (myxathiazol, stigmetallin, oudemansins, strobilurin, or MOA stilbene) are inhibitors at stite III, but in contrast to antimycin A, they bind at the ubiqinone oxidizing (outer side), near the iron-sulfur centre, which itself seems not to be involved in the binding. They hinder the transmission of electrons from ubihydroquinole to the Fe—S-centre without hindering the binding of ubiquinole directly. Probably, a conformational distortion of cytochrome b exhibits such a displacement of ubiquinone that the transfer

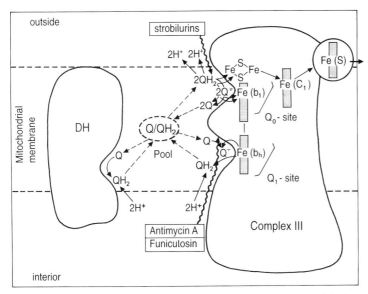

Fig. 19.6 Scheme of the electron flow through the succinodehydrogenase (DH), the ubiquinol/ubiquinone pool (Q/QH$_2$), the bc-complex III, and the sites of interaction of strobilurins, antimycin A and funiculosin. Interior = interior of the mitochondrion, outside = Cytoplasma, Fe—S = iron-sulfur-proteins. (v. JAGOW 1989).

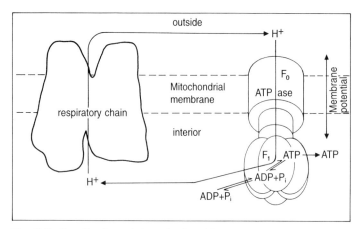

Fig. 19.7 Localisation of the mitochondrial ATP-synthetase (= ATP-ase) within the mitochondrial membrane and the proton driven ATP synthesis from ADP and inorganic phosphate (P_i) via the respiratory chain. (v. JAGOW 1989).

of electrons to the Fe-S-centre is not possible (BRANDT et al. 1988). The exact mechanisms of binding of the fungicides of this type need further investigations. Point mutations can lead to insensitivity of these fungicides in yeast.

In chloroplasts of higher plants MOA-stilbene is an effector of the cytochrome bf-complex at the quinone reduction site. This is in contrast to mitochondria, where the block is at the quinol oxidation site (RICH et al. 1991). The selectivity seems not to be based on differences in mitochondrial binding sites, but on differences of penetration and degradation in various organisms (KÖHLE et al. 1994).

These results demonstrate again that selectivity and low mammalian toxicity can be combined with good control of some fungi.

Fluazinam

Fluazinam is a fungicide type, detected by ISHIHARA SANGYO KAISHA and developed by ICI Agrochemicals. It belongs to the class of phenylpyridinanines (Fig. 19.8). Fluazinam has a selective, but rather broad spectrum of activity against genera, such as *Botrytis, Sclerotinia, Alternaria*, but also for *Phytophthora infestans* in potatoes. For control of P.i. it is used as an EC (500 g a.i./ha). At spraying intervals of 10–14 days even 200–150 g a.i./ha had a long lasting effect, and controlled leaf and tuber blight. It acts protectively, but not curatively or systemically. Some examples for the spectrum of activity are given in table 19.6.

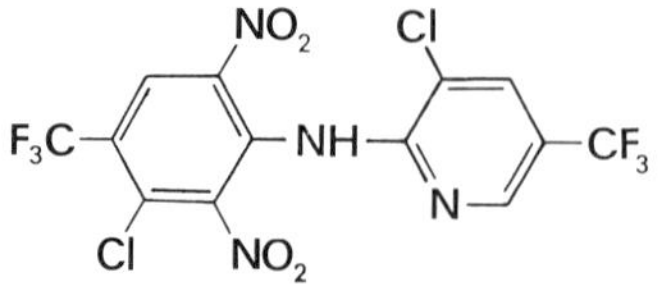

Fig. 19.8 Chemical structure of the fungicide fluazinam.

Table 19.6 Protective activity of fluazinam against several important plant diseases (pot tests) (ANEMA and BOUWMAN 1992)

Disease	MIC (mg/l)
Botrytis cinerea / cucumber	4
Pseudoperonospora cubensis / cucumber	8
Colletotrichum lagenarium / cucumber	8
Venturia inaequalis / apple	8
Sclerotinia sclerotiorum / cucumber	16
Pyricularia oryzae / rice	16
Phytophthora infestans / tomato	16
Thanatephorus cucumerinus / rice	31
Mycosphaerella melonis / cucumber	63
Alternari mali / apple	63
Sphaerotheca fuliginea / cumber	250
Puccinia coronata / oat	250
Gymnosporangium asiaticum / jap. pear	1 000

The mechanism of action is said to be an uncoupling of oxidative phosphorylation, and is distinct from that of other fungicides (HEMMEN and KONRADT 1992). Fluazinam is an extremely potent uncoupler in rat liver mitochondria, but is inactivated by a glutathione transferase at the mitochondrial level, which seems to be related to the Cl atom in the phenyl ring. Lowering of the glutathione level increased its stability (GUO et al. 1991). The uncoupling efficiency was not dependent on the hydrophobicity of the compound, but seems to be governed by the pK value of the deprotonatable secondary amino group. The mechanism of action seems to be a simple protonophoric cycle involving protonation/deprotonation and is primarily governed by the limiting total concentration of the uncoupler species (BRANDT et al. 1992) (Fig. 19.7). (compare chapter 9, Fig. 9.1).

References

ALDRIDGE, W. N., STREET, B. W., and NOLTES, J. G.: The action of 5-coordinate triorganotin compounds on rat liver mitochondria. Chem.-Biol. Interact. **34** (1981): 223–232.

AMMERMANN, E., LORENZ, G., and SCHELBERGER, K.: BAS 490 F – A broadspectrum fungicide with a new mode of action. Brighton Crop Protect. Conf. **5–2** (1992): 403–410.

ANEMA, P. E., and BOUWMAN, J. J.: Fluazinam: A novel fungicide for use against *Phytophthora infestans* in potatoes. Brighton Crop Protect. Conf. **6C–22** (1992): 663–668.

ANKE, T., HECHT, H. J., SCHRAMM, G., and STEGLICH, W.: Antibiotics from basidiomycetes. IX. Oudemansin, an antifungal antibiotic from *Oudemansiella mucida* (Schrader ex Fr.) Hoehnel (Agaricales). J. Antibiotics **XXXII** (1979): 1112–1117.

– OBERWINKLER, F., STEGLICH, W., and SCHRAMM, G.: The strobilurins – new antifungal antibiotics from the basidiomycete *Strobilurus tenacellus* (Pers. ex Fr.) Sing. J. Antibiotics **XXX** (1977): 806–810.

– SCHRAMM, G., SCHWALGE, B., STEFFAN, B., und STEGLICH, W.: Synthese von Strobilurin A und Revision der Stereochemie der natürlichen Strobilurine. Liebigs Ann. Chem. (1984): 1616–1625.

– and STEGLICH, W.: β-Methoxyacrylate antibiotics: from biological activity to synthetic analogues. In: SCHLUNEGGER, U. P., (Ed.): Biologically Active Molecules. Springer, Berlin–Heidelberg 1989, pp. 9–25.

BAUMANN, J.: Untersuchungen über die fungizide Wirksamkeit einiger Organo-Zinnverbindungen insbesondere von Triphenylzinn-acetat, ihren Einfluß auf die Pflanze und ihre Anwendung in der Landwirtschaft. Diss., Landw. Hochschule, Stuttgart-Hohenheim 1958, 74 S.

BEAUTEMENT, K., CLOUGH, J. M., DE FRAINE, P. J., and GODFREY, C. R. A.: Fungicidal β-methoxyacrylates: from natural products to novel synthetic agricultural fungicides. Pestic. Sci. **31** (1991): 499–519.

BECKER, W. F., JAGOW, G. VON, ANKE, T., and STEGLICH, W.: Oudemansin, strobilurin A, strobilurin B and myxathiazol: New inhibitors of the bc segment of the respiratory chain with an E-β-methoxyacrylate system as common structural element. FEBS letters **132** (2) (1981): 329–333.

BOCK, R.: Triphenyl tin compounds and their degradation products. Residue Rev. **79** (1981): 12–225.

BRANDL, F., und HAGMEYER, M.: Neue Möglichkeiten der Saatgutbehandlung bei Getreide auf Basis von Fenpiclonil. Pflanzenschutztagung CIBA-Geigy, Berlin 1990: 7–16.

BRANDT, U., SCHÄGGER, H., and VON JAGOW, G.: Characterisation of binding of the methoxyacrylate (MOA) inhibitors to mitochondrial cytochrome c reductase. Europ. J. Biochem. **173** (1988): 499 ff.

– SCHUBERT, J., GECK, P., and VON JAGOW, G.: Uncoupling activity and physicochemical properties of derivatives of fluazinam. Bioch. Biophys. Acta **1101** (1992): 41–47.

CHRYSAYI-TOKOUSBALIDES, M., and GIANNOPOLITIS, C. N.: Crossresistance in *Cercospora beticola* to triphenyltin and oligomycin: Plant diseases **65** (1981): 267–268.

CLOUGH, J. M., EVANS, D. A., DE FRAINE, P. J., FRASER, T. E. M., GODFREY, CH. R. A., and YOULE, D.: The role of natural products in pesticide discovery: The β-methoxyacrylate fungicides. J. Amer. Chem. Soc. (1993).

ELLNER, H.: Das Glutathion-System in *Phytophthora infestans* (Mont.) de Bary: Charakterisierung ausgewählter Komponenten und Untersuchung ihrer Beeinflußbarkeit. Diss., Math.-Naturw. Fak. Leipzig, 1991.

FARETRA, F., and POLLASTRO, S.: Genetic base of resistance to the phenylpyrrole fungicide CGA 173506 in *Botryotinia fuckeliana* (*Botrytis cinerea*). In: LYR, H., and POLTER, C., (Eds.): Intern. Symposium Modern Fungicides and Antifungal Compounds, May 1992, Reinhardsbrunn. Proceedings. Ulmer Verlag, Stuttgart 1993, pp. 405–409.

GEHMANN, H., NYFELER, R., LEADBEATER, A. J., NEVILL, D., and SOZZI, D.: CGA 173506: A new phenylpyrrole fungicide for broad-spectrum disease control. Brighton Crop Protect. Conf. **5–1** (1990): 399–406.

GODWIN, J. R., ANTHONY, V. M., CLOUGH, J. M., and GODFREY, C. R. A.: ICI A5504: A novel, brad spectrum, systemic β-methoxyacrylate fungicide. Brighton Crop Protect. Conf. **5–6** (1992): 435–442

GUO, Z., MIYOSHI, H., KOMYOJI, T., HAGA, T. and FUJITA, T.: Uncoupling activity of a newly developed fungicide, fluazinam [3-chloro-N-(3-chloro-2, 6-dinitro-4-trifluoromethylphenly)-5-trifluoromethyl-2-pyridinamine]. Biochem. Biophys. Acta **1056** (1991): 89–92.

HAUER, R., UHLEMANN, G., NEUMANN, J., and HÖFER, M.: Proton pumps of the plasmalemma of the yeast *Rhodotorula gracilis*. The coupling to fluxes of potassium and other ions. Bioch. Biophys. Acta **649** (1981): 680–690.

HEMMEN, C., and KONRADT, M.: Fluazinam gegen *Phytophthora infestans* in Kartoffeln. 48. Deutsche Pflanzenschutztagung Göttingen 1992. Mitt. BBA Berlin-Dahlem Heft **283** (1992): 436.

HILBER, U. W.: Comparative studies on genetic variability and fungicide resistance in *Botryotinia fuckeliana* (de Bary) Whetzel against vinclozolin and the phenylpyrrole CGA 173506. Diss. Univ. Basel, 1992, 78 p.

– SCHÜEPP, H., and SCHWINN, F. J.: Resistance to *Botryotina fuckeliana* (de Bary) Whetzel to phenylpyrrole fungicides as compared to dicarboximides. In: LYR, H., and POLTER, C., (Eds.): Intern. Symposium Modern Fungicides and Antifungal Compounds, May 1992 Reinhardsbrunn. Proceedings. Ulmer Verlag, Stuttgart 1993, pp. 105–111.

JAGOW, G. VON: Energiewandlung in der Atmungskette. Forschung Frankfurt. wiss. Magazin d. Univ. Frankfurt/M. **1** (1989): 28–32.
- Inhibitor binding of the mitochondrial bc$_1$ complex and of inhibitor resistant mutant. In: Design of mitochondrial electron transport inhibitors as agrochemicals. Bath Sympos. Biochem. Soc. April 1993. Biochem. Soc. Transactions (in press).
JESPERS, A. B. K.: Mode of action of the phenylpyrrole fungicide fenpiclonil in *Fusarium sulphureum*. Thesis Wageningen Agricult. Univ. N. L. (1994).
- and DEWAARD, M. A.: Biochemical effets of the phenylpyrrole fungicide fenpiclonil in *Fusarium sulphureum* (Schlecht). Proceed. 10th Int. Symp. on Systemic Fungicides and Antifungal compounds. Ulmer Verlag, Stuttgart 1993, pp. 99–104.
KOCH, E., and LEADBEATER, A. J.: Phenylpyrroles – a new class of fungicides for seed treatment. Brighton Crop Protect. Conf. **9B-3** (1992): 1137–1146.
- STECK, B., and MAAREL, H. V. D.: Control of plant pathogenic *Fusarium* species on cereals and non-cereals by seed treatment with fenpiclonil and CGA 173506. **3rd**. European Seminar: *Fusarium* – mycotoxins, taxonomy, pathogenicity and host resistance. IHAR Radzikow, Poland, Proceedings (1992): 1–7.
KÖHLE, H., GOLD, R. E., SAUTER, H., and RÖHL, F.: Biokinetic properties of BAS 490F and some related compounds. In: Design of mitochondrial electron transport inhibitors as agrochemicals. Bath Symposium Biochem. Soc. April 1993. Biochem. Soc. Transactions (in press).
- RADEMACHER, W., and RETZLAFF, G.: Wirkungsmechanismen von Strobilurinen bei Pilzen und Pflanzen. 49. Deutsche Pflanzenschutztagung Heidelberg, Mitt. BBA H. 301 (1994): 396.
KRAUSE, G., KLEPEL, M., JUMAR, A., and FRANKE, R.: Quantitative structure – activity relationships of fungicidal pyrimidines. Tagungsber. Akadem. Landwirtschaftswiss. Berlin **222** (1984): 229–232.
KÜHL, A., and RAUM, J.: CGA 219417 – ein neuer fungizider Wirkstoff mit vielfältigen Einsatzmöglichkeiten. 49. Deutsche Pflanzenschutztagung Heidelberg, Mitt. BBA H. 301 (1994): 392.
LEADBEATER, A. J., and KIRK, W. W.: Control of tuber born diseases of potatoes with fenpiclonil. Brighton Crop Prot. Conf. **6C–21** (1992): 657–668.
- NEVILL, D. J., STECK, B., and NORDMEYER, D.: CGA 173506: A novel fungicide for seed treatment. Brighton Crop Prot. Conf. **7C–27** (1990): 825–830.
LEROUX, P.: Mise en evidence d'une similitude d'action fongicide entre fenpiclonil, l'iprodione et le tolclofos-methyl. Agronomie **11** (1991): 115–117.
MAENO, S., MIURA, I., MASUDA, K., and NAGATA, T.: Mepanipyrim (KIF-3535), a new pyrimidine fungicide. Brighton Crop Prot. Conf. **5–3** (1990): 415–422.
MILLING, R. J., RICHARDSON, C. J., and PILLMOOR, J. B.: The biochemical mode of action of pyrimethanil. 8. JUPAC Congress, Washington 1994.
MIURA, I., KAMURA, T., MAENO, S., HAYASHI, S., and YAMAGUCHI, I.: Mode of action of mepanipyrim (KIF-3535). Abstracts 6th International Congress of Plant Pathology, Montreal/Canada (1993): 3.7.5, p. 89.
NEUMANN, G. L., and WINTER, E. H.: Pyrimethanil: A new fungicide. Brighton Crop Prot. Conf. **5–1** (1992): 395–402.
NEVILL, D., NYFELER, R., and SOZZI, D.: CGA 142705: A novel fungicide for seed treatment. Brighton Crop Prot. Conf. **2–7** (1988): 65–72.
NYFELER, R., EGLI, TH. and NEVILL, D.: Phenylpyrroles, a new class of agricultural fungicides related to the natural antibiotic pyrrolnitrin. 7th JUPAC Congr. of Pestic. Chem. Hamburg, Vol. **1** (1990): 219.
PAQUET, R., and WILKIN, P.: Nouvelle association organo stannique pour le lutte contre de mildioudela pomme de terre (*Phytophthora infestans* De Bary). Mededel. Landb. Wetensch. Gent **36** (1971): 348–354.
RICH, P. R., MADGWICK, S. A., BROWN, S., JAGOW, G. VON, and BRANDT, V.: MOA-Stilben – a new effector of the Q$_i$ site of the chloroplast cytochrome bf complex. Bioch. Soc. Transact. **29** (1991): 263.

RICHARDS, J. C., MILLING, R. J., and PITTIS, J. E.: Achieving biological control of phytopathogenic fungi using β-methoxyacrylates. In: Design of mitochondrial electron transport inhibitors as agrochemicals. Bath Symposium Biochem. Soc. April 1993, Biochem. Soc. Transactions (in press).

RÖHL, F., and SAUTER, H.: Species dependence of mitochondrial respiration inhibition by strobilurin analogues. In: Design of mitochondrial electron transport inhibitors as agrochemicals. Bath Symposium Biochem. Soc., April 1993. Biochem. Soc. Transactions (in press).

SEIDEL, P., LYR, H., and BANASIAR, L.: The influence of some TPTH-derivatives and safener combinations on phytotoxicity and fungitoxicity in comparison to TPTH. Proc. 10th Intern. Symp. Reinhardsbrunn on Modern Fungicides and Antifungal Compounds. Ulmer Verlag, Stuttgart 1993, pp. 431–439.

SELWYN, M. J.: Tinorganic compunds as ionophors and inhibitors of ion translocating ATP-ases. Adv. Chem. Ser. **151** (1976): 204–226.

- DAWSON, A. P., STOCKDALE, M., and GAINS, N.: Chloride-hydroxide exchange across mitochondrial, erythrocyte and artificial lipid membranes mediated by trialkyl- and triphenyl tin compounds. Eur. J. Biochem. **14** (1970): 120–126.

SIJPESTEIJN, A. K.: Biochemical modes of action of agricultural fungicides. World Rev. Pest. Contr. **9** (1970): 85–93.

STACHEWICZ, H.-E., BURTH, U., LEADBEATER, A., MAAREL, H. VAN DER, und ELMSHEUSER, H.: Behandlung von Kartoffeln mit Beret (Fenpiclonil) gegen Lagerkrankheiten und *Rhizoctonia solani*. 47. Pflanzenschutztagung CIBA-GEIGY, Berlin 1990: 17–24.

STOCKDALE, M., DAWSON, A. P., and SELWYN, M. J.: Effects of trikalkyltin and triphenyltin compounds on mitochondrial respriation. Europ. J. Biochem. **15** (1970): 342–351.

VAN DER KERK, G. J. M.: Differentiation between transportable fungicides and sytemic infection inhibiting agents, and their relative perspectives: Acta Acad. Sci. Phytopath. Hungar. **6** (1971): 311–327.

WIGGINS, Th. E.: Inhibitors of quinone mediated electron transport. Biochem. Soc. Transact. **30** (1992): 237.

WOOD, B.: Influence of single applications of fungicides on net photosynthesis in pecan. Plant Disease **68** (1984): 427–428.

WORTHING, C. R.: The pesticide Manual. 6th ed. Brit. Crop. Prot. Counc. 1979.

ZOLLFRANK, G., and LYR, H.: Properties of androprim – a new *Phytophthora fungicide*. Pesticide Newsletters (1988): 40–41.

Chapter 20

Antibiotics as antifungal agents

I. Yamaguchi

The Institute of Physical and Chemical Research (RIKEN), Wako, Japan

Introduction

While pesticides have played an indispensable role in preventing crop losses caused by pests and diseases to meet the increasing population in the world, there is growing concern about pesticides on their side effects on non-target organisms and from environmental viewpoints. This points to the need for selective pesticides with higher degradability in nature. Some older conventional pesticides such as DDT and BHC are reportedly hazardous to mammals through their biological concentration in the foodchain and they remain in the natural environment for a long period. On the contrary, naturally occurring pesticides are generally specific for target organisms and are supposed to be inherently biodegradable since they are synthesized biologically. Thus worldwide interest in pesticides of microbial origin has lately been renewed.

The use of microbial products to protect crops against plant pathogenic bacteria originated in Europe and USA, applying medicinal antibiotics such as streptomycin, tetracycline, or chloramphenicol. Then, cycloheximide and griseofulvin were introduced for the control of fungal plant diseases. The use of these compounds was, however, not extensive. Later, many attempts were made to find microbial products for the prime purpose of plant disease control, but most compounds had no practical promise because of their instability in field use or phytotoxicity. In Japan, however, an epock-making substance named blasticidin S (Fig. 20.1) was discovered (Takeuchi et al. 1958), and it exhibited significant efficacy in controlling rice blast disease, the most serious and damaging of all diseases of rice in temperate and humid climates found in Japan. The success of blasticidin S in practical use inspired further research for new pesticides of microbial origin, leading to the development of excellent antifungal substances such as kasugamycin (Umezawa et al. 1965), polyoxins (Isono et al. 1965 and 1967), validamycin A (Iwasa et al. 1971a), and mildiomycin (Iwasa et al. 1978). In addition, microbial products with insecticidal and herbicidal activity have also been found; polynactins (Ando et al. 1971) and milbemectins (Mishima et al. 1975) as miticides and bialaphos (Tachibana et al. 1982) as a herbicide. If gibberellins, well-known plant growth regulators, are included in this category, microbial products are being used in every sphere of pesticide use. Table 20.1 shows the main pesticides of microbial origin practically used in Japan. It might be surprising that eight compounds out of eleven were discovered in that country.

In recent years, standards for the disposal of industrial wastes have become more and more strict, and consequently the manufacture of synthetic chemicals becomes quite expensive if waste disposal facilities, as well as production plants, must be set up for each

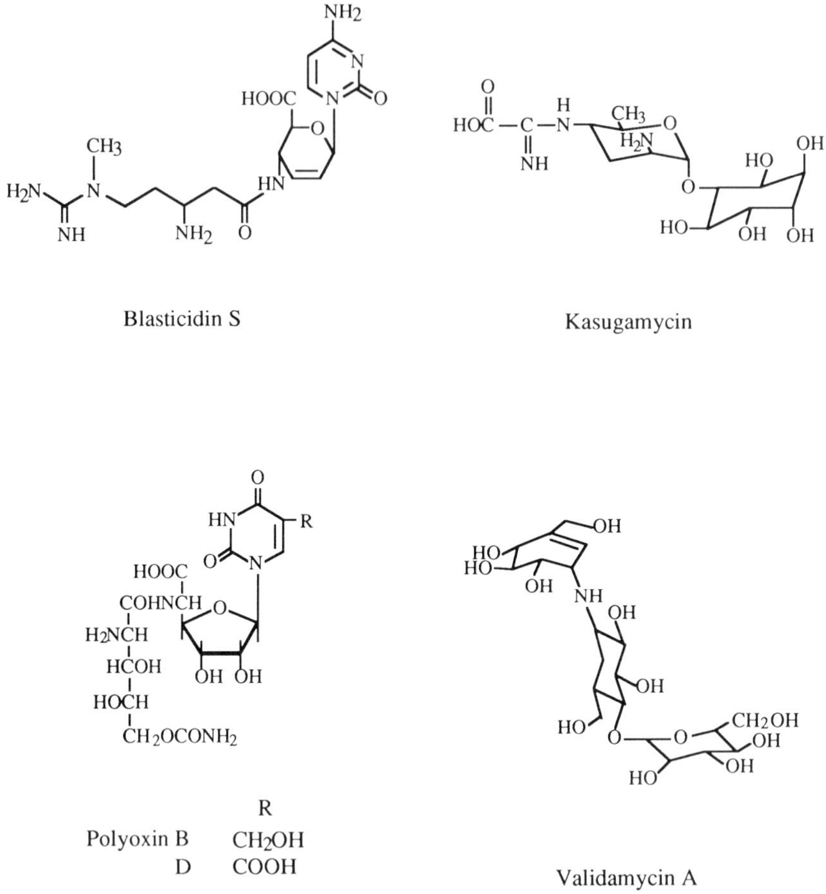

Fig. 20.1 Chemical structures of major antibiotics as antifungal agents.

new product. In this concern, pesticides of microbial origin have an economic advantage, since a variety of substances can be manufactured using one set of equipment and facilities. Additionally, they are produced not from limited fossil resources but from renewable agricultural products through fermentation by microorganisms. From these aspects, pesticides of microbial origin would be more advantageous especially for developing countries, and better results can be expected there in future.

As is true for every scientific technique, their use also has limitations. One disadvantage is the difficulty in their microanalysis, especially when they consist of many components. The second demerit is that, because of their highly specific mechanism of action, they are apt to suffer from the emergence of resistant strains. To cope with this problem, it is sometimes a requisite to combine the agents with other chemicals having different action mechanisms or to use them in rotation. Further, the biggest limitation to the use of microbial products in agriculture would be a concern that their wide use might create resistant strains which could hinder medical treatment of humans. This concern seems particularly noticeable in advanced Western countries, where the use of microbial products started by an application of medical antibiotics as mentioned earlier. Fortunately,

Table 20.1 Microbial products used in agriculture in Japan

Substance	Effective against
Antifungal	
Blasticidin S*)	Rice blast
Kasugamaycin*)	Rice blast
Polyoxins*)	Rice sheath blight, Fungal diseases of fruit trees and vegetables
Validamycin A*)	Rice sheath blight
Mildiomycin*)	Powdery mildew of garden trees
Antibacterial	
Streptomycin	Bacterial diseases of fruit trees and vegetables
Dihydrostreptomycin	Bacterial diseases of fruit trees and vegetables
Oxytetracyclin	Bacterial diseases of fruit trees and vegetables
Insecticidal	
Polynactins*)	Mites of fruit trees and tea plants
Milbemectins*)	Mites of fruit trees and tea plants
Herbicidal	
Bialaphos*)	Weeds in mulberry field and orchard

*) Substances discovered in Japan.

the use of pesticides of microbial origin in Japan for more than 30 years has never met any problem involving cross-resistance with medical antibiotics. Naturally, sufficient precaution must be taken not to cause any resistance in human pathogens in the future. However, this approach should not be limited to pesticides of microbial origin. There would be no difference between microbial products and synthetic chemicals in this respect and the important point is whether or not any pesticides may induce resistance to medicines. At present, the use of microbial products in agriculture is subjected to rigid regulations and restrictions apparently because they are called "antibiotics". This should be reconsidered from scientific points of view and it can be argued that they should be treated in the same way as synthetic chemicals.

Fungicides of Microbial origin

Blasticidin S

Blasticidin S is the first successful agricultural antibiotic developed in Japan. It was isolated from the culture filtrates of *Streptomyces griseochromogenes* (TAKEUCHI *et al.* 1958), and a potent curative effect of blasticidin S on rice blast was found (MISATO *et al.* 1959). Thereafter benzylaminobenzene sulfonate of blasticidin S proved to be least phytotoxic to the host plant without reduction of antifungal activity against *Pyricularia oryzae*, the pathogen of rice blast (ASAKAWA et al. 1963), and this salt has been industrially produced for agricultural use. Three other members of the blasticidin family, blasticidins A, B, and C, have been isolated from the same strain, but these were found to be less effective for practical use.

The chemical structure of blasticidin S has been elucidated to be 1-(1'-cytosinyl)-4-[L-3'-amino-5'-(1"-N-methylguanidino) valerylamino]-1,2,3,4-tetradeoxy-β-D-*erythro*-hex-2-eneuronic acid, as shown in Fig. 20.1 (OTAKE et al. 1966; YONEHARA and OTAKE 1966). The molecule of blasticidin S comprises a novel nucleoside designated cytosinine and a new β-amino acid named blasticid acid (ε-N-methyl-β-arginine). SETO et al. (1966) studied the biosynthesis of blasticidin S by the producing organism using ^{14}C-labeled precursors. The results showed that the pyrimidine ring of the antibiotic arose from cytosine and the 2,3-unsaturated sugar moiety from glucose; arginine served as the precursor for blastidic acid, and the N-methyl group of blastidic acid came from methionine. A metabolic intermediate in the biosynthetic pathway was found to be leucylblasticidin S (SETO et al. 1968).

Blasticidin S has a wide range of biological activities; besides its significant inhibitory effects on the growth of *P. oryzae*, it also exhibits wide antimicrobial and antiviral (HIRAI and SHIMOMURA 1965), as well as antitumor (TANAKA et al. 1961) activities. The acute oral toxicity (LD_{50}) to mice is 53.3 mg/kg. Fortunately, it can be used in the paddy field since toxicity to fish is rather low (LD_{50}/carp > 40 ppm).

MISATO et al. (1959) found the curative effect of blasticidin S on rice blast to be due to its strong inhibitory action on mycelial growth of *P. oryzae*. The antibiotic markedly inhibited the incorporation of ^{14}C-labeled amino acids into protein in the cell-free system of *P. oryzae* (HUANG et al. 1964), while other metabolic pathways including glycolysis, the electron transport, the oxidative phosphorylation and the nucleic acid syntheses were not inhibited by blasticidin S. YAMAGUCHI and TANAKA (1966) reported that blasticidin S remarkably inhibited the incorporation of leucine and phenylalanine into polypeptides in the cell-free proteinsynthesizing system from *Escherichia coli*, and suggested that blasticidin S acts in the step of peptide transfer from peptidyl-tRNA on the donor site to the incoming amino acyl-tRNA on the acceptor site. COUTSOGEORGOPOULOS (1969) observed that the inhibition of polypeptide synthesis by blasticidin S was not reversed by increasing concentration of phenylalanyl-tRNA, and thus blasticidin S did not act as a competitive inhibitor, with respect to phenylalanyl-tRNA, in polypeptide synthesis. Since blasticidin S markedly interfered with the puromycin reaction (YAMAGUCHI and TANAKA 1966), the action of blasticidin S may involve the formation of peptidyl-blasticidin S which remains bound to the ribosomes instead of being released. The mode of action of this antibiotic on the molecular level is still not known with any certainty, but a process related to peptidyl-transferase activity was reported (YUKIOKA et al. 1975).

Efficacy of blasticidin S was evaluated in field trials for rice blast control and it has been put into practical use since 1961. The effective spraying concentration of blasticidin S in the field is 10–40 ppm (1–4 g of blasticidin S/10a), but it occasionally causes chemical injury on rice leaves when sprayed at higher concentrations. Among crops, the tobacco plant is the most susceptible to blasticidin S, followed by eggplant, tomato, and potato; mulberry plants are rather susceptible; grape, pear, and peach plants are resistant, and water melon, cucumber and rice plants are more resistant to the antibiotic.

As for the mammalian toxicity, application of blasticidin S by dusting occasionally causes conjunctivitis if the dust accidentally comes in contact with the eyes. Less accidents were reported for the application of wettable powder or solution. It also induces inflammation of mucous membrane or of injured skin if they are exposed to the antibiotic (OHTA 1963). This toxic effect on mammals are the most unfavorable characteristics of blasticidin S. One of the attempts to remove this defect involves a chemical modification of the antibiotic, but favorable results as expected have not yet been obtained. Biological transformation of blasticidin S was also performed, and a strain of *Aspergillus fumigatus* was found to transform the antibiotic into deaminohydroxyblasticidin S, which is significantly less toxic to mammals but the antimicrobial activity was also fairly reduced. YAMAGUCHI et al. (1975) isolated an enzyme from *Aspergillus terreus* capable of catalyzing the deamination of the cytosine nucleus in blasticidin S. The enzyme was shown to have properties similar to cytidine deaminase in some respects, but quite distinct from the coexisting cytidine deaminase. In addition, the enzyme was proved to act on some derivatives of blasticidin S, but not on cytosine, cytidine, and other well known nucleosides (YAMAGUCHI et al. 1986). The structural gene, *bsr*, of an inactivating enzyme (YAMAGUCHI et al. 1972, 1985, 1986) has been cloned and introduced into tobacco plants using the Ti plasmid vector system, which demonstrated that the *bsr* gene afforded a blasticidin S resistant phenotype to the plants (KAMAKURA et al. 1990). This gives an example for a new approach to the solution of phytotoxic problems with some fungicides.

YONEHARA et al. (1968) discovered a new group of substances designated as detoxins; these were produced by *Streptomyces caespitosus* var. *detoxicus* and by several strains of *Streptomyces mobaraensis*. It exhibited unique biological activities as a selective antagonist blasticidin S, negating the selective inhibitory action of blasticidin S on *Bacillus cerus* but not on *P. oryzae*. Moreover, it has been observed that phytotoxicity to the rice plant was depressed without a loss in curative effect of the antibiotic, and the combined preparation of detoxin complex and blasticidin S produced less eye irritation. Separation of components of the detoxin complex and their characterization were performed (OTAKE et al. 1973), and the chemical structures of the main components were elucidated (KAKINUMA et al. 1974). While troublesome defects of blasticidin S were reduced by the addition of detoxins, they are not yet in use because the difficulties still remain for practical production and cost performance.

A simpler method to alleviate eye irritation caused by blasticidin S was developed; the addition of calcium acetate blasticidin S dust (5% addition) specifically reduced the eye trouble without influencing the antiblast effect, although other mammalian toxicity or phytotoxicity of the antibiotic was not alleviated. This improved dust is now in agricultural use.

The behavior and fate of blasticidin S in the environment were investigated using radioactive compounds prepared biosynthetically from [^{14}C]-cytosine and [^{14}C]-L-methionine (YAMAGUCHI et al. 1972). The sprayed antibiotic was located mostly on the surface of the rice plant, and little was taken up and translocated in the tissue. From the wound or infected part, however, the compound was incorporated and translocated to the apexes. Blasticidin S located on the plant surface was decomposed by sunlight and gave rise to cytosine as the main degradation product. A considerable quantity of the blasticidin S sprayed fell to the ground and was bound onto the soil surface. Significant generation of [^{14}C]-carbon dioxide from the [^{14}C]-blasticidin S-treated soil was observed, and several microbes usually inhabiting the paddy field were found to reduce the biological activity of blasticidin S. From the results obtained, it was suggested that, after application of blasticidin S to the crop field at a low concentration, the antibiotic is easily broken down in the environment, so that there would be no danger of environmental pollution and food contamination. In fact, the residual amount of blasticidin S in the unpolished rice cropped in the field is to be estimated as less than 0.05 ppm by biological assay.

Kasugamycin

Kasugamycin is a water-soluble and basic antibiotic produced by *Streptomyces kasugaensis* (UMEZAWA et al. 1965). This antibiotic controls rice blast disease at a concentration as low as about 20 ppm (ISHIHYAMA et al. 1965), being used as an agricultural antibiotic for rice blast control since 1965. It can be safely used for crop protection without any phytotoxicity and mammalian toxicity. However, kasugamycin-resistant strains in the paddy field raised a serious problem concerning rice blast control by the antibiotic.

The chemical structure of kasugamycin was elucidated by chemical methods (SUHARA et al. 1966) and by with X-ray diffraction analysis (IKEKAWA et al. 1966). As shown in Fig. 20.1, the molecule of kasugamycin is composed of three moieties; D-inositol, kasugamine (2,3,4,6-tetradeoxy-2,4-diaminohexopyranose), and an iminoacetic acid side chain. With regard to biosynthesis of kasugamycin, the incorporation of several sacccharides and amino acids into kasugamycin was examined (FUKAGAWA et al. 1968a); the incorporation of [^{14}C]-glucose (U) and [^{14}C]-mannose (U) into the kasugamine moiety was observed and [^{14}C]-*myo*-inositol and [^{14}C]-glycine were incorporated into kasugamycin mostly in other part of the antibiotic. D-inositol was proved not to be directly incorporated into the D-inositol moiety of kasugamycin, but [^{14}C]-*myo*-inositol was converted to the moiety of kasugamycin (FUKAGAWA et al. 1968b). Further, mass spectroscopic analysis showed that the nitrogen atom of glycine is incorporated into the imino nitrogen of the carboxyformidoyl group (FUKAGAWA et al. 1968c). SUHARA et al. (1968) succeeded in synthesizing kasuganobiosamine and related compounds, and they achieved the total chemical synthesis of kasugamycin.

Kasugamycin selectively inhibits the growth of *P. oryzae* and some bacteria including *Pseudomonas* species, and showed little or no activity against other microorganisms tested. Interestingly, it interferes with the growth of *P. oryzae* in acidic media (pH 5.0), but hardly inhibits it in neutral media (pH 7.0) (HAMADA et al. 1965). Kasugamycin does not inhibit spore germination of *P. oryzae* even at a concentration of 120 µg/ml, but when it is applied to rice plants it shows both protective and curative action. Kasugamycin has been used against rice blast in a large scale. It can control the disease when sprayed at about 20 ppm in aqueous solution. For practical control, kasugamycin is mainly applied as a dust containing 0.3% of active ingredient. No injury was observed to many plants tested. After kasugamycin-resistant strains had been detected in the field, the combined formulations of kasugamycin with chemicals having different action mechanisms came into practical use. Coating the seed with 2% wettable powder protects rice plants from blast disease in the field for a month.

TANAKA et al. (1966) reported that kasugamycin inhibited protein synthesis in a cell-free system of *P. oryzae*. Kasugamycin also inhibits protein synthesis in *Escherichia coli* by interfering with the binding of aminoacyl-tRNA to the mRNA-30 S ribosomal subunit complex. While it is an aminoglycoside antibiotic, it does not cause miscoding in protein biosynthesis. OHMORI (1967) isolated kasugamycin-resistant strains of *P. oryzae* from colonies formed on a agar medium containing kasugamycin *in vitro*. The resistant strains were not different from susceptible parent strains in sporulation and hyphal elongation ability, but infectivity of the resistant strains was remarkably reduced. Resistant strains were not detected in the field for several years after kasugamycin came into practical use, but following the increase of kasugamycin application in the paddy field, the effectiveness of kasugamycin on rice blast gradually declined. Problems arising from virulence of kasugamycin-resistant strains occurred in 1972 at several places in Japan where kasugamycin was mainly used for rice blast control. However, the population of the kasugamycin-resistant strains has rapidly declined when the successive application of kasugamycin was discontinued.

Kasugamycin shows a quite low acute or chronic toxicity to mice, rats, rabbits, dogs, and monkeys; the oral LD_{50} for mice was 2 g/kg. At a concentration of 1000 ppm, no toxicity to fish was observed.

Polyoxins

The polyoxins, a group of peptidylpyrimidine nucleoside antibiotics, are produced by *Streptomyces cacaoi* var. *asoensis* (SUZUKI et al. 1965; ISONO et al. 1965). Polyoxins are composed of fourteen components (A – N) of closely related peptidic nucleosides. They are safely used for the control of many fungal diseases with no toxicity to livestock, fish and plants. Such excellent characteristics are due to the fact that polyoxins selectively inhibit the synthesis of cell-wall chitin in sensitive fungi, as was reported by MISATO and his co-workers (SASAKI et al. 1968a; HORI et al. 1974a). Polyoxins have been widely used for crop protection against such pathogenic fungi as *Rhizoctonia solani*, *Cochliobolus miyabeanus* and *Alternaria alternata* since 1967.

The isolation and chemical elucidation of polyoxins were attributed to the excellent work by ISONO et al. (1965, 1967, 1968a, 1969; Fig. 20.1). Among polyoxins, the C component lacks antifungal activity, but it was a key compound in elucidating the structure of polyoxins, since hydrolytic degradation of all the polyoxins afforded polyoxin C or its analogs. ISONO and SUZUKI (1968b) assigned the structure 1-β-(5'-amino-5'-deoxy-D-allofuranuronosyl)-5-hydroxymethyluracil to polyoxin C on the basis of chemical and physical analysis. The identification of nucleoside with 5-aminohexuronic acid as its sugar group is the first rare example in nature. This prompted the chemical synthesis of polyoxin J by KUZUHARA et al. (1973). The biosynthesis of polyoxins was also attractive since they contain an unique amino acid (polyoximic acid) and 5-substituted pyrimidine ring in their molecules. ISONO and SUHADOLNIK (1976) reported the biosynthesis of the 5-substituted uracil base of the polyoxins from uracil and C-3 of serine by a new enzyme which differs from thymidylate synthetase. Furthermore, they found a new metabolic role for L-isoleucine as a precursor for 3-ethylidene-L-azetidine-2-carboxylic acid (polyoximic acid). The distribution of ^{14}C in this unique cyclic amino acid proved that the intact carbon skeleton of L-isoleucine was utilized directly in its biosynthesis (ISONO et al. 1975).

Polyoxins inhibit the growth of some fungi but are inactive against bacteria and yeasts. All the polyoxins except C and I showed selective antifungal activity against various plant pathogenic fungi (SUZUKI et al. 1966). Among polyoxins, polyoxin D was most effective for rice sheath-blight caused by *Rhizoctonia solani*, whereas polyoxins B was effective for pear black-spot fungus and apple cork-spot fungus at 50–100 ppm. As for its toxicity, oral administration at 15 g/kg and injection at 800 mg/kg to mice did not cause any adverse effect, nor was it toxic to fish during 72 hours of exposure at 10 ppm. Moreover, foliar sprays of 200 ppm of polyoxins have produced no phytotoxicity on most crops, and no injury to the rice plants was observed even at 800 ppm application (SASAKI et al. 1968b).

Trials to control the rice sheath bligh confirmed the efficacy of polyoxins and their persistence in the field. Application could be made at any stage of plant growth without causing phytotoxicity. Besides sheath-blight control, polyoxins showed high efficacy against diseases caused by *Alternaria* spp. in fruit orchards, such as black spot of pear and *Alternaria* leaf spot of apple (EGUCHI et al. 1968). Polyoxin complex has been used in practice in duplicate forms: a polyoxin D-rich fraction for the sheath-blight control, and a B-rich fraction for diseases caused by *Alternaria* spp.

With regards to the mechanism of fungicidal action of polyoxins, a specific physiological reaction of *Alternaria* spp. was observed (EGUCHI *et al.* 1968); polyoxins caused a marked abnormal bulbous phenomenon on germ tubes of spores and hyphal tips of the pathogen, and this abnormally swollen spore became non-infectious. Further, the incorporation of [^{14}C]-glucosamine into cell wall chitin of *Cochliobolus miyabeanus* was shown to be markedly inhibited by polyoxin D without an inhibitory effect on respiration and the synthesis of macromolecules such as protein and nucleic acids (SASAKI *et al.* 1968a). By using a cell-free system of *Neurospora crassa*, ENDO and MISATO (1969) proved that polyoxin D strongly inhibits the incorporation of *N*-acetylglucosamine (GlcNAc) into chitin in a competitive manner between UDP-GlcNAc and polyoxin D. UDP-GlcNAc was accumulated in the polyoxin D-treated mycelia of *C. miyabeanus*, and the relation between polyoxin structure and inhibitory activity on **chitin synthetase** was clarified (HORI *et al.* 1974a). According to the kinetic analysis, the carbamoylpolyoxamic acid moiety of polyoxins was shown to stabilize the polyoxin-enzyme complex, and the pyrimidine nucleoside moiety of the antibiotics to fit into the binding site of the enzyme protein. Therefore, the excellent characteristics of polyoxins can be explained that the antibiotics inhibit the **cell wall synthesis** of sensitive fungi but have no adverse influence on other organisms which have no chitinous cell walls.

Polyoxin resistant strains of *A. alternata* were recognized in some orchards in Japan after exclusive usage of the antibiotics. HORI *et al.* (1974b) suggested that the resistance was caused by a lowered permeability of the antibiotic through the cell membrane into the site of chitin synthesis. The inhibition of mycelial growth of *R. solani* by polyoxins was antagonized by glycyl-L-alanine, glycyl-DL-valine, and DL-alanylglycine (HORI *et al.* 1977). Therefore, polyoxins may be incorporated into the fungal cells through the uptake channels of the dipeptides.

Validamycin A

Validamycin A (VM-A) is an antifungal antibiotic developed for the control of rice-sheath blight (IWASA *et al.* 1971a, b). It was isolated from the culture filtrate of *Streptomyces hygroscopicus* var. *limoneus*, which produced five additional components designated validamycins B to F, together with validoxylamine A (IWASA *et al.*; HORII *et al.* 1972a, b). VM-A has been used for the control of *Rhizoctonia* disease without phytotoxicity and with very low toxicity to mammals (MATSUURA 1983). No toxicity was observed for birds, fish, and insects.

The chemical structure of validamycin A was determined to be 1,5,6-trideoxy-3-O-β-D-glucopyranosyl-5-(hydroxymethyl)-1-[[4,5,6-trihydroxy-3-(hydroxymethyl)-2-cyclohexan-1-yl]amino]-D-chiro-inositol as shown in Fig. 20.1 (SUAMI *et al.* 1980). VM-A can be seen to have two kinds of new hydroxymethyl-branched cyclitols in its molecule, and these cyclitol moieties are quite different from known aminocyclitols, such as streptamine, deoxystreptamine, and actinamine. VM-A is, therefore, an unique aminoglycoside antibiotic and one of the peculiar examples in the field of pseudosugar chemistry.

Validamycins A, C, D, E, and F contain validoxylamine A as a common moiety in their molecules (HORII *et al.* 1972a, b), but they differ from each other at least in one of the following characteristics; the configuration of anomeric center of glucoside, the position of glucosidic linkage, and the number of D-glucose molecules. On the other hand, validamycin B contains validoxylamine B in its molecule, which yielded hydroxyvalidamine instead of validamine derived from VM-A by hydrogenolysis followed by acid hydrolysis (HORII *et al.* 1971). KAMEDA *et al.* (1975) studied microbial transformation of validamycins, and they showed the conversion of validamycin C to

VM-A by *Endomycosis* spp. or *Candida* spp., which has an important significance because validamycin C is about 1 000 times less active than VM-A against *R. solani* in the "dendroid-test" established by IWASA et al. (1971 c).

VM-A is a main component of the validamycin complex and is specifically effective against certain plant diseases caused by *Rhizoctonia* spp., such as web blight, bud rot, damping-off, seed decay, root rot, and black scurf of several crops and southern blight of vegetables as well as sheath blight of rice plant (WAKAE and MATSUURA 1975). While the antibiotic showed neither cidal nor static action on *Rhizoctonia* spp., it caused an abnormal branching at the tips of hyphae of the pathogens, followed by cessation of further development. When it was applied in the early logarithmic phase of lesion expansion on rice plant, sufficient control was achieved by a spray of 30 ppm VM-A solution. VM-A has been commercially used for the control of sheath-blight disease since 1973.

Although validamycin A did not significantly suppress the growth of *R. solani* on a nutritionally rich medium, it caused specifically extensive branching of hyphae and the cessation of colony development on a water agar. Therefore, VM-A appeared to cause morphological changes on the fungus under certain conditions and thus inhibit the growth of the fungus in quantity. Later, NIOH and MIZUSHIMA (1974) found that the antibiotic did not inhibit the fungal growth in mass, nor interfered with the biosynthesis of protein, nucleic acid or cell wall components, but it altered the morphology of the fungus. Meanwhile, it was observed that VM-A had no need of continual contact with the pathogen under nutrientless conditions for the effective control of fungal growth (WAKAE and MATSUURA 1975). This is important in the real state of disease control, since *R. solani* is one of the typical fungi that grows rapidly by transporting nutrients from basal part to hyphal tips through long stretches of hyphae; this type of growth provides the tip part rather a condition of poor nutrition. As for the mode of action of validamycin A, it was shown that VM-A inhibits the **biosynthesis of *myo*-inositol** in *R. solani*, and inositol was supposed to be indispensable for the normal growth and pathogenic activity of the fungus. Later, it was shown that VM-A had potent inhibitory activity against **trehalase** of *R. solani* AG-1, without any significant inhibitory effect on other glycohydrolase enzymes tested. Trehalose is well-known as a storage carbohydrate in the fungus and trehalase is suggested to play an essential role to digest trehalose and to transport D-glucose (SHIGEMOTO et al. 1989).

Antimicrobial activity of VM-A was not detected on about 3 000 species of fungi and bacteria with ordinary methods (IWASA et al. 1971 c), nor was disturbance of microflora on rice plant and crop field observed (WAKAE and MATSUURA 1975). No phytotoxicity was observed for over 150 species of plants sprayed with VM-A even at a concentration of 1 000 ppm. Furthermore, acute and subacute toxicities to mammals were markedly low; in oral administration of VM-A at a dose of 10 g/kg to mice and rats, or in subcutaneous and intravenous administration at the dose of 2 g/kg to mice, all animals examined gave no change for 7 days, and no irritating effects were observed on the skin and the cornea in the rabbit. Oral subacute toxicity tests of VM-A for 4 months in beagle dogs indicated no significant abnormalities of the morphological and biochemical parameters accompanying the daily administration of 200 mg/kg of VM-A (HOSOKAWA et al. 1974). LD_{50} for killifish was found to be greater than 1 000 ppm.

Validamycins have been shown to be susceptible to microbial degradation, and their addition to the soil resulted in complete loss of biological activity by soil microbes. Their half-life in soil was reported to be less than 4 hours. Microbial degradation of VM-A by *Pseudomonas denitrificans* gave rise to D-glucose and validoxylamine A, which was further decomposed into valienamine and validamine (KAMEDA et al. 1975). The metabolic fate of validamycins in animals showed that orally administered VM-A is easily decomposed up to CO_2 by enteric bacteria. VM-A was excreted to feces without being absorbed into the animal body through the intestinal lining, and in the case of intravenous injection intact VM-A was rapidly excreted in the urine. Furthermore, no cross-resistance with medicinal aminoglycoside antibiotics, such as dihydrostreptomycin and kanamycin, was detected in certain human pathogens (MATSUURA 1983).

Validamycin A has been used to protect sheath blight of the rice plant mainly in formulations of 3 – 5% solution or 0.3% dust. Residues in rice grains and straws were less than detectable limit by gas chromatography (KAMEDA and YAMAMOTO 1970). Thus, validamycin A is considered to be one of the ideal chemicals with respect to safety and to environmental pollution.

Mildiomycin

Mildiomycin was isolated from the culture filtrate of *Streptoverticillium rimofaciens* B-98891 (HARADA et al. 1978 a). It is a water soluble and basic antibiotic which belongs to nucleoside antibiotics (HARADA et al. 1978 b). Mildiomycin is highly active against the growth of various pathogens to cause powdery mildews, but weakly active on gram-positive and gram-negative bacteria, most fungi, and some yeast. It shows excellent curative activity on powdery mildew of various plants *in vivo*, and thus it has been practically used for the control of diseases on rose, spindle tree and Indian lilac.

The toxicity of mildiomycin is very low. LD_{50} for acute toxicity in rats and mice is 500 – 1 000 mg/kg by intravenous and subcutaneous injections, and 2.5 – 5.0 g/kg by oral administration. At a concentration of 1 000 ppm there is no irritation to the cornea and skin of rabbits for 10 days. Its toxicity to killifish is not observed at a concentration of 20 ppm for 7 days. Protein synthesis was remarkably inhibited by mildiomycin at concentrations down to 0.02 mM; it interferes with the incorporation of ^{14}C-phenylalanine into polypetides in the cell free system of *E. coli*. The mammalian cell free system from rabbit reticulocytes proved to be less sensitive to mildiomycin in the synthesis of polypeptides than the system from *E. coli* (OM et al. 1984).

Novobiocin, Cycloheximide and Griseofulvin

Novobiocin, also called **cathomycin**, is produced by *Streptomyces spheroides* and other actinomycetes. The occurrence of bacterial canker of tomato is reduced by dipping tomato seedlings in 100 ppm of novobiocin aqueous solution overnight before transplanting. It had been used for the control of tomato cancer since 1968 to 1989, when the registration in Japan was discontinued due to the high cost for safety evaluation.

Cycloheximide was found as a by-product of streptomycin in the culture filtrate of *Streptomyces griseus* and *S. naraensis*. It inhibits the growth of various plant pathogens at low dose levels, but the application was limited because of its high phytotoxicity. Semicarbazone derivatives of cycloheximide showed reduced phytotoxicity while its potential activity against fungi was not changed. The antibiotic had been used for controlling downy mildew of onions for over 20 years in Japan since 1959. This antibiotic had been also used as a fungicide in forests; 3-ppm solution was found to be effective against shoot blight of Japanese larch. When cycloheximide oil formulation was applied to trunks of pines, it was absorbed and translocated to the upper part of tree branches. Its registration was discotinued in 1981.

Utilization of **griseofulvin** for plant protection succeeded in controlling early blight of tomato and *Botrytis* disease of lettuce. Its inhibitory effect against conidial germination of pathogenic fungus of apple blossom blight was observed; by dusting a mixture of apple pollen grains and lycopodium with 25% griseofulvin, blossom blight of apple plants was controlled. When a paste of 50% wettable powder of griseofulvin was applied to the diseased parts of melons, fusarium wilt of melons was controlled. This antibiotic had been used in Japan since 1959, but its use was stopped in 1975 because of the high cost for agricultural use.

Summary and conclusion

Among the pesticides listed in Table 20.1, polyoxins and validamycins are quite safe fungicides; non-phytotoxic, and non-toxic to humans, livestock, and wildlife. Such excellent characteristics seem to be due to their modes of actions. Polyoxins selectively inhibit the synthesis of fungal cell-wall chitin, which does not exist in mammalian cells (HORI et al. 1974a). Validamycin A has an exceptional character, i.e., it is not fungicidal but only deteriorates the normal mycelial growth of the pathogen on plants. Probably because of such moderate activity, no occurrence of resistant strains has been reported in validamycin treatment, despite the largest consumption among pesticides of microbial origin. Further, components of avermectins (DYBAS 1983), insecticidal microbial products developed in USA, were successfully transformed into significantly more active derivatives by chemical modification of the original structures. These findings leave considerable scope for future prospects of microbial products. In addition, new microbial products with novel characters may be found against those pests and diseases such as soil borne diseases and viral diseases which are difficult to be controlled by synthetic chemicals at present. In the near future, crop and pest biochemistry and molecular biology will make rapid further advances, and a new biotechnology will be an important factor for plant protection research and strategy. For example, more efficient production of microbial products will become possible by modern gene engineering. Biorational approaches will also become feasible in the design of new pesticides from microbial products by the extensive use of computer and data processing procedures.

The future shape of agriculture in the world will depend largely on the availability of the proper kinds of pesticides in adequate quantities. The following considerations might be important for the future development of the antibiotics as fungicidal agents.

1. Establishment of screening directions. The development of some new antibiotics were done shortly after establishing a new screening project, e.g., polyoxins against rice-sheath blight.

2. Establishment of new screening test methods. Introduction of a novel screening test method carries with it the possibility for development of new effective antibiotics. Kasugamycin and validamycins were discovered by adopting the new assay systems.

3. Avoidance of pathogens resistance to antibiotics. Kasugamycin and polyoxins have

high selective toxicity to pathogenic fungi. When pathogens resistant to these antibiotics emerged, the alternate or combined application of chemicals with different action mechanisms was effective. Such applications are inevitable in some cases to avoid the development of resistant pathogens.

4. The use of antibiotics as leads for new pesticides. Modifications of existing antibiotics provided new potent chemicals. The best examples of these would be semi-synthetic penicillin and avermectin. More attention should be focused on finding the molecular relationship between chemical structure and biological activity.

References

ANDO, K., OISHI, H., HIRANO, S., OKUTOMI, T., SUZUKI, K., OKAZAKI, H., SAWADA, M., and SAGAWA, T.: Tetranactin, a new miticidal antibiotic. I. Isolation, characterization and properties of tetranactin. J. Antibiot. **24** (1971): 347–352.

ASAKAWA, M., MISATO, T., and FUKUNAGA, K.: Studies on the prevention of the phytotoxicity of blasticidin S. Pestic. Technique. **8** (1963): 24–29.

COUTSOGEORGOPOULOS, C.: Formation of olygophenylalanine in the presence of certain inhibitors of protein synthesis. Federat. Proc. **28** (1969): 844.

DYBAS, R. A.: Avermectins: their chemistry and pesticidal activities. In: MIYAMOTO, J., and KEARNEY, P. C. (Eds.): Pesticide Chemistry; Human Welfare and the Environment. vol. 1., Pergamon Press, Oxford 1983, pp. 83–90.

EGUCHI, J., SASAKI, S., OHTA, N., AKASHIBA, T., TSUCHIYAMA, T., and SUZUKI, S.: Studies on polyoxins, antifungal antibiotics. Mechanism of action on the diseases caused by *Alternaria* spp. Ann. Phytopathol. Soc. Japan **34** (1968): 280–288.

ENDO, A., and MISATO, T.: Polyoxin D, a competitive inhibitor of UDP-N-acetylglucosamine. Chitin N-acetylglucosaminyl-transferase in *Neurospora crassa*. Biochem. Biophys. Res. Comm. **37** (1969): 718–722.

FUKAGAWA, Y., SAWA, T., TAKEUCHI, T., and UMEZAWA, H.: Studies on biosynthesis of kasugamycin. I. Biosynthesis of kasugamycin and the kasugamine moiety. J. Antibioti. **21** (1968a): 50–54.

– – – – Studies on biosynthesis of kasugamycin. III. Biosynthesis of the D-inositol. J. Antibiot. **21** (1968b): 185–188.

– – – – Studies on biosynthesis of kasugamycin. V. Biosynthesis of the amidine group. J. Antibiot. **21** (1968c): 410–412.

HAMADA, M., HASHIMOTO, T., TAKAHASHI, S., YONEYAMA, M., MIYAKE, T., TAKEUCHI, Y., OKAMI, Y., and UMEZAWA, H.: Antimicrobial activity of kasugamycin. J. Antibiot. Ser. A. **18** (1965): 104–106.

HARADA, S., and KISHI, T.: Isolation and characterization of mildiomycin, a new nucleoside antibiotic. J. Antibiot. **31** (1978a): 519–524.

– MIZUTA, E., and KISHI, T.: Structure of mildiomycin, a new antifungal nucleoside antibiotic. J. Am. Chem. Soc. **100** (1978b): 4895–4897.

HIRAI, T., and SHIMOMURA, T.: Blasticidin S, an effective antibiotic against plant virus multiplication. Phytopathol. **55** (1965): 291–295.

HORI, M., EGUCHI, J., KAKIKI, K., and MISATO, T.: Studies on the mode of action of polyoxins. VI. Effect of polyoxin B on chitin synthesis in polyoxin-sensitive and resistant strains of *Alternaria kikuchiana*. J. Antibiot. **27** (1974b): 260–266.

– KAKIKI, K., and MISATO, T.: Studies on the mode of action of polyoxin. Part IV. Further study on the relation of polyoxin structure to chitin synthetase inhibition. Agric. Biol. Chem. **38** (1974a): 691–698.

– – – Antagonistic effect of dipepetides on the uptake of polyxin A by *Alternaria kikuchiana*. J. Pestic. Sci. **2** (1977): 139–149.
Horii, S., Iwasa, T., and Kameda, Y.: Studies on validamycins, new antibiotics. V. Degradation studies. J. Antibiot. **24** (1971): 57–58.
– Kameda, Y.: Structure of validamycin A. J. Chem. Soc. Chem. Comm. (1972a): 747–748.
– – Kawahara, K.: Studies on validamycins, new antibiotics. VIII. Validamycins C, D, E and F. J. Antibiot. **25** (1972b): 48–53.
Hosokawa, S., Ogiwara, S., and Murata, Y.: Oral subacute toxicity of validamycin for 4 months in beagle dogs. J. Takeda Res. Lab. **33** (1974): 119–128.
Huang, K. T., Misato, T., and Asuyama, H.: Effect of blasticidin S on Protein synthesis of *Piricularia oryzae*. J. Antibiot. Ser. A. **17** (1964): 65–70.
Ikekawa, T., Umezawa, H., and Iitaka, Y.: The structure of kasugamycin hydrobromide by X-ray crystallographic analysis. J. Antibiot. Ser. A. **19** (1966): 49–50.
Ishiyama, T., Hara, I., Matsuoka, M., Sato, K., Shimada, S., Izawa, R., Hashimoto, T., Hamada, M., Okami, Y., Takeuchi, T., and Umezawa, H.: Studies on the preventive effect of kasugamycin on rice blast. J. Antibioti. Ser. A. **18** (1965): 115–119.
Isono, K., Asahi, K., and Suzuki, S.: Studies on polyoxins, antifungal antibiotics. XIII. The structures of polyoxins. J. Am. Chem. Soc. **91** (1969): 7490–7505.
– Funayama, S., and Suhadolnik, R. H.: Biosynthesis of the polyoxins, nucleoside peptide antibiotics. A new metabolic role for L-isoleucine as a precursor for 3-ethylidene-ethylidene-L-azetidine-2-carboxylic acid (Polyoximic acid). Biochemistry **14** (1975): 2992–2996.
– Nagatsu, J., Kawashima, Y., and Suzuki, S.: Studies on polyoxins antifungal antibiotics. Part I. Isolation and characterization of polyoxins A and B. Agric. Biol. Chem. **29** (1965): 848–854.
– – Kobinata, K., Saski, K., and Suzuki, S.: Studies on polyoxins, antifungal antibiotics. Part V. Isolation and characterization of polyoxins C, D, E, F, G, H and I. Agric. Biol. Chem. **31** (1967): 190–199.
– Suhadolnik, R. H.: The Biosynthesis of Natural and Unnatural polyoxins by *Streptomyces cacaoi*. Arch. Biochem. Biophy. **173** (1976): 141–153.
– Suzuki, S.: The structures of polyoxins D, E, F, G, H, I, J, K and L. Agric. Biol. Chem. **32** (1968a): 1193–1197.
– – The structure of polyoxin C. Tetrahedron Lett. No. 2 (1968b): 203–208.
Iwasa, T., Higashide, E., and Shibata, M.: Studies on validamycins, new antibiotics. III. Bioassay methods for the determination of validamycin. J. Antibiot. **24** (1971c): 114–118.
– – Yamamoto, H., and Shiba, M.: Studies on validamycins, new antibiotics. II. Production and biological properties of VM-A and B. J. Antibiot. **24** (1971a): 107–113.
– Kameda, Y., Asai, M., Horii, S., and Mizuno, K.: Studies on validamycins, new antibiotics. IV. Isolation and characterization of VM-A and B. J. Antibiot. **24** (1971b): 119–123.
– Suetomi, K., and Kusaka, T.: Taxonomic study and fermentation of producing organism and antimicrobial activity of mildiomycin. J. Antibiot. **31** (1978): 511–518.
Kakinuma, K., Otake, N., and Yohehara, H.: The structure of detoxin D_1. Agr. Biol. Chem. **38** (1974): 2529–2538.
Kamakura, T., Yoneyama, K., and Yamaguchi, I.: Expression of the blasticidin S deaminase gene (*bsr*) in tobacco: Fungicide tolerance and a new selective marker for transgenic plants. Mol. gen. genet. **223** (1990): 332–334.
Kameda, Y., Horii S., and Yamoto, T.: Microbial transformation of validamycins. J. Antibiot. **28** (1975): 298–306.
Kuzuhara, H., Ohrui H., and Emoto, S.: Total synthesis of polyoxin J. Tetrahedron Lett. No. 50 (1973): 5055–5058.
Matsuura, K.: Characteristics of validamycin A in controlling *Rhizoctonia* diseases. IUPAC Pestic. Chem. (1983): 301–308.

MISATO, T., ISHII, I., ASAKAWA, M., OKIMOTO, Y., and FUKUNAGA, K.: Antibiotics as protectant fungicides against rice blast. II. The therapeutic action of blasticidin S. Ann. Phytopathol. Soc. Japan. **24** (1959): 302–306.

– Ko, K., and YAMAGUCHI, I.: Use of antibiotics in agriculture. In: PERLMAN (Ed.): Advances in Applied Microbiology. Academic Press, New York **21** (1977): 53–88.

MISHIMA, H., KURABAYASHI, M., TAMURA, C., SATO, S., KUWANO, H., and SAITO, A.: Structures of milbemycin $\beta 1$, $\beta 2$, and $\beta 3$. Tetrahedron Lett. No. 10 (1975): 711–714.

NIOH, T., and MIZUSHIMA, S.: Effect of validamycin on the general growth and morphology of *Pellicularia sasakii*. J. Gen. App. Microbiol. **20** (1974): 373–383.

OHMORI, K.: Studies on characters of *Piricularia oryzae* made resistant to kasugamycin. J. Antibiot. Ser. A. **20** (1967): 109–114.

OM, Y., YAMAGUCHI, I., and MISATO, T.: Inhibition of protein of synthesis by mildiomycin, an antimildew substance. J. Pesticide Sci. **9** (1984): 317–323.

OTAKE, N., KAKINUMA, K., and YONEHARA, H.: Separation of detoxin complex and characterization of two active principles, detoxin C and D. Agr. Biol. Chem. **37** (1973): 777–2780.

– TAKEUCHI, S., ENDO, T., and YONEHARA, H.: Chemical studies on blasticidin S. III. The structure of blasticidin S. Agric. Biol. Chem. **30** (1966): 132–141.

SASAKI, S., OHTA, N., EGUCHI, J., FURUKAWA, Y., AKASHIBA, T., TSUCHIYAMA, T., and SUZUKI, S.: Studies on polyoxins, antifungal antibiotics. Phytopathol. Soc. Japan. **34** (1968b): 272–279.

– – YAMAGUCHI, I., KURODA, S., and MISATO, T.: Studies on polyoxin action. Part I. Effect on respiration and synthesis of protein, nucleic acids and cell-wall of fungi. Nippon Nogei-Kagaku Kaishi. **42** (1968a): 633–638.

SETO, H., OTAKE, N., and YONEHARA, H.: Studies on the biosynthesis of blasticidin S. Part II. Leucylblasticidin S, a metabolic intermaidate of blastcidin S biosynthesis. Agric. Biol. Chem. **32** (1968): 1299–1305.

– YAMAGUCHI, I., OTAKE, N., and YONEHARA, H.: Biogenesis of blasticidin S. Tetrahedron Lett. No. 22 (1966): 3793–3799.

SHIGEMOTO, R., OKUNO, T., and MATSUURA, K.: Effect of validamycin A on the activity of trehalase of *Rhizoctonia solani* and several sclerotial fungi. Ann. Phytopat. Soc. Japan **55** (1989): 238–241.

SUAMI, T., OGAWA, S., and CHIDA, N.: The revised structure of validamycin A. J. Antibiot. **33** (1980): 98–99.

SUHARA, Y., MAEDA, K., and UMEZAWA, H.: Chemical Studies on kasugamycin. V. The structure of kasugamycin. Tetrahedron Lett. No. 12 (1966): 1239–1244.

– SASAKI, F., MAEDA, K., UMEZAWA, H., and Ohno, M.: The total synthesis of kasugamycin.
J. Am. Chem. Soc. **90** (1968): 6559–6560.

SUZUKI, F., ISONO, K., NAGATSU, J., MIZUTANI, T., KAWASHIMA, Y., and MIZUNO, T.: A new antibiotic. polyoxin A. J. Antibiot. Ser. A. **18** (1965): 131.

– – – KAWASHIMA, Y., YAMAGATA, K., SASAKI, K., and HASHIMOTO, K.: Studies on polyoxins, antifungal antibiotics. Part IV. Isolation of polyoxins C, D, E, F, and G, new components of polyoxin complex. Agric. Biol. Chem. **30** (1966): 817–819.

TACHIBANA, K., WATANABA, T., SEKIZAWA, Y., KONNAI, M., and TAKEMATSU, T.: Finding of the herbicidal activity and the mode of action of bialaphos, L-2-amino-4-[(hydroxy) (methyl) phosphinoyl] butyrylalanyl-alanine. Abstr. 5th. Int. Congr. Pestic. Chem. (1982): IV a – 19.

TANAKA, N., SAKAGAMI, Y., YAMAKI, H., and UMEZAWA, H.: Activity of cytomycin and blasticidin S against transplantable animal tumors. J. Antibiot. Ser. A. **14** (1961): 123–126.

TAKEUCHI, S., HIRAYAMA, K., UEDA, K., SASAKI, H., and YONEHARA, H.: Blasticidin S, a new antibiotic. J. Antibiot. Ser. A. **11** (1958): 1–5.

UMEZAWA, H., OKAMI, Y., HASHIMOTO, T., SUHARA, Y., HAMADA, M., and TAKEUCHI, T.: A new antibiotic, Kasugamycin. J. Antibiot. Ser. A. **18** (1965): 101–103.

WAKAE, O., and MATSUURA, K.: Characteristics of validamycin as a fungicide for *Rhizoctonia* disease control. Rev. Plant Protec. Res. **8** (1975): 81–92.

YAMAGUCHI, I., and MISATO, T.: Active center and mode of reaction of blasticidin S deaminase. Agric. Biol. Chem. **49** (1985): 3355–3361.

— SETO, H., and MISATO, T.: Substrate binding by blasticidin S deaminase, an aminohydrolase for novel 4-aminopyrimidine Nucleosides. Pestic. Biochem. Physiol. **25** (1986): 54–62.

— SHIBATA, H., SETO, H., and MISATO, T.: Isolation and purification of blasticidin S deaminase from *Aspergillus terreus*. J. Antibiot. **28** (1975): 7–14.

— TAKAGI, K., and MISATO, T.: The sites for degradation of blasticidin S. Agric. Biol. Chem. **36** (1972): 1719–1727.

— TANAKA, N.: Inhibition of protein synthesis by blasticidin S. II. Studies on the site of action in *E. coli* polypeptide synthesizing systems. J. Biochem. **60** (1966): 632–642.

YONEHARA, H., and OTAKE, N.: Absolute configuration of blasticidin S. Tetrahedron Lett. 1 (1966): 3785–3791.

— SETO, H., AIZAWA, S., HIDAKA, T., SHIMAZU, A., and OTAKE, N.: The detoxin complex, selective antagonists of blasticidin S. J. Antibiot. **21** (1968): 369–370.

YUKIOKA, M., HATAYAMA, T., and MORISAWA, S.: Affinity labelling of the ribonucleic acid component adjacent to the peptidyl recognition center of peptidyl transferase in *Escherichia coli* ribosomes. Biochem. Biophys. Acta. **390** (1975): 192–208.

Chapter 21

Chemotherapy of human fungal infections

H. VANDEN BOSSCHE

Department of Comparative Biochemistry, Janssen Research Foundation, B2340 Beerse, Belgium

Introduction

The mycoses are among the most ubiquitous of all infections that plague mankind (AJELLO 1991). Fungal infections occur in every climatic zone, afflict all strata of society and are still on the rise. This is related to the increased use of powerful and broad spectrum antibacterials, prosthetics, cytotoxic chemotherapy, transplantation and more aggressive immunosuppression, the increased survival of premature infants in neonatal intensive care units and, finally but not least, the acquired immunodeficiency syndrome (AIDS) (DROUHET and DUPONT 1990a; JACOBS 1990; STEVENS 1992).

This contribution will focus on the antifungals of current use in antifungal chemotherapy. However, before discussing the properties of the antifungal agents available this chapter starts with a brief description of a number of human mycoses.

Fungal infections

Table 21.1 gives an overview of common and rare mycoses and their etiologic agents. *Candida* infections have been found in virtually every tissue of the human body. Oral candidosis (oral thrush) is now the commonest cosmopolitan fungal disease in human immunodeficiency virus (HIV) positive patients (incidence up to 90%) (DROUHET and DUPONT 1990). Oral thrush afflicts approximately 5% of babies in the first week of life and denture stomatitis (denture sore mouth) has been estimated as high as 60% in the elderly (ODDS 1987, 1988). Oral infection often spreads to the esophagus in HIV-infected individuals to produce esophagitis. The prevalence of esophageal candidosis in AIDS patients is between 4 and 14% (DROUHET and DUPONT 1991).

Vulvovaginal candidosis (vaginal thrush) is an increasingly common problem in otherwise healthy women. Seventy-five % of all women are confronted weth vaginal candidosis, at least once in their lifetime (SOBEL 1985). The etiologic agent is *C. albicans* in about 90% of the cases (SOBEL 1989). At specialist hospital clinics over 80% of episodes of fungal vaginitis are due to *C. albicans* 5–15% to other *Candida* species (3–4% *C. tropicalis*) and about 12% to *C. (Torulopsis) glabrata* (SOBEL1989).

Fungi of the genus *Candida* are also frequent causes of skin infections. The most common organism is again *C. albicans*, but *C. tropicalis*, *C. guilliermondi* and *C. parapsilosis* may also cause infection of the skin (HAY and KALTER 1990). The superficial infections caused by *Candida* species also include those affecting keratinized tissues of the body folds (intertrigo), nail folds (paronychia) and nail plates (onychomycosis).

Table 21.1 Mycoses and their etiologic agents[a])

Mycoses	Etiologic agents
Alternaria infection	*Alternaria alternata*
Aspergillosis	*Aspergillus fumigatus, A. flavus, A. niger*
Blastomycosis	*Blastomyces dermatitidis*
Candidosis	*Candida albicans, C. glabrata (Torulopsis glabrata), C. guilliermondi, C. inconspicua, C. krusei, C. lusitaniae, C. parapsilosis, C. tropicalis*
Chromoblastomycosis	*Cladosporium carrionii, Fonsecaea pedrosoi*
Coccidioidomycosis	*Coccidioides immitis*
Cryptococcosis	*Cryptococcus neoformans*
Dermatophytosis	*Trichophyton rubrum, T. mentagrophytes, T. tonsurans, T. verrucosum, Microsporum canis, M. gypseum, Epidermophyton floccosum*
Fusariosis	*Fusarium solani, F. oxysporum, F. moniliforme, F. chlamydosporum, F. anthophilium*
Histoplasmosis	*Histoplasma capsulatum*
Hyalohyphomycosis	*Fusarium proliferatum, Pseudallescheria boydii*
Lobomycosis	*Paracoccidioides loboi*
Onychomycosis	*T. rubrum, T. mentagrophytes, Ep. floccosum, C. albicans, Aspergillus* spp., *Scopulariopsis brevicaulis*
Paracoccidioidomycosis	*Paracoccidioides brasiliensis*
Penicillium infection	*Penicillium marneffei*
Pityriasis versicolor	*Malassezia furfur*
Phaeohyphomycosis	*Exophiala jeanselmei, E. spinifera, Fonsecaea pedrosi, Hendersonula toruloidea, Phialophora richardsiae, P. verrucosa, Scytalidium hyalinum, Wangiella dermatitidis*
Pneumocytosis	*Pneumocystis carinii*
Rhinosporidiosis	*Rhinosporidium seeberi*
Sporotrichosis	*Sporothrix schenckii*
Zygomycosis	*Absidia corymbifera, Basidiobolus haptosporus, Conidiobolus coronatus, Cunninghamella bertholletiae, Mucor* spp., *Rhizopus* spp., *Saksenaea vasiformis*

[a]) Complied from: AJELLO 1991; CHAIPRASERT 1990; DEGREEF 1990; DROUHET and DUPONT 1990a; KAMESWARAN 1991; KOTRAJARAS 1990; KWON-CHUNG 1988; ODDS et al. 1992.

Deep *Candida* infections range from focal infections of, e.g., the peritoneum and the urinary tract, to widespread invasive disease. Disseminated candidosis has been associated with intravenous catheters, thoracic and abdominal surgery, neutropenia and IV drug use (FRASER *et al.* 1992), it occurs rarely in AIDS patients; *C. albicans* and *C. parapsilosis* account for the majority of isolates but *C. krusei* has also been isolated (CHU *et al.* 1990; DROUHET and DUPONT 1991).

Whereas oral candidosis is now the commonest fungal disease in HIV positive patients, cryptococcosis, generally presenting as meningitis, is the most common life-threatening fungal disease affecting AIDS patients. Its prevalence varies from 5% in Western Europe to 20% in tropical Africa (VANDEPITTE 1990). Prior to the AIDS epidemic, cryptococcosis was most commonly reported from patients with e.g. Hodgkins disease, leukemia, sarcoidosis and rheumatoid arthritis (KWON-CHUNG *et al.* 1990). Now, at least 50% of the total cryptococcosis cases reported annually in the U.S.A. are from AIDS patients, ranking as the fourth most life-threatening infection in these patients (KWON-CHUNG *et al.* 1990). The opportunistic fungus *Cryptococcus neoformans* is the only etiologic agent. Inhalation of cryptococci-laden dust by susceptible persons may cause primary pulmonary cryptococcosis, a benign and usually self-limited respiratory disease. However, disseminated cryptococcosis is invariably fatal if untreated.

The dimorphic fungus *Histoplasma capsulatum* is the etiologic agent of histoplasmosis, an important opportunistic infection in patients with impaired cellular immunity, particularly in AIDS patients. Histoplasmosis is diagnosed in all continents but with greatest concentration of cases located in the eastern half of the U.S.A. (RINALDI 1990) where it is at the moment the most common endemic mycosis with over 500,000 cases diagnosed annually (WHEAT 1993). Following inhalation of *H. capsulatum* microaleuriospores most individuals develop asymptomatic or self-limited pulmonary infection (WHEAT 1988). Histoplasmosis cause chronic pulmonary infection in individuals with underlying lung disease, and widespread disseminated infection in those who are immunocompromised (WHEAT 1993).

Another respiratory fungal pathogen of humans is *Coccidioides immitis*. The saprobic phase of this imperfect fungus inhabits alkaline desert soil in regions of the southwestern United States, Mexico, Central and South America (COLE and KIRKLAND 1991). Its airdispersed arthroconidia can initiate a human respiratory disease known as coccidioidomycosis or valley fever. In most patients, *C. immitis* produces relatively minor transient symptoms that may not even need medical attention (GALGIANI et al. 1990). However, coccidioidal infection can also lead to chronic, morbid and even lethal complications. Severe manifestations of infection are common when *Co. immitis* infects immunosuppressed patients such as recipients of organ transplants, patients with lymphoma and patients treated with high doses of corticosteroids. Coccidioidomycosis also poses frequent and diverse problems in patients already infected with HIV (reviewed by GALGIANI et al. 1990).

Blastomycosis is another important endemic systemic mycosis of North America (North American blastomycosis). The etiologic agent is the dimorphic fungus *Blastomyces dermatitidis*. Primary infection in humans follows inhalation of conidia into the lungs. At body temperature, the conidia covert to the yeast form. The acute primary pulmonary infection may be asymptomatic or it may produce an atypical pneumonia syndrome that may resolve spontaneously. However, there is occasionally hematogenous dissemination to extrapulmonary organs e. g. skin, bone and internal genitalia. Skin lesions are frequent and infections of bones also occur frequently (for a review see UTZ 1991).

Paracoccidioidomycosis, also known as South American blastomycosis, is a chronic or subacute systemic mycosis, caused by the dimorphic fungus *Paracoccidioides brasiliensis*. It is the most prevalent of the systemic fungal disorders in Latin America. *P. brasiliensis* grows as a filamentous fungus in soil or in culture at 25 °C, but it becomes yeast-like in tissue or in culture at 37 °C. Although the infection with *P. brasiliensis* is commonly acquired by individuals under the age of 20, the progressive form of the disease occurs most often in the 30−50-year age group, and males are more often affected than females by a ratio of 20 : 1. The disease is generally progressive and necessitates treatment. Once cleared, the disease may leave serious fibrotic sequelae (for rewiews see LONDERO 1991 and RESTREPO-MORENO 1990). It is surprising that both paracoccidioidomycosis and blastomycosis remain rarities in the setting of AIDS (GRAYBILL et al. 1990).

Aspergillus species are very common in the human environment and are responsible for a wide spectrum of human and animal disease, ranging in humans from localized colonization of the skin or ear to life-threatening systemic infections of neutropenic patients. Invasive aspergillosis has become increasingly important as a cause of morbidity and death, initially in patients receiving immunosuppression prior to organ transplantation and latterly in hemotologic patients rendered neutropenic by chemotherapy or underlying disease.

Dermatophytosis (ringworm, tinea) is an infection of the hair, skin, and nails caused by keratinophilic fungi, the dermatophytes. Dermatophytes are classified into three genera, *Trichophyton*, *Microsporum* and *Epidermophyton*. Common dermatophytes are *Trichophyton rubrum*, *T. mentagrophytes*, *T. tonsurans*, *T. verrucosum*, *Microsporum canis*, *M. gypseum* and *Epidermophyton floccosum*. Infections of the feet (tinea pedis) are caused by *T. rubrum* or *T. interdigitale*. The main cause of tinea corporis (ringworm of the body) is *T. rubrum*, but other organims may be implicated. The commonest causes of tinea cruris (ringworm infections of the groin) are *T. rubrum* and *E. floccosum*. Tinea capitis (scalp ringworm) is caused by e. g. *M. canis*, *T. tonsurans* and *T. violaceum*. A distinctive form of scalp ringworm caused by *T. schoenleinii* is called favus (for a review see HAY 1991). Onychomycosis are not only caused by *C. albicans* but also and more often by dermatophytes. The most important species implicated are *T. rubrum* and *T. mentagrophytes* (ANDRÉ and ACHTEN 1991).

Pityriasis versicolor is a chronic superficial fungal disease usually located on the upper trunk, neck and upper arms (FAERGEMANN 1990). This infection is caused by the lipophilic dimorphic yeast *Malassezia furfur*. *M. furfur* is also associated with pityrosporum folliculitis and there seems to be a relationship between *M. furfur* and seborrheic dermatitis (SCHUSTER 1984; FAERGEMANN 1990).

In contrast to this very common and benign infection stands the more severe infection phaeohyphomycosis, caused by dematiaceous fungi (for a review see KOTRAJARAS 1990). Cutaneous phaeohyphomycosis is most commonly caused by *Hendersonula toruloidea* and *Scytalidium hyalinum*. Subcutaneous phaeohyphomycosis involves deep dermis and subcutaneous tissue by traumatic inoculation with the dematiaceous fungi *Exophiala jeanselmei* or *Wangiella dermatitidis*. Other agents include e. g. *Phialophora richardsiae*, *Ph. repens*, *Ex. spinifera*, *Exserophilum rostatum* and *Fonsecaea pedrosoi*. In tissues dark yeast-like or hypha-like elements can be found. Initial lesions usually appear as dark scaly patches, they will gradually become confluent and increase in size and hardness. This may lead to the formation of abscesses. *W. dermatitidis*, *F. pedrosoi* and *Ph. verrucosa* not only can cause localized cutaneous or subcutaneous phaeohyphomycosis, but can disseminate in e. g. immunocompromised patients to, for example, liver, kidney and brain. The most commonly isolated species from phaeohyphomycosis of the brain is *Cladosporium trichoides* (*bantianum*). In brain infections phaeohyphomycosis mortality is usually 100%.

Medical mycologists have traditionally distinguished phaeohyphomycosis and chromoblastomycosis by the absence in tissue of sclerotic bodies [isotropically enlarged, thick-walled, muriform (multicellular) units] in the former and their presence in the latter. Indeed, although agents of phaeohyphomycosis such as *W. dermatitidis* have the genetic potential for multicellular production, this morphology is only infrequently expressed *in vivo* (SZANISZLO et al. 1993). Nevertheless the taxonomy of the agents of chromoblastomycosis is still one of the most confusing areas in the field of medical mycology. Agents of chromoblastomycosis are e. g. *F. pedrosoi* and *Cladosporium carrionii*. Puncture wounds are the main mode of entrance for these agents. Chromoblastomycosis is a localized cutaneous and subcutaneous disease characterized by warty-like lesions that become cauliflower-like in gross pathological appearance.

In Table 21.1 a number of rare mycoses are listed. They are caused by ubiquitous saprophytic fungi, such as *Alternaria*, *Fusarium*, *Penicillium* and zygomycetes. In recent years these fungal infections are encounterd in increasing numbers in immunosuppressed patients (for more information see CHAIPRASERT 1990; VIVIANI and TORTORANO 1990; WARNOCK and JOHNSON 1990; VIVIANI et al. 1993). Another rare mycosis is lobomycosis caused by *Paracoccidioides loboi*. Lobomycosis is an exclusively cutaneos mycosis and can be treated effectively by surgical excision.

Table 21.1 also includes pneumonia due to *Pneumocystis carninii* (pneumocystosis). *P. carinii* became an increasingly important cause of pneumonia in immunosuppressed patients in the 1960 and 1970 and is the commonest opportunistic pathogen in AIDS patients, affecting up to 85% of cases (CHANDLER 1990; MACKENZIE 1990). Up to 1988 there was general agreement that *P. carinii* was a eukaryote and that it must be a protozoan, probably belonging to the Sporozoa. However, there is molecular biological evidence indicating that *P. carinii* is taxonomically related to fungi (for a review see MACKENZIE 1990). *P. carinii* is susceptible to sulfamethoxazole/trimethoprin, pentamidine and other chemotherapeutic agents (e. g. DL-α-difluoromethylornithine) effective against protozoa (MC CANN et al. 1986; ANONYMOUS 1990) whereas antifungals in current use have little or no effect. Therefore, the therapy of pneumocystosis will not be discussed in detail.

Antifungals

An overview of the antifungals in current use or under development is geven by classifying the compounds by their mechanism of action. During recent years considerable advances have been made in the identification of potential targets for antifungal agents. Examples are listed in Table 21.2 (for a review see KERRIDGE and VANDEN BOSSCHE 1990). Although this table gives a variety of potential targets it is surprising that the clinically available antifungals interfere only with a limited number of targets. They either impair the membrane barrier function, inhibit ergosterol synthesis, inhibit macromolecular synthesis or prevent microtubule assembly.

Table 21.2 Examples of targets for antifungal agents

Organelles	Targets	Inhibitors
cell wall	chitinase	allosamidin
	glycosidases	castonospermine, deoxynojirimycin, bromoconduritol
plasma membrane	impairment of barrier function	polyenes (e. g. nystatin and amphotericin B) azole antifungals (high concentrations): miconazole, clotrimazole, propiconazole, penconazole, fluconazole
	chitin synthase(s)	polyoxins, nikkomycins
	β1,3-glucan synthase(s)	papulacandin, echinocandin, cilofungin
	mannoprotein synthesis	tunicamycin
	proton ATPase	miconazole (high concentrations)
endoplasmic reticulum	squalene epoxidase	naftifine, terbinafine, butenafine, tolnaftate, tolciclate
	P450-dependent 14α-demethylase	pyrimidine, pyridine, imidazole and triazole antifungals
	sterol Δ^{14}-reductase	morpholines (amorolfine)
	sterol $\Delta^8 \to \Delta^7$-isomerase	morpholines
nucleus	nucleic acid synthesis	5-fluorocytosine
	microtubules	griseofulvin
cytosol	microtubules	griseofulvin
	ornithine decarboxylase	difluoromethylornithine (DFMO)

Compounds that impair the membrane barrier function

Polyenes

Polyene macrolide antibiotics, produced by species of *Streptomyces*, are characterized by large lactone rings, containing three to eight conjugated double bonds, which are generally combined with one sugar moiety. They can be subdivided into tetraenes (e. g.

natamycin, nystatin), pentaenes (pentamycin) and the heptaenes (e. g. amphotericin B, candicidin). The chemical structures of amphotericin B and nystatin, are given in Fig. 21.1.

Both amphotericin B (AmB) and nystatin contain eight free hydroxyl groups in the macrocyclic lactone ring and possess as glycosidically linked carbohydrate moiety 3-amino-3,6-dideoxy-D-mannopyranose (mycosamine). These polyenes contain a hydrophobic and a hydrophilic chain, the latter contains the free hydroxyl groups, the former

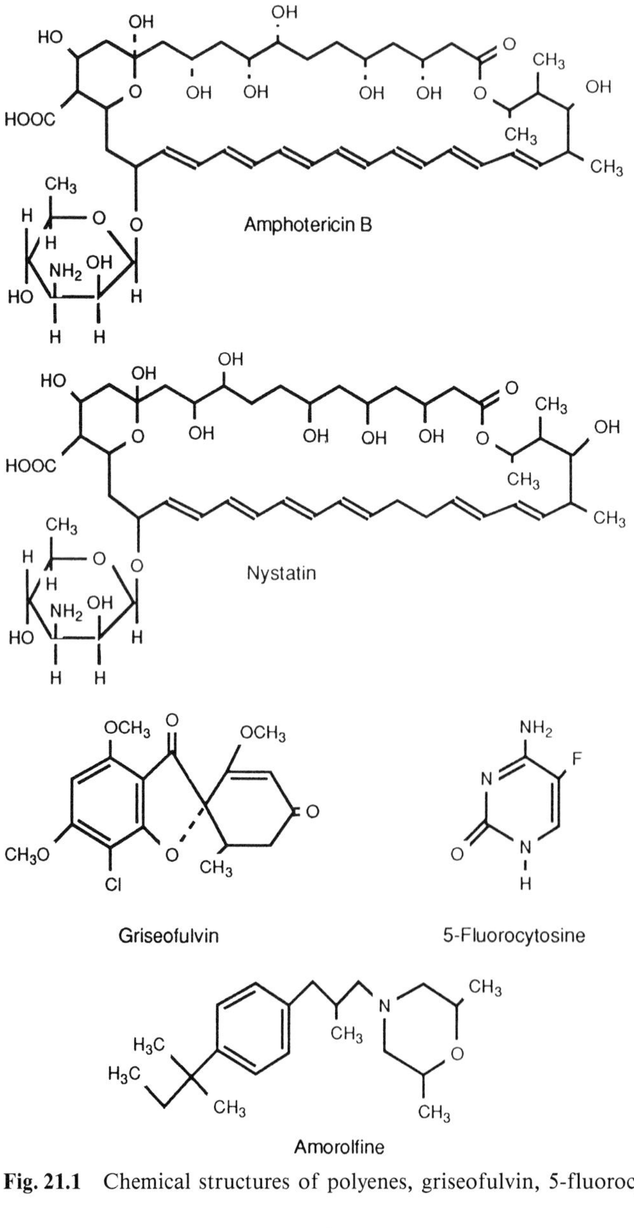

Fig. 21.1 Chemical structures of polyenes, griseofulvin, 5-fluorocytosine and amorolfine.

the polyene chromophore. Mycosamine is bound by a β-glycosidic linkage to the hydroxyl group at carbon 19, it stands together with a carboxyl group at one end of the rod. A hydroxyl group is positioned at the opposite end. AmB is about 24 Å long (ANDREOLI 1974), or about half the thickness of a gel-phase phospholipid bilayer (STORCH and KLEINFELD 1985). At pH 6–7 AmB is almost insoluble in water whereas nystatin is slightly soluble (4 mg/ml). Both are soluble in dimethylsulfoxide (DMSO).

Mechanism of action

The toxicity of AmB to fungal cells is thought to occur as a result of its binding to sterols incorporated in cell membranes changing as such the physical state of the membrane, causing an impairment of membrane functions, resulting in an enhanced permeability to protons and leakage of internal constituents such as K^+, Ca^{2+} and PO_4^{3-} (BOLARD 1986, 1991; VANDEN BOSSCHE et al. 1987a; KERRIDGE et al. 1988; BRAJTBURG et al. 1990, 1992). Although experiments of GRUDA et al. (1991) did not show any difference in AmB interaction with ergosterol or cholesterol, studies by READIO and BITTMAN (1982) indicated that AmB is bound more tightly to ergosterol than to cholesterol, desmosterol, lanosterol, β-sitosterol and stigmasterol. Furthermore, in liposomes, the sensitivity to AmB is increased almost 15 times by the incorporation of ergosterol instead of cholesterol (TEERLINK et al. 1980). Thus, the membrane sterols play a role in toxicity and selectivity, but, other factors may also be involved.

As reviewed by MEDOFF (1988), KERRIDGE (1988), BOLARD (1991) and BRAJTBURG et al. (1990, 1992) the cellular effects of AmB are complex. These effects occur in three dose-dependent stages:
1. At low AmB concentrations (0.01–0.04 µg/ml) an increase of colony-forming units occur. This effect was attributed to an increase in plating efficiency. The same stimulatory effects may be expected to occur on host cells. Indeed, at sublethal doses AmB has stimulating properties, in particular, for cells of the immune system (the *in vitro* and *in vivo* immunomodulating properties are reviewed by MEDOFF 1988, BOLARD 1991, 1992). There is evidence indicating that oxidation-dependent units are involved in AmB-induced stimulation of cells of the immune system (BRAJTBURG et al. 1990).
2. Inhibition of growth and germ-tube formation occurs at intermediate concentrations (0.02–0.10 µg/ml). AmB-induced growth inhibition of for example *C. albicans* can be attributed to several events. One of the earliest indications of changes induced by AmB is K^+ efflux. This efflux could be the consequence of the polyene-induced collapse of the proton gradient. The effects of polyenes on the proton gradient may result from an interaction with the proton-translocating adenosine triphosphatase of the plasma membrane. This interaction may be the result of direct effects of AmB on this ATPase or of AmB-induced alterations in the plasma membrane (SURARIT and SHEPHERD 1987; VANDEN BOSSCHE et al. 1987a). The latter may be related to AmB-ergosterol binding.
3. At higher concentrations (>0.3µg/ml), AmB is lethal to, for example, *C. albicans* and *Cryptococcus neoformans* and is lytic to erythrocytes and protoplasts. Lethality is not a simple consequence of changes in permeability of cell membranes (BRAJTBURG et al. 1990). Evidence for the role of reactive forms of oxygen in the lethal or lytic action of AmB is obtained from studies which showed that AmB injury to cells can be modulated by extracellular scavengers (e.g. 1-phenyl-3-pyrazolidone) and prooxidants (e.g. ascorbic acid). Furthermore, exposure of erythrocytes or protoplasts of *C. albicans* to AmB under hypoxic conditions reduced AmB-induced lysis. In addition, erythrocytes with higher levels of catalase activity are less sensitive to lysis by AmB (reviewed by BRAJTBURG et al. 1990, 1992).

Thus, more than 4 decades after the 1st description of nystatin (HAZEN and BROWN 1950) and 36 years after the isolation of AmB it is clear that the mechanism(s) of action of the polyenes is complex and still not fully understood. The multiplicity of interactions with both fungal and mammalian cells may be the reason why AmB is such an effective, but not a well tolerated antifungal agent.

Antifungal activity

Polyenes are highly effective in the treatment of many mycoses (for their use in oral candidosis see Table 21.8). The most widely used member of the group is amphotericin B. AmB has, depending on the concentration used, fungistatic and fungicidal properties and is active *in vitro* against *Candida* species including *C. glabrata, C. tropicalis* and *C. parapsilosis* as well as *C. albicans*, it is also active against *H. capsulatum, Cr. neoformans, Co. immitis, Rhodoturula, B. dermatitidis, P. brasiliensis, S. schenckii* and strains of *Aspergillus*. It is also active *in vitro* against dermatophytes and some etiologic agents of phaeohyphomycosis (WILSON and RYLEY 1991; KATRAJARAS 1990).

AmB is not absorbed appreciably from the gastrointestinal tract or from sites of muscular injection, and therefore administered almost exclusively by the intravenous route. Because of the insolubility of AmB in water, the conventional formulation is prepared by adding sodium deoxycholate to form a micellar dispersion for parenteral administration (Fungizone-Bristol-Myers Squibb). The history of the clinical use of AmB has been reviewed by BENNETT (1992).

Since AmB is a broad-spectrum antifungal agent, is fungicidal, and has a rapid onset of action it is still indicated for treatment of a life-threatening fungal infection (SAROSI 1990). For example in a great number of infectious disease departments AmB is the treatment of choice in initial therapy of cryptococcal meningitis in patients with AIDS (ARMSTRONG 1988; CHUCK and SANDE 1989; SAROSI 1990). The dose is between 0.5 and 1 mg/kg/day. For discussion of AmB administration schedules see ARMSTRONG and SCHMITT (1990) and SAROSI (1990). If cryptococcosis is diagnosed early, an excellent response can be expected, with more than 70% response rates (ARMSTRONG 1988). A marked improvement is also noted in patients with systemic candidosis (ALY 1990), disseminated coccidioidomycosis or histoplasmosis (ARMSTRONG 1988; SAROSI 1990), and blastomycosis (KLEIN 1990; SAROSI 1990). Intrathecal together with systemic AmB is used in the treatment of meningeal coccidioidomycosis. Intrathecal AmB should be started at 0.1 mg diluted in a small volume of 5% glucose solution and increased gradually to a maximum dose of 1 mg (for details see SAROSI 1990).

Maintenance therapy is indicated in AIDS patients infected with *Cryptococcus*, *Histoplasma* or *Coccidioides*. At present orally absorbed azole antifungals can be used for this purpose (see later).

The clinical use of the conventional AmB formulation is limited by severe side effects (for reviews see SUGAR 1986; ARMSTRONG and SCHMITT 1990; SAROSI 1990). Nephrotoxicity is the most common toxic effect. Signs of nephrotoxicity appear almost immediately after therapy is started. Renal toxicity is mostly reversible, but, permanent changes in renal function and histology have been demonstrated (SAROSI 1990; BENNETT 1992).

Many patients develop fever during the first infusions with AmB. The common local side effect is thrombophloebitis. The list of toxic effects also includes anemia, nausea, vomiting, weight loss and headache. Neurotoxicity is rare, but intrathecal administration may cause significant toxicity (SAROSI 1990).

Attempts to produce a less toxic formulation started with the synthesis of a series of N-acylation products (for reviews see SCHAFFNER 1987; BENNETT 1992). Although their *in vitro* activity was comparable to AmB, they showed reduced chemotherapeutic efficacy. Esterification of the carboxyl group provided the water soluble amphotericin B methylester (AmE) with *in vitro* antifugal activity comparable to AmB and therapeutic efficacy in experimental mycoses. However, although renal toxicity was clearly less, in some patients receiving AmE, neuropsychiatric symptoms (confusion, disorientation, lethargy, depression, anorexia and more rarely, stupor and coma) became evident (SCHAFFNER 1987).

Another and more recent approach is to formulate AmB in liposomes and other lipid complexes. Liposome-encapsulated AmB is less toxic than the usual preparations (LOPEZ-BERESTEIN 1988; MEUNIER et al. 1988) and therefore can be given in higher concentrations (MEUNIER et al. 1988). AmB incorporated into liposomes (L-AmB) composed of dimyristoylphosphatidyl choline and dimyristoylphosphatidyl glycerol (7:3) (JULIANO et al. 1983) shows higher MIC values (RALPH et al. 1991) but it seems to be more active against experimental systemic candidosis in mice than free AmB (LOPEZ-BERESTEIN 1988). Sixteen patients with systemic fungal infections (*Aspergillus* sp.; *Candida* sp. 3; *Histoplasma* sp. 1; *Mucor* sp. 1) who failed to response to fungizone were treated with L-AmB. Of the 16 patients treated, 7 were complete responders, 3 partial and 6 showed no response to treatment (LOPEZ-BERESTEIN 1988). For more information see LOPEZ-BERESTEIN and ROSENBLUM (1992). L-AmB administration was devoid of the severe side effects associated with free AmB. Absence of nepthrotoxicity was also reported by MEUNIER et al. (1988) who used a liposomal preparation composed of egg phosphatidylcholine, cholesterol and stearylamine (4:3:1). The main difficulty at that time was the lack of commercially available liposomal preparations. However, at the moment two formulations are under active clinical evaluation i.e. amphotericin B lipid complex (ABLC, Bristol-Myers Squibb) and amphotericin B colloidal dispersion (ABCD, amphocil™, Liposome technology Inc.) and one was recently launched in the market (Ambisome™, Vestar Ltd).

ABLC is a modified LOPEZ-BERESTEIN formulation containing approximately 33 mol% AmB instead of 5 – 10 mol% (BENNETT 1992). Vestar's ambisome™ contains disteoroylphosphatidylglycerol, hydrogenated phosphatidylcholine, cholesterol and AmB (GONDAL et al. 1989) and is a available as a sterile lyophilized product. Each vial contains 50 mg AmB. A major drawback of this preparation is its price (in U.K. £ 119 per vial). Ambisome and ABLC appear to be less toxic than equal doses of fungizone (BENNETT 1992). Ambisome, while not as potent as free AmB at doses <0.1 mg/kg, is effective in cryptococcal meningitis in mice. Decreased toxicity permitted much higher dosing and at high doses it was superior to free AmB (ALBERT et al. 1992).

The preparation of amphocil™ is based on the specific interaction of AmB with sterols (GUO et al. 1991; LASIC 1992). AmB and cholesteryl sulfate at equimolar ratio form a thermodynamically stable discoidal complex of uniform size. The therapeutic index is 4 – 6-fold improved. In comparison with AmB-deoxycholate at 1 mg/kg/day, ABCD at 5 mg/kg/day prolonged survival of rabbits with invasive pulmonary aspergillosis and demonstrated equivalent microbiological effect. Five mg/kg/day may be the optimal dose for safety and efficacy at least in the system used (ALLENDE et al. 1992). However, GRÜNEWALD et al. (1992) reported treatment failure of ABCD in two HIV-infected patients with cryptococcal meningitis. Both patients had fever and chills during application of ABCD similar to conventional AmB.

Resistance

Despite extensive use of polyene antibiotics for more than 30 years, emergence of acquired or primary resistance seems not to be a significant clinical problem (KERRIDGE 1988; KERRIDGE et al. 1988; ARMSTRONG and SCHMITT 1990; BOLARD 1991). However, acquired resistance has been seen with *C. tropicalis*, *C. parasilosis* and *C. lusitaniae* (BRAJTBURG et al. 1990) and CONLY et al. (1992) reported a disseminated candidosis due to amphotericin B-resistant *C. albicans*.

Compounds that interact with tubulin

Compounds (e. g. benomyl, carbendazim and thiabendazole) that inhibit microtubule assembly are used widely against a broad range of fungal diseases of plants (see chapter 14). One compound of use in medical practice that also interacts with tubulin is griseofulvin (Fig. 21.1).

Griseofulvin

Although griseofulvin had been isolated from *Penicillium griseofulvum* as early as 1939 (OXFORD et al. 1939), its action against experimental ringworm in guinea-pigs was first described in 1956 (GENTLES 1956) and its use as an orally active antifungal against ringworm in man was first described by WILLIAMS et al. in 1958.

Molecular basis of action

Griseofulvin causes major growth abnormalites in sensitive fungal cells, the most characteric effect is "curling" of the hyphae. This distortion of the hyphae is associated with its fungistatic effect, its fungicidal effect results from rupture of the cell wall. In studies with *Basidiobolus ranarum*, the etiological agent of basidiobolomycosis (a subcutaneous zygomycosis known in many tropical countries), GULL and TRINCI (1973) observed that in the presence of griseofulvin daughter nuclei did not separate during cell division. In addition, shortly after treatment of *Aspergillus nidulans* with the antibiotic, spindle distortion was observed (CRACKOWER 1972). Thus, griseofulvin is now considered to be an inhibitor of mitosis (DEKKER 1984). There is evidence that this antibiotic interacts with intact microtubules (Malawista 1975; LANGCAKE et al. 1983). It has been proposed that it may alter a process vital to the sliding of microtubules necessary for the separation of the chromosomes (LANGCAKE et al. 1983). That the effects on the function of microtubules differ from those observed with compounds such as carbendazim, that prevent mitosis by binding to tubulin subunits of the mitotic spindle, is further supported by the lack of cross-resistance (DEKKER 1984).

Griseofulvin also affects the cytoplasmic microtubules (POLAK 1990). Cytoplasmic microtubules are involved in the intracellular transport of endogenous compounds. POLAK (1990) hypothesized that this interaction results in disturbance of the transport of newly synthesized cell wall continents and may lead to the "curling" effect.

Griseofulvin interacts with tubulins from a variety of source (SLOBODA et al. 1982) but it enters sensitive cells only. Indeed, dermatophytes possess a prolonged energy-dependent transport system for this antibiotic, whereas in insensitive cells (e. g. *C. albicans*) this is replaced by a short energy-independent transport system (POLAK 1990).

Clinical use and adverse reactions

Griseofulvin's spectrum of activity is limited to dermatophytes (*Trichophyton* spp., *Epidermophyton flocosum* and *Microsporum canis*). Griseofulvin has become a standard treatment for onychomycosis caused by dermatophytes (DEGREEF 1990). The oral dose is 0.5 – 1 g/day in adults and 10 mg/kg in children (GREER 1990). Contraindications include pregnancy, patients with porphyria, hypersensitivity to griseofulvin, or hepatocellular failure (GREER 1990). Griseofulvin has been shown to induce drug-metabolizing enzymes in humans (OKEY et al. 1986). It has been reported that administration of this antibiotic stimulates markedly the activity of 5-aminolevulinate synthase, the rate-limiting enzyme of porphyrin synthesis, and cause a rapid inhibition of mitochondrial porphyrin-metal chelatase and may thereby inhibit the synthesis of heme in the liver (DE MATTEIS 1982).

Target: nucleic acid synthesis

Although there are innumerable compounds which inhibit macromolecular synthesis in fungi only **5-fluorocytosine** (5-FC, Fig. 21.1) has been exploited as an antifungal agent. This fluorinated pyrimidine has a narrow spectrum of antifungal activity (SCHOLER 1980). It inhibits growth of *Candida* and *Cryptococcus* at low concentrations (MIC's ranging from 0.1 – 1.2 µg/ml), whereas dematiaceous fungi are only moderately sensitive (MIC's: 1 – 25 µg/ml) (KOTRAJARAN 1990; POLAK 1992a). Although it inhibits growth of many *Aspergillus* isolates its clinical efficacy in aspergillosis is equivocal. This difference in sensitivity may results from the fact that in *Aspergilli* 5-FC exerts only fungistatic activity whereas a fungicidal effect is seen in yeast and dematiaceae after prolonged contact (POLAK 1988a, 1990, 1992a).

Molecular basis of action

The mode of action of 5-FC has been discussed in many reviews (see for example VANDEN BOSSCHE 1987a; POLAK 1988a, 1990). Therefor 5-FC's molecular basis of action will be summarized only.

5-FC is taken up by a cytosine permease an enzyme responsible for the uptake of adenine, guanine, hypoxanthine and cytosine. Inside the cell it is rapidly deaminated to 5-fluorouracil (5-FU) by means of a cytosine deaminase. This is a critical enzyme: without the conversion to 5-FU no antifungal activity is observed. The absence of this key-enzyme in mammalian cells has been considered to be the basis for 5-FC's low toxicity for the host (POLAK 1988a). After the formation of 5-FU two pathways are possible: 5-FU can be metabolized via the pyrimidine salvage pathway to

5-fluorouridylic acid, which is incorporated into RNA and leads in *C. albicans*, *Cryptococcus* and *Aspergilli* to a cytostatic effect. The other pathway leads to the formation of 5-fluorodeoxyuridine-monophosphate (5-FDUMP), a potent inhibitor of the thymidylate synthase and hence DNA synthesis. Inhibition of DNA synthesis leads to a fungicidal effect (POLAK 1990).

Resistance

The occurrence of resistant mutants in *C. albicans* and *Cr. neoformans* is an important drawback (POLAK 1992a; IWATA 1992). The most common enzyme deficiency associated with this resistance in clinical isolates of *C. albicans* is a deficiency in the activity of the uridine monophosphate pyrophosphate phosphoribosyl transferase, an enzyme involved in the synthesis of both 5-FdUMP and 5-FUTP (KERRIDGE et al. 1988). The appearance of mutants resistant to 5-FC is drastically reduced when used in combination with amphotericin B (POLAK 1992a).

Synergism and clinical use

MEDOFF et al. (1972) were the first to report that synergy exists between 5-FC and AmB. The most clearcut indication for 5-FC in combination with AmB is in the treatment of cryptococcal meningitis (ARMSTRONG and SCHMITT 1990; POLAK 1992a). POLAK (1992a) also recommends the 5-FC-AmB combination in the treatment of other opportunistic fungal infections, particularly those due to *Candida*.

5-FC is available in 0.5 g tablets and, in some countries in vials containing 2.5 g 5-FC in isotonic saline. The normal daily dose is 150 mg/kg in four divided doses. However, ARMSTRONG and SCHMITT (1990) recommend to initiate therapy at 100 mg 5-FC/kg daily in four divided doses and to follow blood levels. They also recommend not to give 5-FC alone. Blood levels of 5-FC should be monitored in case of impaired renal function, since 5-FC is mainly excreted through renal clearance.

Toxicity

5-FC is generally well tolerated as long as there is no renal insufficiency, the latter may be a problem when 5-FC is used in combination with AmB. The principal side effects are nausea and vomiting. A more serious (but rare) adverse reaction is liver toxicity, and the most serious side effect is bone marrow toxicity (ARMSTRONG and SCHMITT 1990).

Compounds that interact with enzymes associated with the endoplasmic reticulum

An important group of modern antifungal agents of use in medical practice interferes with the biosynthesis of ergosterol. The pathway leading to the synthesis of ergosterol can be divided into 3 segments (Fig. 21.2). The first comprises the synthesis of squalene from acetyl-coenzyme A. The 2nd segment consists of the enzymes involved in the conversion of squalene to lanosterol i.e. the squalene epoxidase and 2,3-oxidosqualene cyclase. In the third segment of the pathway lanosterol is converted into ergosterol. In most fungal cells this segment of the pathway consists of the 24-methyltransferase, 14α-demethylase, 4-demethylase, Δ^{14}-reductase, Δ^{8-7}-isomerase, Δ^5-desaturase, Δ^{22}-desaturase and Δ^{24}-reductase.

All ergosterol biosynthesis inhibitors presently used in medicine inhibit enzymes in the post-squalene segments of the fungal sterol biosynthetic pathway.

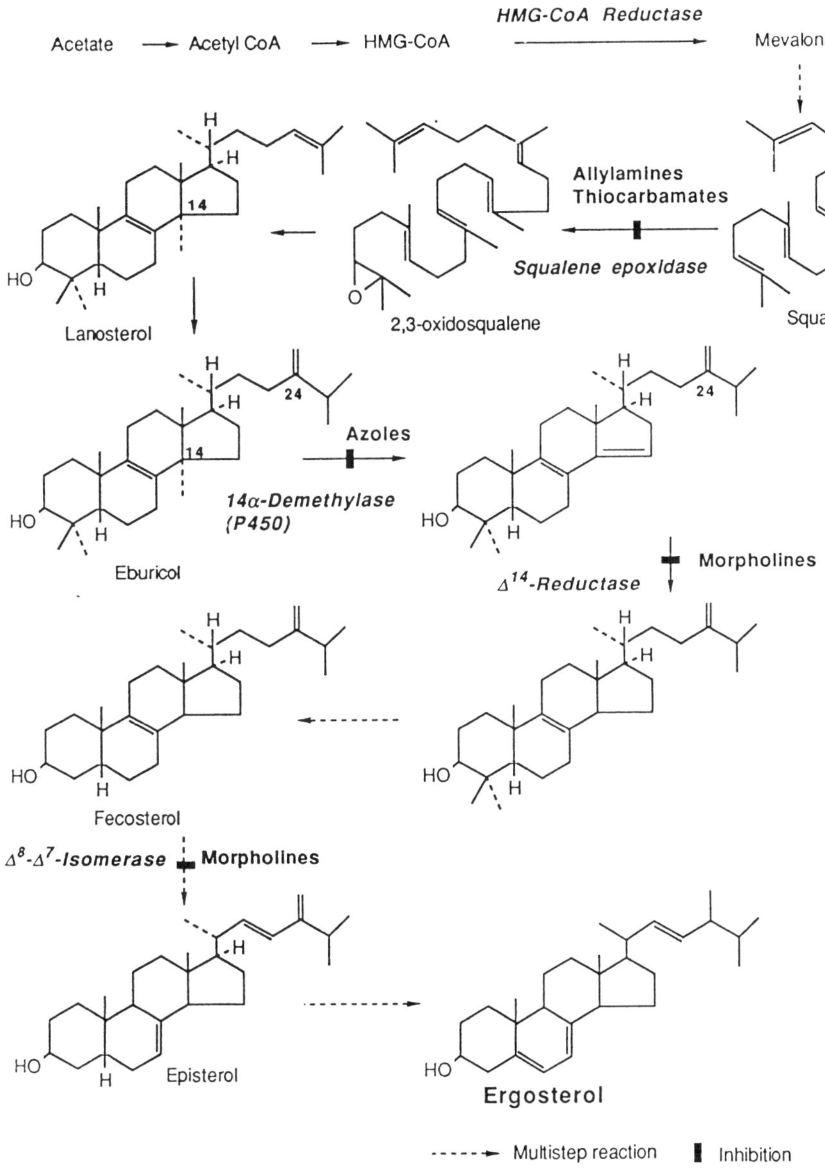

Fig. 21.2 Simplified ergosterol biosynthesis pathway.

Squalene epoxidase inhibitors

The membrane-bound squalene epoxidase (Fig. 21.2) catalyzes the oxidation of squalene to 2,3-oxidosqualene (for reviews see RYDER 1990, 1991; RYDER et al. 1992; MERCER 1991; VANDEN BOSSCHE et al. 1993a).

The squalene epoxidase is composed of a flavin-containing enzyme and a flavoprotein identical to NADPH-cytochrome P450 reductase. This monooxygenase requires next to oxygen, a source of reducing equivalents and phospholipids. In contrast with the other monooxygenases, i.e. cytochromes P450, the squalene epoxidase contains no heme and is not inhibited by carbon monoxide and classical inhibitors of P450 (JANDROSITZ et al. 1991). The purified rat liver squalene epoxidase consists of a single polypeptide chain and has a molecular weight of 51 kDa (ONO et al. 1982). The molecular weight of the S. cerevisiae enzyme is 55 kDa (JANDROSITZ et al. 1991). The rat liver and S. cerevisiae enzymes prefer NADPH as cofactor whilst that from C. albicans prefers NADH. Rat liver epoxidase is inhibited by unsaturated fatty acids whereas that of C. albicans is stimulated (RYDER 1990). The epoxidases from rat liver, C. albicans and S. cerevisiae also differ in their requirements for soluble protein factors. The rat liver enzyme requires these factors for activity, that of S. cerevisiae (M'BAYA and KARST 1987) and C. albicans (RYDER 1990) does not.

Mode of action studies

The first specific inhibitor of fungal squalene epoxidase was **naftifine**. This compound became the prototype of a new class of antifungal agents, the allylamins, of which terbinafine has recently been introduced as an orally and topically active antimycotic agent with *in vitro* fungicidal activity against dermatophytes. Examples of squalene epoxidase inhibitors are shown in Fig. 21.3 (for reviews see RYDER 1990, 1992; RYDER and MIETH 1992; RYDER et al. 1992).

The antifungal activity of the allylamines naftifine (trade name: Naftin), terbinafine (trade name: Lamisil) and SDZ 87-469, of the benzylamine derivative butenafine (RYDER and MIETH 1992) and the thiocarbamates tolnaftate (naphthiomate T) and tolciclate (BARRETT-BEE et al. 1986; RYDER et al. 1986; NOZAWA and MORITA 1992) result from their inhibition of squalene epoxidation causing deficiency of ergosterol and accumulation of squalene. Measuring the effects of naftifine, terbinafine, SDZ 87-469, butenafine and tolnaftate on fungal and mammalian microsomal squalene epoxidation revealed a high degree of selectivity for the fungal enzyme system (RYDER et al. 1992; NOZAWA and MORITA 1992). Naftifine and terbinafine are reversible non-competitive inhibitors of the microsomal squalene epoxidase of C. albicans with IC_{50}-values of 1.1 µM and 0.03 µM, respectively. Butenafine and SDC 87-469 are even more potent inhibitors (IC_{50}-values: 0.045 µM and 0.011 µM) whereas tolnaftate is less active (IC_{50} = 5 µM). Fifty % inhibition of rat liver and guinea-pig microsomal squalene epoxidase is achieved at 144 µM and > 100 µM naftifine, 77 µM and 4 µM terbinafine, and 43 µM and 1.2 µM SDZ 87-469 (RYDER et al. 1992). Fifty % inhibition of the rat liver enzyme is not reached at 0.5 mM tolnaftate (Nozawa and MORITA 1992). The precise mechanism of selectivity is still not claer but may reside in intrinsic differences in the respective epoxidase enzymes.

Recently, RYDER (1991) speculated that the high affinity of terbinafine for the fungal enzyme may result from binding of the molecule to two separate sites on the enzyme, the naphthalene ring to the squalene-binding site, and the side chain to an adjacent lipophilic pocket.

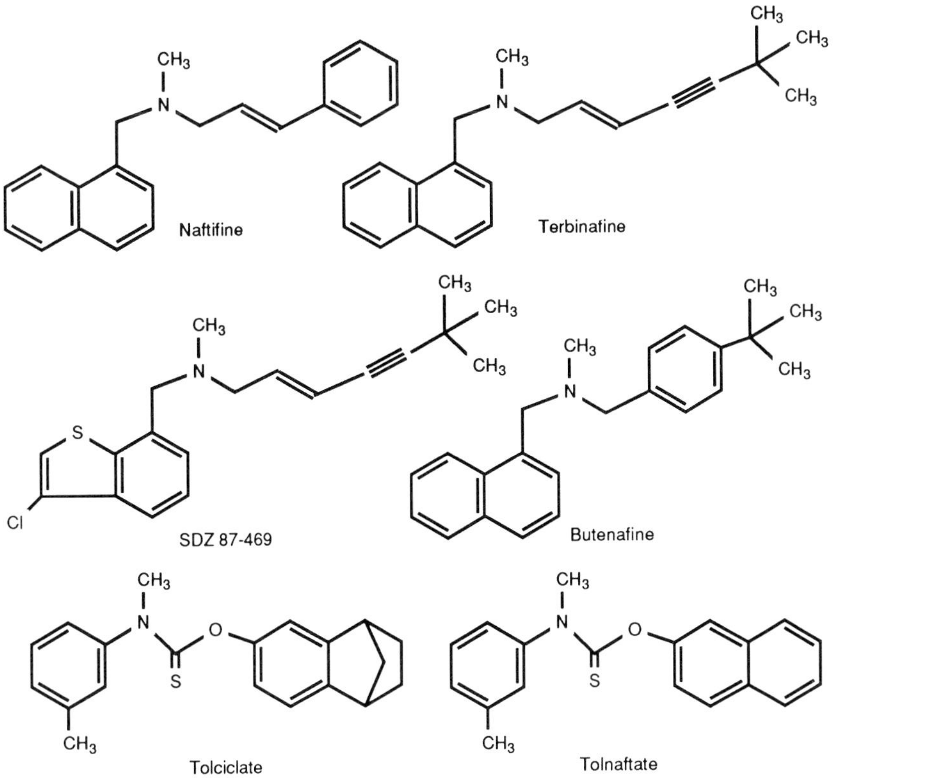

Fig. 21.3 Chemical structures of allylamines, benzylamine derivative and thiocarbamates.

It has been postulated that the antifugal activity of naftifine and terbinafine results solely from their inhibition of fungal ergosterol synthesis and consequent accumulation of squalene (see for example RYDER et al. 1992). Indeed, in dermatophytes (e. g. *T. rubrum* and *T. mentagrophytes*) there is a clear correlation between inhibition of squalene epoxidase, inhibition of growth and fungal cell death (RYDER and MIETH 1992). However, such a clear cut correlation is not seen in *Candida* spp. Although ergosterol synthesis in *C. albicans* (yeast form) and *C. parapsilosis* is almost equisensitive to terbinafine (IC50-values: 8 and 6 µg/ml) the latter species is almost 8 times more susceptible to growth inhibition (RYDER and MIETH 1992). Furthermore, terbinafine is fungicidal against *C. parapsilosis* and fungistatic against *C. albicans* (RYDER and MIETH 1992).

The effects on growth correlate extremely well with inhibition of the residual ergosterol biosynthesis measured by sterol side chain methylation (substrate: [methyl-^{14}C] methionine) in fungal cells incubated for 3 h in the presence of terbinafine. With this method it is possible to measure, in case of a full inhibition of squalene epoxidase, any residual ergosterol biosynthesis occuring distal to the point of inhibition and utilizing endogenous sterol precursors. The extent of this residual biosynthesis varies considerably among the fungi tested. For example, a complete block in ergosterol synthesis is achieved in *T. rubrum* and *T. mentagrophytes* incubated with 0.1 or 1 µg/ml terbinafine. The residual ergosterol synthesis in *C. parapsilosis* and *C. albicans*, incubated in the presence of 1 µg terbinafine/ml is 3.6 and 39% of that measured in cells incubated in the presence of solvent (RYDER 1987).

It is well known that filamentous fungi are much more susceptible to ergosterol biosynthesis inhibition. For example, the hyphal form of *C. albicans* is more susceptible to allylamines than the yeast form. Hypha formation was fully suppressed at a drug concentration causing only 80% inhibition of ergosterol synthesis (RYDER and MIETH 1992). However, this difference in susceptibility to the decreased availability of ergosterol between filamentous fungi and *C. albicans* does not explain the differences observed between *C. albicans* and *C. parapsilosis*. RYDER and MIETH (1992) suggest that rather than ergosterol deficiency, the accumulation of squalene might be the critical effect leading to fungal cell death. Indeed, in *T. mentagrophytes* treated with terbinafine at its MIC (3 ng/ml), cell death coincides with a massive rise in intracellular squalene concentration, while ergosterol content is only slightly decreased (RYDER and MIETH 1992). Furthermore, both *C. parapsilosis* (RYDER 1989) and *T. mentagrophytes* (RYDER et al. 1985) accumulated after 24 h of incubation in the presence of terbinafine (3 µg/ml and 3 ng/ml, respecively) much higher intracellular squalene concentrations than *C. albicans* incubated with 3.1 µg terbinafine/ml (RYDER 1984). Sofar, there is almost no evidence for toxic effects of squalene. A number of studies suggests rather the opposite. For example it has been speculated that squalene may increase membrane fluidity. However, LONG et al. (1988) showed that the squalene that accumulates in yeast grown in the presence of naftifine is located in the cytosol and not in the microsomal fraction. Furthermore, differential scanning calorimetry of multilamellar vesicles of dipalmitoylphosphatidylcholine containing 10 to 35 mole % squalene did not show any effect on the transition temperature or the enthalpy of melting (VANDEN BOSSCHE and MARICHAL 1991). The addition of increasing concentrations of squalene to *T. mentagrophytes* cultures results, after 48 h of growth, in an increased intracellular squalene and ergosterol content. At a squalene concentration of 15 mg/100ml medium the ratio squalene: ergosterol is 6.8 (VANDEN BOSSCHE and MARICHAL, 1993) this is similar to the ratio reached when *T. mentagrophytes* is grown for 48 h in the presence of terbinafine at its MIC value (RYDER 1992). In addition, squalene has been reported to be a safe substance in feeding experiments (HORIE et al. 1990). Therefore, it is speculated than an effect, secondary to the inhibition of the squalene epoxidase may be involved (SISLER and BUCHMAN-ORTH 1990). The epoxidation of squalene requires a reductase that, at least in rat liver, is identical with the NADPH-cytochrome P450 reductase. An inhibition of squalene epoxidase makes the reductase available for e. g. the transfer of an electron to ubiquinone to form semiquinone that can react with oxygen to form superoxide. Superoxide can be converted to oxygen and hydrogen peroxide and the latter reduced to hydroxyl radicals which may abstract H-atoms from polyunsaturated fatty acids (PUFA) to form PUFA radicals. The latter can react with oxygen to form lipid peroxides leading to fungitoxicity (SISLER and BUCHMAN-ORTH 1990). This interesting hypothesis should be evaluated.

Clinical use of naftifine and terbinafine

The lipophilicity and affinity for the skin of both naftifine and terbinafine render them suited to treatment of dermatomycoses (for a recent review see RYDER and MIETH 1992). Topical treatment with naftifine (as 1% cream, Naftin) of infections caused by dermatophytes, including tinea corporis, tinea cruris and tinea pedis led to mycological cure rates of 80 to 100%, and clinical cure rates of 51 to 90% (RYDER and MIETH 1992). Although most *Candida* species and *Malassezia furfur* are *in vitro* less susceptible to naftifine than are dermatophytes (PETRANYI et al. 1987) clinical cure rates of 72 and 76% have been reported in cutaneous candidosis and pityriasis. Naftifine (as a 1% gel formulation used twice daily for 6 months) has also been reported to give cure rates of 42% in patients with onychomycosis of the toes and fingers (see RYDER and MIETH 1992).

In tinea pedis topical terbinafine (as 1% cream, Lamisil) is more active than naftifine giving clinical cure rates of 78% instead of 51%. The activity against other dermatomycoses is similar to that of naftifine with the important exception that the duration of treatment is shorter (results summarized by RYDER and MIETH 1992; MIETH and VILLARS 1992).

Topically appplied 1% terbinafine and naftifine creams are well tolerated. Rare cases of allergic reactions to naftifine and terbinafine have been reported. Side effects such as irritation, redness, burning and dryness have been reported in 2% of the patients treated with terbinafine cream (RYDER and MIETH 1992).

The clinical efficacy of orally administered terbinafine (250 to 500 mg daily) in the treatment of dermatomycoses has recently been reviewed by RYDER and MIETH (1992). In tinea corporis, tinea cruris and tinea pedis mycological cure rates of 90 – 92% and clinical cure rates of 78 – 85% have been reported. In onychomycosis clinical cure rates of 85% are reached after 20 1/2 (finger nails) and 44 weeks (toe nails) of treatment. Cutaneous candidosis responds poorly and pityriasis versicolor does not respond to oral therapy by terbinafine. Although there are some reports suggesting that terbinafine may be useful against some systemic pathogens it is inactive in experimental models of sporotrichosis, phaeohyphomycosis, cryptococcosis and aspergillosis (RYDER and MIETH 1992).

At the recommended dose of 250 mg/day terbinafine has proved to be well tolerated. Adverse effects were noted in 10.4% of patients, most concerned the gastrointestinal tract (RYDER and MIETH 1992; MIETH and VILLARS 1992). Allergic skin reactions have been reported in 1% of the patients (DEGREEF 1990). STRICKER et al. (1992) described eleven patients who developed taste loss after 4 – 8 weeks of treatment with terbinafine. The effect disappears within 3 – 6 weeks after discontinuation of therapy.

14α-Demethylase inhibitors

In the 3rd segment of the ergosterol biosynthesis pathway (Fig. 21.2) lanosterol is first alkylated at C-24 to form eburicol (24-methylenedihydrolanosterol) in filamentous fungi, dimorphic fungi, *Cryptococcus neoformans* and some *C. albicans* strains. Eburicol or lanosterol (*S. cerevisiae*, *C. glabrata* and mammalian cells) from the substrate for the cytochrome P450-dependent 14α-demethylase ($P450_{14DM}$).

$P450_{14DM}$ (product of the *CYP51* gene) purfied to homogeneity form rat liver (TRZASKOS et al. 1986) and *S. cerevisiae* (YOSHIDA 1988; YOSHIDA and AOYAMA et al. 1991) microsomes catalyzes three oxidation steps:

1. the hydrylation of the C-32-methy (14α-methyl) group of lanosterol,
2. 32-hydroxylanosterol is converted to 32-oxolanosta-$\Delta^{8,24}$-dien-3β-ol (3β-hydroxy-8-32-al),
3. in the 3rd step 14α-formyloxy-lanost-8-3β-ol is formed from 3β-hydroxylanost-8-32-al and is further metabolized with the formation of formatic acid and 4,4-dimethyl-$\Delta^{8,14,24}$-cholestatrienol (FISHER et al. 1991). In most fungi the end product of the three monooxygenation reactions is 4,4-dimethyl-$\Delta^{8,14,24(28)}$-ergostatrienol.

The P450-dependent 14α-demethylase is the target enzyme for a long list of modern antifungal agents (cf. chapter 12). The 14-demethylase inhibitors (DMI) of use in medical practice are either imidazole- or triazole derivatives. Their chemical structures are shown in Fig. 21.4; trade names and manufactures are listed in Table 21.3.

Azole antifungal agents

	Generic name	X	R¹	R²	R³
1-Methyl-imidazoles 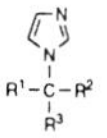	Bifonazole		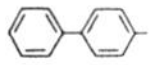 (biphenyl)	(phenyl)	H
	Clotrimazole		(phenyl)	(2-chlorophenyl)	(phenyl)
Ethers of hydroxyethyl or mercaptoethyl-imidazoles 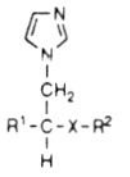	Butoconazole	S	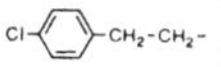 Cl–(phenyl)–CH₂–CH₂–	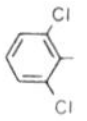 (2,4-dichlorophenyl)	
	Econazole	O	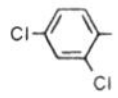 Cl–(phenyl)-Cl	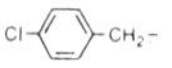 Cl–(phenyl)–CH₂–	
	Fenticonazole	O	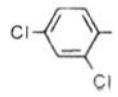 Cl–(phenyl)-Cl	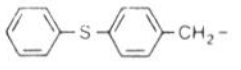 (phenyl)–S–(phenyl)–CH₂–	
	Isoconazole	O	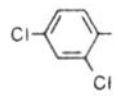 Cl–(phenyl)-Cl	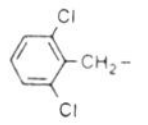 (2,6-dichlorophenyl)–CH₂–	
	Miconazole	O	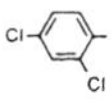 Cl–(phenyl)-Cl	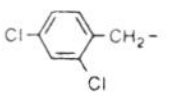 Cl–(phenyl)(Cl)–CH₂–	
	Sulconazole	S	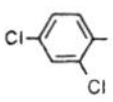 Cl–(phenyl)-Cl	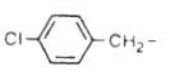 Cl–(phenyl)(Cl)–CH₂–	
	Tioconazole	O	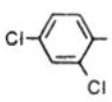 Cl–(phenyl)-Cl	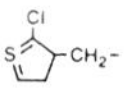 (2-chlorothiophene)–CH₂–	
Ketals of azolyl-acetophenones 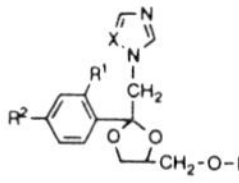	Ketoconazole	CH	Cl	Cl	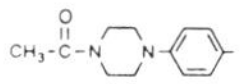 CH₃–C(=O)–N(piperazine)-(phenyl)
	Itraconazole	N	Cl	Cl	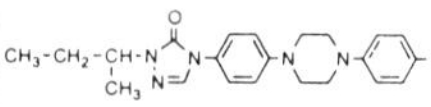 CH₃–CH₂–CH(CH₃)–N(triazolone)-(phenyl)–N(piperazine)-(phenyl)
	Saperconazole	N	F	F	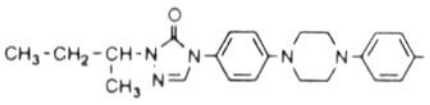 CH₃–CH₂–CH(CH₃)–N(triazolone)-(phenyl)–N(piperazine)-(phenyl)
	Terconazole	N	Cl	Cl	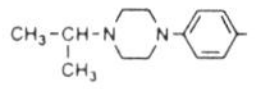 CH₃–CH(CH₃)–N(piperazine)-(phenyl)
Hydroxyethyl-azoles 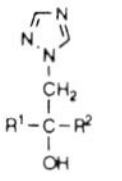	Fluconazole		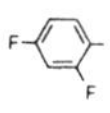 F–(phenyl)–F	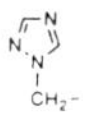 (triazolyl)–CH₂–	
	D 0870 ICI 195,739		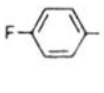 F–(phenyl)	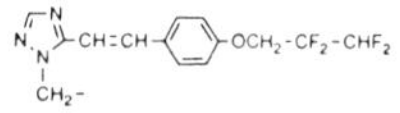 (triazolyl)–CH=CH–(phenyl)–OCH₂–CF₂–CHF₂	

Fig. 21.4 Chemical structures of azole antifungal agents.

Mode of action studies

In 1972, RAGSDALE and SISLER presented the first evidence that the inhibitory site of the pyrimidine antifungal agent triarimol was located in the ergosterol biosynthetic pathway. Later they showed that in *Ustilago maydis* sporidia treated 9 1/2 h with 2 µg triarimol per ml, ergosterol constituted less than 4% of the sterol fraction (RAGSDALE 1975). In treated cells this fraction was composed of eburicol, obtusifoliol and 14-methylfecosterol (structures of sterols are shown in Table 21.6) pinpointing the 14α-demethylase as the target enzyme.

Table 21.3 Azole antifungal agents: origin, trade names

Azole	Origin	Administration	Trade mark
Bifonazole	Bayer	topical	Amycor, Amolmen, Bedriol, Mycospor, Mycosporan
Butoconazole nitrate	Syntex	topical	Exelgyn, Femstat, Gynomyk
Clotrimazole	Bayer	topical	Canesten, Canifug, Expecid, Lotrimon, Mycofug, Pedisafe, Rimazole, Tibatin, Trimysten
Econazole nitrate	Janssen	topical	Econal, Econsole, Epi-pevaryl, Epi-Pevisone, Fitonax, Ifenec, Glyconal, Mycopevaryl, Palavale, Pevaryl, Pevisone, Polycain, Spectazole
Fenticonazole nitrate	Recordati	topical	Falvin, Lamexin
Fluconazole	Pfizer	oral, IV	Diflucan, Triflucan
Isoconazole nitrate	Janssen	topical	Bi-vaspit, Icaden, Nupaten, Travocort, Travogen Tri-nerisona
Itraconazole	Janssen	oral, topical	Fungibet, Itranax, Itrizole, Sempera, Sporal, Sporanox, Trisporal
Ketoconazole	Janssen	oral, topical	Fungarest, Fungoral, Ketazol, Ketoderm, Ketonil, Ketoisdin, Nizoral, Nisral, Orifungal M, Oronazol, Panfungol, Triatop
Miconazole or Miconazole nitrate	Janssen	topical, IV	Albistat, Andergin, Bexicortil, Brentacort, Berntan, Daktacort, Daktar, Daktarin, Daktasol, Daktador, Dermacure, Dermonistat, Dermisdin, Florid, Fungisdin, Micatin, Micolon, Micotef, Mikoderm, Mirasol, Monistat, Mycosolon, Nutragel, Trialone
Sulconazole	Syntex	topical	Exelderm, Myk, Sulcosyn
Terconazole	Janssen	topical	Fungistat, Fungix, Terazol, Tercospor
Tioconazole	Pfizer	topical	Fungibacid, Trosyd, Trosyl, Zoniden

A few years later we were able to prove that the structurally unrelated imidazole derivative miconazole shared with triarimol a common mode of action (VANDEN BOSSCHE et al. 1978). Fifty % inhibition of the acetate incorporation into ergosterol was found after 1 h of incubation of *C. albicans* in the presence of 1 nM miconazole. This inhibition coincided with the accumulation of eburicol, lanosterol, obtusifoliol, 4,14-dimethylzymosterol and 14-methylfecosterol.

The classification of miconazole in the class of DMIs marked the beginning of extensive studies on the interaction of azole antifungal agents with the sterol-biosynthetic pathway in a variety of pathogenic fungi (Table 21.4). The list of azole antifungals presented in Table 21.4 includes, as well as imidazole and triazole derivatives of use in medical practice, also saperconzole and ICI 195, 739 (D0870), both of which have reached the stage of clinical development (for more extented reviews see VANDEN BOSSCHE 1985, 1988; KÖLLER 1992; HITCHCOCK and WHITLLER 1993; VANDEN BOSSCHE et al. 1993c).

Table 21.4 Azole antifungal agents: inhibitors of the 14α-demethylation step in fungal ergosterol biosynthesis

Imidazoles	Species	Ref.[a]	Triazoles	Species	Ref.[a]
Bifonazole	C. albicans	1	Fluconazole	C. albicans	5, 16, 17, 23
	C. glabrata	1		C. glabrata	10, 11, 16, 17
	Ep. floccosum	1		Cr. neoformans	16
	M. canis	1		H. capsulatum	10, 16
	T. mentagrophytes	1, 22			
	T. rubrum	1	Itraconazole	A. fumigatus	6, 7
				C. albicans	6, 10, 16–19
Butoconazole	C. albicans	2		C. glabrata	10, 11, 16, 17
				C. lusitaniae	6, 12
				Cr. neoformans	16, 24
Clotrimazole	C. albicans	1–5		H. capsulatum	10, 16, 17
	C. glabrata	1		P. ovale	6, 13
	Ep. floccosum	1		T. mentagrophytes	20, 22
	M. canis	1			
	T. mentagrophytes	1			
	T. rubrum	1			
			Saperconazole	A. fumigatus	21
				C. albicans	17, 21
Econazole	C. albicans	2–4		C. glabrata	17, 21
				T. mentagrophytes	21
Ketoconazole	A. fumigatus	6, 7			
	A. niger	7			
	C. albicans	2, 4–6, 8, 9	Terconazole	C. albicans	21
	C. glabrata	11		C. glabrata	21
	C. lusitaniae	6, 12			
	H. capsulatum	10	ICI 195, 735	C. albicans	23
	P. ovale	13	(DO870)		
Miconazole	C. albicans	2, 4, 5, 14, 15			
	C. glabrata	10			
	C. krusei	2			
	C. pseudotropicalis	2			
	T. mentagrophytes	15			
Tioconazole	C. albicans	2			

[a] References: 1: BERG et al. (1987); 2: PYE and MARRIOTT (1982); 3: MARRIOTT (1980); 4: GEORGOPAPADAKOU et al. (1987); 5: HITCHCOCK et al. (1990); 6: VANDEN BOSSCHE et al. (1988a); 7: MARICHAL et al. (1985); 8: VANDEN BOSSCHE et al. (1979); 9: VANDEN BOSSCHE et al. (1980); 10: VANDEN BOSSCHE et al. (1990a); 11: VANDEN BOSSCHE et al. (1992b); 12: VANDEN BOSSCHE et al. (1988b); 13: MARICHAL et al. (1986); 14: VANDEN BOSSCHE et al. (1978); 15: MORITA and NOZAWA (1985); 16: VANDEN BOSSCHE et al. (1992a); 17: VANDEN BOSSCHE et al. (1992c); 18: VANDEN BOSSCHE et al. (1984); 19: VANDEN BOSSCHE et al. (1991); 20: VANDEN BOSSCHE et al. (1993a); 21: VANDEN BOSSCHE et al. (1990b); 22: VANDEN BOSSCHE et al. (1990c); 23: BARRETT-BEE et al. (1988) 24: VANDEN BOSSCHE et al. (1993b).

Although all these azole antifungal agents inhibit the fungal 14α-demethylase, the concentration needed to reach 50% or 90% inhibition of ergosterol synthesis are species and drug-dependent. Examples for *C. albicans* and *C. glabrata* are given in Table 21.5. In addition, the accumulating 14-methylsterols also differ in the different species treated with an azole antifungal (Table 21.6). Although there are major differences in the nature of the accumulating sterols, all have a methyl group at C-14, thus proving that the P450-dependent 14α-demethylase is the target enzyme for itraconazole and other imidazole and triazole derivatives.

Since the studies of WILKINSON et al. (1972, 1974) it is known that imidazoles interact with P450, yielding type II binding spectra with a peak at about 430 nm and a trough at about 393 nm. Similar spectral changes were observed when the imidazole antifungals miconazole (VANDEN BOSSCHE et al. 1987b) and ketoconazole (VANDEN BOSSCHE et al. 1984, 1986; YOSHIDA and AOYAMA 1986), and the triazole derivative itraconazole VANDEN BOSSCHE 1987; VANDEN BOSSCHE et al. 1986; YOSHIDA 1988) were added to microsomes from *C. albicans* or *S. cerevisiae* or to purefied P450 $_{14DM}$ from *S. cerevisiae*. These results suggest that the azole groups of these compounds interact with the heme-iron of P450.

Table 21.5 Effects of azole antifungal agents on ergosterol synthesis [a]

Generic name	Inhibition of ergosterol synthesis (μM)			
	C. albicans		*C. glabrata*	
	IC_{50}	IC_{90}	IC_{50}	IC_{90}
Bifonazole	2.3	>10	0.053	0.890
Clotrimazole	0.069	1	0.029	0.420
Econazole	0.084	>1	0.020	0.085
Fluconazole	9.5	>100	2.4	50.1
Itraconazole	0.045	0.220	0.037	0.140
Ketoconazole	0.068	>1	0.052	0.370
Miconazole	0.170	0.300	0.051	3.5
Saperconazole	0.024	0.130	0.028	0.096
Terconazole	0.025	0.160	0.057	0.670

[a] Cells were first grown for 16 h (8 h of which in an orbital shaker) at 30 °C in polypeptone: yeast extract: glucose (10:10:40 g/l) medium, then washed and resuspended in 0.1 M potassium phosphate buffer containing 56 mM glucose (pH 6.5). [^{14}C]-acetate, azole antifungal and/or DMSO were added and the cell suspensions were incubated for 2 h at 30 °C in an orbital shaker (300 rpm). At the end of the incubation period sterols and squalene were extracted and separated by TLC (VANDEN BOSSCHE et al. 1990b). IC_{50}- and IC_{90}-values: concentrations needed to get 50% and 90% inhibition of [^{14}C]-incorporation into egosterol.

It is well known that the reduced form of P450 combines with carbon monoxide and forms the reduced P450-CO-complex having the Soret band at about 450 nm. Miconazole, econazole, clotrimazole, bifonazole, ketoconazole, itraconazole, terconazole, saperconazole, and fluconazole interfere with the appearance of the Soret band at 448–451 nm in the reduced CO-difference spectrum of *S. cerevisiae* (VANDEN BOSSCHE and WILLEMSENS 1982; VANDEN BOSSCHE et al. 1987a), *C. albicans* (VANDEN BOSSCHE and MARICHAL 1991; VANDEN BOSSCHE et al. 1984, 1986, 1987a, 1989), *C. glabrata* (VANDEN BOSSCHE and MARICHAL 1991) or *A. fumigatus* (MARICHAL et al. 1989; BALLARD et al. 1990a) microsomes. Fifty % inhibition of CO-binding to microsomal P450 from *C. albicans* or *C. glabrata* is obtained with nanomolar concentrations of for example miconazole, ketoconazole, terconazole and itraconazole (VANDEN BOSSCHE and MARICHAL 1991). Fluconazole is an 8-times less potent inhibitor of CO-binding. Itraconazole and its imidazole analogue form equistable complexes with *C. albicans* P450, the less hydrophobic ketoconazole forms a less stable P450 complex and the stability of the complex is not increased by replacing the imidazole by a triazole ring (VANDEN BOSSCHE et al. 1988 b.) In addition, nor-ketoconazole (deacylated ketoconazole) has, as compared with ketoconazole, a 3 times lower affinity for the microsomal P450 from *C. albicans* (VANDEN BOSSCHE et al. 1989). Thus, although the binding to the heme iron is a prerequisite for inhibition, the binding of the N-1 substituent to the protein moiety determines the potency.

Studies of YOSHIDA and colleagues (reviewed by YOSHIDA and AOYAMA 1991) prove that the azole antifungal agents interact with $P450_{14DM}$ and inhibit its enzyme activity. Indeed they showed that, for example, ketoconazole inhibited lanosterol demethylation by a reconstituted system consisting of $P450_{14DM}$ and NADPH-cytochrome P450 reductase. Complete inhibition is achieved at a ketoconazole concentration equal to the $P450_{14DM}$ content. This indicates that ketoconazole inhibits the demethylase activity by forming a 1:1 complex.

Ketoconazole and itraconazole are almost equipotent inhibitors of CO-binding to microsomal P450 from *A. fumigatus* (MARICHAL et al. 1989). This does not correspond with itraconazole's greater effect on growth (VAN CUTSEM and JANSSEN 1988) and ergosterol synthesis in intact *A. fumigatus* (MARICHAL et al. 1985; VANDEN BOSSCHE et al. 1988a). It is hypothesized that this difference in activity originates at least partly from the fact that at the intracellular pH (pH < 5.4) measured in the hyphal tip (VANDEN BOSSCHE et al. 1988a), the triazole nitrogen (N-4) of itraconazole is unprotonated whereas the imidazole nitrogen (N-3) of ketoconazole is more than 92% protonated (VANDEN BOSSCHE et al. 1988b). Protonation of the imidazole nitrogen should result in a diminished affinity for the heme iron of $P450_{14DM}$ and thus a lower effect on ergosterol synthesis. This does not mean that all triazole derivatives should show high activity against *Aspergillus*. For example, fluconazole's *in vitro* activity against *Aspergillus* spp. (MIC: 32-64 µg/ml) is much lower than that of itraconazole (MIC: 0.062 – 0.125 µg/ml) (PLEMPEL et al. 1988), as is its activity against systemic aspergillosis in non-predisposed guinea-pigs (VAN CUTSEM and JANSSEN 1988). Itraconazole is more active than amphotericin B (VAN CUTSEM and JANSSEN 1988) whereas fluconazole proved to be much less active against aspergillosis in mice and rats (TROKE 1987). This difference in activity between fluconazole and itraconazole may origate from fluconazole's lower inhibitory effect on ergosterol synthesis. Indeed, studies of BALLARD et al. (1990b) using a cell-free preparation of *A. fumigatus* indicate that fluconazole is about 42-times less active than itraconazole. The low susceptibility of *A. fumigatus* to fluconazole may also be due to low intracellular accumulation of fluconazole. For example, often 1h of incubation *A. fumigatus* accumulated almost 100-fold less fluconazole than itraconazole (VANDEN BOSSCHE et al. 1994).

As already discussed above inhibition of the 14α-demethylase results in a decreased availability of ergosterol and the accumulation of 14-methylsterols and 3-ketosteroids (Table 21.6). In most fungal cells, ergosterol is the sterol best suited to maintain membrane integrity and activity (for reviews see: KÖLLER 1992; MARGALITH 1986; VANDEN BOSSCHE 1990; VANDEN BOSSCHE et al. 1993a). There is evidence that this 24-alkylsterol, at nanomolar concentrations, is also involved in critical non-membrane associated functions in the cell. Indeed, ergosterol can serve as a signal for membrane-associated metabolic events and do so in amounts too small to influence the physical state of the membrane (KAWASAKI et al. 1985; DAHL et al. 1987). Thus, a complete block of ergosterol synthesis as shown in *C. albicans* and *P. ovale* (Table 21.6) should block cell growth and change the physical state of the membrane. However, according to some studies, for its membrane-associated functions ergosterol can be replaced by 14α-methylfecosterol (14-MF). For example, studies of ORTH and SISLER (1990) indicated that *Ustilago maydis* sporidia treated with DMIs, accumulated 14-MF and showed a residual growth rate, as did a 14-demethylase-deficient *U. maydis* mutant (*erg-40*) (SISLER et al. 1983) and a P450-deficient *C. albicans* (strain D10) (BARD et al. 1987). In contrast, the terbinafine-resistant *U. maydis* mutant (AR212) which, upon treatment with the DMI propiconazole showed no accumulation of 14-MF, is much more sensitive to azole antifungals than the

Table 21.6 Sterols present in lipid extracts from fungal cells after incubation in the presence of itraconazole[a])

Species	C. a.	C. g.	Cr. n.	H. c.	P. o.	T. m.	A. f.
Itraconazole (µM)	0.03	0.1	0.03	0.01	0.3	1	1
	% of Total Radioactivity						
(1)	0	14	18	7	0	14	2
(2)	0	0	0	1	0	0	0
(3)	0	0	2	0	0	0	0
(4)	71	57	4	4	0	26	21
(5)	1	1	0	0	0	4	0
(6)	7	0	0	0	0	2	10
(7)	0	6	0	0	0	0	0
(8)	8	0	6	2	7	0	8
(9)	4	0	31	45	93	37	47
(10)	8	21	0	2	0	8	12
(11)	0	0	31	39	0	0	0
(12)	0	1	7	0	0	9	0

[a]) Itraconazole concentrations are those giving more that 80% inhibition of the incorporation from [^{14}C]-acetate into ergosterol. Results are expressed as % of radioactivity incorporated into sterols (I) + 3-ketosteroids (II) + squalene (III). C. a.: *C. albicans*, C. g.: *C. glabrata* both grown for 24 h in CYG-medium; Cr. n.: *Cr. neoformans* grown for 16 h in PYG1-medium; H. c.: *H. capsulatum* grown for 48 h in GY-medium; P.o.: *P. ovale* grown for 48 h in Dixon medium; T. m.: *T. mentagrophytes* grown for 48 h in PYG4-medium; A. f.: *A. fumigatus* grown for 16 h in Sabouraud medium. CYG: casein hydrolysate: yeast extract: glucose (5:5:5 g/l), PYG: polypeptone: yeast extract: glucose (PYG 1:10:10:10 g/l; PYG 4:10:10:40), GY: glucose: yeast extract: cysteine (20:10:0.3 g/l) Dixon medium: malt extract: oxgall: Tween 40: glycerol (40:20 g/l : 10:25 ml/l); Sabouraud: neopeptone: dextrose (10:20 g/l). Except for *T. mentagrophytes* (30 °C), all cells were incubated at 37 °C. (1): ergosterol; (2): brassicasterol; (3): fungisterol; (4): 14α-methyl-ergosta-8,24(28)-dien-3β, 6α-diol; (5): 14α-methyl-ergosta-5,7,22,24(28)-tetraen-3β-ol; (6): 14-methylfecosterol; (7): 4,14-dimethylzymosterol; (8): obtusifoliol; (9): eburicol; (10): lanosterol; (11): obtusifolione; (12): squalene. Data taken from VANDEN BOSSCHE *et al.* (1990a, 1993a, 1993b).

wild-type, which accumulates 14-MF (ORTH and SISLER 1990). It should be noted that the doubling time of the 14-MF accumulating *U. maydis* mutant is 6 1/2 h compared to 2 1/2 h for the wild-type (SISLER et al. 1983). Furthermore, 14-MF is not found in itraconazole-treated *C. glabrata* which is less sensitive than *C. albicans* in which this sterol is accumulating (Table 21.6).

The major accumulating sterol in both *C. albicans* and *C. glabrata* is 14α-methyl-ergosta-8,24(28)-dien-3β, 6α-diol. Studies of KELLY et al. (1991) point to the inability of this 3,6-diol to support growth of *S. cerevisiae* even in the presence of ergosterol.

The only sterols found in itraconazole-treated *P. ovale* are obtusifoliol and eburicol. These sterols are methylated both at position 4 and 14 and are considered equally deleterious for membrane function (LALA et al. 1978; MARICHAL et al. 1990; NES et al. 1978). Both the complete block in ergosterol synthesis and the accumulation of obtusifoliol and eburicol may be at the origin of itraconazole's fungicidal activity against *P. ovale*. [itraconazole is fungicidal at 0.1 μg/ml (VAN CUTSEM 1989)]. High amounts of eburicol are also accumulating in *Cr. neoformans* (VANDEN BOSSCHE et al. 1993b) and the yeast form of *H. capsulatum*. (VANDEN BOSSCHE et al. 1990a) treated with ketoconazole or itraconazole. Next to eburicol, obtusifolione is also accumulating in itraconazole treated *Cr. neoformans* (VANDEN BOSSCHE et al. 1993b) and *H. capsulatum* (VANDEN BOSSCHE et al. 1990a) (Table 21.6) and this 3-ketosteroid is also found in ketoconazole-treated *H. capsulatum* (VANDEN BOSSCHE et al. 1990a). 3-Ketosteroids strongly destabilize the lipid bilayer structure, inhibit the growth of sterol-requiring mycoplasmas and greatly increase the permeability and fragility of erythrocyte membranes (GALLAY and DE KRUYFF 1982). Thus, the interaction (direct or indirect) of itraconazole or ketoconazole with the 3-ketosteroid reductase involved in the 4-demethylation step in the ergosterol biosynthetic pathway might explain the high *in vitro* sensitivity of *Cr. neoformans* and *H. capsulatum* to these antifugal agents. The decreased availability of ergosterol together with the accumulation of eburicol and obtusifolione may be at the origin of itraconazole's excellent *in vivo* activity against *H. capsulatum* and *Cr. neoformans* (VAN CUTSEM 1987, 1992; GRAYBILL et al. 1990; DENNING et al. 1990).

Viable 14α-demethylase-deficient mutants have been described for yeasts only (ZIOGAS et al. 1983; BARUG and KERKENAAR 1984; GUAN et al. 1990). Such a defect seems to be lethal for filamentous fungi and therefore, compounds that completely block ergosterol synthesis and induce the accumulation of 14-methylsterols that disturb membrane functions should be cidal for filamentous fungi. Indeed, itraconazole proved to be cidal for *T. mentagrophytes*, *Microsporum canis* (VAN CUTSEM 1989) and *A. fumigatus* (VAN CUSTEM and JANSSEN 1988). As already mentioned in the section on squalene epoxidase inhibitors, hypha formation is extremely sensitive to ergosterol biosynthesis inhibitors. For example, the addition of 30 nM itraconazole to *C. albicans* cultures in mycelium-promoting medium, resulted in yeast growth only and at 100 nM non-growing clusters of yeast cells were present (VANDEN BOSSCHE and MARICHAL 1993). Analysis of the sterols synthesized from acetate indicated that at 30 nM about 21% of the extract still consisted of ergosterol. Thus, even in the presence of this substantial ergosterol synthesis *C. albicans* is blocked in hypha formation.

The exact mechanism connecting the altered sterol composition with defective morphogenesis is still not really clear. It has been suggested that the tubelike cell structure needs a more stringent condition with regard to physicochemical properties of membranes than the spherical structure and that 14-methylsterols cannot fulfill this requirement (SHIMOKAWA et al. 1986). It is also possible that the ergosterol content is too low to fulfill domain functions that may be more critical for the polarized hypha growth than for yeast budding. Another possibility is that changes in the nature of sterols may affect membrane-bound enzymes that are more needed for hyphal growth. Examples are the chitin synthases. Chitin is an important component of the primary septum in yeast and of the septa and primary wall of hyphae. In *C. albicans* chitin is about three times more abundant in the mycelial than in the yeast cell wall (CHATTAWAY et al. 1968). Mutants of *C. albicans* with a low ergosterol content show increased chitin synthesis (PERTI et al. 1981) and high ergosterol levels inhibit chitin synthase activity (CHIEW et al. 1982). Thus, chitin synthesis should be deregulated in fungal cells treated with ergosterol biosynthesis inhibitors. Indeed, ergosterol biosynthesis inhibitors increase chitin synthesis in for example *C. albicans*, *C. glabrata* and *A. fumigatus* resulting in an irregular distribution of patches of chitin (BARUG et al. 1983; VANDEN BOSSCHE et al. 1985; KERKENAAR 1987; VANDEN BOSSCHE et al. 1988a).

Changes in sterol structure might also lead to an alteration in fatty acid composition. An increase in palmitate was found in *C. albicans* grown in the presence of clotrimazole, econazole, miconazole or ketoconazole (VANDEN BOSSCHE 1985; GEORGOPAPADAKOU et al. 1987). The increased synthesis of saturated fatty acids suggests an effect on the Δ^9 desaturase, a microsomal enzyme whose activity depends on a defined fluidity of the environment. It is thus possible that the azole-induced ergosterol depletion and accompanying accumulation of 14-methylsterols alter the fluidity in such a way that the desaturase is inhibited. From antagonistic effects of unsaturated fatty acids it can be deduced that the decreased availability of unsaturated fatty acids may contribute to the antifungal activity of azole antifungals (Yamaguchi 1977; VANDEN BOSSCHE et al. 1982; GEORGOPAPADAKOU et al. 1987).

Other membrane-bound enzymes whose activities are altered by 14α-demethylase inhibitors are, for example, cytochrome oxidase, peroxidase and ATPases of both mitochondria and plasma membrane (for reviews see VANDEN BOSSCHE 1985, 1988, 1990; KERRIDGE et al. 1988; KELLY et al. 1990; POLAK and HARTMAN 1990). Thus, interaction of azole antifungals with the 14α-demethylase is at the origin of a cascade of perturbations, all together leading to the antifungal activity.

As well as inhibiting the P450-dependent 14α-demethylase some azole antifungal agents affect other targets (for reviews see VANDEN BOSSCHE 1985, 1988). Using differential scanning calorimetry (DSC), it has been shown that high concentrations of miconazole shift the lipid transition temperature of dipalmitoylphosphatidylcholine (DPPC) liposomes from 42 °C to 33 °C without affecting the enthalpy of melting (VANDEN BOSSCHE et al. 1982). Ketoconazole induces a broadening of the main transition peak only. It is suggested that miconazole changes the lipid organization without binding to the lipids, whereas ketoconazole is localized in the multilayer without having an important direct effect on the lipid organization. Itraconazole does also not disturb membrane organization parameters measured by DSC and infrared spectroscopy (BRASSEUR et al. 1991). Conformational analysis studies suggest that the molecular volume and the position of itraconazole in the lipid membrane is similar to that of DPPC. Indeed, the mean molecular area of itraconazole calculated by projecting the molecule on the lipid-water interface is equal to that occupied by DPPC: 60 Å^2/molecule (BRASSEUR et al. 1991). The mean molecular areas of miconazole and ketoconazole are 90 Å^2 and 30 Å^2 (BRASSEUR et al. 1983). Thus, the area occupied per miconazole molecule is much higher than that occupied per DPPC molecule. Such a conformation should result in a destabilizing effect and may explain the direct effects observed with miconazole but not with ketoconazole nor itraconazole. It should be noted that the direct effects of miconazole on membranes are obtained at doses ($\geq 10^{-6}$ M) that can be reached by topical application only (VANDEN BOSSCHE et al. 1982).

Changes in fungal ultrastructure after azole treatment

The ultrastructural changes induced by azole antifungals have been reviewed by BORGERS (1987, 1988), BORGERS et al. (1988), YAMAGUCHI (1988) and YAMAGUCHI et al. (1993). For example, the effects on morphogenesis and cell ultrastructure induced by itraconzole have been studied in *A. fumigatus, C. albicans, Co. immitis, Cr. neoformans, P. ovale, Pa. brasiliensis, S. schenckii* and *T. rubrum* (BORGERS 1987; BORGERS et al. 1981; JANSEN et al. 1991). The primary changes induced by itraconazole and other ergosterol biosynthesis inhibitors are in most species observed at the level of the cell wall and intracellular vacuoles in which lipid-like vesicles assemble. These changes are usually accompanied by a marked increase in cell volume, impaired cell division or abortive hyphal outgrowth. Exposure of *P. ovale* to itraconazole resulted in the disorganization of internal substructures without visibly altering the cell periphery. In contrast with, for example, *C. albicans, Co. immitis, Cr. neoformans* and *T. rubrum*, membranous inclusions were not found in the cell wall of itraconazole-treated *P. ovale* (BORGERS 1987; BORGERS et al. 1981). Cytoplasmic degeneration and plasmolysis predominantly occur in *P. brasiliensis* and *A. fumigatus*. The characteristic changes at the periphery are only occasionally observed (BORGERS 1987). The concentration of azole antifungal necessary to induce irreversible structural degeneration highly depends on the species and morphogenetic form 10^{-10} M (*P. brasiliensis*) to $> 10^{-6}$ M (*C. albicans*) (BORGERS 1987).

Selectivity

Cytochromes P450 are involved in the synthesis and metabolism of endobiotics in mammalian cells. The first candidate to be tested was the $P450_{14DM}$ involved in cholesterol synthesis (VANDEN BOSSCHE et al. 1992a; VANDEN BOSSCHE and JANSSEN 1992). Fifty % inhibition of cholesterol synthesis from mevalonate in subcellular fractions of male rat liver is reached at 2 µM ketoconazole, 6 µM miconazole and 7 µM itraconazole. When macrophages or human hepatoma cells (Hep G2 cells) were incubated in the presence of [^{14}C] acetate and ketoconazole, 50% inhibition of cholesterol synthesis was achieved at 1.5 µM and 0.7 µM, respectively. To reach 50% inhibition in Hep G2 cells 1.3 µM of itraconazole is needed (VANDEN BOSSCHE et al. 1992a). Fluoro substituents on the phenyl ring (saperconazole) provide a lower activity than the corresponding chloro substituents (itraconazole). Indeed, even at 10 µM saperconazole an inhibition of < 40% is found (VANDEN BOSSCHE et al. 1990b). Fluconazole is also a poor inhibitor of the mammalian 14α-demethylase (MARRIOTT and RICHARDSON 1987). These studies indicate that miconazole, ketoconazole, itraconazole, saperconazole and fluconazole inhibit cholesterol synthesis at concentrations much higher than those needed to inhibit ergosterol synthesis in e.g. *A. fumigatus, C. albicans, C. glabrata, Cr. neoformans, H. capsulatum, P. ovale* and *T. mentagrophytes* (Table 21.6).

At concentrations 10 times lower than those needed to inhibit cholesterol synthesis, ketoconazole inhibits the conversion of 17α-hydroxy, 20-dihydroprogesterone into androstenedione by rat testicular microsomes; 50% inhibition is achieved at 0.26 µM indicating that ketoconazole is an inhibitor of the 17,20-lyase (VANDEN BOSSCHE and JANSSEN 1992; VANDEN BOSSCHE and MOEREELS 1994). Similar results were obained with microsomal preparations of pig, dog, bovine and human testes (for reviews see VANDEN BOSSCHE et al. 1987c; VANDEN BOSSCHE 1992; VANDEN BOSSCHE and MOEREELS 1994). In human ovarian cells (mainly theca cells) ketoconazole also inhibits the production of androstenedione from 17α-hydroxyprogesterone. However, 23 µM is needed to achieve

50% inhibition (WEBER et al. 1991). Ketoconazole also interferes with adrenal androgen synthesis (VANDEN BOSSCHE et al. 1987c; AYUB and LEVELL 1989). The effects on adrenal and testicular androgen synthesis have made ketoconazole a good candidate for the treatment of androgen-dependent prostate carcinoma. Indeed, at high doses (400 mg every 8 h instead of 200 mg daily in the treatment of mycoses) ketoconazole is effective both clinically and endocrinologically in the treatment of metastatic prostate carcinoma. However its use is limited by gastric discomfort.

In contrast with ketoconazole, itraconazole (VANDEN BOSSCHE et al. 1986, 1992a), saperconazole (VANDEN BOSSCHE et al. 1990b), fluconazole and ICI 195,739 (BARRET-BEE et al. 1988) are almost devoid of effects on P450-dependent steroid biosynthesis.

Pharmacokinetics of azole compounds

The pharmacokinetic properties of the azole antifungal agents have been reviewed by GASCOIGNE et al. (1981), VAN TYLE (1984), BRAMMER and TARBIT (1987), RITTER (1988), TÄUBER (1988), HEYKANTS et al. (1987, 1989), TARBIT (1990), TARBIT et al. (1990), CAUWENBERGH and HEYKANTS (1990), CAUWENBERGH (1992) and FECZKO (1992). In summary: generally speaking the pharmacokinetics are almost uniform for most topically applied azole derivatives. However, for orally administered azoles, the pharmacokinetics may be significantly different. For example, ketoconazole is predominantly excreted via the sweat, while with itraconazole excretion by the sebaceous glands plays a more important role. Furthermore, the tissue affinity of itraconazole is much greater than that of ketoconazole. In most tissues, itraconazole concentrations are at least 2–3 times the corresponding plasma levels, in adipose tissue even 20 times. Brain concentrations of itraconazole are also higher than plasma concentrations. This explains that, in spite of negligible concentrations in the cerebrospinal fluid (CSF), itraconazole is active in the treatment of cryptococcal meningitis. It should also be noted that, in contrast with ketoconazole, concurrent treatment with the H_2-receptor antagonists cimetidine and ranitidine does not alter single-dose kinetics of itraconazole (HEYKANTS et al. 1989). However, it is recommended that, when itraconazole is administered together with cimetidine or ranitidine, itraconazole blood levels should be regularly monitored.

The bioavailability of ketoconazole is improved when it is given with a meal (GASCOIGNE et al. 1981). To ensure optimal oral absorption, itraconazole may be administered either in capsules shortly after a meal or in solution, the absorption of which is not influenced by the presence of food in the stomach (HEYKANTS et al. 1989).

The pharmacokinetics of fluconazole differ in nearly every respect. This is related to the fact that fluconazole is much less lipophilic than other azole antifungal agents. Furthermore, in contrast to the other azole derivatives, fluconazole's protein binding is only 12%. Fifty to 90% of plasma concentrations are reached in CSF. Fluconazole is eliminated predominantly unchanged via the kidney. Fluconazole's pharmacokinetic profile compensates for its low affinity for the target enzyme in fungal cells. Indeed, as already discussed, fluconazole is a less potent inhibitor of ergosterol synthesis than, for example, itraconazole and ketoconazole. This lower inhibitory potency may originate from fluconazole's lower protein-binding capacity.

At fluconazole dosages of 100 mg daily or more, drug interactions of known or potential clinical consequence may occur with warfarin, hypoglycemics and phenytoin (FECZKO 1992). Cimetidine and rifampin enhance the metabolism of fluconazole (FECZKO 1992). Fluconazole is a poor inhibitor of cyclosporine metabolism by human liver microsomes (BACK and TJIA 1991). Clinical interaction studies are somewhat equivocal. The balance of evidence suggests that fluconazole at doses of 200 mg/day will cause an elevation of cyclosporine blood levels (BACK and TJIA 1991).

Ketoconazole causes marked inhibition of the P450-dependent cyclosporine hydroxylation (P450 3A4) by human liver microsomes (IC_{50} value $= 0.24$ μM), while itraconazole is 10 times less potent (BACK and TJIA 1991). A cyclosporine-ketaconazole interaction was first reported by LOKIEC et al. (1982) in 5 bone marrow transplant patients. A study of SCHROEDER et al. (1987) showed that cyclosporine and ketoconazole can be administered together safely, providing that there is an appropriate reduction in the dosage of cyclosporine. This results in maintenance of adequate immunosuppression without development of nephrotoxicity. This drug interaction also provides a significant reduction in the costs associated with organ transplantation (FIRST et al. 1991).

As with fluconazole, clinical itraconazole-cyclosporine interaction studies are equivocal. Nevertheless, on the basis of their experience in heart and lung transplant recipients, KRAMER et al. (1990) recommend that the dose of cyclosporine be immediately reduced by 50% when itraconazole is instituted at daily doses of 400 mg in order to maintain nontoxic cyclosporine levels. Thereafter, cyclosporine levels must be measured frequently until a new steady state is reached. When itraconazole therapy is stopped, the cyclosporine dose should not be increased until cyclosporine levels start decreasing.

Rifampin and phenytoin reduce serum levels of ketoconazole and itraconazole. Itraconazole does not affect the metabolism of warfarin and phenytoin (VAN TYLE 1984; BENNETT 1990).

Spectrum of activity

The *in vitro* and *in vivo* activities of the azole antifungal agents have been reviewed several times (see for example: VAN CUTSEM and THIENPONT 1972; HEEL et al. 1978, 1980, 1982; VAN TYLE 1984; FROMTLING 1984; CAUWENBERGH and DE DONCKER 1987; VAN CUTSEM et al. 1987, 1989; TROKE 1987; ODDS 1988; PLEMPEL et al. 1988; MENDLING 1988; VAN CUTSEM and JANSSEN 1988; TUCKER et al. 1988; HAY 1988; WILSON and RYLEY 1990; HAY and KALTER 1990; VAN CUTSEM 1989, 1990, 1992; GREER 1990; BENNETT 1990; BLATCHFORD 1990; GRAYBILL 1990; VIVIANI et al. 1990; JUST et al. 1990; DENNING et al. 1990 a, b; HAY 1992; FECZKO 1992; VAN CUTSEM and CAUWENBERGH 1992; STEVENS 1992; DISMUKES 1992; DUPONT 1992; SUGAR et al. 1992). The reader is referred to these articles for a detailed account of both the imidazole and triazole antifungal agents.

Candidosis

Nowadays there are several rapidly acting topical and oral treatments for vaginal candidosis. Azole antifungals are available for local use as creams, lotions, vaginal tablets, ovules or coated tampons. Not only many formulations are available, nowadays physicians are also faced with a great number of different azole derivatives. Examples are listed in Table 21.7. Therapeutic regimens vary from 1 day up to 14 days of treatment (Table 21.7).

Table 21.7 Azole antifungal agents for the treatment of vaginal candidosis[a]

Azole	Formulation	Duration of treatment[b]
Butoconazole local	2% cream	once daily, 3 days
Clotrimazole local	1% cream	once daily, 6–14 days
	10% cream	once single application
	100 mg vaginal tablet	1 tablet/day, 6–7 days
		2 tablets/day, 3 days
	500 mg vaginal tablet	1 tablet
Miconazole nitrate local	2% cream	once daily, 10–14 days
	100 mg ovule	1 ovule/day, 7–14 days
	200 mg ovule	1 ovule/day, 3–7 days
	1 200 mg ovule	one single application
Econazole local	50 mg ovule	once daily, 6 days
	150 mg ovule	once daily, 3 days
	depot ovule	one single application
Terconazole local	0.8% cream	once daily, 3 days
	80 mg ovule	once daily, 3 days
Ketoconazole local	400 mg ovules	once daily, 5 days
oral[c]	200 mg tablets	2 times 1 tablet daily for 5 days
	suspension (20 mg/ml)	2 times daily 10 ml for 5 days
Itraconazole[d] oral	100 mg capsules	2 times 200 mg (1 day treatment)
Fluconazole oral	50 mg capsules	150 mg, single dose
	100 mg tablets	

[a]) Adopted from MENDLING 1988 and SOBEL 1990

[b]) The vaginal ovules, tablets or cream should be inserted at bed time for best results.

[c]) The two tablets can be taken together or one in the morning and one at night. They should be taken during a meal.

[d]) Two 100 mg capsules should be taken in the morning, the other 2 at night.

Oral treatment of vaginal candidosis achieves as effective cure rates (80%–95%) as local treatment (84%–93%) (MENDLING 1988; VANDER PAS 1989; SOBEL 1990; WESEL 1990). With the vast number of different treatments available for vulvovaginal candidosis the choice between topical and oral therapy is not straightforward. However, in a U.K. patient survey on the treatment of vaginal candidosis (TOOLY 1990) oral therapy was the most popular choice.

Although compliance is an important factor in selecting a medication, the advantage of oral therapy must be weighed against the potential side effects and toxicity associated with systemic treatment. For example, it is clear that oral therapy is contraindicated during pregnancy. The use of fluconazole may encourage the emergence of *C. glabrata* and *C. krusei*, both to be much less sensitive to fluconazole than *C. albicans* (MARRIOTT and RICHARDSON 1989; ODDS 1993). *C. glabrata* is also less susceptible than *C. albicans* to ketoconazole whereas both show similar sensitivity to itraconazole (PLEMPEL et al. 1988). Therefore, itraconazole may be indicated in infections caused by *C. albicans*, *C. glabrata* and *C. krusei*.

In mild cases of *Candida* oropharyngitis, local application of miconazole gel or clotrimazole can be prescribed (HAY and KALTER 1990; JUST et al. 1990). Treatment schedules are listed in Table 21.8. The efficacy of the topical antifungals may be insufficient in the treatment of more chronic forms of candidosis or if there is accompanying esophagitis. For these forms an orally absorbed drug should be used. The first developed orally active azole ketoconazole (200 – 400 mg daily) appears to be a suitable alternative to oral nystatin. In chronic infections, it usually takes between 3 to 4 weeks to achieve remission (HAY and KALTER 1990). Ketoconazole at a dose of twice daily 200 mg produced after 28 days of treatment a mycological cure rate of 92% in AIDS patients (GAZZARD and SMITH 1990). JUST et al. (1990) used the ketoconazole suspension at a dosage of 400 – 600 mg/day in patients in advanced stages of HIV infection. This treatment led to a rapid improvement and disappearance of clinical symptoms. The compound was well tolerated without severe side effects. It should be noted that ketoconazole is contraindicated in patients with hepatic failure or in patients in whom liver function tests show significant abnormalities, or those recovering from hepatitis (HAY 1991). The importance of acid for absorption was also found critically important in AIDS patients, who frequently poorly absorbed ketoconazole because of achlorhydria (LAKE-BAKAAR et al. 1988).

SMITH et al. (1990) compared itraconazole 200 mg once daily with ketoconazole 200 mg twice daily in 111 HIV-infected patients with oral candidosis. After 1 week of treatment, 75% and 82% of the patients on itraconazole and ketoconazole respectively had responded clinically. After 4 weeks, the clinical response rate was 93% in each group, and mycological cure rates were 83% for itraconazole and 85% for ketoconazole. One patient discontinued itraconazole because of toxicity (rash), five patients discontinued ketoconazole (2 nausea, 2 hepatotoxicity and 1 rash). All these adverse reactions resolved on discontinuing medication. Dispite successful clinical and mycological clearance, 80% of patients had a further episode of candidosis within the next 3 months.

Initial results obtained with an itraconazole solution (10 mg/ml) at a dosage of 100 mg twice daily in 39 patients with AIDS were encouraging (DESMET et al. 1989). All patients were mycologically cured after 7 days of treatment. Because of the fairly fast recurrence

Table 21.8 Treatment regimens for oropharyngeal and esophageal candidosis[a])

Compounds	Dosage per day	Duration
Topical		
Nystatin	10^6 units $\times 5$[b])	
Amphotericin B	200 – 400 mg $\times 5$	2 – 3 weeks
Miconazole (gel)	2.5 ml (= 62 mg) $\times 4$	
Clotrimazole (troches)	5 – 10 mg $\times 3 – 5$	
Systemic		
Ketaconazole (oral)	200 – 400 mg	
Fluconazole (oral)	50 – 100 mg	10 – 15 days
Itraconazole (oral)	200 mg	
Amphotericin B (parenteral)	15 – 20 mg	

[a]) Adopted from DUPONT (1992)
[b]) times daily

in the degree of colonization, a maintenance therapy of 200 mg itraconazole twice daily once a week was instituted.

The oral itraconazole solution (100 mg twice daily) has been shown to be as effective as fluconazole (capsules 50 mg/day) for the treatment of oropharyngeal and esophageal candidosis in 20 HIV-positive patients (SOUBRY et al. 1991). Fluconazole (50–100 mg/day) seems slightly more effective and better tolerated but much more expensive than ketoconazole (DUPONT 1992). Both ketoconazole and fluconazole can be used daily when relapses are frequent (DUPONT 1992). However, the fact that the use of fluconazole may encourage the emergence of *C. glabrata* and *C. krusei* may oblige practitioners to ascertain the identity of the etiologic agents.

Resistance of *C. albicans* to azole antifungals is known to be exceptional. However, since the introduction of fluconazole in clinical practice an increasing number of patients with clinical resistance to fluconazole with persisting thrush or esophagitis is observed (KITCHEN et al. 1991; WILLOCKS et al. 1991; DUPOUY-CAMET et al. 1991; DUPONT 1992; DUPONT et al. 1992; RODRIGUEZ-TUDELA et al. 1992; ODDS 1993; BART-DELABESSE et al. 1993; JOHNSON et al. 1993; VANDEN BOSSCHE et al. 1994). Resistance is mostly observed after long periods of treatment. The major reason why resistance in *C. albicans* is rare may be the diploid nature of this species and the absence of a known sexual cycle (KERRIDGE et al. 1988). WARNOCK et al. (1988) reported a *C. glabrata* strain that had become resistant to fluconazole after 9 days of treatment of a patient infected with the strain with 400 mg once daily. Such fast appearance of resistance may be related to the haploid nature of *C. glabrata*. Our studies have revealed that both lower fluconazole uptake and increased ergosterol synthesis (due to higher squalene epoxidase and P450-dependent 14α-demethylase activities) are involved in the mechanism of fluconazole resistance. The fluconazole-induced increase in ergosterol synthesis is at the origin of cross-resistance with ketoconazole, itraconazole and the decreased sensitivity to amphotericin B (VANDEN BOSSCHE et al. 1992b). Although it is not established whether low-dose fluconazole (50–100 mg/day) may predispose to emergence of resistance during therapy, it is possible that the fluconazole dose used should be higher to avoid resistance.

Infection of the nail fold with *Candida* spp. (paronychia) can be treated with a topical azole lotion such as miconazole, econazole or clotrimazole (HAY and KALTER 1991). In onychomycosis caused by *Candida*, treatment with oral ketoconazole or itraconazole can be of help (see: Onychomycosis). It should be noted that therapy with ketoconazole necessitates regular monitoring of liver function. Ketoconazole (100–400 mg daily) and itraconazole (100 mg daily) (HAY and CLAYTON 1987; HAY 1991) have been used with success in chronic mucocutaneous candidosis.

Cryptococcosis

A limited number of antifungal agents is suitable to treat cryptococcal meningitis in both non-immunocompromised and immunocompromised patients. Next to amphotericin B and AmB +5-FC, the triazoles fluconazole and itraconazole provide new approaches to the treatment of cryptococcosis, the primary fungal cause of death in AIDS patients.

VIVIANI et al. (1990) treated patients with cryptococcosis with itraconazole (200–400 mg/day) for active infection and/or prevention of relapse (200 mg/day).

Active infection was treated successfully with itraconazole alone in 9 of 12 patients and with itraconazole plus flucytosine (150 to 200 mg/day) in 8 of 10 patients. Failure occured in 2 patients. One of them, after initial clinical and mycological improvement, received rifampicin and had a relapse. Of the 31 patients who received itraconazole maintenance therapy (200 mg/day) for up to 27 months 4 (13%) relapsed.

In a Californian study therapy with itraconazole (200 mg twice daily) was monitored by clinical response, culture and cryptococcal antigen testing in 57 AIDS patients (DENNING et al. 1990a, b). Of the 19 evaluable patients with cryptococcemia at the start of therapy, 17 had negative blood cultures on therapy. Of the 36 evaluable patients with cryptococcal meningitis, 23 (64%) had complete response, 8 (22%) a partial response and 5 (14%) failed therapy. Partial responses or failures were associated with failure of previous therapy, severe disease, low serum concentrations, or a resistant organism. In several instances, low serum concentrations were attributable to concurrent rifampicin therapy. Among the 9 evaluable patients with pulmonary cryptococcosis, 8 responded (DENNING et al. 1990a). Cryptococcuria was abolished in 4 of 10 evaluable patients.

The data collected by VIVANI et al. (1990) and DENNING et al. (1990a, b) indicate that itraconazole is much less toxic and is much better tolerated than amphotericin B.

DUPONT et al. (1990, see also DROUHET and DUPONT 1990b and a review by DUPONT 1992) found after 45–60 days of treatment with 200 mg fluconazole daily that 50% of the 14 patients treated had negative cultures of the cerebrospinal fluid. Negative cultures were found in CSF of 27 of the 31 patients treated with 400 mg fluconazole per day. Relapse rate under maintenance therapy was 23%. In a group of 64 African patients with AIDS and cryptococcosis, treatment with a daily dose of 400 mg fluconazole during the acute phase showed a clinical cure in 63% of the evaluable patients. Mycological responponse to treatment with negative culture was found in 76% of the patients (at day 60–90) (LAROCHE et al. 1992). SAAG et al. (1992) compared intravenous amphotericin B (mean daily dose: 0.4 mg/kg in patients with successful treatment and 0.5 mg/kg in patients with treatment failure) with oral fluconazole (200 mg per day) as primary therapy for AIDS-associated acute cryptococcocal meningitis in a randomized multicenter trial. Treatment was successful in 25 of the 63 AmB recipients and in 44 of the 131 fluconazole recipients. There was no significant difference between the groups in overall mortality due to cryptococcosis; however, mortality during the first two weeks of therapy was higher in the fluconazole group (15% vs. 8%). Thus, fluconazole is certainly an effective alternative to AmB as primary treatment of cryptococcal meningitis when a daily dose of 400 mg is used. As indicated 50 to 90% of fluconazole's plasma concentrations are reached in CSF whereas itraconazole's CSF concentrations are negligible. Nevertheless itraconazole's efficacy in cryptococcal meningitis is at least similar to that of fluconazole. This suggests that meningeal and parenchymal penetration are critical for activity.

Histoplasmosis

According to DISMUKES (1992) AmB is still the drug of choice for AIDS patients with disseminated histoplasmosis. However, because of some failures in patients on AmB, and because of the inconvenience and complications of AmB therapy, alternative new approaches to treatment have been sought (GRAYBILL et al. 1990).

The first candidate was ketoconazole, which *in vitro* is a potent inhibitor of ergosterol synthesis and growth of *H. capsulatum* (VANDEN BOSSCHE *et al.* 1990a). However, the results obtained so far in AIDS patients are not encouraging (GRAYBILL *et al.* 1990). A potential explanation relates to reduced of ketoconazole in patients with AIDS (GRAYBILL *et al.* 1990).

Itraconazole is more potent *in vitro* than ketoconazole (VANDEN BOSSCHE *et al.* 1990a) and in non-AIDS patients it has been found to be highly effective for treatment of histoplasmosis (SAAG and DISMUKES 1988). Patients with widely disseminated disease associated with HIV infection also responded well to itraconazole 400 mg per day (GRAYBILL and SHARKEY 1993). In a prospective multicenter non-comparative trial WHEAT *et al.* (1992) administered itraconazole at a dose of 300 mg twice daily for 3 days then 200 mg twice daily for 12 weeks. Responding patients were then maintained on 200 mg twice daily for at least 52 weeks. Of the 61 patients with AIDS enrolled, 51 responded to therapy. Clinical improvement and clearance of fungemia occurred by week 8 in all responding cases. This study showed that itraconazole is a safe and effective alternative to AmB for treatment of mild and moderately-severe histoplasmosis in patients with AIDS (WHEAT *et al.* 1992).

Fluconazole at daily doses of 100 – 400 mg may also be an effective alternative to AmB for maintenance treatment of histoplasmosis in patients with AIDS (NORRIS *et al.* 1992).

Blastomycosis

Pulmonary and disseminated blastomycosis respond to ketoconazole (for review see KLEIN 1990). Treatment is initiated with a dosage of 400 mg/day. In patients who tolerate the initial dose and have no evidence of clinical progression of disease, the 400 mg/day dosage should be maintained for a minimum of 6 months. In patients whose disease progresses the dosage can be advanced by increments of 200 mg/day up to 800 mg/day. Ketoconazole is generally well tolerated in the dosage range of 400 mg/day. Higher dosages (800 mg/day) may be associated with gastrointestinal and endocrinological effects. Liver function tests should be monitored before starting treatment and at frequent intervals during treatment.

Although ketoconazole is much better tolerated than amphotericin B it is less well tolerated than the triazoles fluconazole and itraconazole. Fluconazole at doses up to 400 mg/day is not highly effective in blastomycosis (SUGAR *et al.* 1992; GRAYBILL and SHARKEY 1993). On the basis of the Mycoses Study Group trials, itraconazole (400 mg/day) appears to be the optimal drug from viewpoints of both efficacy and tolerability (GRAYBILL 1990; GRAYBILL and SHARKEY 1993).

Paracoccidioidomycosis

Ketoconazole at oral doses of 200 mg/day (children: 5 mg/kg/day) is highly effective in the treatment of paracoccidioidomycosis (South American blastomycosis) (RESTREPO-MORENO 1990). In rare cases (for example impaired absorption) the dosage may be increased to 400 mg/day. It is recommended that the course of ketoconazole therapy be maintained for a minimum of 6 to 12 months. Relapses are less frequent (8 – 10%) than those occuring with amphotericin B (30%) (RESTREPO-MORENO 1990). Ketoconazole became the drug of choice for paracoccidioidomycosis by 1983 (SUGAR et al. 1992).

Fluconazole at doses of 200–400 mg/day was effective in 34 of the 37 (92%) patients treated. The six months treatment period appeared to be adequate for avoiding relapses (NEGRONI 1993). Itraconazole at doses of 100 mg/day given for 6 months resulted in the clinical cure of over 99% of the patients treated. The relapse rate was low (2.1%) and all patients who relapsed responded to a renewed treatment with itraconazole (NEGRONI 1993).

Coccidioidomycosis

Primary coccidioidal pneumonia can be treated with ketoconazole at a dose of 400 mg/day given for a period of 2 to 6 months (GALGIANI 1993). For the treatment of patients with progressive forms of non-meningeal coccidioidomycosis, especially those that are immunocompromised, amphotericin B is usually selected (GALGIANI 1993). However, the toxicity of amphotericin B and the high rate of disease relapse have emphasized the need for effective, much less toxic alternatives (TUCKER et al. 1990).

Miconazole IV has been used for some time but despite its effectiveness it has never rivaled amphotericin B (GALGIANI 1993). Toxicity, including nausea, hyponatremia, anemia, pruritus, phlebitis and thrombocytosis, occurred in nearly half of patients (STEVENS 1983).

Initial reports on the efficacy of ketoconazole were more promising (GRAYBILL 1990). However, in the Mycoses Study Group trial, only 23% of 56 patients achieved remission at 400 mg per day, and 32% of 56 patients at 800 mg/day (GALGIANI et al. 1988). Of those who achieved remission, 38% relapsed (GRAYBILL 1990). Even when dosing was increased up to 1600 mg per day, less than 50% of patients achieved remission. At doses above 800 mg/day as many as half of the patients developed nausea and vomiting. Other toxicities included rash, pruritus, liver abnormalities and suppression of adrenal and testicular steroid synthesis (GRAYBILL 1990).

In a study by TUCKER et al. (1990) 79 patients with refractory coccidioidomycosis were enrolled at an initial dose of 50 mg itraconazole per day. This dosage was progressively increased to 200 mg twice a day with meals. Responses were seen in 42 of the 58 assessable patients. Among responders, 17 showed a response within 3 months of initiation, 16 within 3 to 6 Months, and 9 between 6 and 9 months. Itraconazole was well tolerated; mild gastrointestinal intolerance was the most frequently noted side effect. In a study by the NIAID Mycoses Study Group itraconazole was used as long a year at a dose of 400 mg/day. Fifty-seven % of 51 patients responded. The relapse rate after discontinuing treatment was 16% (results summarized by GALGIANI 1993). The same group also studied the efficacy of fluconazole in non-meningeal coccidioidomycosis. At 400 mg/day 61% of the 25 patients responded. The relapse rate with fluconazole was 36% (see GALGIANI 1993).

Sporotrichosis

Treatment of disseminated cutaneous sporotrichosis with oral ketoconazole, 200 mg daily for 10 days, resulted in considerable clinical improvement. Because of a few persistent skin lesions, the dose of ketoconazole was increased for a short interval to 600 mg daily and then reduced to a 400 mg daily dose. The patient tolerated treatment well and his lesions healed with both hypertrophic and atrophic hyperpigmented scars. (CULLEN et al. 1992).

Itraconazole and saperconazole, both at doses of 100 mg/day, appear highly effective in the treatment of lymphocutaneous sporotrichosis (GRAYBILL and SHARKEY 1993). Treatment duration is variable, but should be extended at least until complete mycological cure (VAN CUTSEM and CAUWENBERGH 1992). In osteoarticular disease itraconazole at doses up to 400 mg/day gives response rates comparable to amphotericin B and is much better tolerated (GRAYBILL and SHARKEY 1993).

Disseminated fungal disease due to *Penicillium marneffei*

Infections with the dimorphic fungus *P. marneffei* have been successfully treated with ketoconazole (VIVIANI and TORTORANO 1990) or itraconazole (DUPONT 1992; VIVIANI et al. 1993).

Aspergillosis

Among the azole antifungal agents with possible activity against aspergillosis, itraconazole appears promising. Indeed, *in vitro* and in animal models itraconazole is more active against *Aspergillus* spp. than ketoconazole and fluconazole (VAN CUTSEM and JANSSEN 1988; VAN CUTSEM and CAUWENBERGH 1992). DUPONT (1990) administered daily 200 to 400 mg itraconazole to 49 patients with pulmonary aspergilloma (14), chronic necrotizing pulmonary aspergillosis (14), and invasive aspergillosis (21). Itraconazole had a partial effect on pulmonary aspergilloma. Chronic necrotizing pulmonary aspergillosis appears to be one of the best indications for itraconazole. All patients but one responded to treatment. Invasive aspergillosis is difficult to treat. Osseous and pulmonary involvement were cured in six patients with itraconazole alone and in 9 patients with itraconazole plus initial treatment with amphotericin B and flucytosine. In a study of DENNING et al. (1990) 12 of 15 patients with aspergillosis responded to itraconazole (400 mg/day). Ten of the patients were immunocompromised, 8 of them responded to therapy. Others have confirmed the value of itraconazole in treatment of innvasive aspergillosis (for a review see GRAYBILL 1992).

Fluconazole has been approved for the treatment of invasive aspergillosis in Japan. However, in doses less than 600 mg/day, fluconazole is unlikely to be effective (SUGAR et al. 1992).

Dermatophyte infections

The introduction of the azole antifungal agents for the treatment of dermatophytoses has opened the way for consistent and therefore ultimately effective therapy. The following azole antifungals are used topically: bifonazole, clotrimazole, econazole, isoconazole, miconazole, sulconazole and tioconazole. Ketoconazole is used both topically and orally and itraconazole orally.

For the treatment of superficial mycoses, topical antifungal therapy has proven to be efficacious. However, for patients with infections on large skin aeras topical treatment creates problems: a continuous application of a drug once or twice daily for a minimum 3 weeks on large body surface is difficult and considerable amounts of cream or lotion are needed. Furthermore, application of a topical antimycotic in hairy regions is mostly not homogenous. Therefore, although focal lesions of the glabrous skin (tinea corporis, tinea faciei, tinea cruris) respond well to topical azole antifungals, chronic, recalcitrant cutaneous ringworm lesions, covering extensive body areas, and tinesa capitis require systemic antifungal treatment. A list of treatments of tinea corporis, cruris, pedis, manuum and capitis is given in Table 21.9.

Table 21.9 Azole antifungal agents for dermatophytosis[a])

Tinea	Common agents	Azole antifungal	Use
Tinea capitis barbae	T. tonsurans M. canis T. verrucosum	**Oral:** Ketoconazole[b])	adults: 200 – 400 mg daily children: 50 – 100 mg daily minimum duration: 4 – 8 weeks
Tinea corporis cruris	T. rubrum T. verrucosum M. canis M. gypseum T. rubrum E. floccosum	**Topical:** Ketoconazole 2% Econazole 1% Miconazole 2% Bifonazole 1% **Oral:** Ketoconazole[b]) Itraconazole	once daily (2 – 3 weeks) twice daily (2 – 3 weeks) twice daily (2 – 3 weeks) once daily (2 – 3 weeks) adults: 200 mg daily (4 weeks) children: 50 – 100 mg daily 100 mg daily (15 – 30 days)
Tinea pedis[c]) manuum	T. mentagrophytes E. floccosum T. rubrum	**Topical:** Ketoconazole 2% Econazole 1% Miconazole 2% Bifonazole 1% **Oral:** Ketoconazole Itraconazole	once daily (4 – 6 weeks) twice daily (4 – 6 weeks) twice daily (4 – 6 weeks) once daily (4 weeks) 200 mg daily (4 weeks) 100 mg daily (30 days)

[a]) Partly adopted from GREER (1990)
[b]) Oral ketoconazole is contraindicated in patients with hepatic failure or in patients whom liver functions tests show significant abnormalities.
[c]) Chronic hyperkeratotic, moccasin-type tinea pedis requires 6 – 8 weeks of topical treatment. For recalcitrant tinea pedis oral treatment may be used simultaneously with topical.

Onychomycosis

Onychomycosis is caused by dermatophytes, yeasts and moulds. The most common dermatophytes are *Trichophyton rubrum*, *T. mentagrophytes* and *Epidermophyton floccosum*. *C. albicans* is the primary cause of yeast onychomycosis and *Scopulariopsis brevicaulis* and *Aspergillus* spp. are common causes of mould onychomycosis (DEGREEF 1990).

Onychomycosis and fungal infections of the surrounding nail are difficult to treat. Because fingernails have a linear growth rate 2 to 3 times that of toenails, fingernail mycoses respond better to therapy than toenail mycoses. Three forms of treatment are possible: (1) systemic treatment; (2) local treatment; (3) removal of the affected keratin. Combination of oral and topical therapy plus chemical and surgical procedures (for example use of urea 40% to dissolve keratin and surgical nail avulsion) will often shorten the treatment time (DEGREEF 1990). Onychomycosis caused by yeasts can be treated orally with ketoconazole (200–400 mg/day taken with meals). For *Candida* infections of the fingernails the duration is about 6–7 months.

Itraconazole rapidly reaches the nail plate (within 7 days of starting oral therapy) and high itraconazole concentrations have been detected in nail plates 6 months after stopping treatment (HEYKANTS et al. 1989). Since itraconazole is very active against *Candida* spp. and dermatophytes, has a broader spectrum of activity (e.g. *Aspergillus*) than ketoconazole and certainly than griseofulvin and is well tolerated future studies are likely to confirm itraconazole's clinical superiority to ketoconazole and griseofulvin. At present, the recommended treatment schedule in onychomycosis is 200 mg itraconazole daily for 3 months (CAUWENBERGH 1993).

For the local treatment of onychomycosis partial removal of the nail plate is necessary. It should be noted that when topical agents are used alone treatment can last for up to 18 months (DEGREEF 1990). Azole antifungal agents used for local application are e.g. miconazole, tioconazole and bifonazole.

Pityriasis

Pityriasis versicolor is a mild superficial fungal infection of the stratum corneum caused by *Malassezia furfur*. A number of different topical imidazole antifungal agents are available. They usually need to be applied twice a day for at least 3 weeks. Ketoconazole and bifonazole have to be applied once daily only. A short course of oral ketoconazole (200 mg daily for 10 days) offers a convenient alternative to topical treatments (HAY 1991).

The available clinical data (reviewed by HAY 1991) show that topical ketoconazole is effective for the management of sebborrheic dermatitis (ketoconazole cream once daily 2–6 weeks or ketoconazole shampoo 2%, 2 times weekly during 2–4 weeks and further once weekly) and of pityriasis capitis (dandruff) a mild form of sebborrheic dermatitis linked with the presence of *Pityrosporum ovale* (ketoconazole shampoo).

Chromoblastomycosis and Phaeohyphomycosis

Cutaneous phaeohyphomycosis is difficult to treat (for a review see KOTRAJARAS 1990). *Hendersonula toruloidea* and *Scytalidium hyalinum* are insensitive to ketoconazole. Topical clotrimazole and miconazole for 6 weeks gave no responsive results. *S. hyalinum* is sensitive to itraconazole, however, *H. toruloidea* isolates are at least 100-times less sensitive (VAN CUTSEM et al. 1987).

Ketoconazole is moderately effective in the treatment of subcutaneous phaeohyphomycosis. Ketoconazole plus 5-fluorocytosine can produce additive or synergistic effects (KOTRAJARAS 1990). The most promising azole for treatment of phaeohyphomycosis seems to be itraconazole. Indeed, a study of (SHARKEY et al. (1990) showed clinical improvement or remission in 9 patients (dosages ranging from 50 to 600 mg/day for 1 to 48 months), two patients have had stabilization of disease and 6 patients failed treatment. Although itraconazole appears to be highly effective in some patients, treatment of phaeohyphomycosis remains a significant problem, treatment courses are long and the illness may not completely resolve.

Itraconazole improved the prognosis of chromoblatomycosis. It should be noted that infections by *Cladosporium carrionii* are more sensitive to itraconazole than infections caused by *Fonsecaea predrosoi* (VAN CUTSEM and CAUWENBERGH 1992).

Δ^{14}-Reductase and $\Delta^8 \to \Delta^7$-Isomerase inhibitors

N-substituted morphiline fungicides such as fenpropimorph and **amorolfine** are active against fungi pathogenic to plants and humans. Amorolfine (Fig. 21.1; Ro 14-4767/002, Loceryl) is the only morpholine derivative in clinical use (for reviews see POLAK 1988b; 1990, 1992b; POLAK and HARTMAN 1991; MERCER 1988). Incubation of *C. albicans* for 24 h in the presence of a 31.5 µM concentration of amorolfine resulted in an accumulation of 5α-ergosta-8,14-dienol (ignosterol). This indicates that amorolfine is an inhibator of the sterol Δ^{14}-reductase (Fig. 21.2). Although no accumulation of Δ^8-sterols (fecosterol) was detected amorolfine might share with fenpropimorph its inhibitory effects on the $\Delta^8\to\Delta^7$-isomerase (MERCER 1988). Indeed, such an inhibition could only be detected if sufficient sterol has leaked through the Δ^{14}-reductase catalyzed step. In crude extracts of *S. cerevisiae* the IC_{50} value of amorolfine for the Δ^{14}-reductase is 2.93 µM and for the $\Delta^8\to\Delta^7$-isomerase 0.0018 µM. (POLAK 1992b). Inhibition of both enzymes resulted in a decreased ergosterol synthesis. Studies of MARCIREAU et al. (1990) on the effects of the amorolfine analog fenpropimorph suggest that the effects of this morpholine are not due to the accumulation of abnormal sterols (e. g. ignosterol) in treated cells, but are linked to the decreased ergosterol content.

Fungi, grown in inhibitory concentrations of fenpropimorph and of 14-demethylase inhibitors show the same morphological alterations. They cause an irregular deposition of $\beta 1,3$ and $\beta 1,4$-polysaccharides, probably chitin (VANDEN BOSSCHE 1985; POLAK 1992b). This suggests that the decreased availability of ergosterol results in an uncoordinated synthesis of chitin.

Spectrum of activity and therapeutic efficiency

Amorolfine is efficacious *in vitro* against dermatophytes, *Candida* species, dimorphic and dematiaceous fungi. It is less active against moulds. Growth of aspergilli and most zygomycetes are inhibited at high concentrations only (POLAK 1988b). Ninety-nine % of the cells are killed after 48 h contact with amorolfine at concentrations of 0.003 – 0.01 µM for dermatophytes, 0.25 µM/ml for dematiaceae, 1 µM/ml for *C. albicans* and 1.7 µM/ml for *H. capsulatum* (POLAK 1992b).

The therapeutic efficacy of this morpholine derivative is limited to superficial fungal infections. In clinical trials amorolfine is efficaceous in vaginal candidosis (50 and 100 mg vaginal tablets) and in dermatomycosis (0.5% cream, 2% spray) (POLAK 1992b). After 6 months of twice-weekly treatment with a 5% nail lacquer 54.2% of the 142 patients were mycologically and clinically cured, 19.7% showed significant improvement and 26% did not respond to terapy. This 5% nail lacquer is a available in a number of countries.

Acknowledgements: The author is grateful to Drs. G. CAUWENBERGH, M. JANSSEN and F. C. ODDS for their helpful comments, and to Cindie MICHIELSEN for typing the manuscript.

References

AJELLO, H.: The public health importance of human mycoses. Health cooperation papers. **13** (1991): 19–24.

ALBERT, M., STAHL-CARROLL, T., and GRAYBILL, J.: Comparsion of Ambisome (Abs) to Amphotericin B(AMB) in cryptococcal meningitis (CM). Abstract 178. Program and Abstracts of the 32nd Interscience Conference on Antimicrobial Agents and Chemotherapy (Anaheim Oct. 11–14, 1992).

ALLENDE, M., FRANCIES, P., BERENGUER, J., LEE, J., GARRETT, K., DOLLENBERG, H., PIZZO P., and WALSH, T.: Efficacy of amphotericin B colloidal dispersion (ABCD) in the treatment of invasive pulmonary aspergillosis in rabbits. Abstract 180. Program and Abstracts of the 32nd Interscience Conference on Anticrobial Agents and Chemotherapy (Anaheim Oct. 11–14, 1992).

ALY, R.: Systemic candidiasis (candidosis). In: JACOBS, P.H., and NAHL, L. (Eds): Antifungal Drug Therapy. Marcel Dekker Inc., New York 1990, pp. 165–172.

ANDRÉ, J., and ACHTEN, G.: Onychomycosis. Health cooperation papers **13** (1991): 83–90.

ANDREOLI, T. E.: The structure and function of amphotericin B-cholesterol pores in lipid bilayer membranes. Ann. N.Y. Acad. Sci. **235** (1974): 448–468.

Anonymous: WHO Model Prescribing Information. Drug Used in Parasitic Diseases. World Health Organization, Geneva 1990, pp. 48–49.

ARMSTRONG, D.: Life-treatening opportunistic fungal infections in patients with the acquired immunodeficiency syndrome. Ann. N.Y. Acad. Sci. **544** (1988): 443 450.

— SCHMITT, H. J.: Older drugs. In: RYLEY, J. F. (Ed.): Chemotherapy of Fungal Diseases. Springer-Verlag, Berlin 1990, pp. 439–454.

AYUB, M., and LEVELL, M. J.: The inhibition of human adrenal steroidogenic enzymes *in vitro* by imidazole drugs including ketoconazole. J. Steroid. Biochem. **32** (1989): 515–524.

BACK, D. J., and TJIA, J. F.: Comparative effects of the antimycotic drugs ketoconazole, fluconazole, itraconazole and terbinafine on the metabolism of cyclosporine by human liver microsomes. Br. J. Pharmac. **32** (1992): 624–626.

BALLARD, S. A., KELLY, S. L., ELLIS, S. W., and TROKE, P. F.: Interaction of microsomal cytochrome P450 isolated from *Aspergillus fumigatus* with fluconazole and itraconazole. J. Med. Ved. Mycol. **28** (1990a): 327–334.

— ELLIS, S. W., KELLY, S. L., and TROKE, P. F.: A novel method for studying ergosterol biosynthesis by a cell-free preparation of *Aspergillus fumigatus* and its inhibition by azole antifungal agents. J. Med. Vet. Mycol. **28** (1990b): 335–344.

BARD, M., LEES, R. J., BARBUCH, R. J., and SANGLARD, D.: Characterization of a cytochrome P450 deficient mutant of *Candida albicans*. Biochem. Biophys. Res. Commun. **147** (1987): 794–800.

BARRETT-BEE, K. J., LANE, A. C., and TURNER, R. W.: The mode of action of tolnaftate J. Med. Vet. Mycol. **24** (1986): 155–160.
- LEES, J., PINDER, P., CAMPBELL, J., and NEWBOULT, L.: Biochemical studies with a novel antifungal agent, ICI 195,739. Ann. N. Y. Acad. Sci. **554** (1988): 231–244.

BART-DELABESSE, E., BOIRON, P., CARLOTTI, A., and DUPONT, B.: *Candida albicans* genotyping in studies with patients with AIDS developing resistance to fluconazole. J. Clin. Microbiol. **31** (1993): 2933–2937.

BARUG, D., and KERKENAAR, A.: Resistance in mutagen-induced mutants of *Ustilago maydis* to fungicides which inhibit ergosterol biosynthesis. Pestic. Sci. **15** (1984): 78–84.
- SAMSON, R. A., and KERKENAAR, A.: Microscopic studies of *Candida albicans* and *Torulopsis glabrata* after *in vitro* treatment with bifonazole. Arzneim. Forsch. **33** (1983): 528–537.

BENNETT, J. E.: Current status and perspectives of antifungal therapy. In: VANDEN BOSSCHE, H., MACKENZIE, D. W. R., CAUWENBERGH, G., VAN CUTSEM, J., DROUHET, E., and DUPONT, B. (Eds.): Mycoses in AIDS Patients. Plenum Press, New York 1990, pp. 199–206.
- Developing drugs for the deep mycoses: a short history. In: BENNETT, J. E., HAY, R. J., and PETERSON, P. K. (Eds): New Strategies in fungal disease. Churchill Livingstone, Edinburgh 1992, pp. 3–12.

BERG, D., BÜCHEL, K.-H., PLEMPEL, M., and REGEL, E.: Biochemical characteristics of bifonazole. In: FROMTLING, R. A. (Ed.): Recent Trends in the Discovery, Development and Evaluation of Antifungal Agents. J. R. Prous Science Publishers, S. A., Barcelona 1987, pp. 313–334.

BLATCHFORD, N. R.: Treatment of oral candidosis with itraconazole. A review. J. Am. Acad. Dermatol. **23** (1990): 565–567.

BOLARD, J.: How do polyene macrolide antibiotics affect the cellular membrane properties. Biochim. Biophys. Acta **864** (1986): 257–304.
- Mechanism of action of an Anti-*Candida* drug. amphotericin B and its derivatives. In: PRASSAD, R.(Ed.): *Candida albicans*. Cellular and Molecular Biology. Springer-Verlag, Berlin 1991, pp. 214–228.
- Modulation of the immune defences against fungi by amphotericin B and its derivatives. In: YAMAGUCHI, H., KOBAYASHI, G. S., and TAKAHASHI, H. (Eds.): Recent Progress in Antifungal Chemotherapy. Marcel Dekker Inc., New York 1992, pp. 293–304.

BORGERS, M.: Changes in fungal ultrastructure after itraconazole treatment. In: FROMTLING, R. A. (Ed.): Recent Trends in the Discovery, Development and Evaluation of Antifungal Agents. J. R. Prous Science Publisher, S. A., Barcelona 1987, pp. 193–206.
- Ultrastructural correlations of antimycotic treatment. Curr. Top. Med. Mycol. **2** (1988): 1–38.
- LEVINE, H. B., and COBBS, J. M.: Ultrastructure of *Coccidioides immitis* after exposure to the imidazole antifungals miconazole and ketoconazole. Sabouraudia **19** (1981): 27–38.
- VAN DE VEN, M. A., and VAN CUTSEM, J.: An ultrastructural study of *Aspergillus fumigatus*: effects of azoles. In: VANDEN BOSSCHE, H., MACKENZIE, D. W. R., and CAUWENBERGH, G. (Eds.): *Aspergillus* and Aspergillosis. Plenum Press, New York 1988, pp. 199–211.

BRAJTBURG, J., POWDERLY, W. G., KOBAYASHI, G. S., and MEDOFF, G.: Amphotericin B: Current understanding of mechanisms of action. Antimicrob. Ag. Chemother. **34** (1990): 183–188.
- GRUDA, I., and KOBAYASHI, G. S.: Amphotericin B: Studies focused on improving its therapeutic efficacy. In: YAMAGUCHI, H., KOBAYASHI, G. S., and TAKAHASHI, H. (Eds.): Recent Progress in Antifungal Chemotherapy. Marcel Dekker Inc., New York 1992, pp. 65–76.

BRAMMER K. W., and TARBIT, M. H.: A review of the pharamocokinetics of fluconazole (UK-49,858) in laboratory animals and man. In: FROMTLING, R. A. (Ed): Recent Trends in the Discovery, Development and Ealuation of Antifungal Agents. J. R. Prous Science Plublishers, S. A., Barcelona 1987, pp. 141–156.

BRASSEUR, R., GOORMAGTIGH, E., RUYSSCHAERT, J. M., DUQUENOY, P. H., MARICHAL, P., and VANDEN BOSSCHE, H.: Lipid-itraconazole interaction in lipid model membranes. J. Pharm. Pharmacol. **43** (1991): 167–171.

- VANDENBOSCH, C., VANDEN BOSSCHE, H., and RUYSSCHAERT, J. M.: Mode of insertion of miconazole, ketoconazole and deacylated ketoconazole in lipid layers. Biochem. Pharmacol. **32** (1983): 2175–2180.
CAUWENBERGH, G.: Skin kinetics of azole antifungal drugs. Curr. Top. Med. Mycol. **4** (1992): 88–136.
- Clinical efficacy of itraconazole: focus on cutaneous infections. In: RIPPON, J. W., and FROMTLING, R. A. (Eds.): Cutaneous Antifungal Agents. Marcel Dekker, Inc., New York 1993, pp. 295–308.
- and DE DONCKER, P.: The clinical use of itraconazole in superficial and deep mycoses. In: FROMTLING, R. A. (Ed.): Recent Trends in the Discovery, Development and Evaluation of Antifungal Agents. J. R. Prous Science Publishers, S. A., Barcelona 1987, pp. 273–284.
- and HEYKANTS, J.: Pharmacokinetics of antifungals. In: VANDEN BOSSCHE, H., MACKENZIE, D. W. R., CAUWENBERGH, G., VAN CUTSEM, J., DROUHET, E., and DUPONT, B. (Eds.): Mycoses in AIDS Patients. Plenum Press, New York 1990, pp. 245–254.
CHAIPRASERT, A.: Rare and unusual fungal diseases. In: JACOBS, P. H., and NAHL, L. (Eds.): Antifungal Drug Therapy. Marcel Dekker, New York 1990, pp. 279–296.
CHANDLER, F. W.: Epidemiology of AIDS and its opportunistic infections. In: VANDEN BOSSCHE, H., MACKENZIE, D. W. R., CAUWENBERGH, G., VAN CUTSEM, J., DROUHET, E., and DUPONT, B. (Eds.): Mycoses in AIDS Patients. Plenum Press, New York 1990, pp. 3–12.
CHATAWAY, F. W., HOLMES, M. R., and BARLOW, A. J. E.: Cell wall composition of the mycelial and blastospore forms of *Candida albicans*. J. Gen. Microbiol. **51** (1968): 367–376.
CHIEW, Y. Y., SULLIVAN, P. A., and SHEPHERD, M. G.: The effects of ergosterol content and alcohols on germ-tube formation and chitin synthase activity in *Candida albicans*. Can. J. Biochem. **60** (1984): 15–20.
CHU, F. E., CARROW, M., BLEVINS, A., and ARMSTRONG, D.: Candidemia in patients with acquired immunodeficiency syndrome. In: VANDEN BOSSCHE, H., MACKENZIE, D. W. R., CAUWENBERGH, G., VAN CUTSEM, J., DROUHET, E., and DUPONT, B. (Eds.): Mycoses in AIDS Patients. Plenum Press, New York 1990, pp. 75–82.
CHUCK, S. L., and SANDE, M. A.: Infections with *Cryptococcus neoformans* in the acquired immunodeficiency syndrome. N. Eng. J. Med. **321** (1989): 794–799.
COLE, G. T., and KIRKLAND, T. N.: Conidia of *Coccidioides immitis*. Their significance in disease inhibition. In: COLE, G. T., and HOCH, H. C. (Eds.): The fungal Spore and Disease Inhibition in Plants and Animals. Plenum Press, New York 1991, pp. 403–443.
CONLY, J., RENNIE, R., JOHNSON, J., FARAH, S., and HELLMAN, L.: Disseminated candidiasis due to amphotericin B-resistant *Candida albicans*. J. Infect. Dis. **165** (1992): 761–764.
CRACKOWER, S. H. B.: The effect of griseofulvin on mitosis in *Aspergillus nidulans*. Can. J. Microbiol. **18** (1972): 683–687.
CULLEN, S. I., MAUCERI, A. A., and WARNER, N.: Successful treatment of disseminated cutaneous sporotrichosis with ketoconazole. J. Am. Acad. Dermatol. **27** (1992): 463–464.
DAHL, C., BIEMANN, H.-P., and DAHL, J.: A protein kinase antigenically related to pp. $60^{v\text{-}src}$ possibly involved in yeast cell cycle control: positive *in vivo* regulation by sterol. Proc. Natl. Acad. Sci. USA **84** (1987): 4012–4016.
DE MATTEIS, F.: Loss of microsomal components in drug-induced liver damage, in cholestasis and after administration of chemicals which stimulate heme catabolism. In: SCHENKMAN, J. B., and KUPFER, D. (Eds.): Hepatic Cytochrome P450 Monooxygenase System. Pergamon Press, Oxford 1982, pp. 307–340.
DEGREEF, H.: Onychomycosis. Br. J. Clin. Pract. **44** (Suppl. 71) (1990): 91–97.
DEKKER, J.: Development of resistance to antifungal agents. In: TRINCI, A. P. J., and RYLEY, J. F. (Eds.): Mode of Action of Antifungal Agents. Cambridge University Press, Cambridge 1984, pp. 89–111.
DENNING, D. W., TUCKER, R. M., HOSTETLER, J. S., GILL, S., and STEVENS, D. A.: Oral itraconazole therapy of cryptococcal meningitis and cryptococcosis in patients with AIDS. In: VANDEN

Bossche, H., Mackenzie, D. W. R., Cauwenbergh, G., van Cutsem, J., Drouhet, E., and Dupont, B. (Eds.): Mycoses in AIDS patients. Plenum Press, New York 1990a, pp. 305–324.

– – Hanson, L. H., and Stevens, D. A.: Itraconazole in opportunistic mycoses: cryptococcosis and aspergillosis. J. Am. Acad. Dermatol. **23** (1990b): 602–607.

Desmet, P., Kayembe, K., Stoffels, P., Mulumba, M. P., de Beule, K., and Cauwenbergh, G.: Treatment of oral candidosis in AIDS patients with itraconazole oral solution. Abstract book 3rd Symposium Topics in Mycoses in AIDS Patients. Paris, November 20–23, 1989, pp. 175–176.

Dismukes, W. E.: Treatment of systemic fungal diseases in patients with AIDS. In: Yamaguchi, H., Kobayashi, G. S., and Takahashi, H. (Eds.): Recent Progress in Antifungal Chemotherapy. Marcel Dekker Inc., New York 1992, pp. 227–238.

Drouhet, E., and Dupont, B.: Mycoses in AIDS patients: an overview. In: vanden Bossche, H., Mackenzie, D. W. R., Cauwenbergh, G., van Cutsem, J., Drouhet, E., and Dupont, B. (Eds.): Mycoses in AIDS Patients. Plenum Press, New York 1990a, pp. 27–53.

– – Cryptococcasis. In: Jacobs, P. H., and Nahl, L. (Eds.): Antifungal Drug Therapy. Marcel Dekker Inc., New York 1990b, pp. 143–164.

– – Candidosis in heroin addicts and AIDS: new immunologic data on chronic mucocutaneous candidosis. In: Tümbay, E., Seeliger, H. P. R., and Ang, Ö. (Eds.): *Candida* and Candidamycosis. Plenum Press, New York 1991, pp. 61–72.

Dupont, B.: Itraconazole therapy in aspergillosis: study in 40 patients. J. Am. Acad. Dermatol. **23** (1990): 607–614.

– Antifungal therapy in AIDS patients. In: Bennett, J. E., Hay, R. J., and Peterson, P. K. (Eds.): New Strategies in Fungal Disease. Churchill Livingstone, Edinburgh 1992, pp. 290–300.

– Hilmarsdottir, I., Datry, A., Gentilini, P., Dellamonica, P., Bernard, E., Lefort, S., Frottier, J., Choutet, P., Vilde, J. L., and the French study group on fluconazole in cryptococcal meningitis in AIDS patients: Cryptococcal meningitis in AIDS patients. A pilot study of fluconazole therapy in 52 patients. In: vanden Bossche, H., Mackenzie, D. W. R., Cauwenbergh, G., van Cutsem, J., Drouhet, B., and Dupont B. (Eds.): Mycoses in AIDS Patients. Plenum Press, N. Y. 1990, pp. 287–303.

– Improvisi, L., Eliaszewicz, M., Pialoux, G., and Gemo, T.: Resistance of *Candida albicans* to fluconazole (FCZ) in AIDS patients. Abstract 1203. Program and Abstracts of the 32nd Interscience Conference on Antimicrobial Agents and Chemotherapy (Anaheim Oct. 11–14, 1992).

Dupouy-Camet, J., Paugam, A., di Donato, C., Viguie, C., Vicens, I., Volle, P. J., and Tourte-Schaefer, C.: Résistance au fluconazole en milieu hospitalier. Concordance entre la résistance de *Candida albicans in vitro* et l'échec thérapeutique. Press Med. **20** (1991): 1341.

Faergemann, J. N.: Pityriasis (tinea) versicolor, tinea nigra and piedra. In: Jacobs, P. H., and Nahl, L. (Eds.): Antifungal Drug Therapy. Marcel Dekker Inc., New York 1990, pp. 23–30.

Feczko, J. M.: Overview of fluconazole. In: Yamaguchi, H., Kobayashi, G. S., and Takahashi, H. (Eds.): Recent Progress in Antifungal Chemotherapy. Marcel Dekker Inc., New York 1992, pp. 191–201.

First, M. R., Schroeder, T. J., Alexander, J. W., Stephens, G. W., Weiskittel, P., Myre, S. A., and Pesce, A. J.: Cyclosporine dose reduction by ketoconazole administration in renal transplant recipients. Transplantation. **51** (1991): 365–370.

Fisher, R. T., Trzaskos, J. M., Magolda, R. L., Ko, S. S., Brosz, C. S., and Larsen, B.: Lanosterol 14α-demethylase. Isolation and characterization of the third metabolically generated oxidative demethylation intermediate. J. Biol. Chem. **266** (1991): 6124–6132.

Fraser, V., Jones, M., Dunkel, J., Sturfer, S., Medoff, G., and Dunagan, W. C.: Candidemia in a tertiary care hospital, epidemiology, risk factors, and predictors of mortality. Clin. Infec. Dis. **15** (1992): 412–421.

Fromtling, R. A.: Imidazoles as medically important antifungal agents: an overview. Drugs of Today. **20** (1984): 325–349.

GALGIANI, J. N.: Current therapy for coccidioidomycosis. In: VANDEN BOSSCHE, H., ODDS, F. C., and KERRIDGE, D. (Eds.): Dimorphic Fungi in Biology and Medicine. Plenum Press, New York 1993, pp. 397–403.
– AMPEL, N. M., DOLS, C. L., and FISH, D.: *Coccidioides immitis* in AIDS patients. In: VANDEN BOSSCHE, H., MACKENZIE, D. W. R., CAUWENBERGH, G., VAN CUTSEM, J., DROUHET, E., and DUPONT, B. (Eds.): Mycoses in AIDS Patients. Plenum Press, New York 1990, pp. 171–178.
– STEVENS, D. A., GRAYBILL, J. R., DISMUKES, W. E., and CLOUD, G. A.: Ketoconazole therapy of progressive coccidioidomycosis. Comparison of 400- and 800 mg doses and observations at higher doses. Am. J. Med. **84** (1988): 603–610.
GALLEY, J., and DE KRUIJFF, B.: Correlation between molecular shape and hexagonal H_{II} phase promoting ability of sterols. FEBS Lett. **143** (1982): 133–143.
GASCOIGNE, E. W., BARTON, G. L., MICHAELS, M., MEULDERMANS, W., and HEYKANTS, J.: The kinetics of ketoconazole in animals and man. Clin. Res. Rev. **1** (1981): 177–187.
GAZZARD, B. G., and SMITH, D.: Oral candidosis in HIV-infected patients. Brit. J. Clin. Pract. **44** (Suppl. 71) (1990): 103–108.
GENTLES, J. C.: Experimental ringworm in guinea-pigs: oral treatment with griseofulvin. Nature **182** (1956): 476–477.
GEORGOPAPADAKOU, N. H., DIX, B. A., SMITH, S. A., FREUDENBERGER, J., and FUNKE, P. T.: Effect of antifungal agents on lipid biosynthesis and membrane integrity in *Candida albicans*. Antimicrob. Ag. Chem. **31** (1987): 46–51.
GOLD, W., STOUT, H. A., PAGANO, J. F., and DONOVICK, R.: Amphotericin A and B, antifungal antibiotics produced by a streptomycete. I. *In vitro* studies. Antibiotics Annual 1956: pp. 576–586.
GONDAL, J. A., SWARTZ, R. P., and RAHMAN, A.: Therapeutic evaluation of free and liposome-encapsulated amphotericin B in the treatment of systemic candidiasis in mice. Antimicrob. Ag. Chemother. **33** (1989): 1544–1548.
GRAYBILL, J. R.: Azole therapy of systemic fungal infection. In: BERG, D., and PLEMPEL, M. (Eds.): Sterol Biosynthesis Inhibitors. Pharmaceutical and Agrochemical Aspects. Ellis Horwood Ltd., Chichester 1988, pp. 520–533.
– The modern revolution in antifungal drug therapy: In: VANDEN BOSSCHE, H., MACKENZIE, D. W. R., CAUWENBERGH, G., VAN CUTSEM, J., DROUHET, E., and DUPONT, B. (Eds.): Mycoses in AIDS Patients. Plenum Press, New York 1990, pp. 265–277.
– Antifungal therapy in the non-compromised host. In: BENNETT, J. E., HAY, R. J., and PETERSON, P. K.: New Strategies in Fungal Disease. Churchill Livingstone, Edinburgh 1992, pp. 271–289.
– and SHARKEY, P. K.: Treatment of sporotrichosis, blastomycosis and histoplasmosis. In: VANDEN BOSSCHE, H., ODDS, F. C., and KERRIDGE, D. (Eds.): Dimorphic Fungi in Biology and Medicine. Plenum Press, New York 1993, pp. 381–389.
– – JOHNSON, P., and NIGHTINGALE, S.: The major endemic mycoses in the setting of AIDS: clinical manifestations. In: VANDEN BOSSCHE, H., MACKENZIE, D. W. R., CAUWENBERGH, G., VAN CUTSEM, J., DROUHET, E., and DUPONT, B. (Eds.): Mycoses in AIDS Patients. Plenum Press, New York 1990, pp. 179–190.
GREER, D. L.: Dermatophytosis (Ringworm). In: JACOBS, P. H., and NAHL, L. (Eds.): Antifungal Drug Therapy. Marcel Dekker Inc., New York 1990, pp. 5–30.
GRUDA, I., MILETTE, D., BROTHER, M., KOBAYASHI, G. S., MEDOFF, G., and BRAJTBURG, J.: Structure-activity study of inhibition of amphotericin B (fungizone) binding to sterols, toxicity to cells and lethality to mice by esters of sucrose. Antimicrob. Ag. Chemother. **35** (1991): 24–28.
GRÜNEWALD, TH., DORMANN, A., and RUF, B.: Treatment failure of amphotericin B colloidal dispersion (ABCD) in AIDS-associated cryptococcal meningitis – report of two cases. Abstract 1209. Program and Abstracts of the 32nd Interscience Conference on Antimicrobial Agents and Chemotherapy (Anaheim Oct. 11–14, 1992).

GUAN, J., KERKENAAR, A., and DE WAARD, M. A.: Studies on mechanism of resistance to imazalil in *Penicillium italicum*. Tag.-Ber. Akad. Landwirtsch. Wiss. Berlin **291** (1990): 115–125.

GULL, K., and TRINCI, A. P. J.: Griseofulvin inhibits fungal mitosis. Nature **244** (1973): 292–294.

GUO, L. S. S., FIELDING, R. M., LASIC, D. D., HAMILTON, R. L., and MUFSON, D.: Novel antifungal drug delivery: stable amphotericin B-cholesteryl sulfate discs. Int. J. Pharm. **75** (1991): 45–54.

HAY, R. J.: Azoles in the treatment of systemic opportunistic mycoses. In: BERG, D., and PLEMPEL, M. (Eds.): Sterol Biosynthesis Inhibitors. Pharmaceutical and Agrochemical Aspects. Ellis Horwood Ltd., Chichester 1988, pp. 507–519.

– Ketoconazole in Perspective. Adis International Ltd., Chester 1991.

– Ringworm. Health cooperation papers **13** (1991): 71–82.

– Historical perspectives of and projected needs for systemic azole antifungals. In: YAMAGUCHI, H., KOBAYASHI, G. S., and TAKAHASHI, H. (Eds.): Recent Progress in Antifungal Chemotherapy. Marcel Dekker Inc., New York 1992, pp. 173–182.

– and KALTER, D.: Superficial *Candida* infections. In: JACOBS, P. H., and NAHL, L. (Eds.): Antifungal Drug Therapy. Marcel Dekker Inc., New York 1990, pp. 31–51.

HAZEN, E. L., and BROWN, R. H.: Two antibiotics produced by a soil actinomycete. Science **112** (1950): 423.

HEEL, R. C., BROGDEN, R. N., SPEIGHT, T. M., and AVERY, G. S.: Econazole: a review of tis antifungal activity and therapeutic efficacy. Drugs. **16** (1978): 177–201.

– – PAKES, G. E., SPEIGHT, T. M., and AVERY, G. S.: Miconazole: a preliminary review of its therapeutic efficacy in systemic fungal infections. Drugs. **19** (1980): 7–30.

– – CARMINE, A., MORLEY, P. A., SPEIGHT, T. M., and AVERY, G. S.: Ketoconazole: a review of its therapeutic efficacy in superficial and systemic fungal infections. Drugs. **23** (1982): 1–36.

HEYKATNS, J., MICHIELS, M., MEULDERMANS, W., MONBALIU, J., LAVRIJSEN, K., VAN PEER, A., LEVRON, J. C., WOESTENBORGHS, R., and CAUWENBERGH, G.: The pharmacokinetics of itraconazole in animals and man: an overview. In: FROMTLING, R. A. (Ed.): Recent Trends in the Discovery, Development and Evaluation of Antifungal Agents. J. R. Prous Science Publishers, S. A., Barcelona 1987, pp. 223–249.

– VAN PEER, A., VAN DE VEN, V., VAN ROOY, P., MEULDERMANS, W., LAVRIJSEN, K., WOESTENBORGHS, R., VAN CUTSEM, J., and CAUWENBERGH, G.: The clinical pharmacokinetics of itraconazole: an overview. Mycoses. **32** (Suppl. 1) (1989): 67–87.

HITCHCOCK, C. A., and WHITTLE, P. J.: Chemistry and mode of action of fluconazole. In: RIPPON, J. W., and FROMTLING, R. A. (Eds.): Cutaneous Antifungal Agents. Marcel Dekker Inc., New York 1993, pp. 183–197.

– DICKINSON, K., BROWN, S. B., and EVANS, E. G. V.: Interaction of azole antifungal antibiotics with cytochrome P450-dependent 14α-demethylase purified from *Candida albicans*. Biochem. J. **266** (1990): 475–480.

HORIE, M., TSUCHIYA, Y., HAYASHI, M., IIDA, Y., IWASAWA, Y., NAGATA, Y., SAWASAKI, Y., FUKUZUMI, H., KITANI, K., and KAMEI, T.: NB-598: a potent competitive inhibitor of squalene epoxidase. J. Biol. Chem. **265** (1990). 18075–18078.

IWATA, K.: Drug resistance in human pathogenic fungi. Europ. J. Epidemiol. **8** (1992): 407–421.

JACOBS, P. H.: Antifungal therapy. In: JACOBS, P. H., and NAHL, L. (Eds.): Antifungal Drug Therapy. Marcel Dekker Inc., New York 1990, pp. 1–4.

JANDROSITZ, A., TURNOWSKI, F., and HÖGENAUER, G.: The gene encoding squalene epoxidase from *Saccharomyces cerevisiae*: cloning and characterization. Gene **107** (1991): 155–160.

JANSEN, T., VAN DE VEN, M.-A. A., BORGERS, M. J., ODDS, F. C., and VAN CUTSEM, J. M. P.: Fungal morphology after treatment with itraconazole as a single oral dose in experimental vaginal candidosis in rats. Am. J. Obstet. Gynecol. **165** (1991): 1552–1557.

JOHNSON, E. M., LUKER, J., SCULLY, C., and WARNOCK, D. W.: Azole resistant *Candida* species among HIV-positive patients given long-term fluconazole for oral candidosis Abstract 741. Program and Abstracts of the 33rd Interscience Conference on Antimicrobial Agents and Chemotherapy (New Orleans, Oct. 17–20, 1993).

Juliano, R., Lopez-Berestein, G., Mehta, R., Hopfer, R., Mehta, K., and Kasi, L.: Pharmacokinetic and therapeutic consequences of liposomal drug delivery: fluorodeoxyuridine and amphotericin B as examples. Biol. Cell. **47** (1983): 39–46.

Just, G., Steinheimer, D., Schnellbach, M., Böttinger, C., Helm, E. B., and Stille, W.: Treatment of candidosis in AIDS patients. In: vanden Bossche, H., Mackenzie, D. W. R., Cauwenbergh, G., van Cutsem, J., Drouhet, E., and Dupont, B. (Eds.): Mycoses in AIDS Patients. Plenum Press, New York 1990, pp. 279–285.

Kameswaran, S.: Rhinosporidiosis. Health Cooperation Papers **13** (1991): 151–157.

Kawasaki, S., Ramgopal, M., Chin, J., and Bloch, K.: Sterol control of the phosphatidylethanolamine-phosphatidylcholine conversion in yeast mutant-GL7. Proc. Natl. Acad. Sci. USA **82** (1985): 5716–5719.

Kelly, S. L., Kenna, S., Bligh, H. F. J., Watson, P. F., Stansfield, I., Ellis, S. W., and Kelly, D. E. (1990): Lanosterol to ergosterol-enzymology, inhibition and genetics. In: Kuhn, P. J., Trinci, A. P., Jung, M. J., Goosey, M. W., and Copping, L. G. (Eds.): Biochemistry of Cell Wall and Membranes in Fungi. Springer-Verlag, Berlin 1990, pp. 223–243.

– Rowe, J., and Watson, P. F.: Molecular genetic studies on the mode of action of azole antifungal agents. Biochem. Soc. Transact. **19** (1991): 796–798.

Kerkenaar, A.: The mode of action of dimethylmorpholines. In: Fromtling, R. A. (Ed.): Discovery, Development and Evaluation of Antifungal Agents. J. R. Prous Science Publishers, S. A., Barcelona 1987, pp. 523–542.

Kerridge, D.: Antifungal drugs. Drugs of Today **24** (1988): 705–715.

– Fasoli, M., and Wayman, F. J.: Drug resistance in *Candida albicans* and *Candida glabrata*. Ann. N. Y. Acad. Sci. **544** (1988): 245–259.

– and vanden Bossche, H.: Drug discovery: a biochemist's approach. In: Ryley, J. F. (Ed.): Chemotherapy of Fungal Diseases. Springer-Verlag, Berlin 1990, pp. 31–76.

Kitchen, V. S., Savage, M., and Harris, J. R. W.: *Candida albicans* resistance in AIDS. J. Infect. **22** (1991): 204–205.

Klein, B. S.: North American blastomycosis. In: Jacobs, P. H., and Nahl, L. (Eds.): Antifungal Drug Therapy. Marcel Dekker Inc., New York 1990, pp. 173–179.

Köller, W.: Antifungal agents with target sites in sterol functions and biosynthesis. In: Köller, W. (Ed.): Target Sites of Fungicide Action. CRC Press, Boca Raton 1992, pp. 119–206.

Kotrajaras, R.: Phaeohyphomycosis. In: Jacobs, P. H., and Nahl, L. (Eds.): Antifungal Drug Therapy. Marcel Dekker Inc.., New York 1990, pp. 71–86.

Kramer, M. R., Marchall, S. E., Denning, D. W., Keogh, A. M., Tucker, R. M., Galgiani, J. N., Lewiston, N. J., Stevens, D. A., and Theodore, J.: Cyclosporine and itraconazole interaction in heart and lung transplant recipients. Ann. Inter. Med. **113** (1990): 327–329.

Kwon-Chung, K. J.: *Aspergillus*: diagnosis and description of the genes. In: vanden Bossche, H., Mackenzie, D. W. R., and Cauwenbergh, G. (Eds.): *Aspergillus* and Aspergillosis. Plenum Press, New York 1988, pp. 11–21.

– Varma, A., and Howard, D. H.: Ecology of *Cryprococcus neoformans* and prevalence of its two varities in AIDS and non-AIDS associated cryptococcus. In: vanden Bossche, H., Mackezie, D. W. R., Cauwenbergh, G., van Cutsem, J., Drouhet, E., and Dupont, B. (Eds.): Mycoses in AIDS Patients. Plenum Press, New York 1990, pp. 103–113.

Lake-Bakaar, G., Tom, W., Lake-Bakaar, D., Gupta, N., Beidas, S., Elsakr, M., and Straus, E.: Gastropathy and ketoconazole malabsorption in the acquired immunodeficiency syndrome (AIDS). Ann. Intern. Med. **109** (1988): 471–473.

Lala, A. K., Lin, H. K., and Bloch, K.: The effect of some alkyl derivatives of cholesterol on the permeability properties and microviscosities of model membranes. Bioorganic Chem. **7** (1978): 437–445.

Langcake, P., Kuhn, P. J., and Wade, M.: The mode of action of systemic fungicides. In: Hutson, D. H., and Roberts, T. R. (Eds.): Progress in Pesticide Biochemistry and Toxicology. Vol. 3. Wiley, Chichester 1983, pp. 1–109.

LAROCHE, R., DUPONT, B., TOUZE, J. E., TAELMAN, H., BOGAERTS, J., KADIO, A., M'PELE, P., LATIF, A., AUBRY, P., DURBEC, J. P., and SAUNIERE, J. F.: Cryptococcal meningitis associated with acquired immunodeficiency syndrome (AIDS) in African patients: treatment with fluconazole. J. Med. Vet. Mycol. **30** (1992): 71–78.

LASIC, D. D.: Mixed micelles in drug delivery. Nature **355** (1992): 279–280.

LOKEIC, F., POLRIER, O., GLUCKMAN, E., DEVERGIE, A., and ARCESE, W.: A pharmacokinetic study of cyclosporine A: preliminary results. In: Bone marrow transplantation in Europe. Voll. 2. Proceedings of the 5th European Symposium on bone marrow transplantation. Courchevel, France, March 16–18, 1981, pp. 160–164.

LONDERO, A. T.: Paracoccidioidomycosis. Health cooperation papers **13** (1991): 187–196.

LONG, M. I., STEEL, C. C., and MERCIER, E. I.: Location of squalene accumulation and physiological effects of ergosterol depletion in naftifine-grown yeast. Biochem. Soc. Transact. **16** (1988): 1044–1045.

LOPEZ-BERESTEIN, G.: Liposomes as carriers of antifungal drugs. Ann. N. Y. Acad. Sci. **544** (1988): 590–597.

– and ROSENBLUM, M. G.: In: GOLDSTEIN, A. L., and GARACI, E. (Eds.): Treatment and Pharmacokinetics of Liposomal-amphotericin B patients with Systemic Fungal Infections. Combinations Therapies. Biological Response Modifiers in the Treatment of Cancer and Infectious Diseases. Plenum Press, New York 1992, pp. 105–112.

M'BAYA, B., and KARST, F.: *In vitro* assay of squalene epoxidase of *Saccharomyces cerevisiae*. Biochem. Biophys. Res. Commun. **147** (1987): 556–564.

MACCANN, P. P., BACCHI, C. J., CLARKSON, A. B., BEY, P., SJOERDSMA, A., SCHESTER, P. J., WALZER, P. D., and BARLOW, J. L. R.: Inhibition of polyamine synthesis by a difluoromethylornithine in African trypanosomes and *Pneumocystis carinii* as a basis of chemotherapy: biochemical and clinical aspects. Am. J. Trop. Med. Hyg. **35** (1986): 1153–1156.

MACKENZIE, D. W. R.: *Pneumocystis carinii*: a nomadic taxon. In: VANDEN BOSSCHE, H., MACKENZIE, D. W. R., CAUWENBERGH, G., VAN CUTSEM, J., DROUHET, E., and DUPONT, B. (Eds.): Mycoses in AIDS Patients. Plenum Press, New York 1990, pp. 55–63.

MALAWISTA, S. E.: Microtubules and the movement of melanin granules in frog dermal melanocytes. Ann. N. Y. Acad. Sci. **253** (1975): 702–710.

MARCIREAU, C., GUILLOTON, M., and KARST, F.: *In vivo* effects of fenpropimorph on the yeast *Saccharomyces cerevisiae* and determination of the molecular basis of the antifungal property. Antimicrob. Ag. Chemother. **43** (1990): 989–993.

MARGALITH, P. Z.: Steroid Microbiology. Charles C. Thomas publisher. Springfield 1986.

MARICHAL, P., GORRENS, J., and VANDEN BOSSCHE, H.: The action of itraconazole and ketoconazole on growth and sterol synthesis in *Aspergillus fumigatus* and *Aspergillus niger*. J. Med. Vet. Mycol. **23** (1985): 13–21.

– – VAN CUTSEM, J., VAN GERVEN, F., and VANDEN BOSSCHE, H.: Effects of ketoconazole and itraconazole on growth and sterol synthesis in *Pityrosporum ovale*. J. Med. Vet. Mycol. **24** (1986): 487–489.

– VANDEN BOSSCHE, H., GORRENS, J., BELLENS, D., and JANSSEN, P. A. J.: Cytochrome P450 of *Aspergillus fumigatus* – Effects of itraconazole and ketoconazole. In: SCHUSTER, I. (Ed.): Cytochrome P450: Biochemistry and Biophysics. Taylor and Francis, London 1989, pp. 177–180.

– – MOEREELS, H., and BRASSEUR, R.: Mode of insertion of azole antifungals and sterols in membranes. In: BRASSEUR, R. (Ed.): Molecular Description of Biological Membranes by Computer added Conformational Analysis. Vol. 2. CRC Press, Boca Raton 1990, pp. 27–42.

MARRIOTT, M. S.: Inhibition of sterol biosynthesis in *Candida albicans* by imidazole-containing antifungals. J. Gen. Microbiol. **117** (1980): 253–255.

– and RICHARDSON, K.: The discovery and mode of action of fluconazole. In: FROMTLING, R. A. (Ed.): Recent Trends in the Discovery, Development and Evaluation of Antifungal Agents. J. R. Prous Science Publishers, S. A., Barcelona 1987, pp. 81–92.

MEDOFF, G.: The mechanism of action of amphotericin. In: VANDEN BOSSCHE, H., MACKENZIE, D. W. R., and CAUWENBERGH, G. (Eds.): *Aspergillus* and Aspergillosis. Plenum Press, New York 1988, pp. 161–164.
- COMFORT, M., and KOBAYASHI, G. S.: Synergystic action of amphotericin B and 5-fluorocytosine against yeast-like organisms. Proc. Soc. Exp. Biol. Med. **138** (1971): 571–574.

MENDLING, W.: Azoles in the therapy of vaginal mycoses. In: BERG, D., and PLEMPEL, M. (Eds.): Sterol Biosynthesis Inhibitors. Pharmaceutical and Agrochemical Aspects. Ellis Horwood Ltd., Chichester 1988, pp. 480–506.

MERCER, E. I.: The mode of action of morpholines. In: BERG, D., and PLEMPEL, M. (Eds.): Sterol Biosynthesis Inhibitors. Pharmaceutical and Agrochemical Aspects. Ellis Horwood Ltd., Chichester 1988, pp. 120–150.
- Sterol biosynthesis inhibitors: their current status and modes of action. Lipids **26** (1991): 584–597.

MIETH, H., and VILLARS, V.: Terbinafine: clinical efficacy and development. In: YAMAGUCHI, H., KOBAYASHI, G. S., and TAKAHASHI, H. (Eds.): Recent Progress in Antifungal Chemotherapy. Marcel Dekker Inc., New York 1992, pp. 135–146.

MORITA, T., and NOZAWA, Y.: Effects of antifungal agents on ergosterol biosynthesis in *Candida albicans* and *Trichophyton mentagrophytes*: differential inhibitory sites of naphtiomate and miconazole. J. Invest. Dermatol. **85** (1985): 434–437.

NEGRONI, R.: Azole compounds in the treatment of paracoccidioidomycosis. In: VANDEN BOSSCHE, H., ODDS, F. C., and KERRIDGE, D. (Eds.): Fungal Dimorphism. Plenum Press, New York 1993, 391–396.

NES, W. R., SEKULA, B. C., NES, W. D. and ADLER, J. H.: The functional importance of structural features of ergosterol in yeast. J. Biol. Chem. **253** (1978): 6218–6225.

NORRIS, S., MC KINSEY, D., LANCASTER, D., and WHEAT,. J.: Retrospective evaluation of fluconazole maintenance therapy for disseminated histoplasmosis in AIDS. Abstract No. 1207. Program and Abstracts of the 32nd Interscience Conference on Antimicrobial Agents and Chemotherapy (Anaheim, Oct. 11–14, 1992).

NOZAWA, Y., and MORITA, T.: Biochemical aspects of squalene epoxidase inhibition by a thiocarbamate derivative, naphthiomate T. In: YAMAGUCHI, H., KOBAYASHI, G. S., and TAKAHASHI, H. (Eds.): Recent Progress in Antifungal Chemotherapy. Marcel Dekker Inc., New York 1992, pp. 53–64.

ODDS, F. C.: *Candida* infections: an overview. Crit. Rev. Microbiol. **15** (1987): 1–5.
- *Candida* and Candidosis. A Review and Bibliography. Baillière Tindall, London 1988.
- Resistance of yeasts to azole-derivative antifungals. J. Antimicrob. Chemother. **31** (1993): 463–471.
- ARAI, T., DISALVO, A. F., EVANS, E. G. V., HAY, R. J., RANDHAWA, H. S., RINALDI, M. G., and WALSH, T. J.: Nomenclature of fungal diseases: a report and recommendation from a sub-committee of the international society for human and animal mycology (ISHAM). J. Med. Vet. Mycol. **30** (1992): 1–10.

OKEY, A. B., ROBERTS, E. A., HARPER, P. A., and DENISON, M. S.: Induction of drugmetabolizing enzymes: mechanisms and consequences. Clin. Biochem. **19** (1986): 132–141.

ONO, T., NAKAZONO, K., and KOSAKA, H.: Purification and partial characterization of squalene epoxidase from rat liver microsomes. Biochim. Biophys. Acta **709** (1982): 84–90.

ORTH, A. B., and SISLER, H. D.: Mode of action of terbinafine in *Ustilago maydis* and characterization of resistant mutants. Pestic. Biochem. Physiol. **37** (1990): 53–63.

OXFORD, A. E., RAISTRICK, M., and SIMONART, P.: Studies in biochemistry of microorganisms LX Griseofulvin $C_{17}H_{16}O_6Cl$ metabolic product of *Penicillium griseofulvin*. Dierckx. Biochem. J. **33** (1939): 240–248.

PESTY, M., CAMPBELL, J. M., and PEBERDY, J. F.: Alteration of ergosterol content and chitin synthase activity in *Candida albicans*. Current Microbiol. **5** (1981): 187–190.

PETRANYI, G., STÜTZ, A., RYDER, N. S., MEINGASSNER, J. G., and MIETH, H.: Experimental antimycotic activity of naftifine and terbinafine. In: FROMTLING, R. A. (Ed.): Recent Trends in the Discovery, Development and Evaluation of Antifungal Agents. J. R. Prous Science Publishers, S. A., Barcelona 1987, pp. 441–459.

PLEMPEL, M., BERG, D., BÜCHEL, K.-H., and RITTER, W.: Experimental antimycotic activity of azole derivatives – experience, knowledge and questions. In: BERG, D., and PLEMPEL, M. (Eds.): Sterol Biosynthesis Inhibitors. Pharmaceutical and Agrochemical Aspects. Ellis Horwood Ltd., Chichester 1988, pp. 349–396.

POLAK, A.: Mode of action of 5-fluorocytosine in *Aspergillus fumigatus*. In: VANDEN BOSSCHE, H., MACKENZIE, D. W. R., and CAUWENBERGH, G. (Eds.): *Aspergillus* and Aspergillosis. Plenum Press, New York 1988a, pp. 165–170.

– Morpholines in clinical use. In: BERG, D., and PLEMPEL, M. (Eds.): Sterol Biosynthesis Inhibitors. Pharmaceutical and Agrochemical Aspects. Ellis Horwood Ltd., Chichester 1988b, pp. 430–448.

– Mode of action studies. In: RYLEY, J. F. (Ed.): Chemotherapy of Fungal Diseases. Springer-Verlag, Berlin 1990, pp. 153–182.

– 5-Fluorocytosine and its combination with other antifungal agents. In: YAMAGUCHI, H., KOBAYASHI, G. S., and TAKAHASHI, H. (Eds.): Recent Progress in Antifungal Chemotherapy. Marcel Dekker Inc., New York 1992a, pp. 77–85.

– Amorolfine, Ro 14-4767/002, Loceryl. In: YAMAGUCHI, H., KOBAYASHI, G. S., and TAKAHASHI, H. (Eds.): Recent Progress in Antifungal Chemotherapy. Marcel Dekker Inc., New York 1992b, pp. 125–134.

– and HARTMANN, P. G.: Antifungal chemotherapy – are we winning? Progress in Drug Research **37** (1991): 191–269.

PYE, G. W., and MARRIOTT, M. S.: Inhibition of sterol C14 demethylation by imidazole-containing antifungals. Sabouraudia. **20** (1982): 325–329.

RAGSDALE, N. N.: Specific effect of triarimol on sterol biosynthesis in *Ustilago maydis*. Biochim. Biophys. Acta **380** (1975): 81–96.

– and SISLER, H. D.: Inhibition of ergosterol synthesis in *Ustilago maydis* by the fungicide triarimol. Biochem. Biophys. Res. Commun. **46** (1972): 81–96.

RALPH, E. D., KHAZINDAR, A. M., BARBER, K. R., and GRANT, C. W. M: Comparative *in vitro* effects of liposomal amphotericin B, amphotericin B-deoxycholate, and free amphotericin B against fungal strains determined by using MIC and minimal lethal concentration susceptibility studies and time-kill curves. Antimicrob. Ag. Chemother. **35** (1991): 188–191.

READIO, J. D., and BITTMAN, R.: Equilibrium binding of amphotericin B and its methyl ester and borate complex to sterols. Biochim. Biophys. Acta **685** (1982): 219–224.

RESTREPO-MORENO, A.: Paracoccidioidomycosis (South American blastomycosis). In: JACOBS, P. H., and NAHL, L. (Eds.): Antifungal Drug Therapy. Marcel Dekker Inc., New York 1990, pp. 181–205.

RINALDI, M. G.: *Histoplasma* in AIDS patients. In: VANDEN BOSSCHE, H., MACKENZIE, D. W. R., CAUWENBERGH, G., VAN CUTSEM, J., DROUHET, E., and DUPONT, B. (Eds.): Mycoses in AIDS Patients. Plenum Press, New York 1990, pp. 163–169.

RITTER, W.: Pharmacokinetics of azole compounds. In: BERG, D., and PLEMPEL, M. (Eds.): Sterol Biosynthesis Inhibitors. Pharmaceutical and Agrochemical Aspects. Ellis Horwood, Chichester 1988, pp. 397–429.

RODRIGUEZ-TUDELA, J. L., LA GUNA, F., MARTINEZ-SUAREZ, J. V., CHAVES, F., and DRONDA, F.: Fluconazole resistance of *Candida albicans* isolates from AIDS patients receiving prolonged antifungal therapy. Abstract 1204. Program and Abstracts of the 32nd Interscience Conference on Antimicrobial Agents and Chemotherapy (Anaheim, Oct. 11–14, 1992).

RYDER, N. S.: Selective inhibition of squalene epoxidation by allylamine antimycotic agents. In: NOMBELA, C. (Ed.): Microbial Cell Wall Synthesis and Autolysis. Elsevier Science Publishers, Amsterdam 1984, pp. 313–321.

- Biochemical mode of action of the allylamine antimycotic agents naftifine and SF 86-327. In: IWATA, K., and VANDEN BOSSCHE, H. (Eds.): *In vitro* and *in vivo* Evaluation of Antifungal Agents. Elsevier Science Publishers, Amsterdam 1986, pp. 89–99.
- Mechanism of action of the allylamine antimycotics. In: FROMTLING, R. A. (Ed.): Recent Trends in the Discovery, Development and Evaluation of Antifungal Agents. J. R. Prous Science Publishers, S. A., Barcelona 1987, pp. 451–459.
- Squalene epoxidase-Enzymology and inhibition. In: KUHN, P. J., TRINCI, A. P. J., JUNG, M. J., GOOSEY, M. W., and COPPING, L. G. (Eds.): Biochemistry of Cell Walls and Membranes in Fungi. Springer-Verlag, Berlin 1990, pp. 189–203.
- Squalene epoxidase as a target for the allylamines. Biochem. Soc. Transact. **19** (1991): 769–774.
- Terbinafine: Mode of action and properties of the squalene epoxidase inhibition. Br. J. Dermatol. **126** (Suppl. 39) (1992): 2–7.
- and MIETH, H.: Allylamine antifungal drugs. Curr. Top. Med. Mycol. **4** (1992): 158–188.
- SEIDL, G., PETRANYI, G., and STÜTZ, A.: Mechanism of fungicidal action of SF 86-327, a new allylamine antimycotic agent. In: ISHIGAMI, J. (Ed.): Recent Advances in Chemotherapy. University of Tokyo Press, Tokyo 1985, pp. 2558–2559.
- FRANK, I., and DUPONT, M. C.: Ergosterol biosynthesis inhibition by the thiocarbamate antifungal agents, tolnaftate and tolciclate. Antimicrob. Ag. Chemother. **29** (1986): 858–860.
- STUETZ, A., and NUSSBAUMER, P.: Squalene epoxidase inhibitors. Structural determinants for activity and selectivity of allylamines and related compounds. In: NES, W. D., PARISH, E. J., and TRZASKOS, J. M. (Eds.): Regulation of Isopentenoid Metabolism. American Chemical Society, Washington D. C. 1992, pp. 192–204.

SAAG, M. S., and DISMUKES, W. E.: Treatment of histoplasmosis and blastomycosis. Chest **93** (1988): 848–851.
- POWDERLY, W. G., CLOUD, G. A., ROBINSON, P., GRIECO, M. H., SHARKEY, P. K., THOMPSON, S. E., SUGAR, A. M., TUAZON, G. U., FISHER, J. F., HYSLOP, N., JACOBSON, J. M., HAFNER, R., DISMUKES, W. E., and the Niaid Mycoses Study Group and the AIDS clinical trials group: Comparison of amphotericin B with fluconazole in the treatment of acute AIDS-associated cryptococcal meningitis. N. Eng. J. Med. **326** (1992): 83–89.

SAROSI, G. A.: Amphotericin B still the 'gold standard' for antifungal therapy. Postgrad. Med. **88** (1990): 151–166.

SCHAFFNER, C. P.: Amphotericin B derivates. In: FROMTLING, R. A. (Ed.): Evaluation of Antifungal Agents. Prous Science Publishers S. A., Barcelona 1987, pp. 595–618.

SCHOLER, H. J.: Flucytosine. In: SPELLER, D. C. E. (Ed.): Antifungal Chemotherapy. John Wiley & Sons Ltd., Chichester 1980, pp. 35–106.

SCHROEDER, T. J., MELVIN, D. B., CLARDY, W., WADHWA, N. K., MYRE, S. A., REISING, J. M., COLLINS, W. J. A., PESCE, A. J., and FIRST, M. R.: Use of cyclosporine and ketoconazole without nephrotoxicity in two heart transplants recipients. J. Heart Transplant. **6** (1987): 84–89.

SHARKEY, P. K., GRAYBILL, J. R., RINALDI, M. G., STEVENS, D. A., TUCKER, R. M., PETERIE, J. D., HOEPRICH, P. D., GREER, D. L., FRENKEL, L., COUNTS, G. W. et al.: Itraconazole treatment of phaeohyphomycosis. J. Am. Ac. Dermat. **23** (1990): 577–586.

SHIMOKAWA, O., KATO, Y., and NAKAYAMA, H.: Accumulation of 14-methylsterols and defective hyphal growth in *Candida albicans*. J. Med. Vet. Mycol. **24** (1986): 327–336.

SHUSTER, S.: The aetiology of dandruff and the mode of action of therapeutic agents. Br. J. Dermatol. **111** (1984): 235–242.

SISLER, H. D., and BUCHMAN-ORTH, A. M.: Oxygen radical generation as a basis of fungitoxicity. Tagungsber. Akad. Landwirtschaftswiss. **291** (1990): 17–23.
- WALSH, R. C., and ZIOGAS, B. N.: Ergosterol biosynthesis: a target of fungitoxic action. In: MATSUNAKA, S., HUTSON, D. H., and MURPHY, S. D. (Eds.): Pesticide Chemistry: human welfare and the environment. Vol 3. Mode of action, metabolism and toxicology. Pergamon Press, New York 1983, pp. 129–134.

SLOBODA, R. D., VAN BLARICOM, G., CREASEY, W. A., ROSENBAUM, J. L., and MALAWISTA, S. E.: Griseofulvin: association with tubulin and inhibition of *in vitro* microtubule assembly. Biochem. Biophys. Res. Commun. **105** (1982): 882–888.

SMITH, D. E., MIDGLEY, J., ALLEN, M., CONNOLLY, G. M., and GAZZARD, B. G.: Itraconazole versus ketoconazole in oral and oesophageal candidosis in patients with HIV. AIDS. **5** (1991): 1367–1371.

SOBEL, J. D.: Epidemiology and pathogenesis of recurrent vulvovaginal candidiaisis. Am. J. Obstet. Gynaecol. **152** (1985): 924–935.

– New insights into the pathogenesis of vaginal candidosis. Int. J. Exptl. Clin. Chemother. **2** (Suppl. 1) (1989): 9–18.

– Therapeutic considerations in fungal vaginitis. In: RYLEY, J. F. (Ed.): Chemotherapy of Fungal Diseases. Springer-Verlag, Berlin 1990, pp. 365–383.

SOUBRY, R., CLERINX, J., BANYANAGILIKI, V., VAN DE PERRE, P., and TAELMAN, H.: Comparison of itraconazole oral solution and fluconazole capsules in the treatment of oral and esophageal candidiasis in HIV-infectes patients. Preliminary results. Abstract M. B. 2201. 7th International Conference on AIDS, 1991, p. 232.

STEVENS, D. A.: Historical perspectives and state-of-the-art treatment of systemic mycoses compared to newer problems in management of fungal infections in debilitated and immunosuppressed hosts. In: YAMAGUCHI, H., KOBAYASHI, G. S., and TAKAHASHI, H. (Eds.): Recent Progress in Antifungal Chemotherapy. Marcel Dekker Inc., New York 1992, pp. 215–226.

– Miconazole in the treatment of coccidioidomycosis. Drugs. **26** (1983): 347–354.

STORCH, J., and KLEINFELD, A. M.: The lipid structure of biological membranes, TIBS, November 1985: 418–421.

STRICKER, B. H. CH., DE JONG, P. A. C., SCHREUDER, F., BIJLMER-IEST, J. C., HERMANN, W. A., and VAN ULSEN, J.: Loss of sense of taste during terbinafine therapy. Ned. Tijdschr. Geneeskd. **136** (1992): 2238–2240.

SUGAR, A. M.: The polyene macrolide antifungal drugs. In: PETERSON, P. K., and VERHOEF, J. (Eds.): Antimicrobial Agents Annual/1. Elsevier, Amsterdam 1986, pp. 229–244.

– ANAISSIE, E. J., GRAYBILL, J. R., and PATTERSON, T. F.: Fluconazole. J. Med. Vet. Mycol. **30** (Suppl. 1) (1992): 201–212.

SURARIT, R., and SHEPHERD, M. G.: The effects of azole and polyene antifungals on the plasma membrane enzymes of *Candida albicans*. J. Med. Vet. Mycol. **25** (1987): 403–413.

SZANISZLO, P. J., MENDOZA, L., and KARUPPAYIL, S. M.: Clues about chromoblastomycotic and other dematiaceous fungal pathogens using *Wangiella* as a model. In: VANDEN BOSSCHE, H., ODDS, F., and KERRIDGE, D. (Eds.): Dimorphic Fungi in Biolgy and Medicine. Plenum Press, New York 1993, pp. 241–255.

TARBIT, M. H.: Pharmacokinetic aspects of antifungal therapy. In: RYLEY, J. F. (Ed.): Chemotherapy of Fungal Diseases. Springer-Verlag, Berlin 1990, pp. 183–204.

– ROBERTSON, W. R., and LAMBERT, A.: Hepatic and endocrine effect of azole antifungal agents. In: RYLEY, J. F. (Ed.): Chemotherapy of Fungal Diseases. Springer-Verlag, Berlin 1990, pp. 205–229.

TÄUBER, U.: Pharmacokinetics of antimycotics with emphasis on local treatment. Ann. N. Y. Acad. Sc. **544** (1988): 415–426.

TEERLINK, T., DE KRUYFF, B. and DEMEL, R. A.: The action of pimaricin, etrescomycin and amphotericin B on liposomes with varying sterol content. Biochim. Biophys. Acta **599** (1980): 484–492.

TOOLEY, P. J. H.: Treatment of vaginal candidosis – a UK patients survey 1989. Brit. J. Clin. Pract. **44** (Suppl. 71) (1990): 73–76.

TROKE, P. F.: Efficacy of fluconazole in animal models of superficial and opportunistic systemic fungal infections. In: FROMTLING, R. A. (Ed.): Recent Trends in the Discovery, Development and Evaluation of Antifungal Agents. J. R. Prous Science Publishers, S. A. Barcelona 1987, pp. 103–112.

TUCKER, R. M., WILLIAMS, P. L., ARATHOON, E. G., and STEVENS, D. A.: Treatment of mycoses with itraconazole. Ann. N. Y. Acad. Sci. **544** (1988): 451–470.
- DENNING, D. W., ARATHOON, E. G., RINALDI, M. G., and STEVENS, D. A.: Itraconazole therapy for nonmeningeal coccidioidomycosis: clinical and laboratory observations. J. Am. Acad. Dermatol. **23** (1990): 593–601.

UTZ, J. P.: Blastomycosis. Health cooperation papers **13** (1991): 181–186.

VAN CUTSEM, J.: The antifungal activity of ketoconazole. Am. J. Med. **74** (1983): 9–15.
- The in vitro antifungal spectrum of itraconazole. Mycoses **32** (Suppl. 1) (1989): 7–13.
- Fungal models in immunocompromised animals. In: VANDEN BOSSCHE, H., MACKENZIE, D. W. R., CAUWENBERGH, G., VAN CUTSEM, J., DROUHET, E., and DUPONT, B. (Eds.): Mycoses in AIDS Patients. Plenum Press, New York 1990, pp. 207–222.
- In vitro antifungal spectrum of itraconazole and treatment of systemic mycoses with old and new antimycotic agents. Chemotherapy **38** (Suppl. 1) (1992): 3–11.
- and CAUWENBERGH, G.: Results of itraconazole treatment in systemic mycoses in animals and man. In: YAMAGUCHI, H., KOBAYASHI, G. S., and TAKAHASHI, H. (Eds.): Recent Progress in Antifungal Chemotherapy. Marcel Dekker Inc., New York 1992, pp. 203–214.
- and JANSSEN, P. A. J.: In vitro and in vivo models to study the activity of antifungals against Aspergillus. In: VANDEN BOSSCHE, H., MACKENZIE, D. W. R., and CAUWENBERGH, G. (Eds.): Aspergillus and Aspergillosis. Plenum Press, New York 1988, pp. 215–227.
- and THIENPONT, D.: Miconazole, a broad-spectrum antimycotic agent with antibacterial activity. Chemotherapy. **17** (1972): 392–404.
- VAN GERVEN, F., and JANSSEN, P. A. J.: Activity of orally, topically and parenterally administered itraconazole in the treatment of superficial and deep mycoses: animal models. Rev. Infect. Dis. **9** (Suppl. 1) (1987): 515–532.
- – – Saperconazole, a new potent antifungal triazole: in vitro activity spectrum and therapeutic efficacy. Drugs of the Future. **14** (1989): 1187–1209.

VANDEN BOSSCHE, H.: Biochemical targets for antifungal azole derivatives: hypothesis on the mode of action. Curr. Top. Med. Mycol. **1** (1985): 313–351.
- Itraconazole a selective inhibitor of the cytochrome P450 dependent ergosterol biosynthesis. In: FROMTLING, R. A. (Ed.): Recent Trends in the Discovery, Development and Evaluation of Antifungal Agents. J. R. Prous Science Publishers, S. A., Barcelona 1987, pp. 207–221.
- Mode of action of pyridine, pyrimidine and azole antifungals. In: BERG, D., and PLEMPEL, M. (Eds.): Sterol Biosynthesis Inhibitors. Pharmaceutical and Agrochemical Aspects. Chichester 1988, pp. 79–119.
- Importance and role of sterols in fungal membranes. In: KUHN, P. J., TRINCI, A. P. J., JUNG, M. J., GOOSEY, M. W., and COPPING, L. G. (Eds.): Biochemistry of Cell Walls and Membranes in Fungi. Springer-Verlag, Berlin 1990, pp. 135–157.
- Ergosterol biosynthesis inhibitors. In: PRASAD, R. (Ed.): *Candida albicans*. Cellular and Molecular Biology. Springer-Verlag, Berlin 1991, pp. 239–257.
- Inhibitors of P450-dependent steroid biosynthesis: from research to medical treatment. J. Steroid Biochem. Molec. Biol. **43** (1992): 1003–1021.
- and MARICHAL, P.: Mode of action of anti-*Candida* drugs: focus on terconazole and other ergosterol biosynthesis inhibitors. Am. J. Obstet. Gynecol. **165** (1991): 1193–1199.
- – Azole antifungals: mode of action. In: YAMAGUCHI, H., KOBAYASHI, G. S., and TAKAHASHI, H. (Eds.): Recent Progress in Antifungal Chemotherapy. Marcel Dekker Inc., New York 1992, pp. 25–40.
- – Is there a role for sterols and steroids in fungal growth and transition from yeast to hyphal-form and vice versa? An overview. In: VANDEN BOSSCHE, H., ODDS, F., and KERRIDGE, D. (Eds.): Dimorphic Fungi in Biology and Medicine. Plenum Press, New York 1993, pp. 177–190.
- and MOEREELS, H.: 17α-Hydroxylase/17,20-lyase. In: SANDLER, M., and SMITH, J. (Eds.): Design of Enzyme Inhibitors as Drugs. Vol. 2., Oxford University Press, Oxford 1994, pp. 438–461.

- and WILLEMSENS, G.: Effects of the antimycotics, miconazole and ketoconazole, on cytochrome P450 in yeast microsomes and rat liver microsomes. Archiv. Int. Physiol. Biochem. **90** (1982): B218–B219.
- – COOLS, W., LAUWERS, W. F. J., and LE JEUNE, L.: Biochemical effects of miconazole in fungi: II. Inhibition of ergosterol biosynthesis in *Candida albicans*. Chem. Biol. Interact. **21** (1978): 59–78.
- – – CORNELISSEN, F., LAUWERS, W. F., and VAN CUTSEM, J.: Inhibition of ergosterol synthesis in *Candida albicans* by ketoconazole. Arch. Int. Physiol. Biochem. **87** (1979): 849–850.
- – – – – – In vitro and in vivo effects of the antimycotic drug ketoconazole on sterol synthesis. Antimicrob. Ag. Chemother. **17** (1980): 922–928.
- RUYSSCHAERT, J. M., DEFRISE-QUERTAIN, F., WILLENSENS, G., CORNELISSEN, F., MARICHAL, P., COOLS, W., and VAN CUTSEM, J.: The interaction of miconazole and ketoconazole with lipids. Biochem. Pharmacol. **31** (1982): 2609–2617.
- WILLEMSENS, G., MARICHAL, P., COOLS, W., and LAUWERS, W.: The molecular basis for the antifungal activities of *n*-substituted azole derivatives. Focus on R51211. In: TRINCI, A. P. J., and RYLEY, J. F. (Eds.): Mode of Action of Antifungal Agents. Cambridge University Press, Cambridge 1984, pp. 321–341.
- BELLENS, D., COOLS, W., GORRENS, J., MARICHAL, P., VERHOEVEN, H., WILLEMSENS, G., DE COSTER, R., BEERENS, D., HAELTERMAN, C., COENE, M.-C., LAUWERS, W., and LE JEUNE, L.: Cytochrome P450: target for itraconazole. Drug. Develop. Res. **8** (1986): 287–298.
- WILLEMSENS, G., and MARICHAL, P.: Anti-*Candida* drug-the biochemical basis for their activity. Crit. Rev. Microbiol. **15** (1987a): 57–72.
- MARICHAL, P., GORRENS, J., BELLENS, D., VERHOEVEN, H., COENE, M.-C., LAUWERS, W., and JANSSEN, P. A. J.: Interaction of azole derivatives with cytochrome P450 isozymes in yeast, fungi, plants and mammalian cells. Pestic. Sci. **21** (1987b): 289–306.
- DE COSTER, R. and AMERY, W.: Pharmacological and clinical uses of ketoconazole. In: FURR, B. J. A., and WAKELING, A. E. (Eds.): Pharmacology and Clinical uses of Inhibitors of Hormone Secretion and Action. Ballière Tindall, London 1987c, pp. 288–307.
- MARICHAL, P., GEERTS, H., and JANSSEN, P. A. J.: The molecular basis for itraconazole's activity against *Aspergillus fumigatus*. In: VANDEN BOSSCHE, H., MACKENZIE, D. W. R., and CAUWENBERGH, G. (Eds.): *Aspergillus* and Aspergillosis. Plenum Press, New York 1988a, pp. 171–197.
- – GORRENS, J., GEERTS, H., and JANSSEN, P. A. J.: Mode of action studies. Basis for the search of new antifungal drugs. Ann. N. Y. Acad. Sci. **544** (1988b): 191–207.
- – – – COENE, M.-C., WILLEMSENS, G., BELLENS, D., ROELS, I., MOEREELS, H., and JANSSEN, P. A. J.: Biochemical approaches to selective antifungal activity, focus on azole antifungals. Mycoses **32** (Suppl. 1) (1989): 35–52.
- – – – BELLENS, D., COENE, M.-C., LAUWERS, W., LE JEUNE, L., MOEREELS, H., and JANSSEN, P. A. J.: Mode of action of antifungals of use in immunocompromised patients. Focus on *Candida glabrata* and *Histoplasma capsulatum*. In: VANDEN BOSSCHE, H., MACKENZIE, D. W. R., CAUWENBERGH, G., VAN CUTSEM, J., DROUHET, E., and DUPONT, B. (Eds.): Mycoses in AIDS Patients. Plenum Press. N. Y. 1990a, pp. 223–243.
- – – WILLEMSENS, G., BELLENS, D., GORRENS, J., ROELS, I., COENE, M.-C., LE JEUNE, L., and JANNSEN, P. A. J.: Saperconazole: a selective inhibitor of the cytochrome P450-dependent ergosterol synthesis in *Candida albicans*, *Aspergillus fumigatus* and *Trichophyton mentagrophytes*. Mycoses **33** (1990b): 335–352.
- – – GORRENS, J., and COENE, M.-C.: Biochemical basis for the activity of oral antifungal drugs. Brit. J. Clin. Pract. **33** (Suppl. 71) (1990c): 41–46.
- – – COENE, M.-C., WILLEMSENS, G., LE JEUNE, L., COOLS, W., and VERHOEVEN, H.: Cytochrome P450-dependent 14α-demethylase. Target for antifungal agents and herbicides. In: NES, W. D., PARISH, E. J., and TRZASKOS, J. M. (Eds.): Regulation of Isopentenoid Metabolism. American Chemical Society, Washington D. C. 1992a, pp. 219–230.

- - and ODDS, F. C.: Molecular mechanisms of drug resistance in fungi. Trends Microbiol. **2** (1994): 393–400.
- - - LE JEUNE, L., and COENE, M.-C.: Characterization of an azole-resistant *Candida glabrata* isolate. Antimicrob. Ag. Chemother. **36** (1992b): 2602–2610.
- - - and MOEREELS, H.: Molecular mechanisms of antifungal activity and fungal resistance: focus on inhibitors of ergosterol biosynthesis. In: MARESCA, B., KOBAYASHI, G., and YAMAGUCHI, H. (Eds.): Molecular Biology and its Application to Medical Mycology. NATO Series book. Springer-Verlag, Berlin 1993a, pp. 179–197.
- - LE JEUNE, L., COENE, M.-C., GORRENS, J., and COOLS, W.: Effect of itraconazole on cytochrome P-450-dependent sterol 14α-demethylation and reduction of 3-ketosteroids in *Cryptococcus neoformans*. Antimicrob. Ag. Chemother. **37** (1993b): 2101–2105.
- HEERES, J., BACKX, L. J. J., MARICHAL, P., and WILLEMSENS, G.: Discovery, chemistry, mode of action, and selectivity of itraconazole. In: RIPPON, J. W., and FROMTLING, R. A. (Eds.): Cutaneous Antifungal Agents. Marcel Dekker Inc., New York 1993c, pp. 263–283.
VANDEPITTE, J.: Clinical aspects of cryptococcosis in patients with AIDS. In: VANDEN BOSSCHE, H., MACKENZIE, D. W. R., CAUWENBERGH, G., VAN CUTSEM, J., DROUHET, E., and DUPONT, B. (Eds.): Mycoses in AIDS Patients. Plenum Press, New York 1990, pp. 115–122.
VAN DER PAS, H.: International clinical experience with itraconazole in the treatment of vaginal candidosis. Int. J. Exptl. Clin. Chemother. **2** (Suppl. 1) (1989): 43–46.
VAN TYLE, J. H.: Ketoconazole. Mechanism of action. spectrum of activity, pharmacokinetics, drug interactions, adverse reactions and therapeutic use. Pharmacotherapy. **4** (1984): 343–373.
VIVIANI, M. A., and TORTORANO, A. M.: Unusual mycoses in AIDS patients. In: VANDEN BOSSCHE, H., MACKENZIE, D. W. R., CAUWENBERGH, G., VAN CUTSEM, J., DROUHET, B., and DUPONT, B. (Eds.): Mycoses in AIDS Patients. Plenum Press, N. Y. 1990, pp. 147–155.
- - PAGANO, A., VIGEVANI, G. M., GUBERTINI, G., CRISTINA, S., ASSAISSO, M. L., SUTTER, F., FARINA, C., MINETTI, B., FAGGIAN, G., CARETTA, M., DI FABRIZIO, N., and VAGLIA, A.: European experience with itraconazole in systemic mycoses. J. Am. Acad. Dermatol. **23** (1990): 587–593.
- HILL, J. O., and DIXON, D. M.: *Penicillium marneffei*: dimorphism and treatment. In: VANDEN BOSSCHE, H., ODDS, F. C., and KERRIDGE, D. (Eds.): Dimorphic Fungi in Biology and Medicine. Plenum Press, New York 1993, pp. 413–422.
WARNOCK, D. W.: Immunological aspects of Candidosis in AIDS patients. In: VANDEN BOSSCHE, H., MACKENZIE, D. W. R., CAUWENBERGH, G., VAN CUTSEM, J., DROUHET, E., and DUPONT, B. (Eds.): Mycoses in AIDS Patients. Plenum Press, New York 1990, pp. 83–91.
- and JOHNSON, E. M.: Clinical manifestations and management of hyalohyphomycosis and other uncommon forms of fungal infections in the compromised patient. In: WARNOCK, D. W., and RICHARDSON, M. D. (Eds.): Fungal Infections in the Compromised Patient. John Wiley & Sons, Chichester 1990, pp. 247–310.
- BURKE, J., COPE, N. J., JOHNSON, N. A., VON FRAUENHOFER, N. A., and WILLIAMS, E. W.: Fluconazole resistance in *Candida glabrata*. Lancet ii (1988): 1310.
WEBER, M. M., WILL, A., ADELMAN, B., and ENGELHARDT, D.: Effect of ketoconazole on human ovarian C17,20-desmolase and aromatase. J. Steroid Biochem. Molec. Biol. **38** (1991): 213–218.
WESEL, S.: Itraconazole: a single-day oral treatment for acute vulvovaginal candidosis. Brit. J. Clin. Pract. **44** (Suppl. 71) (1990): 77–80.
WHEAT, J.: Histoplasmosis. Infect. Dis. Clin. NA. **2** (1988): 841–859.
- Diagnosis of histoplasmosis-Review of current methods. In: VANDEN BOSSCHE, H., ODDS, F., and KERRIDGE, D. (Eds.): Dimorphic Fungi in Biology and Medicine. Plenum Press, New York 1933, pp. 333–340.
- HAFNER, R. E., RITCHIE, M., and SCHNEIDER, D.: Itraconazole (Itra) is effective treatment for histoplasmosis in AIDS: Prospective multicenter non-comparative trial. Abstract No. 1206. Program and Abstracts of the 32nd Interscience Conference on Antimicrobial Agents and Chemotherapy (Anaheim, Oct. 11–14, 1992).

WILKINSON, C. F., HETNARSKI, K., and YELLIN, T. O.: Imidazole derivatives. A new class of microsomal enzyme inhibitors. Biochem. Pharmacol. **21** (1972): 3187–3192.

— — and HICKS, L. J.: Substituted imidazoles as inhibitors of microsomal oxidation and insecticide synergists. Pestic. Biochem. Physiol. **4** (1974): 299–312.

WILLIAMS, D. J., MARTEN, R. M., and SARKANY, I.: Oral treatment of ringworm with griseofulvin. Lancet **2** (1958): 1212.

WILLOCKS, L., LEEN, C. L. S., BRETTLE, R. P., URQUHART, D., RUSSELL, T. B., and MILNE, L. J. R.: Fluconazole resistance in AIDS patients. J. Antimicrob. Chemother. **28** (1991): 937–939.

WILSON, R. G., and RYLEY, J. F.: Screening and evaluation *in vitro*. In: RYLEY, J. F. (Ed.): Chemotherapy of Fungal Diseases. Springer-Verlag, Berlin 1990, pp. 111–128.

YAMAGUCHI, H.: Antagonistic action of lipid components of membranes from *C. albicans* and various other lipids on two imidazole antimycotics. Antimicrob. Ag. Chem. **12** (1977): 16–25.

— Morphological aspects of azole action. In: BERG, D., and PLEMPEL, M. (Eds.): Sterol Biosynthesis Inhibitors. Pharmaceutical and Agrochemical Apects. Ellis Horwood Ltd., Chichester 1988, pp. 56–78.

— NISHIYAMA, Y., and AOKI, Y.: Morphological aspects of antifungal action. In: VANDEN BOSSCHE, H., ODDS, F., and KERRIDGE, D. (Eds.): Dimorphic Fungi in Biology and Medicine. Plenum Press, New York 1993, pp. 361–372.

YOSHIDA, Y.: Cytochrome P450 of fungi: primary target for azole antifungal agents. Curr. Top. Mycol. **2** (1988): 388–418.

— and AOYAMA, Y.: Interaction of azole fungicides with yeast cytochrome P450 which catalyzes lanosterol 14α-demethylation. In: IWATA, K., and VANDEN BOSSCHE, H. (Eds.): *In vitro* and *in vivo* Evaluation of Antifungal Agents. Elsevier Science Publishers, Amsterdam 1986, pp. 123–134.

— — Cytochromes P450 in ergosterol biosynthesis. In: RUCKPAUL, K., and REIN, H. (Eds.): Frontiers in Biotransformation Vol. 4. Microbial and Plant Cytochromes P450: biochemical characteristics, genetic engineering and practical applications. Akademie Verlag, Berlin 1991, pp. 127–148.

ZIOGAS, B. N., SISLER, H. D., and LUSBY, W. R.: Sterol content and other characteristics of pimaricin-resistant mutants of *Aspergillus nidulans*. Pestic. Biochem. Physiol. **20** (1983): 320–329.

Chapter 22

Fungicides for wood preservation

E.-H. POMMER

Limburgerhof, Germany

Introduction

Under favourable environmental conditions, many fungi can cause decay and discolouration of wood. They cause a variety of visible changes to the timber that is attacked. This may occur in freshly felled logs, freshly sawn timber, wood in service becoming moist, or wood in service in contact with the ground. In moist conditions the white rot and brown rot fungi, which belong to the *Basidiomycetes*, are the most important destroyers of wood. These fungi produce different types of enzymes, which can cause decomposition either of all the woody substances, with the cellulose part remaining (white rot), or vice versa (brown rot). The fungal deterioration of wood in contact with the ground or water is mainly due to attack by soft rot fungi, the hyphae of which grow inside the cell wall, forming long tunnels by consuming cellulose. If the timber is moist, its surface becomes soft. These fungi belong to the *Ascomycetes*. Apart from these wood-destroying fungi, there are two other groups which can cause discolouration of timber to a greater or lesser degree: bluestain fungi and moulds. The bluestain fungi develop in the surface of timber, and the pigmented hyphae can penetrate into the sapwood, consuming the carbohydrates. As a result of discolouration, joinery becomes shabby and the quality of the timber is diminished. Moulds are able to grow mostly on freshly sawn timber, producing masses of coloured hyphae and spores. They do not attack the wood, but there is a loss of quality as a result of the discolouration. The spores may be harmful to health. Moulds belong either to *Ascomycetes* or to *Deuteromycetes*.

Depending on the use of the wood, which has no or insufficient resistance to fungal attack, it must be protected by wood preservatives. For example, this protection may be temporary on freshly felled logs or sawn timber in order to prevent the growth of moulds and bluestain fungi. Wood in the ground, in soil contact, or in water must be treated against wood-rotting fungi. Timbers with a risk of wetting, and external joinery, both coated and uncoated, are susceptible to infestation with moulds, blue-staining fungi, or wood-destroying fungi.

Among the wood preservatives that are effective against wood-destroying fungi, fixing salt types based on inorganic arsenic, chromium, copper, boron and fluorine salts, creosotes, pentachlorophenol (PCP) and its sodium salt (PCP-Na), tributyltin oxide (TBTO) and other TBT formulations are common. However, toxicological problems and environmental hazards may arise if preservatives containing arsenic or chromium are incorrectly handled, and this includes burning of treated wood. Damage to health can result from PCP and PCP-Na on account of dioxin impurities or dioxin formation when wood treated with PCP is burned.

On account of this situation, the industry makes every effort to detect and develop compounds that are less toxic and less damaging to the environment. In addition, the new preservatives should have a broad spectrum of fungicidal activity as well as a long period of effectiveness. A problem that ist hard to solve is the demand for rapid degradation of the chemical outside the wood in order to permit simple waste management. Furthermore, the properties of wood, such as strength and the capacity to be painted or glued, should not be influenced.

Fixing salts

The term "fixing salts" is used for water-borne preservatives which become insoluble after the wood is treated. Water-borne preservatives are mostly inorganic salts; at present only a few organic chemicals are available.

The majority of fixing salts consist of several components, of which chromium has the function of being a fixing agent; copper is effective against soft rot fungi, and a third subtance controls brown rot and white rot. These combinations are also effective against insects, and some also against termites. Table 22.1 presents some of the conventional fixing salt combinations and a few of the water-borne organic preservatives. It must be emphasised that AAC (quats) are effective only in wood not in contact with the ground.

Table 22.1 Some fixing salt types in wood preservatives

CCA:	Chromium – copper – arsenic. There are various formulations with different types of chemicals such as sodium dichromate, copper sulphate, arsenic pentoxide and variable ratios of the constituents. CCAs were first introduced in the thirties. They exhibit a long period of good effectiveness against most of the wood-destroying organisms, including termites. In hardwood with ground contact they are less effective.
CCB, CCF, CCFB:	Chromium – copper – boron, chromium – copper – fluorine, chromium – copper – fluoroborate. In these combinations the arsenic is replaced by boron. Compared with CCA, these salts are to some extent less effective.
CC:	Chromium – copper. This mixture shows high efficacy against soft rot fungi, which is of interest principally for cooling towers.
AAC:	Alkyl ammonium compounds (quats): their proportion and applications will be described in the appropriate section.
Cu – HDO:	This preservative is an organic copper compound which is made water-soluble by forming a complex with an amine. It is a new type of preservative, containing no chromine. It is discussed in more detail in the section "N-cyclohexyldiazenium-dioxy-chelates".

Tributyltin (TBT) compounds

For health and safety reasons, the use of tri-n-butyltin oxide (TBTO) (Fig. 22.1 a) and other TBT formulations is now restricted in some countries such as Japan despite of the fact that they are proven effective wood preservatives and agents for controlling wood-infesting insects.

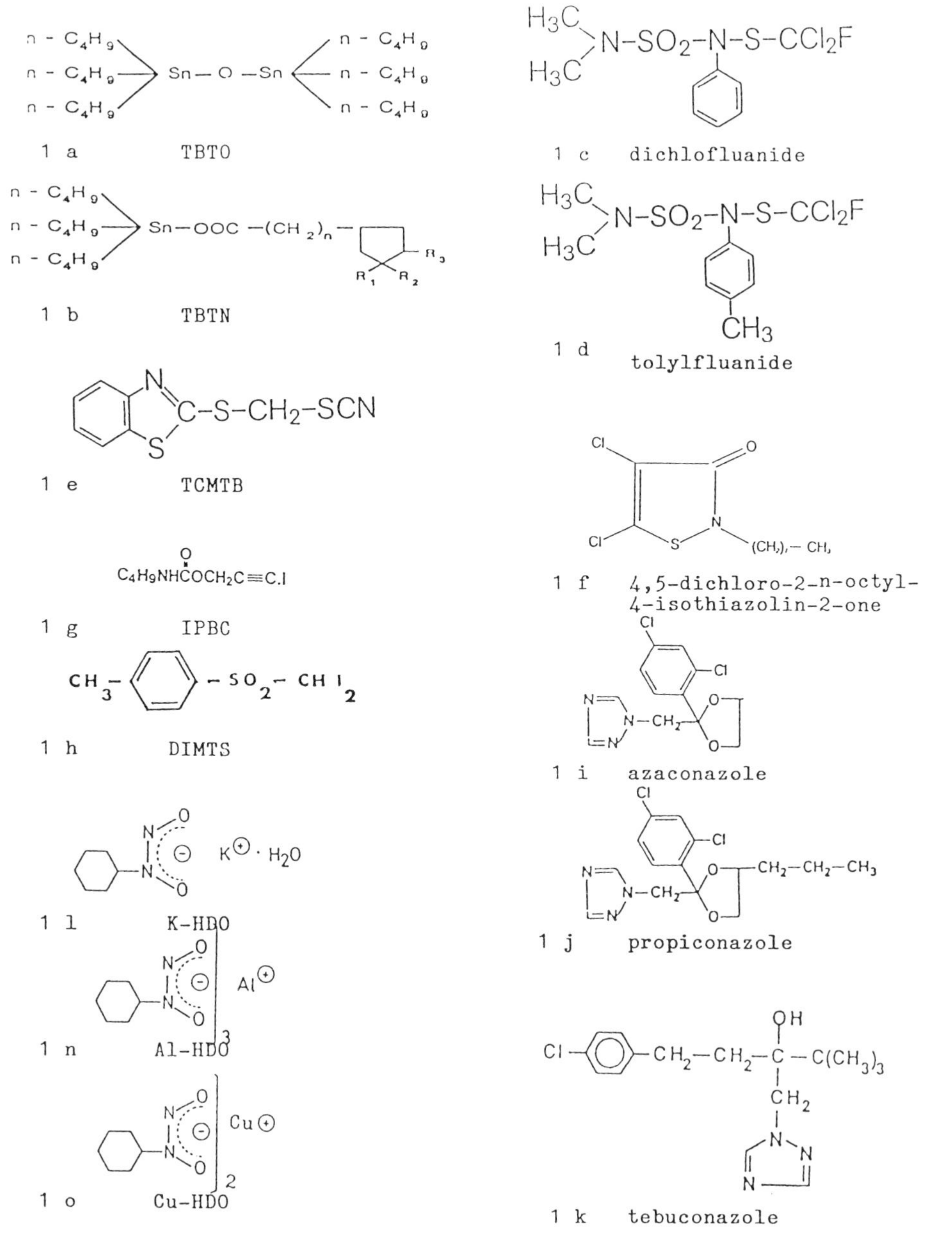

Fig. 22.1 Structures of compounds to be used as fungicides for the preservation of wood.

The effectiveness of TBT compounds as wood preservatives has been known since 1958, and was described by HOF and LUIJTEN (1959). In particular, they are very effective against wood-destroying *Basidiomycetes*. The best known substance is TBTO. However, because of their low volatility and low leaching rates from treated timber, other TBT compounds such as TBT naphthenate (TBTN) (Fig. 22.1 b) have gained some interest.

The oral LD_{50}(rat) for TBTO is 127 mg/kg and for TBTN 224 mg/kg. TBTO is sparingly soluble in water: 33 ppm; its vapour pressure is 10^{-5} mbar at 20°C. The corresponding values for TBTN are 1.3 ppm and 10^{-3} mbar respectively (BECKER 1984). LANDSIEDEL and PLUM (1981) investigated the volatility of TBT compounds at 65°C, and were able to show that after 144 hours 26% of TBTO and only 6.5% of TBTN were lost through volatilisation.

The fungicidal effects of TBT compounds against the *Basidiomycetes* which cause bown rot are clear. Much higher concentrations sometimes have to be used in order to protect timber against white rots such as *Coriolus versicolor* and soft rot fungi. The toxic limits obtained by different investigators vary considerably. However, it must be taken into account that methods such as DIN, BS, EN, AWPA and ASTM test procedures were used. In a review article by BECKER (1987, 1988), a very large number of results were put together.

In Scandinavia and Great Britain TBTO and TBTN are mainly used for the protection of joinery. It is usual to impregnate the timber by means of a double vacuum process. In Britain, softwood windows treated with 1% TBTO preservatives are said to be protected against decay for 30 years (CONNELL et al. 1990).

It has been shown that in both aged and freshly treated timber TBTO and TBTN can be degraded to products such as dibutyltin and monobutyltin with a reduced protective effect against wood-destroying fungi (HENSHAW et al. 1978; EDLUND et al. 1982; EDLUND et al. 1985). It is assumed that several factors cause this degradation, e. g. UV-light, heat, oxidation, microorganisms, and the wood substance itself. JERMER et al. (1983) and JERMER and EDLUND (1985) reported that there was considerable breakdown of TBTO when treated samples were heated to 50°C and 70°C respectively. It corresponded to the degradation obtained in TBTO-treated windows that had been in service for about 5 and 7 years. The most commonly occurring fungus that causes bluestain, *Aureobasidium pullulans*, is specifically tolerant to TBTO and seems to be able to detoxify it (SAVORY and CAREY 1979). Quite a number of fungi have an enzyme system to dealkylate TBTO. The degradation products are tri-, di- and monobutyltin compounds (DUDLEY-BRENDELL and DICKENSON 1982). DUDLEY (1981) reported the degradation of TBTO by some gramnegative bacteria in the presence of suitable sources of carbohydrates.

Little is known about the interactions between TBT compounds and wood. It must be assumed that there are reactions such as the addition of the TBT group onto the hydroxyl groups of cellulose or the formation of TBT carbonate as a result of a reaction with CO_2, which seems to be responsible for the durability of TBTO in the wood (DUDLEY-BRENDELL and DICKINSON 1982). There are very few reports of the failure of TBTO-treated timber in service (HENSHAW et al. 1978; IRMSGARD et al. 1985).

Sulphamides

About 30 years ago, dichlofluanide (Fig. 22.1 c) and tolylfluanide (Fig. 22.1 d) were developed simultaneously as fungicides for plant protection (see Chapter 20) and as agents for the protection of materials (see Chapter 23); their chemistry and launch details are given in Table 22.2. Some details of the physical, chemical and toxicological properties are presented in Table 22.3.

Table 22.2 Dichlofluanide and tolylfluanide as fungicides for wood preservatives

	Dichlofluanide	Tolylfluanide
Chemical name	N,N-dimethyl-N'-phenyl-N'-(dichloro-fluoromethylthio)-sulphamide	N,N-dimethyl-N'-tolyl-N'-(dichloro-fluoromethyl-thio)-sulphamide
Common name	dichlofluanide	tolylfluanide
Code Number	Ue 5 932	Ue 5 933
Trade name	Preventol A 4-S	Preventol A 5
Year of introduction by	1964, Bayer	1964, Bayer
Patent Number	GB 1 056 642	GB 1 056 642
		EP 7 910 377

Table 22.3 Some chemical, physical and toxicological properties of dichlofluanide and tolylfluanide (Source: Bayer Product Information 1988)

	Dichlofluanide	Tolylfluanide
Vapour pressure (mbar)	10^{-6} at 20 °C	10^{-7} at 20 °C
Solubility at 20 °C (g/100 ml)	10^{-5} at 45 °C	10^{-4} at 20 °C
Water	0.0002	0.0002
Acetone	30	46
ethylacetate	11.7	30
Xylene	6.5	22
Solvesso 100	7.5	18
Shellsol AB		12
White spirit	1.2	1
The active ingredients hydrolyse in aqueous solutions, are incompatible with alkalis and amines, are light-stable, non-volatile, and non-leaching.		
LD_{50} (mg/kg rat) oral	5 000	5 000
LD_{50} (mg/kg rat) dermal	5 000	5 000
LC_{50} inhalation rat (mg/m^3; 4 h)	300	265

These products are effective against moulds, blue-staining fungi and wood-destroying *Basidiomycetes*. In the field of wood preservation they were developed as fungicides for preventing bluestain on timber in service. METZNER (1977) quotes an experiment in which the surface of planed timber was treated with 0.5 g dichlofluanide/m^2. After outdoor weathering for 6 months a laboratory trial was carried out with *Aureobasidium pullulans* and *Sclerophoma pityophila* as the test organisms. Unlike with PCP (3.5 g/m^2) and TBTO (2.5 g/m^2) there was no development of bluestain. Depending on the formulation, dichlofluanide penetrates more or less deeply into the treated wood, and there will be a stain-free zone of several millimetres. This was confirmed by VALCKE (1989) and LOVE (1991). Primers and wood stains with a low content of binders generally contain 0.6 – 0.8% of either dichlofluanide or tolylfluanide. In repeated co-operative studies using a formulation containing 1% dichlofluanide, it was shown that on varnished wood panels this compound is excellent in controlling bluestain after artificial as well as natural weathering followed by laboratory trials with *Aureobasidium pullulans* and *Sclerophoma pityophila* (BRAVERY and DICKINSON 1985).

Both products show remarkable efficacy against wood-destroying fungi (METZNER 1977; PAULUS and KÜHLE 1984). Due to their favourable behaviour in the wood against leaching and evaporative ageing, dichlofluanide (METZNER 1977) and tolylfluanide (PAULUS and KÜHLE 1984) can be used to prevent wood being attacked by wood-destroying fungi such as *Coniophora puteana*, *Coriolus versicolor* and *Gloeophyllum abietinum*. The mean toxic limits are generally said to lie in the range of 1.3 kg a.i./m^3 to 3.7 kg a.i./m^3.

2-(Thiocyanomethylthio)benzthiazole (TCMTB)

In 1970 TCMTB (Fig. 22.1e) was introduced by Buckman Laboratories as a seed dressing to control diseases caused by *Rhizoctonia* and *Helminthosporium*; in about 1975 the company began the development of this compound for the field of wood preservation.

TCMTB has a low vapour pressure: 3.57×10^{-6} torr (25°C), its water solubility is reported to be 33 mg/l (VAN DEN ENDE 1990). According to BUTCHER and DRYSDALE (1978) the oral LD$_{50}$ (rat) is 1590 mg/kg. The commercially available product Busan 30 contains 30% a.i. in a solvent-borne formulation.

TCMTB is mainly used to prevent the development of sapstain on freshly sawn timber. From the results published, it appears that protection lasts a period of 3 to 5 months (PLACKETT 1982; DRYSDALE 1986; DRYSDALE and KEIRLE 1986; ROSE 1987). CZERJESEI et al. (1984) reported that 0.04% TCMTB (1.2% Busan 30) protected softwood in Canada for 16 months. A distinct improvement in anti-sapstain effectiveness of TCMTB could be maintained by combining this active compound with methylene-bis-thiocyanate (MBT), which is also effective against sapstain and moulds (DRYSDALE 1987; LEWIS et al. 1985; NOMURA 1991). This can be attributed to the fact that MBT shows a rapid movement and high initial fungotoxicity, whereas the immobile TCMTB would remain on the surface to improve the residual surface toxicity (WILLIAMS and LEWIS 1991).

In cooperative studies investigating the influence of different weathering systems on the effectiveness of fungicides preventing bluestain fungi on varnished wood panels, a formulation containing 1% showed variable performance. Compared with dichlofluanide and IPBC at the same level of active compound, TCMTB was less effective (BRAVERY and DICKENSON 1985).

GREAVES (1977), using a filter paper assay technique, found TCMTB to be effective against the soft rot fungus *Chaetomium globosum*. In other investigations carried out by THORNTON (1977), HEDLEY (1979) and VERBECKE et al. (1989), similar results were obtained. KONABE (1987) determined the toxic limits of TCMTB on birch blocks exposed to soil and found a value of 1.5 kg a.i./m^3.

TCMTB is effective against wood-destroying fungi; however, the determined toxic limits against brown rot and white rot fungi differ, depending on the type of formulation or diluent employed (BLOW 1980; BRAVERY and CAREY 1985; VAN DEN ENDE 1990).

On the basis of laboratory experiments. WILLIAMS (1990) suggests that under favourable conditions active bacterial biodetoxification of TCMTB could occur within a short time during storage of timber treated with anti-sapstain.

N-substituted isothiazolones (ITA's)

Investigating the efficacy of N-substituted isothiazolones for the control of wood-destroying fungi, PRESTON et al. (1984) stated that in general the ITA's were more active against brown and soft rot fungi than against white rot fungi. The authors compared a series of N-substituted dichloro-isothiazolinones with n-octyl-isothiazolin-3-one, which has been shown to be effective against both decay fungi and moulds (HEDLEY et al. 1979). Of the compounds tested, 4,5-dichloro-2-n-octyl-4-isothiazolin-3-one (Fig. 22.1 f) showed the most superior activity. This new wood preservative was introduced by ROHM and HAAS in 1987 under the trade name Kathon 925; an experimental formulation was tested as RH 287. The commercial product contains 25% a..i. and 75% aromatic solvent as the carrier. The active compound is soluble in a wide range of commercial solvents and can be formulated as a water-miscible emulsion. Vapour pressure and water solubility are low: 7.4×10^{-6} torr and 10 ppm respectively. According to NICHOLAS et al. (1984), the oral LD_{50} (rat) for RH 287 is 2.56 g/kg.

Kathon 925 was incorporated into a solvent-based varnish and applied to *Pinus sylvestris* panels which were exposed for 18 months. Despite heavy falls of rain, panels treated with 0.3 – 0.4% a.i. were completely free of staining fungi and moulds. The fungicidal effectiveness was superior to that of 4% PCP (ROHM and HAAS 1989; ANON. 1989).

The toxicity values of the a.i. of Kathon 925 are given in Table 22.4.

Table 22.4 Efficacy of Kathon 925 against wood-destroying *Basidiomycetes* (GREENLEY and HEGARTY 1988)

Fungi	Toxicity values*) (kg a.i./m³)		
	without ageing	after leaching	after evaporative ageing
Coniophora puteana	0.64 – 1.26	0.34 – 0.67	1.39 – 1.88
Gloeosporium trabeum	0.65 – 1.29	0.33 – 0.67	0.55 – 1.01
Coriolus versicolor	0.65 – 1.27	0.33 – 0.67	0.54 – 1.03
Poria placenta	0.32 – 0.64	–	0.55 – 1.08

*) Test procedure according to EN 113, EN 83, EN 73

On account of its favourable behaviour in wood, this fungicide is recommended for the protection of timber of hazard classes 2 and 3 (3 = timber directly exposed to weather, not in ground contact). GRENNLEY et al. (1988) observed in a stake test after 48 months' exposure that stakes treated with 4.6 kg/m³ showed only a very low degree of deterioration. A non-sterile soil block-test was used to determine the efficacy of Kathon 925 against soft rot fungi. NICHOLAS et al. (1984) determined a toxicity threshold value of 0.5 kg a.i./m³.

Kathon 925 is also highly effective against wood-infesting insects (ROHM and HAAS 1989; Anon. 1989).

According to CROW and LEONARD (1965) and MILLER et al. (1975), the mode of action of isothiazolinone biocides is based on a reaction of the isothiazolone ring with nucleophilic cell entities such as enzymes, proteins and amino-acids. The reaction begins with opening of the ring, and this initiates the biocidal activity.

Organic iodo compounds

Looking for an alternative fungicide to PCP, 3-iodo-3-propynyl butyl carbamate (IPBC), (Fig. 22.1 g), which was originally developed by the Troy Chemical Corp. as a microbicide for the protection of paints, was found to be effective against wood-destroying and discolouring fungi. Troysan Polyphase is the trade name of a formulation containing 17% IPBC; other names are Woodbrite NTX and Argosy.

Its solubility in water is relatively high: 156 ppm at 20°C; in aromatic and polar solvents solubility is high. Vapour pressure is $1,8 \times 10^{-6}$ mm Hg at 20°C. The acute oral LD_{50} (rat) is 1580 mg/kg.

HANSEN (1984) reported on the fungicidal efficacy of IPBC. In test procedures according to EN 113 without leaching and after leaching, he determined the toxicity values for some important *Basidiomycetes* that cause brown and white rot. For better distribution of the a.i. in the timber he employd a binder-containing formulation. The toxicity values found for treated timber were 0.09 kg a.i./m^3 unleached and 0.2 – 0.8 kg a.i./m^3 after leaching. TSUNODA (1992) found that organo-iodine compounds performed differently as fungicides in the amended JWPA decay test (Standard 1989) when applied as surface treatment to timber and followed by frequent drying and leaching cycles. For beech sapwood IPBC provided no satisfactory protection against attack by *Coriolus versicolor*, whereas it was effective against the brown rot fungi *Tyromyces palustris* and *Serpula lacrymans*.

LEWIS *et al.* (1985) examined the effectiveness of anti-sapstain chemicals. In laboratory tests, sapwood slats of fresh unseasoned wood were immersed in formulations containing 0.25% and 1.0% a.i. and artificially inoculated. After incubation for 5 weeks the slats were assessed for the percentage area affected by stain and mould. Compared with methylen-bis-thionate (MBT), the level of stain control was lower. In a 5-month field test with various anti-sapstain formulations in which hemfir and fir were used, PLACKETT (1982) showed that 1% IPBC-based treatments provided protection of hemfir timber comparable to that of treatments containing 0.65% PCP; at the 0.5% level IPBC was ineffective.

BRAVERY and DICKINSON (1985) reported on co-operative studies to assess the effects of different artificial weathering systems on the performance of fungicides to control bluestain on varnished panels. After weathering, the panels were inoculated with a mixed spore suspension of *Aureobasidium pullulans* and *Sclerophoma pityophila*, followed by a six-week incubation period. 1% IPBC was generally effective in preventing blue staining on the surfaces of the panels; the average stain-free zones were several milimetres.

Di-iodomethyl-b-tolylsulfone (DIMTS)

STAMM *et al.* (1987) reported on DIMTS (Fig. 22.1 h) as a fungicide for wood protection. The compound was originally developed by Abbott Laboratories as a preservative for materials under the trade name Amical 48. Solubility in water is 0.1 mg/l, and the compound is stable at pH 2.0 – 10.5. The oral LD_{50} (mouse) is 10,000 mg/kg. At levels of 0.5% to 1.0% the substance performs well against sapstain-causing fungi and moulds, including *Trichoderma* sp. (CASSENS and ENSYN 1981). An improvement in effectiveness could be obtained by a combination of DIMTS and didecyldimethyl-ammonium chloride (HAYWARD *et al.* 1984). THOMPSON and TOOLE (1974) determined the effectiveness of DIMTS against wood-destroying *Basiodiomycetes* in a soil block test. In a concentration of 0.08% DIMTS provided almost complete protection as measured by average weight loss of the wood blocks. NAKAMURA (1983) carried out a field test with pressure-treated stakes of six wood species over a period of six years. All stakes, above and below ground, that were treated with 0.5% DIMTS were largely resistant to decay and termite attack for six years, with the exception of two wood species which were protected for four years below ground.

Triazoles

Since 1973, numerous fungicides for plant protection have been developed from the triazole class (cf. chapter 12). However, it took approximately another ten years before a triazole derivative was introduced as a wood preservative: azaconazole (Fig. 22.1 i). Two other triazoles, which have alrcady proved successful as fungicides in the field of plant protection, have been developed as wood preservatives in recent years. On account of their spectra of activity, especially against wood-decaying fungi, their low toxicity, and their favourable behaviour in the environment, they are of special interest: propiconazole and tebuconazole (Fig. 22.1 j and 22.1 k).

Table 22.5 presents an orverview of these compounds, which have been introduced as commercial products.

Table 22.5 Triazole derivatives as fungicides for wood preservation

Chemical name	1-(2-(2′,4′-dichloro-phenyl)-1,3-dioxolan-2-yl-methyl)-1 H-1,2,4-triazole	($\pm$)-(cis+transl)-1-((2-(2,4-dichlorophenyl)-4-propyl-1,3-dioxolan-2-yl-methyl-1 H-1,2,4-triazole	($\pm$)-α-(2-(4-chlorophenyl)-α(1,1-dimethylethyl)-1 H-1,2,4-triazole-1-ethanol
Common name	Azaconazole	Propiconazole	Tebuconazole
Code number	R 28 644	R 49 362	–
Trade name	Madurox	WOLCOSEN	Preventol A 8
Year of introduction by	1982, Janssen	1989, Janssen	1988, Bayer
Patent Number	EP 0038 109 US 4 542 146 EP 0 148 526 US 4 648 988	EP 0038 109 US 542 146	DE 3 018 866

Table 22.6 gives some details of the physico-chemical and toxicological properties of the three triazole fungicides. It must be emphasised that tebuconazole has both very low water solubility and very low vapour pressure, which are important preconditions for use in wood preservatives. Propiconazole, as a liquid, has the best solubility, which is advantageous when formulating products.

Depending on the area of application, for example dipping tanks, vacuum pressure or brushing, different formulations are available: (light) organic solvent- and water-borne types. An emulsifiable concentrate of azaconazole was developed as a fungicidal additive to glue for the manufacture of chipboard.

The three triazoles are effective against wood-destroying *Basidiomycetes*, Table 22.7 gives the toxicity values. They are based on the results from a number of tests.

From the published results it may be concluded that in treated timber both propiconazole and tebuconazole are highly resistant to leaching and evaporation. It was shown by VALCKE and STEVENS (1991) that in addition to this favourable behaviour the concentration gradient of propiconazole in the outermost layers of the wood specimens was not altered, and that the active ingredient was evenly distributed in the wood.

EXNER (1991) reported on vacuum pressure treatment of southern pine wood with solutions of tebuconazole followed by a soil block test. Attack by *Gloeophyllum trabeum* was prevented by a very low fungicide concentration. It should be emphasised that propiconazole displays strong activity against *Poria* (VALCKE 1989).

In a stain test with *Aureobasidium pullulans* the inhibitory concentrations were found to be in the range from 0.15 to 0.31%; finishing stains containing 0.33% azoconazole showed good anti-bluestain efficacy (VALCKE 1985).

VALCKE and GOODWINE (1989) demonstrated that 0.6% propiconazole gave satisfactory protection agains bluestain on timber in service.

Table 22.6 Some details of the physico-chemical and toxicological properties of triazole fungicides (LECLERCQ 1983; VALCKE and GOODWINE 1985; VALCKE 1989; EXNER 1991; PAULUS 1992)

	Azaconazole	Propiconazole	Tebuconazole
Vapour pressure (mbar)	4×10^{-9} at 20 °C	1.3×10^{-6} at 20 °C	7.2×10^{-9} at 20 °C
	2×10^{-6} at 70 °C	1.5×10^{-5} at 40 °C	4.5×10^{-6} at 60 °C
Solubility (g/100 ml)			
Water	0.03	0.011	0.0032
Ethylene glycol	3.5	very soluble	27
Propylene glycol	6.6	very soluble	5
White spirit	?	very soluble	0.3
Xylene	4.5	very soluble	4
Stability		stable in acid and alkaline solutions	
LD_{50} (mg/kg rat) oral	308	1514	3900
LD_{50} (mg/kg rat) dermal	2560	4000	5000
LD_{50} inhalation rat (mg/m^3; 4 h)	?	5800	5000

Table 22.7 Toxicity values*) of triazole fungicides against *Basidiomycetes* (VALCKE and GOODWINE 1985; VALCKE 1989; Bayer Product Information on Preventol A 8, 1990; VALCKE and STEVENS 1991)

Fungi	Toxicity values (kg a.i./m^3)		
	Azaconazole	Propiconazole	Tebuconazole
Coniophora puteana	0.68 – 1.41	0.27 – 0.39	0.05 – 0.2
Gloeophyllum trabeum	0.66 – 1.35	0.4 – 0.75	0.15
Poria placenta	0.67 – 1.38	0.22	0.3 – 0.5
Coriolus versicolor	0.51 – 0.99	0.25 – 0.33	0.2 – 0.4

*) Test procedure according to EN 73 (1978), EN 113 (1986), EN 84 (1987)

Despite the fact that good results in preventing the development of *Aureobasidium pullulans* and *Sclerophoma pityophila* were obtained on test specimens (WÜSTENHÖFER *et al.* 1990), it is advisable to apply tebuconazole in combination with diclofluanide in order to prevent the growth of bluestain fungi (Bayer Product Information on Preventol A 8 (1990)).

The mode of action of triazole derivatives is primarily based on the inhibition of ergosterol biosynthesis as a result of biosynthetic reactions in the conversion of lanosterol to ergosterol being blocked (see Chapter 13).

N-Cyclohexyl-diazeniumdioxy-chelates

HICKMANN *et al.* (1979) proved by X-ray structure analysis that the compounds initially described as potassium, aluminium and copper salts of N-cyclohexyl-hydroxylamine are the corresponding N-cyclohexyl-diazeniumdioxy-chelates; The chemical structures are presented in Figures 22.1 l, n, o. The chelates used in wood preservation are summarised in Table 22.8.

Table 22.8 N-Cyclohexyldiazeniumdioxy-chelates as fungicides for wood preservatives

Chemical name	N-cyclohexyl-diazeniumdioxy-potassium hydrate	Tris-(N-cyclohexyl-diazeniumdioxy)-aluminium	Bis-(N-cyclohexyl-diazenium-dioxy)-copper
Code number	KR 4091	–	–
Trade name	Xyligen K; Xyligen 25 F; 30 F	Xyligen Al	Cu-HDO
Year of introduction by	1967, BASF	1970, BASF/Wolman	1989, BASF, Wolman
Patent Number	DE 102 474	DE 102 474	DE 2 410 603

The potassium chelate, Xyligen K, was the first to be introduced as a preservative for wood-based panels against wood-attacking *Basidiomycetes*, especially brown rot fungi such as *Coniophora puteana* or *Gloeophyllum trabeum*. As a result of its solubility in water, Xyligen K is freely miscible with the adhesives used in the production of chipboard and plywood made from *conifers*. The oral LD_{50} (Rat) ist 440 mg/kg for the aqueous solution.

Toxicity values for AL-HDO and CU-HDO are presented in Table 22.9.

Table 22.9 Efficacy of AL-HDO and Cu-HDO against wood-destroying fungi (GÖTTSCHE and MARX 1989)

Fungi	Toxicity values*) (kg a.i./m^3)			
	AL-HDO		Cu-HDO	
	without ageing	after leaching	without ageing	after leaching
Coniophora puteana	0.2 – 0.5	0.4 – 0.8	0.5 – 0.9	0.5 – 0.8
Poria monticola	0.1 – 0.3	0.3 – 0.4	0.2 – 0.4	0.2 – 0.4
Coriolus versicolor	3.8 – 6.0	= 6.8	0.9 – 2.0	0.4 – 1.0
*Chaetomium**) globosum* (Beech)		12.0		2.0 – 3.8

*) Test procedure according to DIN 52 176; DIN 52 172
**) Vermiculite burial test procedure (KERNER-GANG, GERSONDE 1972)

It can be seen clearly that AL-HDO shows a lack of effectiveness against the white rot fungus *Coriolus versicolor* — comparable values were determined by METZNER (1977) and TSUNODA (1991) — and the causative fungus of soft rot, *Chaetomium globosum*. In contrast to AL-HDO, the copper chelate is effective against these fungi. AL-HDO in amounts of approx. 2% ist an ingredient in primers for the treatment of windows and other joinery, which are applied either in a spraying tunnel or by the double vacuum procedure. Two of the remarkable features of AL-HDO are its excellent penetration into the treated timber and its good stability even after several years of outdoor weathering (GÖTTSCHE and MARX 1989). The active compound has a very low vapour pressure of 1×10^{-7} mbar/20°C and an oral LD_{50} (rat) of 5600 mg (GÖTTSCHE and MARX 1989). In the framework of their investigations of the light stability of low-toxicity wood preservatives, FUSE and TANA (1981) found that AL-HDO is essentially stable on exposure to light.

Since REUTHER et al. (1974) were successful in converting the practically water-insoluble HDO copper chelate into a water-soluble form by means of complex formation, it was possible to develop new types of wood preservatives for vacuum pressure processes (GÖTTSCHE and MARX 1989). First a water-borne combination was commercially introduced, containing 12.5% Cu-HDO and 6% boric acid (Wolmanit CX-50). As stated by GÖTTSCHE and MARX (1989), the purpose of adding boric acid was to obtain protection of heartwood by diffusion. It was possible to improve the effectiveness of this product against soft rot by adding cupric (II) hydroxide carbonate. This new formulation (Wolmanit CX-S), containing 6.1% Cu-HDO, 8.1% cupric (II) hydroxide carbonate and 4% boric acid, is highly effective against wood-destroying fungi. Moreover, there is also insecticidal activity against the larvae of the house longhorne beetle. Table 22.10 gives the toxicity values of Wolmanit CX-50 and Wolmanit CX-S.

Table 22.10 Cu-HDO formulations/toxicity values*) (kg/m³) (Dr. Wolman GmbH, Anon. 1992)

Fungi/test procedure	Wolmanit CX-S	Wolmanit CX-50
*Basidiomycetes***) (after leaching)	1.9 – 3.0	1.4 – 2.4
Soft rot (vermiculite/pine)	3.0 – 4.7	4.7 – 7.5

*) Test procedures according to EN 84 (1979) and EN 113 (1986)
**) *Coniophora, Poria, Gloeophyllum trabeum*

On account of its excellent fixation in pressure-treated wood, the Cu-HDO formulation can be used for wood protection in hazard classes 1 – 4. Hazard class 4: timber in contact with ground or fresh water; permanently wet and with a moisture content permanently of 20%.

The oral LD_{50} of Cu-HDO is approximately 860 mg/kg.
With regard to corrosion, experiments were carried out in accordance with DIN 52 168-A with the result that neither formulation promotes corrosion (Dr. Wolman GmBH; personal communication). Results form a combustion test with a preservative containing Cu-HDO showed that the combustion gases CO, CO_2, HCN and NO_x practically corresponded to the concentrations from untreated timber that has been burnt (HETTLER et al. 1992).

Alkyl ammonium compounds (AAC)

After extensive research by BUTCHER and DRYSDALE (1977, 1978), BUTCHER et al. (1977), BUTCHER and GREAVES (1982), and PRESTON (1983), wood preservatives based on alkyl ammonium compounds (ACC) were introduced into commercial use in New Zealand in 1978. Many tests have been carried out to determine the efficacy of different groups of AACs against wood-attacking *Basidiomycetes*, soft rot, or sapstain-causing fungi. Substances originating from primary, secondary and tertiary amines were of less interest than quaternary ammonium compounds, the "quats". BECKER (1983, 1989) has reviewed the types of AACs that were investigated. Quats have shown a very good performance against quite a number of fungi, including *Basidiomycetes* and soft rot fungi. Of the compounds tested, alkyldimethylbenzylammonium chloride, dialkyldimethylammonium chloride, and to some extent trimethylalkylammonium chloride types (Fig. 22.2 a, b, c) containing alkyl carbons with chain lengths of C_8 to C_{12}/C_{14} were most promising. Against wood-destroying *Basidiomycetes* (NICHOLAS and PRESTON 1980) gave a generalised toxic limit of $1.0-6.2$ kg a.i./m^3 for unleached and leached timber treated with dialkyldimethylammonium chloride. Toxic limits have been determined by various authors for benzalkonium chloride in tests simulating exposure to ground contact; the calculated mean toxic limit was 4 kg a.i./m^3, *Coniophora puteana* had a toxic threshold estimated to be of the order of 6 kg a.i./m^3 pine sapwood (BUTCHER 1985), and could be regarded as tolerant of AAC. RUDDICK (1987) reported on the field performance of pressure-treated pine sapwood stakes in Canada. He observed that stakes treated with dialkyldimethylammonium chloride and alkyltrimethylammonium choride respectively failed totally to decay within the first two or three years. Stakes treated with dimethyldidecylammonium chloride, which has been regarded as the most successful AAC, showed substantial decay in six years of testing. There was no improvement with increased preservative retention. In New Zealand, as a result of the observation of decay in battens after long-term storage, the retentions of alkyldimethylbenzalkonium chloride (benzalkonium chloride) for moderate decay hazards were increased from 2.5 kg/a.i./m^3

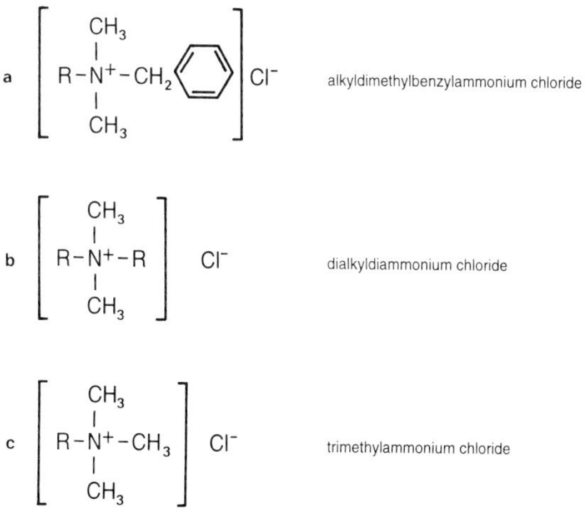

Fig. 22.2 Types of alkylammonium (AACs) to be used in wood preservation.

originally to 3.75 kg a. i./m^3 (BUTCHER 1985). BLOW (1985) investigated the influence of different types of formulations on the toxicity values of benzalkonium chloride, and found a solvent-based formulation to be superior to a water-based one.

In commercial use there arose a number of cases of the non-performance of AACs. The problems and possible reasons for failure have been discussed in detail by BUTCHER (1985). These have been ascribed to substandard treatment, poor distribution of the preservative, misuse in service, depletion of the active ingredient during long-term storage, and attack by AAC-tolerant *Coniophora* spp. Despite investigations to find the reasons, no clear explanations were obtained (RUDDICK and SAM 1982; RUDDICK 1984; CONRAD and PIZZI 1986). JIN and PRESTON (1991) found that didecyldimethylammonium chloride had a greater affinity for lignin than for cellulose. This may lead to insufficient absorption of the preservative on cellulose, and therefore provide little protection against cellulose degradation by brown rot fungi.

On account of biological problems that may occur with AAC's, there are various publications that deal with mixtures of AACs and other fungicides such as copper salts (SUNDMAN 1984; RUDDICK 1987; HENNINSON and JERMER 1988; TSUNODA and NICHIMOTO 1987), IPBC (LEIGHTLEY 1986) and guazatine (EDLUND et al. 1988) for improving long-term performance and broadening the spectrum of activity.

Others

In addition to copper-naphthenate, which has been known for more than a hundred years, in recent times there have been a few new developments in the field of metal carboxylates (metal soaps) as wood preservatives: Acypetacs-copper (2.75% copper), Acypetacs-zinc (3% zinc) (HILDITCH et al. 1983; HILDITCH 1991), and zinc 2-ethyl hexanoate (LOVE 1991). These preservatives have a low toxicity for mammals, and are used for industrial and DIY remedial treatments of joinery with a low risk of wetting.

Methyl benzimidazole-2-yl carbamate (carbendazim, BCM), introduced in 1973 as a fungicide for plant protection (see Chapter 14), is of lesser importance as a wood preservative. BCM is effective against bluestain fungi. On account of its very low solubility in water as well as in organic solvents, the substance does not penetrate sufficiently into the timber to prevent internal bluestain.

References

BAYER: Produkt Information Preventol A4–S, 1988.
BAYER: Produkt Information Preventol A5, 1988.
BAYER: Produkt Information Preventol A8, 1990.
BECKER, H.: Organozinnverbindungen als Holzschutzmittel. Seifen-Öle-Fette-Wachse **113** (1987): 773–776.
– Organozinnverbindungen als Holzschutzmittel, Seifen-Öle-Fette-Wachse **114** (1988): 61–63, 99–100.
– Neue Holzschutzmittel-Gruppe: Alkylammoniumverbindungen. Seifen-Öle-Fette-Wachse **109** (1983): 603–606.

- Neue Holzschutzmittel-Gruppe: Alkylammoniumverbindungen. Seifen-Öle-Fette-Wachse **110** (1984): 15–17.
- Alkylammoniumverbindungen als Holzschutzmittel. Seifen-Öle-Fette-Wachse **115** (1989): 681–684.

BLOW, D.: Alkyl dimethyl benzyl ammonium chloride: toxicity to *Coniophora puteana* when formulated in water and organic solvent. Int. Res. Group on Wood Preservation. Document No.: IRG/WP/2250 (1986).

BRAVERY, A. F., and CAREY, J. K.: Some data on the activity of alternative fungicides for wood preservation. Int. Res. Group on Wood Preservation. Document No.: IRG/WP/3333 (1985).

- and DICKINSON, D. J.: Artificial wheathering as an aid to assessing the effectiveness of chemicals for preventing blue stain in service – a co-operative study. Int. Res. Group on Wood Preservation. Document No.: IRG/WP/2215 (1984).

BUTCHER, J.A.: Benzalkonium chloride (an AAC preservative): Criteria for approval, performance in service, and implications for the future. Int. Res. Group on Wood Preservation. Document No. IRG/WG/3328 (1985).

- DRYSDALE, J.: Relative tolerance of seven wood-destroying basidiomycetes to quaternary ammonium compounds and copper-chrome-arsenate preservatives. Material und Organismen **12** (1977): 271–277.
- Laboratory screening trials with chemicals for protection of sawn timber against mould, sapstain & decay. Int. Biodeterior. Bull. **14** (1978 a): 11–19.
- Efficacy of acidic and alkaline solutions of alkylammonium compounds as wood preservatives. N. Z. Forest Sci. **8** (1978 b): 403–409.
- GREAVES, H.: Alkylammonium compound preservatives: Recent New Zealand and Australian Experiences. Int. Res. Group on Wood Preservation. Document No.: IRG/WP/3188 (1982).
- PRESTON, A., and DRYSDALE, J.: Initial screening trials of some quaternary ammonium compounds and amine salts as wood preservatives. For. Prod. J., **27** (1977): 19–22.

CASSENS, D. L., and ESLYN, W. E.: Field trials of chemicals to control sapstain and mold on yellow polar and southern yellow pine timber. For. Prod. J. **33** (1983): 52–56.

CONNELL, M., CORNFIELD, J. A., and WILLIAMS, G. R.: A new preservative – a double edged swort. Int. Res. Group on Wood Preservation. Document No.: IRG/WP/3573 (1990).

CONRADIE, W. E., and PIZZI, A.: Tannin as a cause of failure of AAC preservatives. Holz als Roh- und Werkstoff **44** (1986): 328–332.

CROW, D. W., and LEONARD, N. J.: 3-isothiazolone-cis-thiocyanoacrylamide equilibria. J. Organ. Chem. **30** (1965): 2660–2665.

CZERJESI, A. J., BYRNE, A., and JOHNSON, E. L.: Long-term protection of stored lumber against mould, stain, and specifically decay: a comparative field test of fungicidal formulations. Int. Res. Group on Wood Preservation. Document No.: IRG/WP/3281 (1984).

Dr. WOLMAN GmbH: Product information Wolmanit CX-S, 1992.

DRYSDALE, J. A.: A field trial to assess the potential of antisapstain chemicals for long-term protection of sawn radiata pine. Intern. Res. Group on Wood Preservation. Document No.: IRG/WP/3375 (1986).

- Commercially available anti-sapstain chemicals in New Zealand – an update. Int. Res. Group on Wood Preservation. Document No.: IRG/WP/3416 (1987).
- and KEIRLE, R. M.: A comparative field test of the effective-ness of antisapstain treatments on radiata pine roundwood. Int. Res. Group on Wood Preservation. Document No.: IRG/WP/3376 (1986).

DUDLEY-BRENDELL, T. E., and DICKINSON, D.: Detoxification of preservatives: Tri-n-butyltin oxide as a biocide. Int. Res. Group on Wood Preservation. Document No.: IRG/WP/1156 (1982).

EDLUND, M., and HENNINGSSON, B.: Field and laboratory studies on antisapstain preservatives. Int. Res. Group on Wood Preservation. Document No.: IRG/WP/3205 (1982).

- JERMER, J., HENNINGSSON, B., and HINTZE, W.: Chemical and biological investigations of double-vacuum treated windows after $7^1/_2$ years in service. Int. Res. Group on Wood Preservation. Document No.: IRG/WP/3329 (1985).

– and HENNIGSSON, B.: Chemical and biological studies of organotin treated and painted wood stakes after outdoor exposure. Int. Res Group on Wood Preservation. Document No.: IRG/WP/3419 (1987).

EXNER, O.: Tebuconazole – a new triazole fungicide for wood preservation. Int. Res. Group on Wood Preservation. Document No.: IRG/WP/3680 (1991).

FUSE, G., and TANA, H.: On the stability against light of low toxic wood-fungicide. J. Antibact. Antifung. Agents **9** (1981): 273–284.

GÖTTSCHE, R., and MARX, H.-N.: Kupfer-HDO – ein vielseitiger Wirkstoff im Holzschutz. Holz als Roh- und Werkstoff **47** (1989): 509–513.

GREAVES, H.: Progress towards controlling soft-rot of treated hardwood poles in Australia. Int. Res. Group on Wood Preservation. Document IRG/WP/289 (1977).

GREENLEY, D. E., and HEGARTY, B. M.: Report on field test results for dichloro-n-octyl-isothiazolone; a potential new wood preservative. Int. Res. Group on Wood Preservation. Document No.: IRG/WP/3495 (1988).

HANSEN, J.: IPBC – a new fungicide for wood protection. Int. Res. Group on Wood Preservation. Document No.: IRG/WP/3295 (1984).

HAYWARD, P., RAE, W., and DUFF, J.: Mixtures of fungicides screened for the control of sapstain on *Pinus radiata*. Int. Res. Group on Wood Preservation. Document No.: IRG/WP/3237 (1984).

HEDLEY, M. E., PRESTON, A. F., CROSS, D. J., and BUTCHER, J. B.: Screening of agricultural and industrial chemicals as wood preservatives. Int. Biodeter. Bull. **15** (1979): 9–18.

HENSHAW, W., LAIDLAW, R. A., ORSLER, R. J., CAREY, J. K., and SAVORY, J. G.: The permanence of tributyltin oxide in timber. Ann. Convent. B.W.P.A., 1978.

HETTLER, W., BREYNE, ST., and MAIER, M.: Gesundheits- und Umweltaspekte bei der Anwendung von Cu-HDO-haltigen Holzschutzmitteln im Kesseldruckverfahren. 19. Holzschutz-Tagung in Rosenheim, Oktober 1992: 217–239.

HICKMANN, E., HÄDICKE, E., and REUTHER, W.: "Isonitramines": Nitrosohydroxylamines or hydroxidiazenium oxides? Tetrahydron Letters No. 216 (1979): 2457–2460.

HILDITCH, E. A.: Organic solvent preservatives: application and composition. In: THOMPSON, R. (Ed.): The Chemistry of Wood Preservation. Royal Society of Chemistry, Cambridge 1991, pp. 69–87.

– SPARKS, C. R., and WORRINGHAM, J. H. M.: Further developments in metallic soap based wood preservatives. Ann. Convent. B.W.P.A. (1983).

HOF, T., and LUIJTEN, J. G. A.: Organotin compounds as wood preservatives. Timber Technol. **67** (1959): 83–84.

IRMSGARD, F., JENSEN, B., LANDSIEDEL, H., and PLUM, H.: Stability and performance of tributyltin compounds. Int. Res. Group on Wood Preservation. Document No.: IRG/WP/3275 (1984).

JERMER, J., EDLUND, M.-L., HENNINGSSON, B., HINTZE, W., and OHLSEN, S.: Chemical and biological investigations of double-vacuum treated windows after 5 years in service. Int. Res. Group on Wood Preservation. Document No.: IRG/WP/3219 (1983).

– EDLUND, M. L., HENNINGSSON, B., and HINTZE, W.: Studies on accelerated ageing procedures with TBTO-treated wood. Int. Res. Group on Wood Preservation. Document No.: IRG/WP/2244 (1985).

JIN, L., and PRESTON, A. F.: The interaction of wood preservation with lignocellulosic substrates. Holzforschung **45** (1991): 455–459.

KERNER-GANG, W., and GERSONDE, M.: Untersuchungen zur Prüfung mit Moderfäulepilzen im Vermiculit-Eingrabe-Verfahren. Material und Organismen **7** (1972): 241–258.

KONABE, H. C.: Effectiveness of Busan 30 treated birch blocks in a soil medium. Int. Res. Group on Wood Preservation. Document No.: IRG/WP/3409 (1987).

LANDSIEDEL, H., and PLUM, H.: Untersuchungen über die Flüchtigkeit von Tributylzinn-Verbindungen. Holz als Roh- und Werkstoff **39** (1981): 261–264.

LECLERCQ, A.: Azaconazole, a potential new wood preservative. I: Preliminary results concerning white and brown rot fungi. Material und Organismen **18** (1983): 65–77.

LEIGHTLEY, L. E.: An evaluation of anti-sapstain chemicals in Queensland, Australia. Int. Res. Group on Wood Preservation. Document No.: IRG/WP/3374 (1986).

– and NICHOLAS, D. D.: In ground performance of wood treated with a substituted isothiazolone. Int. Res. Group on Wood Preservation. Document No.: IRG/WP/3612 (1990).

LEWIS, D. A., WILLIAMS, G. R., and EATON, R. A.: The development of prophylatic chemicals for the treatment of green timber. B. W. P. A. Conventin 1985.

LOVE, D. J.: Some applications of boron and zinc organic compounds in timber preservation. In: THOMPSON, R. (Ed.): The Chemistry of Wood Preservation. Royal Society of Chemistry, Cambridge 1991, pp. 282–293.

METZNER, W., BUCHWALD, G., CYMOREK, S., and HINTERBERGER, H.: Neuere Entwicklungen auf dem Gebiet der Insektizide und Fungizide für ölartige Holzschutzmittel. Holz als Roh- und Werkstoff **35** (1977): 233–237.

MILLER, G. A., LEWIS, N. S., and WEILER, E. D.: Acrylamides derivatives of -isothiazolones. USP. 3914301 (1975).

NAKAMURA, Y.: Laboratory evaluation of the recent fungicides for sapstain and lumber mould by J. P. W. A. test methods. Wood Preservation **12** (1986): 167–179.

NICHOLAS, D. D., and PRESTON, A. F.: Evaluation of alkylammonium compounds as potential wood preservatives. Proc. Americ. Wood Preserv. Assoc. 1980.

– – GREENLEY, D. E., and PARIKH, S.: Evaluation of a substituted isothiazolone as a potential new wood preservative. Intern. Res. Group on Wood Preservation. Document No.: IRG/WP/3306 (1983).

NOMURA, Y.: Laboratory and field evaluations of a novel formulation, BAM, as an anti-sapstain agent. Intern. Res. Group on Wood Preservation. Document No.: IRG/WP/3639 (1991).

PAULUS, W.: Microbicides for the protection of materials: yesterday, today and tomorrow. In: ROSSMORE, H. (Ed.): Biodeterioration 8, Elsevier Science Publishers, Amsterdam 1991, pp. 36–52.

– and KÜHLE, E.: Tailoring of microbicides for material protection – the trihalogenmethylthio group in various microbicides. VI. Int. Biodet. Symp. Washington (1984).

PLACKETT, D. V.: Field evaluation of alternative antisapstain chemicals. Int. Res. Group on Wood Preservation. Document No.: IRG/WG/3198 (1982).

PRESTON, A. F.: Dialkyldimethylammonium halides as wood preservatives. J. A. O. C. S. **60** (1983): 567–570.

– NICHOLAS, D. D., GREENLEY, D. E., WALCHESKI, P. J., and MCKRAIG, P. A.: Efficacy of N-substituted isothiazolones for the control of wood decay fungi. 6th Internat. Biodeteriorat. Symp. Washington, August 1984, pp. 100–107.

REUTHER, W., POMMER, E.-H., RAFF, P., and GÖTTSCHE, R.: Fungizid für den Holzschutz, DE-PS 24 10 603, 06. 03. 1974.

ROHM and HAAS: Kathon 925 biocide for wood preservation. Product Information (1989).

Rose, M., and BEDOYA, A.: Field evaluation of antisapstain products. Intern. Res. Group on Wood Preservation. Document No.: IRG/WP/3402 (1987).

RUDDICK, J. N.: Field testing of alkylammonium wood preservatives. Int. Res. Group on Wood Preservation. Document No.: IRG/WP/3248 (1983).

– The influence of staining fungi on the decay resistance of wood treated with alkylammonium compounds. Material und Organismen **21** (1984): 139–149.

– A field evaluation of modified and unmodified alkylammonium compounds. Int. Res. Group on Wood Preservation. Document No.: IRG/WP/3436 (1987).

- and SAM, A. R. H.: Didecyldimethylammonium chloride – a quaternary ammonium wood preservative. Material und Organismen **17** (1982): 299–313.
SAVORY, J. G., and CAREY, J. K.: Decay in external framed joinery in the United Kingdom. J. Inst. Wood Sci. **8** (1979): 176–180.
STAMM, J. M., LITTEL, K. J., CASATI, F. M. H., and FRIEDMAN, M. B.: Discussion of diiodomethyl p-tolyl sulfone (Amical® 48) as a fungicide for wood preservation. Int. Res. Group on Wood Preservation. Document No.: IRG/WP/3425 (1987).
SUNDMAN, C. E.: Tests with ammonical copper and alkylammonium compounds as wood preservatives. Int. Res. Group on Wood Preservation. Document No.: IRG/WP/3299 (1984).
THOMSON, W. S., and TOOLE, E. R.: Project CR-811-73-3601, Evaluation of Amical® 48 as wood preservative, research sponsored by Abbott Laboratories (1974). Quoted from Int. Res. Group on Wood Preservation. Document IRG/WP/3425 (1987).
THORNTON, J. D.: Potential toxicants for controlling soft rot in preservative treated hardwoods. II. Laboratory screening tests using sawdust. Material und Organismen **12** (1977): 201–210.
TSDUNODA, K.: JWPA method for testing effectiveness of surface coatings with preservatives against decay fungi. Intern. Res. Group on Wood Preservation. Document No.: IRG/WP2164 (1981).
- and NICHIMOTO, K.: Fungicidal effectiveness of amended alkyl ammonium compounds. Int. Res. Group on Wood Preservation. Document No.: IRG/WP/3421 (1987).
VALCKE, A. R.: Azaconazole, a new wood preservative. American Wood Preservers' Association (1985).
- Suitability of propiconazole (R 49362) as a new-generation wood-preserving fungicide. Int. Res. Group on Wood Preservation. Document No.: IRG/WP/3529 (1989).
- and GOODWINE, W. R.: Azaconazole, a new wood preservative. American Wood Preservers' Association, 1985.
- and STEVENS, M.: Stability, performance and distribution of propiconazole (R 49362) in acceleratedly aged wood. Int. Res. Group on Wood Preservation. Document No.: IRG/WP/3647 (1991).
VAN DEN ENDE, R.: The use of TCMTB in applications other than sapstain prevention: a review. Int. Res. Group on Wood Preservation. Document No.: IRG/WP/3606 (1990).
VERBECKE, W., VAN ACKER, J., and STEVENS, M.: Evaluation of laboratory soft rot test on basis of weight loss and residual strength. Int. Res. Group on Wood Preservation. Document No.: IRG/WP/2236 (1989).
WILLIAMS, G. R.: Observations on the failure of antisapstain treated timber under non-drying conditions. Int. Res. Group on Wood Preservation. Document No.: IRG/WP/1437 (1990).
WÜSTENHÖFER, B., WEGEN, H.-W., and METZNER, W.: Tebuconazole, a new wood preserving fungicide. Int. Res. Group on Wood preservation. Document No.: IRG/WP/3634 (1990).

Chapter 23

Fungicides for the preservation of materials

E.-H. POMMER

Limburgerhof, Germany

Introduction

Preservatives for the protection of materials are variously referred to as material protectants, microbicides, biocides, mildewcides, fungicides, and fungistats. The term "preservative" indicates that a material is preserved in a suitable manner for short- or long-term in-service use or to combat microorganisms that cause biodeterioration or spoilage in industrial products. This chapter will deal with fungicides in this area of application. However, it must be taken into account that bacteria as well as fungi are frequently involved in damaging processes. Many fungicides also exhibit bactericidal or bacteriostatic effects, and vice versa. Such compounds are used as microbicides or biocides.

A considerable number of different fungi and bacteria can be involved in the biodeterioration of materials or industrial products. Because the spectrum of antimicrobial activity does not always provide the protection required, it is necessary in such cases to use combinations of preservatives.

Several countries have introduced legislation to check or re-check extensively microbicides (and preservatives in general) for their effects on the environment, including their toxicological properties. As a result of this, the development of new compounds is stagnating due to the high costs. On the other hand, there is a tendency for effective compounds such as mercurials and higher chlorinated phenols to decline in importance, and microbicides which are suspected of being mutagenic, teratogenic or carcinogenic will be withdrawn.

Areas of application

It is impossible to review the use of fungicides for material preservation in detail in this chapter. Consequently, what I have reviewed is their use in different areas of application, the chemistry of the active ingredients, and, as far as possible, the manufacturers and marketing companies, the LD_{50} for rat (Table 23.1; Fig. 23.1), and sources. In addition, more detailed information on several selected areas of application which may be of general interest is given: fungicides for paint films, water-based paints and surface coatings, polymer emulsions, plastics, and leather.

Table 23.1 Fungicides for the preservation of materials

Active ingredient (Common name or abbreviation)	Trade name(s)	Producer or supplier	LD$_{50}$ rat (oral) mg/kg	Main areas of application when used as fungicides	References
tetramethyl thiuram disulfide (thiram, TMTD)	various	Du Pont Bayer	800–3000	latex paints (paint films)	Wallhäusser und Fink (1985) Grant et al. (1986) Allsopp and Allsopp (1983)
methylene bisthiocyanate (MBT)	MBT Chemviron T-9 Biocide N-948 Cytox 3522	Tenneco Chemviron Stauffer American Cyanamid	79	leather slimicide (paper-mill)	
1,2-dibromo-2,4-dicyano-butane	Tektamer 38	Merck and Co.	720	adhesives, binders, paints, polymer emulsion	Wallhäusser (1988) Gillatt (1990)
diiodomethyl p-tolyl sulphone (DIMTS)	Amical 48	Abott Angus	1580	paints leather (wetblue) adhesives	Wallhäusser und Fink (1976) Allsopp and Allsopp (1983) Abbott Technical Information (1986)
3-iodo-2-propynyl butyl carbamate (IPBC)	Troysan Polyphase	Troy	10000	paints	Rose (1981) Sharpell (1989)
trans-1,2-bis(n-propyl-sulfonyl)-ethylene	Vancide PA	Vanderbilt	2730	paints	Trotz and Pitts (1981)
2,4,5,6-tetrachloroisophthalo-nitrile (chlorothalonil)	Nopcocide N-96	Diamond Shamrock	3000	latex paints	Grant et al. (1986)
N,N-dimethyl-N'-phenyl-N'-(dichloro-fluormethylthio)-sulphamide (dichlofluanide)	Preventol A 4-S	Bayer	5000	solvent based paints and varnishes (coatings)	Pauli (1972) Grant et al. (1986) Wallhäusser (1988)
N,N-dimethyl-N'-tolyl-N'-(dichloro-fluormethylthio)-sulphamide (tolylfluanide)	Preventol A 5	Bayer	5000	solvent based paints and varnishes (coatings)	Bayer Product Information Preventol A 5 (1988)
di-n-decyl-dimethyl-ammoniumchloride	Bardac 22	Lonza	530/650	textiles	Yeager (1977) Trotz and Pitts (1981) Wallhäusser (1988)

Chemical name	Trade name	Manufacturer		Application	Reference
2,6-dimethyl, 1,3-dioxanol-4-acetate	Giv-Gard DXN	Givaudan	1900	paints	Sharpell (1989)
sodium 2-pyridinethio-N-oxide	Sodium-Omadine Pyrion-Na	Olin Pyrion	875	polymer emulsions latex paints metal working fluids	Wallhäusser (1988)
2,3,5,6-tetrachloro-4-methylsulphonyl pyridine	Dowicil S 13 Densil S	Dow ZENECA (ICI)	2000	adhesives, sealants, polymer emulsions, latex paints, plastics	Pauli (1972, Pauli (1973) Dyckhoff and Pommer (1987) Denham (1988)
N-(trichloromethane-sulphenyl)-cyclohex-4-ene-1,2-dicarboximide (captan)	Vancide 89 Captan Orthocide	Vanderbilt Chevron	9000	solvent based paints (coatings), plastics	Allsopp and Allsopp (1983)
N-(trichloromethyl-thio)-phthalimide (folpet)	Phaltan Fungitrol 11 Cosan P Preventol A 3	Chevron Cosan Bayer	10000	solvent based paints plastics	Pauli (1973) Allsopp and Allsopp (1983) Paulus (1992)
methyl benzimidazole-2-yl carbamate (carbendazim, BCM)	various Mergal BCM	Du Pont Hoechst BASF	15000	paints (paint films) (preferred use in combination with other fungicides)	Allsopp and Allsopp (1983) Diehl (1987) Wallhäusser (1988)
2-(4'-thiazolyl)-benzimidazole (Thiabendazole, TBZ)	Metasol TK 100 Metasol TK 50	Merck and Co.	3300	paints paper manufacture	Diehl (1987) Wallhäusser (1988)
Tetrahydro-3,5-dimethyl-1,3,5-thiadiazine-2-thione (dazomet)	Chemviron TK-100 various	Chemviron Stauffer Chemviron Cosan Chem. BASF	500	paints, adhesives, leather slimicide (paper mills)	Wallhäusser (1988)
2-n-octyl-4-isothiazolin-3-one (OITZ)	Kathon 893 C (45%) Skane M 8 Kathon LM (5%) Kathon LP Kathon 893 MW	Rohm & Haas	550/650 (20% w/v) 6000	paint films, fabric mildewcide leather (wetblue) metal working fluids	Rohm and Haas Product Informations Skane M-8 (1988) and Kathon 893 MW (1988) Wallhäusser (1988) Lindner and Neuber (1990)
4,5-dichloro-2-n-octyl-4-isothiazolin-3-one	Kathon 925	Rohm & Haas	1560	silicone sealants	Heaton et al. (1991)
5-chloro-2-methyl-3(2 H)-isothiazolone/2-methyl-3(2H)-isothiazolone	Kathon LX (13.9%) Kathon LX (1.5%) Acticid SPX	Rohm & Haas Thor	467/561 3811	polymer emulsions, latex paints adhesives, fuels	Hettige and Sheridan (1989) Pitcher (1989) Gillatt (1990)

Table 23.1 continued

Active ingredient (Common name or abbreviation)	Trade name(s)	Producer or supplier	LD_{50} rat (oral) mg/kg	Main areas of application when used as fungicides	References
1,2-benzoiso-thiazolin-3-one (BIT)	Proxel XL 2 (9.5%) Proxel GXL	ZENECA (ICI)	2000	Polymer emulsions, latex paints adhesives, paper coatings	Wallhäusser (1988) Gillatt (1990)
2-methyl-4,5-trimethyl-ene-4-isothiazolin-3-one (MTI)	Promexal (5%)	ZENECA (ICI)	1831/2359	polymer emulsions, latex paints	Eacott (1991) Pommer and Eacott (1993)
2-(thiocyanomethylthio) benzothiazole (TCMTB)	Busan Butrol 30 Tolcide C 80	Buckman Buckman Albright and Wilson	1590	leather, paint films polymer emulsions paper manufacture	van Deren and Weiss (1978) Tasker (1987) Lindner and Neuber (1990)
dithio-2,2'-bis (benzylmethylamide)	Densil P	ZENECA (ICI)	15000	adhesives, sealants, paint films latex paints, leather (wetblue)	ICI Product Information Densil P (1992)
2,2'-oxybis-(4,4,6-trimethyl-1,3,2-dioxaborinane)/2,2'-(1-methyl-trimethylenedioxy)-bis (4-methyl-1,3,2-dioxaborinane)	Biobor JF	US Borax		hydrocarbon fuels, ship fuels home heating fuels	Elphick and Hunter (1968) Hettige and Sheridan (1989) Pitcher (1989)
copper-8-quinolate	Cunilate Quindex Mergal HS 14	Ventron Nuodex Riedel	10000	textiles, plastics paint films	Singh et al. (1972) Wallhäusser and Fink (1976) Yeager (1977)
copper-naphthenate	Noudex Copper Naphthenate	Nuodex	6000	textiles (tents, cordage)	Yeager (1977)
zinc 2-pyridinethiol-1-oxide	Zinc Omadine Zinc Pyrion	Olin Pyrion	200	plastics	Olin Product Data Zinc Omadine (1992)
10,10'-oxybisphenoxy-arsine (OBPA)	Vinycene SB-1 (5%) Vinycene BP-5 Estabex ABF	Thiokol/ Ventron Akzo	20000	plastics	Wolf and Rily (1965) Yeager (1977) Stühlen and Pommer (1983)
tributylin oxide (TBTO)	TBTO Advacide FC Advacide ATO Metatin Fundex bioMeT Cotin	Schering various Acima	127	latex paints, paint films plastics, textiles	Enninga and Bordes (1968) Miller (1972) McCarthy and Greaves (1988)

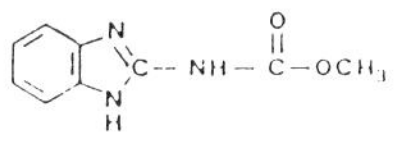

Fungicides for the preservation of materials 507

carbendazim, BCM

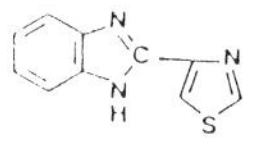

Thiabendazole, TBZ

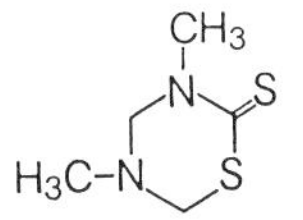

dazomet

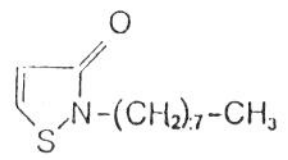

OITZ

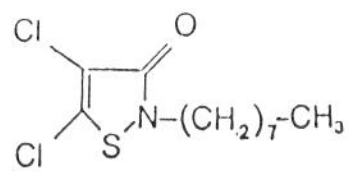

4,5-dichloro-2-n-octyl-4-isothiazolin-3-one

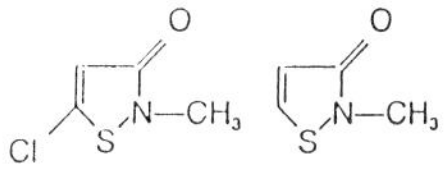

5-chloro-2-methyl-3(2H)-isothiazolone/ 2-methyl-3(2H)-isothiazolone

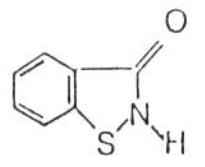

BIT

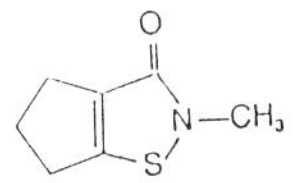

MTI

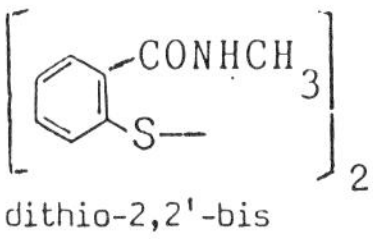

dithio-2,2'-bis(benzylmethylamide)

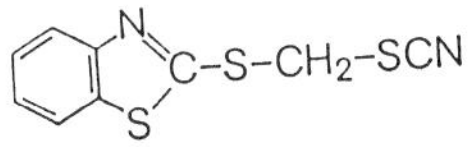

TCMTB

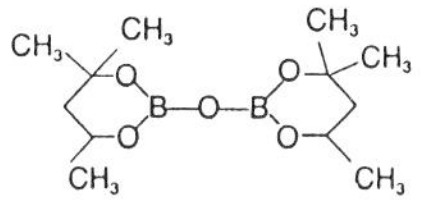

copper-8-quinolate

2,2'-oxybis-(4,4,6-trimethyl-1,3,2-dioxaborinane) / 2,2'-(1-methyltrimethylenedioxy)-bis(4-methyl-1,3,2,-dioxaborinane)

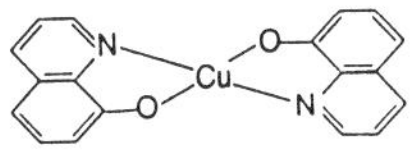

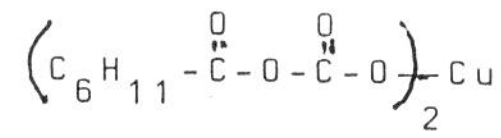

copper-naphthenate

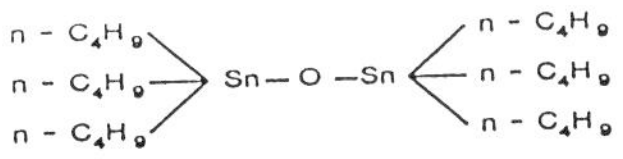

TBTO

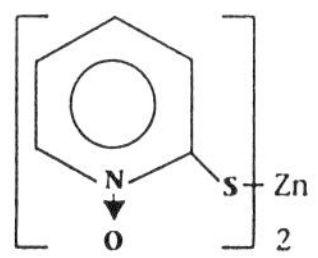

zinc 2-pyridinethiol-1-oxide

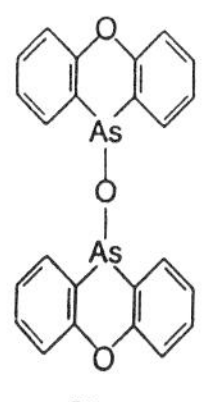

OBPA

Fig. 23.1

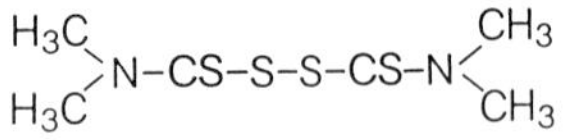

TMTD

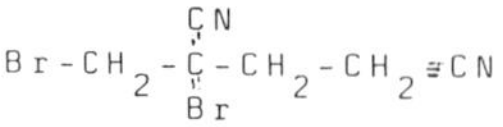

1,2-dibromo-2,4-dicyano-butane

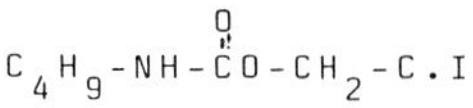

IPBC

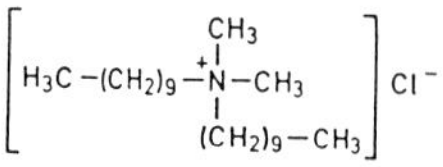

di-n-decyl-dimethyl-ammoniumchloride

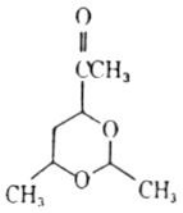

2,6-dimethyl, 1,3-dioxanol-4-acetate

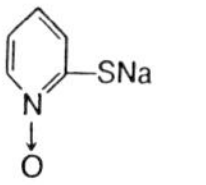

sodium 2-pyridin-ethio-N-oxide

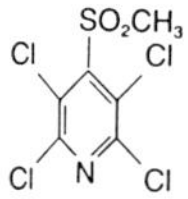

2,3,5,6-tetrachloro-4 (methylsulphonyl) pyridine

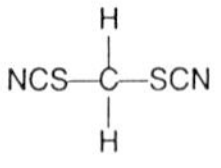

MBT

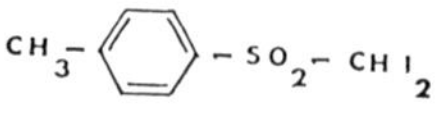

DIMTS

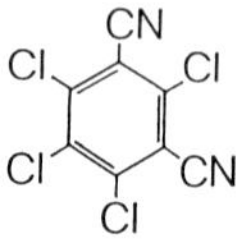

chlorothalonil

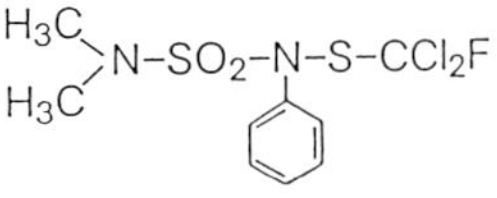

dichlofluanide

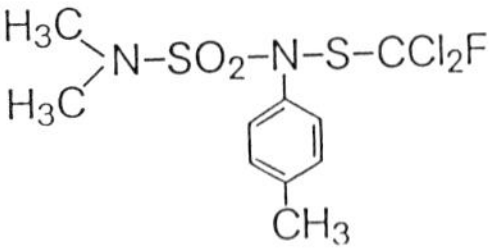

tolylfluanide

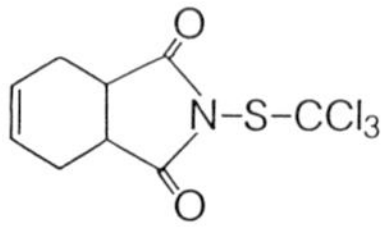

captan

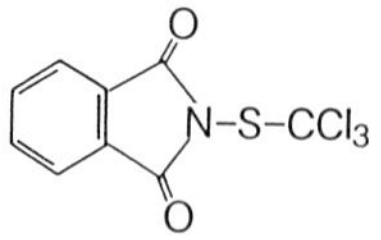

folpet

Fig. 23.1 Chemical structures of fungicides for the preservation of materials.

Fungicides for paint films, surface coatings, water-based paints and polymer emulsions

Interior and exterior paint films of both oil- and water-based types can be colonised by fungi if sufficient moisture is present. The growth of fungi is frequently supported by contaminating nutrients. On interior paint films, emulsion paints are generally affected to a greater degree than gloss paints, with moulds developing on damp walls and ceilings. Examples of fungi encountered on painted surfaces are: *Alternaria alternata, Alternaria dianthicola, Aspergillus fumigatus, Aspergillus ustus, Aspergillus versicolor, Aureobasidium pullulans, Cladosporium cladosporioides, Cladosporium sphaerospermum, Curvularia lunata, Fusarium chlamydosporum, Fusarium oxysporum, Penicillum* spp., *Phoma glomerata, Ulocladium atrum, Rhodotorula* sp. (Ross *et al.* 1968; LUMPKINS *et al.* 1973; SOLOMON 1975; Grant *et al.*, 1986; LIM *et al.*, 1989; HEATON *et al.*, 1991).

On exterior paints films and surface coatings *Aureobasidium pullulans* is said to be the most important filmdegrading organism. There are indications that bacterial precolonisation of the surfaces, especially by *Pseudomonas aeruginosa* promotes fungal development. Furthermore, green algae and cyanobacteria may grow on exterior paint films and surface coatings.

Of the numerous paint fungicides, mention should be made of the following compounds:

2-n-octyl-4-isothiazolin-3-one, OITZ (Fig. 23.1),
4,5-dichloro-2-n-octyl-4-isothiazolin-3-one (Fig. 23.1),
tetramethyl thiuram disulphide (TMTD) (Fig. 23.1),
N,N-dimethyl-N'-phenyl-N'-(dichloro-fluoromethylthio)-sulphamide, dichlofluanide (Fig. 23.1),
tributylin oxide, TBTO (Fig. 23.1),
diiodomethyl p-tolyl sulphone, DIMTS (Fig. 23.1),
3-iodo-2-propynyl butyl carbamate, IPBC (Fig. 23.1),
2,4,5,6-tetrachloroisophthalonitrile, cholorothalonil (Fig. 23.1),
dithio-2,2'-bis(benzmethylamide) (Fig. 23.1),
2,3,5,6-tetrachloro-4(methylsulphonyl)pyridine (Fig. 23.1),
2-(thiocyanomethylthio)benzothiazole, TCMTB (Fig. 23.1).

Within the framework of evaluating fungicidal emulsion paints, GRANT *et al.* (1986) tested eight commercially available paint fungicides by means of laboratory tests, site trials, and a high humidity test chamber. Although the results did not permit prediction of the actual period over which the tested fungicides in paint films are effective, it could be stated that TMTD performed extremely well in the tests. In a solvent-based gloss paint there were indications that dichlofluanide was effective as a film protectant. In outdoor exposure trials with fungicidal latex house paints carried out in Florida and Puerto Rico, MARK (1981) observed inadequate performance of chlorothalonil, IPBC, DIMTS, and OITZ. LIM *et al.* (1989) reported on fungal problems in buildings subject to humid climatic conditions in the tropics. The authors tested commercial paints containing as fungicides *inter alia* benzothiazole, butyl carbamate, a halogenated aromatic compound, a heterocyclic and halogenated aromatic nitrogen compound, an N-S-heterocyclic compound, and 8-hydroxyquinoline. As a result of their investigations, the authors stated that only 8-hydroxyquinoline was effective over a period of two years, two other compounds permitted only slight growth, and the other five paint films were infected by six months. HEATON *et al.* (1991) determined the efficacy of 4,5-dichloro-2-n-octyl-4-isothiazolin-3-one in urethane oil based paint in laboratory experiments and field experiments in the UK and in Singapore. After 18 months' exposure of painted panels there was no mould growth on the fungicide-containing paint films, whilst untreated paint films developed extensive mould growth after only a few months.

Water-based emulsion paints contain a wide variety of ingredients. Apart from water and polymer emulsion (also described as latices, binders, polymer dispersions), which can be derived from polyvinyl acetate (PVA), PVA/acrylic, acrylic, vinylpropionate/acrylic, styrene acrylic, styrene butadiene, mention must be made of the following: thickeners, defoamers, dispersing agents, fillers, pigments, etc. Microbial contamination resulting either from the raw materials used in emulsion paints or during the course of manufacture can create problems with bacterial and/or fungal growth both in the wet stage and in-can. Both aerobic and anaerobic bacteria are essential for microbial spoilage. However, moulds can grow on the paint surfaces, and, for example, by producing cellulases, which cause a breakdown of the cellulosic thickeners, or pigmented moulds, can cause discolourations. In emulsion paints JACUBOWSKI et al. (1983) found species such as *Aspergillus*, *Penicillium*, *Fusarium*, *Scopulariopsis*, *Geotrichum*, *Torula*, *Rhodotorula*, *Sporobolomyces*, and *Saccharomyces*. Polymer emulsions themselves consist of numerous components, some of which are susceptible to attack by bacteria, yeasts, and moulds (GILLAT 1990). CARTER (1982), JACUBOWSKI et al. (1982), ELSOM (1988), EVERTSEN (1988) isolated yeasts and moulds such as *Candida boidinii*, *Pichia* sp., *Rhodotorula rubra*, *Saccharomyces* sp., *Torula* sp., *Aspergillus* sp., *Cladosporium* sp., *Fusarium* sp., *Geotrichum candidum*, and *Penicillium* sp. from spoiled polymer emulsions.

Because microorganisms grow in the aqueous phase of an emulsion paint or polymer emulsion, it is important for the microbicide not only to be water-soluble, but also to remain in the aqueous phase.

Microbicides for the preservation of both water-based emulsion paints and polymer emulsions often do not have a sufficiently broad spectrum of antimicrobial activity. Combinations of two or more active agents are therefore common – but few of them show synergistic effects.

There are a considerable number of microbicides suitable for use as preservatives. However, for reasons of environmental damage the use of a number of active ingredients has recently been restricted.

From the chemical class of substituted isothiazolinones, which has been known for 15 years, three preservatives have been developed which are highly active against bacteria, moulds and fungi which cause microbial spoilage:

A combination of 5-chloro-2-methyl-3(2H)-isothiazolinone and 2-methyl-3(2H)isothiazolinone in a ratio of approx. 3:1, stabilised with magnesium salts; pH optimum 4–12 (Fig. 23.1),

1,2-benzoisothiazolinon-3-one(BIT); pH optimum 4–12 (Fig. 23.1),

2-methyl-4,5-trimethylene-4-isothiazolin-3-one(MTI); pH optimum 4–10 (Fig. 23.1).

Most emulsion paints and many polymer emulsions are adjusted to alkaline pH-values in order to improve their stability, the pH-value generally being between 8.0 and 9.5. WALLHÄUSSER (1988) reports that the chemical stability of a mixture of 5-chloro-2-methyl-3(2H)-isothiazolone and 2-methyl-3(2H)isothiazolone decreases at a pH higher than 7.5 EACOTT (1991) showed that in an MTI-containing solution, the pH of which was raised to 9.0 with ammonium hydroxide, 84% of the active ingredient remained after 6 months' storage. BTIs are to some extent less active against yeasts and moulds than the two other isothiazolinones. POMMER and EACOTT (1993) ascertained in their tests that the onset of fungicidal (microbicidal) or fungistatic (microbistatic) efficacy of MTI is comparatively slow after contamination with microorganisms, and may take several days. With regard to the mode of action of isothiazolinones, see Chapter 22; however, whether MTI has yet another mode of action is something that should be checked further.

Further preservatives are listed in Table 23.1.

Fungicides for plastic (plasticised polyvinyl chloride and polyester polyurethane)

Plastics may be defined as materials consisting mainly of macromolecular organic compounds obtained synthetically or by modification of natural products. These materials have replaced classical natural products in many areas. Two groups are of major importance with regard to microbial degradation if used under conditions which encourage the development of microorganisms:

Polyvinylchloride (PVC) itself is known to be resistant to microbial degradation. However, plasticised PVC is susceptible to a greater or lesser degree, depending on the additives used (with softeners in first place). Damage may become visible as discolouration, embrittlement, or destruction of the plastic material. Fungi rather than bacteria are mainly involved in these processes.

Examples of fungi that change soft PVC are: *Alternaria alternata*, Alternaria sp., *Aureobasidium pullulans*, *Chaetomium globosum*, *Cladosporium sphaerospermum*, *Curvularia* sp., *Fusarium semitectum*, *Monilia* sp., and *Pestalotia odorata* (HAMILTON 1983; INOUE 1987).

Thermoplastic polyurethane elastomers are widely used on account of their favourable product properties. DARBY and KAPLAN (1968) and KAPLAN et al. (1968) showed in comprehensive trials that polyestertype polyurethanes were generally susceptible to microorganisms, especially to fungi, whereas polyether polyurethanes could be regarded as morderately to highly resistant. From the studies by EVANS and LEVISOHN (1968), FILIP (1978), GRIFFIN (1980), PATHIRANA and SEAL (1983), and POMMER and Lorenz (1985) it may be concluded that enzymatic attack plays an important role in the decomposition of polyester polyurethanes. Examples of PUR-degrading fungi are: *Alternaria alternata*, *Aspergillus niger*, *Aspergillus ustus*, *Fusarium oxysporum*, *Gliocladium roseum*, *Paecilomyces lilacinus*, *Penicillium* spp., *Scopulariopsis fusca* and *Trichoderma viride* (PATHIRANA and SEAL 1983; POMMER and LORENZ 1985; WALES and SAGAR 1985).

There are few preservatives available for the protection of plastics against biodeterioration on account of the demands that are made of the active ingredients. The preservative must be compatible with the manufacturing process; in particular, it must not react chemically with the systems. Durability, resistance to leaching, no or very low volatility, UV-stability, relative colourlessness, and environmental acceptability of the active ingredient are essential requirements.

For approximately 30 years, 10,10''-oxybisphenoxarsine (OBPA) (Fig. 23.1) has proved to be the most effective plastic fungicide. It is widely used as a protectant of plasticised (soft) PVC which is processed for use for shower curtains, floor and wall coverings, coated fabrics for upholstery, mattress covers, ditch and pool liners, marine upholstery, awnings, etc. STÜHLEN and POMMER (1983) investigated *inter alia* the fungistatic effect of

OBPA in samples of plasticised PVC containing different types of both plasticiser and pigments, after four weeks' weathering, four weeks' leaching, and untreated. The authors confirmed the efficacy of OBPA; however, they concluded from the results that the components of a PVC formulation can have a negative influence on the microbicide, which manifests itself, for example, in a higher leaching rate from the final product. It was also able to be demonstrated that OBPA-containing polyester PUR-coated fabrics remained unaffected in a 120-day soil burial test. In contrast to untreated samples, samples with a coating containing OBPA showed neither fungal attack nor pink staining, nor physical degradation (hydrolytic cracking) (Anon 1986). Similar results were obtained with polyester PUR shoe-soles treated with OBPA (Anon 1985).

Another active ingredient for preventing fungal problems with wall coverings in bathrooms and kitchens is 2,3,5,6-tetrachloro-4(methylsulphonyl)pyridine (Fig. 23.1). DENHAM (1988) reported the fungicidal efficacy of the compound in treated plasticised PVC sheets without weathering and after weathering, compared with a microbicide-free control. After inoculation with six fungi known to be contaminants of flexible PVC and an incubation period of 28 days, only slight growth could be detected on the weathered samples.

DYKHOFF and POMMER (1987) found in a soil burial test with thermoplastic polyester PUR samples, protected with the pyridine fungicide, that fungal attack was prevented for a period of 16 weeks; the formation of hydrolytic cracking was retarded for approximately 10 weeks.

Fungicides for leather

Raw hides are highly susceptible to bacterial decomposition, and short-term preservation is therefore imperative. After soaking and further operations such as liming and chairing, the skins are processed by chrome tanning into wetblues. Moulds are the main problem with stored pickled pelts, wetblues and vegetable-tanned leather. Wetblues are quite often stored or shipped in the wet stage. At a low pH and in the presence of proteins and fats, the growth of moulds is favoured. GATTNER et al. (1988) and LINDNER and NEUBER (1990) describe the following species that are primarily responsible for damage to leather: *Aspergillus* sp., *Mucor* sp., *Paecilomyces variotii*, *Penicillium* sp., *Rhizopus nigricans* and *Trichoderma viride*, which is found most frequently. Finished leather and leather products can be damaged by moulds of the genera *Aspergillus* and *Penicillium* (REISS 1983).

The use of PCP or other higher chlorinated phenols as fungicides for wetblue preservation was cut back in favour of products showing lower environmental toxicity. LINDNER and NEUBER (1990) reported on four fungicides which are in commercial use:
2-(thiocyanomethyltheio)benzothiazole(TCMTB), Fig. 23.1,
methylene bisthiocyanate (MBT), Fig. 23.1,
diiodomethyl-p-tolyl sulphone (DIMTS), Fig. 23.1,
2-n-octyl-isothiazolin-3-one(OITZ), Fig. 23.1.

TCMTB (VAN DEREN and WEISS 1978) and MBT are highly active against moulds. Traces of isothiocyanate as an impurity from manufacture may, together with iron ions, cause red discolouration of wetblues. DIMTS is unstable under UV-light; yellowing of wetblues and leather manufactured from them is possible. The use of OITZ can be limited by its high reactivity. In the presence of sodium sulphite there is an opening of the isothiazolone ring followed by inactivation of the fungicide. The authors came to the conclusion that the ideal fungicide for wetblue preservation is not available so far.

Conclusion

The presented examples dealing with the use of fungicides as preservatives for materials clearly show that the ideal antifungal agent which meets all requirements does not as yet exist. This may be attributable to the fact that the range of fungi that can be involved in the spoilage or deterioration of a specific material is so extensive that the fungicidal or fungistatic effectiveness of an active ingredients is often inadequate. Moreover, the physical and chemical properties of an active ingredient and its possible interactions with the components of the product to be protected must be taken into account. As a result of the new regulatory measures, it is to be expected that in the foreseeable future the number of available fungicides and microbicides for material preservation will decrease.

References

ALLSOPP, C., and ALLSOPP, D.: An updated survey of commercial products used to protect materials against biodeterioration. Internat. Biodeteriorat. Bull. **19** (1983): 99–144.
Anonym: The resistance of vinycene treated polyester polyurethane shoe soles to micro-organisms. Morton Thiokol/Ventron Division (1985).
Anonym: Vinycene prevents delamination and cracking of polyurethane coated fabrics. Morton Thiokol/Ventron Division (1986).
Bayer Product Information 'Preventol A 5' (1988).
CARTER, G.: The preservation of latex emulsions. Proc. Plastic Rubber Inst. Conf. Emulsion Polymers. Paper 12 (1982): 1–9.
DARBY, R. T., and KAPLAN, A. M.: Fungal susceptibility of polyurethanes. Appl. Microbiol. **16** (1968): 900–905.
DENHAM, K.: Densil S: an unique fungicide/algicide from ICI Specialty Chemicals. ICI Europe Limited (1988).
DIEHL, K.-H.: Biozide gegen mikrobielle Materialzerstörung. farbe+lack **93** (1987): 267–271.
DYCKHOFF, H., and POMMER, E.-H.: Beständigkeit von thermoplastischen Polyurethanen gegenüber Mikroorganismen. Kunststoffe **77** (1977): 597–601.
EACOTT, C.: A new biocide for the preservation of aqueous-based paints. Surface Coatings International (Jocca) **74** (1991): 322–323.
ELPHICK, J. J., and HUNTER, S. P. K.: Evaluating biocidal fuel additives for intermittent use in aircraft fuel systems. In: WALTERS, A. H., and ELPHICK, J. J. (Eds.): Biodeterioration of Materials. Elsevier Publishing Comp., Amsterdam–London–New York 1968, pp. 364–370.
ELSOM, S. J.: The biodeterioration of polymer emulsions – a synopsis. In: Proc. Internat. Biodeteriorat. Res. Group meeting. Document No.: IBRG/P89/04A (1988).
ENNINGA, R., and BORDES, W. J.: Fungicides in latex paints. In: WALTERS, A. H., and ELPHICK, J. J. (Eds.): Biodeterioration of materials. Elsevier Publishing Comp., Amsterdam–London–New York 1968, pp. 326–332.

EVANS, D. M., and LEVISOHN, I.: Biodeterioration of polyester-based polyurethane. Internat. Biodeterioat. Bull. **4** (1968): 89–92.

EVERTSEN, P.: Biodeteroration of polymer emulsions/dispersions. In: Proc. Internat. Biodeteriorat. Res. Group Meeting. Document No.: IBRG/P89/04B (1988).

FILIP, Z.: Decomposition of polyurethane in a garbage landfill leakage water and by soil microorganisms. Europ. Appl. Microbiol. Biotechnol. **5** (1978): 399–407.

GATTNER, H., LINDNER, W., and NEUBER, H.-U.: Mikrobieller Befall bei der Lederherstellung und seine Kontrolle mit modernen Konservierungsmitteln. Das Leder **39** (1988): 66–73.

GILLATT, J.: The biodeterioration of polymer emulsions and its prevention with biocides. Internat. Biodeteriorat. **26** (1990): 205–216.

GRANT, C., BRAVERY, A. F., SPRINGLE, W. R., and WORLEY, W.: Evaluation of fungicidal paints. Internat. Biodeteriorat. Bull. **22** (1986): 179–194.

GRIFFIN, G. J. L.: Synthetic polymers and the living environment. Pure Appl. Chem. **52** (1980): 399–407.

HAMILTON, N. F.: Biodeterioration of flexible polyvinylchloride films by fungal orgnisms. In: OXLEY, T. A., and BARRY, S. (Eds.): Biodeterioration 5, John Wiley and Sons, Chichester–New York–Brisbane–Toronto–Singapore 1983, pp. 663–678.

HEATON, P. E., CALLOW, M. E., and BUTLER, G. M.: Control of mould growth by anti-fungal paints. Internat. Biodeteriorat. **27** (1991): 163–173.

HETTIGE, G. E. G., and SHERIDAN, J. E.: Effects of biocides on microbiological growth in middle destillate fuel. Internat. Biodeteriorat. **25** (1989): 175–189.

ICI Product Information 'Densil P' (1992).

INOUE, M.: The study of fungal contamination in the field of electronics. In: HOUGHTON, D. R., SMITH, R. N., and EGGINS, H. O. W. (Eds.): Biodeterioration 7. Elsevier Applied Science, London–New York 1987, pp. 580–584.

JACUBOWSKI, J. A., SIMPSON, S. L., and GYURIS, J.: Microbiological spoilage of latex emulsions: causes and prevention. J. Coat. Technol. **54** (1982): 39–44.

– GYURIS, J., and SIMPSON, S. L.: Microbiology of modern coating systems. JCT. **55** (1983): 49–54.

KAPLAN, A. M., DARBY, R. T., GREENBERGER, M., and ROGERS, M. R.: Microbial deterioration of polyurethane systems. Developments Industr. Microbiol. **9** (1968): 201–217.

LIM, G., TAN, T. K., and TOH, A.: The fungal problem in buildings in the humid tropics. Internat. Biodeteriorat. **25** (1989): 27–37.

LINDNER, W., and NEUBER, H.-U.: Preservation in tannery. Internat. Biodeteriorat. **26** (1990): 195–203.

LUMPKINS, E. D., CORBIT, S. L., and TIEDEMANN, G. M.: Airborne fungi survey. 1. Culture-plate survey of the home environment. Ann. Allergy **31** (1973): 361–370.

MARK, S.: Protecting latex house paints from mildew. Modern Paints and Coatings (July 1981): 32–35.

MCCARHY, B. J., and GREAVES, P. H.: Mildew – causes, detection methods and prevention. Wool Sci. Rev. **65** (1988): 27–48.

Olin Product Data 'Zinc Omadine' (1992).

PATHIRANA, R. A., and SEAL, K. J.: *Gliocladium roseum* (Bainier), a potential biodeteriogen of polyester polyurethane elastomers. In: OXLEY, T. A., and BARRY, S. (Eds.): Biodeterioration 5. Academic Press, New York 1983, pp. 679–689.

PAULI, O.: Paint fungicides – a review. In: WALTERS, A. H., and HUECK-VAN DER PLAS (Eds.): Biodeterioration of Materials Vol. 2. Applied Science Publishers, London 1972, pp. 355–359.

– Anti-mildew coatings. I. Review. J. Oil Col. Chem. Assoc. **56** (1973): 285–288.

PAULUS, W.: Microbicides for material protection with special regard to the paint industry. Färg och lack No. 9 (1992): 161–166.

PITCHER, D. G.: Histories of microbiological fuel contamination – cause, effect and treatment. Internat. Biodeteriorat. **25** (1989): 207–218.

POHLMANN, J. L.: Fungicidal efficacy of a diiodomethyl-p-tolylsulfone emulsion in metalworking fluid. In: ROSSMORE, H. W. (ed.): Biodeterioration and biodegradation 8. Elsevier Science Publishers Ltd., Barking, Essex 1991, pp. 502–503.

POMMER, E.-H., and EACOTT, C.: Ein wasserlösliches Isothiazolinon-Derivat zur Konservierung von Dispersionen, farbe+lack **99** (1993): 105–108.

– and LORENZ, G.: The behaviour of polyester and polyether polyurethanes towards microorganisms. In: SEAL, K.-J. (Ed.): Biodeterioration and Biodegradation of Plastics and Polymers. Biodeterioration Society, Cranfield, 1985, pp. 77–86.

REISS, J.: Schimmelpilze – Lebensweise, Nutzen, Schaden, Bekämpfung. Springer-Verlag, Berlin–Heidelberg–New York–Tokyo 1986.

Rohm and Haas Product Information 'Skane M-8' Mildewcide (1988).

Rohm and Haas Product Information 'Kathon 893 MW' Metalworking Fluid Fungicide (1988).

ROSE, A. H.: History and scientific basis of microbial biodeterioration of materials. In: ROSE, H. (Ed.): Economic Microbiology Vol. 6, Microbial Biodeterioration. Academic Press, London-New York 1981, pp. 1–18.

ROSS, T.: Biodeterioration of paint and paint films. J. paint technol. **41** (1969): 266–274.

SHARPELL, F. H.: Biocides in speciality products. Special. Chemicals **9** (1989): 233–236.

SINGH, I. D., PERTI, S. L., and TANDON, R. N.: Anti-cockroach and anti-fungal surface coatings. In: WALTERS, A. H., and HUECK-VAN DER PLAS, E. H. (Eds.): Biodeterioration of Materials Vol. 2. Applied Science Publishers, London 1972, pp. 301–310.

SOLOMON, W. R.: Assessing fungus prevalence in domestic interiors. J. Allergy Clinic. Immunol. **56** (1975): 235–242.

STÜHLEN, F., and POMMER, E.-H.: Fungistatische Ausrüstung von Weich-PVC. Kunststoffe **73** (1983): 32–35.

TASKER, N.: Formulations for the use of TCMTB as a preservative and biocide in paint and other materials. Special. Chemicals **7** (April 1987).

TROTZ, S. I., and PITTS, J. J.: Industrial antimicrobial agents. In: KIRK-OTHMER (Ed.): Encyclopedia of Chemical Technology, Vol. 13, Third Edit. John Wiley & Sons Inc., New York 1981, pp. 223–253.

VAN DEREN, J. M., and WEISS, E. F.: Controlling fungal growth on leather. Correlation of TCMTB uptake and duration of mold resistance. JALCA **73** (1978): 498–507.

VIGO, T. L.: Protection of textiles from biological attack. In: LEWIN-SELLO (Ed.): Chemical Processing of Fibers and Fabrics, Functional Finishes, Part A. Marcel Dekker, New York 1983, pp. 367–426.

WALLHÄUSSER, K. H., and FINK, W.: Konservierung von Dispersionen und Dispersionsfarben. Farbe und Lack **82** (1976): 108–125.

– – Die Konservierung von wäßrigen Kunststoffdispersionen. farbe+lack **91** (1985): 277–286.

– Praxis der Sterilisation, Desinfektion – Konservierung, 4. Auflage. Georg Thieme Verlag, Stuttgart–New York 1988.

WOLF, P. F., and RILEY, W. H.: Fungistatic performance of 10, 10′-oxybisphenoxyarsine in exterior latex and asphalt coatings. Appl. Microbiol. **13** (1965): 28–33.

YEAGER, C. C.: Fungicides in Industry. In: SIEGEL, M. R., and SISLER, H. D. (Eds.): Antifungal Compounds. Marcel Dekker, New York 1977, pp. 371–396.

Chapter 24

Computer aided discovery and optimization of fungicides

James J. STEFFENS and DANIEL A. KLEIER

DuPont de Nemours and Company, Agricultural Products, Newark, Delaware 19714, USA

Introduction

The discovery of fungicides has historically been approached either as a game of chance or as a process of imitating successful products, whether they be natural or synthetic. Modern database technologies and molecular modeling can improve the odds of success for both of these traditional approaches. More recent target site directed approaches are also enriched by computer aided molecular modeling. Once discovered, lead structure optimization is facilitated by a variety of computer aided analyses of structure-activity relationships. Even many aspects of field performance can now be addressed using computer modeling.

In this chapter we shall provide an overview of these methodologies, highlighting examples where they have been used in fungicide discovery and optimization. We shall detail examples demonstrating application of molecular modeling to the design of fungicides active as inhibitors at two different steps in the ergosterol biosynthetic pathway.

Computerwares for discovery and optimization of fungicides:
A   Molecular graphics

Comparative molecular structure analysis

Imitation has been practiced as a form of fungicide design since long before the advent of computers. However, modern computer graphics has considerably improved the ability of a designer to compare visually two structures, one known to possess a desirable attribute, the other designed to be an imitation. Structures can be superimposed in a manner not possible with physical models, and molecular surfaces that have been color coded for a variety of physical properties can be compared. Several outstanding examples of this approach to fungicide design have been described. For example, flutriafol was designed as an inhibitor of the C-14 demethylase step of ergosterol biosynthesis by comparing its structure to that of the natural substrate, lanosterol, as well as to the common conformation thought to be realized at the active site by the class of azole fungicides to which flutriafol (Fig. 24.2) belongs (MARCHINGTON and LAMBROS 1987). Several examples using computer graphics to design imitations of high energy intermediates of reactions catalyzed by fungal enzymes have been described (URCH 1991; BASARAB et al. 1992) and one will be featured below.

Examination of protein binding sites

Today a fungicide designer, blessed with the crystal structure of a target enzyme or receptor, can examine his target site with modern high quality computer graphics in ways never before possible. Much as an artist might experience a landscape in a variety of lighting situations, he can experience the target site by viewing it in a variety of coloring schemes, each highlighting different structural features. Highly abstracted principles of complementarity (e.g., square pegs don't fit in round holes, oil and water don't mix, opposites attract, etc.) guide the design of structures that snugly fit into the target. Molecular models of structures so designed can be manually docked into the protein binding site to gain an impression of fit. At this level, the design process is as much art as science. The spirit of this approach is illustrated by a pair of computer graphics studies of the fit of the triazole fungicides triflumizole (NAKAYAMA et al. 1989) and mylcobutanil (FUJIMOTO et al. 1988) into the active site of a bacterial cytochrome P-450. Opportunities for applying this approach will increase as the number of crystal structures relevant to fungal biochemistry become available. We anticipate that the odds of success will be improved by a more scientific approach that includes conformational analysis of both protein and effectors as well as quantification of the degree of complementarity between them. These calculations are within the domain of molecular mechanics and electronic structure theory.

B Molecular mechanics and electronic structure

Where are the electrons in a molecule and what are the molecular conformations and intermolecular interactions that they determine? These are the questions addressed by electronic structure and molecular mechanics calculations, questions whose answers may be of considerable value in understanding the action of fungicides as well as in their design.

Electronic structure

Electronic structure calculations provide a host of descriptors of ground state charge distributions that can be useful for understanding structure-fungicidal activity relationships. For example, partial atomic charges are significant correlates of fungicidal activity for the phenethyl 1,2,4-triazole class of fungicides (FUJIMOTO et al. 1988) exemplified by myclobutanil (Systhane). Conformational analyses performed with electronic structure programs have also served to elucidate the relationship of structure to activity for several classes of fungicides (KATAGI et al. 1987; FUJIMOTO et al. 1988).

The nature of transition states and high energy intermediates are experimentally elusive. Although most molecular mechanics force fields are not parameterized for the unusual bonding situations that characterize many of these transient species, some electronic structure methods can produce plausible models of them. Later in this chapter we describe electronic structure calculations on the high energy carbocation intermediates of the $\Delta^8 \to \Delta^7$-isomerase and $\Delta^{8,14}$-reductase steps of ergosterol biosynthesis. These calculations were performed using the semi-empirical AM1 method (DEWAR 1985) which is well suited for treatment of carbocations (STEWART 1989). These structures have been used to rationalize the activity of experimental cyclopentylamine fungicides (BASARAB et al. 1992) and to define a hypothetical pharmacophore used as a query to search 3D databases for transition state analogues.

In addition to addressing questions concerning ground state properties of fungicides, electronic structure calculations can also shed light on the nature of electronic excited states. Such calculations can elucidate the nature of the transitions that may initiate photochemical decomposition under field conditions, and can suggest molecular modifications that can enhance photochemical stability.

Molecular mechanics

A collection of balls connected by springs is the molecular mechanics representation of a molecule. The spring force constants are determined either empirically or can be estimated from electronic structure calculations. Molecular mechanics calculations can be conveniently performed on large collections of atoms including proteins (WEINER et al. 1984). Molecular mechanics force fields are also being used increasingly in computer aided ligand-docking experiments.

Protein modeling and mechanical docking

To build a realistic model of a protein from sequence information alone has been a long standing and elusive goal for computational chemists. The problem becomes tractable only when theoretical approaches are supplemented by experimental data. For example, plausible models for the active site of a mammalian cytochrome P-450 (FERENCZY and MORRIS 1989) and for the entirety of two yeast Cyt P-450's (BOSCOTT et al., personal communication; MORRIS and RICHARDS 1991) have been built starting with sequence information for the modeled enzymes and the x-ray crystal structure of a bacterial homologoue (*vide infra*). The crystal structure of a highly homologous protein is but one piece of experimental data that can be used to build a plausible protein model.

Homology building also uses homology on a smaller scale to extract short segments from crystal structures of globally unrelated proteins as starting points for insert or other poorly characterized segments in the model built protein. In a sense the entire protein crystal structure data bank is brought to bear on the model building process when empirical rules for predicting secondary structure from sequence (GARNIER 1978; EISENBERG et al. 1984; CHOU and FASMAN 1974) are used. Structures built by using a combination of empirical rules and prefabricated elements from homologous proteins can then be refined by molecular mechanics.

This kind of a multi-faceted approach has been used to build models of membrane bound Cytochrome P450's present in yeast such as *Saccharomyces cerevisiae* (MORRIS and RICHARDS 1991) and *Candida albicans* (BOSCOTT et al. personal communication) starting with the crystal structure of a homologous soluble Cyt P450 found in the bacterium *Pseudomonas putida* (Cyt P450$_{cam}$). In the case of the *C. albicans* enzyme (CytP450$_{CA}$) the sequences of the two enzymes were first aligned using both sequence correlation, as well as the calculated propensity of the *Candida* sequence to adopt a secondary configuration similar to that of Cyt P450$_{cam}$. Secondary structure features of Cyt P450$_{CA}$ were then assigned to be those of corresponding sequences of CytP450$_{cam}$ where homologies exist, and according to empirical rules for inserts. The secondary structure of CytP450$_{CA}$ was then folded around the template structure of CytP450$_{cam}$. Next residues in the CytP450$_{cam}$ tertiary sructure were mutated where dictated by the alignment to the corresponding residues appearing in the CytP450$_{CA}$ sequence. Next short coil inserts were placed in the CytP450$_{cam}$ structure and regularized, and larger inserts were sought in a fragment database of proteins. Two especially long inserts with predicted helical structure were placed on the surface of the nascent protein by searching its surface for regions of sufficient hydrophobicity. The final structure was subjected to constrained minimization and molecular dynamics to relax the new backbone structure and remove sidechain clashes.

In many instances other experimentally derived constraints can also be used to refine a model built protein structure. Spectroscopically determined interatomic distances and angles can be useful input for protein model building. Distance geometry (WUTHRICH 1989) and systematic search techniques (MARSHALL 1984) generate 3-dimensional structures that are consistent with these geometric parameters. Additional constraints on models come from experiments such as site-directed mutagenesis, which can establish the functional role of a given residue in a sequence and provide circumstantial evidence for the spatial location of the residue with respect to active sites and effector sites in a protein. Similar information can be drawn from labeling experiments such as those described below for lanosterol 14-demethylase.

Mechanical docking of ligands is facilitated by energy evaluations for large receptor-ligand complexes. These calculations have been enabled by molecular mechanics programs that are specially parameterized for treatment of proteins (WEINER *et al.* 1984). The effect of solvent, especially hydrophobic effects, are often difficult to assess in these calculations, but simulations using thermodynamic perturbation theory (LYBRAND 1986) have much to promise in this regard. Some methods actually enable models for moderators of biological action to be grown into a binding site while monitoring the energetics of interaction (MOON and HOWE 1991).

Mechanical docking has been used effectively to study the binding of triazole fungicides to the model built active site of the yeast Cyt P450 from *C. albicans* (BOSCOTT *et al.*, personal communication) described above. Fortunately, the active site of the model built enzyme exhibits secondary structure elements that are little changed from those in Cyt P450$_{cam}$ even though not all residues are conserved in this region. In order to rationalize inhibition by the fungicides, the substrate (24-methylene-24,25-dihydrolanosterol) was first manually docked into the model built active site. The degree to which azole inhibitors could be made to maximize overlap with the docked substrate − subject to the constraint that the N-4 nitrogen atom remain chelated to the iron − was found to be reasonably well correlated with IC$_{50}$ values for inhibition of the Cyt P450$_{CA}$ enzyme. Thus, this model provides a way for rationalizing activity, but more importantly may provide an invaluable framework for designing more selective inhibitors active against *Candida* infections in humans. A similar approach starting with sequence information for a target Cyt P450 from a plant disease agent could provide an invaluable model for the design of novel plant disease control agents.

C Statistical modeling

Quantitative structure-activity relationships (QSAR's)

The members of an analogue series are variations on a theme, the theme being defined by a basic molecular framework that is common to all molecules in a series. If fungicidal assay results are available for a set of analogues of sufficient diversity, quantitative relationships can be established between the measured activity and various descriptors that characterize the diversity. QSAR's so established can be of great value for optimizing the activity within an analogue series of fungicides (FUJIMOTO 1987). If QSAR's are established at various levels of removal from the intrinsic target site (e. g., *in vitro*, fungal cell, *in vivo*, *in situ*), it may be possible to factor out contributions of transport and metabolism to the higer levels of activity. This feature of QSAR analysis is illustrated by a recent study on phenylcarbamate fungicides, wherein increasing lipophilicity of substituents on the phenyl ring was related to specific and favorable interactions at the active site but to adverse effects on transport (KAMOSHITA 1992). In this same QSAR study branching of alkyl groups on the ester was related to the prevention of metabolic degradation. Due to multiple sources of variability, assays for fungicidal activity often yield categorical ratings rather than ratings on a continuous scale. Thus, standard QSAR techniques of multiple linear regression (HANSCH and FUJITA 1964) may need to be replaced by pattern recognition techniques (TAKAHASHI *et al.* 1987), cluster analysis (MCFARLAND and GANS 1986), or variants of discriminant analysis (MORIGUCHI *et al.* 1980; KAMOSHITA *et al.* 1992).

With increasing awareness of environmental hazards, the safety of experimental fungicides needs to be considered earlier in discovery programs. In this regard QSAR models (ENSLEIN 1988) and expert systems (SANDERSON and EARNSHAW 1991) can be of value in designing compounds which minimize toxic effects on non-target organisms. For example, QSAR's have been developed to predict a variety of toxic endpoints (ENSLEIN 1988) including carcinogenesis, rat LD_{50}, fathead minnow LC_{50}, etc- Use of these models can be of value in prioritizing fungicide leads for early toxicity testing.

Although QSAR analysis is usually performed retrospectively, QSAR principles can be of great value in designing an analogue program once a new lead is discovered (PLUMMER 1990). What is desired is a representative sampling of all variations that can be built from the common theme that characterizes an analogue series. The variations often take the form of functional groups (substituents) that replace hydrogens at various locations on the common framework. The fractional factorial design approach (Box *et al.* 1978; WOLD *et al.* 1986) is one of several procedures for selecting a representative sampling of structurally diverse analogues for synthesis and testing. It can be performed either by hand (AUSTEL 1982) or with the assistance of computer (MARSILI 1988). Alternative methods include cluster analysis (HANSCH and LEO 1979) and difficulty information approaches (BORTH and MCKAY 1985).

Quantitative structure-property relationships (QSPR's)

The physical properties of a fungicide candidate will effect not only its intrinsic activity, but also its uptake, transport, metabolism and environmental fate. Thus, the ability to estimate these parameters may be critical to the success of a fungicide discovery program.

Quantitative structure-property relationships (QSPR's) can play a major role in making these estimates. Among the most often used QSPR's are those that relate measures of lipophilicity such as octanol-water partition coefficients to structure (LEO 1984), but models have also been developed for estimating acid dissociation constants, water solubilities and vapor pressures (LYMAN et al. 1982).

D Pharmacokinetic modeling

To protect crops from disease a fungicide must move from the site of application to the site of infection. Barriers to this movement often limit effectiveness. Pharmacokinetic modeling can be of value in designing compounds that can overcome these barriers. For example, a pharmacokinetic model of the plant vascular system accounts well for the phloem mobility of many pesticides (GRAYSON and KLEIER 1990) in terms of their physical properties and should be of value in designing phloem mobile fugicides (BROMILOW and CHAMBERLAIN 1989) for which there is considerable demand (JACOB and NEUMANN 1987). As another example, a statistical model for the acropetal mobility of fungicides in rice plants has been of value in elucidating the role of physical properties in the performance of isoprothiolane fungicides on rice diseases (UCHIDA and SUZUKI 1982).

Pharmacokinetic models for analyzing environmental exposure to chemicals can be an aid in enhancing effectiveness and improving safety. For example, exposure of aquatic organisms to fungicides can be anticipated using models of aquatic ecosystems that account for partitioning, metabolism, hydrolysis and photochemistry (BURNS et al. 1982). The sensitivity of the predicted exposure to the physical and chemical properties of the compound modeled can suggest chemical modifications that may increase both safety and effectiveness.

E Molecular databases

2D-Databases and computer assisted random screening

Large collections of retained chemical samples can be a rich source of leads for applications (e.g. plant disease control) other than those for which they were originally prepared. Unless the new application suggests a reason for screening compounds from specific structural families, exploration of a wide range of structural types should improve chances for success in a limited sampling of the collection. Screening every Nth chemical does not guarantee broad exploration unless the collection is already structurally

---

Fig. 24.1 Top: AM1 optimized structure of carbocation intermediate. Side chain has been appended in conformation generated by CONCORD (PEARLMAN et al.). Middle: Hypothetical pharmacophore for inhibition of isomerase generated from structure of carbocation intermediate. Distance ranges bracket measured distances in carbocation intermediate (C8—O = 6.45 Å; C8—C24 = 7.52 Å; O—C24 = 13.86 Å). Bottom: Sample structure found by searching MACCS FCD3D database.

Computer aided discovery and optimization of fungicides 523

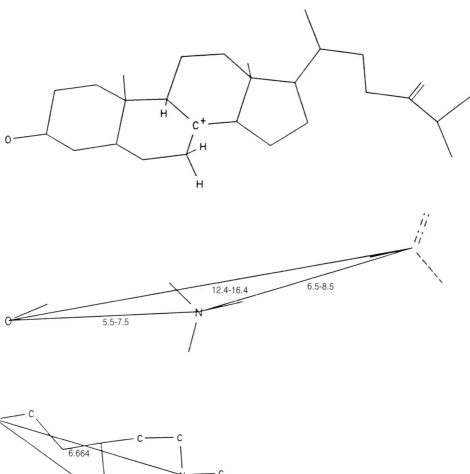

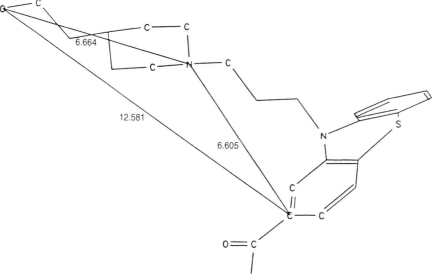

Fig. 24.1

diverse. A better approach would be to cluster the compounds into families based upon chemical similarity and then to sample a few compounds from each family. If a 2D structural database for the collection is available, this process can be facilitated by computer, but requires some metric of similarity upon which to base family assignment. Similarity metrics may focus on topological or substructural features and algorithms for clustering based upon these similarity indices are numerous (WILLETT 1987). Once instructed in the metric, the computer can be an efficient judge of familial resemblance. Sampling across families then ensures structural diversity in selection lists for bioassays.

3D-Databases and pharmacophore searching

The recognition that carbon is tetrahedral stimulated a revolution in the development of chemistry. Adding a third dimension to chemical databases is stimulating a similar revolution (BORMAN 1992; MARTIN 1992) in chemical information management. 3D-databases can be a rich source of ideas for novel fungicides if the requirements for activity can be specified in the form of a 3D-pharmacophore query. Pharmacophore hypotheses can be defined by analyzing the shape of a receptor (DIXON and KUNTZ 1988), by determining a common arrangement of functional groups that can be reasonably realized by a set of ligands thought to bind to a common receptor, or by analyzing the structure of a transition state or high energy intermediate for an enzyme catalyzed reaction. For example, Fig. 24.1 illustrates a hypothetical pharmacophore that may specify requirements for inhibition of the $\Delta^8 \rightarrow \Delta^7$-isomerase reaction of the ergosterol biosynthesis pathway (see below). This hypothetical pharmacophore has been defined by reference to the structure of the modeled high energy carbocation intermediate for this reaction, and is phrased as a MACCS-3D query (CHRISTIE 1990). The carbocation center of the modeled intermediate is replaced by a tertiary amine which is likely to be protonated at physiological pH's. Typically structures found by the search, such as the CONCORD generated structure (PEARLMAN et al.) shown from the MDL's FCD3D database (CHRISTIE 1990), are not expected to be active themselves, but serve as starting points for further structural elaboration or pruning.

Bio-isostere databases

The equivalence of functional groups for the purpose of evoking a given biological response is captured in the notion of bio-isosteres. For example, it is often observed that an imidazole ring can replace a triazole without compromising fungicidal activity. EMIL is a database of bio-isosteric transformations (FUJITA 1993) that is presently under construction by a consortium of pharmaceutical and agricultural chemical companies. Given the structure of a bioactive compound, the associated EMIL software will parse the structure for the presence of functional groups and replace them with bio-isosteres. This list of structures should be a rich source of ideas for novel compounds with reasonable odds of activity at the same or similar sites as the bioactive structure used as input.

The application of computer modeling to selected classes of fungicides: survey of recent modeling efforts around triazole fungicides

Because of the potent and economically important disease control spectrum demonstrated by the triazole class of fungicides, the large array of structural types of triazoles exhibiting commercial levels of activity, and the known mode of action of these materials, there has been considerable attention devoted to understanding the structure-activity relationships in this class. Studies involving several triazole and related fungicides, and several different fungal and yeast organisms, strongly corroborate the concept that the compounds act by inhibiting the 14a-demethylation of lanosterol, an early and critical step in ergosterol biosynthesis in filamentous fungi (reviewed in KÖLLER 1992 and in chapter 13 and 21 of this volume). We shall compare several modeling studies which have focused on how triazoles might compare with lanosterol, the substrate of the target enzyme. These studies have usually assumed that the hydrophobic groups of the fungicide somehow conform to the bulk of the sterol molecule, while the triazole plays the role of a ligand in the sixth coordination position of the iron porphyrin system (Fig. 24.3).

One of the first serious modeling efforts along these lines (MARCHINGTON 1984; MARCHINGTON and LAMBROS 1987) consisted of computing an energy-minimized structure of *RR*-diclobutrazol; confirming that the computed conformation of this molecule was comparable to the conformation determined by X-ray crystallography; retrieving the X-ray crystal structure of lanosterol from a crystal structure database; and the actual overlaying of the two structures. A key assumption made for the overlay was that the N-4 nitrogen of the triazole ring should be superimposed on the oxygen of the initial hydroxylated product of lanosterol, where this oxygen would also be positioned as the sixth ligand of the heme system. A second assumption was that, in the bound conformation of diclobutrazol, the hydroxyl group should not be intramolecularly hydrogen-bonded to the triazole ring. Although in nonpolar solvents there is infrared evidence for this hydrogen bonding, such bonding does not occur in the crystal structure of the compound. Furthermore, hydrogen bonding is not observed in infrared dilution experiments with triadimenol. For these reasons the conformational refinement procedure applied to diclobutrazol apparently ignored intramolecularly hydrogen-bonded structures. The point here is that, at least for diclobutrazol, minimally two low energy conformations are available to the molecule, and a choice was made to ignore the one involving intramolecular hydrogen bonding.

In the computer-derived comparison of lanosterol and *RR*-diclobutrazol (Fig. 24.3), the *t*-butylcarbinol group lies in a plane with the B ring of the sterol backbone, such that one of the inhibitor methyl groups is close to the angular methyl group at the A,B-ring junction. The 2,4-dichlorobenzyl moiety is largely in the space occupied by the extended sterol sidechain. It was argued that the hydroxyl group of the inhibitor would be in a position to form a hydrogen bond to a propionate sidechain of the heme, or a carboxylate group of the protein thought at the time to be important in the oxidative reaction. A very similar model was proposed for the binding of paclobutrazol to lanosterol 14-demethylase (SUGAVANAM 1984; WORTHINGTON 1987).

In the modeling of flutriafol binding to the demethylase, a conformation different from that of diclobutrazol was arrived at (MARCHINGTON and LAMBROS 1987). Conformational analysis, presumably neglecting conformations with intramolecular hydrogen bonding, again resulted in a conformation of the inhibitor closely resembling that found in the crystal structure. Computer overlap of both flutriafol and *RR*-paclobutrazol with lanosterol led to a somewhat different conformation of the sterol (WORTHINGTON 1987), where the sidechain is now no longer in the approximate plane of the ring system. One of the fluorophenyl rings of flutriafol occupies the space of the *t*-butylcarbinol group of

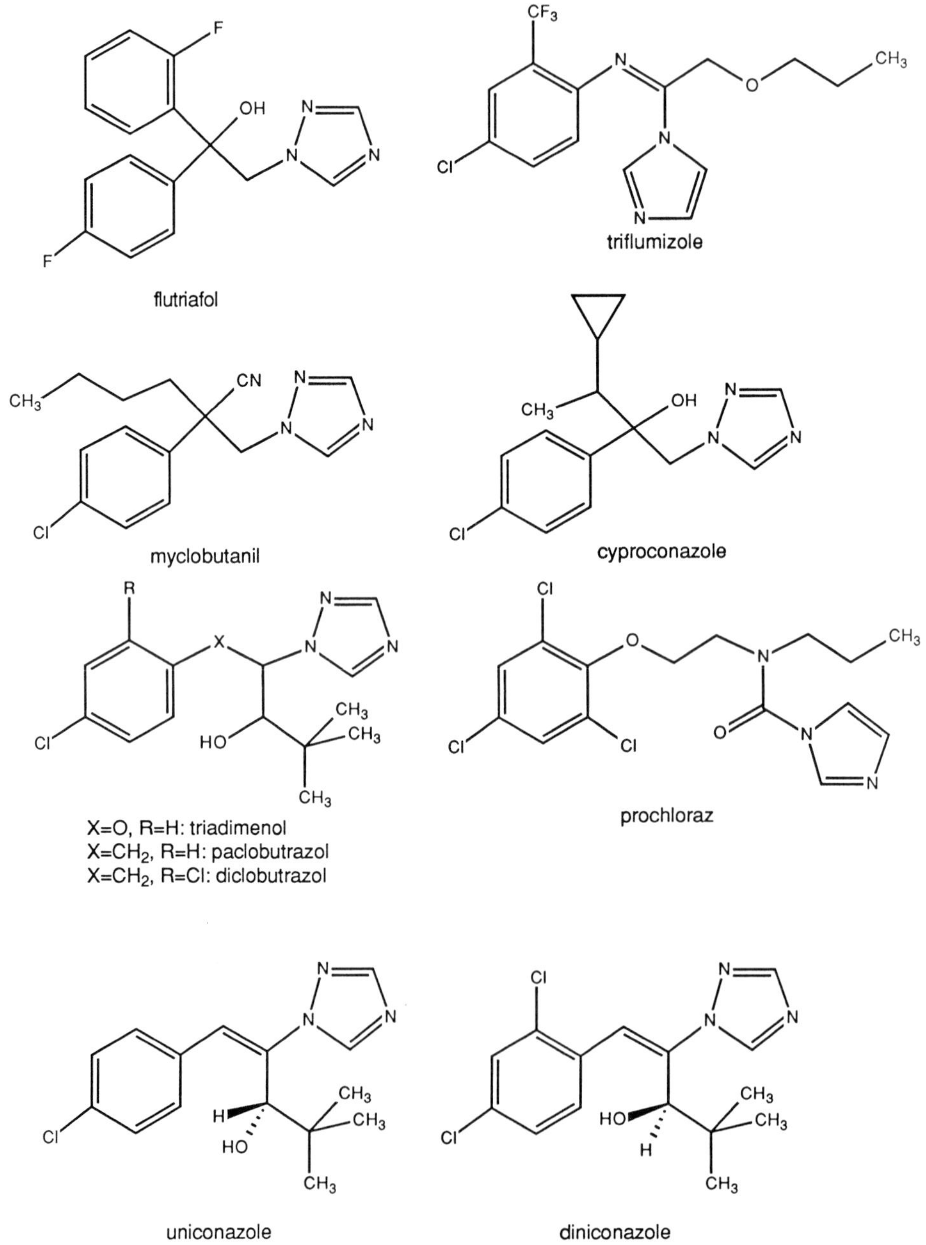

Fig. 24.2 Chemical structures of imidazole and triazole fungicides discussed in this chapter.

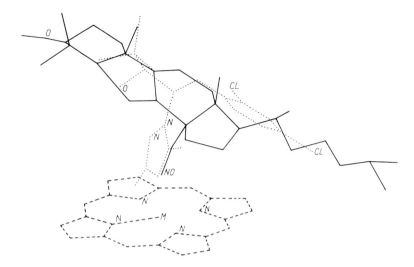

Fig. 24.3 Comparison of the optimized structure of *RR*-diclobutrazol with the crystal structure of 32-hydroxylanosterol, reproduced from MARCHINGTON and LAMBROS 1987.

paclobutrazol, and the other fluorophenyl ring is in the vicinity of sterol sidechain and the chlorophenyl ring of paclobutrazol. The result is that the HO−CH−CH-triazole linkage of paclobutrazol, and the HO−C−CH$_2$-triazole linkage of flutriafol are in orientations offset from one another. There is no reason *a priori* why these particular atoms should map on top of one another in optimized binding modes. However, if hydrogen bonding involving the hydroxyl group is important for the binding of one inhibitor, it is difficult to invoke the same system of hydrogen bonding for minimum energy binding of the other. This was the first study which attempted to rationalize the binding of two structurally divergent inhibitor types with the same model.

Among inhibitors of 14-demethylase, prochloraz is structurally unusual in being an N,N-dialkylcarbamoylimidazole (the corresponding triazoles are much less active), and unusual biologically, having particularly high efficacy against cereal eyespot (*Pseudocercosporella*). Computer modeling was performed on this class of compounds in an effort to interpret structure-activity trends (BAILLIE 1987). The assumptions used in this modeling effort have not been discussed in detail. However, the model compared prochloraz with an extended conformation of lanosterol, where the propyl sidechain mapped to the sterol ring system, and the phenoxyethyl group conformed to the conformation of the lanosterol sidechain. It is interesting in this model that lanosterol is positioned somewhat further above the heme than in the models discussed above, so that the 14a-methyl group corresponds in position to the carbonyl group of the inhibitor. In this conformation, there appears to be excellent overlap between the phenoxyethyl group and the sterol sidechain.

An interesting case is represented by diniconazole, which is a 14-demethylase inhibiting fungicide, and uniconazole, which is a plant growth regulator by virtue of inhibiting 19-demethylation of *ent*-kaurene. The enzyme specificity difference is apparently due to the absolute stereochemistry at the carbinol carbon. The corresponding Z-isomer of diniconazole is much less active as a fungicide and, in fact, gives a P-450 difference spectrum atypical of azole fungicides (KATAGI *et al.* 1987). The authors reasoned that conformations of these molecules in solution, rather than crystal structure conformations, would best represent the binding conformations. By carrying out infrared experiments in solvents approximating the polarity of the P-450 enzyme active site, and using model compounds lacking either the aryl or triazole rings, they were able to demonstrate that diniconazole exists in a solution conformation in which the hydroxyl group is intramolecularly hydrogen bonded to the N-2 nitrogen of the triazole. Thus, their initial conformation for modeling purposes was the one rejected by MARCHINGTON and WORTHINGTON. Nevertheless, the structure of lanosterol was based on crystal structure data. The actual computer modeling began with the clearly stated assumption that the triazole N-4 nitrogen and the 14a-methyl group should map to approximately the same position above the heme iron. The olefinic linkage and the *t*-butylcarbinol map in the vicinity of the C and D rings and the angular methyl group between them, and the dichlorophenyl ring occupies the space of the sterol sidechain.

The published structure-activity studies around myclobutanil represent an intellectually satisfying and self-consistent interplay among QSAR analysis, computer modeling, and conformational analysis (FUJIMOTO et al. 1988). It is also the first study to employ the crystal structure of the *Pseudomonas putida* camphor hydroxylase as an aid in positioning the lanosterol substrate with respect to the heme. Myclobutanil represents one of a series of 2-arylethyltriazoles optionally − but not necessarily − substituted with a cyano group on the benzylic carbon. Detailed quantitative structure-activity studies were carried out with several fungal organisms, using both *in vitro* tests and whole plant assays, as well as a yeast ^{14}C 14-demethylase assay. The correlations varied somewhat with the organism; however, the size of the optimal alkyl substituent on the benzylic carbon was largely constant, as was the finding that *ortho*-substituted phenyl compounds were most active when they lacked the benzylic cyano group, and compounds without *ortho*-substitution were most active with the cyano. To rationalize this empirical finding, conformational analysis was performed on a series of analogs with increasing substitution at the *ortho*- and benzylic positions. Clear conformational preferences were observed in those compounds bearing a single substituent at either one or the other position. Computer modeling was then carried out, beginning with the lanosterol 14a-methyl group positioned with respect to the heme iron similarly to the oxidized methylene group of camphor in the crystal structure of the camphor hydroxylase enzyme-substrate complex: that is, with the reactive C−H bond above the heme and offset from the orthogonal axis passing through the heme iron. When the triazole inhibitors were mapped onto the lanosterol skeleton, it was clear that the best alignment was obtained when the inhibitor was in a "folded" conformation, *i.e.*, with the planes of the aryl and triazole rings perpendicular to one another. More specifically, this conformation positioned the aryl ring perpendicular to the heme iron-triazole bond. The compounds which favored or allowed this conformation were those with either an *ortho*-phenyl substituent or a benzylic cyano group, and these were, satisfyingly, the most active fungicides.

In the foregoing discussion one of the aims has been to highlight the assumptions which various research groups have used in their computer modeling efforts. A critical assumption is the initial conformation of the demethylase inhibitor which was chosen. Whereas in one case (KATAGI et al. 1987) a conformation was selected which predominates in solution, in another case (MARCHINGTON 1984) such a conformation was specifically rejected in favor of one derived from electronic structure calculations, but which closely resembles the conformation found in the crystal structure. As was clearly shown in the modeling efforts around myclobutanil and its analogues, there exists for many structures a population of low-energy conformations available to any given compound. The relative probability of any one conformational state's being populated will depend on its relative free energy, which is in turn dependent on non-bonded and hydrogen-bonding interactions, as well as solvation forces. Upon binding of a small molecule to a biopolymer, solvation forces are largely or entirely eliminated and are replaced by a series of specific interactions characteristic of the binding domain. These may act to select one conformation from those which are accessible, one which is not necessarily the most stable solution conformation. As crystal structure and nuclear magnetic evidence accumulates on protein ligand interactions, it is becoming clear that, in some cases, the bound ligand may assume more than one conformation, with a dynamic equilibrium existing between them (CLORE et al. 1984); or that the protein conformation itself (BIRDSALL et al. 1984), and/or number of water molecules entrapped, may vary depending on the specific ligand. Molecular mechanics calculations can now take into consideration the possibility of intra- and intermolecular hydrogen bonding. However, the subtle influence of hydrogen bonding on coformational changes associated with desolvation and binding are very difficult to account for, even with the most sophisticated of modern force fields. None of these considerations makes the choice of an inhibitor conformation any simpler. The best practice is likely to be aware, through molecular mechanics calculations, of the population of conformations likely to be accessible at physiological temperatures, and then to choose the one with best overlaps to the natural substrate. The work of FUJIMOTO et al. on myclobutanil is an elegant example of this approach. As a mental image, Emil Fischer's "lock and key" hypothesis (FISCHER 1894), while revolutionary in its time, confers a conformational rigidity to both the docking ligand and the biopolymer receptor, which we must now begin to unlearn.

Inputs into a model for inhibitor binding to the active site of lanosterol 14-demethylase: biochemical and mechanistic information

A model which attempts to account for the binding of a fairly diverse group of inhibitor structures as well as that of a substrate must be consistent with what is known about three-dimensional structure of the binding domain, and the enzymology of the catalyzed reaction. These consideration are briefly discussed here with respect to lanosterol 14-demethylase. Reference will be made to specialized reviews of these issues. A comprehensive review of the literature on P-450 enzymes will be found in ORTIZ DE MONTELLANO 1986a.

The essential oxidative steps catalyzed by lanosterol 14-demethylase are becoming understood in increasing detail. The reaction proceeds in sequential steps involving the intermediacy of 14-hydroxymethyl and 14-formyl compounds. Not only is there now solid evidence for these intermediates, but they have been shown to have lower K_m values and to be more reactive than the initial lanosterol substrate (AOYAMA et al. 1987; AOYAMA et al. 1989). Therefore, a model for the binding of lanosterol should also account for the tighter binding, as well as the greater reactivity, of its oxidation intermediates.

For a long time a three dimensional representation of any cytochrome P-450 was limited to the crystal structure of the camphor hydroxylase, P-450$_{cam}$ of *Pseudomonas putida* (POULOS 1986; POULOS et al. 1987). The P-450 enzymes from eukaryotic organisms, including yeasts and other fungi, are membrane-bound, and they have so far resisted crystallization. However, there is sufficient sequence homology between the bacterial enzyme and those from higher organisms to allow computer modeling of the substrate binding domain of other P-450 enzymes. This has been carried out with a mammalian enzyme, termed nifedipine oxidase, P-450$_{NF}$, which catalyzes the hydroxylation of several xenobiotic compounds as well as steroid hormones (FERENCZY and MORRIS

1989). Empirical rules were used to align the known sequence of the mammalian P-450$_{NF}$ with that of the bacterial P-450$_{CAM}$. Then, a crude model of P-450$_{NF}$ was built from that of P-450$_{CAM}$ by replacing amino acid sidechains of the latter structure with those suggested by the alignment. The mutated structure was then refined using molecular mechanics to yield a plausible model for the active site of P-450$_{NF}$. The resultant model led to the favorable computer docking of several substrates of this enzyme. With the determination of the amino acid sequence of P-450 lanosterol demethylase from yeast (KALB et al. 1987), it is possible to carry out similar computer modeling experiments. Work along these lines is well advanced (MORRIS and RICHARDS 1991).

In the absence of information based on crystal structure, one must depend on chemical probes of the P-450 active site. Beginning with heme alkylations with olefins and acetylenes (KUNZE et al. 1983; reviewed in MIWA and LU 1986) and culminating most recently with arylations of lanosterol 14-demethylase using aryldiazenes (TUCK et al. 1992 and references therein), ORTIZ DE MONTELLANO and coworkers have probed the size of the substrate binding domain and the accessibility of the heme group. In the early work it became clear that only a portion of the tetrapyrrole ring system is, in general, accessible to alkylation, while similar reactions performed on free porphyrins yield products arising from alkylation of each of the four pyrroles. In the most recent work on lanosterol 14-demethylase, only pyrrole ring C is arylated when the intermediate aryl-iron complex is oxidized. That pyrrole arylation proceeds with biphenyldiazene suggests that the substrate must bind specifically over the C ring of the porphyrin, and that the binding pocket must be at least 10 Å high in this region. These studies strongly suggest a model for lanosterol binding represented by Figure 24.4 Top.

Considerable progress has been made in understanding the mechanisms of cytochrome P-450 catalyzed reactions (for a review see MCMURRY and GROVES 1986; ORTIZ DE MONTELLANO 1986b). This work has culminated in the isolation of a chemical model for Compound I, the ultimate oxidant in P-450 reactions (GROVES et al. 1981). The characterization of horseradish peroxidase Compound I as well as the model system allows this ultimate oxidant to be drawn either as an oxo-iron (V) or as an iron (IV) radical cation, in which the iron-oxygen bond is 1.6 Å (Fig. 24.5). Electronic structure considerations (GROVES and NEMO 1983; summarized in ORTIZ DE MONTELLANO 1986b) suggest that the reactive ferryl oxygen orbital is orthogonal to the iron-oxygen bond, that is, parallel to the porphyrin ring system. If the hydroxylation reaction is viewed as taking place via hydrogen atom abstraction followed by hydroxyl transfer to the resultant carbon radical (Fig. 24.5), then hydrogen atom trasfer must take place parallel to the porphyrin plane (MCMURRY and GROVES 1986). The crystal structure of the P-450$_{cam}$ camphor complex is in accord with this prediction (POULOS 1986), with the two C−H bonds of the carbon of camphor undergoing oxidation aligned in the proper orientation. Applying this model to the active site of lanosterol 14-demethylase, one would anticipate the 14α-methyl group to be displaced from the axis of the iron-oxygen bond by some distance (Fig. 24.4 bottom). In considering the interaction of a triazole inhibitor with the heme, one should recognize that the triazole will likely bind as a sixth ligand in a semi-eclipsed conformation with respect to a *trans* pair of porphyrin nitrogens, in order to maximize overlap with orbitals on iron (NYFELER and HUXLEY 1986). This is especially important with respect to binding inhibitors possessing a hydroxyl substituent capable of intramolecular hydrogen bonding: if the internal hydrogen bond exists in the bound conformation, then the orientation of the triazole with respect to the porphyrin ring system will affect the entire orientation of the bound inhibitor. Conceivably the orientation of the triazole ring with respect to the heme system could also affect the conformation of the rest of the inhibitor through nonbonded interactions.

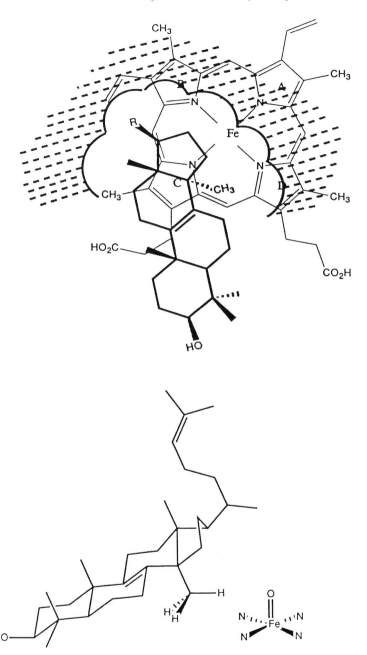

Fig. 24.4 Two hypothetical views of lanosterol bound to the active site of lanosterol demethylase. Top: View illustrating the binding of the substrate relative to pyrrole ring C. Bottom: View illustrating the orientation of the reacting C—H bond of the 14-methyl group with respect to the ferryl system.

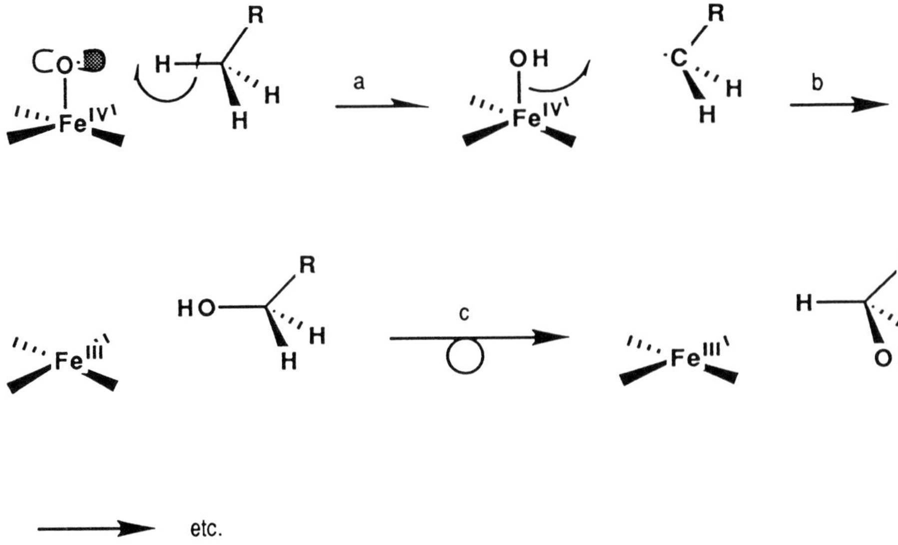

Fig. 24.5 Hypothetical mechanism of the lanosterol demethylase reaction illustrating. a) alignment of the C—H bond with the reactive oxygen orbital and hydrogen atom abstraction, b) oxygen transfer, and c) rotation of the resulting hydroxymethyl group of the substrate prior to the second hydroxylation reaction.

We must now account for the subsequent steps in the reaction. If the 32-hydroxylanosterol intermediate were to remain complexed to the iron as a ligand in the sixth position, either as a hydroxy species or as a alkoxide, then it would have to be displaced by molecular oxygen for the second oxygenation step to occur. It is unlikely that a geometry favorable for further reaction would serve to stabilize the intermediates as ligands in the sixth coordination position of the iron. Rather, the mechanistic considerations discussed above would suggest that the 32-hydroxy intermediate is somewhat displaced from the position from which it would optimally ligate the iron. The hydroxymethyl group would then merely have to rotate 120° to present the second hydrogen atom for abstraction by the ferryl system. Such a rotation would be favored further, if to do so would set up a hydrogen bond with a suitable donor or acceptor. A third and final oxidation step must still occur, mechanistic details of which have been suggested (FISCHER *et al.* 1991). However, the discussion to this point serves to put geometric constraints on the binding of lanosterol and its oxygenated intermediates.

Inputs into a computer model derived from the structure-activity of known inhibitor classes

The information available from inhibitor studies must now be taken into consideration. Some of this will derive from optimal structural parameters of azole fungicides. Other details will come from a consideration of the relative efficacy of enantio- and diastereomeric inhibitors. In considering structure-activity data, it is important to consider the type of biological activity on which correlations are based. It is well recognized in theory, but often ignored in practice, that the biological data used in a correlation should be aligned with the ultimate uses of the study. With the ultimate aim of identifying the best possible fungicide, the activity of a range of candidates in whole plant tests must be determined against a range of pathogens, with the ultimately intended market in view: how, for example, to balance cereal rust *versus* eyespot activity while maintaining high levels of mildew control. Computer modeling of inhibitor binding to an enzyme active site should not be guided by whole plant data, because confounding factors such as transport and metabolism obscure the trends in intrinsic activity. For modeling studies of this type it is far superior to use *in vitro* data, derived preferably from enzyme assays or, in the present case, cytochrome P-450 difference spectra (for a discussion of possible dangers in relying on data derived from P-450 difference spectra, see KAPTEYN et al. 1992). The best inhibitor may not be the best fungicidal candidate. From a full and complete knowledge of the binding domain, however, it is possible that novel structures, not readily derived from random synthesis, but possessing the physical properties necessary for high *in vivo* activity, will be obtained.

Of the azole fungicides developed recently, a 1-substituted imidazole or triazole is generally separated from an aromatic system by two atoms (prochloraz and tebuconazole are the most notable exceptions among the agricultural fungicides) and from a heteroatom also by two atoms (again, prochloraz stands out, as well as flusilazole, penconazole, and myclobutanil as deviating from this pattern).

Much of the modeling work referred to earlier attempted to rationalize by model building the dependence of activity on absolute stereochemistry among enantiomeric or diastereomeric isomers. From this work several patterns were apparent, namely that all activity does not reside with one diastereomer, but that, in molecules such as triadimenol and diclobutrazole, the absolute stereochemistry of the *t*-butylcarbinol portion is most critical (WIGGINS and BALDWIN 1984; DEAS et al. 1986; BURDEN et al. 1987; YOSHIDA and AOYAMA 1991). In cyproconazole, which has a chiral center within the alkyl group and at the carbinol carbon, both *R* and *S* carbinols appear equally active (GISI et al. 1986). In this case it was argued that both the active *RR* and *SS* diastereomers can map equally well to the A, B, and C rings of lanosterol substrate. Among the diastereomers of etaconazole and propiconazole, the *S* configuration at the 2-position of the dioxolane ring is important for activity. However, the 4-position, bearing the alkyl substituent, can have either the *S* or *R* stereochemistry with little effect on activity (EBERT et al. 1989). Thus the 14-demethylase binding domain for these inhibitors must be able to accomodate the 4-alkyl substituent either in a *cis* or *trans* configuration relative to the triazolylmethyl group. Alternatively the *SS* and *SR* isomers must bind in two entirely different orientations in the active site, but with approximately equal affinity.

Modeling sterol intermediates for the design of isomerase-reductase inhibitors

Both in order to forestall the appearance of resistance to azole fungicides (SCHEINPFLUG 1988; DEKKER, chapter 2 of this volume), as well as to combat such resistance once it appears (KÖLLER 1988; HEANEY 1988), companion fungicides with activity against cereal powdery mildew and rust, used either alternately or in combination with azoles, are extremely useful. The tertiary amine fungicides such as fenpropimorph, fenpropidin, and tridemorph (Fig. 24.6) fulfil this role extremely well. The fungicidal mode of action of these compounds is generally assumed to be inhibition of sterol biosynthetic steps occurring subsequent to 14-demethylation, namely reduction of the 14,15-double bond and isomerization of the 8,9-double bond to the 7,8-position (for reviews, see MERCER 1991; KÖLLER 1992; KERKENAAR, Chapter 11 of this volume). It has been proposed that, in their protonated form, the amine fungicides are able to mimic carbocationic intermediates thought to occur in both the isomerase- and reductase-catalyzed reactions (BALOCH and MERCER 1987; TATON et al. 1989).

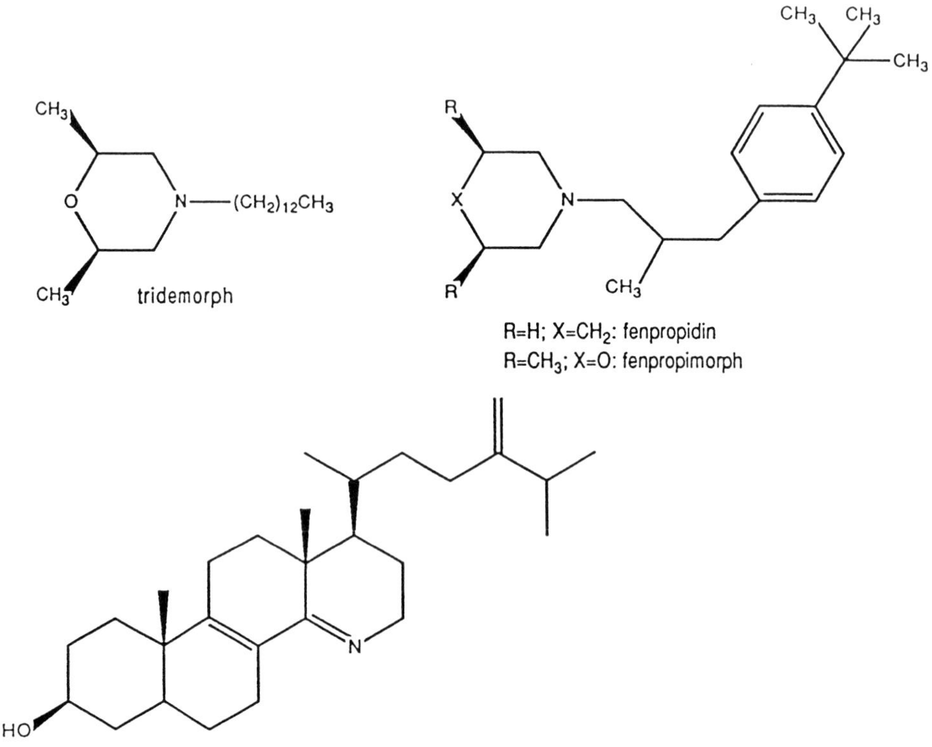

Fig. 24.6 Structures of commercial and natural product sterol isomerase and reductase inhibitors discussed in this chapter.

According to a popular theory (PAULING 1946) the emulation of transition state structures and high energy intermediates should be a fruitful way to design competitive enzyme inhibitors. Using an hypothesis for the mechanisms of these reactions and for the mode of inhibition by the amine fungicides, several groups have attempted to design inhibitors more active than the commercial compounds, by hoping to capture additional binding energy with structures more closely resembling the natural substrates. An example of this approach can be found in nature: the natural product A25822B, obtained from cultures of *Geotrichum flavobruneum*, is highly active against fungi and yeasts. Through determination of sterol intermediates (WOLOSHUK et al. 1979) and enzyme inhibition data (BOTTEMA and PARKS 1978), this compound has been shown to be a highly specific inhibitor of the $\Delta^{8,14}$-reductase. The synthetic efforts, however, to design partial sterol structures with appropriately positioned amine functionalities, have met with limited success. Although it has been possible to design compounds which are up to ten times better inhibitors (measured as I_{50} values) than the commercial compounds (RAHIER et al. 1986; RAHIER et al. 1990; and AKERS et al. 1991), this success has not translated into more active fungicides in the field.

Three groups have published on a similar concept ot applying stereochemical constraints to the fenpropidin structure (URCH 1991; HUXLEY-TENCER et al. 1992; BASARAB et al. 1992). The strategy was based on the hypothesis that, if the 2-methylpropylene chain of fenpropidin and fenpropimorph represents the 14, 13, 17, and 18 carbon atoms of the sterol ring system, then by cyclizing this linking group, several degrees of freedom would be removed, and resemblance to the sterol ring D would be enhanced. In doing so, both *cis* and *trans*, as well as optical isomers are generated. Both HUXLEY-TENCER et al. and BASARAB et al. separated the stereoisomers, while the former group also separated optical isomers of each. The former group assayed both isomerase and reductase enzyme activities from *Ustilago maydis* but reported only isomerase data; the latter group employed a substrate which allowed them to look specifically at reductase activity derived from *Saccharomyces cerevisiae*. HUXLEY-TENCER and coworkers found the appropriate isomer of 1,3-disubstituted cyclopentane was up to 20-fold more active as a isomerase inhibitor compared to fenpropimorph, while BASARAB et al. found all 1,3-disubstituted and 1,2,3-trisubstituted cyclopentanes to be weaker than fenpropimorph as reductase inhibitors.

There exist four diastereomeric pairs of isomers of 1,2,3-trisubstituted cyclopentanes. BASARAB et al. obtained crystal structures of the two most active of these, the *rel-R,S,S* and *rel-R,R,S* isomers (Fig. 24.8). These were compared with the energy-minimized structure for the carbocationic intermediate in the reductase reaction (for a discussion of the methodology applied to ionic species, vide supra). The crystal structure conformation of the *rel-R,S,S* isomer, with the aryl and morpholine rings in *pseudo*-equatorial conformations and the methyl group *pseudo*-axially disposed, overlays particularly well with the D-ring and the angular methyl group. The crystal structure of the *rel-R,R,S* isomer has the morpholine ring in a *pseudo*-axial orientation. It is this latter isomer, however, which is the most active as a reductase inhibitor. To rationalize this, molecular mechanics calculations were performed on the inhibitors. A conformation exists for the *rel-R,R,S* isomer which places both the *t*-butylphenyl and morpholine rings in planar arrangement with the cyclopentane (Fig. 24.7) and is computed to be of comparable stability to the molecular mechanics refined crystal structure. This result again demonstrates that crystal structure conformations should not be used uncritically for molecular modeling purposes.

Reasoning that excessive flexibility still remains in the cyclopentane compounds, BASARAB et al. provided additional rigidity to the system by incorporating an epoxy linkage at the carbons bearing the methyl and aryl substituents, thereby fixing them *cis* to one another (Fig. 24.8). The compound bearing the 3,5-*cis*-dimethylpiperidine ring *trans* to the *t*-butylphenyl ring is forty times more active as a reductase inhibitor than the *cis* isomer and 2.5 times more active than fenpropimorph. Here again, the *trans* analogue has the phenyl and heterocyclic rings disposed *pseudo*-equatorially, and thus maps best to the planar conformation of the sterol ring system.

The lack of improvement in fungicidal efficacy in all of this tertiary amine chemistry is vexing. Detailed correlations will have to await a complete set of isomerase and reductase enzyme data and *in vitro* fungal growth measurements on the same set of compounds. Then a correlation of fungicidal activity with *in vitro* data, perhaps using the

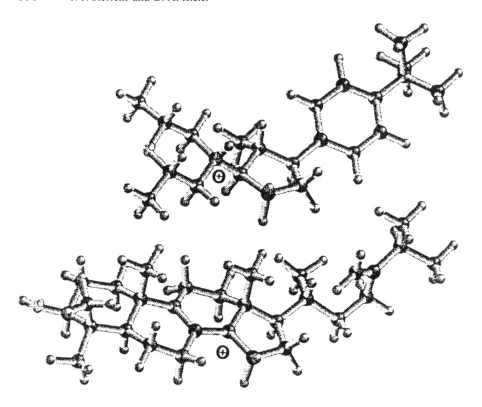

Fig. 24.7 Comparison of the optimized *rel-R,R,S*-isomer (top) with the substrate for the $\Delta^{8,14}$-reductase reaction. Molecular graphic by SYBYL, Tripos Associates, St. Louis, Missouri.

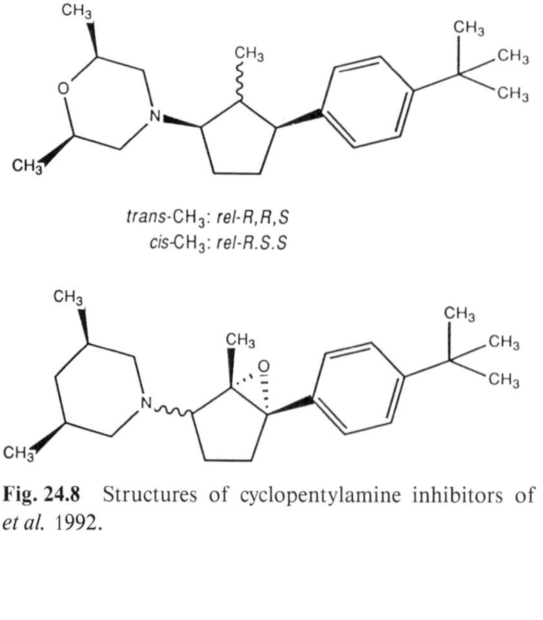

trans-CH$_3$: rel-R,R,S
cis-CH$_3$: rel-R,S,S

Fig. 24.8 Structures of cyclopentylamine inhibitors of $\Delta^{8,14}$-reductase synthesized by BASARAB *et al.* 1992.

HANSCH-FUJITA method of correlation (KAMOSHITA et al. 1992) will allow identification of those plant factors controlling the expression of fungicidal activity. The geometry at the carbocation centers discussed above are planar, a geometry that is not realized by the protonated nitrogen of morpholine-type fungicides. Furthermore, the electronic structure of the protonated morpholine is not characterized by a low lying unoccupied orbital as are the carbocation intermediates. Thus, structures that better emulate the nature of these carbocations may have improved inhibitory capability for the isomerase and reductase steps in ergosterol biosynthesis.

Conclusion

In order for a compound to succeed as a commercial plant disease control agent it must possess many attributes including a) intrinsic activity at some target site, usually an enzyme or receptor; b) mobility from the site of application to the site of action; c) suitable stability under field conditions; and d) safety. The availability of crystal structures for proteins validated as target sites for fungicidal action well enable rapid advances in computer aided discovery of compounds with intrinsic activity. In this chapter we have highlighted computerwares available for this purpose and given an overview of the computer tools that are available for discovering and/or designing compounds with the other requisite attributes as well.

Acknowledgements: The authors wish to express their gratitude to Kevin T. KRANIS for useful discussions as well as some of the modeling work presented in this chapter; to Gregory S. BASARAB and Toshio FUJITA for a careful and critical readings of the manuscript; to Anthony F. MARCHINGTON for permission to reproduce Figure 24.3; and to Paul E. BOSCOTT and W. Graham RICHARDS for a preprint of their manuscript on modeling the *Candida albicans* Cyt P-450, prior to publication.

References

AKERS, A., AMMERMANN, E., BUSCHMANN, E., GÖTZ, N., HIMMELE, W., LORENZ, G., POMMER, E.-H., RENTZEA, C., RÖHL, F., SIEGEL, H., ZIPPERER, B., SAUTER, H., and ZIPPLIES, M.: Chemistry and biology of novel amine fungicides: attempts to improve the antifungal activity of fenpropimorph. Pestic. Sci. **31** (1991): 521–538.

AOYAMA, Y., YOSHIDA, Y., SONODA, Y., and SATO, Y.: Metabolism of 32-hydroxy 24,25-dihydrolanosterol by purified cytochrome P-450$_{cam}$ from yeast. J. Biol. Chem. **262** (1987): 1239–1243.

– – – – Deformylation of 32-oxo-24,25-dihydrolanosterol by the purified cytochrome P-450 14DM (lanosterol 14a-demethylase) from yeast. Evidence confirming the intermediate step of lanosterol 14a-demethylation. J. Biol. Chem. **264** (1989): 18502–18505.

AUSTEL, V.: A manual method for systematic drug design: Eur. J. Med Chem., Chem. Ther. **17** (1982): 9–16.

BAILLIE, A. C.: Prochloraz and its analogs. In: BAKER, D. R., FENYES, J. G., and MOBERG, W.L. (Eds.): Synthesis and Chemistry of Agrochemicals. ACS Symposium Series 355. American Chemical Society, Washington, D. C. 1987, pp. 328–339.

BALOCH, R. I., and MERCER, E. I.: Inhibition of sterol $\Delta^8 \to \Delta^7$-isomerase and Δ^{14}-reductase by fenpropimorph, tridemorph and fenpropidin in cell-free enzyme systems from *Saccharomyces cerevisiae*. Phytochem. **26** (1987): 663–668.

BASARAB, G. S., LIVINGSTON, R. S., VOLLMER, S. J., JOHNSON, C. B., and KRANIS, K. T.: Design of sterol reductase inhibitors: insights into the binding conformation of tertiary amine fungicides. In: BAKER, D. R., FENYES, J. G., and STEFFENS, J. J. (Eds.): Synthesis and Chemistry of Agrochemicals III, ACS Symposium Series No. 504. American Chemical Society, Washington, D. C. 1992, pp. 414–427.

BIRDSALL, B., BEVAN, A. W., PASCUAL, C., ROBERTS, G. C. K., FEENEY, J., GRONENBORN, A., and CLORE, G. M.: Multinuclear NMR characterization of two coexisting conformational states of the *Lactobacillus casei* dihydrofolate reductase-trimethoprim-NADP$^+$ complex. Biochem. **23** (1984): 4733–4742.

BORMAN, S.: New 3-D search and de novo design techniques aid drug development. Chem. and Eng. News, Aug 10 (1992): 18–26.

BORTH, D. M., and MCKAY, R. J.: A difficulty information approach to substituent selection in QSAR studies. Technometrics **27** (1985): 25–35.

BOSCOTT, P. E., GRANT, G. H., and RICHARDS, W. G.: Modelling cytrochrome P450 14a demethylase (*Candida albicans*) from P450 Cam. Personal communication.

BOTTEMA, C. K., and PARKS, L. W.: Δ^{14}-Sterol reductase in *Saccharomyces cerevisiae*. Biochim. Biophys. Acta **531** (1978): 301–307.

BOX, G. E. P., HUNTER, W. G., and HUNTER, J. S.: Statistics for Experimenters. John Wiley & Sons, New York 1978.

BROMILOW, R. H., and CHAMBERLAIN, K.: Designing molecules for systemicity. In: Monogr.-Brit. Plant Growth Regul. Group **18** (1989): 113–128.

BURDEN, R. S., CARTER, G. A., CLARK, T., COOKE, D. T., CROKER, S. J., DEAS, A. H. B., HEDDEN, P., JAMES, C. S., and LENTON, J. R.: Comparative activity of the enantiomers of triadimenol and paclobutrazol as inhibitors of fungal growth and plant sterol and gibberellin biosynthesis. Pestic. Sci. **21** (1987): 253–267.

BURNS, L. A., CLINE, D. M., and LASSITER, R. R.: Exposure Analysis Modeling System (EXAMS). U. S. Department of Commerce, National Technical Information Service, Springfield, Virginia 1982.

CHOU, P. Y., and FASMAN, G. D.: Prediction of protein conformation. Biochem. **13** (1974): 222–245.

CHRISTIE, B. D., HENRY, D. R., GUNER, O. F., and MOOCK, T. E.: MACCS-3D: a tool for three-dimensional drug design. Online Information 90, Dec. (1990): 137–167. MACCS-3D is a software product from Molecular Design Ltd., San Leandro, California 94577.

CLORE, G. M., GRONENBORN, A. M., BIRDSALL, B., FEENEY, J., and ROBERTS, G. C. K.: ^{19}F-n.m.r. studies of 3′,5′-difluoromethotrexate binding to *Lactobacillus casei* dihydrofolate reductase. Biochem. J. **217** (1984): 659–666.

DEAS, A. H. B., CARTER, G. A., CLARK, T., CLIFFORD, D. R., and JAMES, C. S.: The enantiomeric composition of triadimenol produced during metabolism of triadimefon by fungi. Pestic. Biochem. Physiol. **26** (1986): 10–21.

DEWAR, M. J. S., ZOEBISCH, E. G., HEALY, E. F., and STEWART, J. J. P.: AM1: A new general purpose quantum mechanical molecular model. J. Am. Chem. Soc. **107** (1985): 3902–3909.

DIXON, J. S. and KUNTZ, I. D.: Using shape complementarity as an initial screen in designing ligands for a receptor binding site of known three-dimensional structure. J. Med. Chem. **31** (1988): 722–729.

EBERT, E., ECKHART, W., JÄKEL, K., MOSER, P., SOZZI, D., and VOGEL, C.: Quantitative structure activity relationships of fungicidally active triazoles: analogs and stereoisomers of propiconazole and etaconazole. Z. Naturforsch. **44 c** (1989): 85–96.

EISENBERG, D., SCHWARZ, E., KOMAROMY, M., and WALL, R.: Analysis of membrane and surface protein sequences with the hydrophobic moment plot. J. Mol. Biol. **179** (1984): 125–142.

ENSLEIN, K.: An overview of structure-activity relationships as an alternative to testing in animals for carcinogenicity, mutagenicity, dermal and eye irritation, and acute oral toxicity. Toxicol. Ind. Health **4** (1988): 479–498.

FERENCZY, G. G., and MORRIS, G. M.: The active site of cytochrome P-450 nifedipine oxidase: a model-building study. J. Mol. Graphics **7** (1989): 206–211.

FISCHER, E.: Einfluß der Configuration auf die Wirkung der Enzyme. Ber. deut. chem. Ges. **27** (1894): 2985–2993.

FISCHER, R. T., TRZASKOS, J. M., MAGOLDA, R. L., KO, S. S., BROSZ, C. S., and LARSEN, B.: Lanosterol 14 a-demethylase-isolation and characterization of the third metabolically generated oxidative demethylation intermediate. J. Biol. Chem. **266** (1991): 6124–6132.

FUJIMOTO, T. T., QUINN, J. A., EGAN, A. R., SHABER, S. H., and ROSS, R. R.: Quantitative structure-activity studies of the fungitoxic properties of phenethyl 1,2,4-triazoles. Pestic. Biochem. Physiol. **30** (1988): 199–213.

— SHABER, S. H., CHAN, H. F., QUINN, J. A., and CARLSON, G. R.: Successful exploitation of 2-cyanoarylethyltriazoles as agricultural fungicides. In: BAKER, D. R., MOBERG, W. K., and CROSS, B. (Eds): ACS Symposium Series 355, American Chemical Society, Washington, D. C. 1987, pp. 318–327.

FUJITA, T.: EMIL: Concept and Features of EMIL, a system for lead evolution of bioactive compounds. In: WERMUTH, C. G. (Ed.): Trends in QSAR and Molecular Modelling '92, ESCOM Science Publishers B. V., Leiden, 1993.

GARNIER, J., OSGUTHORPE, D. G., and ROBSON, B.: Analysis of the accuracy and implications of simple methods for predicting the secondary structure of globular proteins. J. Mol. Biol. **20** (1978): 97–120.

GISI, U., SCHAUB, R., WIEDMER, H., and UMHEL, E.: SAN 619 F, a new triazole fungicide. In: 1986 British Crop Protection Conference, Pests and Diseases, Vol. 1. BCPC Publications, Thornton Heath Surry 1986, pp. 33–40.

GRAYSON, B. T., and KLEIER, D. A.: Phloem mobility of xenobiotics IV. Modeling of pesticide movement in plants. Pestic. Sci. **30** (1990): 67–79.

GROVES, J. T., HAUSHALTER, R. C., NAKAMURA, M., NEMO, T., and EVANS, B. J.: High valent iron-porphyrin complexes related to peroxidase and cytochrome P-450. J. Am. Chem. Soc. **103** (1981): 2884–2886.

— and NEMO, T. E.: Aliphatic hydroxylation catalyzed by iron porphyrin complexes. J. Am. Soc. **105** (1983): 6243–6248.

HANSCH, C., and FUJITA, T.: A method for the correlation of biological activity and chemical structure. J. Am. Chem. Soc. **86** (1964): 1616–1626.

— and LEO, A.: Substituent Constants for Correlation Analysis in Chemistry and Biology. John Wiley & Sons, New York 1979.

HEANEY, S. P.: Population dynamics of DMI fungicide sensitivity changes in barley powdery mildew. In: DELP, C. J. (Ed.): Fungicide Resistance in North America. APS Press, St. Paul 1988, pp. 89–92.

HUXLEY-TENCER, A., FRANCOTTE, E., and BLADOCHA-MOREAU, M.: 1(R)-(2,6-cis-dimethylmorpholino)-3(S)-(p-tert-butylphenyl)cyclopentane: a representative of a novel, potent class of biorationally designed fungicides. Pestic. Sci. **34** (1992): 65–74.

JACOB, F., and NEUMANN, S.: Principles of uptake and systemic transport of fungicides within the plant. In: LYR, H. (Ed.): Modern Selective Fungicides. Gustav Fischer Verlag, Jena 1987, pp. 13–29.

KALB, V. F., WOODS, C. W., TURI, T. G., DEY, C. R., SUTTER, T. R., and LOPER, J. C.: Primary structure of the P450 lanosterol demethylase gene from *Saccharomyces cerevisiae*. DNA **6** (1987): 529–537.

KAMOSHITA, K., TAKAYAMA, C., TAKAHASHI, J., and FUJINAMI, A.: Application of the Hansch-Fujita method to the design of imide and carbamate fungicides. In: FUJITA, T., and DRABER, W. (Eds.):

Development of Agrochemicals: Structure-Activity Relationships. CRC Press, Boca-Raton 1992, pp. 429–444.
KAPTEYN, J. C., PILLMOOR, J. B., and DE WAARD, M. A.: Isolation of microsomal cytochrome-P450 isozymes from *Ustilago maydis* and their interaction with sterol demethylation inhibitors. Pestic. Sci. **34** (1992): 37–43.
KATAGI, T., MIKAMI, N., MATSUDA, T., and MIYAMOTO, J.: Application of molecular orbital calculations to estimate the active conformation of azole compounds. In: BAKER, D. R., FENYES, J. G., and MOBERG, W. K. (Eds.): Synthesis and Chemistry of Agrochemicals. ACS Symposium Series 355. American Chemical Society, Washington, D. C. 1987, pp. 340–352.
KÖLLER, W.: Sterol demethylation inhibitors: mechanism of action and resistance. In: DELP, C. J. (Ed.): Fungicide Resistance in North America. APS Press, St. Paul 1988, pp. 79–88.
– Antifungal agents with target sites in sterol functions and biosynthesis. In: KÖLLER, W. (Ed.): Target Sites of Fungicide Action. CRC Press, Boca Raton 1992, pp. 119–206.
KUNZE, K. L., MANGOLD, B. L. K., WHEELER, C., BEILAN, H. S., and ORITZ DE MONTELLANO, P. R.: The cytochrome P-450 active site: regiospecificity of prosthetic heme alkylation by olefins and acetylenes. J. Biol. Chem. **258** (1983): 4202–4207.
LEO, A.: Partitioning in pesticide mode of action and environmental problems. In: MAGEE, P. S., KOHN, G. K., and MENN, J. J. (Eds.): ACS Symposium Series 255. American Chemical Society, Washington, D. C. 1984, pp. 213–224.
LYBRAND, T. P., MCCAMMON, J. A., and WIPFF, G.: Theoretical calculation of relative binding affinity in host-guest systems. Proc. Nat. Acad. Sci. (USA) **83** (1986): 833–835.
LYMAN, W. J., REEHL, W. F., and ROSENBLATT, D. H.: Handbook of chemical and physical property estimation methods: Environmental behavior of organic compounds. McGraw-Hill, New York 1982.
MARCHINGTON, A. F., and LAMBROS, S. S.: Computer design of fungicides. In: LYR, H. (Ed.): Modern Selective Fungicides. Gustav Fischer Verlag, Jena 1987, pp. 325–336.
– The design of triazole fungicides. In: MAGEE, P. S., KOHN, G. K., and MENN, J. J. (Eds.): Presticide Synthesis through Rational Approaches. ACS Symposium Series 255. American Chemical Society, Washington, D. C. 1984, pp. 173–183.
MARSHALL, G. R., VAN OPDENBOSCH, N., and FONT, J.: Systematic search of conformational space: use and visualization. In: EMMETT, J. C. (Ed.): Proc. 2nd SCI-RSC Medicinal Chem. Symposium, Spec. Publ. – Royal Chem. Soc. (1984): 96–108.
MARSILI, M.: An expert system for chemometrics-based optimization in chemistry: Tetrahedron Comput. Methodol. **1** (1988): 71–80.
MARTIN, Y.: 3D database searching in drug design. J. Med. Chem. **35** (1992): 2145–2154.
MCFARLAND, J. W., and GANS, D. J.: On the significance of clusters in the graphical display of structure-activity data. J. Med. Chem. **29** (1986): 505–514.
MCMURRY, T. J., and GROVES, J. T.: Metalloporphyrin models for cytochrome P-450. In: ORTIZ DE MONTELLANO, P. R. (Ed.): Cytochrome P-450. Plenum Press, New York–London 1986, pp. 1–28.
MERCER, E. I.: Morpholine antifungals and their mode of action. Biochem. Soc. Trans. **19** (1991): 788–793.
MIWA, G. T., and LU, A. Y. H.: The topology of the mammalian cytochrome P-450 active site. In: ORTIZ DE MONTELLANO, P. R. (Ed.): Cytochrome P-450. Plenum Press, New York–London 1986, pp. 77–88.
MOON, J. B., and HOWE, W. J.: Computer design of bioactive molecules: a method for receptor based *de novo* ligand design. Proteins: Struct. Funct. Genet. **11** (1991): 314–328.
MORIGUCHI, I., KOMATSU, K., and MATSUSHITA, Y.: Adaptive least squares methods applied to structure-activity correlation of hypotensive N-alkyl-N″-cyano-N′-pyridylguanidines. J. Med. Chem. **23** (1980): 20–26.
MORRIS, G. M., and RICHARDS, W. G.: Molecular modeling of the sterol C-14 demethylase of *Saccharomyces cerevisiae*. Biochem. Soc. Trans. **19** (1991): 793–795.

NAKAYAMA, A., IKURA, K., KATSUURA, K., HASHIMOTO, S., and NAKATA, A.: Quantitative structure activity relationships, conformational analyses and computer graphics study of triflumizole analogs, fungicidal N-(1-imidazol-1-ylalkylidene)-anilines. J. Pestic. Sci. **14** (1989): 23–37.

NYFELER, R., and HUXLEY, P.: The application of modern methods in chemical fungicide research. In: 1986 Brit. Crop Protect. Con., Pests and Diseases, Vol. 1. BCPC Publications, Thornton Heath, Surrey 1986, pp. 207–215.

ORTIZ DE MONTELLANO, P. R. (Ed.): Cytochrome P-450. Plenum Press, New York–London 1986 a.

– Oxygen activation and transfer. In: ORTIZ DE MONTELLANO, P. R.: Cytochrome P-450. Plenum Press, New York – London 1986 b, pp. 217–271.

PAULING, L.: Molecular architecture and biological reactions. Chem. Eng. News **24** (1946): 1375–1377.

PEARLMAN, R. S., RUSINKO, A., III, SKELL, J. M., BALDUCCI, R., and McGARITY, C. M.: CONCORD, distributed by Tripos Associates, Inc., 1699 South Hanley Rd., Suite 303, St. Louis, Missouri 63944.

PLUMMER, E. L.: The application of quantitative design strategies in pesticide discovery. In: LIPKOWITZ, K. B., and BOYD, D. B. (Eds.): Reviews in Computational Chemistry. VCH Publishers, New York 1990, pp. 119–168.

POULOS, T. L.: The crystal structure of cytochrome P-450 cam. In: ORITZ DE MONTELLANO, P. R.: Cytochrome P-450. Plenum Press, New York–London 1986, pp. 505–524.

– FINZEL, B. C., and HOWARD, A. J.: High-resolution crystal structure of P-450$_{cam}$. J. Molec. Biol. **195** (1987): 687–700.

RAHIER, A., TATON, M., BOUVIER-NAVÉ, P., SCHMITT, P., BENVENISTE, P., SCHUBER, F., NARULA, A. S., CATTEL, L., ANDING, C., and PLACE, P.: Design of high energy intermediate analogues to study sterol biosynthesis in higher plants. Lipids **21** (1986): 52–62.

– – and BENVENISTE, P.: Inhibition of sterol biosynthesis enzymes *in vitro* by analogues of high-energy carbocationic intermediates. Bioch. Soc. Trans **18** (1990): 48–52.

SANDERSON, D. M., and EARNSHAW, C. G.: Computer prediction of possible toxic action from chemical structure: the DEREK system. Human Experiment. Toxicol. **10** (1991): 261–273.

SCHEINPFLUG, H.: Resistance management strategies for using DMI fungicides. In: DELP, C. J., (Ed.): Fungicide Resistance in North America. APS Press, St. Paul 1988, pp 93–94.

STEWART, J. J. P.: Optimization of parameters for semiempirical methods II. Applications. J. Comput. Chem. **10** (1989): 221–264.

SUGAVANAM, B.: Diastereoisomers and enantiomers of paclobutrazol: Their preparation and biological activity. Pestic. Sci. **15** (1984): 296–302.

TAKAHASHI, Y., SASKI, S., TAMARU, M., SHIMASAKI, I., ITO, S., KAWADA, S., and SUDA, Y.: Application of the ORMUCCS method to structure-activity studies of the fungicidal activity of mepronil derivatives. Quant. Struct.-Act. Relat. **6** (1987): 17–21.

TATON, M., BENVENISTE, P., and RAHIER, A.: Microsomal $\Delta^{8,14}$-sterol $\Delta^{8,14}$-reductase in higher plants. Characterization and inhibition by analogues of a presumptive carbocationic intermediate of the reduction reaction. Eur. J. Biochem. **185** (1989): 605–614.

TUCK, S. F., AOYAMA, Y., YOSHIDA, Y., and ORTIZ DE MONTELLANO, P. R.: Active site topology of *Saccharomyces cereviseae* lanosterol 14 a-demethylase (CYP51) and its G310D mutant (cytochrome P-450 SG1). J. Biol. Chem. **267** (1992): 13175–13179.

UCHIDA, M., and SUZUKI, T.: Affinity and mobility of fungicides in soils and rice plants. In: MIYAMOTO, J., and KEARNEY, P. C., (Eds.): IUPAC Pesticide chemistry: Human welfare and the envirnoment. Pergamon, Oxford 1982, pp. 371–376.

URCH, C. J.: Synthesis and fungicidal properties of cyclic tertiary amines. In: BAKER, D. R., FENYES, J. G., and MOBERG, W. K. (Eds.): Synthesis and Chemistry of Agrochemicals II. ACS Symposium Series No. 443. American Chemical Society, Washington, D. C. 1991, pp. 515–526.

WEINER, S. J., KOLLMAN, P. A., CASE, D. A., CHANDRA SINGH, U., GHIO, C., ALAGONA, G., PROFETA, S., and WEINER P.: A new force field for molecular mechanics simulation of nucleic acids and proteins, J. Am. Chem. Soc. **106** (1984): 765–784.

WIGGINS, T. E., and BALDWIN, B. C.: Binding of azole fungicides related to diclobutrazol to cytochrome P-450. Pestic. Sci. **15** (1984): 206–209.

WILLETT, P.: Similarity and Clustering in Chemical Information Systems. Wiley Interscience, New York 1987.

WOLD, S., SJOSTROM, M., CARLSON R., LUNDSTEDT, T., HELBERG, S., SKAGERBEERG, B., WIKSTORM, C., and OHMAN, J.: Multivariate design. Analyt. Chem. Acta **191** (1986): 17–32.

WOLOSHUK, C. P., SISLER, H. D., and DUTKY, S. R.: Mode of action of the azasteroid antibiotic 15-aza-24-methylene-D-homocholesta-8,14-dien-3 b-ol in *Ustilago maydis*. Antimicrob. Agents Chemother. **16** (1979): 81–86.

WORTHINGTON, P. A.: Synthesis and fungicidal activity of triazole tertiary alcohols. In: BAKER, D. R., FENYES, J. G., and MOBERG, W. K. (Eds.): Synthesis and Chemistry of Agrochemicals. ACS Symposium Series 355. American Chemical Society, Washington, D. C. 1987, pp. 302–317.

WUTHRICH, K.: The development of nuclear magnetic resonance spectroscopy as a technique for protein structure determination. Accounts Chem. Research **22** (1989): 36–46.

YOSHIDA, Y., and AOYAMA, Y.: Sterol 14 a-demethylase and its inhibition: structural considerations on interaction of azole antifungal agents with lanosterol 14 a-demethylase (P-450_{14DM}) of yeast. Biochem. Soc. Trans. **19** (1991): 778–782.

Chapter 25

Disease control by nonfungitoxic compounds

Hugh D. Sisler*) and Nancy N. Ragsdale**)

* Department of Botany, University of Maryland, USA
** National Agricultural Pesticide Impact Assessment Program, USDA, Washington, USA

Introduction

Nonfugitoxic disease control compounds are chemicals which have little or no effect on the rate of growth of the pathogen *in vitro* at concentrations which control pathogenic activity *in vivo*. The antifungal activity of these compounds, therefore, is based on actions which lead to inhibition of growth and reproduction of the pathogen in its parasitic phases. The compound may act directly on the pathogen to prevent it from becoming established in plant tissue or from causing disease once it has become established. On the other hand, it may affect the host-parasite interaction in such a way that host defense mechanisms kill or halt encroachment of the fungal pathogen.

Compounds which are converted to a derivative *in vivo* that is directly toxic to the pathogen do not belong in the group of compounds under consideration here. However, with systemic fungicidal compounds, disease control may involve both direct fungitoxicity and enhanced host resistance mechanisms. These combination effects are discussed later in connection with the fungicides metalaxyl and fosetyl-Al.

Nonfungitoxic disease control chemicals offer several advantages over conventional fungitoxic compounds. First, these compounds are more likely to affect target sites specific to fungi or higher plants than conventional fungicides and would therefore constitute less of an environmental hazard. Second, nonfungitoxic compounds which regulate host resistance are likely, in many cases, to be active at extremely low levels. Third, compounds that enhance host defense activity are less likely to encounter fungal resistance than are many conventional systemic fungicides.

Various aspects of plant disease control by nonfungitoxic chemicals have been discussed in other publications. The reader is referred to the following references for information or points of view that may differ in some respect from those presented here (Sisler 1977; Langcake 1981; Dekker 1983; Wade 1984).

Action on host defense systems

Higher plants have evolved passive as well as active defense systems to ward off potential pathogenic fungi and other microorganisms in the environment. Only those fungi which have developed mechanisms to overcome these defense systems are able to cause disease.

Gaps in host defense systems which allow these fungi to become established as pathogens have created the need to develop antifungal compounds for use in disease control. Thus far, the chemicals used for this purpose have been mainly fungitoxic compounds that act directly on the pathogen to prevent or eradicate infections. While this simple and straightforward approach will almost certainly remain in use, an attractive alternative is to chemically modify the host-parasite interaction so that the host defense mechanisms provide the antifungal activity needed to control the pathogen.

Concepts of host-parasite interactions as they relate to host susceptibility and resistance responses have been discussed by HEATH (1981), BUSHNELL and ROWELL (1981) and TEPPER and ANDERSON (1984). It is believed by some investigators that resistance gene products act as receptors for signal molecules (elicitors) produced by challenging microorganisms. When recognition does not occur or if the elicitor is not produced, the plant does not respond in a resistant manner. Elicitor recognition may trigger the expression of many structural genes which leads to a multitude of events that constitute a hypersensitive or resistance reaction. These events may include phytoalexin production, phenol production and lignification as well as a marked increase in chitinase, β-1,3-glucanase and proteinase-inhibitor activities (LEGRAND et al. 1991; SCHEEL et al. 1991). The consequence may be detrimental to the pathogen as well as to the higher plant cells in the local vicinity. It seems reasonable that resistance in some cases might result from triggering of only one or a few of the aforementioned types of events.

In a chemical regulation of resistance, the compound applied could induce pathogen (or elicitor) recognition, or might act as a selective activator of specific resistance events, but in the absence of the pathogen, should not produce an effect that is injurious to the host plant. The compound (nonfungitoxic disease control agent) might act by changing the conformation of the elicitor binding site (receptor), by preventing the pathogen from producing a suppressor, by blocking the binding of a suppressor to the receptor or to the elicitors, or by inducing elicitor production by the pathogen in cases where production is weak or absent. KUC (1984) has discussed the regulation of resistance by substances which systemically immunize the plant so that it responds in a resistant manner when challenged by the pathogen.

Compounds such as probenazole (SEKIZAWA and MASE 1981), 2,2-dichloro-3,3-dimethylcyclopropane carboxylic acid (LANGCAKE 1981) and isonicotinic acids (MÉTRAUX et al. 1991) which will be discussed later, are reputed to increase host resistance possibly by one of the mechanisms discussed above.

Action on pathogenic mechanisms

In addition to the effects on the pathogen that accentuate host resistance responses, nonfungitoxic compounds may act on the pathogen in a vareity of ways to block pathogenicity. They could block the induction or action of enzymes involved in penetration of the host or spread of the pathogen within the host tissue. Ruffianic acid, for example, is a compound reported to inhibit pectolytic and cellulolytic enzymes of *Fusarium* and *Verticillium* species (GROSSMAN 1968). The action of cutinase inhibitors as antipenetrants is described in a later section of this article.

Other examples of antipathogenic action are the interference with appressorial development or function as has been observed for tricyclazole, pyroquilon and other compounds described in the section on antipenetrants or with adhesion of spores to the plant surfaces as described for *Pyricularia oryzae* spores by HAMER *et al.* (1988).

Phytotoxins produced by fungal pathogens play a critical role in some cases in pathogenesis or severity of a disease. The role of toxins in disease development is firmly established in the case of the "host specific toxins" (SCHEFFER 1976), but is less clear in the case of many nonspecific fungal toxins. Nevertheless, some of the latter toxins seem necessary for infection of the host while others play a role in determining disease severity (RUDOLPH 1976). Blocking production or counteracting effects of a phytotoxin, therefore, can be a disease control mechanism. As pointed out by DEKKER (1983) there are few cases where this has been done. On the other hand, there are studies which indicate that it might not be advisable to interfere with the production or action of some phytotoxins because they may play a role in eliciting host resistance responses. For example, LANGCAKE *et al*, (1983) showed that picolinic acid, which is a nonspecific phytotoxin produced by *P. oryzae*, causes a hypersensitive response much like that produced by *P. oryzae* in rice plants treated with the sensitizing agent 2,2-dichloro-3,3-dimethylcyclopropane carboxylic acid. Moreover, tenuazonic acid, another toxin produced by *P. oryzae* is reported to elicit defense reactions in rice leaves (LEBRUN *et al.* 1984). The reactions were more intense in those varieties with a high level of general resistance to *P. oryzae* than in those with a low level.

One possible mechanism for fungal pathogens to overcome plant resistance is through metabolic detoxication of phytoalexins (VAN ETTEN 1982). A number of cases of phytoalexin degradation by plant pathogenic fungi have been reported (SISLER 1977). The extent to which this mechanism is used by fungi to overcome plant resistance is unclear; however, there are indications that detoxication of the phytoalexin pisatin is required for pathogenicity by *Nectria haematococca* on peas (VAN ETTEN 1982). More recently, direct evidence for this has been obtained by transforming nonpathogenic isolates of *haematococca* to pathogens with a gene for pisatin demethylating activity (CIUFETTI *et al.* 1988). Blocking induction or action of enzymes involved in phytoalexin degradation may prove to be a useful method of plant disease control by nonfungitoxic compounds.

In the following section, the action of several chemicals on fungal pathogenicity or host resistance will be examined in some detail.

Mode of action of various chemicals

Cutinase inhibitors

The plant epidermis and the cuticle in particular, are formidable barriers which many fungal pathogens penetrate in order to gain access to plant tissue. Penetration is believed to be accomplished by mechanical forces, enzymatic action or a combination of both. There is now conclusive evidence that infection can be prevented by compounds or agents which act specifically on the penetration process but are nonfungitoxic to growth of the pathogens *in vitro*.

Experimental control of certain diseases has been obtained by compounds which block fungal cutinase activity. Specific cutinase antiserum or nonfungitoxic concentrations of the potent cutinase inhibitor diisopropylfluorophosphate (Fig. 25.1) protect pea epicotyls from infection by *Fusarium solani* (MAITI and KOLATTUKUDY 1979) and papaya fruit from infection by *Colletotrichum gloeosporioides* (DICKMAN et al. 1982).

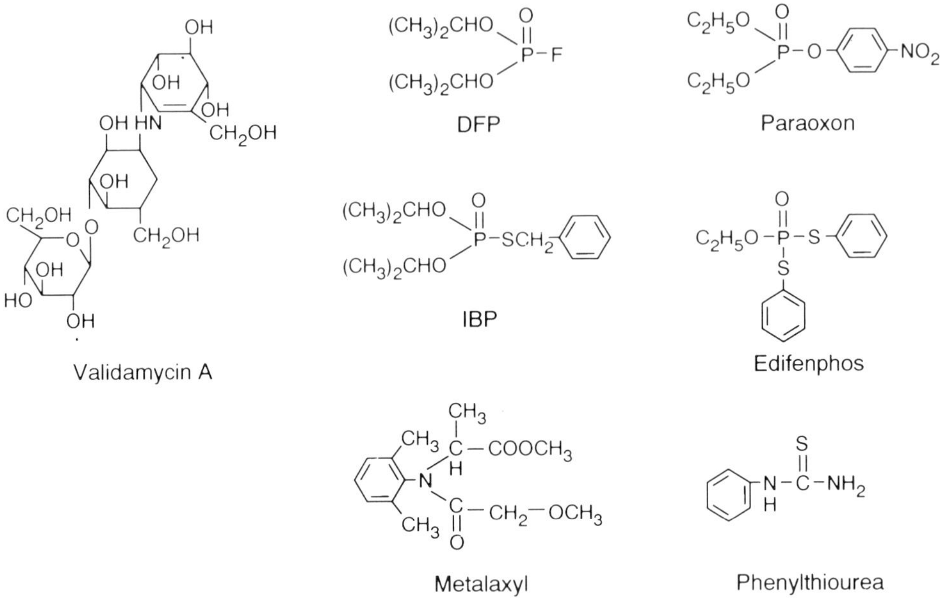

Fig. 25.1 Structures of several plant disease control chemicals discussed in the text. DFP represents diisopropylfluorophosphate, IBP represents S-benzyl-0,0-diisopropylphosphorothiolate.

Several organic phosphorus pesticides (insecticides and fungicides) inhibit cutin esterase of *F. solani* (KÖLLER et al. 1982a) and *C. gloeosporioides* (DICKMAN et al. 1983) and protect plant tissue from infection at concentrations which are not fungitoxic *in vitro*. Infection is prevented only when the tissue is not wounded, which suggests that protection results from an antipenetrant action. There is, in fact, a general degree of correspondence between effectiveness of various compounds as inhibitors of cutinase of *F. solani* and protection from infection by this pathogen. Among the highly effective compounds in both types of activity are the insecticides paraoxon (Fig. 25.1), 0,0-dimethyl-0-(2,4,5-trichlorophenyl)phosphate and 0,0-diethyl-0-(3,5,6-trichloro-2-pyridyl)-phosphate. The I_{50} for each of these 3 compounds as cutinase inhibitors is well below 1 µM while a high degree of protection against infection is afforded by a concentration of 50 nM or less. The latter two compounds are potent inhibitors of cutinase of *C. gloeosporioides*, and effectively inhibit infection of papaya fruit by this pathogen at nM concentrations.

The organic phosphorus fungicides Kitazin (IBP) and Hinosan (edifenphos) (Fig. 25.1) are appreciably less effective than the aforementioned insecticides both as cutinase inhibitors and as protectants of pea tissue from infection by *F. solani*. They do, however, give good protection at concentrations which show little or no toxicity to growth of

F. solani in vitro. IBP (S-benzyl-0,0-diisopropylphosphorothiolate), and edifenphos (0-ethyl-S,S-diphenylphosphorodithiolate) are used primarily to control rice blast disease caused by *P. oryzae*. In this organism they inhibit phospholipid N-methyltransferase (AKATSUKA et al. 1977; KODAMA et al. 1980). This mechanism is presumed to be the basis for their fungitoxicity and plant protective action. The role that cutinase inhibition and antipenetrant action plays in their protective activity against *P. oryzae* has apparently not been determined. In antipenetrant tests made with rice sheaths and Cellophane film by ARAKI and MIYAGI (1977), 10 µg/ml of IBP proved to be quite inhibitory to spore germination and appressorial formation by *P. oryzae*; however, appressorial penetration of both rice sheaths and Cellophane film was strongly inhibited by the compound at 2 µg/ml. Since the antipenetrant action in the case of Cellophane is not dependent on the inhibition of cutinase, it is doubtful that control of rice blast disease by IBP is due specifically to cutinase inhibition. It would seem that highly potent cutinase inhibitors such as paraoxon would be more effective than the fungicides edifenphos or IBP for rice blast control if antipenetrant activity based on cutinase inhibition were the primary mode of protection.

In other studies, it has been shown that the fungicide benomyl but not its fungitoxic degradation product carbendazim (MBC), protects pea stems from infection by *F. solani* (KÖLLER et al. 1982b). Benomyl breaks down to yield the transient toxicant butylisocyanate in addition to carbendazim, and since the toxicity of carbendazim to growth of *F. solani in vitro* is essentially the same as that of benomyl (KÖLLER et al. 1982b), it was suggested that inhibition of cutinase by butylisocyanate is responsible for preventing infection by *F. solani*. The authors also suggested that inhibition of cutinase by butylisocyanate may explain why benomyl is superior to MBC for the control of certain other fungal diseases.

Ebelactones A and B produced by actinomycetes, are potent inhibitors of the cutinases of *Venturia inaequalis* and *Rhizoctonia solani* (KÖLLER et al. 1990). No tests were reported for plant protective activity of these cutinase inhibitors.

Melanin biosynthesis inhibitors

Reductase inhibitors. Among the most interesting and successful nonfungitoxic disease control chemicals are melanin biosynthesis inhibitors (Fig. 25.2). While members of this group block melanin biosynthesis in a variety of Ascomycetes and imperfect fungi (TOUKOUSBALIDES and SISLER 1978; WHEELER 1983) they give practical control only of rice blast disease caused by *Pyricularia oryzae* and experimental control of certain diseases caused by *Colletotrichum* species.

Melanin biosynthesis inhibitors (MBI) are antipenetrants that act on the pathogen to prevent it from piercing the plant epidermis. The antipenetrant action of pentachlorobenzyl alcohol (PCBA) (ISHIDA et al. 1969; ARAKI and MIYAGI 1977) and of fthalide (ARAKI and MIYAGI 1977) was known before the compounds were recognized as melanin biosynthesis inhibitors (WOLOSHUK et al. 1982; YAMAGUCHI et al. 1982). MBI compounds at concentrations which control diseases, do not inhibit spore germination or appressorial formation on epidermal surfaces (WOLOSHUK and SISLER 1982; YAMAGUCHI et al. 1982; INOUE et al. 1984) or on the surface of barriers such as Cellophane (ARAKI and MIYAKI 1977; YAMAGUCHI et al. 1982), nitrocellulose (KUBO et al. 1982b) or Formvar plastic

(WOLOSHUK et al. 1983; WOLKOW et al. 1983). These compounds do, however, prevent penetration of epidermal or other barriers by appressoria of *P. oryzae* (ISHIDA et al. 1969; ARAKE and MIYAGI 1977; WOLOSHUK and SISLER 1982; CHIDA et al. 1982; OKUNO et al. 1983; INOUE et al. 1984) or by appressoria of *C. lagenarium* (KUBO et al. 1982b) and *C. lindemuthianum* (WOLKOW 1982). The fact that these compounds do not control disease if the epidermal wall is punctured (WOLOSHUK et al. 1983; WOLKOW 1982; INOUE et al. 1984) or if their application is delayed for more than 8 hr after inoculation (INOUE et al. 1984) is consistent with their specific mechanism of antipenetrant action.

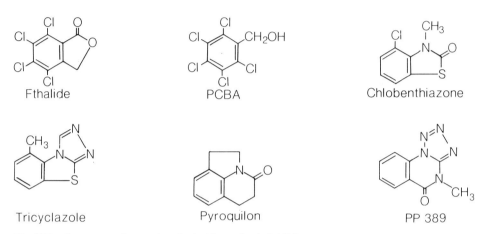

Fig. 25.2 Structures of several melanin biosynthesis inhibitors.

The antipathogenic activity of tricyclazole parallels the ability of the compound to block the polyketide pathway (Fig. 25.3) leading to melanin biosynthesis (TOKOUSBALIDES and SISLER 1978; WOLOSHUK et al. 1980; YAMAGUCHI et al. 1982). This tricyclazole sensitive pathway of melanin biosynthesis is widely distributed among Ascomycetes and imperfect fungi, but apparently does not occur in Basidiomycetes (WHEELER 1983). At concentrations which control rice blast disease, these compounds appear to affect only melanin biosynthesis in the plant pathogen, *P. oryzae*. Tricyclazole, for example, completely blocks melanin biosynthesis at 0.1 µg/ml but does not inhibit growth at concentrations as high as 20 µg/ml (TOKOUSBALIDES and SISLER 1978). Tricyclazole blocks the polyketide pathway to melanin (Fig. 25.3) between 1,3,6,8-THN and scytalone and between 1,3,8-THN and vermelone (TOKOUSBALIDES and SISLER 1979; WHEELER and STIPANOVIC 1979; WOLOSHUK et al. 1980). The inhibited steps are NADPH dependent reduction reactions (WHEELER 1982). The conversion of 1,3,8-THN to vermelone appears to be most sensitive to tricyclazole in *P. oryzae* and *V. dahliae* (TOKOUSBALIDES and SISLER 1979; WOLOSHUK et al. 1980; CHIDA and SISLER 1987b). This conversion in *C. lagenarium* is also very sensitive to tricyclazole (KUBO et al. 1985).

One may raise the question of how the blocking of melanin biosynthesis is related to antipenetrant action. *P. oryzae* as well as *C. lindemuthianum* and *C. lagenarium* develop an appressorial structure for penetrating host epidermal cell walls. There is a specific melanization of the appressorial walls of these fungi prior to penetration. In the presence

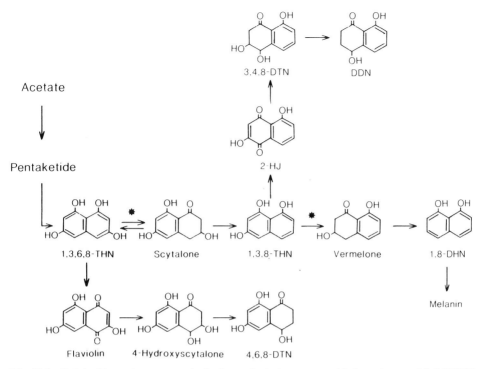

Fig. 25.3 Polyketide pathway to melanin in *pyricularia oryzae* with branches at 1,3,6,8-THN (1,3,6,8-tetrahydroxynaphthalene) and 1,3,8-THN (1,3,8-trihydroxynaphthalene). Other abbreviations indicate the following compounds: 2HJ (2-hydroxyjuglone); 3,4,8-DTN (3,4-dihydro-3,4,8-trihydroxy-1-(2H)naphthalenone); DDN (3,4-dihydro-4,8-dihydroxy-1(2H)naphthalenone); 4,6,8-DTN (3,4-dihydro-4,6,8-trihydroxy-1-(2H)naphthalenone; 1,8-DHN (1,8-dihydroxy-naphthalene). The asterisks indicate main sites of action of reductase inhibitors such as tricyclazole. Cerulenin blocks the pathway between acetate and pentaketide.

of an MBI such as tricyclazole, or if the fungus is genetically deficient in melanin biosynthesis, penetration of epidermal or other barriers does not occur. For example, buff mutants of *P. oryzae* which are essentially identical with the tricyclazole treated wild type, do not form melanized appressoria and are nonpathogenic (WOLOSHUK *et al.* 1980) because they cannot penetrate the host epidermis. Treatment of *C. lindemuthianum* (WOLKOW 1982) or *C. lagenarium* (KUBO *et al.* 1982b) with tricyclazole or pyroquilon blocks appressorial melanization and also appressorial penetration of barriers such as epidermal walls or nitrocellulose membranes. These appressoria germinate to produce hyphae which grow laterally along the barrier surface. Albino mutants of *C. lagenarium* form appressoria without melanized walls that behave essentially the same as those developed from the tricyclazole treated wild type fungus (KUBO *et al.* 1982a; 1982b). These albino mutants are not pathogenic. Two albino mutants of *Pyricularia* isolated by CHUMLEY and VALENT (1990) were tested and found to be incapable of penetrating plant epidermal barriers unless an exogenous source of a melanin precursor such as scytalone was used to restore appressorial melanization (BUSTAMAM and SISLER 1987; CHIDA and SISLER 1987a, b). This behavior is consistent with the observation of CHUMLEY and

VALENT (1990) that albino mutants of *Pyricularia* are not pathogenic on unwounded plants, but are virulent pathogens if scytalone is provided in the spore suspension used for inoculation. As with *Pyricularia*, the addition of scytalone restores melanization and barrier penetration capacity to albino mutants of *C. lagenarium* (YAMAGUCHI and KUBO 1992).

Cerulenin. The antibiotic, cerulenin, is produced by the fungus *Cephalosporium caerulens* and is moderately toxic to growth of *P. oryzae*. It shows varying degrees of toxicity to many other fungi. Cerulenin blocks the malonyl-ACP condensation reaction in fatty acid and polyketide biosynthesis by inhibiting b-ketoacyl-thioester synthetase and also blocks sterol biosynthesis by inhibiting the enzyme hydroxymethyl CoA-synthetase (OMURA 1976). Even though the antibiotic inhibits primary fungal metabolism, it is useful as a selective inhibitor of secondary polyketide metabolism in cells such as those of mature *Colletotrichum* and *Pyricularia* spores that have already synthesized a reserve of fatty acids and sterols. Cerulenin selectively inhibits pentaketide melanin biosynthesis (Fig. 25.3) and thus blocks the barrier penetration capacity of appressoria of *C. lagenarium* (KUBO et al. 1986). and *P. oryzae* (CHIDA and SISLER 1987a) with little or no effect on spore germination of appressorial formation. Appressoria formed in the presence of cerulenin resemble those produced by albino strains of *P. oryzae* and *C. lagenarium*. Melanization and penetrating capacity can be restored in these appressoria by supplying an exogenous source of scytalone, vermelone or 1,8-DHN (CHIDA and SISLER 1987a).

Even though it is well established that blocking the melanin biosynthetic pathway with reductase inhibitors or cerulenin leads to the loss of epidermal-penetration capacity of appressoria of *P. oryzae* and *Colletotrichum* species, the reason for this is not entirely clear. The following have been proposed as explanations:

(1) Melanization of the appressorial wall is necessary for the architecture and rigidity needed to support and focus the mechanical forces involved in the penetration process (WOLOSHUK et al. 1983; WOLKOW et al. 1983). Increased wall rigidity and focus might result from simple deposition of melanin beween wall fibrils or through cross-linking of wall polymers by an oxidation product of an immediate melanin precursor (SISLER et al. 1984).

(2) Melanin seals the appressorial wall of *P. oryzae* which then serves as a differential permeability barrier to water and solutes, thus allowing appressoria to establish and maintain the high internal solute concentration required to produce adequate hydrostatic pressure for penetration of epidermal barriers (HOWARD and FERRARI 1989). These investigators were able to establish that higher solute concentrations are required to plasmolyze melanized than unmelanized appressoria. WOLOSHUK et al., (1983) had earlier recognized the differential permeability of melanized appressorial walls as compared to unmelanized walls and the possibility that hydrostatic pressure differences might exist between appressoria with the two types of walls.

(3) Toxic pentaketide melanin precursors such as 2-hydroxyjuglone accumulate when reductase reactions are blocked by inhibitors in wild-type appressoria of *P. oryzae* and these toxins interfere with the barrier penetration capacity (YAMAGUCHI et al. 1983). This explanation cannot account for the lack of barrier penetrating capability of albino or cerulenin treated appressoria because they do not accumulate these toxins. Nervertheless, the fact that the penetration capacity of wild-type appressoria treated with reductase inhibitors is more fully restored by 1,8-DHN when toxin accumulation is prevented by

cerulenin (CHIDA and SISLER 1987a) suggests that accumulating toxins play a role in antipenetrant action, particulary when the reductase inhibitors are used at a concentration above the minimum effective dose.

(4) Lateral germination could adequately explain the inability of unmelanized appressoria of *Colletotrichum* species to penetrate epidermal or other barriers (YAMAGUCHI and KUBO 1992), but it has not been determined whether lateral germination is the consequence or the cause of the penetration failure. This explanation does not account for barrier penetration failures by unmelanized appressoria of *P. oryzae* because they rarely germinate in a lateral direction (CHIDA and SISLER 1987a; HOWARD and FERRARI 1989).

(5) Lack of melanization leads to inadequate adhesion of *P. oryzae* appressoria to the barrier surface (INOUE et al. 1987); however, HOWARD and FERRARI (1989) found adhesive capacity of melanized and unmelanized appressoria to be similar. It would be expected that lateral germination on barrier surfaces would be associated with adhesion failure of unmelanized appressoria of *P. oryzae* but, as mentioned above, lateral germination of unmelanized appressoria of this fungus rarely occurs.

Validamycin

Validamycin (Fig. 25.1) is a water soluble, weakly basic compound classified as an aminoglucoside antibiotic (SUAMI et al. 1980; TRINCI 1985) which is active primarily for control of diseases caused by *Rhizoctonia* type fungi (WAKAE and MATSUURA 1975). It is widely used in Japan to control sheath blight of rice caused by *Rhizoctonia solani* = (*Pellicularia sasakii*). Under poor nutritional conditions, the compound inhibits radial growth of *R. solani* on agar media. NIOH and MIZUSHIMA (1974) observed, however, that the antibiotic does not reduce the mycelial mass when the fungus is grown in liquid culture, even though it alters the morphology. TRINCI (1985) made similar observations for the action of validamycin A on growth of *Rhizoctonia cerealis*. When grown on Vogel's solid medium containing 50 mM glucose, the radial growth rate is markedly inhibited by 10 µM validamycin, but the doubling time in Vogel's liquid medium is not affected by this concentration of the antibiotic. The antibiotic induces excessive branching which results in denser colonies that expand in diameter more slowly than control colonies (NIOH and MIZUSHIMA 1974; TRINCI 1985). There are indications that validamycin A inhibits formation of a hyphal extension factor in *R. solani* (SHIBATA et al. 1982). The pathogenicity of *P. sasakii* is markedly reduced on rice plants treated with validamycin. The addition of meso-inositol partially restores pathogenicity and also prevents the morphological effect produced in *R. solani in vitro*. It was suggested that the antibiotic interferes with the biosynthesis of mesoinositol which is necessary for pathogenicity of *P. sasakii* (WAKAE and MATSUURA 1975). Meso-inositol, however, did not counteract the inhibitory effect of validamycin on radial growth of *R. cerealis in vitro* (TRINCI 1985). On the other hand, the onset of inhibition of radial growth of both *R. solani* and *R. cerealis* by 0.2 µM validamycin is delayed in direct proportion to the glucose concentration present between 1.5 and 10 mM, but glucose concentration does not affect the ultimate growth rate attained in antibiotic treated cultures. The mechanism underlying this antagonism by glucose is not understood, but it might be explained by observations that validamycins are potent inhibitors of trehalase activity in *R. solani* and other fungi

(ASANO et al. 1987; SHIGEMOTO et al. 1989). In nutritionally poor environments, blocking the action of trehalase on the storage sugar, trehalose, may deprive the fungus of a source of glucose needed for hyphal tip extension and infection cushion formation. This type of action would be consistent with the suggestion of TRINCI (1985) that validamycins may control disease by preventing penetration of the host by the pathogen and by reducing the rate of spread so that enough time is allowed for the host to mobilize its defense system.

Probenazole

Probenazole (Oryzemate), a relative of saccharin, is a systematic compound that controls rice blast disease caused by *P. oryzae* and bacterial leaf blight caused by *Xanthomonas oryzae*. Probenazole (Fig. 25.4) is ordinarily applied as a subermerged treatment to the

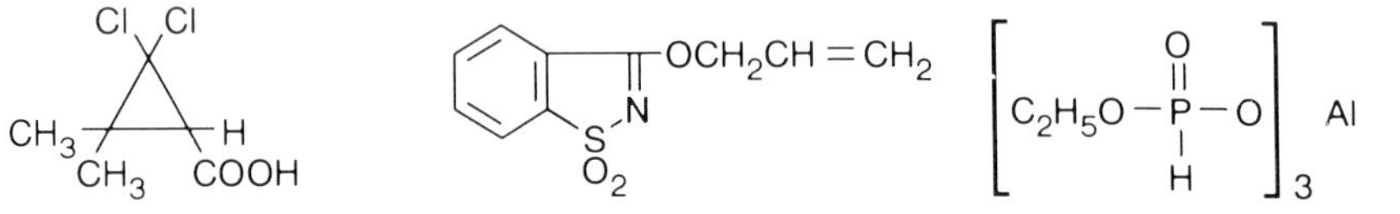

2,2-Dichloro-3,3-dimethyl cyclopropane carboxylic acid

Probenazole

Aluminium tris-O-ethyl phosphonate (fosetyl-Al)

Fig. 25.4 Structures of three compounds reported to increase host resistance responses.

roots of rice plants. The compound does not show appreciable toxicity to hyphal growth of *P. oryzae* on agar media or to conidial germination on glass slides; however, conidial germination on rice leaf sheaths as well as appressorial formation and penetration are quite sensitive to the compound (WATANABE 1977). While these observations suggest that probenazole may act prior to penetration, the compound is known to suppress lesion spread after the fungus has penetrated into the plant tissue (WATANABE 1977). Evidence indicates that root application of probenazole leads to blast control through enhancement of the resistance response of the rice plant rather than through direct fungitoxicity of the applied chemical (WATANABE et al. 1979; SEKIZAWA and MASE 1981). Analysis of probanazole treated plants inoculated with *P. oryzae* reveals an accumulation of a-linolenic acid and three other fungitoxicants with similar properties. These toxicants are believed to form a chemical barrier to the invading pathogen. There is also an augmentation of peroxidase, phenylalanine ammonia lyase and catechol-O-methyltransferase in treated plants inoculated with *P. oryzae* which is believed to facilitate the formation of a lignoid barrier around invaded cells (SEKIZAWA and MASE 1981).

Isonicotinic acid

The experimental disease control compounds 2,6-dichloro-isonicotinic acid (CGA 41396) and its methyl ester derivative (CGA 41397) (Fig. 25.5) give local and systemic protection of cucumber plants from infection by *Colletotrichum lagenarium* (MÉTRAUX et al. 1991).

Protection cannot be explained on the basis of direct fungitoxicity of these compounds or of fungitoxic conversion products, but appears to be due to a modification of host plant physiology that leads to disease resistance (MÉTRAUX et al. 1991). This is indicated by the strong induction of the antifungal enzyme chitinase in cucumber plants following treatment with CGA 41396 or CGA 41397. The local and systematic protection produced resembles that resulting from localized infections with necrotrophic pathogens described by KUC (1984). The compounds also protect cucumber from infection by the bacterial pathogen, *Pseudomonas lachrymans*.

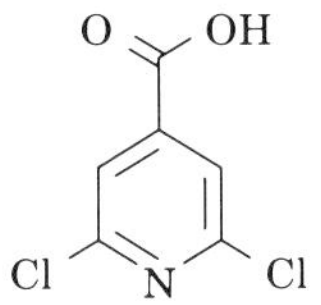

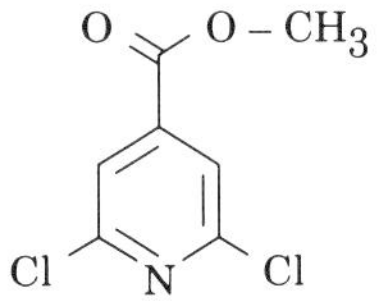

Fig. 25.5 2,6-dichloroisonicotinic acid (CGA 41396) and its methyl ester derivative (CGA 41397).

Activity of these chemicals is not limited to control of cucumber diseases. For example, treatment with CGA 41396 gives protection from diseases caused by *Erwinia amylovora* on pear, *Xanthomonas vesicatoria* on pepper, *Peronospora tabacina* on tobacco and *Pyricularia oryzae* as well as *Xanthomonas oryzae* on rice. Isonicotinic acid derivatives resemble probenazole in regard to their capacity to control a fungal as well as a bacterial disease of rice. WARD et al. (1991) showed that the onset of systemic acquired resistance (SAR) induced in tobacco by the methyl ester of 2,6-dichloro-isonicotinic acid or salicylic acid is correlated with the expression of a number of resistance genes. Salicylic acid is believed to be an endogenous signal compound involved in triggering expression of SAR genes.

A recent publication (MORTON and NYFELER 1993) indicates that phytotoxicity of foliar sprays preclude the commercial development of the insonicotinic acid derivatives as disease control agents. While the potential of compounds of this type to control a broad spectrum of pathogens on many plant species is very exciting, the extent to which prohibitive phytotoxicity will limit their practical use remains an interesting question.

DDCC

At nonfungitoxic concentrations, the compound 2,2-dichloro-3,3-dimethylcyclopropane carboxylic acid (DDCC) specifically controls rice blast disease caused by *P. oryzae* (LANGCAKE and WICKINS 1975). While it is an interesting experimental chemical, DDCC (Fig. 25.4) is not used for the practical control of rice blast disease. In contrast to tricyclazole, DDCC does not prevent epidermal penetration by *P. oryzae*, but promotes a markedly enhanced host response once penetration has occurred (CARTWRIGHT et al. 1980). Rice plants treated with DDCC respond in a rapid hypersensitive fashion to penetration by *P. oryzae* whereas untreated plants respond in a delayed and mild fashion. There is a rapid development of intracellular hyphae in cells of untreated plants, but hyphal development is halted soon after penetration in cells of treated plants (CART-

WRIGHT et al. 1980). The suppression of hyphal development in DDCC treated plants apparently results from the accumulation of the fungitoxic substances, momilactones A and B, in the tissue surrounding the invasion site. The accumulation of these phytoalexins is more rapid and far greater in magnitude in leaves of DDCC treated plants than in leaves of untreated plants.

The data concerning control of rice blast disease by DDCC clearly indicate that host defense systems are the source of antifungal substances which halt the invasion of *P. oryzae*. However, the mechanism leading to the DDCC enhancement of the resistance response is unknown. There are several lines of evidence to suggest that the primary target of DDCC action is in the host plant rather than in the pathogen (LANGCAKE et al. 1983). First, picolinic acid applied to wounds on DDCC treated plants causes an intense hypersensitive type reaction but a much less intensive response when applied to wounds on untreated plants. Picolinic acid is reputed to be a phytotoxin produced by *P. oryzae*. Second, greater momilactone accumulation occurs in response to mycelial extracts of *P. oryzae* in treated plants than in untreated. This suggests that prior interaction with the plant is necessary for DDCC activity (LANGCAKE et al. 1983)

The DDCC-*P. oryzae*-rice plant system is an interesting experimental model which can contribute further to an understanding of the chemical regulation of host/parasite interactions. Perhaps the most critical bit of information needed in this systems is the identity of the primary target of DDCC which triggers the enhanced resistance response.

Phenylthiourea

Phenylthiourea (PTU) prevents disease development in certain plants at concentrations exhibiting little or no fungitoxicity *in vitro*. The compound (Fig. 25.1), however, is primarily of research interest since it has not been adopted for practical use.

When cucumber seedlings are allowed to take up PTU by root absorption from solutions containing 50 µg/ml, they are protected from infection by *Cladosporium cucumerinum* (KAARS SIJPESTEIJN 1969). The sap of these seedlings contains 10 to 20 µg/ml of PTU, but is not toxic to growth of the pathogen *in vitro*. Lignification is enhanced in PTU treated plants around sites of penetration of *C. cucumerinum*, and this is believed to be the basis of protection (KAARS SIJPESTEIJN and SISLER 1968; KAARS SIJPESTEIJN 1969). PTU is a potent inhibitor of polyphenol oxidase (tyrosinase) activity. According to the hypothesis of KAARS SIJPESTEIJN (1969), PTU inhibits polyphenol oxidase in the plant tissue as well as in the invading pathogen. As a consequence, phenolic precursors of lignin accumulate and are rapidly converted to lignin by the elevated levels of peroxidase found in the PTU treated tissue.

PTU controls rice blast disease when used at 100 to 200 µg/ml. The ED_{95} values of PTU for inhibition of mycelial growth, spore germination and appressorial formation by the pathogen, *P. oryzae*, are greater than 1,000 µg/ml. PTU acts as an antipenetrant toward *P. oryzae* on both Cellophane film and rice sheaths (ARAKI and MIYAGI 1977). The compound inhibits penetration of Cellophane film by both appressoria and hyphae at 25 µg/ml. This is in contrast to the action of fthalide (Rabcide) which results only in the inhibition of appressorial penetration. These results indicate that PTU can act directly on the pathogen as an antipenetrant.

Fosetyl-Al

Aluminium tris-O-ethylphosphonate (fosetyl-Al) known as Aliette is a systemic, organic phosphite compound used to control certain diseases caused by Peronosporales. The compound (Fig. 25.4) is degraded in buffer or plant tissue to phosphorous acid (BOMPEIX et al. 1980), a product on which protective activity is apparently based. Fosetyl-Al controls plant diseases at concentrations which have little or no effect on growth of the pathogen *in vitro* under certain conditions. The low *in vitro* toxicity has been interpreted as an indication that antifungal activity of the compound is indirect and mediated through host defense systems (BOMPEIX et al. 1980).

Fosetyl-Al treatment enhances phenol accumulation and produces necrotic blocking (defense) reactions around infection sites on tomato leaves inoculated with *Phytophthora capsici* (BOMPEIX et al. 1980; DURAND and SALLÉ 1981). Induction of these defense reactions is blocked by phosphate ions. Fosetyl-Al also enhances production of antifungal stilbenes and flavonoids and reduces disease symptoms in grape leaves inoculated with *Plasmopara viticola* (LANGCAKE 1981). These observations, in addition to its low toxicity to mycelial growth *in vitro*, support the idea that fosetyl-Al acts by stimulating the natural defense system of the host plant.

FARIH et al. (1981) have observed, however, that fosetyl-Al at a concentration of 10 µg/ml strongly inhibits sporangial formation or zoospore release in some *Phytophthora* species. Moreover, a study by FENN and COFFEY (1984) indicates that a direct fungitoxic effect of fosetyl-Al (or of HPO_3) *in vivo* is involved in plant protective activity. The fungitoxicity of fosetyl-Al is affected by the amount of phosphate present (BOMPEIX et al. 1980; FENN and COFFEY 1984) and probably also by the rate of release of H_3PO_3. The high level of phosphate present in culture medium used for *in vitro* tests and possibly a more rapid conversion of the compound to H_3PO_3 *in vivo* than *in vitro* may have led some investigators to underestimate the potential direct fungitoxicity of fosetyl-Al *in vivo*.

FENN and COFFEY (1984) found H_3PO_3 is quite fungitoxic *in vitro* to *Phytophthora cinnamomi*, being appreciably more so than fosetyl-Al. The EC_{50} values of these compounds for this fungus growing on low phosphate medium were 0.05 PO_3 milliequivalents of H_3PO_3 (4 µg/ml) and 0.45 PO_3 milliequivalents of fosetyl-Al (54 µg/ml), FENN and COFFEY (1984) found similar *in vitro* and *in vivo* activity for H_3PO_3, but lower activity for fosetyl-Al *in vitro* than *in vivo*. However, efficacy of the two compounds *in vivo* was similar for control of seedling root rot of *Persea indica* caused by *P. cinnamomi*.

The observations of FENN and COFFEY (1984) clearly indicate that plant protection by fosetyl-Al and H_3PO_3 is initiated by a direct interaction between the chemicals and the pathogen, a conclusion that is supported by more recent studies (DOLAN and COFFEY 1988; FENN and COFFEY 1989; SMILLIE et al. 1989). However, the protection that results from this interaction is believed to be due to a combination of direct fungitoxicity of the chemicals and an enhanced response of the host resistance system to the fosetyl-Al or H_3PO_3 modified pathogen (AFEK and SZTEJNBERG 1989; DERCKS and CREASY 1989; GUEST and BOMPEIX 1990; NEMESTOTHY and GUEST 1990; SAINDRENAN et al. 1988; SMILLIE et al. 1989). The mode of action of these phosphonate inhibitors on *Phytophthora* pathogens appears to produce changes in the fungi that are particularly favorable for enhancement of host resistance reactions. Even low phosphite concentrations that do not reduce growth rate of *Phytophthora palmivora* in culture lead to changes in lipid and cell wall composition that possibly cause reduced virulence of the pathogen (DUSTIN et al. 1990). At these low inhibitor concentrations, there is a decrease of macromolecular materials that can be washed from the surface of mycelia grown on agar plates. These materials may contain suppressors that normally act to delay expression of host defense reactions (GRANT et al.). Various aspects of enhanced host resistance resulting from the action of phosphonates on fungal pathogens are discussed by GRIFFITH et al. (1992).

Metalaxyl

Metalaxyl, an acylalanine fungicide (Fig. 25.1) is an important control agent for pathogenic fungi in the order Peronosporales, such as soil-borne *Pythium* and *Phytophthora* spp., downy mildews and potato late blight (DAVIDSE and DE WAARD 1984). Metalaxyl interferes with a template-bound RNA polymerase (DAVIDSE 1984) and is quite toxic *in vitro* to the aforementioned types of fungi. Nevertheless, the antifungal activity of metalaxyl in plants sometimes appears accentuated, promoting the thought that the fungicide also affects pathogenicity or resistance mechanisms (WADE 1984). This idea is supported by studies showing that potato tubers treated with metalaxyl are resistant to decay caused by *Fusarium sambucinum* and *Alternaria solani*, fungi to which metalaxyl shows no direct toxicity (BARAK and EDGINGTON 1983).

Metalaxyl promoted glyceollin production in soybean seedlings infected with a compatible race of *Phytophthora megasperma*. The phenomenon was believed to contribute to disease control (WARD et al. 1980). Further study (STÖSSEL et al. 1982) indicated that metalaxyl acted directly on the fungus and not on the host and that necrosis and enhanced glyceollin production in plant tissue are secondary effects, possibly due to release of elicitors from metalaxyl damaged hyphae, failure to produce a suppressor or a reduction in growth rate thus allowing host cells enough time to develop a resistance response. Another investigation (LAZAROVITS and WARD 1982) led to the the conclusion that metalaxyl alone and not glyceollin restricted spread of the pathogen in tissue, even though appreciable accumulation of the latter had occurred. BÖRNER et al. (1983), working also with a compatible race of *P. megasperma*, found that while metalaxyl did not increase glyceollin production per treated hypocotyl, it did affect glyceollin distribution in the tissue so that very high levels accumulated around infection sites. BÖRNER et al. (1983) suggested that an increased release of elicitor from fungal cell walls together with growth inhibition of hyphae can explain high and localized glyceollin accumulation. Thus, it seems that metalaxyl may increase host resistance through its toxic action on a compatible fungal pathogen within plant tissue. WARD (1984) provided further evidence for this. Using glyphosate to block the shikimic acid pathway in tissue infected with a compatible race of *P. megasperma*, the effectiveness of metalaxyl was reduced when it was present in the tissue at concentrations marginally inhibitory to the fungus. In other studies, CAHILL and WARD (1989) showed that metalaxyl treatment resulted in an increase of elicitor activity in culture fluids of a metalaxyl sensitive isolate of *P. megasperma* while HWANG and SUNG (1989) showed that production of the phytoalexin, capsidiol, was increased when pepper plants infected with *Phytophthora capsici* were treated with metalaxyl. In the latter study, metalaxyl did not elicit capsidiol production in uninfected plants which indicates that enhanced production is mediated through the action of the fungicide on the pathogen.

Miscellaneous compounds

There is an extensive literature concerning the effect of plant hormones, growth regulator chemicals and various metabolites such as amino acids on disease development, but space permits only a few comments to be made about such compounds. This subject has been discussed by DEKKER (1983).

The use of plant hormones and plant growth regulators has been studied extensively, particularly in respect to the control or reduction of severity of vascular wilt diseases caused by *Fusarium*, *Verticillum* and *Ceratocystis* species. The literature on this subject has been reviewed by ERWIN (1977). Among the mechanisms proposed to explain vascular wilt control by plant hormones or growth regulators are restriction of propagule spread and increased levels of antifungal compounds in the infected tissue. In no case is the molecular basis of the disease control mechanism known. Some of the most promising results have been obtained with growth retardants exhibiting antigibberellin activity. For example, a marked reduction of *Verticillium* propagules occurred in cotton plants treated with 2-chloroethyl trimethylammonium chloride (CCC). Yield of cotton seed and lint was increased 16 and 12 percent by 10 and 25 g/ha doses of CCC respectively. Yields were unaffected by 50 g/ha and were reduced 6 percent by 75 g/ha (ERWIN 1977). In another case, promising results were obtained by using the non-bactericidal growth retardant 3,5-dichlorophenoxy acetic acid for control of potato scab caused by *Streptomyces scabies* (MCINTOSH et al. 1968).

When growth regulators affect plant growth but do not reduce crop yield, there may be promise for their use in disease control; however, as DEKKER (1983) has pointed out, limited disease control and adverse side effects produced by chemicals of this type greatly restrict their value as disease control agents. One practical application of a plant growth regulator for disease control is the use of the isopropyl ester of 2,4-dichlorophenoxyacetic acid to delay senescence and thereby maintain tissue resistance of citrus fruits to stem-end infection by *Alternaria* (ECKERT 1977).

Because of highly specific associations with the host, a modification of host or pathogen physiology might be expected to succeed more often for control of obligate parasites than for other types of fungal pathogens. A number of cases of experimental control of powdery mildews and rusts have, in fact, been described. The growth regulator kinetin, for example, has proven to be active against powdery mildews of cucumber (DEKKER 1963) and tobacco (COLE and FERNANDES 1970). Other natural metabolites such as methionine control powdery mildew on cucumber (DEKKER 1969) and barley (AKUTSU 1977). Since folic acid reversed methionine activity toward powdery mildew of cucumber, DEKKER (1969) suggested that amino acid interfered with folic acid metabolism. In another case, soybean lecithin controlled strawberry and cucumber powdery mildew but was not toxic to the fungi *in vitro* (MISATO et al. 1977). Disease control was believed to result from an action of lecithin on the host plants.

Synthetic chemicals with no apparent fungitoxicity also control powdery mildews and rusts. One of these is the compound Indar (4-n-butyl-1,2,4-triazole) which is specific for control only of brown rust of wheat caused by *Puccinia recondita* (VON MEYER et al. 1970). The mechanism of action of this compound remains obscure, although WATKINS et al. (1977) suggest from ultrastructural studies that the compound acts on the pathogen rather than through defense mechanisms of the host.

Treatment of tomato or eggplant seedlings with dinitroaniline herbicides results in markedly increased resistance to vascular wilts caused by *Fusarium* and *Verticillium* species (GRINSTEIN et al. 1984). Tomato plants susceptible to *Fusarium oxysporium* f. sp. *lycopersici* when pretreated with the dinitroaniline herbicide, trifluralin, and inoculated with the aforementioned pathogen, accumulate fungitoxic compounds. These toxins do not accumulate in comparable plants not pretreated with trifluralin. However, toxins do accumulate in inoculated, monogenic resistant plants without treatment with trifluralin. This herbicide is regarded as a sensitizer which conditions the plant to produce toxins, which are probably phytoalexins, upon challenge by the pathogen (GRINSTEIN et al. 1984). It thus appears to correct some deficiency in the host/parasite interaction necessary for a resistance response by the host. In this respect, the action of trifluralin resembles that of a dichlorocyclopropane carboxylic acid in the rice/*P. oryzae* system (LANGCAKE 1981).

Summary and Conclusions

Although practical control of fungal diseases by nonfungitoxic compounds has been achieved in several instances, the area still remains one with few successes. Among the successful compounds are the melanin biosynthesis inhibitors, which directly block pathogenicity of *P. oryzae* and probenazole, which enhances the resistance response in the rice plant.

Validamycin, like the melanin biosynthesis inhibitors, also acts on the pathogen, but its success may depend in part on host resistance responses.

Disease control by probenazole and isonicotinic acids is mediated through host defense systems which actually produce the antifungal action. The isonicotinic acids appear to act as signal molecules for the activation of plant disease resistance genes.

In the case of fosetyl-Al and H_3PO_3, a question remains concerning the role of direct fungitoxicity and host resistance responses in the ultimate antifungal action. Disease control obtained with these compounds seems to be due to an enhanced host resistance response resulting from a mild fungitoxic action of the chemicals on the pathogen.

The operation of a disease control mechanism of this type occurs in metalaxyl treated soybean tissue infected with *P. megasperma*. In this case, the contribution of the host resistance system is probably important only at marginally toxic concentrations of the fungicide. The extent to which the fungitoxic action of various systemic fungicides is supported by enhanced host resistance mechanisms remains an interesting question.

There are good prospects that chemical regulation of host resistance or of fungal pathogenicity by nonfungitoxic compounds will increase in significance as mechanisms of disease control. Progress in this area is presently hinderd by lack of fundamental knowledge concerning factors involved in fungal pathogenicity and host parasite interactions which lead to resistance or susceptibility.

References

AFEK, U., and SZTEJNBERG, A.: Effects of fosetyl-Al and phosphorus acid on scoparone, a phytoalexin associated with resistance of citrus to *Phytophthora citrophthora*. Phytopathology **79** (1989): 736–739.

AKATSUKA, T., KODAMA, O., and YAMADA, H.: A novel mode of action of Kitazin P in *Pyricularia oryzae*. Agric. Biol. Chem. **41** (1977): 2111–2112.

AKUTSU, K., AMANO, K., and OGASAWARA, N.: Inhibitory action of methionine upon barley powdery mildew (*Erysiphe graminis* f. sp. *hordei*). I. microscopic observation of development of the fungus on barley leaves treated with mehionine. Ann. Phytopath. Soc. Japan. **43** (1977): 33–39.

ARAKE, F., and MIYAGI, Y.: Effects of fungicides on penetration by *Pyricularia oryzae* as evaluated by an improved cellophane method. J. Pesticide Sci. **2** (1977): 457–461.

ASANO, N., YAMAGUCHI, T., KAMEDA, Y., and MATSUI, K.: Effect of validamycins on glycohydrolases of *Rhizoctonia solani*. J. Antibiotics **40** (1987): 526–532.

BARAK, E., and EDGINGTON, L. V.: Bioactivity of the fungicide metalaxyl in potato tubers against *Phytophthora infestans* and other fungi. Can. J. Plant Pathol. **5** (1983): 200.

BÖRNER, H., SCHATZ, G., and GISEBACH, H.: Influence of the systemic fungicide metalaxyl on glyceollin accumulation in soybean infected with *Phytophthora megasperma* f. sp. *glycinea*. Physiol. Plant Pathol. **23** (1983): 145–152.

BOMPEIX, G., RAVISÉ, A., RAYNAL, G., FETTOUCHE, F., and DURAND, M. C.: Modalités de l'obtention des nécroses bloquantes sur feuilles détachees de tomate par l'action du tris-O-éthyl phosphonate d'aluminium (Phoséthyl d'aluminium) hypothéses sur son mode d'action *in vivo*. Ann. Phytopathol. **12** (1980): 337–351.

BUSHNELL, W. R., and ROWELL, J. B.: Suppressors of defense reactions; a model for roles in specificity. Phytopathology **71** (1981): 1012–1014.

BUSTAMAM, M., and SISLER, H. D.: Effect of pentachloronitrobenzene, pentachloroaniline and albinism on epidermal penetration by appressoria of *Pyricularia*. Pestic. Biochem. Physiol. **28** (1987): 29–37.

CARTWRIGHT, D. W., LANGCAKE, P., and RIDE, J. P.: Phytoalexin production in rice and its enhancement by a dichlorocyclopropane fungicide. Physiol. Plant. Pathol. **17** (1980): 259–267.

CAHILL, D. M., and WARD, E. W. B.: Effects of metalaxyl on elicitor activity, stimulation of glyceollin production and growth of sensitive and tolerant isolates of *Phytophthora megasperma* f. sp. *glycinea*. Physiol. Mol. Plant Pathol. **35** (1989): 97–112.

CHIDA, T., and SISLER, H. D.: Restoration of appressorial penetration ability by melanin precursors in *Pyricularia oryzae* treated with antipenetrants and in melanin-deficient mutants. J. Pestic. Sci. **12** (1987a): 49–55.

– – Effect of inhibitors of melanin biosynthesis on appressorial penetration and reductive reactions in *Pyricularia oryzae* and *Pyricularia grisea*. Pestic. Biochem. Physiol. **29** (1987b): 244–251.

– UEKITA, T., SATAKE, K., HIRANO, K., AOKI, K., and NOGUCHI, T.: Effect of fthalide on infection process of *Pyricularia oryzae* with special observation of penetration site of appressoria. Ann. Phytopath. Soc. Japan. **48** (1982): 58–62.

CHUMLEY, F. G., and VALENT, B.: Genetic analysis of melanin-deficient, nonpathogenic mutants of *Magnoporthe grisea*. Mol. Plant-Microbe Interactions **3** (1990): 135–143.

CIUFETTI, L. M., WELTRING, K. M., TURGEON, B. G., YODER, O. C., and VAN ETTEN, H. D.: Transformation of *Nectria hematococca* with a gene for pisatin demethylating activity and the role of pisatin detoxification in virulence. J. Cellular Biochem., Suppl. **12 C** (1988): 278.

COLE, J. C., and FERNANDES, D. L.: Changes in the resistance of tobacco leaf to *Erysiphe cichoracearum* D.C. induced by topping, cytokinins and antibiotics. Ann. Appl. Biol. **66** (1970): 239–243.

DARVAS, J. M., TOERIEN, J. C., and MILNE, C. L.: Control of avocado root rot by trunk injection with phosethyl-Al. Plant Dis. **68** (1984): 691–693.

DAVIDSE, L. C.: Antifungal activity of acylalanine fungicides and related chloroacetanilide herbicides. In: TRINCI, A. P. J., and RILEY, J. F. (Eds.): Mode of Action of Antifungal Agents. Cambridge Univ. Press, London 1984, pp. 239–255.

— and DE WAARD, M. A.: Systemic fungicides. Adv. Plant. Pathology **2** (1984): 191–257.

DEKKER, J.: Effect of kinetin on powdery mildew. Nature (London) **197** (1963): 1027–1028.

— L-Methionine induced inhibition of powdery mildew and its reversal by folic acid. Netherlands J. Plant Pathol. **75** (1969): 182–185.

— Non-fungicidal compounds which prevent disease development. In: Plant Protection for Human Welfare. Proc. 10th Int. Congr. Plant Protect. **1** (1983): 237–248.

DERECKS, W., and CREASY, L. L.: Influence of fosetyl-Al on phytoalexin accumulation in the *Plasmopara viticola*-grapevine interaction. Physiol. Mol. Plant Pathol. **34** (1989): 203–213.

DICKMAN, M. B., PATIL, S. S., and KOLATTUKUDY, P. E.: Purification characterization and role in infection of an extracellular cutinolytic enzyme from *Colletotrichum gloeosporioides*. Penz. on *Carica papaya* L. Physiol. Plant. Pathol. **20** (1982): 333–347.

— — Effects of organophosphorous pesticides on cutinase activity and infection of papayas by *Colletotrichum gloeosporioides*. Phytopathology **73** (1983): 1209–1214.

DOLAN, T. E., and COFFEY, M. D.: Correlative *in vitro* and *in vivo* behavior of mutant strains of *Phytophthora palmivora* expressing different resistances to phosphorous acid and fosetyl-Na. Phytopathology **78** (1988): 974–978.

DURAND, M. C., and SALLÉ, G.: Effet du tris-O-éthyl phosphonate d'aluminium sur le couple *Lycopersicum esculentum* Mill. – *Phytophthora capsici* Leon. Etude cytologique et cytochimique Agronomie **9** (1981): 723–732.

DUSTIN, R. H., SMILLIE, R. H., and GRANT, B. R.: The effect of sub-toxic levels of phosphonate on the metabolism and potential virulence factors of *Phytophthora palmivora*. Physiol. Mol. Plant Pathol. **36** (1990): 205–220.

ECKERT, J. W.: Control of postharvest diseases. In: SIEGEL, M. R., and SISLER, H. D. (Eds.): Antifungal Compounds. V. 1. Marcel Dekker, New York, 1977, pp. 269–352.

ERWIN, D.C.: Control of vascular pathogens. In: SIEGEL, M. R., and SISLER, H. D. (Eds.): Antifungal Compounds. V. 1. Marcel Dekker, New York 1977, pp. 163–224.

FARIH, A., TSAO, P. H., and MENGE, J. A.: Fungitoxic activity of efosite aluminium on growth, sporulation and germination of *Phytophthora parasitica* and *P. citrophthora*. Phytopathology **71** (1981): 934–936.

FENN, M. E., and COFFEY, M. D.: Studies on the *in vitro* and *in vivo* antifungal activity of fosetyl-Al and phosphorus acid. Phytopathology **74** (1984): 606–611.

— — Quantification of phosphonate and ethyl phosphonate in tobacco and tomato tissues and significance for the mode of action of two phosphonate fungicides. Phytopathology **79** (1989): 76–82.

GRANT, B. R., DUSTAN, R. H., GRIFFITH, J. M., NIERE, J. O., and SMILLIE, R. H.: The mechanism of phosphonic (phosphorous) acid action in *Phytophthora*. Australasian Plant Pathol. **19** (1990): 115–121.

GRIFFITH, J. M., DAVIS, A. J., and GRAND, B. R.: Target sites of fungicides to control Oomycetes. In: KÖLLER, W. (Ed.): Target Sites of Fungicide Action. CRC Press, Boca Raton 1992, pp. 69–100.

GRINSTEIN, A., LISKER, N., KATAN, J., and ESHEL, Y.: Herbicide induced resistance to plant wilt diseases. Physiol. Plant Pathol. **24** (1984): 347–356.

GROSSMANN, F.: Studies on the therapeutic effects of pectolytic enzyme inhibitors. Netherland J. Plant Pathol. **74**, Supplement 1 (1968): 91–103.

GUEST, D. I., and BOMPEIX, G.: The complex mode of action of phosphonates. Australasian Plant Pathol. **19** (1990): 113–115.

HAMER, J. E., HOWARD, R. J., CHUMLEY, F. G., and VALENT, B.: A mechanism for surface attachment in spores of a plant pathogenic fungus. Science **239** (1988): 288–290.

HEATH, M. C.: A generalized concept of hos-parasite specificity. Phytopathology **71** (1981): 1121–1123.

HOWARD, R. J., and FERRARI, M. A.: Role of melanin in appressorium function. Exp. Mycol. **13** (1989): 403–418.

HWANG, B. K., and SUNG, N. K.: Effect of metalaxyl on capsidiol production in stems of pepper plants infected with *Phytophthora capsici*. Plant Disease **73** (1989): 748–751.

INOUE, S., KATO, T., JORDON, V. W. L., and BRENT, K.: Inhibition of appressorial adhesion of *Pyricularia oryzae* to barley leaves by fungicides. Pestic. Sci. **19** (1987): 145–152.

– UEMATSU, T., and KATO, T.: Effects of chlobenthiazone on the infection process by *Pyricularia oryzae*. J. Pesticide Sci. **9** (1984): 689–695.

ISHIDA, M., SUMI, H., and OKU, H.: Pentachlorobenzyl alcohol, a rice blast control agent. Residue Rev. **25** (1969): 139–148.

KAARS SIJPESTEIJN, A.: Mode of action of phenylthiourea, a therapeutic agent for cucumber scab. J. Sci. Fd. Agric. **20** (1969): 403–405.

– and SISLER, H. D.: Studies on the mode of action of phenylthiourea, a chemotherapeutant for cucumber scab. Neth. J. Plant Pathol. **74** (1968), Supplement 1: 121–126.

KODAMA, O., YAMASHITA, K., and AKATSUKA, T.: Edifenphos, inhibitor of phosphatidylcholine biosynthesis in *Pyricularia oryzae*. Agric. Biol. Chem. **44** (1980): 1015–1021.

KÖLLER, W., ALLAN, C. R., and KOLATTUKUDY, P. E.: Protection of *Pimus sativum* from *Fusarium solani* f. sp. *pisi* by inhibition of cutinase with organophosphorus pesticides. Phytopathology **72** (1982a): 1425–1430.

– – – Inhibition of cutinase and prevention of fungal penetration into plants by benomyl-A possible protective mode of action. Pestic. Biochem. Physiol. **18** (1982b): 15–25.

– TRAIL, F., and PARKER, D. M.: Ebelactones inhibit cutinases produced by fungal plant pathogens. J. Antibiotics **43** (1990): 734–735.

KUBO, Y., KATOH, M., FURUSAWA, I., and SHISHIYAMA, J.: Inhibition of melanin biosynthesis by cerulenin in appressoria of *Colletotrichum lagenarium*. Exp. Mycol. **10** (1986): 301–306.

– SUZUKI, K., FURUSAWA, L., ISHIDA, N., and YAMAMOTO, M.: Relation of appressorium pigmentation and penetration of nitrocellulose membranes by *Colletotrichum lagenarium*. Phytopathology **72** (1982a): 498–501.

– – – and YAMAMOTO, M.: Effect of tricyclazole on appressorial pigmentation and penetration from appressoria of *Colletotrichum lagenarium*. Phytopathology **72** (1982b): 1198–1200.

– – – – Melanin biosynthesis as a prerequisite for penetration by appressoria of *Colletotrichum lagenarium*: Site of inhibition by melanin-inhibiting fungicides and their action on appressoria. Pestic. Biochem. Physiol. **23** (1985): 47–55.

KUC, J.: Systemic plant immunization. In: LYR, H., and POLTER, C. (Eds.): Systemic Fungicides and Antifungal Compounds. Tagungsber. Akad. Landwirtschaftswiss. **222**. DDR, Berlin (1984): 189–198.

LANGCAKE, P.: Alternative chemical agents for controlling plant disease. Phil. Trans. Royal Soc. London B **295** (1981): 83–101.

– CARTWRIGHT, D. W., and RIDE, J. P.: The dichlorocyclopropanes and other fungicides with indirect mode of action. In: LYR, H., and POLTER, C. (Eds.): Systemische Fungizide and antifungale Verbindungen. Akademie-Verlag, Berlin 1983, S. 199–210.

– and WICKINGS, S. G. A.: Studies on the action of the dichlorocyclopropanes on the host-parasite relationship in the rice blast disease. Physiol. Plant Pathol. **7** (1975): 113–126.

LAZAROVITS, G., and WARD, E. W. B.: Relationship between localized glyceollin accumulation and metalaxyl treatment in the control of *Phytophthora* rot in soybean hypocotyls. Phytopathology **72** (1982): 1217–1221.

LEBRUN, M. H., ORCIVAL, J., and DUCHARTRE, C.: Resistance of rice to tenuazonic acid, a toxin from *Pyricularia oryzae*. Rev. Cytol. Biol. Végét-Bot. **7** (1984): 249–250.

LEGRAND, M., STINTZI, A., HEITZ, T., GEOFFORY, P., KAUFFMAN, S., and FRITIG, B.: Biological activity of PR-proteins from tobacco; characterization of a proteinase inhibitor In: HENNECKE, H., and VERMA, D. (Eds.): Advances in Molecular Genetics of Plant-Microbe Interactions. V. 1. Kluwer Academic, Dordrecht 1991, pp. 399–402.

MAITI, I. B., and KOLATTUKUDY, P. E.: Prevention of fungal infection of plants by specific inhibition of cutinase. Science **205** (1979): 507–508.

MCINTOSH, A. H., BATEMAN, G. L., CHAMBERLIN, K., DAWSON, G. W., and BURRELL, M. M.: Decreased severity of potato common scale after foliar sprays of 3,5-dichlorophenoxy acetic acid, a possible antipathogenic agent. Ann. Appl. Biol. **99** (1968): 275–281.

MÉTRAUX, J. P., GOY, P. A., STAUB, T., SPEICH, J., STEINEMANN, A., RYALS, J., and WARD, E.: Induced systemic resitance in cucumber in response to 2,6-dichloro-isonicotinic acid and pathogens. In: HENNECKE, H., and VERMA, D. (Eds.): Advances in Molecular Genetics of Plant-Microbe Interactions. V. 1. Kluwer Academic, Dordrecht 1991, pp. 432–439.

MISATO, T., HOMMA, Y., and KO, K.: The development of a natural fungicide, soybean lecithin. Netherl. J. Plant Pathol. **83**, Supplement 1 (1977): 395–402.

MORTON, H. V., and NYFELER, R.: Utilizing derivatives of microbial metabolites and plant defenses to control diseases. In: DUKE, S. O., MENN, J. J., and PLIMMER, J. R. (Eds.): Newer Pest Control Agents and Technology With Reduced Environmental Impact. Amer. Chem. Soc., Washington, D. C. 1993, 316–322.

NEMESTOTHY, G. N., and GUEST, D. I.: Phytoalexin accumulation, phenylalanine ammonia lyase activity and ethylene biosynthesis in fosetyl-Al treated resistant and susceptible tobacco cultivars infected with *Phytophthora nicotianae* var. *nicotiane*. Physiol. and Molecular Plant Pathol. **37** (1990): 207–219.

NIOH, T., and MIZUSHIMA, S.: Effect of validamycin on the growth and morphology of *Pellicularia sasakii*. J. Gen. appl. Microbiol. **20** (1974): 373–383.

OKUNO, T., MATSUURA, K., and FURUSAWA, I.: Recovery of appressorial penetration by some melanin precursors in *Pyricularia oryzae* treated with tricyclazole and in a melanin deficient mutant. J. Pesticide Sci. **8** (1983): 357–360.

OMURA, S.: The antibiotic cerulenin, a novel tool for biochemistry as an inhibitor of fatty acid synthesis. Bacterial. Rev **40** (1976): 681–697.

RUDOLPH, K.: Nonspecific toxins. In: HEITEFUSS, R., and WILLIAMS, P. H. (Eds.): Physiological Plant Pathology. Springer-Verlag, Berlin 1976, pp. 270–315.

SAINDRENAN, P., BARCHIETTO, T., AVELINO, J., and BOMPEIX, G.: Effects of phosphite on phytoalexin accumulation in leaves of cowpea infected with *Phytophthora cryptogea*. Physiol. Molec. Plant Pathol. **32** (1988): 425–435.

SCHEEL, D., COOLING, C., HEDRICH, R., KAWALLECK, P., PARKER, J. E., SACKS, W. R., SOMSICH, I. E., and HALBROCK, K.: Signals in plant defense gene activation. In: HENNECKE, H., and VERMA, D. (Eds.): Advances in Molecular Genetics of Plant-Microbe Interactions. V. 1. Kluwer Academic, Dordrecht 1991, pp. 373–380.

SCHEFFER, R. P.: Host-specific toxins in relation to pathogenesis and disease resistance. In: HEITEFUSS, R., and WILLIAMS, P. H. (Eds.): Physiological Plant Pathology. Springer-Verlag Berlin 1976, pp. 247–269.

SEKIZAWA, Y. and MASE, S.: Mode of controlling action of probenazole against rice blast disease with reference to the induced resistance mechanism in rice plant. J. Pesticide Sci. **6** (1981): 91–94.

SHIBATA, M., MORI, K., and HAMASHIMA, M.: Inhibition of hyphal extension factor formation by validamycin in *Rhizoctonia solani*. J. Antibiot. **35** (1982): 1422–1423.

SHIGEMOTO, R., OKUNO, T., and MATSUURA, K.: Effect of validamycin A on the activity of trehalase of *Rhizoctonia solani* and several sclerotial fungi. Ann. Phytopathol. Soc. Japan **55** (1989): 238–241.

SISLER, H. D.: Fungicides: Problems and prospects. In: SIEGEL, M. R., and SISLER, H. D. (Eds.): Antifungal Compounds. V. 1. Marcel Dekker, New York 1977, pp. 531–547.

- WOLOSHUK, C. P., and WOLKOW, P. M.: Specific regulation of appressorial function. In: LYR, H., and POLTER, C. (Eds.): Systemic Fungicides and Antifungal Compounds. Tagungsber. Akad. Landwirtschaftswiss. 222 DDR, Berlin (1984): 17–28.
- SMILLIE, R., GRANT, B. R., and GUEST, D.: The mode of action of phosphite: Evidence for both direct and indirect modes of action on three *Phytophthora* spp. in plants. Phytopathology **79** (1989): 921–926.
- STÖSSEL, P., LAZAROVITS, G., and WARD, E. W. B.: Light and electron microscopy of *Phytophthora* rot in soybeans treated with metalaxyl. Phytopathology **72** (1982): 106–111.
- SUAMI, T., OGAWA, S., and CHIDA, N.: The revised structure of validamycin A. J. Antibiot. **33** (1980): 98–99.
- TEPPER, C. S., and ANDERSON, A. J.: The genetic basis of plant-pathogen interaction. Phytopathology **74** (1984): 1143–1145.
- TOKOUSBALIDES, M. C., and SISLER, H. D.: Effect of tricyclazole on growth and secondary metabolism in *Pyricularia oryzae*. Pestic. Biochem. Physiol. **8** (1978): 26–32.
- – – Site of inhibition by tricyclazole in the melanin biosynthetic pathway of *Verticillium dahliae*. Pestic. Biochem. Physiol. **11** (1979): 64–73.
- TRINCI, A. P. J.: Effect of validamycin A and L-sorbose on the growth and morphology of *Rhizoctonia cerealis* and *Rhizoctonia solani*. Exp. Mycol. **9** (1985): 20–27.
- VAN ETTEN, H. D.: Phytoalexin detoxification by monooxygenases and its importance for pathogenicity. In: ASADA, Y., BUSHNELL, W. R., OUCHI, S., and VANCE, C. P. (Eds.): Plant Infection – The Physiological and Biochemical. Basis. Japan Sci. Soc. Press, Tokyo 1982, pp. 315–327.
- VON MEYER, W. C., GREENFIELD, S. A., and SEIDEL, M. C.: Wheat leaf rust control by 4n-butyl-1,2,4-triazole, a systemic fungicide. Science **169** (1970): 997–998.
- WADE, M.: Antifungal agents with an indirect mode of action. In. TRINCI, A. P. J., and RILEY, J. F. (Eds.): Mode of Action of Antifungal Agents. Cambridge Univ. Press. Cambridge 1984, pp. 283–298.
- WAKAE, O., and MATSUURA, K.: Characteristics of validamycin as a fungicide for *Rhizoctonia* disease control. Rev. Plant Protec. Res. **8** (1975): 81–92.
- WARD, E. R., UKNES, S. J., WILLIAMS, S. C., DINCHER, S. S., WIEDERHOLD, D. L., ALEXANDER, D. C., AHI-GOY, P., MÉTRAUX, J. P., and RYALS, J. A.: Coordinate gene activity in response to agents that induce systemic acquired resistance. Plant Cell **3** (1991): 1085–1094.
- WARD, E. W. B.: Suppression of metalaxyl activity by glyphosate: evidence that host defense mechanisms contribute to metalaxyl inhibition of *Phytophthora megasperma* f. sp. *glycinea* in soybeans. Physiol. Plant Pathol. **25** (1984): 381–386.
- – LAZAROVITS, G., STÖSSEL, P., BARRIE, S. D., and UNWIN, C. H.: Glyceollin production associated with control of *Phytophthora* rot of soybeans by the systemic fungicide, metalaxyl. Phytopathology **70** (1980): 738–740.
- WATANABE, T.: Effects of probenazole (Oryzemate) on each stage of rice blast fungus (*Pyricularia oryzae* Cavara) in its life cycle. J. Pesticide Sci. **2** (1977): 395–404.
- – SEKIZAWA, Y., SHIMURA, M., SUZUKI, Y., MATSUMOTO, K., IWATA, M., and MASE, S.: Effects of probenazole (Oryzemate) on rice plants with reference to controlling rice blast. J. Pesticide Sci. **4** (1979): 53–59.
- WATKINS, J. E., LITTLEFIELD, L. J., and STATLER, G. D.: The effect of the systemic fungicide 4-n-butyl-1,2,4-triazole on the development of *Puccinia recondita* f. sp. *tritici* in wheat. Phytopathology **67** (1977): 985–989.
- WHEELER, M. H.: Melanin biosynthesis in *Verticillium dahliae*: Dehydration and reduction reactions in cell-free homogenates. Exp. Mycol. **6** (1982): 171–179.
- – Comparsions of fungal melanin biosynthesis in Ascomycetous, imperfect and Basidiomycetous fungi. Trans. Br. Mycol. Soc. **81** (1983): 26–36.

- and GREENBLATT, G. A.: The inhibition of melanin biosynthetic reactions in *Pyricularia oryzae* by compounds that prevent rice blast disease. Exp. Mycol. **12** (1988): 151–160.
WOLKOW, P. M.: Compounds which specifically block epidermal penetration by *Colletotrichum lindemuthianum*. M. S. Thesis, Univ. of Maryland, 1982.
- SISLER, H. D., and VIGIL, E. L.: Effect of inhibitors of melanin biosynthesis on structure and function of appressoria of *Colletotrichum lindemuthianum*. Physiol. Plant Pathol. **23** (1983): 55–71.
WOLOSHUK, C. P., and SISLER, H. D.: Tricyclazole, pyroquilon, tetrachlorophthalide, PCBA, coumarin, and related compounds inhibit melanization and epidermal penetration by *Pyricularia oryzae*. J. Pesticide Sci. **7** (1982): 161–166.
- - TOKOUSBALIDES, M. C., and DUTKY, S. R.: Melanin biosynthesis in *Pyricularia oryzae*: Site of tricyclazole inhibition and pathogenicity of melanin-deficient mutants. Pestic. Biochem. Physiol. **14** (1980): 256–264.
- - and VIGIL, E. L.: Action of the antipenetrant, tricyclazole, on appressoria of *Pyricularia oryzae*. Physiol. Plant Pathol. **22** (1983): 245–259.
YAMAGUCHI, I., and KUBO, Y.: Target sites of melanin biosynthesis inhibitors. In: KÖLLER, W. (ed.): Target Sites of Melanin Biosynthesis Inhibitors. CRC Press, Boca Raton 1992, pp. 101–118.
- SEKIDO, S., MISATO, T.: The effect of non-fungicidal anti-blast chemicals on the melanin biosynthesis and infection by *Pyricularia oryzae*. J. Pesticide Sci. **7** (1982): 523–529.
- - SETO, H., and MISATO, T.: Cytotoxic effect of 2-hydroxyjuglone, a metabolite in the branched pathway of melanin biosynthesis in *Pyricularia oryzae*. J. Pesticide Sci. **8** (1983): 545–550.

Chapter 26

Synergism and antagonism in fungicides

M A. DE WAARD* and U. GISI**

* Department Phytopathology, Wageningen, Agricultural University, Wageningen, The Netherlands
** Sandoz Agro Ltd., Agrobiological Research Station, CH-4108 Witterswil. Switzerland

Introduction

Fungicides are often combined in mixtures to extend the spectrum of antifungal activity and to counteract development of fungicide resistance. Additional advantages of mixtures may be due to synergistic interactions by which the efficiency of the individual components can be increased or the amount of active ingredients can be reduced. Lower amounts of active ingredients may be important for compounds with some toxicological or environmental concerns as well as for phytotoxic or expensive materials. The latter consideration stimulated DIMOND and HORSFALL (1944) to develop synergistic mixtures of cuprous oxides and sulphur in order to conserve the war-short material copper. Up till now, the intentional use of fungicidal mixtures with a synergistic interaction has remained in general rather limited. This contrasts to the commercial development of synergistic mixtures of insecticides (O'BRIEN 1967; WILKINSON 1976) and drugs (KERRIDGE and WHELAN 1984). Due to recent developments the prospects for use of fungicide mixtures showing synergistic interactions may improve, e. g. as part of an antiresistance strategy. The purpose of this chapter is to review the concepts of joint action in mixtures and the biochemical and physiological mechanisms which may underlie the synergistic action. Also, possible interactions of fungicide-insecticide and fungicide-herbicide mixtures will be discussed briefly.

Definitions and test methods

The expected response of a mixture of two chemicals is the sum of the effects of the components separately. This is usually referred to as additive action. It may occur between chemicals with an identical or a different mode of action and has been designated by BLISS (1939) and FINNEY (1947) as similar and independent joint action, respectively. In case of similar joint action, dose-response curves of the chemicals are often parallel, so that one component can be substituted at a constant proportion for the other, without altering the toxicity of the mixture. The toxicity of the mixture is directly predictable from the relative proportions of the constituents. With independent joint action dose-response curves are often not parallel and the additive toxicity of a mixture can be different in size on various efficacy levels. It can only be predicted from the dose-response curves for each component applied alone and in mixture and the correlation in sensitivity to the two toxicants.

In case of synergism the effectiveness of the mixture can not be computed from that of the individual ingredients. Synergism is defined as the simultaneous action of two or more compounds in which the total response of an organism to the pesticide combination is greater than the sum of the individual components (NASH 1981). Antagonism is present if the total response is smaller than the sum of the components.

In order to test whether an observed response can be considered as synergistic or antagonistic, the expected response in the absence of synergism or antagonism should be well defined. It requires a null hypothesis or reference model (MORSE 1978). This presents no great difficulty if only one component of the mixture affects the test organism when applied on its own, but if more components are active, the reference model is sometimes hard to define. In studies on joint action of herbicides MORSE (1978) and NASH (1981) refer to two types: the additive dose model (ADM) and the multiplicative survival model (MSM). These models are also valid for other pesticide combinations.

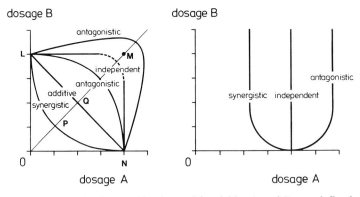

Fig. 26.1 Isoboles for combinations of fungicides A and B at a defined response, e.g. the EC_{50}. Left: both chemicals A and B fungitoxic; right: only chemical A fungitoxic. Explanation of letters in text.

A common ADM is the isobole method in which curves or isoboles represent lines of equal response, e.g. the EC_{50} (TAMMES 1964). With mixtures of two toxic components the line of equal response representing additive action is a straight line LN (Fig. 26.1: left). This reference isobole is valid for most cases of similar joint action. Isoboles drawn from observed responses to mixtures indicate synergism when they are located below this reference and antagonism or independent interaction when above. The interaction can be quantified by the ratio OQ/OP (synergism) and OQ/OM (antagonism). An independent interaction represents the exceptional situation that the response is that of the most potent ingredient of the mixture. Interactions for mixtures with only one toxicant are much more simple (Fig. 26.1: right).

The MSM holds for mixtures of which the constituents produce their effect in entirely different ways. It is, therefore, a reference model for chemicals displaying independent joint action. The reference isobole for additive action of this model does not coincide with the straight line LN but lies within triangle LMN for most response levels; for high response levels it can even lie within triangle LON (MORSE 1978). The recognition that the location of the reference isobole may vary, often fails. Methods to assess accurately the location of the reference isobole in the MSM are, however, not conclusive. Improved analytic procedures to do so are described by Morse (1978) and Nash (1981).

A simple mathematical method to test the additivity of a pesticide combination is based on the equation used by Abbott (1925)

$$E = X + Y - (XY/100)$$

in which X and Y represent percent inhibition by toxicant A (dosage p) and B (dosage q), respectively. E is the expected percent inhibition by the mixture of A and B (dosage p+q). The equation has been modified by Colby (1967) in

$$E_1 = \frac{X_1 Y_1}{100}$$

in which X_1 and Y_1 represent growth as a percentage of control with toxicant A (dosage p) and toxicant B (dosage q), respectively. E_1 is the expected growth as a percentage of control with the mixture A and B (dosage p + q). The observed response is obtained with the mixture containing the same rate of the constituents as applied singly. A deviation of the expected response from the observed response (E_{obs}/E_{exp}) would indicate synergism (E>1) or antagonism (E<1). The method can be considered as a MSM. The method is subject to serious criticism (Morse 1978; Nash 1981) and should, therefore, be used with care. The accuracy of the method is especially doubtful at high response levels of the components (Levy et al. 1986).

Another method has been described by Wadley (1945, 1967) as cited by Gisi et al. (1985). Dose-response curves of the single components and the mixture are constructed and converted to linear regressions using a logit-log transformation; EC-values ("equally effective concentrations") for different levels of control are calculated. The expected response $EC_{90(exp)}$ to a two component mixture is calculated according to the formula

$$EC_{90(exp)} = \frac{a + b}{\dfrac{a}{EC_{90(A)}} + \dfrac{b}{EC_{90(B)}}}$$

A and B in the formula represent the two components and a and b the ratio or the amounts of the components in the mixture. The level of interaction R can be calculated as $R = EC_{90(exp)}/EC_{90(obs)}$ in which $EC_{90(obs)}$ is the observed EC_{90} value of the specific mixture. Unlike the Abbott method, the Wadley approach allows estimation of the interactions at any fungicide concentration and can be used to define the optimum ratio of the components in the mixture providing highest control levels with minimum amounts of the components (Levy et al. 1986).

All methods described above can be used for a wide variety of test conditions and for various types of pesticides. Interactions of pesticides can be studied in vitro (e. g. incorporation of fungicide in agar) and on the plant under laboratory as well as field conditions, as long as the components are tested individually and in the mixture at identical concentrations and intervals (e. g. for field studies, Gisi 1991).

The methods described below only apply to in vitro tests for fungicides or bactericides. A rapid, qualitative method is the crossed-paper strip bioassay, originally described by Bonifas (1952) and modified by others (Corbaz 1963; Katagiri and Uesugi 1977; de

Waard and van Nistelrooy 1982). Filter paper strips impregnated with test chemicals are placed cross-wise on agar seeded with a fungus. Measurement and visual assessments of the inhibition zones at the crossing of the strips indicate the type of the interaction. Disadvantages of the method are the somewhat inaccurate rating method and the difficulty to handle water-insoluble chemicals or chemicals with a high water solubility (differences in diffusion gradient).

A semi-quantitative method is the paper disk bioassay, in which disks impregnated with a chemical are placed on agar seeded with the test organism. Interactions with another chemical can be assessed by measuring the inhibition zone around disks on agar amended or not with a second chemical at a sublethal concentration. The method can be modified in various ways (e. g. Goss and Marshall 1985) and has similar drawbacks as the crossed-paper method.

Possible explanations of synergism and antagonism

Synergism and antagonism may be related to different mechanisms. A most obvious one is based on chemical interactions between components of mixtures before of during uptake. Interactions of this type have already been described by Horsfall (1945) under the name "potentiated" synergism or antagonism.

Mixtures of sulphur and metallic oxides like those of cuprous, cupric, zinc and lead are, for instance, synergistic due to formation of metallic sulphides. Antagonism based on chemical detoxification has been described for several conventional fungicides (Torgeson 1967). Thiols, for instance, reduce the toxicity of phthalimides and of chlorothalonil (Lukens and Sisler 1958; Vincent and Sisler 1968). Another example of synergistic interaction is the complex formation of dithiocarbamates with copper ions, resulting in a bimodal dose-response curve (Goksøyr 1955). Complex formation between polyene macrolides and certain sterols is also the reason why sterols antagonize the toxicity of filippin and related antibiotics (Dekker 1969).

Mechanisms of synergism and antagonism may also relate to interactions with physiological processes in the target organism. From a theoretical point of view interaction with one of the following processes can be involved (de Waard 1985). 1. Non-mediated diffusion (influx) across the plasma membrane. 2. Carrier-mediated transport (influx) across the plasma membrane. 3. Energy-dependent efflux from the fungal cell. 4. Transport to the target site. 5. Activation. 6. Detoxification (biodegradation). 7. Affinity for the target site. 8. Circumvention of the target site. 9. Compensation of the target site. The companion chemical may affect these processes in such a way that the efficacy of the fungicide is enhanced. Such specific physiological interactions may particularly explain synergism or antagonism with site-specific fungicides (examples described under SBI and phenylamide fungicides).

Formulation agents may enhance fungicide efficacy; they should not be regarded as synergists since the intrinsic toxicity of the active ingredients is normally not changed. Formulation agents merely improve leaf coverage, prevent run-off from leaves and enhance penetration or redistribution of a fungicide in a plant. However, they can also behave as true synergists. Detergents, for instance, enhance the in vitro toxicity of sulphur to *B. cinerea*, possibly by changing its membrane permeability to sulphur (PEZET and PONT 1984). Many surfactants are weak fungicides, e. g. against powdery mildews and may, therefore, add to the activity of a mixture with a true fungicide.

Effects of companion chemicals may also result in better field performance of a mixture if dealing with a disease complex. Pathogens of a disease complex may have a differential sensitivity to fungicides, by which the action of single fungicides may be less effective when used alone but more effective when combined (BERGER and WOLF 1974; EISA and BARAKAT 1978; FUSHTEY 1975; PAPAVIZAS and LEWIS 1977). Similarly, a disease may be simultaneously incited by fungicide-sensitive and fungicide-resistant strains of the same pathogen. Again, the single fungicide may fail to control the disease but a mixture with a companion fungicide to which all strains are equally sensitive or to which the resistant strains display negatively-correlated cross-resistance may control the population successfully. The only example for negative cross resistance is the control of sensitive and benzimidazole-resistant *Botrytis* in vineyards by a mixture of carbendazim and isopropyl *N*-(3,4-diethoxyphenyl)-carbamate (LEROUX et al. 1985).

Examples of antagonism

The action of antagonistic compounds is often based on circumvention or compensation of the target site of the fungicide in a biosynthetic pathway. In consequence, knowledge of such compounds often forms a part of mode of action studies. An example of compensation has been demonstrated for the experimental fungicide 6-azauracil (AzU). AzU is metabolized to 6-azauridine-5'-phosphate which inhibits the decarboxylation of orotidine-5'-phosphate to form uridine-5'-phosphate. Uracil and other precursors of uridine-5'-phosphate reverse AzU toxicity because of compensation (DEKKER and OORT 1964).

Toxicity of 2-aminopyrimidines towards powdery mildew can be reduced by adenine, guanidine, folic acid, pyridoxal-5-phosphate and adenosine. A correlation between the activity of these chemicals and the mode of action of 2-aminopyrimidines, i. e. the inhibition of adenosine deaminase activity, has not been established (c. f. HOLLOMON 1984). The reversal by riboflavin was caused by direct photo-inactivation of the fungicides (BENT 1970).

The toxicity of carboxamide fungicides depends on the carbon source for the fungus in the nutrient medium. Toxicity of the fungicide is higher if an acetate substrate instead of glucose is available for the fungus. This behaviour relates to the inhibitory action of carboxamides on succinate dehydrogenase activity in the Krebs cycle. Inhibition of energy generation is, to a certain extent, circumvented in the presence of glucose (RAGSDALE and SISLER 1970; GEORGOPOULOS et al. 1972; RITTER et al. 1973).

In *Rhizoctonia* spp. validamycin A causes various morphological effects on mycelial growth like a decrease in the maximum rate of hyphal extension and an increase in hyphal branching, whereas growth yield is not inhibited. Validamycin A is, therefore, regarded as an antifungal agent with an indirect mode of action (TRINCI 1984). Its effectiveness in preventing infection of rice by *Rhizoctonia solani* is reduced by the presence of meso-inositol. In liquid cultures meso-inositol formation in the mycelium is inhibited by validamycin (WAKAE and MATSURA 1975). The antagonism may, therefore, be based on compensation of a reduced meso-inositol content of the mycelium.

The mode of action of etridiazole is probably based on stimulation of phospholipase activity and induced lipid peroxidation (chapter 6). Procaine is an inhibitor of phospholipase activity and antagonizes the toxicity of etridiazole to *Mucor mucedo*. Calcium ions have a similar effect. The antagonistic activity of tocopherol in mixture with etridiazole may be the protection of cell membranes from lipid peroxidation and formation of oxygen radicals (LYR et al. 1977; RADZUHN and LYR 1984).

The toxicity of fungicides which inhibit sterol biosynthesis (SBIs) can be alleviated in certain fungi by a variety of lipophilic compounds like ergosterol, fatty acids, phospholipids, triglycerides and non-ionic detergents (c. f. SISLER and RAGSDALE 1984). This is another example of compensation based on a substitute for ergosterol in fungal membranes. The antagonism may also result from complex formation or partitioning of the SBIs into undissolved residues of the chemicals. However, these physico-chemical mechanisms would not explain the specificity of antagonism in different fungi. Other mechanisms, which might be involved are decreased affinity of SBIs to the membrane-bound target enzyme or a reduction in membrane permeability for the fungicide (SHERALD et al. 1973; DE WAARD and VAN NISTELROOY 1982). There is no explanation available for the antagonistic action of calcium chloride, tetracaine, carboxin and dialkyldithiocarbamates on the fungitoxicity of certain SBIs (DE WAARD and VAN NISTELROOY 1982).

Examples of synergism

Numerous examples of synergism among fungicides are mentioned in reviews and papers by DIMOND and HORSFALL (1944), CORBAZ (1963) and SCARDAVI (1966). Cases of synergism between multisite fungicides are listed in Tab. 26.1; virtually no information on the mechanisms involved is available. In contrast, synergism among organophosphorus compounds, SBIs and phenylamides are well-documented and will be discussed in more detail.

Table 26.1 Cases of fungicides synergism

Components in mixture	Pathogen	Reference
Anilazine/zinc or copper	*Botrytis cinerea*	GOSS and MARSHALL 1985
	Colletrotrichum coccodes	
Zineb/polygam	*Plasmopara viticola*	FLIEG and POMMER 1965
Carboxin/mancozeb	various	BAICU and NÄGLER 1974
Chloroneb/thiram	*Pythium ultimum*	RICHARDSON 1973
Copper/zineb	*Plasmopara viticola*	THIOLLIERE 1985
Dimethyldithiocarbamates/complex-forming agents	*Botrytis cinerea*	MATOLCSY et al. 1971
Dodine/captan	*Venturia inaequalis*	POWELL 1960
	Xanthomonas prunei	DIENER and CARLTON 1960
Ethazole/pentachloronitro-benzene	*Pythium aphanidermatum*	RAHIMIAN and BANIHASHEMI 1982

Organophosphorus compounds

Crossed-paper strip bioassays have revealed synergistic interactions between organophosphorus thiolate fungicides (PLTs) and phosphoramidates like dibutyl N-methyl N-phenylphosphoramidate (BPA). The toxic action of PTLs in *Pyricularia oryzae* is caused by inhibition of phosphatidylcholine biosynthesis. The toxic action of BPA is probably reduced by rapid fungal hydroxylation and N-demethylation resulting in the formation of non-toxic products. PTLs like diisopropyl S-benzyl phosphorothiolate (IBP) inhibit the metabolism of BPA, thereby enhancing BPA toxicity. BPA detoxification by certain PTL-resistant mutants was much slower than by the wild type. This probably explains the higher sensitivity of these mutants to BPA and the absence of synergism (UESUGI and SISLER 1978). Isoprothiolane has a similar mode of action as PTLs and also displays synergism with BPA (KATAGIRI and UESUGI 1977).

Sterol biosynthesis inhibitors (SBIs)

Crossed-paper strip bioassays revealed synergistic interactions in wild-type and SBI-resistant strains of *Aspergillus nidulans* between the SBI fenarimol and a variety of other chemicals (Tab. 26.2; DE WAARD and VAN NISTELROOY 1982, 1984a). In a similar test, fenarimol and sodium orthovanadate showed synergism to wild-type and SBI-resistant strains of *Penicillium italicum* (DE WAARD and VAN NISTELROOY 1984b). The synergism with the inorganic chemicals and the detergents might be due to increased solubility of fenarimol in the medium. However, for the respiratory inhibitors, phthalimides and sodium orthovanadate a mechanism based on interference with the accumulation of fenarimol in the mycelium has been proposed. Accumulation by both fungi is the sum of passive influx by non-mediated diffusion and energy-dependent efflux. The sensitive and resistant strains differ from each other in efflux activity, being inducible and constitutive, respectively. The constitutive efflux activity in resistant strains results in a permanent operation of a permeability barrier by which fenarimol does not accumulate in

Table 26.2 Chemicals which show synergism with fenarimol to wild-type and SBI-resistant strains of *Aspergillus nidulans*

Anionic agents	sodium lauryl sulphate
	sodium tetradecyl sulphate
Cationic agents	dodecylamine
	dodine
Inorganic chemicals	hydrochloric acid
	sodium hydroxide
Phthalimide fungicides	captafol
	captan
	folpet
Respiratory inhibitors	carbonyl cyanide 3-chlorophenylhydrazone (CCCP)
	dicyclohexylcarbodi-imide (DCCD)

mycelium, as is the case in the sensitive strain. This is probably the mechanism of resistance to fenarimol (DE WAARD and VAN NISTELROOY 1979, 1980, 1981, 1984b). Phthalimide fungicides, respiratory inhibitors and sodium orthovanadate do not interfere with passive influx of fenarimol but do inhibit its energy-dependent efflux. In consequence, the permeability barrier for fenarimol is annihilated and the fungicide accumulates to similar levels in sensitive and resistant strains. The annihilation of fenarimol efflux probably explains the synergistic interaction, being highest in resistant isolates. It is suggested that fenarimol efflux is a membrane process, related with maintenance of an electrochemical proton gradient over membranes by plasma membrane ATPase (DE WAARD and VAN NISTELROOY 1984b). In this concept, the synergism is due to interference with this enzyme, either directly by inhibition of its activity (orthovanadate) or indirectly by depletion of its substrate (mitochondrial respiratory inhibitors, phthalimides). Fenarimol accumulation by various other pathogens is also energy-dependent. This also applies to accumulation of imidazole and triazole SBIs by *P. italicum* (DE WAARD 1988). Therefore, the phenomenon seems to be of general relevance and useful for the design of synergistic mixtures (DE WAARD 1987). Many SBIs are marketed in mixture with multisite fungicides which inhibit respiration. Synergism has been oberved for a mixture of the SBI fenpropimorph with chlorothalonil in control of *Pyrenophora teres* on barley (HAMEL and LARTAUD 1983). Another case of synergistic interaction has been described for the fungicide tridemorph which enhanced the inhibitory action of its formulating product Nekanil LN on oxygen consumption by several fungi (KERKENAAR and KAARS SIJPESTEIJN 1979). The interaction between tridemorph and several other respiratory inhibitors (BERGMANN 1979) seems to have a similar character. An explanation for the synergism may be that tridemorph enhances the transport of the companion chemicals to their site of action. Recently, it was shown that mixtures of pyrazophos and propiconazole were synergistic in the control of *Erysiphe graminis* f. sp. *hordei* and *Drechslera teres*. The underlying mechanism is not known (ZEUN and BUCHENAUER 1991a, 1991b).

Phenylamides

Synergistic interactions in mixtures of phenylamides (oxadixyl, metalaxyl, benalaxyl) with many other fungicides have been described by several authors. In *Phytophthora infestans* on potato and *Plasmopara viticola* on grape vines synergism was observed when oxadixyl was combined with mancozeb, cymoxanil or phosetyl-Al (GISI et al. 1983, 1985). Synergistic activity of the same mixtures was also observed with phenylamide-resistant isolates of *P. infestans* (GRABSKI and GISI 1985). A mixture of metalaxyl and mancozeb was synergistic in control of downy mildew of cucumber (SAMOUCHA and COHEN 1984); similarly, synergistic interactions occurred between benalaxyl and mancozeb in control of *Phytophthora capsici* in vitro (GARIBALDI et al. 1985). Maximum levels of synergism were found in the three-component mixture oxadixyl plus mancozeb plus cymoxanil at ratios of 1 + 7 + 0.4 (GRABSKI and GISI 1987). Disease control by the three-component mixture was little influenced by the phenylamide-sensitivity of the pathogen strains, whereas oxadixyl/mancozeb mixtures were much less effective against resistant than against sensitive strains. By comparing sensitive and resistant strains, synergy levels were higher for all cymoxanil-containing mixtures (oxadixyl + cymoxanil, mancozeb + cymoxanil, oxadixyl + mancozeb + cymoxanil), but lower for oxadixyl plus mancozeb (GRABSKI and GISI 1987). Mathematical models suggest that selection for resistance is slow if synergistic interactions in a mixture are equally high or higher for resistant compared to sensitive subpopulations (LEVY and STRIZYK 1985). Therefore, two-component mixtures of phenylamides plus multisite fungicides may not delay build-up of resistant subpopulations as effectively as three-component mixtures containing cymoxanil. This assumption was confirmed by greenhouse and field experiments (SAMOUCHA and GISI 1987b; COHEN and SAMOUCHA 1990). Synergism of mixtures containing a phenylamide and a multisite fungicide is of general significance for this class of chemicals and has important implications for practice. Several hypotheses of the physiological and biochemical mechanism(s) of synergism are available. No chemical interactions are involved, since a split application, i.e. drench application with the systemic component oxadixyl and a foliar spray with the contact fungicide resulted in significant synergism in control of *P. infestans* on tomato (SAMOUCHA and GISI 1987a). As in other classes of fungicides, a mixture of components with different sites of action may result in an increased uptake into the host and pathogen cells; preliminary results support this theory (COHEN and GISI, unpublished). Fungicides in mixtures may be degraded in host or fungal cells less rapidly than when applied alone. Some evidence was found to support this theory by Gozzo *et al.* (1988). During the infection process, the pathogen may be faced with several consecutive sites of inhibition if the fungicides are applied together. The pathogen may be weakened by the contact fungicide at the germination stage decreasing its aggressiveness and reducing the likelihood of successful infections. The systemic fungicide can then inhibit the pathogen at concentrations which are normally sublethal (SAMOUCHA and COHEN 1986). A short exposure of sporangia of *P. infestans* to sublethal dosages of mancozeb or other respiration inhibitors reduced their aggressiveness to the plant tissue treated with oxadixyl (BASHAN *et al.* 1991).

Fungicide/insecticide and fungicide/herbicide mixtures

As fungicide mixtures are used to broaden the spectrum of activity and to minimise the process of selection for resistant subpopulations, fungicide/insecticide and fungicide/herbicide mixtures may control diseases and pests or diseases and weeds simultaneously in the same field and at the same time. Combinations of fungicides and insecticides is common practice in seed treatment, whereas mixtures of fungicides and herbicides are used in specific situations only, e.g. in rice and cereals to control early weed emergence and damping-off or stem diseases. Surprisingly little is known about interactions for fungicide/herbicide mixtures, whereas many studies are available for fungicide/insecticide combinations, as was summarised recently by KATARIA and GISI (1990) and KATARIA *et al.* (1989), respectively. The inhibition of mycelial growth in vitro and of seedling infections by *Rhizoctonia solani* was for most fungicide/insecticide mixtures additive and sometimes even synergistic, e.g. pencycuron with different insecticides in vitro and carboxin applied to the seed already coated with phosphamidon, monocrotophos, endosulfan or dimethoate. When the SBI fungicide cyproconazole was mixed with one of the herbicides DNOC, dicamba, ioxynil or bromoxynil, the control of *Rhizoctonia cerealis* and *Pseudocercosporella herpotrichoides* was synergistic both in vitro and against the disease. The mixtures were at best additive in their activity against the two weeds *Avena fatua* and *Sinapis alba*. The mechanism(s) of synergism are still speculative. Some of the herbicides and insecticides are weakly fungitoxic on their own (DNOC, bromoxynil, parathion-methyl). They may accumulate in the fungal cell and stimulate the passive uptake of the fungicide by increasing the permeability of membranes and interfering with mitochondrial electron transport as well as uncoupling phosphorylation for ATP generation (MORELAND et al. 1982). Synergism may occur if sublethal concentrations of herbicides (or insecticides) affect the fungus to an extent that sublethal concentrations of the fungicide become detrimental.

Conclusions

Optimisation of the field performance of fungicide mixtures implies minimisation of antagonistic and exploitation of synergistic interactions. Synergistic mixtures allow a decrease of the amount of chemicals without reducing the overall fungicidal activity. Such mixtures provide more reliable disease control in most cases and offer a longer duration of activity as compared to single components alone. Synergistic interactions have been demonstrated for various compounds and existing commercial fungicides. Synergistic activity in vitro does not necessarily imply synergism in disease control since fungicidal components of the mixture may have a different residual activity or may become spatially separated in the plant by systemic activity. Problems which may be involved in the development of synergistic mixtures are: (1) difficulty in finding safe compounds, (2) the costs for two fungicides, (3) the ratio of the fungicides necessary in the combination, (4) the simultaneous formulation of the fungicides, (5) variation in the spectrum of antifungal activity. Despite these difficulties it is believed that synergistic mixtures of fungicides may become more important in future. Fundamental studies on fungicide action and resistance may provide rational leads. First of all synergism in mixtures of already labelled fungicides should be explored. Synergistic mixtures may also extend the antifungal spectrum and counteract development of resistance.

References

ABBOTT, W.. S.: A method of computing the effectiveness of an insecticide. J. Econ. Entomology **18** (1925): 265–267.

BAICU, T., and NÄGLER, M.: Die Wirkung der Fungizidmischung Carboxin und Mancozeb bei der Bekämpfung einiger Weizen- und Gerstenkrankheiten. Arch. Phytopathol. Pflanzensch., Berlin **10** (1974): 395–404.

BASHAN, B., LEVY, Y., and COHEN, Y.: Synergistic interactions between phenylamide fungicides and respiration inhibitors in controlling phenylamide-resistant populations of *Phytophthora infestans* on potato tuber disks. Phytoparasitica **19** (1991): 73–77.

BENT, K. J.: Fungitoxic action of dimethirimol and ethirimol. Ann. appl. Biol. **66** (1970): 103–113.

BERGER, R. D., and WOLF, E. A.: Control of seedborne and soilborne mycoses of "Florida Sweet" corn by seed tratment. Plant Dis. Rep. **58** (1974): 922–923.

BERGMANN, H.: Wirkung von Tridemorph auf *Torulopsis candida* Berlese. Z. Allg. Mikrobiol. **19** (1979): 155–162.

BLISS, C. I.: The toxicity of poisons applied jointly. Ann. appl. Biol. **26** (1939): 585–615.

BONIFAS, V.: Détermination de l'association synergique binaire d'antibiotes et de sulfamides. Experienta **8** (1952): 234–235.

COHEN, Y., and SAMOUCHA, Y.: Competition between oxadixyl-sensitive and -resistant field isolates of *Phytophthora infestans* on fungicide-treated potato crops. Crop Protection **9** (1990): 15–20.

COLBY, S. R.: Calculating synergistic and antagonistic responses of herbicide combinations. Weeds **15** (1967): 20–22.

CORBAZ, R.: Recherches en laboratoire concernant l'association binaire de fongicides: synergie et antagonisme, Phytopathol. Z. **48** (1963): 337–347.

DEKKER, J.: Antibiotics. In: TORGESON, D. C. (Ed.): Fungicides, an Advanced Treatise. Academic Press, New York 1969.

– and OORT, A. J. P.: Mode of action of 6-azauracil against powdery mildew. Phytopathology **54** (1964): 815–818.

DE WAARD, M. A.: Negatively correlated cross-resistance and synergism as strategies in coping with fungicide resistance. 1984 Brit. Crop Protect. Conf. – Pests and Diseases **2** (1984): 573–584.

– Fungicide synergism and antagonism. 1985 Fungicides for Crop Protection. BCPC monograph **31** (1985): 89–95.

– Synergism in fungicides. Tag. Ber. Akad. Landwirtsch. Wiss. DDR, Berlin **253** (1987): 197–203.

– Accumulation of SBI fungicides in wild-type and fenarimol-resistant isolates of *Penicillium italicum*. Pesticide Science **22** (1988): 371–382.

– and VAN NISTELROOY, J. G. M.: Mechanism of resistance to fenarimol in *Aspergillus nidulans*. Pesticide Biochem. Physiol. **10** (1979): 219–229.

– – An energy-dependent efflux mechanism for fenarimol in a wild-type strain and fenarimol-resistant mutants of *Aspergillus nidulans*. Pesticide Biochem. Physiol. **13** (1980): 255–266.

– – Induction of fenarimol-efflux activity in *Aspergillus nidulans* by fungicides inhibiting sterol biosynthesis. J. gen. Microbiol. **126** (1981): 483–489.

– – Antagonistic and synergistic activities of various chemicals on the toxicity of fenarimol to *Aspergillus nidulans*. Pesticide Sci. **13** (1982): 279–286.

– – Effects of phthalimide fungicides on the accumulation of fenarimol by *Aspergillus nidulans*. Pesticide Sci. **15** (1984a): 56–62.

– – Differential accumulation of fenarimol by a wild-type isolate and fenarimol-resistant isolates of *Penicillium italicum*. Neth. J. Pl. Path. **90** (1984b): 143–153.

DIENER, U. L., and CARLTON, C. C.: Dodine-captan combination controls bacterial spot of peach. Plant Dis. Rep. **44** (1960): 136–138.

DIMOND, A. E., and HORSFALL, G.: Synergism as a tool in the conservation of fungicides. Phytopathology **34** (1944): 136–139.

EISA, N. A., and BARAKAT, F. M.: Relative efficiency of fungicides in the control of damping-off and *Stemphylium* leafspot of broad bean (*Vicia faba*). Plant Dis. Rep. **62** (1978): 114–118.

FINNEY, D. J.: Probit Analysis. A statistical treatment of the sigmoid response curve. Cambridge Univ. Press, Cambridge 1947.

FLIEG, O., and POMMER, E.-H.: Polyram Combi. Die Landwirtschaftliche Versuchsstation Limburgerhof 1914–1964. BASF AG, Ludwigshafen 1965, 319–328.

FUSHTEY, S. G.: The nature and control of snow mold of fine turfgrass in Southern Ontario. Can. Plant Dis. Survey **55** (1975): 87–96.

GARIBALDI, A., ROMANO, M. L., and GULLINO, M. L.: Synergism between fungicides with different mechanisms of action against acylalanine resistant strains of *Phytophthora* spp. EPPO Bulletin **15** (1985): 545–551.

GEORGOPOULOS, S. G., ALEXANDRI, E., and CHRYSAYI, M.: Genetic evidence for the action of oxathiin and thiazole derivatives on the succinic dehydrogenase system of *Ustilago maydis* mitochondria. J. Bacteriol. **110** (1972): 809–817.

GISI, U.: Synergism between fungicides for control of *Phytophthora*. In: LUCAS, J. A., SHATTOCK, R. C., SHAW, D. S., and COOKE, L. R. (Eds.): Phytophthora. Cambridge University Press, Cambridge 1991, pp. 361–372.

– BINDER, H., and RIMBACH, E.: Synergistic interactions of fungicides with different modes of action. Trans. Br. mycol. Soc. **85** (1985): 299–306.

– HARR, J., SANDMEYER, R., and WIEDMER, H.: A new systemic oxazolidinone fungicide (SAN 371 F) against diseases caused by *Peronosporales*. Meded. Fac. Landbouww. Rijksuniv. Gent. **48** (1983): 541–549.

GOKSØYR, J.: The effect of some dithiocarbamyl compounds on the metabolism of fungi. Physiol. Plant. **8** (1955): 719–727.

GOSS, V., and MARSHALL, W. D.: Synergistic interactions of zinc or copper with anilazine. Pesticide Sci. **16** (1985): 163–171.

GOZZO, F., PIZZIGRILLI, G., and VALCAMONICA, C.: Chemical evidence of the effects of mancozeb on benalaxyl in grape plants as possible rationale for their synergistic interaction. Pesticide Biochem. Physiol. **30** (1988): 136–141.

GRABSKI, C., and GISI, U.: Mixtures of fungicides with synergistic interactions for protection against phenylamide resistance in *Phytophthora*. 1985 Fungicides for Crop Protection. BCPC monograph **31** (1985): 315–318.

– – Quantification of synergistic interactions of fungicides against *Plasmopara* and *Phytophthora*. Crop Protection **6** (1987): 64–71.

HAMPEL, H., and LARTAUD, G.: The control of important cereal diseases using fenpropimorph mixtures. Proc. 10th Intern. Congr. Plant Protect., Brighton, **3** (1983): 926.

HOLLOMON, D. W.: Antifungal activity of substituted 2-aminopyrimidines. In: TRINCI, A. P. J., and RYLEY, J. F. (Eds.): Mode of action of Antifungal Agents. Cambridge Univ. Press, Cambridge 1984.

HORSFALL, J. G.: Synergism and antagonism. Plant Dis. Rep., Suppl. **157** (1945): 162–166.

KATAGIRI, M., and UESUGI, Y.: Similarities between the fungicidal action of isoprothiolane and organophosphorus thiolate fungicides. Phytopathology **67** (1977): 1415–1417.

KATARIA, H. R., and GISI, U.: Interactions of fungicide-herbicide combinations against plant pathogens and weeds. Crop Protection **9** (1990): 403–409.

– SINGH, H., and GISI, U.: Interactions of fungicide-insecticide combinations against *Rhizoctonia solani* in vitro and in soil. Crop Protection **8** (1989): 399–404.

KERKENAAR, A., and KAARS SIJPESTEIJN, A.: On a difference in the antifungal activity of tridemorph and its formulated product Calixin. Pesticide Biochem. Physiol. **12** (1979): 124–129.

KERRIDGE, D. and WHELAN, W. L.: The polyene macrolide antibiotics and 5-fluorocytosine: molecular action and interaction. In: TRINCI, A. P. J., and RYLEY, J. F. (Eds.): Mode of Action of Antifungal Agents. Cambridge Univ. Press, Cambridge 1984.

LEROUX, P., GREDT, M., MASSENOT, F., and KATO, T.: Activité du phenylcarbamate S32165 sur *Botrytis cinerea*, agent de pourriture grise de la vigne. 1985 Fungicides for Crop Protection. BCPC monograph **31** (1985): 443–446.

LEVY, Y., BENDERLY, M., COHEN, Y., GISI, U., and BASSAND, D.: The joint action of fungicides in mixtures: comparison of two methods for synergy calculation. EPPO Bulletin **16** (1986): 651–657.

– and STRIZKY, S.: Une modélisation du développement de souches résistantes aux fongicides systémiques compte tenu du mode d'application, de l'adaptabilité génétique du champignon et de la synergie entre fongicides. EPPO Bulletin **15** (1985): 519–525.

LUKENS, R. J., and SISLER, H. D.: Chemical reactions involved in the fungitoxicity of captan. Phytopathology **48** (1958): 235–244.

Lyr, H., Casperson, G., and Laussmann, B.: Wirkungsmechanismus von Terrazol bei *Mucor mucedo*. Z. Allg. Mikrobiol. **17** (1977): 117−129.

Matolcsy, G., Hamrán, M., and Bordás, B.: Increased antifungal action of zinc dimethyldithiocarbamate in the presence of complex forming compounds. Pesticide Sci. **2** (1971): 229−231.

Moreland, D. E., Huber, S. C., and Novitzky, W. P.: Interaction of herbicides with cellular and liposome membranes. ACS Symposium Series **182** (1982): 79−96.

Morse, P. M.: Some comments on the assessment of joint action in herbicide mixtures. Weed Sci. **26** (1978): 58−71.

Nash, R. G.: Phytotoxic interaction studies − Techniques for evaluation and presentation of results. Weed Sci. **29** (1981): 147−155.

O'Brien, R. D.: Insecticides, Action and Metabolism. Academic Press, New York 1967.

Papavizas, G. S., and Lewis, J. A.: Effects of cottonseed treatment with systemic fungicides on seedling disease. Plant Dis. Rep. **61** (1977): 538−542.

Pezet, P. and Pont, V.: Sensitivity of *Botrytis cinerea* Pers. to elemental sulfur in the presence of surfactants. Experienta **40** (1984): 354−356.

Powell, D.: The inhibitory effects of certain fungicide formulations to apple scab conidia. Plant Dis. Rep. **44** (1960): 176−178.

Radzuhn, B., and Lyr, H.: On the mode of action of the fungicide etridiazole. Pesticide Biochem. Physiol. **22** (1984): 14−23.

Ragsdale, N. N., and Sisler, H. D.: Metabolic effects related to fungitoxicity of carboxin. Phytopathology **60** (1970): 1422−1427.

Rahimian, M. K., and Banihashemi, Z.: Synergistic effects of ethazole and pentachloronitrobenzene on inhibition of growth and reproduction of *Pythium aphanidermatum*. Plant Disease **66** (1982): 26−27.

Richardson, L. T.: Synergism between chloroneb and thiram applied to peas to control seed rot and damping-off by *Pythium ultimum*. Plant Dis. Rep. **57** (1973): 3−6.

Ritter, G., Kluge, E., and Lyr, H.: Beziehungen zwischen Carboxin-Resistenz und glykolytischer Potenz bei Pilzen. Z. Allg. Mikrobiol. **13** (1973): 243−250.

Samoucha, Y., and Cohen, Y.: Synergy between metalaxyl and mancozeb in controlling downy mildew in cucumbers. Phytopathology **74** (1984): 1434−1439.

− − Efficacy of systemic and contact fungicide mixtures in controlling late blight in potatoes. Phytopathology **76** (1986): 855−859.

− and Gisi, U.: Possible explanations of synergism in fungicide mixtures against *Phytophthora infestans*. Ann. appl. Biol. **110** (1987 a): 303−311.

− − Use of two- and three-way mixtures to prevent buildup of resistance to phenylamide fungicides in *Phytophthora* and *Plasmopara*. Phytopathology **77** (1987 b): 1405−1409.

Scardavi, A.: Synergism among fungicides. Ann. Review Phytopathol. **4** (1966): 335−348.

Sherald, J. L., Ragsdale, N. N., and Sisler, H. D.: Similarities between the systemic fungicides triforine and triarimol. Pesticide Sci. **4** (1973): 719−727.

Sisler, H. D., and Ragsdale, N. N.: Biochemical and cellular aspects of the antifungal action of ergosterol biosynthesis inhibitors. In: Trinci, A. P. J., and Ryley, F. J. (Eds.): Mode of Action of Antifungal Agents. Cambridge Univ. Press, Cambridge 1984, pp. 257−282.

Tammes, P. M. L.: Isoboles, a graphic representation of synergism in pesticides. Neth. J. Plant Path. **70** (1974): 73−80.

Thiolliere, J.: Progres apportes par cuprosan − Première association synergique organo − cuprique. 1985 Fungicides for Crop Protection. BCPC monograph **31** (1985): 227−230.

Thorgeson, D. C. (Ed.): Fungicides, an Advanced Treatise, vol. 2. Academic Press, New York 1967.

Trinci, A. P. J.: Antifungal agents which effect hyphal extension and hyphal branching. In: Trinci, A. P. J., and Ryley, J. F. (Eds.): Mode of Action of Antifungal Agents. Cambridge Univ. Press, Cambridge 1984, pp. 113−134.

UESUGI, Y., and SISLER, H. D.: Metabolism of a phosphoramidate by *Pyricularia oryzae* in relation to tolerance and synergism by a phosphorothiolate and isoprothiolane. Pesticide Biochem. Physiol. **9** (1978): 247–254.

VINCENT, P. G., and SISLER, H. D.: Mechanism of antifungal action of 2,4,5,6-tetrachloroisophthalonitril. Physiol. Plant. **21** (1968): 1249–1264.

WADLEY, F. M.: The evidence required to show synergistic action of insecticides and a short cut in analyses. ET-223, U. S. Department of Agriculture, Washington 1945.

– Experimental Statistics in Entomology. Graduate School Press, U. S. Department of Agriculture, Washington 1967.

WAKAE, O., and MATSUURA, K.: Characteristics of validamycin as a fungicide for *Rhizoctonia* disease control. Rev. Plant Protect. Res. **8** (1975): 81–92.

WILKINSON, C. F.: Insecticide synergism. In: METCALF, R. L., and MCKELVEY, J. J. (Eds.): The Future for Insecticides. John Wiley & Sons, New York 1976.

ZEUN, R., and BUCHENAUER, H.: Synergistic interactions of the fungicide mixture pyrazophos-propiconazole against barley powdery mildew. Z. Pflanzenkr. Pflanzensch. **98** (1991 a): 526–538.

– – Synergistic effects of pyrazophos and propiconazole against *Pyrenophora teres*. Z. Pflanzenkr. Pflanzensch. **98** (1991 b): 661–668.

Chapter 27

Outlooks

H. LYR

Institut für Integrierten Pflanzenschutz der Biologischen Bundesanstalt Kleinmachnow, Germany

Broad usage of fungicides in agriculture has increased crop yields world wide and this has contributed to a decrease of shortage of food in many regions of the world. Because of the high economic effect lay here the centres of efforts of fungicide research and development with great benefits for other areas of application. In recent years many new and more effective compounds have been developed and introduced to the market.

Moreover, several other fungicidal or biocidal compounds are in use to treat timber, various organic materials, such as leather, textiles, plastics, paints a. o., in order to prevent a deterioration of organic materials by fungi (cf. chapters 22 and 23).

Some antifungal compounds have been taken from the market because of toxic side effects or other undesireable properties in the environment.

Also in medicine, where mycoses have had a tendency to increase (cf. chapter 21), new compounds with improved properties allow a more effective curing of fungal diseases in humans.

The amount of fungicides produced world wide is represented by the sum of 5 Mrd. US $ in 1992. The total economic benefit of their positive effects can be estimated up to 20 to 50 Mrd. US $. But, one must admit that progress in the world is unequal.

Development of new compounds is concentrated because of the high costs for research and development mainly in large firms f. e. in USA, Germany, Switzerland, France, Japan and the United Kingdom. The effective use of modern fungicides requires a high level of education of farmers and other users to bring the desired economic benefit. Many developing or other countries are not able to pay for urgently needed new fungicidal products to improve their agricultural crop yields. Richer countries have achieved by a combination of good agricultural practice, use of high yield varieties and application of modern pesticides a high level of agricultural production or even an overproduction in most crops. In these countries peoples are more concerned about a possible contamination of the environment by "pesticides" or toxic residues in the food, problems which are in the public opinion very often overestimated.

These considerations lead partly to a more extensive production mainly in field crops, such as grains, to the use of lower doses and fewer appliations, which brought economic problems for fungicide producers and might slow down the efforts for development of new compounds, which is no desireable goal.

In contrast to routine applications with fixed spraying programs, intelligent use of modern fungicides requires a more disease directed (integrated) approach, which allows f. e. in wheat a complete control of most diseases by 1 – 3 applications of fungicides.

In any case, in spite of higher costs of modern fungicides, proper usage guarantees a good economic benefit, and in addition higher qualities of the products (for example, free of highly toxic mycotoxins).

New Toxophores

Numerous compounds with fungicidal or antifungal activity do already exist, but many of them are merely chemical analogues of earlier invented compounds, and have similar or identical properties. Therefore, the real number of fungicides types, which differ in their mechanism of action, their spectrum of activity and their relationships regarding fungal cross resistance patterns is much smaller. Therefore is still a need for new groups of compounds for integration into antiresistance strategies and for better control of some important fungi, which are still difficult to control. As described in chapter 19 and others, several new types of fungicides, such as phenylpyrroles, strobilurins (β-methoxyacrylates), anilinopyrimidines, dimethomorph, fluazinam and some others have been recently introduced. They have a new mechanism of action, different from that of older groups of fungicides. This means that still unexplored targets within fungal cells do exist, which can be used for antifungal effects by newly synthesized compounds.

The probability of finding desirable new toxophores is decreasing (RYLEY and RATHMELL 1984), but it seems quite certain that still many other targets can be attacked by new compounds to be found in the future.

The main route of finding new toxophores are still chemical screening programs. Their relative success in the past did not stimulate alternative programs (GEISSBÜHLER 1984). It would seem that a more directed, or even a biorational approach could be an attractive alternative. The limitations of such approaches are mainly insufficient biochemical analyses of targets for a better description of essential target structures on a molecular level, which are useable for chemists to construct new molecules (SCHWINN and GEISSBÜHLER 1984).

But the modelling of secondary and tertiary protein structures is complicated and therefore it is difficult to design chemical structures with high affinity for specific target sites (cf. chapter 24). Therefore, for the foreseeable future a pure biorational design *sensu stricto* will be only a desirable aim, but not yet a practical accomplishment. This does not mean, that this is absolutely impossible, but more detailed information from genetic engineering methods is needed. The computer will be an unavoidable tool for design of molecule structures or to elucidate certain correlations between chemical structures and biological effects (cf. chapter 24). More directed biochemical investigations on new targets in fungi are still needed (LYR 1979; BALDWIN 1984). Also a deeper understanding of host parasite interactions on a biochemical level are necessary. Unfortunately efforts are declining rather than increasing because of an inadequate support for various reasons. Very often the complicated situation, which demands long term experiences and intensive, detailed studies, is underestimated.

New or better quality of fungicides

Although we usually speak of "systemic" fungicides, it should be kept in mind that with very few exceptions only a xylem transport of fungicides with the water stream is realized (cf. chapter 3). Only Aliette (phosphite) has a phloem mobility, but this molecule is very specific for *Oomycetes*. To control root pathogenic organisms or wilt diseases a break through towards new highly effective molecules with phloem mobility is necessary. New ambimobile herbicides (glyphosate, gluphosinate) demonstrate that such a goal is not too fantastic. The mass flow through the phloem is quantitatively much inferior to that of the xylem and specific conditions regulate loading and unloading of the phloem sap. Therefore, the demands for the molecule which meets all these conditions is much higher than that for a xylem transport. But only through the phloem can roots, bulbs, new leaf growth, and fruits be reached by a leaf application of a fungicide. Soil applications are profitable only in exceptional cases.

Of course a minimum risk for resistance development is also a highly desired feature for new compounds.

For several important pathogens (such as *Fusaria*, *Botrytis*, *Ophiobolus*, *Penicillia*, *Oomycetes* a. o.) new solutions for an effective control are needed.

Low mammalian toxicity, no disturbance of the environment and a moderate price are nowadays quite normal conditions.

To bring together all desirable features in one molecule is surely still a dream, but a nice one.

Induced Resistance

More information has accumulated that indicates the possibility of a biological or chemical induction of resistance in plants. Several chemicals, structurally unrelated, have been detected as inducers (cf. chapter 25). The biochemical background is only partly understood, but much progress has been made recently.

The practical significance of this phenomenon for disease control is still an open question, but the possibility is highly attractive because problems of resistance are not expected to occur.

A handicap seems to be the dependence of intensity of induction on environmental conditions. Whether other plants react as strongly as the models, tobacco and cucumber, needs to be demonstrated. The specificy of some compounds for host/parasite combinations seems to be rather high.

It may be that this principle is already operating in nature and perhaps we get only aware of the outbreak of epidemics, where the induced resistance may have been overrun by high infection levels or specific weather constellations.

If special physiological conditions are required to bring into operation an induced resistance, it would be hard to manage it in practice. This remains to be elucidated.

Synergism and antagonism in fungicides

These phenomena (chapter 26) need further attention from a therotetical as well as from a practical point of view. The directed use of synergists can be very favourable for increasing antifungal spectrum or decreasing the danger of resistance formation in the fungal population. Synergistic interactions with simultaneously applied herbicides and insecticides should not be neglected (YEGEN and HEITEFUSS 1970; HEITEFUSS 1970; HEITEFUSS 1972), as well as possible antagonistic interactions, which diminish the fungicidal effect (KATARIA and DODAN 1983).

Biological control of plant diseases

Due to some drawbacks of chemical control of plant diseases and pests, the possibilities for biological control have been widely discussed and tested. Biological control mentioned here means the inoculation or application of antagonistic fungi or bacteria. Fungi of the genus *Trichoderma* have long been used for biological control experiments, more or less successfully. Of course other fungal organism such as *Peniophora gigantea* (against *Fomes annosus*) or bacteria such as *Pseudomonas* spp. or sporulating genera (*Bacillus*) have also been used. It seems that there are possibilities for successful applications especially against soil borne pathogens (dressing, inoculation of seeds, combinations with soils disinfectants). Whether vascular diseases such as Dutch elm disease can be controlled by *Pseudomonas* spp. (STROBEL and LANIER 1981) or other bacteria remains to be observed. A disadvantage of working with living organisms is their dependence on the actual ecological conditions which are hard to predict or to control. Therefore, both positive and negative results have been described. The directed genetic manipulation of certain isolates can perhaps increase the effect of biological control measures, but due to their inherent limitations they can solve only a very small part of the overall problem.

Breeding of resistant plant varieties

We can look back on several decades of intensive efforts to select cultivars resistant to fungal diseases. In spite of many successes in the past, a breakthrough of new aggressive strains or new pathogens could not be avoided. The successes did not lead to such a stabilization of plant production that the application of chemicals could be eliminated. Recently there is an increased optimism for solving the problem of fungal disease resistance with the help of genetic engineering methods and new selection methods using cell cultures. But at present, the genetic base of resistance in plants is very obscure and even biochemically only scarcely understood. If one considers that each pathogen-host-interaction probably has different causes of resistance and eventually also various plant species or cultivars differ in this respect, the difficulties for the gene technique to construct R-genes and to insert them with high stability into a productive plant variety are evident. Therefore very quick progress can not be expected (CHILTON 1984). Regarding horizontal resistance, breeders can be confronted with toxicological problems (accumulation of defense substances such a alkaloids, glycosides, phytoalexins and other secondary plant products which may even have carcinogenic or estrogenic properties) that require residue analysis and toxicological evaluations equal to those of pesticides.

Because many other problems such as high productivity, stress tolerance, product qualities, special demand from a technological point of view must be realized by breeding programs (SPAAR and LYR 1984) application of fungicides (and other pesticides and growth regulating substances) will be required to stabilize plant production for a long time. The possibility of breeding varieties resistant to toxigenic, common molds such as *Fusarium, Penicillium, Aspergillus* and other generea which are known as producers of highly toxic mycotoxins is nearly on unsolvable problem (LYR 1985). The long life span of trees makes breeding for resistance very ineffective because of the high adaptibility of fungal species. The main line of progress will probably be the selection of more tolerant varieties with a broad genetic variability (BREMERMANN 1983).

Conclusions

The system of fungal diseases control in the future will be more sophisticated and contain more options than those presently available. In some cases, epidemiological information may be used to guide biological/chemical control strategies in good agricultural practise, or integrated systems of plant protection.

There will still be a requirement for new compounds with new mechanisms of action, low susceptibility to resistance, low mammalian toxicity and some new features (for example phloem mobility). These compounds will contribute to a still higher efficiency of fungal disease control.

New toxophores will de developed either by recent screening methods or by an advanced biorational design. In any case, advances will depend very much on a deeper understanding of the biology and biochemistry of at least the main diseases and the main crop plants and their varieties. Efforts of plant protection science in cooperation with those of other disciplines of science all over the world should find solutions to many of the unsolved problems of plant protection in the coming decades.

References

BALDWIN, B. C.: Potential targets for the selective inhibition of fungal growth. In: TRINCI, A. P. J., and RYLEY, J. F. (Eds.): Mode of Action of Antifungal Agents. Brit. Mykol. Soc. (1984): 43–62.

BREMERMANN, H. J.: Theory of catastrophic diseases of cultivated plants. J. theor. Biol. **100** (1983): 255–274.

CHILTON, MARY-DELL: Genetic engineering-prospects for use in crop management. Brit. Crop Protect. Conf. (1984): 1 A, 3–9.

GEISSBÜHLER, H.: Biorational reflections in agricultural chemical research. Chimia **38** (1984): 307–316.

HEITEFUSS, R.: Nebenwirkungen von Herbiziden auf Pflanzenkrankheiten und deren Erreger. Z. Pflanzenkrankh. u. Pflanzenschutz, Sonderheft V (1970): 117–127.

— Ursachen der Nebenwirkungen von Herbiziden auf Pflanzenkrankheiten. Z. Pflanzenkrankh. u. Pflanzenschutz, Sonderheft VI (1972): 80–87.
KATARIA, H. R., and DODAN, D. S.: Impact of two soil-applied herbicides on damping-off of cowpea caused by *Rhizoctonia solani*. Plant and Soil **73** (1983): 275–283.
KUĆ, J., and TUZUN, S.: Immunization for disease resistance in tobacco. Recent Adv. in Tobacco Sci. **9** (1983): 174–213.
— Induced systemic resistance to plant disease and phytointerferons — are they compatible? Fitopathologia Brasilia (1985): 17–40.
LYR, H.: Differentialmerkmale zwischen Pilzen und höheren Pflanzen als Basis für eine selektive Wirkung systemischer Fungizide. In: LYR, H., and POLTER, C. (Eds.): Systemic Fungicides. Abh. Akad. Wiss. DDR, 2 N. Akademie-Verlag, Berlin 1979, pp. 7–13.
— Mykotoxine — eine vermeidbare Gefahr. Biol. Rdsch. **23** (1985): 285–293.
RATHMELL, W. G.: The discovery of new methods of chemical disease control: current developments, future prospects and the role of biochemical and physiological research. Advanc. in Plant Pathol. **2** (1983): 259–288.
RYLEY, J. F., and RATHMELL, W. G.: Discovery of antifungal agents: *in vitro* and *in vivo* testing. In: TRINCI, A. P. J., and RYLEY, J. F. (Eds.): Mode of Action of Antifungal Agents. Brit. Mykol. Soc. (1984): 63–87.
SALT, ST. D., and KUĆ, J.: Elicitation of disease resistance in plants by the expression of latent genetic information. In: HEDIN, P. A. (Ed.): Bioregulators for Pest Control. ACS Symposium, Series **276** (1985): 47–68.
SCHWINN, F. and H. GEISSBÜHLER: Towards a more rational approach to fungicide design. Crop. protection (1986).
SPAAR, D., and LYR, H.: The use of fungicides and resistant crop varieties as basis of modern plant protection strategies. In: LYR, H., and POLTER, C. (Eds.): Systemic Fungicides and Antifungal Compounds. Tagungsber. Akad. Landwirtschaftwiss. DDR, **222** (1984): 7–16.
STROBEL, G. A., and LANIER, G. N.: Dutch elm disease. Scientific American **245** (1981): 56–66.
YEGEN, O., and HEITEFUSS, R.: Nebenwirkungen von Natriumtrichloracetat (NaTA) auf den Wurzelbrand der Rüben und das antiphytopathogene Potential des Bodens. Zucker **23** (1970): 694–700; 723–729.

Subject Index

(underlined pages show chemical structures)

AAC 497
Acylalanines 327 f
— genetics of resistance 43
— mechanism of action 336
— properties 331
— uptake, transport 335
— usage 333
AFUGAN (= pyrazophos) 374
Aldimorph (FALIMORPH) 163, 164
— degradation 169
— mechanism of action 191
— properties, usage 170
— uptake, transport 167
ALIETTE (Al-fosethyl) 327
Alkylammonium compounds (AAC) 497
Allylamines 444
Al-fosethyl (ALIETTE) 327
— spectrum of activity 16, 331
— transport 64
α-Amanitin 346, 347
Ambimobility (transport) 54
2-Aminopyrimidine fungicides 355 ff
— cross resistance 365
— in vitro effects 361
— mechanism of action 362
— mixtures 366
— practical application 356 f
— resistance 364
— uptake, metabolism 358 f
Amitrole, transport 63
Amorolfine 436, 468 f
— mechanism of action 190
Amphothericin B 436, 460
Ancymidol 211, 259 f
— cell division 265
— effect on fungi 264
— gibberellin biosynthesis 262

— plant growth 259
— polyamine level 268
Andoprim 330
— properties 402 f
— spectrum of activity 17
Anilazine 390
— properties, usage 395
Anilino-Pyrimidines 402 f
— mechanism of action 404, 405
— usage 403, 404
Antibiotics 415 ff
— usage 417
Antifungal agents 435
Antimycin A
— site of attack 150
Antipenetrants 545 ff
Apoplastic transport 53
Aromatic hydrocarbon fungicides (AHF) 75 ff
— genetic of resistance 42
— mechanism of action 82 ff
— resistance 93
— spectrum of activity 77
Aspergillosis 433, 465
Azaconazole 487
— wood preservation 494
Azide
— site of attack 150
Azole fungicides 448
— for human chemotherapy 448
— trade names 449

BAS 111 W
— abscisic acid 269
— auxin level 268
— cytokinin level 269
— growth retardation 264
BAS 45 406 F 220, 227, 238
— properties 227, 238

BAS 490 F (Kresoxim-methyl) 329, 331, 408
– mechanism of action 336, 339, 410
– spectrum of activity 16, 332, 409
– translocation 335
– usage 333
BASITAC (= mepronil)
BAVISTIN (= carbendazim)
BAYCOR (= bitertanol)
BAYFIDAN (= triadimenol)
BAYLETON (= triadimefon)
Benalaxyl (GALBEN) 328, 331
– mode of action 349 f
– resistance 349 f
– usage 333
BENLATE (= benomyl)
Benodanil (CALIRUS) 134
– practical application 140
Benomyl (BENLATE) 306
– cutinase inhibition 547
– mechanism of action 307 ff
– resistance 307 ff
– spectrum of activity 16
– transport 64
Benzimidazole fungicides 291 ff, 292
– genetics of resistance 41
– mechanism of action 305 ff
– resistance 307 f
– selectivity 310
– usage 293 f
BERET (= fenpiclonil)
Bialophos
– usage 417
Bifonazole 448 f, 450
Binapacryl 389
Biphenyl (Diphenyl) 75 f
– spectrum of activity 77
– usage 76
Bitertanol (BAYCOR) 220
– properties, usage 223, 234
Blasticidin S 415, 416
– degradation 419
– mechanism of action 418
– properties 419
– usage 417, 418
Blastomycosis 433, 463
Bordeaux mixture 324
BOTRYSAN (= anilazine)
BRASSICOL (= QUINTOCENE)
BRAVO (= chlorothalonile)
Bromuconazole 221
– properties, usage 231, 246

Bupirimate (NIMROD) 355 f
– metabolism 362
– practical application 359
– resistance 365
– uptake, transport 362
Butenafine 445
Buthiobate 210, 211, 212
– properties 211
– spectrum of activity 212
Butoconazole 448 f, 450, 459

CALIRUS (= benodanil)
CALIXIN (= tridemorph)
Candidosis 431, 458 f
CANESTEN (= clotrimazole)
Captan
– spectrum of activity 13
Carbendazim 29, 296, 306
– mechanism of action 305 ff
– residues 300
– resistance 25, 299, 307
– transport 64
– usage 293 ff
Carboxamide fungicides 133 ff
Carboxin (VITAVAX) 133
– genetic of resistance 43
– spectrum of activity 16
– transport 64
– usage 136 f
Carboxin fungicides 133 ff
– alternative respiration, pathway 156
– application 137 f
– degradation 141
– growth stimulation 142
– inhibition respiration 150
– mechanism of action 149 ff
– ozone protection 143
– receptor interaction 151 f, 155
– resistance 142
– side effects 158
– site of interaction 150
– structure-activity relation 137 f
– systemic movement 135
– transport 64
CCC
– fungal control 557
CDPF 315
CERCOBIN (= thiophanate-methyl)
Cerulenin
– melanin biosynthesis 550
Chlobenthiazone 548

- melanin biosynthesis 548
Chloramphenicol 415
Chloroneb (DEMOSAN) 75 f
- cross resistance 89, 130
- lipid peroxidation 89
- mechanism of action 83 f
- properties, usage 78 f
- resistance 79
- spectrum of activity 16, 17, 77, 80
Chlorothalonile 390
- properties, usage 396
Chlozolinate (SERINAL) 100
Chromoblastomycosis 467 f
Clotrimazole 448, 450, 459
Coccidioidomycosis 433, 464
COLBY equation 567
Colchicine
- inhibition of mitosis 306 f
Computer modelling 517 ff
- of triazoles 525
Copper
- genetic of resistance 44
CORBEL (= fenpropimorph)
Crop protection, worldmarket 14
Cross resistance 24
Cryptococcosis 432, 461 f
CURZATE (= cymoxanil)
Cutinase inhibitors 545
Cyanide
- site of attack 150
Cyclafuramid 136
Cyclic amine fungicides
- effects on enzymes 195
- effects on growth 186
- lipid synthesis 187
- mechanism of action 185 ff
- sites of inhibition 188, 197
Cycloheximide 415
- genetic of resistance 47
- usage 424
N-Cyclohexyl-diazeniumdioxy-chelates 487
- wood preservation 495
Cymoxanil (CURZATE) 327
- mechanism of action 336, 339
- mixtures 334
- resistance 339
- spectrum of activity 16, 332
- translocation 335
- usage 333
CYPREX (= dodine)
Cyproconazole 526, 221

- properties 229
- usage 242
Cyprodinil (CGA 219 417) 405
Cyprofuram 328
- mode of action 349 f
Cytochrom P 450
- binding of triazoles 520
- catalysis 530
- crystal structure 529 f
- inhibition 263, 283, 435, 444, 447

DACONIL (= chlorothalonil)
DCPF 315
DDCC (2,2-dichloro-3,3-dimethylcyclopropane)
- mechanism of action 552
DELSENE (= carbendazim)
14α-Demethylase inhibitors 447 ff, 451
DEMOSAN (= chloroneb)
DENMERT (= buthiobate)
Dermophytosis 433, 465 ff
DESMEL (= propiconazole)
DEXON (= fenaminosulf) 331
- site of attack 150
Diazinon 355
Dicarboximide fungicides (DCOF) 99 ff, 100, 101
- behaviour in soil and plants 104
- cross resistance 106 ff, 127
- effect on flavin enzymes 125
- effect on fungal cells 120 ff
- genetic of resistance 42
- lipid peroxidation 123
- mechanism of action 119 ff
- mixtures 101
- osmotic sensitivity 110
- resistance 106 ff, 127
- selectivity 127
- spectrum of activity 102
- toxicity 101
- uptake 103
- usage 102
Dichlofluanid (EUPAREN) 390, 487
- properties, usage 394
- spectrum of activity 16
- wood preservation 488 f
Dichlone
- spectrum of activity 13
2,4-Dichloro-3-methoxyphenol (DCMP) 75 f
- lipid peroxidation 89
- spectrum of activity 80

4,5-Dichloro-2-n-octyl-4-isothiazolin-2-one
  (ITAs) 487
 − wood preservation 491
Dichlozoline (SCLEX) 100
Diclobutrazole 220
 − lanosterol interaction 525, 527
 − properties, usage 225, 237
Dicloran 75 f
 − cross resistance 89, 130
 − properties, usage 79
 − spectrum of activity 16, 77
Diethofencarb 296, 315
 − mechanism of action 315
Difenoconazole 221
 − properties, usage 229, 242
Diisopropylfluorophosphate (DFP)
 − cutinase inhibitor 546
Dimethachlor (OHRIC) 100
Dimethirimol (MILCURB) 355 f
 − mechanism of action 363
 − practical appliction 358
 − resistance 364
Dimethomorph 329, 331
 − mechanism of action 336, 339
 − mixtures 334
 − spectrum of activity 332
 − translocation 335
 − usage 333
DIMTS (di-iodomethyl-o-tolylsulfone)
 − wood preservation 492 f
Diniconazole 220, 526
 − interaction 32-hydroxylanosterol 527
 − interaction 14-C demethylase 527
 − properties, usage 227, 239
Dinocap 389
Diphenyl (= Biphenyl) 75 f
 − cross resistance 89
 − properties, usage 81
 − spectrum of activity 77
Diphenylamine (DPA) 296
 − benzimidazole resistance 315
Ditalimfos (PLONDREL) 374
DMI-fungicides 205 ff
 − abscisic acid biosynthesis 269
 − antisenescence 268
 − chlorophyll content 271
 − cross resistance 280
 − effects on phytohormones 268 f
 − ethylen biosynthesis 268
 − genetic of resistance 45

 − gibberellic acid biosynthesis 261, 262
 − resistance 208 ff, 280 ff
 − reversal of toxicity 286
 − side effects on plants 206 f, 259 ff
 − spectrum of activity 206 ff
 − systemic properties 206 f
Dodemorph 163, 164
 − mechanism of action 190
 − properties, usage 170
Dodine (CYPREX) 390
 − genetic of resistance 45
 − properties, usage 391
DOWCO 269 (= pyroxychlor)
DOWICIDE (= 2-phenylphenol)
DRAWIFOL (= metomeclan)
Drazoxolon 389
DYRENE (= anilazine)

Ebelactones A, B
 − cutinase inhibitors 547
EBP (= kitazin) 374
Econazole 448 f, 450, 459
Edifenphos (= HINOSAN) 374
 − cutinase inhibition 546
 − mechanism of action 376
 − practical application 376
Electronic structures
 − triazoles 518
Epoxyconazole 221
 − properties, usage 231, 245
Ergosterol biosynthesis 443
ESBP (= inezin, inejin) 373
Etaconazole 220
 − properties, usage 224 f, 236
Ethirimol (= MILGOE) 355 f
 − genetic of resistance 47
 − mechanism of action 363
 − metabolism 362
 − practical application 356
 − resistance 364
Etridiazole (= TERRAZOL) 75 f
 − cross resistance 89
 − effects on fungal cells 90
 − mechanism of action 91, 92
 − properties, usage 79
 − spectrum of activity 16, 17, 77
EUPAREN (= dichlofluanid)

F 427 (carboxin derivative) 136
 − spectrum of activity 136
FADEMORPH (= trimorphamide)

Subject Index   589

FALIMORPH (= aldimorph)
Fenaminosulf (DEXON) 150, 331, 389
Fenapanil 215, 216
– usage 218
Fenarimol 210
– properties 211
– resistance 282
– synergism 286
– usage 213
Fenbucanozole 221
– properties, usage 230, 244
Fenfuram (PANORAM) 135
– mixtures 140
Fenitropan 389
Fenpiclonil (BERET) 406
– cross resistance 407
– properties, usage 406
– spectrum of activity 406
Fenpropidine (PATROL) 163, 164, 534
– degradation 169
– effect on plant growth 186
– genetic of resistance 46
– mechanism of action 191
– resistance 178 f
– side effects 176
– spectrum of activity 175, 535
– translocation 167
Fenpropimorph (CORBEL) 163, 164, 468, 534, 535
– degradation 169
– effect on cell wall 186
– effect on fungal cells 186
– effect on fungal growth 186
– effect on plants 194
– genetic of resistance 46
– mechanism of action 189
– resistance 178 f
– sites of inhibition 188
– spectrum of activity 173
– translocation 167
Fenticonazole 448 f
Fentin fungicides 400 f
– fentin acetate 400 f
– fentin chloride 400
– fentin hydroxide 400
– genetic of resistance 47
– mechanism of action 401
Fixing salts 485
– wood preservation 486
Fluazinam 411
– mechanism of action 336, 339, 412

– properties, usage 333, 411
– site of attack 150
– spectrum of activity 332, 411
– translocation 335
Fluconazole 448 f, 450, 459
Fludioxonil
– properties, usage 407
5-Fluorocytosine 436, 441
Fluotrimazole 219, 220
– properties, usage 219, 233
Fluquinconazole 221
– properties, usage 232, 247
Flurprimidol
– effect on plant growth 259
Flusilazole 220
– properties, usage 226, 238
Flutolanil (MONCUT) 134
– application 140
Flutriafol 220, 526
– properties, usage 226, 237
– C14-demethylase 517
FOLICUR (= tebuconazole)
FOLOSAN (= quintocene)
FONGARID (= furalaxyl)
Fosethyl-Al (ALIETTE) 327, 552
Fosethylphosphonates 327
– mechanism of action 336, 337
– mixtures 334
– resistance 339
– spectrum of activity 332
– translocation 335
– usage 333
Fthalide 548
– melanin biosynthesis 547
FUNDAZIL (= imazalil)
FUNGAFLOR (= imazalil)
Fungicide resistance 23 ff
– avoidance 31
– build-up 27
– definition 24
– fitness 28
– genetic basis 27, 39 ff
– major genes 40 f
– mechanism 25
– negative cross resistance 34
– polygenic control 45
– resistance factor 33
– selection pressure 29
Fungicides
– blue staining fungi 485 f
– causes of selectivity 19 f

- detoxification 26
- new toxophores 580
- outlooks 579 ff
- selectivity 13 ff
- spectra of activity 13, 15 f
- synergism 35, 565 ff
- translocation 53 ff
- uptake 26
- wood preservation 485 ff
- world market 14

Furalaxyl 327
- properties, usage 331, 333

FURAVAX (= methfuroxam)
Furcarbanil 135
Furconazole 221
- properties, usage 230, 244

Furmecyclox (XYLIGEN B) 136
- application 140
- wood preservation 140

GALBEN (= benalaxyl)
Gibberellin biosynthesis
- cytochrom P450 263
- effects of DMI enantiomers 264

Glyphosate, transport 63
Griseofulvin 436, 415
- usage 425, 440 f

Growth regulators
- fungal control 557

Growth retardants 259 ff
Guanidines 389 f
Guazatine 390
- properties, usage 392

Herbicides
- fungal control 557 f

Hexachlorobenzene 75 f
- properties, usage 77

Hexaconazole 221
- properties, usage 228, 241

Histoplasmosis 433, 462 f
Host defense systems 543
Hymexazole (TACHIGAREN) 327
- mechanism of action 336
- spectrum of activity 17, 332
- translocation 335
- usage 333

IBP (isobenfos = KITAZIN P) 374
- cutinase inhibition 546

- mechanism of action 378
- practical application 378

ICADEN (= isoconazole nitrate) 449
ICI 195, 448, 739
ICI A 5504 330, 331, 408
- mechanism of action 336, 339
- spectrum of activity 332
- translocation 335
- usage 333

Imazalil 215
- properties, usage 217
- resistance 282

Imibenconazole 221
- properties, usage 230, 243

Imidazole fungicides 215
Iminoctadine 390
- properties, usage 393

Inabenfide, growth regulator 260
INDAR
- fungal control 557

Induced resistance 553, 581
IPBC (3-iodo-3-propynyl-butylcarbamate) 487
- wood preservation 492

Iprobenfos (KITAZIN P) 374
- cutinase inhibition 379
- mechanism of action 378
- practical application 376

Iprodione (ROVRAL) 101
- cross resistance 130
- mixtures 101

Isoconazole 448 f
Isomerase-reductase inhibitors 533 f, 534
$\Delta^8 \to \Delta^7$ Isomerase inhibition 188
- fenpropidin 196
- fenpropimorph 196

Isonicotinic acid 553
- induced resistance 552

Itraconazole 448, 450, 453, 459

Kasugamycin 415, 416
- genetic of resistance 43
- mechanism of action 420
- properties 420
- resistance 25
- usage 417, 420

KETAZOL (= ketoconazole) 449
Ketoconazole 448 f, 450, 459
KITAZIN (EBP) 373
- resistance 26

Kresoxim-methyl (BAS 490 F) 329

Subject Index

Lanosterol demethylase <u>531</u>, <u>532</u>

Maneb
— spectrum of activity 13
MDCP <u>296</u>, 314, <u>315</u>
Mebenil <u>134</u>
Melanin biosynthesis inhibitors 547 f
Melanization inhibitors <u>548</u> ff
MELPREX (= dodine)
Menapyrim <u>402</u>
— mechanism of action 403
— properties, usage 403 f
— spectrum of activity 404
Mepronil (BASITAC) <u>134</u>
— application 140
Metalaxyl <u>328</u>, 331 f
— indirect effects 556
— mechanisms of action 346 f, 349
— mixtures 334
— resistance 347 f
— spectrum of activity 17
— usage 333
Metconazole <u>221</u>
— properties, usage 232, 246
Methfuroxam (FURAVAX) <u>135</u>
— application 140
Methionine
— fungal control 557
Metolachlor 349, 350
— mode of action 349
Metomeclan (DRAWIFOL) <u>100</u>
MICODERM (= miconazole) 449
Miconazole <u>448</u> f, 450, 459
— growth retardant 260
MILCURB (= dimethirimol)
Mildiomycin 415
— mechanism of action 424
— properties 424
— usage 417
MILGO (E) (= ethirimol)
MILSTEM (= ethirimol)
Molecular databases 522
Molecular graphics 517 f
MONCEREN (= pencycuron) 390, 397
MONCUT (= flutolanil)
Morpholine fungicides 163 ff, 185 ff
— effects on bacteria 196 f
— effects on NADH-oxidase 196
— effects on organelles and enzymes 195
— effects on plants 191
— genetic of resistance 46

— mechanism of action 185 ff
— properties, degradation 169
— resistance 177
— site of inhibition 188, 197
— uptake and transport 166
Multiple resistance 25
Myclobutanil <u>520</u>, 526
— conformation 528
— cytochrome c interaction 518
— properties, usage 227, 240
Myclozoline <u>100</u>
Mycoses, human 431 ff
— fungal species 432 f
MYCOSPOR (= bifonazole) 449
MYR (sulconazole) 449
Myxothiazol <u>408</u>

Naftifine 444
NIMROD (= bupirimate)
Nitrophthal-isopropyl 389
Nocodazole <u>306</u>
— resistance 311
— selectivity 311
Non fungitoxic compounds 543
Novobiocin (CATHOMYCIN)
— usage 424
Nuarimol <u>210</u>
— properties, usage 214
Nystatin <u>436</u>, 460
— mechanism of action 437
— usage 438 f

Ofurace <u>328</u>, 331
— mode of action 349 f
— usage 333
OHRIC (= dimethachlor)
OLPISAN (= 1,2,4-trichloro-3,5-dinitro-benzene)
Onychomycosis 466 f
Oomycetes fungicides 323 ff
— diseases 324, 325
— mechanism of action 336 f
— resistance 339 f
— selectivity 331
— taxonomy *Oomycetes* 223
— uptake, transport 335
— usage 333
Organophosphorus fungicides 373 ff, <u>374</u>
— mechanism of action and resistance 378 f
— practical application 375 f
Oudemansin <u>408</u>

Oxadixyl <u>328</u>, 331
- mode of action 349 f
- usage 333
Oxamyl, transport 63
Oxycarboxin (PLANTVAX) <u>133</u>
- application 137
- systemic movement 135 f

Paclobutrazol 260, <u>526</u>, 527
- abscisic acid 269
- antisenescence 268
- cell division 265
- chlorophyll content 271
- gibberellin biosynthesis 262
- lanosterol interaction 527
- plant growth 260 ff
- stress tolerance 272
PANOCTINE (= guazatine)
PANORAM (= fenfuram)
Paracoccidioidomycosis 433, 463
Paraoxon
- cutinase inhibition 546
PATAFOL (= ofurace)
Pathogenic interactions 544
PCNB (= quintocene)
- cross resistance 130
Penconazole <u>220</u>
- properties, usage 225, 236
Pencycuron (MONCEREN) <u>390</u>
- properties, usage 397
Pentachlorobenzyl alkohol <u>548</u>
- melanin biosynthesis 547
Pentachloronitrobenzene (PCNB, QUINTOCENE) <u>75</u>
- cross resistance 89
- mechanism of action 87
- properties, usage 78
- spectrum of activity 77
Pentachlorophenol (PCP) 485
- cross resistance 89
PERSULON (= fluotrimazol)
PEVISONE (= econazole nitrate) 449
Phaeohypomycosis 434, 467 f
Pharmacokinetic modeling 522
Phenylamide fungicides <u>327</u>
- genetics of resistance 43
- mechanism of action 336, 346 f, 348 f
- resistance 339
- spectrum of activity 332
- structure/activity 348
- translocation 335

- usage, properties 333, 334
N-Phenylcarbamate <u>315</u>
- mechanism of action 314 ff
N-Phenylformamidoxime <u>315</u>
o-Phenylphenol (OPP) <u>75</u> f
- genetics of resistance 42
- properties, usage 80
- spectrum of activity 80
Phenylpyrroles 405 ff
- mechanism of action 407
Phenylthiourea <u>546</u>
- mechanism of action 554
Phorat 373
Phosphorothiolates 373
- cytochrom P450 379
- metabolites 379
Piericidin
- site of attack 150
Piperalin
- effect on growth 186
- mechanism of action 191
Piperazine fungicides 209, <u>210</u>
Piperidine fungicides
- mechanism of action 185 ff
Pityriasis 434, 467
PLANTVAX (= oxycarboxin) 133
Polyenes 435 ff, <u>436</u>
Polyketide pathway
- melanin biosynthesis inhibitors <u>549</u>
Polynactines
- usage 417
Polyoxin antibiotics 415, <u>416</u>
- biosynthesis 421
- mechanism of action 422
- properties 421
- resistance 26, 422
- usage 417, 421
PP 389
- melanin biosynthesis <u>548</u>
PP 969 <u>220</u>
- properties, usage 227, 239
PREVICUR (= prothiocarb)
Probenazole
- mechanism of action <u>552</u>
Prochloraz (SPORTAK) <u>215</u>, <u>526</u>
- lanosterol interaction 527
- properties, usage 216, 218
- resistance 281 f, 285
Procymidone (= SUMILEX) <u>100</u>
- mixtures 101
Propachlor 349

- mechanism of action 349
Propamocarb <u>327</u>, 331
- mechanism of action 336
- spectrum of activity 332
- translocation 335
- usage 333
Propiconazole (TILT, DESMEL) <u>220</u>, <u>487</u>
- cross resistance 285
- field resistance 285
- properties, usage 224, 235
- wood preservation 493 f
Protein modeling (P450) 519
Prothiocarb (PREVICUR) <u>327</u>, 332
Pyracarbolid (SICAROL) <u>133</u>
- application 140
Pyrazophos (AFUGAN) <u>374</u>
- mechanism of action 381
- practical application 375 f
- resistance 26, 383
- synergism 383
Pyridine fungicides <u>210</u>
Pyrifenox <u>210</u>
- properties 211
- usage 212
Pyrimethanil <u>402</u>
- properties, usage 404 f
Pyrimidine fungicides <u>210</u>
Pyroquilon
- melanin biosynthesis <u>548</u>
Pyroxychlor, transport 63, 389

QSAR 521
QSPR 521
Quinomethionate 389
QUINTOCENE (PCNB)
- resistance 40

Reductase inhibitors
- melanin biosynthesis 547 f
Δ^8, Δ^{14}-Reductase inhibitors <u>536</u>
Δ^{14}-Reductase
- fenpropidin 195
- fenpropimorph 195
- inhibition 188
- tridemorph 195
Resistance factor 24
RIDOMIL (= metalaxyl)
RIZOLEX (= tolclofos-methyl)
RONILAN (= vinclozolin)
Rotenone, site of attack 150
ROVRAL (= iprodione)
RUBIGAN (= fenarimol)

Salicylic acid
- induced resistance 553
Salizylanilid (SHIRLAN) <u>133</u>
- application 140
SANDOFAN (= oxadixyl)
Saperconazole <u>448</u>, 450
SAPROL (= triforine)
SCLEX (= dichlozoline)
SDZ-87-469 <u>445</u>
Sec-butylamine 389
Selectivity
- antimycotica 456
- fungicides 13 ff
SERINAL (= chlozolinate)
SHIRLAN (= salizylanilide)
SICAROL (= pyracarbolid)
SISTHANE (= fenapanil)
SPORAL (= itraconazole) 449
Sporotrichosis 464 f
SPORTAK (= prochloraz)
SSF-109 <u>221</u>
- properties, usage 231, 245
Statistical modeling 521
Sterol biosynthesis
- inhibition fenpropimorph 189
- inhibition tridemorph 187, 189
Streptomycin 415
- genetic of resistance 44
- usage 417
Strobilurin A+E <u>408</u>
Strobilurin analogues <u>329</u>
Strobilurins 407 ff
- mechanism of action 339, 409 f
- site of attack 150
- spectrum of activity 409
Structure activity relationship 517 f
Sulconazole <u>448</u> f
Sulphamides 393 ff
SUMILEX (= procymidone)
Systemic transport 53 ff

TACHIGAREN (= hymexazole) 327
TCMTB (2-[thiocyanomethylthio]-benzthiazole <u>487</u>
- wood preservation 490
Tebuconazole <u>221</u>
- properties, usage 228, 240
- wood preservation 493 f
Tecto (= thiabendazole)
TERAZOL (= terconazole) 449
Terbinafine 444 ff

Terconazole <u>448</u>, 450, 459
TERRAZOL (= etridiazole)
Tetcyclasis, growth retardant 260, 261, 262
- abscisic acid 269
- cell division 265
- cytokinin level 269
- sterol synthesis 265
Tetrachloronitrobenzene 75 f
Tetraconazole <u>221</u>
- properties, usage 230, 243
Tetracyclines 415
- usage 417
Thiabendazole <u>306</u>
- resistance 307, 308
- selectivity 310
Thiophanate-methyl <u>292</u>, <u>296</u>
- spectrum of activity 16
Thioquinox 389
Thiram
- spectrum of activity 13
TILT (= propiconazole)
Tioconazole 449, 450
Tolciclate <u>445</u>
Tolclofos-methyl (RIZOLEX) <u>75</u>, <u>374</u>
- activity to *Rhizoctonia* 17
- mechanism of action 380
- practical application 375, 377
- properties, usage 76, 81
- spectrum of activity 16, 17, 77
Tolerance to fungicides 24
Tolnaftate 445 f
Tolylfluanide <u>390</u>, <u>487</u>
- properties, usage 394
- wood preservation 488 f
TOPAS (= penconazole)
Translocation of fungicides 53 ff
- phloem 59
- terminology 53, 54
- tests 65
- uptake by leaves 60
- uptake by roots 57
- uptake by the phloem 59
- xylem 55
Triadimefon (BAYLETON) <u>220</u>
- properties, usage 222, 233
- resistance 284
- spectrum of activity 16
- transport 64
Triadimenol <u>220</u>, <u>526</u>
- cross resistance 281
- field resistance 285

- metabolism in plants 261
- synthesis gibberellic acid 262
- properties, usage 223, 233
Triamiphos (WEPSYN) 374
Triapenthenol 260
- abscisic acid 269
- cross resistance 281
- growth retardant 260
- sterol, gibberellin biosynthesis 266
Triarimol
- gibberellin biosynthesis 262
Triazole fungicides 219 ff, <u>220</u>, <u>221</u>
- abscisic acid biosynthesis 269 f
- antisenescence 268
- effect on fungal gibberellin synthesis 265
- growth retardation 260
- properties, usage 221 ff
- resistance 280 ff
- stress tolerance 272
Triazole herbicides 267
Tributyltins (TBT) 486 f
- tributyltin-oxide (TBTO) 487
- tributyltin-naphthenate (TBTN) 487
1,3,5-Trichloro-2,4,6-trinitrobenzene (PHOMASAN) <u>75</u> f, 78
1,2,4-Trichloro-3,5-dinitrobenzene (OLPISAN) 75 f, 78
Tricyclazole
- melanin biosynthesis <u>548</u>
Tridemorph (CALIXIN) <u>164</u>
- effect on fungal cells 186
- effect on growth 186
- effect on organelles and enzymes 195
- effect on plants 192
- genetic of resistance 46
- lipid biosynthesis 187
- properties, usage 171
- protein synthesis 187
- resistance 178
- side effects 178
- sites of inhibition 188
- spectrum of activity 16
- sterol biosynthesis 189
- uptake, translocation 166
TRIFLUCAN (= fluconazole) 449
Triflumizole <u>215</u>
- computer graphics 518
- properties 219, 526
Triforine <u>210</u>
- properties 209

Subject Index 595

– resistance 284
– spectrum of activity 16, 212
– transport 64
Trimorphamide (FADEMORPH) 163, <u>164</u>
– properties 177
– translocation 167
Triphenyltins 400 f
Triticonazole <u>221</u>
– properties, usage 232, 247
TRITISAN (= QUINTOCENE)
TROSYL (= tioconazole) 449
β-Tubulin 305 f
– fungicide interaction 305
– mutation 308 f

Uniconazole <u>526</u>
– amylase induction 262
– cytokinin level 268
– ethylen biosynthesis 268
– growth regulator 261

Validamycin A <u>416</u>, <u>546</u>
– biosynthesis 422
– degradation 423
– mechanism of action 423, 551
– usage 417, 422
VIGIL (= diclobutrazole)
Vinclozolin (RONILAN) <u>100</u>
– antidots 124
– cross resistance 130
– degradation 104
– effects on flavin enzymes 124
– effects on fungal cells 122
– mixtures 101
– NADPH oxidation 124
– properties, usage 102, 105 f
– spectrum of activity 16
VITAVAX (= carboxin)

Wadley equation 567

Zarilamide <u>330</u>

Nutritional Disorders of Plants
- Development, Visual and Analytical Diagnosis -

Edited by Prof. Dr. Dr. Werner BERGMANN, Jena.
In cooperation with 70 donors of slides.

In Englisch. Translator: Brian Patchett, Rostock

1992. 742 pp., 945 colour pictures, 15 fig., 111 tab., hard cover DM 298,-/ÖS 2325,-/ SFr 286,50/US-$ 214,-
ISBN 3-334-60422-5;   US-ISBN 1-56081-357-1

This book comprises a most valuable data base for the study of all problems of plant nutrition and plant damage and will meet the requirements of a wide range of users, especially students in agricultural colleges, agronomists, botanists, physiologists, ecologists, and environmentalists. The most impressive and informative part of the book are the 945 colour pictures (with captions in English, French and Spanish) showing changes in growth and development of plants under nutrient stress (from deficiency to toxicity). The imbalance in the supply of following elements is presented in colour pictures: N, P, S, K, Ca, Mg, B, Mo, Cu, Fe, Mn, and Zn; in addition, there are illustrations of heavy metal toxicities, gaseous compounds (SO_2, NH_4, Cl, HF, and NO_3), herbicides, and salts (NACl). The symptoms are given for various groups of plants such as crop and ornamental plants, forest trees, tropical and subtropical vegetation. In a special chapter the background and reasons for using plant analysis and the evaluation of analytical data by "computerised nutrient element charts" are discussed. Thirteen tables (with the text in English, French, and Spanish) are presented showing "adequate ranges" of the mineral content of many plants, including cereals, root crops, vegetables, flowers, fruit and forest trees and some other plants, in order to permit the interpretation of analytical data.

Colour Atlas
Nutritional Disorders of Plants
- Visual and Analytical Diagnosis -

Three language edition: English, French, Spanish.
Edited by Prof. Dr. Dr. Werner BERGMANN, Jena.
In cooperation with 70 donors of slides. Revision of the English text by
V. M. Shorrocks, Wigginton/Hertfordshire/ GB

1992. 386 pp., 945 colour pictures, 5 fig., 14 tab., hard cover DM 189,-/ÖS1474,-/ SFr 181,50/US-$ 135,-
ISBN 3-334-60423-3;   US-ISBN 1-56081-358-X

This book with the French title "Pertubations dans la nutrition des plantes - Diagnostics visuels et analytiques" and the Spanish title "Trastornos de la nutrición de la plantas - Diagnostico visual y analítico" is a shortened edition of the a. m. book. Besides the 945 colour pictures of deficiency and toxicity symptoms of plant, the three language edition includes the following chapters: A "key" for the identification of nutrient deficiency symptoms, a short survey on the recognition of toxicity symptoms, and a chapter concerning plant analysis and the evaluation of analytical data including 12 tables with "adequate ranges" of plant nutrients of more than 90 plant species. This edition is for scientists, students and practitioners of agriculture and forestry, pomology, horticulture, viticulture, phytopathology, ecology, and environmental control and for the extension service in English, French, and Spanish speaking countries.